ALAN RAYBURN

OXFORD
UNIVERSITY PRESS

70 Wynford Drive, Don Mills, Ontario M3C 1J9
www.oupcan.com

Oxford New York
Athens Auckland Bangkok Bogotá
Buenos Aires Calcutta Cape Town Chennai
Dar es Salaam Delhi Florence Hong Kong
Istanbul Karachi Kuala Lumpur Madrid
Melbourne Mexico City Mumbai Nairobi
Paris São Paulo Singapore Taipei
Tokyo Toronto Warsaw

and associated companies in
Berlin Ibadan

Canadian Cataloguing in Publication Data

Rayburn, Alan
Dictionary of Canadian place names

Includes bibliographical references and index.

ISBN 0-19-541086-6 (bound) ISBN 0-19-541470-5 (pbk.)

1. Names, Geographical - Canada. 2. Canada - History, local.
I. Title.

FC36.R38 1999 917.1'003 C97-931286-8
F1004.R39 1999

Cover & Text Design: Brett Miller
Cover Illustration: Paul Fleming, Creative House

1 2 3 4 — 02 01 00 99
This book is printed on permanent (acid-free) paper ∞.
Printed in Canada

PREFACE

Place names are a significant reflection of a nation's cultural and linguistic heritage. They are ever-present on road signs and maps, in correspondence, magazines, and newspapers, and in all kinds of official and unofficial records and documents.

The National Toponymic Data Base, maintained by Geomatics Canada for the Canadian Permanent Committee on Geographical Names, comprises some 450,000 official place names from Cape Spear in the east to Demarcation Point in the west, and Middle Island in the south to Cape Columbia in the north. Each has its own specific detail as to location and jurisdiction, and a brief account of origins and meanings is available for about 25 per cent of them. These data are supplemented by a huge collection of paper records in files and cards managed by the Committee's secretariat. In addition, each of the provinces and territories, with authority over place names within their jurisdiction while sharing authority over names within federal crown lands, has developed its own computerized and paper records on the locations, origins, and meanings of its toponyms.

Considering the richness and variety of Canada's toponymic tapestry, there have been very few place-name books written about the country as a whole. The first was George Armstrong's *The Origin and Meaning of Place Names in Canada*, published by Macmillan of Canada in 1930. Most of his choices were drawn from Ontario, with a very small number selected from the other provinces and territories. In 1972 Macmillan republished the book without correcting its errors and misinterpretations. Names given after 1930 by the numerous expeditions to the North and to the Cordillera in Alberta, British Columbia, and Yukon were ignored, and no information from Canada's newest province, Newfoundland, was provided.

In the mid-1970s Macmillan invited William B. Hamilton, then a professor of education at the University of Western Ontario, to revise the Armstrong volume, but he recommended a completely new book. In January 1978, after he had become the director of the Atlantic Institute of Education, *The Macmillan Book of Canadian Place Names* was published to much acclaim. It provided brief details on the background and significance of more than 2,500 place names, and listed an extensive bibliography of toponymic sources. Five years later a second edition of the book was published.

A third book that deals with Canadian place names is not well known in Canada, even among place-name researchers. It is the *Illustrated Dictionary of Place Names: United States and Canada*, edited by Kelsie B. Harder of the State University College in Potsdam, NY. It was first published by Van Nostrand Reinhold in New York City in 1976, and reprinted in 1985 by Facts on File Publications of New York City and Oxford, England. The Canadian names had been compiled for the book by Robert O'Brien of Morningside Editorial Associates, in New York City. Although only five of the book's 100 references refer directly to Canadian toponymic sources, there are interesting albeit brief details on Canadian places that could only have come from a large number of published Canadian sources. Errors in spelling (e.g., St. Catherines for St. Catharines, Louisville for Louiseville), spurious origins and meanings (e.g., Milton, Moncton, Peterborough, Taber), and curious omissions of cities (e.g., Corner Brook, Red Deer, Woodstock) render the book a questionable source for reliable information on Canada's names. A good feature of the book, however, is the straight alphabetical arrangement of both countries together, even listing Mac- and Mc- names in their appropriate alphabetical order, rather than bundling them together.

There is a youthful vigour in Canada's toponymic character, although documentary evidence for some of its place names may be traced back over 500 years. Change has been a constant, however, with hundreds of new names and redesignations having taken place since 1990. The cities of Halifax and Dartmouth in Nova Scotia are now 'metropolitan areas' in the regional municipality of Halifax. The city of Miramichi in New Brunswick embraces the former towns of Chatham and Newcastle and some adjacent villages. In Prince Edward Island the new town of Stratford comprises four former municipal communities. In Québec the Communauté urbaine de l'Outaouais has been divided, and its outlying municipalities brought together in the new regional county of Les Collines-de-l'Outaouais. Fort McMurray in Alberta is no longer a city, having become an urban service centre in the regional municipality of Wood Buffalo. The new city of Abbotsford in British Columbia has swallowed the district municipality of Matsqui. The Ontario municipality of Clarington, with over 50,000 people, has emerged from the redesignation of the so-called town of Newcastle, and the municipal township of Flos, recalling one of the three pet dogs (the others, Tay and Tiny) of Lady Sarah Maitland, has disappeared into the blandly named municipal township of Springwater.

Change will continue to mark Canada's municipal structures. At the end of 1996 the Ontario government peremptorily announced that come 1 January 1998 the city of Toronto would be enlarged to take over the municipality of Metropolitan Toronto, the cities of Etobicoke, North York, Scarborough, and York, and the borough of East York to form the new city of about 2,300,000 people. There may be other dramatic changes in Ontario's administrative structure in the next few years, resulting in its 815 municipalities being trimmed to 450 or less.

On 1 April 1999 the misnamed Northwest Territories (a single territory since 1905) will be divided in two. The new territory of Nunavut, extending from the border of Manitoba to Cape Columbia at the north end of Ellesmere Island, will have a larger land area than any of Canada's provinces. The rest of the Northwest Territories, based on the Mackenzie River valley, has not yet decided on a suitable name for itself, although there seems to be a consensus to continue using the name Northwest Territories.

Alan Rayburn
Nepean, Ontario
1 June 1997

ACKNOWLEDGMENTS

This book on Canadian toponymy could not have been produced without the guidance and help provided by many people from across Canada and in the United States. The idea of producing an Oxford dictionary of Canada's place names was first proposed in the fall of 1993 by Brian Henderson, who was then with Oxford University Press Canada. The project was continued by Ric Kitowski, Anne Erickson, and Phyllis Wilson, who each provided encouragement and support. I also acknowledge the critical reading of the text by the Press's copy-editors, who resolved a number of discrepancies and clarified the intent of many passages.

A special debt of gratitude is extended to Helen Kerfoot, who succeeded me as the executive secretary of the Canadian Permanent Committee on Geographical Names (CPCGN) in 1988. She permitted full access to the Committee's files and its geographic names data base, developed and maintained by Geomatics Canada in the federal Department of Natural Resources. Considerable help was also provided by Kathleen O'Brien, Jocelyne Revie, Paul O'Blenes, Peter Revie, and Debbie Poupart of the CPCGN secretariat's staff.

The reference staff of the National Library are specially thanked for their guidance to the library's vast collection of published histories and reference materials. The assistance of Thomas Rooney of the Ottawa Public Library's Ottawa Room, and the encouragement of Marlene MacLean, of the Nepean Central Library, are acknowledged with thanks.

I am especially indebted to the suggestions and corrections provided by the following in the provincial and territorial offices responsible for mapping and naming: Jeffrey Ball and Allan Day of the Surveys, Mapping and Remote Sensing Branch of Ontario's Ministry of Natural Resources; Merrily Aubrey, co-ordinator of Alberta's geographic names program in the Department of Community Development; Janet Mason, British Columbia's geographical names officer in the Ministry of Environment, Lands and Parks; Gerald Holm, Manitoba's provincial toponymist in the Department of Natural Resources; David Wills, Nova Scotia's geographical names research officer in the Department of Natural Resources; David Arthur, secretary of the Saskatchewan Geographic Names Board in the Saskatchewan Property Management Corporation; Dan Rae, New Brunswick's geographical names officer in the Geographic Information Corporation; Louise Profeit-LeBlanc of the Yukon Geographical Names Program in the Department of Tourism; and Alain Vallières, of the Commission de toponymie du Québec.

Valuable assistance was provided by Roger Payne, the executive secretary of the domestic names committee of the United States Geographic Names Board. Frank Hamlin, of Richmond, BC, offered several useful suggestions and corrections. Gordon Handcock helped resolve a number of problems relating to Newfoundland place names.

I express my sincere appreciation to André Lapierre, of the University of Ottawa's Canadian Studies Program, for his extensive and scholarly knowledge of geographical names of French origin. Thomas Hillman, of the Canadian Postal Archives in the National Archives of Canada, clarified several postal names and dates. Thanks are extended to Luc Soucy of the Archives of the Oblates of Mary Immaculate in Ottawa. Finally, I acknowledge the considerable patience and understanding of my wife Mary, while I laboured for many months in researching and writing the stories about Canadian place names for this work.

This book is dedicated to the Canadian Permanent Committee on Geographical Names, whose antecedent, the Geographic Board of Canada, was founded on 18 December 1897. It commemorates the initiative of the pioneers who perceived the need to standardize and record Canada's toponymic heritage. And it honours the scholars, scientists, and other researchers who have recognized over the past 100 years the significance of place names to the culture and character of each province and territory as well to the country as whole.

INTRODUCTION

Name Selection

There are approximately 450,000 official geographical names in Canada. This dictionary has 5075 entries, with 1150 associative names listed within the entries for a total of 6225 names, which is less than 1.5 per cent of the total number of official names. Even so, that total is more than twice as many as in the previous books written on Canada's toponymy. Associative names, such as Niagara Escarpment within the entry for Niagara, and Nepisiguit Bay within the entry for Nepisiguit River, are displayed in bold type. If all associative names listed in the most recent volumes of the *Gazetteer of Canada* and the *Répertoire toponymique du Québec*, such as Niagara Gorge and Nepisiguit Falls, had been included in the entries, the total number of names in this book could easily have surpassed 12,000.

Precedence is given to incorporated cities, towns, villages, and similar urban centres designated by other municipal terms with populations of 150 or more. Added to these are unincorporated places with at least 150 people, and well-known urban neighbourhoods. Also included are regional, district, and county municipalities. All incorporated rural municipalities in Ontario and Québec with populations over 1500 are included, as well as many of them with fewer people. The names of Canada's national parks and many of the provincial parks are noted.

All lakes and islands with an area of 350 square kilometres, and many even smaller, are included. All rivers with lengths of 75 kilometres, and some well-known rivers shorter than 75 kilometres, are listed. The names of falls with heights of 200 metres or more, and several with lesser heights, are noted. All mountains with elevations of 2900 metres and more, and some other well-known mountains and peaks with lower elevations, are listed. The better-known names of capes, points, straits, channels, bays, and inlets are also included.

In most instances the year of naming populated centres is either the opening of postal service or the creation of a railway stop, although many originated before either of these important services were established. For each municipality the incorporation year of its current status (i.e., city, town, etc.) is provided in parentheses. Terms such as town and village are used only for municipalities incorporated as such, avoiding the common practice of the media of using them for any grouping of people.

For each populated place and physical feature named after an individual, the person's years of birth and death are given, when readily available. Such dates serve two useful purposes: first, to differentiate between two persons with identical names, such as two called Sir James Outram, one the grandson of the other, each of whom has a mountain in Canada named for him; and second, to guide readers to biographical resources to learn more about the careers of the persons honoured.

Italics identify precise forms of previous names shown on early maps and in documents, such as *cap de Lorraine* for Cape St. Lawrence, at the northwestern end of Cape Breton Island. They are also used for non-English words (*alameda*), for the names of ships (HMS *Ajax*), and for former names of places, such as *Bytown*, which preceded Ottawa. As well, historical Aboriginal names of places and Aboriginal words from which official names have been derived are shown in italics. Efforts have been made to ensure the appropriate Aboriginal language (e.g., Cree, Gwitch'in, Inuvialuktun, Mi'kmaq, Ojibwa) is used to identify the sources of names, rather than using the general word Indian, which is about as useful as describing Paris as a name of European origin.

Considerable care has been taken to ensure the official forms of current names as established by the appropriate federal, provincial, and territorial names authority are used throughout the book, such as Montréal and Îles de la Madeleine in Québec, and Sault Ste. Marie and St. Thomas in Ontario, retaining the precise forms of their names as published in the statutes and in the gazetteers. Québec is accented for the city and the present province (although the latter in the English version of the Canadian Constitution is spelled without an accent), but in references to the province known as Quebec before 1791, which extended beyond the Great Lakes, it is spelled without an accent. Before 1905, North-West Territories was the correct spelling, but since then the name of this single territory has been spelled Northwest Territories.

The names of physical features whose generic terminology precedes the specific (e.g., Mount Assiniboine, Point Pelee, Île d'Orléans) are alphabetized under the specific words, that is, Assiniboine, Pelee, and Orléans. At least one good toponymic reference, William B. Hamilton's *Place Names of Atlantic Canada*, lists such names under the generic, without cross-references, so that Bay d'Espoir, Cape Race, and Mount Caubvick are listed under the letters B, C, and M respectively.

Sources for Toponymic Information

General. To compile the entries in this book several sources, mainly toponymic, but also biographical, historical, and geographical, have been consulted. The voluminous files and card records of the Canadian Permanent Committee on Geographical Names (CPCGN), dating from the 1890s, are the basis for much of the information provided. Especially important are the 2850 letters received from postmasters by geographer James White in the period 1905–9, which few writers of other books on Canadian toponymy have consulted. The 13 volumes of the *Dictionary of Canadian Biography* were essential in resolving several points about the careers of noted individuals involved in the development of Canada before 1911. I also relied extensively on William F. Ganong's *Crucial Maps in the Early Cartography and Place-Nomenclature of the Atlantic Coast of Canada* (1964), John Robert Colombo's *Canadian Literary Landmarks* (1984), and Alan Cooke and Clive Holland's *The Exploration of Northern Canada* (1978).

Alberta. No province has had more toponymic studies published about its places than Alberta. The first, *Place-Names of Alberta* (1928), was written by Robert Douglas, who was then the secretary of the Geographic Board of Canada. It was the basis of most of the information compiled in *2000 Place Names of Alberta* (1972) by Eric and Patricia Holmgren, which they supplemented in a subsequent volume before publishing *Over 2000 Place Names of Alberta* in 1976. Meanwhile in 1973 Ernest Mardon produced *Community Names of Alberta*. Between 1991 and 1996 the four volumes of *Place Names of Alberta*, compiled by three successive geographical names program co-ordinators of Alberta's Historic Sites Service, have resulted in a comprehensive and reliable overview of the province's toponymy.

British Columbia. In 1906 the Geographic Board of Canada published Capt. John

Walbran's excellent *British Columbia Coast Names*, which remains a masterpiece for both its research and elaborate details. In 1969 Philip and Helen Akrigg published the first edition of *1001 British Columbia Place Names*, followed by revised editions in 1970 and 1973. Thirteen years later they produced their much enlarged and well-written *British Columbia Place Names. Place Names of the Canadian Alps*, an excellent study of the names of mountains, valleys, and populated places in the Rocky Mountains (in both British Columbia and Alberta) and the Columbia Mountains, both as far north as Prince George, was published in 1990 by climbers William L. Putnam, Glen W. Boles, and Roger W. Laurilla. Because very little information has been published on the names of peaks in the Coast Mountains northwest of Vancouver, and in the Cassiar Mountains in the northern part of the province, the files and card records of the CPCGN's secretariat were consulted to furnish the necessary details on the names of mountains and other physical features in these areas of the province.

Manitoba. The Geographic Board of Canada published Robert Douglas's *Place-Names of Manitoba* in 1933, three years after he had died. This remained the definitive work on the province's toponymy until J.B. Rudnyckyj published in 1970 his *Manitoba Mosaic of Place Names*, which relied almost entirely on the Douglas work. Ten years later a much better study by Penny Ham, entitled *Place Names of Manitoba*, was published. It was based on the records of the provincial names office and extensive correspondence with knowledgeable individuals across the province.

New Brunswick. In 1896 naturalist William F. Ganong produced a huge study on New Brunswick's place names, and supplemented it 10 years later with additions and corrections. His published toponymic works, including several on names of Aboriginal origin, and unpublished notes at the New Brunswick Museum formed the basis of my *Geographical Names of New Brunswick* (1975).

Newfoundland. Without the *Encyclopedia of Newfoundland* (1981–94) and William B. Hamilton's *Place Names of Atlantic Canada* (1996) it would not have been possible to give Newfoundland justice in the treatment of its geographical names. Before Newfoundland entered Confederation in 1949 no records for the province's names had been collected by the names office in Ottawa. Other than the scholarly *Place Names of the Avalon Peninsula* (1971) by Edgar R. Seary, there had been very little published toponymic material on the rest of the province before Joseph Smallwood's five-volume *Book of Newfoundland*, completed in 1975, and his five-volume *Encyclopedia of Newfoundland* were published.

Northwest Territories. In 1910, geographer James White published *Place Names — Northern Canada*, which provided brief details on the 450 years of documented naming in the North. Subsequently his fine work was added to by staff member Frank Stevenson in the 1960s and early 1970s, and supplemented since then by other staff members, and, after the mid-1980s, by Randolph Freeman, the territorial toponymist.

Nova Scotia. In 1922 Thomas Brown wrote the slim *Nova Scotia Place Names*, which remained the only general study of the province's toponymy until 1967, when provincial archivist Bruce Fergusson published *Place Names and Places of Nova Scotia.* William B. Hamilton's recent book on the names of Atlantic Canada gives a better reflection of the toponymy of the province.

Ontario. The entries for this province are based extensively on my *Place Names of Ontario* (1997). That work was produced after extensive research in the records of the CPCGN, some 250 local histories, Herbert F. Gardiner's *Nothing But Names* (1899), which is a superb study of Ontario's county and township names, the comprehensive review of the province's places by Nick and Helma Mika in their three-volume *Places of Ontario* (1977–83), Floreen Carter's two-volume *Place Names of Ontario* (1984), and her *Ghost & Post Offices of Ontario* (1986).

Prince Edward Island. In 1925 Robert Douglas produced *Place-Names of Prince Edward Island.* Using this study as a base, I wrote *Geographical Names of Prince Edward Island*, which was published in 1973, the centennial of the province's entry into Confederation. The records of new names and name changing in Prince Edward Island have been kept up to date by the CPCGN's secretariat.

Québec. This study has relied almost entirely on one of the best toponymic studies ever produced in the world, the well-documented and beautifully illustrated *Noms et lieux du Québec* (1994). The letters from postmasters received by geographer James White, 1905–9, were the only other significant sources used to verify the backgrounds of some of the approximate 500 Québec place names included in this book.

Saskatchewan. In 1968 E.T. Russell, a public school principal in Saskatoon, published *What's in a Name?*, a compilation of 679 names of the province's populated places. The material was essentially gathered by students, and verified through extensive correspondence with knowledgeable individuals throughout the province. A second edition in 1973 with 1604 names and a third in 1980 with 1837 followed. Although much of the historical details on the places is fascinating, the various editions suffered from a lack of consultation with the official names records in Regina and Ottawa.

Yukon. One of the better books on Canadian toponymy is Robert Coutts's *Yukon Places & Names*, published in 1980. Relying on the excellent records of the CPCGN's secretariat, he was able to supplement his research with other territorial sources. An important reference for the names of features on the Alaska border from Stewart, BC to the Beaufort Sea has been Donald J. Orth's *Dictionary of Alaska Place Names* (1971).

PROVINCIAL AND TERRITORIAL ADMINISTRATIVE TERMINOLOGY

Each of the provinces and territories has a different set of municipal and territorial designations. For example, Ontario stands apart from the rest by being the only jurisdiction with municipalities known as a borough, a separated town, and a metropolitan municipality, and an administrative unit known as a police village. Only Québec uses regional county municipality, Cree village municipality, and urban community. While all incorporated places in Canada are in fact municipalities, both Ontario and Québec have permitted the use of 'municipality' alone for some places without qualifying it. 'Community' is generally used in this book, outside Prince Edward Island and Newfoundland, for any grouping of people, whether located within a municipality or not, and I have used it interchangeably with 'place' outside these two provinces to describe *unincorporated* centres of population. I have distinguished the *incorporated* communities in Prince Edward Island and Newfoundland as 'municipal communities'.

Alberta. All of the province's 15 cities and 112 towns, as well as 96 of its 110 villages, have separate entries in the dictionary. The remaining 14 villages, having populations of less than 150, are excluded. The municipality of Crowsnest Pass is included, but none of the 37 municipal districts, 29 counties, and 53 summer villages is listed, because their names are relatively less well known.

British Columbia. Described in the book are British Columbia's 43 cities, 50 district municipalities, 14 towns, 120 villages, and 1 resort municipality. Excluded are the names of the 29 regional districts and the 33 Indian government districts, although many of their names are noted under other municipal categories.

Manitoba. This province's 5 cities, 36 towns, and 38 villages are described. Its 106 rural municipalities are excluded, but the names of many of them are noted under other municipal entries.

New Brunswick. Included are New Brunswick's 7 cities, 29 towns, and 76 villages. Also listed are its 14 counties, which are no longer organized as municipalities, but their names are well known.

Newfoundland. Listed for this province are its 3 cities, 157 towns, 99 of its 131 municipal communities, and 5 of its 168 local service districts. There are no villages, counties, or regional municipalities in Newfoundland.

Northwest Territories. Incorporated municipalities in this jurisdiction include 1 city, 5 towns, 1 village, 35 hamlets, 2 settlement corporations, and 2 settlements. Twelve places identified in territorial records as communities have no legal community status, but their names have been included.

Nova Scotia. Entries have been compiled for this province's 9 municipal counties, 12 municipal districts, 3 regional municipalities, 31 towns, and 22 villages. The former cities of Halifax, Sydney, and Dartmouth are now metropolitan areas within regional municipalities.

Ontario. This province has 1 metropolitan municipality, 10 regional municipalities, 1 district municipality, 24 counties, 3 united counties, 51 cities, 1 borough, 1 municipality, 143 towns, 4 separated towns, and 107 villages, and all their names are included in the book. There are 468 municipal townships, but only those with populations of 1,000 or more are listed. The 11 unincorporated districts in Northern Ontario are included because their names are well known, but almost all the 2000-plus unorganized and mostly unpopulated townships in that part of the province are excluded. A separated town is a municipality within a county, but does not participate as part of a county's administration. There are some 62 police villages in the province, which are not municipalities but exist within municipal townships to manage such amenities as street lighting and sidewalk paving. Many of them are inactive, and no new police villages may be created. There used to be several local improvement districts in Northern Ontario, which managed urban-style amenities within prescribed areas, but all of them have been redesignated by the province as municipal townships.

Prince Edward Island. Incorporated municipalities in this province include 2 cities, 7 towns, and 64 municipal communities. Incorporated villages were converted to municipal communities in 1983. The province's 3 counties have never been organized along municipal lines, but their well-known names are included.

Québec. Québec's municipal structure is the most complex in the country. It has 3 urban communities, 96 regional county municipalities, 2 cities, 261 *villes*, 160 villages, 90 township municipalities, 8 united township municipalities, 346 parish municipalities, 564 municipalities, 14 northern villages, 8 Cree villages, and 1 Naskapi village. All their names are included except those of *villes* with less than 150 people, township and united township municipalities with less than 1,000 people, and municipalities and parish municipalities with less than 1,500 people. The two cities are Dorval and Côte-Saint-Luc, both on the west side of Montréal. Otherwise *ville* is used for urban centres ranging from Montréal (population 1,018,000) to L'Île-Dorval (population 6). Although *ville* normally translates as 'town', it would not be appropriate to describe Montréal, Sherbrooke, and Hull as towns, or to call Beauceville, Témiscaming, and Lennoxville cities. Therefore, the standard used in Ontario that towns must have 15,000 people to apply for city status has been used to distinguish Québec's cities from its towns. Gaspé is the only exception; it is described in the book as a town, although it has over 17,000 people, but they are widely distributed over 950 km^2.

Saskatchewan. This province has a wide variety of incorporated municipalities. It has 12 cities, 147 towns, 324 villages, 40 resort villages, 299 rural municipalities, 166 organized hamlets, 13 northern hamlets, 13 northern villages, and 10 northern settlements. The names of all the towns and cities are included in this book, but only those 118 villages with populations exceeding 150 are listed. None of the rural municipalities and resort villages is given a separate entry, and only some of the larger organized hamlets and northern municipalities are listed separately.

Yukon. This territory has 1 city, 3 towns, 4 villages, and 3 hamlets. All are included except the hamlet of Elsa, which has a population of fewer than 10.

ABBREVIATIONS

Provinces and territories

The standard two-letter abbreviations are used for the provinces and territories, with QC, rather than PQ, being used for Québec, as it is preferred by the province. For postal purposes, Labrador has LB as its abbreviation. However, for this book, the abbreviation of NF has been used for the entire province of Newfoundland.

AB	Alberta	NT	Northwest Territories
BC	British Columbia	ON	Ontario
MB	Manitoba	QC	Québec
NB	New Brunswick	PE	Prince Edward Island
NF	Newfoundland	SK	Saskatchewan
NS	Nova Scotia	YT	Yukon

Other abbreviations

Adm.	Admiral	MN	Minnesota
AL	Alaska	MO	Missouri
Bro.	Brother	MT	Montana
CA	California	ND	North Dakota
CO	Colorado	NE	Nebraska
Comm.	Commodore	NH	New Hampshire
Cpl	Corporal	NJ	New Jersey
F/O	Flight Officer	NY	New York
Fr	Father	OH	Ohio
FSgt	Flight Sergeant	PA	Pennsylvania
Gen.	General	PO	Petty Officer
IA	Iowa	Pvt.	Private
ID	Idaho	Rear-Adm.	Rear Admiral
IL	Illinois	SD	South Dakota
Insp.	Inspector	Sub-Lt	Sub-Lieutenant
Lt-Cdr	Lieutenant-Commander	Tp	Trooper
Lt-Col	Lieutenant-Colonel	Vice-Pres.	Vice President
MA	Massachusetts	Visc.	Viscount
Maj.-Gen.	Major-General	VT	Vermont
ME	Maine	W/Cdr	Wing Commander
Mgr	Monsignor	WA	Washington
MI	Michigan	WI	Wisconsin

SELECTED REFERENCES

Akrigg, G.P.V., and Helen B. *British Columbia Place Names.* Victoria: Sono Nis Press 1986

Armstrong, G.H. *The Origin and Meaning of Place Names in Canada.* Toronto: Macmillan of Canada 1930, 1972

Aubrey, Merrily K. *Place Names of Alberta,* Vol. 4, Northern Alberta. Calgary: University of Calgary Press 1996

[Barkham, Selma Huxley]. *The Basque Coast of Newfoundland.* [Plum Point, NF]: Great Northern Peninsula Development Corporation 1989

Carter, Floreen. *Ghost & Post Offices of Ontario.* Oakville, ON: Personal Impressions Publishing 1986

———. *Place Names of Ontario.* 2 vols. London: Phelps Publishing Company 1984

Colombo, John Robert. *Canadian Literary Landmarks.* Willowdale, ON: Hounslow Press 1984

Commission de toponymie du Québec. *Noms et lieux du Québec: Dictionnaire illustré.* Québec: Les Publications du Québec 1994

Cooke, Alan, and Clive Holland. *The Exploration of Northern Canada, 500 to 1920.* Toronto: The Arctic History Press 1978

Coutts, R. *Yukon Places & Names.* Sidney, BC: Gray's Publishing Limited 1980

Dawber, Michael. *Where the Heck is Balaheck? Unusual Place-Names from Eastern Ontario.* Burnstown, ON: General Store Publishing House 1995

Dictionary of Canadian Biography. Vols. 1–13. Toronto: University of Toronto Press 1967–94

[Douglas, Robert]. *Place-Names of Manitoba.* Ottawa: Geographic Board of Canada 1933

———. *Place-Names of Alberta.* Ottawa: Geographic Board of Canada 1928

———. *Place-Names of Prince Edward Island, with Meanings.* Ottawa: Geographic Board of Canada 1925

Encyclopedia of Newfoundland and Labrador. St. John's: Newfoundland Book Publishers (1967) Ltd. Joseph R. Smallwood, ed. in chief. Vol. 1 1981, Vol. 2 1984. Harry Cuff Publications Ltd, Vol. 3 1991. Cyril R. Poole, ed. in chief. Harry Cuff Publications Ltd, Vol. 4 1993; Vol. 5 1994

Fergusson, Charles Bruce, ed. *Place-Names and Places of Nova Scotia.* Halifax: The Public Archives of Nova Scotia 1967

Ganong, W.F. *Crucial Maps in the Early Cartography and Place-Nomenclature of the Atlantic Coast of Canada.* Toronto: University of Toronto Press 1964

———. *A Monograph of the Place-Nomenclature of the Province of New Brunswick.* Transactions of the Royal Society of Canada, Ser. 3, Sect. 2, Vol. 2, pages 216–89, 1896; Additions and Corrections, ibid., Ser. 3, Sect. 1, Vol. 2, pages 4–57, 1906

———. *An Organization of the Scientific Investigation of the Indian Place-Nomenclature of the Maritime Provinces of Canada.* Transactions of the Royal Society of Canada, Ser. 3, Sect. 2, Vol. 5, pages 179–93, 1912; Ser. 3, Sect. 2, Vol. 6, pages 179–96, 1912; Ser. 3, Sect. 2, Vol. 7, pages 81–106, 1913; Ser. 3, Sect. 2, Vol. 8, pages 259–93, 1914; Ser. 3, Sect. 2, Vol. 9, pages 375–448, 1915

Gardiner, Herbert Fairbairn. *Nothing but Names: An Inquiry into the Origin of the Names of the Counties and Townships of Ontario.* Toronto: George N. Morang 1899

Ham, Penny. *Place Names of Manitoba.* Saskatoon: Western Producer Prairie Books 1980

Hamilton, William B. *Place Names of Atlantic Canada.* Toronto: University of Toronto Press 1996

———. *The Macmillan Book of Canadian Place Names.* Toronto: Macmillan of Canada 1978

Harder, Kelsie, ed. *Illustrated Dictionary of Place Names: United States and Canada.* New York: Facts on File Publications 1985

Harrison, Tracey. *Place Names of Alberta,* Vol. 3, Central Alberta. Calgary: University of Calgary Press 1994

Holmgren, Eric J. and Patricia M. Holmgren. *Over 2000 Place Names of Alberta,* 3rd ed.

Saskatoon: Western Producer Prairie Books 1976

Jackson, John N. *Names Across Niagara*. St. Catharines: Vanwell Publishing Limited 1989

Karamitsanis, Aphrodite. *Place Names of Alberta*, Vol. 2, Southern Alberta. Calgary: University of Calgary Press, 1992

———. *Place Names of Alberta*, Vol. 1, Mountains, Mountain Parks and Foothills. Calgary: University of Calgary Press, 1991

Lapierre, André. *Toponymie française en Ontario*. Montréal: Éditions Études Vivantes 1981

Macpherson, L.B. *Nova Scotia Postal History, Vol. 1: Post Offices* (1754–1981). Halifax: Petheric Press Ltd 1982

Mardon, Ernest G. *Community Names of Alberta*. Lethbridge: University of Lethbridge 1973

McManus, George. *Post Offices of New Brunswick, 1783–1930*. Toronto: Jim A. Hennock Ltd 1984

Melvin, George H. *The Post Offices of British Columbia, 1858–1970*. Vernon, BC: Vernon Press 1972

Mika, Nick and Helma. *Places in Ontario, Their Name Origins and History*. Belleville: Mika Publishing Company, 1977–83. Part 1, A-E, Part 2, F-M, Part 3, N-Z

———. *Community Spotlight: Leeds, Frontenac, Lennox and Addington, and Prince Edward Counties*. Belleville: Mika Publishing 1974

Murray, G. Douglas. *The Post Office on Prince Edward Island (1787–1990)*. Bridgetown, NS: Mailman Publishing 1990

National Atlas Information Service. *Facts from Canadian Maps: a Geographical Handbook*, 2nd ed. Ottawa: Geographical Services Division, Canada Centre for Mapping, Energy, Mines and Resources Canada, 1989

Orth, Donald J. *Dictionary of Alaska Place Names*. Geological Survey Professional Paper 567. Washington: United States Government Printing Office 1967, with minor revisions, 1971

Phillips, James W. *Washington State Place Names*. Seattle and London: University of Washington Press 1971

Putnam, William, Glenn W. Boles, and Roger W. Laurilla. *Place Names of the Canadian Alps*. Revelstoke, BC: Footprint Publishing 1990

Rayburn, Alan. *Place Names of Ontario*. Toronto, University of Toronto Press 1997

———. *Naming Canada: Stories About Place Names From Canadian Geographic*. Toronto: University of Toronto Press 1994

———. *Geographical Names of New Brunswick*. Ottawa: Surveys and Mapping Branch, Energy, Mines and Resources Canada 1975

———. *Geographical Names of Prince Edward Island*. Ottawa: Surveys and Mapping Branch, Energy, Mines and Resources Canada 1973

Robinson, William G., ed. *Manitoba Post Offices*. Vancouver: William Topping 1988

———. *Saskatchewan Post Offices*. Vancouver: William Topping 1987

Rudnyckyj, J.B. *Manitoba Mosaic of Place Names*. Winnipeg: Canadian Institute of Onomastic Sciences 1970

Russell, E.T. *What's In A Name: The Story Behind Saskatchewan Place Names*, 3rd ed. Saskatoon: Western Producer Prairie Books 1973

Scott, David E. *Ontario Place Names: The Historical, Offbeat or Humorous Origins of Close to 1,000 Communities*. Vancouver and Toronto: Whitecap Books 1993

Seary, E.R. *Family Names of the Island of Newfoundland*. St. John's: Memorial University of Newfoundland 1977

———. *Place Names of the Avalon Peninsula of the Island of Newfoundland*. Toronto: University of Toronto Press 1971

Smith, Robert C. *Ontario Post Offices, Volume 1: An Alphabetical Listing*. Toronto: The Unigrade Press 1988

Voorhis, Ernest. *Historic Forts of the French Regime and of the English Fur Trading Companies*. Typescript. Ottawa: Department of the Interior 1930.

Walbran, Captain John T. *British Columbia Coast Names, 1592–1906*. Vancouver: J.J. Douglas Ltd for the Vancouver Public Library 1971. (Originally published by the Geographic Board of Canada in 1906)

White, James, ed. *Handbook of Indians of Canada*. Toronto: Coles Publishing Company

1974 (Facsimile ed. of the 1913 edition by the Geographic Board of Canada)
———. 'Place-names in Georgian Bay', *Papers and Records*. Ontario Historical Society, Vol. 11, 1913
———. 'Place-names — Northern Canada', *Ninth Report*. Part 4. Ottawa, Geographic Board of Canada, 1910, pp. 231–455

CANADA

Its Origin, Meaning, and Application

Explorer Jacques Cartier approached Île d'Anticosti in August 1535, and learned from his two young Aboriginal companions, who had returned from Gaspé to France with him the previous year, that the *chemin de Canada* (route to **Canada**) lay beyond. Cartier knew that their father, Donnacona, was a chief of an Iroquoian-speaking tribe at Stadaconé, the present-day site of the city of Québec, and found out that it was also called **Canada**. In the report of his voyage Cartier appended a list of words, noting that *kanata* meant 'town', interpreted as a cluster of dwellings. By the mid-1540s French maps portrayed the territory north of Île d'Anticosti as **Canada**, and maps in the 1550s had the name inscribed on both sides of what became known as the *grande rivière du Canada*. In 1791, **Canada** gained official status when the two jurisdictions of **Upper Canada** and **Lower Canada** were created, comprising the southern parts of present-day Ontario and Québec. Together they became the Province of **Canada** in 1841, with the two regions being called **Canada West** and **Canada East**. When union with the Atlantic provinces was proposed in the 1860s more than 30 suggestions were made for the new country's name, including Acadia, Borealia, Cabotia, Hochelaga, Mesapelagia, Tuponia, and Transatlantica. Father of Confederation Thomas D'Arcy McGee settled the choice by remarking: 'Now I would ask any Hon. Member of the House how he would feel if he woke up some fine morning and found himself, instead of a Canadian, a Tuponian or Hochelagander?' With the implementation of the British North America Act on 1 July 1867, **Canada** became the name of the new dominion.

Abbey SK Northwest of Swift Current, the post office serving the area of this village (1913) was called *Longworth* in 1910. In 1913 the Canadian Pacific Railway named its station after Abbey in County Galway, Ireland, with the post office being renamed that year.

Abbotsford BC A city (1995) on the south side of the Fraser River, southeast of Vancouver, it comprises the former district municipalities of Abbotsford and Matsqui. The original townsite was chosen in 1889 by J.C. Maclure for the Canadian Pacific Railway, and named after Harry Braithwaite Abbott (1829-1915), general superintendent of the Pacific Division of the CPR, and brother of Sir John J.C. Abbott, prime minister of Canada, 1891-92. Its post office (first *Abbottsford*, then Abbotsford) opened in 1892, and replicated a post office named after his parents in Québec (now Saint-Paul-d'Abbotsford), and Sir Walter Scott's mansion near Melrose, Roxburghshire, Scotland. **Abbott Peak**, northwest of the head of Duncan Lake, was also named for Harry Abbott.

Abegweit Passage NB, PE This passage, at the narrowest point of Northumberland Strait, where the Confederation Bridge crosses from the mainland, was named in 1962 after MV *Abegweit*, which was put in service to the island in 1947. The vessel's name was derived from the Mi'kmaq name for Prince Edward Island, *Abahquit*, 'lying parallel with the land', often rendered loosely by 'cradled by the waves'.

Abercorn QC A village (1929) in the regional county of Brome-Missisquoi and on the Vermont border south of Cowansville, it was named in 1848, possibly after James Hamilton, 1st Duke of Abercorn (1811-85), a member of the household of the Prince of Wales. His title was derived from a place west of Edinburgh, Scotland.

Abercrombie NS Northwest of New Glasgow, this place's post office was named in 1882 after **Abercrombie Point**. The point was likely called after Gen. James Abercrombie, who was killed in 1775 at Bunker Hill, opposite Boston, during the American Revolution.

Aberdeen SK A town (1988) northeast of Saskatoon, it was named in 1905 by the Canadian Northern Railway after John Campbell Gordon, 1st Marquess of Aberdeen and Temair (1847-1934), governor general of Canada, 1893-8.

Aberdeen Lake NT A widening of the Thelon River, 140 km west of Baker Lake, this lake (1095 km^2) was named in 1893 by Joseph Burr Tyrrell after John Campbell Gordon, 7th Earl of Aberdeen, governor general of Canada, 1893-8.

Aberfoyle ON Located in Wellington County, near Guelph, this place's post office was named in 1851 after a town in Perthshire, Scotland.

Abernethy SK Located southwest of Melville, the post office in this village (1904) was named in 1884 by the Revd Alexander Robson after a place in Perthshire, Scotland, southeast of Perth.

Abitibi QC A regional county in the area of Amos in northwestern Québec, it was formed in 1983 from the former county of Abitibi. At the same time, the regional county of **Abitibi-Ouest**, in the area of La Sarre and Macamic, was created. The area of these regional counties, formerly part of the district of Ungava, became part of the province of Québec in 1898.

Abitibi, Lake ON, QC Situated midway between James Bay and Lake Timiskaming, this lake's name means 'waters of the middle' in both Cree and Algonquin. With an area of 932 km^2, the lake is about four-fifths in Ontario and one-fifth in Québec. The **Abitibi River** (547 km) drains the lake into the Moose River, which flows north into James Bay.

Abraham, Plaines d' QC This famous landmark in the city of Québec was named after Abraham Martin (1589-1664), an early pioneer, who acquired the property in 1635 and 1645. It was the site of the historic battle in September 1759, between the British forces led by Maj.-Gen. James Wolfe and the French forces of the Marquis de Montcalm. Known at that time as *Hauteurs d'Abraham* and *Heights of Abraham*, it was called *Plains of Abraham* in 1759 by British Capt. John Knox. The French form was introduced by Surveyor General Joseph Bouchette in 1815. The area is also known as Battlefields National Park and Parc des Champs-de-Bataille.

Abrams Village PE West of Summerside, this municipal community (1974) was named after Abraham Arsenault, the first settler in the area. Noted as early as 1829 as *Abraham's Village*, Abrams Village's post office was open from 1880 to 1918.

Abruzzi, Mount BC In the Rocky Mountains, north of Sparwood, this mountain (3267 m) was named in 1920 after Luigi Amadeo di Savoia-Aosta, Duke of the Abruzzi (1873-1933), who was the first to climb Mount St. Elias (on the British Columbia-Alaska border) in 1899.

Acadieville NB A community west of Richibucto, its post office was named *Acadie* in 1879 after **Acadieville Parish** and was changed to Acadieville in 1955. The parish was named in 1876 after Acadian settlers had been given free land grants, and for the early French name for that area of the Maritime provinces. Acadie and Acadia have their roots in *Archadia*, a name given to the area of the present-day Atlantic seaboard from Virginia to Nova Scotia in 1524 by explorer Giovanni da Verrazzano, who was exploring on behalf of the King of France.

Acamac NB A neighbourhood in the city of Saint John and on the west side of the Saint John River, its name was proposed by naturalist William F. Ganong in 1902 to replace the Canadian Pacific Railway's *Stevens* station. Ganong believed that the name in Maliseet meant 'south bay', for the nearby bay, but later determined it meant either 'white ash ridge', or 'opposite side of the water'.

Acme AB A village (1910) northeast of Calgary, its post office was called *Tapscott* from 1905 to 1909, when the Canadian Pacific Railway had built as far north as that point, choosing a name derived from Greek, and meaning 'highest point'.

Actinolite ON In Hastings County, north of Belleville, this place was known as *Troy* and *Bridgewater* until 1895, when it was named for the talc-like substance extracted from open-pit mines.

Acton ON Settled about 1820, and named *Danville* and later *Adamsville* for early settlers, Acton was named in 1846 by Robert Swan after his former home town of Acton, on the west side of Greater London, England. It was united in 1974 with the town of Georgetown and most of the township of Esquesing to form the town of Halton Hills in the regional municipality of Halton.

Acton QC A regional county south of Drummondville, it was formed in 1982 from the county of Bagot and part of the county of Shefford. Its main centre is the town of **Acton Vale**, which was established in 1861 and named after the township of Acton. The township, proclaimed in 1806, was named in 1795 after the town of Acton in England's Greater London, west of the centre of the city of London.

Adamant Mountain BC In the Selkirk Mountains, north of Revelstoke, this mountain (3365 m) anchors the northwest corner of the **Adamant Range**. The mountain was named in 1909 by climber Howard Palmer after its precipitous south face.

Adams Lake BC Located northeast of Kamloops, this lake is drained south by the **Adams River** into the west end of Shuswap Lake. It took its name from Chief Sel-howt-ken of the Shuswap, who had been baptized Adam by Fr John Nobili in 1849. It was called *Sl-hes-tal-len* by the Shuswap and identified on an 1827 map as *choo-choo-ach*.

Adamsville QC In the town of Bromont and the regional county of La Haute-Yamaska, it was named after miller George Adams (1813-85), a native of Vermont, who settled here in 1847 and became the postmaster 20 years later.

Adelaide ON A township west of Strathroy in Middlesex County, Adelaide was named in 1830 for Amelia Adelaide Louisa Theresa Carolina (1792-1849), wife of William IV, who became King that year.

Adelaide Peninsula NT Separating Queen Maud Gulf from Rasmussen Basin, this peninsula was named in 1839 by Hudson's Bay Company explorer Thomas Simpson after Queen Adelaide, the wife of William IV.

Adjala ON Part of the municipal township of Adjala-Tosorontio in Simcoe County, Adjala has a name of uncertain origin. Although it is claimed to be named after a daughter of Shawnee chief Tecumseh, he is not known to have had a daughter. With nearby townships named after noted Aboriginal chiefs, Adjala may have been named after an Ojibwa family group called *Ad-je-jawk,* noted by traveller John Tanner in 1830.

Admaston ON This township in Renfrew County was named in 1843 after a village in Staffordshire, near Blithfield Hall and Bagot's Bromley, the home of Sir Charles Bagot (1781-1843), governor general of Canada, 1841-3. The village is some 38 km north of Birmingham. An 1836 map identifies the township as *Kanmore.*

Admirals Beach NF This municipal community (1968) in St. Mary's Bay, southwest of St. John's, received its name from the practice in the late 1700s and throughout the 1800s of calling the first captain of a vessel arriving in a harbour 'the admiral'.

Admiralty Inlet NT At the north end of Baffin Island, and separating Brodeur and Borden peninsulas, this 280-km inlet was named in 1820 by Lt William (later Sir William) E. Parry after the board of the British Admiralty.

Adolphustown ON A township in Lennox and Addington County, Adolphustown was named in 1784 for Prince Adolphus Frederick (1774-1850), seventh son of George III. **Adolphustown** post office was opened in 1822.

Advocate Harbour NS On the north side of Minas Channel, this place is located west of Parrsboro. As there is a reference to it on Surveyor General Charles Morris's map of 1767, it predates Loyalist settlement. It is reported a lawyer was shipwrecked there and drowned. During the French regime, it was identified on maps as *Havre à l'Avocat.*

Afton PE A municipal community (1974), it embraces the localities of Cumberland, Fairview, New Dominion, Nine Mile Creek, Rice Point, and Rocky Point. The name was chosen in 1974, probably after the Robbie Burns poem, *Flow Gently, Sweet Afton.*

Agassiz BC A community in the district municipality of Kent, northeast of Chilliwack, its post office was named in 1888 after Capt. Lewis A. Agassiz, who settled here in 1862. After serving in the Royal Welsh Engineers, he had arrived in Victoria in 1858, and joined the gold rush in the Cariboo.

Agassiz Ice Cap NT In central Ellesmere Island, this ice cap was named *The Chinese Wall* in 1883 by Lt James B. Lockwood, who explored the northwest coast of Kalaallit Nunaat (Greenland) the previous year. Adolphus Greely, who organized expeditions, 1881-4, named it *Mer de Glace Agassiz* after Swiss-American naturalist Jean Louis Rodolphe Agassiz. It was subsequently called *Agassiz Glacier* before 1910, and amended to Agassiz Ice Cap in 1967.

Agawa Bay ON Located on the east side of Lake Superior, 95 km north of Sault Ste. Marie, this bay is the site of famous pictographs on **Agawa Rock**, rediscovered in 1851 by American ethnologist Henry Rowe Schoolcraft. **Agawa Canyon** is a popular tourist destination for the changing of the leaf colours in autumn. In Ojibwa *agawa* may mean 'sheltered harbour'.

Agincourt ON Situated near the centre of the city of Scarborough, this place was named in 1858 by merchant John Hill in honour of the British defeat of the French in 1415 at Agincourt (now Azincourt) in France, south of Calais. Hill had petitioned a childhood friend, Joseph-Élie Thibaudeau, Liberal member of the Assembly of the Province of Canada, to intervene on his behalf for a post office. Thibaudeau was surprised with Hill's choice, but thinking it a pretty name, endorsed it.

Aguanish QC A municipality (1957) in the regional county of Minganie and on the north shore of the Gulf of St. Lawrence, it was settled in the mid-1800s. It was named after the Rivière Aguanus, whose name in Montagnais is believed to mean 'little shelter'.

Ailsa Craig ON In Middlesex County, northwest of London, this village (1874) was named in 1858 after David Craig, and after the rocky island of Ailsa Craig at the entrance to the Firth of Clyde in Scotland.

Ainslie, Lake NS Southeast of Inverness, this lake was named about 1820 after George Robert Ainslie (1776-1839), the last lieutenant-governor of Cape Breton Island, 1816-20.

Airdrie AB Directly north of Calgary, this city (1985) was founded by the Canadian Pacific Railway in 1889, and named by engineer William MacKenzie after Airdrie, a town in Lanarkshire, Scotland, east of Glasgow.

Air Force Island NT In Foxe Basin, and west of Baffin, this island (1720 km^2) was discovered in 1948 by a Royal Canadian Air Force aerial survey. It was officially named the following year. It was one of the last major islands found in the Canadian North.

Ajax ON The site of an armed forces munitions factory during World War II, this town (1955) was named in 1941 after the British cruiser HMS *Ajax*. In December 1939 the cruiser forced the German battleship *Admiral Graf Spee* to scuttle itself in the mouth of the Rio de la Plata, off Montevideo, Uruguay.

Akimiski Island NT On the west side of James Bay, and opposite the mouth of Attawapiskat River, this island (3001 km^2) was named *Viner's Island* in 1674 by Hudson's Bay Company governor Charles Bayley after Sir Robert Viner (Vyner), a charter member of the company, and a London financier. It was identified on a 1901 geological map as Akimiski Island, after the Cree word for 'land across', a name officially endorsed in 1948.

Aklavik NT Located on the west bank of Peel Channel and in the Mackenzie Delta, this hamlet (1974) was named *Aklavuk* in 1910 by Col G.L. Jennings, of the Royal Canadian Mounted Police, after the Inuvialuktun word for 'place where they hunt brown bears'. The present spelling was authorized in 1921. **Aklavik Channel** unites the East and West channels of the Mackenzie River.

Akulivik QC A northern village (1979) north of Povungnituk and on the shore of Hudson Bay, it was founded in 1922. Its name in Inuktitut means either 'middle point of a trident', or 'point between two bays'.

Akwesasne ON, QC An Indian reserve, opposite Cornwall, it embraces lands not only in Ontario and Québec, but also extends into New York State. The name in Mohawk means 'place where the partridge beats its wings'.

Alameda SK A town (1907) east of Estevan, its post office was named in 1883 by Christian Troyer after the city of Alameda, CA, on San Francisco Bay, adjacent to Oakland. He had gone there in the 1850s, and moved back to Canada in 1882. In Spanish, *alameda* describes the poplar or cottonwood tree.

Alaska Highway BC, YT Beginning in Dawson Creek, BC, this highway extends for 2451 km to Fairbanks, Alaska, with 1965 km in Canada. It was built in 1942-3 to provide a better connection with the Soviet Union during the Second World War. First called the Alaska Military Highway, it became known as the *Alcan Highway* among the engineers and workers who built it. United States Congressman Anthony J. Dimond of Alaska proposed the

Alaska Highway, which was adopted by both the Canadian and American governments on 19 July 1943.

Albanel QC A municipality (1990) northwest of Lac Saint-Jean and in the regional county of Maria-Chapdelaine, it was named after the township of Albanel, created in 1883 in honour of the Jesuit missionary Charles Albanel (1616-96).

Albany PE Northeast of Borden, this place's post office was named in 1867. Said to have been named after Prince Leopold, Duke of Albany (1853-84), Queen Victoria's youngest son, although he did not succeed to the dukedom until 1881. The title was previously held by Prince Frederick (1763-1827), second son of George III.

Albany River ON Named in 1683 by the Hudson's Bay Company for James, Duke of York and Albany (1633-1701), who became King James II in 1685, the Albany is the largest river (982 km to the head of Cat River) entirely within the province.

Albemarle ON A township in Bruce County, it was named in 1855 by William Coutts Keppel, Viscount Bury, superintendent of Indian affairs in Canada, after his father George Thomas Keppel, 6th Earl of Albemarle, a military officer and diplomat.

Albert NB A county in the southeastern part of the province, it was named in 1845 after Prince Albert (1819-61), who had married Queen Victoria in 1840. *See also* Riverside-Albert.

Alberta In 1882 the Marquess of Lorne named the provisional district of Alberta after his wife, Princess Louise Caroline Alberta (1848-1939), daughter of Queen Victoria. The district embraced that part of the present province (1905) south of Grande Prairie and west of Medicine Hat. The province was established in 1905. **Mount Alberta** (3619 m), on the continental divide northwest of Calgary, was named in 1898 by British climber Norman Collie. The summer village (1920) of **Alberta Beach**, on the east side of Lac Ste. Anne, west of Edmonton, was named by the Canadian Northern Railway, with its post office opening in 1917.

Alberta, Mount YT In the Centennial Range of the St. Elias Mountains, this mountain (3348 m) was named in 1967 after the province of Alberta, and in honour of the centennial of Confederation.

Albert, Mont QC Rising to 1151 m in the Monts Chic-Chocs and west of Gaspé, it was named in 1845 by geologist Alexander Murray after Prince Albert, the consort of Queen Victoria.

Alberton PE A town (1913), in the northwest part of the province, its post office was named in 1868 by Benjamin Rogers, after Albert Edward, the Prince of Wales, who visited Charlottetown in 1860. Its post office had been called *Cascumpec* in 1839 after the adjacent Cascumpec Bay.

Albion ON Named in 1819 after an ancient poetical name of England, in allusion to the cliffs of Dover, this township was united in 1974 with Caledon Township and half of Chinguacousy Township to become part of the town of Caledon in the regional municipality of Peel.

Albreda, Mount BC In the Monashee Mountains, southeast of Valemount, this mountain (3075 m) was named in 1863 by William Wentworth Fitzwilliam, Viscount Milton after his aunt, Lady Albreda Elizabeth Wentworth Fitzwilliam. The viscount accompanied Dr Walter Cheadle during an expedition through the West, 1862-3.

Aldborough ON A township in Elgin County, it was named in 1792 after the town of Aldeburgh, on the North Sea coast of Suffolk, England.

Aldergrove BC In the west side of the city of Abbotsford, this place's post office was named in 1885 after the local abundance of second-growth alder trees.

Aldershot NS This name was first used after 1867 for a militia-drilling site at Aylesford, taking its name from the military camp southwest of London, England, founded in 1854. In 1904 it was relocated to the north side of Kentville, and developed into a suburban residential area.

Aldershot ON An urban centre in the city of Burlington, this place was named *Aldershott* in 1856 after Aldershot in Hampshire, England, the site of a noted military camp. The spelling was corrected to Aldershot about 1886.

Alderville ON An Aboriginal community in Alnwick Township, Northumberland County, northeast of Cobourg, this place was first named *Alnwick* in 1841. In 1860 it was renamed after the Revd Robert Alder (1796-1873), a noted English preacher, who visited the Mississauga on the Alderville Reserve about 1839.

Aldrich, Cape NT On Ellesmere Island, east of Cape Columbia, this cape was named in 1875-6 by British explorer George (later Sir George) Nares for his first lieutenant, Pelham Aldrich. He established that the area of the two capes was the most northerly land of the Arctic Archipelago.

Alert NT The name of this weather station and military base, near the most northerly point of Canada, was proposed in 1948 by W/Cdr G.H. Newcome to the interdepartmental meteorological committee. Established the following year, the station took its name from HMS *Alert*, the flagship of the proposed British Admiralty expedition to the North Pole in 1875-6. Led by George (later Sir George) S. Nares, the expedition was the first to pass by the site of Alert.

Alert Bay BC A village (1946) on Cormorant Island, on the north side of Vancouver Island, where Johnstone Strait joins Queen Charlotte Strait, it took its post office name from the bay in 1885. The bay was named in 1860 by Capt. George H. Richards after HMS *Alert*, which served on the Pacific station, 1858-61 and 1865-8.

Alexander MB West of Brandon, this place's post office was named *Pultney* in 1882. It was renamed *Alexander Station* in 1885, probably for pioneer Alexander Speers, and was shortened to Alexander in 1892.

Alexander Island NT In the Queen Elizabeth Islands and between Bathurst and Melville islands, this island (484 km^2) was named in 1962 after Harold Rupert Leofric George, 1st Earl Alexander of Tunis (1891-1969), governor general of Canada, 1946-52.

Alexandra PE Located southeast of Charlottetown, this municipal community (1972) was named about 1870 after Princess Alexandra Caroline Maria Charlotte Louise Julia (1844-1925), who married the Prince of Wales, the future Edward VII, in 1863. Its post office was open from 1884 to 1915.

Alexandra Falls NT Located 42 km upriver from the mouth of Hay River, this feature (32 m) was named in 1872 by Anglican bishop William C. Bompas after the Princess of Wales, who became Queen Alexandra in 1901, when Edward VII succeeded Queen Victoria.

Alexandra, Mount AB, BC This summit (3338 m) on the continental divide, northeast of Calgary, was named in 1902 by British climber Norman Collie after Queen Alexandra, consort of Edward VII.

Alexandria ON This town (1903) was called *Priest's Mill* in the early 1800s after a grist mill built by the Most Revd Alexander McDonell (1762-1840), the first Catholic bishop of Upper Canada. He resettled many disbanded soldiers of the Scottish Glengarry Regiment in Glengarry County. In 1825 the name was changed to Alexandria in his honour.

Alfred ON The township in the united counties of Prescott and Russell was named in 1798 for Prince Alfred (1780-82), a son of George III. The village (1952) was first known as *Bradyville* for hotelkeeper Thomas Brady, but became **Alfred** on the opening of the post office in 1842.

Algoma ON Created as a district in 1859 in Northern Ontario, its name may have been devised by American ethnologist Henry Rowe Schoolcraft from Ojibwa *ah-ga-mic*, 'the other side of the water', describing the north shore of Lake Huron in relation to Manitoulin Island.

Algonquin ON In the united counties of Leeds and Grenville, north of Brockville, this place was named by postmaster Silas Wright in 1861 after the Algonquin nation. The nation's name was given by the Mi'kmaq, interpreted as 'at the place of spearing fish from the bow of a canoe'.

Algonquin Provincial Park ON This park was created in 1893, and now covers an area of of 7536 km^2. It was named after the Algonquin nation, which has occupied the Ottawa Valley since at least the early 1600s. The Mi'kmaq named the nation, interpreted as 'at the place of spearing fish from the bow of a canoe'.

Alice ON A township in Renfrew County, it was named in 1855 for Princess Alice Maud Mary (1843-78), the second daughter of Queen Victoria.

Alida SK A terminus of a Canadian Pacific Railway line, northeast of Estevan, the post office of this village (1926) was opened in 1913. It was named by CPR president Lord Thomas Shaughnessy after Lady Alida, wife of Sir Harry Brittain, an English author who participated in Imperial conferences.

Alix AB Northeast of Red Deer, the post office in this village (1907) was established in 1905, and named after Alexia ('Alix') Westhead, the wife of a pioneer rancher.

Allan SK Founded as a Canadian Northern Railway station southeast of Saskatoon, both the post office and the station of this town (1965) were established in 1908, and named after a Grand Trunk Pacific Railway foreman. The name also reflects the practice of the railway of identifying its stations in alphabetical order, with Zelma to the southeast and Bradwell to the northwest. *Curzon* post office served the area from 1904 to 1909.

Allardville NB A community southeast of Bathurst, it was named in 1934 after Mgr Jean-Joseph-Auguste Allard (1884-1971), who led a group of unemployed Bathurst families to undeveloped land on the road leading to the Miramichi River. **Allardville Parish** was named in 1946.

Allenford ON A community in Bruce County, west of Owen Sound, Allenford was named in 1868 for James Allen (1826-95), who had settled at a ford of Sauble River 11 years earlier.

Allen, Mount AB, BC Near Lake Louise, this mountain (3310 m) was named in 1924 after American alpinist Samuel Evans Stokes Allen (1874-1945). Allen had named it *Mount Shappee* after the Stoney word for 'sixth', as it was the sixth of the Ten Peaks (now Wenkchemna Peaks).

Alliance AB Southeast of Camrose, this village (1918) was named in 1916 by Canadian Northern Railway land buyer Tom Edwards after his hometown of Alliance, OH, southeast of Cleveland. Its post office had been called *Galahad* in 1907, but this was transferred in 1916 to the next village to the northwest.

Alliston ON An urban community in the town of New Tecumseth, Simcoe County, Alliston was founded in 1847 by William Fletcher, naming it after his birthplace in Yorkshire, likely Allerston, west of the east coast city of Scarborough, England.

Allumettes, Île aux QC In the Ottawa River, opposite Pembroke, it was called *Isle des Algoumequins* by Samuel de Champlain in 1613. In the second half of the seventeenth century, explorer Nicolas Perrot referred to it as *Isle du Borgne autrement ditte l'Isle des Allumettes.* Le Borgne was a one-eyed Aboriginal leader on the island. The name is attributed to reeds that reminded travellers of matches.

Allumette Lake ON, QC A 30-km-long widening of the Ottawa River, Allumette Lake extends from Chalk River to **Allumette**

Rapids at Morrison Island, below Pembroke. **Lower Allumette Lake** extends for 20 km from there to the rapids below Westmeath. The first explorers found reeds there suitable for the making of *allumettes*, 'matches'.

Alma NB A village (1966), 45 km southwest of Moncton, its post office was called *Salmon River* in 1848 and was renamed after **Alma Parish** in 1873. The parish had been named in 1855 after the victory in 1854 of British and French forces on the heights of the Alma River in the Crimea.

Alma NS West of Stellarton, this place's post office was named in 1855 after the British-French victory over the Russians a year earlier, on the heights of the Alma River in the Crimea. Earlier it was called *Middle River*, after the Middle River of Pictou.

Alma ON This place in Wellington County was named in 1855 after the battle the previous year between the British/French allies and the Russian forces on the heights of the Alma River in the Crimea.

Alma QC A city (1924) on the Saguenay River and in the regional county of Saint-Jean-Est, it was called after **Île d'Alma**, named about 1861 by surveyor Edmond-A. Duberger after the 1854 Crimean War battle. Created first as the municipality of *Saint-Joseph-d'Alma* in 1879, the city's name was shortened to Alma in 1954.

Almonte ON A town (1880) in Lanark County, Almonte replaced *Ramsay* and *Waterford* in 1856. It was named for Juan Nepomucene Almonte (1804-69), Mexican ambassador to the United States. His name was presumably chosen because he strongly opposed the American government's aggressive trading practices, which were affecting Mexico as well as the British provinces to the north. Before 1856 it was *Shepherd's Falls*, for David Shepherd, *Shipman's Mills*, for Daniel Shipman, and *Ramsayville*, after Ramsay Township.

Alnwick ON A township in Northumberland County, Alnwick was named in 1798 after a town in Northumberland, England, north of Newcastle upon Tyne.

Alsask SK On the Alberta border, west of Kindersley, the station in this village (1947) was named in 1909 by the Canadian Northern Railway by joining the first syllables of Alberta and Saskatchewan. Its post office was opened two years later.

Alsek River YT Rising in the Kluane Ranges of the St. Elias Mountains, this river (386 km) flows south across the Alaska border into Dry Bay, part of Yakutut Bay, an inlet of the Pacific Ocean. Although noted as *R Alsehk* as early as 1825 by Russian Capt. Tebenkov, other names (*Behring, Jones, Harrison*) were proposed before Alsek River was officially recognized in 1891 by the Canadian and American governments.

Alton ON In the town of Caledon, south of Orangeville, Alton was named in 1855 by its first postmaster, John Meek, after Alton, IL, a city on the Mississippi, 30 km north of St. Louis. Meek, who was American-born, had a copy of a newspaper from Alton, IL.

Altona MB A town (1956) northwest of Emerson, its Canadian Pacific Railway station was named in 1894 after a suburb of Hamburg, Germany. Its post office was opened the following year.

Altruist Mountain BC In the Coast Mountains, northwest of Pemberton, this mountain (3124 m) was named in the 1950s by Alpine Club of Canada climbers. Because they had been lured to ascend peaks that did not lead to the top of the mountain, this 'delightfully cynical name' was proposed.

Alverstone, Mount YT In the St. Elias Mountains and on the Yukon-Alaska boundary, this mountain (4439 m) was named in 1908 by the Canadian government after Lord Richard Everard Webster Alverstone (1842-1915). A British commissioner in 1903 on the northwestern Canada-United States boundary dispute, he voted in favour of the American claims.

Alvinston ON Located in Lambton County, southeast of Sarnia, this village (1880) was named in 1854 after Alverstone on England's Isle of Wight.

Amabel ON A township in Bruce County, it was named in 1856 by William Coutts Keppel, Viscount Bury, superintendent of Indian affairs of Canada, after Henrietta Amabel Yorke, sister-in-law of Sir Edmund Head, who was then governor general of Canada.

Amadjuak Lake NT On Baffin Island and northwest of Iqaluit, this lake (3058 km^2) was named *Amaxdjuak* in 1888 by anthropologist Franz Boas. The present spelling was used in 1897 by geologist Robert Bell. The name means 'great water' in Inuktitut.

Amaranth MB Located just west of Lake Manitoba and northeast of Neepawa, this place's post office was named by Robert Johnson in 1911 after his native Amaranth Township, Dufferin County, Ontario, northwest of Orangeville.

Amaranth ON A township in Dufferin County, Amaranth was named in 1821 after the never-fading flower of the poets, as well as the common plant called pigweed.

Ameliasburgh ON A township in Prince Edward County, Ameliasburgh was named in 1787 for Princess Amelia (1783-1810), fifth daughter and youngest child of George III. The community of **Ameliasburg** (without the terminal 'h'), in the township south of Belleville, was named in 1832. The two forms of the name continue to be used.

Amery, Mount AB This mountain (3329 m) was named by order in council in 1927 after Leopold Charles Morris Stenet Amery (1873-1955), who was then secretary for the dominions, 1925-9.

Amet Sound NS On the south side of Northumberland Strait, this name was derived from a former French designation of **Amet Island**, *Isle Darmet,* from *ormet*, 'helmet island'. Pierre Jumeau in 1685 referred to it as *Isle porképic*, other map-makers showed it in the 1750s as *Isle Remedy*, and in 1775 chart-maker J.F.W. DesBarres called it *Frederick Bay*.

Amherst NS A town (1889), near the New Brunswick border, its first post office was named *Cumberland* in 1810 after the county. It was renamed Amherst in 1841 after Amherst Township. The township was given its name after 1759 for Jeffery Amherst, 1st Baron Amherst (1717-97), who, with Adm. Edward Boscawen, had captured the French fortress of Louisbourg in 1758 and Montréal in 1760. He was the commander-in-chief of the British forces in North America, 1760-3.

Amherstburg ON A town (1878) in Essex County, south of Windsor, Amherstburg was the site in the late 1790s of *Fort Amherstburg* (later called Fort Malden). The place and fort were named for Lord Jeffery Amherst, commander-in-chief of the British forces in North America, 1760-3.

Amherst Island ON The island and township in Lennox and Addington County were both named in 1792 for Lord Jeffery Amherst, who led the British troops in the capture of Louisbourg in 1758 and of Montréal in 1760.

Amherstview ON A community in Lennox and Addington County, west of Kingston, Amherstview was named in 1965 for the view of Amherst Island, almost 1 km offshore.

Amiraults Hill NS East of Yarmouth, this place was founded in the early 1800s by Jacques Amirault, a native of the area of Pubnico. *Amirault Hill* post office served the community from 1904 to 1970, after it had been called *Amiros Hill* in 1898.

Amisk AB The post office in this village (1956), southwest of Wainwright, was named in 1908, just before the Canadian Pacific Railway arrived. The name was derived from the Cree word for 'beaver'.

Amos QC A town in the regional county of Abitibi, its name originated in 1914 after Alice

Amos, the wife of Sir Lomer Gouin, who was then premier of Québec.

Amqui QC A town (1961) in the valley of the Rivière Matapédia, and in the regional county of La Matapédia, it replaced *Saint-Benôit-Joseph-Labre* in 1948. Amqui post office was opened in 1879, taking its name from the Rivière Humqui, derived from a Mi'kmaq word possibly meaning 'place where it plays', perhaps in reference to the turbulent waters of the rapids here. Or it may refer to joyous gatherings here by the Mi'kmaq.

Amund Ringnes Island NT One of the Sverdrup Islands and in the central part of the Queen Elizabeth Islands, this island (5255 km^2) was named *Amund Ringnes's Land* in 1900 by Otto Sverdrup, during his 1898-1902 Norwegian Arctic expedition, after Amund Ringnes, a Norwegian patron.

Amundsen Gulf NT Separating Victoria and Banks islands from the mainland, this extension of the Beaufort Sea was named before 1910 by the British Admiralty after Norwegian explorer Roald Amundsen. In 1906 he became the first person to sail a ship through the famous Northwest Passage from the Atlantic to the Pacific.

Anahim Lake BC In the Chilcotin country, between Williams Lake and Bella Coola, this lake was named after Chilcotin chief Anahim. In 1861 explorers Ranald Macdonald and John Barnston passed by *Lake Anawhim*. *Anaham Lake* post office was opened at the south end of the lake in 1907 and was changed to **Anahim Lake** in 1939.

Anarchist Mountain BC East of Osoyoos, this mountain (1499 m) is adjacent to the 900-m climb of steep hairpin turns on Highway 3 from the Okanagan Valley to the Okanagan Highlands. It was named after Richard G. Sidley, who arrived in the district in 1889, and was appointed a justice of the peace and customs officer at Sidley, on the Washington State border, southeast of the mountain. His extreme anarchistic political views resulted in his dismissal.

Ancaster ON A town (1974) in Hamilton-Wentworth Region, Ancaster was created as a township in 1792 in Lincoln County, taking its name from Ancaster in England's Lincolnshire, south of the city of Lincoln. It had been earlier called *Wilson's Mills*, after miller James Wilson, and *Rousseaux's Mills*, after noted interpreter and trader Jean-Baptiste Rousseau.

Anchor Point NF On the Strait of Belle Isle, west of St. Anthony, this municipal community (1974) may have been named after an adjacent safe anchorage for travellers crossing from the southern Labrador coast.

Anderdon ON A township in Essex County, Anderdon was created in 1837. It may have named after an Aboriginal clan called *Arendharenon*, 'rock people', probably a euphemism for 'snakes', or after an officer stationed at Fort Malden.

Anderson River NT Rising north of Great Bear Lake, this river (692 km) flows northwest into Wood Bay of the Beaufort Sea. It was named in 1857 by Hudson's Bay Company explorer Roderick MacFarlane after HBC chief factor James Anderson. Anderson, who was in charge of the HBC's Mackenzie district, referred to it two years earlier as *Inconnue or Beghulatesse*.

Andrew AB Northeast of Edmonton, this village (1930) was named in 1902, when the post office was opened. It was named after pioneer settler Andrew Whitford. Whitford, the next Canadian Pacific station to the east, and nearby Whitford Lake, were also named after him.

Andromeda, Mount AB This mountain (3450 m), near the head of Athabasca River, was named in 1938 by alpinist Rex Gibson after the mythological wife of Perseus.

Angikuni Lake NT A widening of the Kazan River, and southeast of Dubawnt Lake, this lake (437 km^2) was named before 1910 after the Chipewyan word for 'big'.

Anguille, Cape NF The most westerly point of the island of Newfoundland, its name in

French means 'eel'. The **Anguille Mountains**, rising to 526 m, extend for 50 km along the coast of St. George's Bay.

Angus ON In Simcoe County, west of Barrie, it was named in 1857 by the Canadian Northern Railway for Angus Morrison (1822-82), a director of the company, a member of the Legislative Assembly, 1854-67, and of the House of Commons, 1867-74.

Angusville MB Northwest of Brandon, near Riding Mountain National Park, this community's post office was called *Snake Creek* in 1909, but was renamed in 1911 after pioneer settlers John, Frank, William, Sam, and Adam Angus.

Anjou QC In the Communauté urbaine de Montréal, northeast of the Montréal city centre, this city was created in 1956 from the parish of Saint-Léonard-de-Port-Maurice. It was named after the ancient province in west central France. The city is often informally called Ville d'Anjou.

Anmore BC A village (1987) north of Port Moody, its post office was named in 1951 after resident Ann Moore.

Annaheim SK Located northeast of Humboldt, the post office in this village (1977) was named in 1904 after St. Ann's Parish, established 26 July 1903, the feast day of the mother of the Virgin Mary. The name literally means 'home of Ann'.

Annan ON In Grey County, northeast of Owen Sound, this place was named in 1874 after Annan, Dumfriesshire, Scotland.

Annandale-Little Pond-Howe Bay PE This municipal community (1975) embraces the localities of Annandale, Howe Bay, and Little Pond. Annandale was named about 1868 by James Johnston, whose parents came from Annan, Dumfriesshire, Scotland in 1840. About 1880 the school district of Howe Bay was called after the adjacent bay, named in 1765 by Samuel Holland for Sir William Howe (1729-1814), commander-in-chief of British forces in North America in 1755. The post office at Howe Bay was called Sailors Hope from 1884 to 1913. Little Pond had a post office from 1886 to 1913.

Annapolis Royal NS A town (1893) on the **Annapolis River**, it is located at the site where Samuel de Champlain established *Port-Royal* in 1604. After its capture in 1710 by Col Francis Nicholson, it was renamed after Queen Anne. *Annapolis* post office was opened in 1785 and the word 'Royal' was added to it in 1903 after it had long been called that. The Annapolis River rises south of Berwick, and flows southwest into the **Annapolis Basin**. The **Annapolis Valley** embraces the valleys of the Annapolis and Cornwallis rivers, from Digby in the southwest to Grand Pré in the northeast.

Annieopsquotch Mountains NF In the central part of the island of Newfoundland, these mountains, between Victoria Lake on the east, and Red Indian Lake on its west, rise to 687 m. In Mi'kmaq the name means 'rocky mountains'.

Anticosti, Île d' QC Quebec's largest island, it lies in the estuary of the St. Lawrence River. Named *isle de l'Assumption* by Jacques Cartier in 1535, it was called Anticosti by historian Marc Lescarbot in 1609. The Mi'kmaq at that time called it *Natigôsteg*, 'forward land', whereas the Montagnais on the north shore still know it as *Natashquan*, 'where bears are hunted'.

Antigonish NS In **Antigonish County**, and at the head of **Antigonish Harbour**, the area of this town (1889) was settled in 1784 by Col Timothy Hierlihy and officers and men of the Nova Scotia Volunteers. Its post office was called Antigonish in 1816. The name, noted in 1672 by Acadian governor Nicolas Denys as *Articougnesche*, likely means 'flowing through broken marsh', from Mi'kmaq *nalegitkoonech*, while the town's site had another Mi'kmaq name that meant 'where the bears tear the branches off trees'. What remained of *Sydney County*, established in 1784, after Guysborough County was separated from it in 1836, became Antigonish County in 1863.

Apohaqui NB Located west of Sussex, this community was founded in 1858 by the European and North American Railway. The name was devised from the Maliseet word for 'junction of two streams'. The post offices *Mouth of Millstream* and *Studholm* served the area until 1868, when Apohaqui post office was opened.

Appin ON In Ekfrid Township, Middlesex County and southwest of London, this place was first called *Ekfrid Centre* in 1854. Seven years later Dr D.S. McKellar named it after his birthplace in the Appin district of Scotland, near Fort William, northwest of Glasgow.

Apple Hill ON In the united counties of Stormont, Dundas and Glengarry, north of Cornwall, this place was named in 1882 when the Canadian Pacific Railway was built through Sandy Kennedy's apple orchard.

Appleton ON In Lanark County, east of Carleton Place, this place was first called *Teskeyville* and *Appletree Falls*, after some apple trees at the foot of falls on the Mississippi River. It was named Appleton when the post office was opened in 1857.

Apsley ON In Peterborough County, northeast of Peterborough, Apsley was named in 1865, possibly after Apsley House, the Duke of Wellington's home on Hyde Park Corner in London, England.

Aquaforte NF This municipal community (1972) is located on the east coast of the Avalon Peninsula. Its name, derived from Portuguese *aguaforte*, 'freshwater', appeared as *R. da aguea* on a 1519 map, and as *Agoforte* in the early 1600s. The change to Aquaforte may have been done by Sir Robert Robinson in 1669.

Arborfield SK Northeast of Tisdale, the post office in this town (1950) was named in 1910 by the post office department after Arbor Day, when public tree planting is undertaken in the spring. *Fairfield* had been requested, but was rejected because it was already in use in the province.

Arborg MB A village (1964) northwest of Gimli and on the Icelandic River, its post office was called *Ardal*, Icelandic for 'river dale', in 1909. It was renamed in 1911 after the Icelandic for 'river town'.

Arcadia NS East of Yarmouth, this place's post office was first called *Upper Chebogue* in 1860 after its location on the Chebogue River. It was renamed in 1875 for the brig *Arcadia*, built here in 1817. The brig may have been named after the regional name Acadia, which originated from *Archadia*, given in the early 1500s by explorer Giovanni da Verrazzano to the present-day Atlantic seaboard from Virginia to Nova Scotia after the classical land of rustic pleasure.

Archerwill SK When the Canadian Pacific Railway built a line north from Wadena to Tisdale in 1924, a station was established midway between the two. William Pierce, secretary-treasurer of the rural municipality of Barrier Valley, met with councillors Archie Campbell and Ernie Hanson, and devised a name from the first syllables of their three given names. Its post office was opened in 1925, and it became an incorporated village in 1947.

Archibald, Mount YT In the St. Elias Mountains, west of Haines Junction, this mountain (2560 m) was named in 1971 after Dr E.S. Archibald, director of the department of agriculture's experimental farm service.

Arcola SK Located northeast of Estevan, the post office in this town (1903) was named in 1889 by postmaster Peter McLellan after a place in Italy where Napoleon had defeated the Austrians in November 1797.

Arctic Bay NT Located on **Arctic Bay**, a small arm of Admiralty Inlet and on the northwestern side of Baffin Island, this hamlet (1976) was officially named in 1951. In 1926 it had been called *Tukik*.

Arctic Red River NT Rising in the Mackenzie Mountains, this river flows north to join the Mackenzie River at the charter community of Tsiigehtchic. The community was the site of the Hudson's Bay Company post of *Arctic Red River* about 1900. The river's name appeared in a 1906 report by Charles Camsell and was approved in 1945. *See also* Tsiigehtchic.

Arden MB Located east of Neepawa, this place's Manitoba and North Western Railway station name was suggested before 1883 by Frederick W. Stobart, a Winnipeg dry goods merchant and promoter of the railway. He named it after his father's birthplace in North Yorkshire, England, northeast of Thirsk. It had earlier been called *Beautiful Plains.* Its post office opened as *Arden Station* in 1884, and was changed to Arden in 1897.

Arden ON In Frontenac County, northwest of Kingston, this place was named in 1865 after Alfred Lord Tennyson's poem *Enoch Arden*, published the previous year.

Argenteuil QC A regional county west of Montréal, its name was derived from a seigneury granted in 1682 to Charles-Joseph d'Ailleboust Des Muceaux (*c.* 1623-1700), and transferred in 1697 to his son, Pierre d'Ailleboust d'Argenteuil (1659-1711). The name has its roots in the town of Argenteuil, a northwestern suburb of Paris.

Argentia NF On the east coast of Placentia Bay, this neighbourhood in the town of Placentia was first known as *Little Placentia.* In the early part of the twentieth century its present name was proposed by Fr St John, after the mining of silver (*argentum* in Latin) in an adjacent cliff. Although the mine failed, the name survived.

Argonaut Mountain BC Located in the Selkirk Mountains, north of Revelstoke, this mountain (2976 m) was named in 1913 by mountaineer Howard Palmer after the some 200 Overlanders who trekked from Eastern Canada in 1862. They had also acquired the nickname Argonauts of '62. In 1912 Palmer had proposed *Dentiform Mountain*, and the next year he suggested *Big Tooth* before the present name was adopted by the Geographic Board.

Argyle NS A municipal district in Yarmouth County, it was established in 1879. It was called after the settlement of Argyle, named in 1766 by Capt. Ranald MacKinnon, who received a grant there a year earlier. He named it after Argyllshire, Scotland. Argyle's post office was opened in 1852 and a string of communities around **Argyle Sound** that received post offices included **Lower Argyle** (1865), **Central Argyle** (1878), **Argyle Sound** (1879), **Argyle Head** (1884), and **Argyle South** (1913).

Arichat NS Located on the south side of Isle Madame, this place's name was derived from Mi'kmaq *neliksaak*, possibly meaning 'camp ground' or 'split rocks'. Referred to as *Nerichat* and *Nerichac* on early maps, Arichat's post office was established in 1824 and West Arichat's was opened in 1867.

Arkell, Mount YT Southwest of Whitehorse, this mountain (2209 m) was named about 1926 after Arkell Creek. The creek was named in 1890 by E.J. Glave and Jack Dalton after W.J. Arkell, owner of the *Frank Leslie Illustrated News of New York.* Arkell organized the Glave-Dalton expedition to the rich goldfields on the Fortymile River in 1890 and 1891. When news of the fabulous wealth of the Yukon reached Arkell in 1897, he tried to lay claim to the whole of the Klondike area.

Arkona ON Surrounded by the town of Bosanquet, Lambton County, northeast of Sarnia, this village (1876) was first called *Eastman's Corners* and *Bosanquet.* In 1857 it was named Arkona after the German cape that extends into the Baltic Sea. Nial Eastman was among the community's first German settlers in the 1830s.

Armagh QC A village (1950) in the regional county of Bellechasse, east of the city of Québec, it was named after the township of Armagh, named in 1795, and proclaimed in 1799. The township's name was given after the city of Armagh, the principal centre of County Armagh, Ireland, and the primatial capital of all Ireland for both the Catholic Church and the Church of Ireland (Anglican).

Armstrong BC Near the north end of the Okanagan Valley, this city (1913) was named in 1891-2 by the Shuswap and Okanagan Railway (a subsidiary of the Canadian Pacific), after London, England financier William Charles Heaton-Armstrong (1853-1917), who visited the place in 1892.

Armstrong ON A divisional point on the transcontinental line of the Canadian National Railways, northwest of Lake Nipigon, Armstrong was established in the early 1900s. It was likely named after Hector Heaton-Armstrong, a railway bond financier in London, England.

Armstrong, Mount YT This mountain (2157 m) in the Russell Range, north of Macmillan River, was named in 1926 by member of Parliament George Black and gold commissioner Percy Reid after Neville Alexander Drummond Armstrong (1874-1954), who managed the Yukon Goldfields Company in the Klondike, 1898-1903. He remained in the Yukon searching for gold until 1928, except for service overseas with the Canadian Expeditionary Forces during the First World War.

Arnaud, Rivière QC A river (150 km) in the region of Kativik, it flows east into Ungava Bay. It was earlier named *Rivière Payne*, after its source in a lake named in the 1880s after meteorologist Frank F. Payne. It was renamed in 1968 after Oblate Fr Charles-André Arnaud (1826-1914), who had served in the 1870s at Fort Chimo (Kuujjuaq).

Arnold's Cove NF A town (1972) on Placentia Bay, and near the narrowest part of the Isthmus of Avalon, its name occurred in population returns of 1836. Its name may have been given after the forename of a fisherman.

Arnprior ON A town (1892) in Renfrew County, Arnprior was named in 1831 by brothers George and Andrew Buchanan after their birthplace in Stirlingshire, Scotland, north of Glasgow. Their cousin, Archibald McNab, opened McNab Township, and persuaded them to build mills at the mouth of the Madawaska River.

Aroostook NB A village (1966) on the west bank of the Saint John River and at the mouth of the **Aroostook River**, south of Grand Falls, its post office was first called *Arestook* in 1852, and renamed Aroostook two years later. The river, which rises in the state of Maine, derived its name from Maliseet *woolahstook*, 'good river for everything', the same description they used for the Saint John River above tide. The river's name took several forms from 1699, when it was spelled *Arassatuk*, to 1832, when map-maker Joseph Bouchette called it *Ristook or Aroostook River*. Surveyor General Thomas Baillie settled on the present form in the 1850s. In 1839 a boundary dispute called the Aroostook War broke out between New Brunswick and Maine in the river's valley, with each side calling out its troops. The route of the boundary was resolved three years later by the Webster-Ashburton Treaty.

Arrowsmith, Mount BC East of Port Alberni, on Vancouver Island, this mountain (1817 m) was named about 1853 after British map-maker Aaron Arrowsmith (1750-1823), and his nephew John Arrowsmith (1790-1873), who continued the Arrowsmith reputation for excellence in cartography.

Artemesia ON A township in Grey County, Artemesia was named in 1822 after the Greek word *artemisia*, a genus of plants that includes 'mugwort' and 'wormwood'. A Greek festival of the goddess Artemis (also Artemisia) involved feasting and amusements.

Arthabaska QC A regional county in the area of Victoriaville, southeast of the city of Québec, it was named in 1982 after the township of Arthabaska. The township was proclaimed in 1802, and was named after the Cree *ayabaskaw*, 'place of bulrushes and reeds'. The former town of **Arthabaska** has been part of the city of Victoriaville since 1994.

Arthur ON A township in Wellington County, it was named in 1835 for Arthur Wellesley, the Duke of Wellington (1769-1852). The village (1871) of **Arthur** was named in 1847.

Arthurette NB This place on the Tobique River, southeast of Grand Falls, was named in 1864 by Lt-Gov. Sir Arthur Hamilton-Gordon (1829-1912) after Arthuret, near Carlisle, England, where his friend Sir James Graham, a former first lord of the Admiralty, was buried three years earlier.

Arthur Meighen, Mount BC In the Premier Range, west of Valemount, this mountain (3150 m) was named in 1962 after Arthur Meighen (1874-1960), prime minister of Canada, 1920-1 and 1926.

Arthur Wheeler, Mount YT In the St. Elias Mountains, this mountain was named in 1970 by the Alpine Club of Canada after surveyor Arthur Oliver Wheeler (1860-1945), who was the founder and first president of the club.

Artillery Lake NT Northeast of Great Slave Lake, this lake (535 km^2) was named in 1834 by explorer George (later Sir George) Back after the Royal Artillery. Some of Back's crew were members of it.

Arva ON In Middlesex County, north of London, this place was named in 1852 after Arvagh, County Cavan, Ireland, southwest of the town of Cavan.

Arviat NT On the west coast of Hudson Bay and near the centre of Canada, this hamlet (1977) received its name in 1989, replacing *Eskimo Point*. The name means 'place of the bowhead whale' in Inuktitut. The hamlet was first settled in 1946 and *Eskimo Point* post office opened here two years later.

Arvida QC Now in the city of Jonquière, it was an incorporated town from 1926 to 1975. Its name was devised from the first two letters of each name of Arthur Vining Davis. He was then the president of Alcoa, the Aluminum Company of America, whose subsidiary, the Aluminum Company of Canada, built an aluminum plant here in 1925.

Asbestos QC A town (1939) in the regional county of **Asbestos**, north of Sherbrooke, its name first occurred in 1884, five years after prospector Evan William found asbestos here. The regional county had been known as *L'Or-Blanc* from 1982 to 1990, when it was changed to Asbestos.

Ascot QC In the regional county of Sherbrooke, east of the city of Sherbrooke, this municipality was named in 1989, deriving its name from the township of Ascot. The township was named in 1803 after a village in England's Berkshire, west of London. The municipality of **Ascot Corner** was named in 1901.

Ashburn ON In the town of Whitby, Durham Region, this place north of the town centre was named in 1852 after Ashbourne, Derbyshire, England, northwest of the city of Derby.

Ashcroft BC A village (1952) on the Thompson River, west of Kamloops, it was named in 1865 after the Ashcroft Ranch, established in 1862 by brothers Clement Francis and Henry Pennant Cornwall. The ranch was named after their family home in England. The Canadian Pacific Railway laid out the present site in 1880. Its post office was called *Ashcroft Station* from 1886 to 1899, when it was changed to Ashcroft.

Ashern MB A community east of Lake Manitoba, it was established by the Canadian Northern Railway in 1911, and named after the company's timekeeper, A.S. Hern.

Ashfield ON A township in Huron County, it was named in 1840 after Ashfield in Suffolk, England, the home of Lord Thomas John Thurlow, the son-in-law of Lord Elgin, governor general of Canada, 1847-54.

Ashton ON On the border of Lanark County and Ottawa-Carleton Region, Ashton was named by John Sumner in 1851 after Ashton-under-Lyne, an eastern suburb of Manchester, England.

Asperity Mountain BC In the Waddington Range of the Coast Mountains, this mountain

(3716 m) was named in 1928 by W.A.D. (Don) Munday. He chose the name because its sharp features established a mood of danger among climbers.

Asphodel ON A township in Peterborough County, Asphodel was named in 1821 after *asphodelus*, a Greek word for the lily family and the daffodil. In classical mythology, dead heroes wandered sadly in the asphodel fields.

Aspy Bay NS On the northeast side of Cape Breton Island, just south of Cape North, this name likely has its roots in Mi'kmaq *kespoogwit*, 'end place'. A Basque map of the early 1600s displays *Aispe* in the area. Pierre Jumeau's map of 1685 has *B. d'Aspé* for it, while earlier maps locate *C. Gaspa* at Cape North. There would appear to be a connection between Gaspé, in Québec, and the name of this bay.

Asquith SK West of Saskatoon, the station and post office in this town (1908) were named in 1907 by the the Canadian Pacific Railway after Herbert Henry Asquith, 1st Earl of Asquith (1852-1928), Liberal prime minister of Great Britain, 1908-16.

Assiginack ON A township in Manitoulin District, it was named in 1864 for Jean-Baptiste Assiginack (*c.* 1768-1866), a chief of the Ottawa nation, and the chief interpreter for the Manitoulin Island region in the early 1800s. His son, Francis Assikinack (1824-63) was also an interpreter and teacher.

Assiniboia SK Situated southwest of Moose Jaw, this town (1913) was named in 1912 by the Canadian Pacific Railway after the district of *Assiniboia*, which was the southern part of present Saskatchewan and a part of southern Alberta between 1882 and 1905. The district was named after an earlier territory of *Rupert's Land* from 1811 to 1869, itself named after the Siouan tribe called the Assiniboine. Louis Riel first proposed Assiniboia as the name of Canada's fifth province in 1870, but subsequently substituted Manitoba. The first post office in the area was called *Leeville* from 1908 to 1912.

Assiniboine River MB, SK The principal western tributary of the Red River, it rises north of Yorkton, SK, and flows southeast, then east through Brandon, Portage la Prairie, and Winnipeg. The name is derived from the Ojibwa description of the practice of the large Siouan tribe, first encountered in the area of the Lake of the Woods and Lake Winnipeg, of cooking their food by using heated stones. Gov. James Knight, of the Hudson's Bay Company post at York Factory, called them Stone Indians in 1715, and Pierre Gaultier de La Vérendrye referred to them as Assiniboils in 1730. In his 1738-9 journal, La Vérendrye called the river *rivière des Assibiboiles*, and its junction with the Red as *la fourche de Assiliboiles.*

Assiniboine, Mount AB, BC Southwest of Banff, this mountain (3618 m) was named in 1884 by geologist George Dawson after the Assiniboine, a variant name of the Stoney tribe of the Siouan nation.

Aston-Jonction QC A village (1951) in the regional county of Nicolet-Yamaska, northwest of Victoriaville, it was named after the township of Aston, created in 1806, and the crossing of two rail lines. Aston was derived from Ashton-under-Lyne, on the northeast side of Manchester, England.

Astorville ON In Nipissing District, southeast of North Bay, this place was called *Levesqueville* in 1895 after settler Joseph-Alphonse Lévesque. In 1904 Fr Joseph-Antonin Astor moved the post office next door to the priest's residence, and renamed it after himself.

Athabasca AB The Hudson's Bay Company post of *Athabasca Landing* was established north of Edmonton in 1877. Located on the **Athabasca River**, its post office was opened in 1901. Incorporated as the town of *Athabasca Landing* in 1911, it became Athabasca two years later. At the beginning of the nineteenth century both geographer David Thompson and fur trader Peter Fidler called the river Athabasca, likely adapted from *athabaska* 'where there are reeds', the Cree description of the mouth of the river on the southwestern side of **Lake Athabasca**. Peter Pond identified Lake Athabasca (7936 km^2, with about three-quarters in Saskatchewan) as *Great Araubaska* on a 1790 map. **Athabasca Pass** (1748 m) was discovered by geographer David Thompson in 1810. **Mount Athabasca** (3491 m) was named in 1898 by British climber Norman Collie. In 1882 the district of Athabasca was created as part of the North-West Territories, but in 1902 the Geographic Board of Canada respelled it, the town and all names with Athabasca as *Athabaska*. In 1948 the board restored the preferred local spelling.

Athalmer BC Adjacent to Invermere, and at the north end of Windermere Lake in the upper Columbia River valley, it was named in 1899 by engineer Frederick W. Aylmer. He chose the roots of his surname, *athel*, 'noble', and *mere*, 'lake', to create it.

Athens ON A village (1888) in the united counties of Leeds and Grenville, northwest of Brockville, it was first known as *Dixon's Corners* and *Bates' Mills* before it was named *Farmersville* in 1836. In 1888, after a new high school joined a grammar school and a model school for teachers, merchant Arza Parish was inspired to rename the educational centre for the Greek classical city of education and learning.

Atherley ON At The Narrows, where Lake Simcoe joins Lake Couchiching, this place in Simcoe County was named in 1851 after Azerley, north of Leeds in North Yorkshire, England.

Athol ON A township in Prince Edward County, it was named in 1835 after the Atholl district of Perthshire, Scotland.

Atholville NB A village (1966) on the south shore of the Restigouche River and on the west side of Campbellton, its post office was called *Shives Athol* in 1906 after Athol House. The house was built in the early 1800s by Robert Ferguson, a native of Blair Atholl, Scotland, who settled here in 1796. Another post office called *Ferguson Manor* was opened in 1916 and was renamed Atholville in 1922.

Atikokan ON A township in Thunder Bay District, between Thunder Bay and Fort Frances, Atikokan was a divisional point of the Canadian Northern Railway in 1899, taking its name from **Atikokan River**, whose name means 'caribou bones' in Ojibwa. An open-pit mine began operating in 1939, and a planned community was laid out three years later.

Atikonak Lake NF The name of this lake (358 km^2), south of Churchill Falls, was derived from the Montagnais word for 'whitefish lake'.

Atlin BC On the east shore of **Atlin Lake** (589 km^2), and in the northwestern part of the province, its post office was named in 1899. The name of the 95-km-long lake was derived from the Inland Tlingit *ahklen* or *aht'lah*, 'big lake'.

Atna Peak BC In the Coast Mountains, southeast of Kitimat, this mountain (2755 m) was named about 1907 after the Carrier word for 'strangers' or 'other people'. **Atna Lake** lies to the northeast of the peak.

Attawapiskat ON Located on the **Attawapiskat River**, northwest of Moosonee, the Hudson's Bay Company trading post *Attawapiscat House* was established in 1900. Attawapiskat post office was opened in 1961. The name was possibly derived from Cree *atawao*, 'he trades' and *pisket*, 'divided', suggesting a 'trading post beside a divided outlet'.

Atwood ON In Elma Township, Perth County, southwest of Listowel, this place was settled in 1854 and named *Elma Centre*. In 1876 a railway station was called Newry Station after a nearby community. It was renamed in 1883 by William Dunn, whose niece, Eliza Gray of Detroit, observed the new place was in the shadow of a surrounding grove of trees.

Auburn NS In the Annapolis Valley, southwest of Kentville, this place's post office was called *Palmer Road* in 1864. It was renamed in 1889 after Oliver Goldsmith's 'Sweet Auburn', in his *Deserted Village* (1770).

Auburn ON A community in Huron County, east of Goderich, it was named in 1862 when a post office was opened on the west side of the Maitland River, across from *Manchester*. It was moved to the police village of *Manchester* about 1900, with both names coexisting until 1978, when the police village was called Auburn.

Augusta ON A township in the united counties of Leeds and Grenville, it was named in 1787 for Princess Augusta Sophia (1768-1840), second daughter of George III.

Augusta, Mount YT On the Yukon-Alaska boundary, and in the St. Elias Mountains, this mountain (4289 m) was named in 1891 by American geologist Israel C. Russell after his wife J. Augusta (Olmsted) Russell.

Augustine Cove PE Situated southeast of Borden-Carleton, this place had a post office from 1854 to 1913. It was called after the adjacent cove, named in 1765 by Surveyor General Samuel Holland after Lt-Col Augustin Prévost, the father of Sir George Prevost (1767-1816), the future governor-in-chief of British North America, 1811-15. The father, a native of Switzerland, had joined the British Army and took part in the capture of Québec in 1759.

Aulac NB Located east of Sackville, this place's post office was first named *Westmorland Point* in 1853. It was renamed Aulac in 1919, taking its name from the **Aulac River**, which flows into the Tantramar River. As early as 1768, map-maker John Montresor noted *au Lac* as a populated place. Nova Scotia Surveyor General Charles Morris called the river *R du Lac* in 1756, with the form of the present name being settled in 1901 by Surveyor General Thomas Loggie.

Aulavik National Park NT Located on the north side of Banks Island, this national park (12,275 km^2) was established in 1992 to protect the sensitive environment in the area of Thomsen River. Its name in Inuvialuktun means 'place where people travel'.

Aulds Cove NS On the mainland side of the Canso Causeway, this place is named after Alexander Auld, who operated grist and sawmills here before 1833. There was postal service here as early as 1841, with Aulds Cove post office opening in 1877.

Aurora ON A town (1888) in York Region, it was named in 1854 on the suggestion of merchant and first postmaster Charles Doan. He may have been prompted to suggest the name of the Greek goddess of the dawn as a reflection of a new dawn of prosperity on that part of Yonge Street.

Ausable River ON Called by the French in the 1700s *Rivière aux Sables* (meaning 'sandy river'), this river is characterized by a long sandy beach in the area of Grand Bend, where it enters Lake Huron. Before 1892, it made a sharp bend at the village, and flowed south for 16 km to enter the lake at present Port Franks.

Austerity, Mount BC In the Selkirk Mountains, this mountain (3347 m) was named in 1911 by climber Howard Palmer, who found the climb up its west side to be quite difficult.

Austin MB This place's Canadian Pacific Railway station was named in 1881 by the Marquess of Lorne after Sidney Austin, a correspondent for the *London Graphic*, who followed the progress of track being laid west. Its post office opened two years later. It had been earlier called *Three Creeks*. *See also* Sidney.

Austin QC A municipality in the regional county of Memphrémagog, southwest of Magog, it was named in 1938 after Loyalist Nicholas Austin, who settled here in the 1790s.

Australia Mountain YT Southeast of Dawson, this mountain (1593 m) was named after Australian miners, who prospected the nearby creeks for gold after 1891.

Auyuittuq National Park NT On the east side of Baffin Island, and north of Pangnirtung, this park reserve (21,471 km^2) was established as *Baffin Island National Park* in 1972. It was renamed three years later after the Inuktitut word *auyuittuq*, 'land of the big ice', or literally 'place that does not melt'.

Avalanche Mountain BC In the Selkirk Mountains, this mountain (2864 m) was named in 1881 by Maj. Albert Rogers after his Shuswap assistants encountered a snowslide on its west side. Subsequently, the Canadian Pacific Railway built snowsheds after avalanches had disrupted traffic through Rogers Pass.

Avalanche Peak YT In the St. Elias Mountains, this peak (4212 m) was named in 1958 by George Wallerstein, who had led an expedition from southern California to the St. Elias Mountains the previous year. His party observed a huge avalanche sweep down from the peak.

Avalon Peninsula NF Joined to the main part of the island of Newfoundland by the **Isthmus of Avalon**, this peninsula (which comprises four distinct peninsulas) was named in 1623 'the Province of Avalon' by Sir George Calvert, Baron Baltimore, after the original site of Avalon, the island of the blessed in the time of King Arthur, in Somerset, England. By 1813 it was known as a peninsula.

Avignon QC A regional county, extending from the mouth of the Rivière Matapédia in the west to Maria in the east, it was established in 1981. The name was probably derived from a former post office (1886-1922) near present-day Saint-Alexis-de-Matapédia, west of the mouth of the river. It was likely named after the famous city in the south of France.

Avondale NF Near the head of Conception Bay, this town (1974) was first known as *Salmon Cove*. It was renamed in 1906 after Avondale, County Wicklow, Ireland.

Avonlea SK This village (1912) was established in 1911 by the Canadian Northern Railway on its Moose Jaw-Radville line, southeast of Moose Jaw. It was named after a fictional place in Lucy Maud Montgomery's *Anne of Green Gables* (1908). She had relatives who had settled in the area. The post office had been called *New Warren* in 1904, and was renamed Avonlea in 1912.

Avonmore ON In the united counties of Stormont, Dundas and Glengarry, northwest of Cornwall, it was named in 1864 after the Gaelic words meaning 'big river'. There are places called Avonmore in County Wicklow, Ireland and Warwickshire, England.

Avonport NS At the mouth of Gaspereaux River, east of Wolfville, this community was first called *Horton Point* after the original township. Its post office was opened in 1864 and called after the **Avon River**, whose mouth is just to the east.

Avon River ON A tributary of the North Thames River, it was first known as the *Little Thames River*. It was renamed about 1833 in association with the naming of Stratford.

Axel Heiberg Island NT One of the Sverdrup Islands and in the Queen Elizabeth Islands, this island (43,178 km^2) was named *Heiberg Land* in 1900 by Otto Sverdrup, during his 1898-1902 Norwegian Arctic expedition, after Consul Axel Heiberg, a Norwegian patron.

Ayer's Cliff QC A village (1909) in the regional county of Memphrémagog and at the south end of Lac Massawippi, southeast of Magog, it was named *Ayer's Flat* in 1836 after Thomas Ayer, the first settler here. Believing that 'Flat' suggested marshy and low land, it was renamed in 1904 to stimulate the sale of shoreline lots.

Aylen Lake ON Situated on the south border of Algonquin Park, it was known as *Little Opeongo Lake* before being renamed about 1850 for Ottawa Valley lumberman Peter Aylen (1799-1868). Aylen lived in Bytown (Ottawa) in the 1830s, and was later a businessman in Aylmer, QC.

Aylesford NS In the Annapolis Valley, west of Kentville, the post office in this village (1968) was named in 1836 after the former *Aylesford Township*, created in 1786. The township was named after the 4th Earl of Aylesford (1751-1812), George III's lord steward of the household.

Aylmer ON A town (1887) in Elgin County, it was first known as *Hodgkinson's Corners* after long-time town clerk and postmaster Philip Hodgkinson. Some Americans from Troy, NY, wanted to call it *Troy*, but this connection suggested disloyalty to the Crown. It was named in 1838 for Matthew Whitworth-Aylmer, 5th Baron Aylmer (1775-1850), governor-in-chief of British North America, 1830-5. To distinguish it from Aylmer, QC, *Aylmer West* was often added to postmarks.

Aylmer QC A city (1975) in the Communauté urbaine de l'Outaouais, west of Hull, it was first known in 1816 as *Symmes Landing* after Charles Symmes, a nephew of Hull founder Philemon Wright. It was renamed in 1848 after Lord Aylmer, governor-in-chief of British North America, 1830-5.

Aylmer Lake NT Northeast of Great Slave Lake, this lake (809 km^2) was named in 1834 by explorer George (later Sir George) Back after Lord Aylmer, governor-in-chief of British North America, 1830-5.

Ayr ON In Waterloo Region, southwest of Cambridge, Ayr was named in 1840 by James Jackson, a native of Ayr, in Ayrshire, Scotland.

Azilda ON In the town of Rayside-Balfour, Sudbury Region, this place was named *St. Azilda* in 1891 for Azilda Brisebois, the wife of Joseph Bélanger, who had settled here in the 1880s. When it was found no such saint existed, it was renamed *Rayside* in 1900 after the township. Residents wanted *St. Azilda* reinstated, but it was restored as plain Azilda in 1901.

B

Babbage River YT Flowing northeast into Beaufort Sea, southeast of Herschel Island, this river was named in 1826 by John (later Sir John) Franklin after Charles Babbage (1792–1871), British mathematician and computer scientist. In 1834 he conceived the principles of the modern computer.

Babine Lake BC North of Burns Lake, and near the centre of the province, the name of this lake (479 km^2) in French means 'a large or pendulous lip', which described the distended lip of young Carrier women. Fort Babine (also called *Kilmar's Fort*) was located on the lake by the Hudson's Bay Company in 1822.

Baccaro Point NS In the southwestern part of the province, this name may have a French origin. A 1631 map by Jean Guérard identified a bay there as *B. Careaux*, perhaps describing the right-angled bend in present Barrington Passage. Jean-Baptiste-Louis Franquelin referred to the point as *Pte de Baccaro* in 1686. The name is less likely associated with Portuguese *baccalau*, 'codfish'. The communities of **West Baccaro** and **East Baccaro** are located at the point.

Back River NT Rising northeast of Great Slave Lake, this river (974 km) flows northeast into Chantrey Inlet, an inlet of the Arctic Ocean. It was named *Backs River* in 1902 after Capt. George (later Sir George) Back (1796–1878), who explored the river in 1834. The name was changed to its present form by the Geographic Board in 1926. Back had called it *Thlew-ee-choh or Great Fish River. Thlew-ee-choh* is possibly Dogrib for 'great fish river'.

Baddeck NS A village (1950) on Cape Breton Island, its name was derived from Mi'kmaq *petecook*, 'lying on the backward turn', in reference to the entrance to the southwest-flowing **Baddeck River**. Postal service at Baddeck dates from 1837, with *Bedeque* being commonly used in the early years. *See also* Bedeque, PE.

Baden ON In Waterloo Region, west of Kitchener, Baden was named in 1854 by Jacob Beck, a native of the Grand Duchy of Baden, in southwestern Germany.

Badger NF At the confluence of **Badger Brook** and the Exploits River, west of Grand Falls-Windsor, this town (1963) was organized as a sawmilling centre in 1909 by the Anglo-Newfoundland Development Company. The source of the name is not recorded in published references.

Badger's Quay NF An urban centre in the town of New-Wes-Valley and on the northwest side of Bonavista Bay, it was settled in the mid–1800s. The source of its name, first recorded as *Badger's Key* in 1891, is not indicated in published references.

Badham, Mount YT In the St. Elias Mountains, and at the head of Donjek Glacier, this mountain (3848 m) was named in 1919 after boundary surveyor Francis Molyneaux Badham. He was killed during the First World War, while serving in the Canadian Army.

Baffin Island NT Canada's largest island (507,451 km^2), and the fifth largest in the world, it was named in 1819 by William (later Sir William) E. Parry after William Baffin (d. 1622). With Robert Bylot, Baffin explored the waters of Davis Strait and **Baffin Bay** in 1615–6. Although they carefully explored the coasts and inlets of Devon and Baffin islands, they failed to investigate Lancaster Sound, the main eastern entrance to the Northwest Passage.

Baie-Comeau QC Located in the regional county of Manicouagan and on the north shore of the St. Lawrence River, this city's post office was named *Comeau Bay* in 1929 after noted writer Napoléon-Alexandre Comeau (1848–1923). The present form was adopted in 1936, when a pulp and paper mill was built by the *Chicago Tribune*'s Col Robert McCormick; the city was incorporated the next year.

Baie-d'Urfé QC A town in the Communauté urbaine de Montréal, between Sainte-Anne-de-Bellevue and Beaconsfield, it was named *Baie-d'Urfée* in 1911 (corrected 1960) after the adjacent bay of Lac Saint-Louis. The bay's name was given after pioneer missionary François-Saturnin Lascaris d'Urfé (1641–1701), who was the first resident priest here in 1686.

Baie-James QC A municipality in west central Québec, embracing an area of 333,000 km^2, it was named in 1972 after James Bay. The bay had been named in honour of navigator Thomas James (*c.* 1593–*c.* 1635), who explored it and Hudson Bay in 1631–2.

Baie-Saint-Paul QC Located in the regional county of Charlevoix, northeast of the city of Québec, the town and parish municipality were named after the adjoining bay, whose name occurred in references before the mid-1600s. The parish was called *Saint-Pierre-et-Saint-Paul-de-la-Baie-Saint-Paul* in 1855, and was shortened to the current name in 1964. The town was incorporated in 1913, having been made a village in 1893.

Baie-Sainte-Anne NB On the south side of Miramichi Bay, this place's post office was named in 1917, taking its name from the adjacent bay. Settled in the 1760s by Acadians, the bay was first identified on maps as *French Bay* and *Lower Bay du Vin*.

Baie-Trinité QC A village (1955) in the regional county of Manicouagan, northeast of the city of Baie-Comeau, it is at the mouth of the **Rivière de la Trinité**, noted on maps as early as 1682.

Baie Verte NB, NS At the head of **Baie Verte**, a 27-km extension of Northumberland Strait, this community was first settled in 1700 by Acadians. After their expulsion in 1755, it was resettled in 1761 by New Englanders, who retained the French name. The name was given for the green salt-water grasses at the head of the bay. French map-maker Pierre Jumeau called the bay *B. Verte ou de S. Claude* in 1685. Although English map-maker Herman Moll referred to it as *Green Bay* in 1715, English chart-maker J.F.W. DesBarres used *Bay Verte* in 1779, and the French form subsequently prevailed.

Baie Verte NF On the north coast of the island of Newfoundand and at the head of **Baie Verte** (meaning 'green bay'), the names of this town (1958) and the bay relate to the period of French occupation on the French Shore both before 1763, the date of the transfer of the island to British title, and by French migratory fishermen until 1904.

Bailieboro ON In Peterborough County, south of Peterborough, Bailieboro was named in 1861 by James Aiken after his home town of Bailieborough, in County Cavan, Ireland.

Baker Brook NB A village (1967) on the Saint John River, and at the mouth of **Rivière Baker-Brook**, 16 km southwest of Edmundston, its post office was opened as *Baker's Brook* in 1851. It was called *Gagnon*, after postmaster Jean Gagnon, from 1882 to 1893, when it was renamed Baker Brook. The river was identified in 1792 by surveyor Isaac Hedden as *Marienequactacook*, adapted from its Maliseet name, meaning 'turtle river'. Maps in the mid-1800s referred to it as *Bakers R* and *Baker's Brook*, after John Baker (1796–1868), who opposed British sovereignty in the 1820s and '30s, touching off a boundary dispute between New Brunswick and Maine. The route of the boundary was settled by the Webster-Ashburton Treaty in 1842. *See also* Aroostook *and* Lac-Baker.

Baker Lake NT A hamlet (1977) on the north shore of Baker Lake, it was founded after the Second World War as an Inuit settlement. The lake (1783 km^2) was named in 1761 by Capt. William Christopher after Sir William and Richard Baker, brothers associated with the Hudson's Bay Company. Christopher had entered Chesterfield Inlet and explored it for 150 km in search of the Northwest Passage.

Bala ON In the district municipality of Muskoka, northwest of Gravenhurst, this place was named in 1870 by postmaster Thomas W. Burgess, who had pleasant memories of the Bala area in northern Wales.

Balcarres SK A town (1951) east of Fort Qu'Appelle, its post office was established in 1884. It had been settled about 1880, with settlers having to haul their grain some 20 km south to Indian Head, whose postmaster was Balcarres Crawford. There is a Balcarres House, south of St. Andrews, Fifeshire, Scotland.

Balderson ON John Balderson, a native of Lincolnshire, England, settled at the site of this community in Lanark County in 1816. Its post office was opened in 1858.

Baldur MB Southeast of Brandon, this place's Canadian Northern Railway station was named in 1888 by S. Christopherson, then living in nearby Grund, after Baldur, the Norse god of sun and light, and son of Odin (supreme god and creator), and of Freyja (goddess of married love and the hearth). The post office was opened in 1891.

Baleine, Grande rivière de la QC With a length of 727 km, and draining an area of 44,735 km^2, this river flows west into Hudson Bay. It was first called *Great Whale River* in 1744, with the French form *R. de la Grande Baleine* occurring on a map in 1914. The present form was adopted in 1962.

Balfour, Mount AB Northwest of Banff, this mountain (3272 m) was named in 1858 by Dr James (later Sir James) Hector, after Scottish botanist John Hutton Balfour (1808–84).

Balgonie SK Located 25 km east of Regina, this town (1907) was named in 1882 by the Canadian Pacific Railway after the Balgonie estate in Fifeshire, Scotland. Its post office was opened in 1883.

Ballinafad ON On the boundary of Halton Region and Wellington County, north of Georgetown, it was named in 1842 after Ballinafad in County Sligo, Ireland, southeast of the town of Sligo.

Balmertown ON In Kenora District, northeast of Red Lake, this mining centre was named in 1948 after Balmer Township, which had been named earlier for Balmer Niely, an official of McIntyre-Porcupine Gold Mines of Timmins. In 1950 the township and parts of three adjoining townships became the improvement district of *Balmertown*, which was changed to the municipal township of Golden in 1985.

Balmoral NB A village (1972) east of Campbellton, it was laid out in 1856 and settled after 1874 under the free grants provisions. Its post office was named in 1875 after the sovereign's residence in Perthshire, Scotland, acquired by Prince Albert in 1847. **Balmoral Parish** was established in 1876.

Bamfield BC On the east side of Barkley Sound, southwest of Port Alberni, this place's post office was named in 1903 after trader and Indian agent William Eddy Banfield, who had come to Vancouver Island in 1846, and retired here. The circumstances surrounding his death (1862) were mysterious. Postal authorities misspelled his name.

Bancroft ON A town (1996) in Hastings County, it was first known as *York Mills* and *York River*. It was renamed in 1879 by Senator Billa Flint after his mother-in-law, Elizabeth Ann Bancroft Clement, a New Hampshire native, who died in Belleville in 1851.

Banff AB A town (1992) in **Banff National Park**, it took its name from a nearby station on the Canadian Pacific Railway, established in 1883. It was possibly proposed by Harry Sandison of Winnipeg, a member of the National Parks Board, after his birthplace of Banff, Banffshire, Scotland, near where both Donald Smith, Lord Strathcona, and George Stephen, Lord Mount Stephen, were born. The national park (6641 km^2) was called the *Rocky Mountains Park Reserve* in 1887 and it was renamed after the town in 1930.

Banks Island BC On the east side of Hecate Strait, south of Prince Rupert, this island was named in 1788 by Capt. Charles Duncan after Sir Joseph Banks (1743–1820), the noted English botanist and president of the Royal Society, 1778-1820.

Banks Island NT The most westerly of the islands of the Arctic Archipelago, this island (70,028 km^2) was named in 1820 by William (later Sir William) Parry after Sir Joseph Banks.

Bannockburn ON In Hastings County, north of Madoc, this place was named in 1862 to celebrate the famous fourteenth century victory of the Scots over the English at Bannockburn, near Stirling, Scotland.

Baptiste Lake ON Located in Hastings County, northwest of Bancroft, it may have been named after an Ojibwa chief called St Jean Baptiste, who was met there by surveyor John Snow at the outlet of the lake in 1854–5.

Barachois NB A community east of Shediac, its post office was named in 1853. A 'barachois' is a pond enclosed by a bar at sea level. The term was introduced into East-Coast geographical terminology by the Basques and the Acadians, with its original application applying to a gravel bar where fishing boats could be beached.

Barachois QC A community in the town of Percé and the regional county of Pabok, its origin may be traced to a map of 1689 by Basque pilot Pierres Detchevery in the form of *Barrachoa.* The word *barachois* means 'lagoon', a deformation of French *barre échouée* or *barre-à-cheoir*, an apt description of **Barachois de Malbaie** adjacent to the community.

Barbeau Peak NT The highest peak east of the Rocky Mountains, it (2616 m) was named in 1969 by the Canadian Permanent Committee on Geographical Names after Marius Barbeau (1883–1969), a folklorist and ethnologist who studied the music and traditional songs of the Québécois and Aboriginals. He collected an immense archive of folklore materials, while working from 1911 to the mid-1960s at the predecessor of the Museum of Civilization.

Barkers Point NB In the city of Fredericton, and on the east side of the Nashwaak River, this place's post office was named in 1912 after the point. It was an incorporated village from 1966 to 1973, when it was annexed by the city. Andrew Barker settled at the point in 1825. His father, Thomas, was a Loyalist from New York, who received a grant of 202 ha (500 a) in the area of Maugerville, downriver from Barkers Point.

Barkerville BC A restored urban site and provincial historic park in the Cariboo, it was the site where William 'Billy' Barker (d. 1894) made a huge strike in 1862, and launched the fabled Cariboo gold rush. After 1900 the place fell into decline, becoming a ghost town in the 1950s.

Barkley Sound BC Southwest of Port Alberni and on the west coast of Vancouver Island, this wide bay comprising many islands and channels was named by Capt. Charles William Barkley (1759–1832), while sailing up the west coast in 1787.

Barnard, Mount AB, BC Located on the continental divide, southwest of Howse Pass, this mountain (3339 m) was named in 1924 after Sir Frank Stillman Barnard (1856–1936), lieutenant-governor of British Columbia, 1914-19.

Barnes Ice Cap NT In central Baffin Island, this ice cap was named in 1950 by Patrick Baird of the Arctic Institute of North America after H.T. Barnes of McGill University, a pioneer in the study of ice.

Barnston QC A township municipality (1855) in the regional county of Coaticook, southwest of the town of Coaticook, it was named after the township. The township was named in 1795, and proclaimed in 1801, after a village in Essex, England, northeast of London.

Barnwell AB This place east of Lethbridge was founded by Mormon settlers in 1902 and first named *Woodpecker*, after Woodpecker Island in the nearby Oldman River, and then *Bountiful*, possibly after the city in Utah, itself called after the Land of Bountiful in the Book of Mormon. It was named in 1909 by the

Canadian Pacific Railway after Richard Barnwell, a CPR purchasing agent in Winnipeg.

Barons AB A village (1910) northwest of Lethbridge, it was first called *Blayney* in 1907 by Jack Warnock after the town of Castleblayney in County Monaghan, Ireland. Two years later Canadian Pacific Railway purchasing agent C.S. Noble named the station after an official of the CPR, either George Stephen, 1st *Baron* Mount Stephen, or Donald Smith, 1st *Baron* Strathcona and Mount Royal. When the Union Bank opened, it called itself Baron's Bank. Other businesses followed suit, and the post office and station adopted Barons.

Barraute QC Located in the regional county of Abitibi, north of Val-d'Or, the village was named in 1918 after the township, whose name had been proclaimed two years earlier. It was given after Pierre-Jean Bachoie *dit* Barraute (1723–60), an officer of the Béarn Regiment, who died during the Battle of Sainte-Foy.

Barrhaven ON An urban centre in the city of Nepean, Ottawa-Carleton Region, its original 81 ha (200 a) was acquired in the 1960s by Melville (Mel) Barr (1909–93) for a harness raceway. When another raceway was built nearby, he sold the property to residential developers.

Barrhead AB Located northwest of Edmonton, the post office in this town (1946) was named in 1914 by the Paddle River and District Co-operative after the home town of James McGuire. He was born in Barrhead, Renfrewshire, Scotland, a southwestern suburb of Glasgow.

Barrie ON The city (1959) of Barrie was named by surveyor William Hawkins in 1832 after Robert (later Sir Robert) Barrie (1774–1841), the senior naval officer in Upper and Lower Canada, and commissioner of the dockyard at Kingston, 1819-34. His wife Julia expressed an interest in settling here, but he rejected the idea, and returned to England in 1834. A township in Frontenac County and an island in Manitoulin District were also named after him.

Barrière BC On the North Thompson River, and north of Kamloops, this community is at the mouth of Barrière River. The river was likely named after a French word for fence by a French-speaking traveller, who observed the practice of the Shuswap of erecting weirs across the river to catch fish. The post office was named in 1914.

Barrington NS At the head of **Barrington Bay** and in the southwestern part of the province, this place was settled in 1761 by the families of New England fishermen from Cape Cod and Nantucket Island. Barrington Township was created in 1767, and named after William Wildman Barrington, 2nd Viscount Barrington (1717–93), British secretary of war, 1755-61 and 1765–78, and chancellor of the exchequer, 1761–2. Barrington post office was opened in 1843. The community of **Barrington Passage**, on the passage separating Cape Sable Island from the mainland, received a post office in 1855.

Barrow Strait NT Part of Parry Channel, with Devon and Cornwallis islands on the north, and Somerset Island on the south, this strait was named in 1819 by William (later Sir William) Parry, after Sir John Barrow (1766–1848).

Barry's Bay ON The adjacent bay was probably named for lumberman William Byers (*c.* 1810–85). In 1847 surveyor John Haslett noted Byers's hay farm nearby, but wrote Barry's Bay for the 8-km arm of Kaminiskeg Lake, and this mistake was repeated by subsequent surveyors and map-makers. The village was incorporated in 1932.

Barryville NB Located northeast of Chatham (Miramichi), this place's post office was named in 1904 after postmaster James Barry. A land grant was received here by Edward Barry in the early 1800s.

Barss Corner NS In Lunenburg County, northwest of Bridgewater, this place was first known as *Centreville*. The site of *New Germany* post office in 1854, it was renamed Barss Corners in 1882 after an early settler.

Bartibog Bridge NB Located at the mouth of **Bartibog River**, 13 km northeast of Chatham (Miramichi), this community's post office was named in 1878, after the bridge over the river. The river, identified on a 1785 map by its present name, may have been called after a Mi'kmaq chief, called Bartabogue by both the French and English.

Bas-Caraquet NB A village on the shore of Chaleur Bay and east of the town of Caraquet, its post office was called *Lower Caraquet* in 1877. It was renamed Bas-Caraquet in 1955. The village was first incorporated as *Lower Caraquet* in 1966, but was changed to Bas-Caraquet five years later.

Bashaw AB South of Camrose, the post office in this town (1964) was named in 1910 after settler Eugene Bashaw, who owned the townsite.

Baskatong, Réservoir QC Situated near the head of the Rivière Gatineau, northwest of the town of Mont-Laurier, it was created in 1926-7. It was named in 1962 after the township of Baskatong, given about 1870 after an Algonquin word for 'place where sand holds back the water'.

Bassano AB On the main line of the Canadian Pacific Railway, the station of this town (1911) was named in 1884 after the Marquis de Bassano, a native of Italy, who was a CPR shareholder.

Bass River NS On the north shore of Cobequid Bay, west of Truro, this community was founded in 1765. Located on a river known for its bass fishing, its post office was opened in 1869.

Bastard ON Part of the municipal township of Bastard and South Burgess in the united counties of Leeds and Grenville, Bastard was named in 1798 for John Pollexfen Bastard, a British member of Parliament. His home was at Kitley, near Plymouth, Devon. *See also* Kitley.

Batawa ON In Hastings County, north of Trenton, it was developed in 1939 as a company village by Czech-born shoe manufacturer Thomas John Bata, who added the last syllable to emulate names like Ottawa and Oshawa.

Batchawana Bay ON On the east shore of Lake Superior, north of Sault Ste. Marie, its name may be derived from Ojibwa *obatchiwanang*, 'where the flowing water rises to a boil through the force of its current'.

Bath NB A village (1966) on the east bank of the Saint John River, north of Woodstock, its post office was opened in 1853 as *Munquart*, after the Munquart River. It was renamed Bath in 1868, possibly after the city of Bath, in Somerset, England, or after Bath at the mouth of the Kennebec River, in Maine.

Bath ON In Lennox and Addington County, west of Kingston, this village (1859) was named in 1819 after the famous English city on the River Avon, east of Bristol.

Bathurst NB This city (1966) in the northern part of the province was named in 1826 by Lt-Gov. Sir Howard Douglas after Henry Bathurst, 3rd Earl Bathurst (1762–1834), who was then Britain's secretary of war and for the colonies. Acadian governor Nicolas Denys referred to it as *Nipisiguit*, after the Nepisiguit River, when he wrote *Description géographique et historique des costes de l'Amérique septentrionale* here in 1672. His son Richard Denys de Fronsac called it *St-Pierre*, when he lived there from 1671 to 1687. It was subsequently called *St. Peters* by the early English settlers.

Bathurst ON A township in Lanark County, it was named in 1816 for Henry, 3rd Earl Bathurst (1762–1834), then Britain's secretary of war and the colonies. He was the son-in-law of the Duke of Richmond, governor-in-chief of British North America, 1818-9, and a brother-in-law of Lady Sarah Maitland, whose husband, Sir Peregrine Maitland, was lieutenant-governor of Upper Canada, 1818-28.

Bathurst, Cape NT Northeast of Tuktoyaktuk, this cape extends into the Arctic Ocean, with Beaufort Sea on the west, and Amundsen Gulf on the east. It was likely named in 1826 by

John (later Sir John) Richardson, after Henry, 3rd Earl of Bathurst (1762–1834), secretary of war and for the colonies, 1812–27 and president of the council, 1828–30.

Bathurst Inlet NT A long inlet of the Arctic Ocean, south of Victoria Island, it was named in 1821 by John (later Sir John) Franklin, after Henry, 3rd Earl of Bathurst (1762–1834). The locality of Bathurst Inlet, on the west side of the inlet and at the mouth of the Burnside River, was the site of a Hudson's Bay Company post from 1944 to 1964, when it was moved to the present site of Umingmaktok, on the east side of the inlet.

Bathurst Island NT One of the Queen Elizabeth Islands, this island (16,042 km^2) was named in 1819 by William (later Sir William) Parry, after Henry, 3rd Earl of Bathurst. In 1851 Capt. William Penny named it *Queen Victoria Island*, but Parry's earlier name was adopted by the Geographic Board in 1905.

Batiscan QC A municipality at the mouth of Rivière Batiscan, and in the regional county of Francheville, northeast of Trois-Rivières, it was named in 1986 after the parish of Saint-François-Xavier-de-Batiscan, which had been erected in 1845. Samuel de Champlain mentioned the river in 1603, and referred to an Aboriginal captain called Batiscan in 1610. Among suggested meanings of the name are 'mist', 'light cloud', and 'reeds at the mouth'.

Batoche SK Located on the east bank of the South Saskatchewan River, northeast of Saskatoon, this historic locality was founded in 1871 by ferry operator Xavier Letendre *dit* Batoche. In 1885 it was Louis Riel's headquarters, and the site where Gen. Sir Frederick Middleton defeated the Métis leader Gabriel Dumont in May of that year. Its post office was opened in 1884.

Battersea ON In Frontenac County, northeast of Kingston, this place was named in 1857 by miller Henry Van Luven after Battersea in London, England, on the south side of the River Thames, opposite Westminster and Chelsea.

Battleford SK On the south bank of the North Saskatchewan River and at the mouth of the **Battle River**, this town (1904) was chosen in 1875 as the capital of the North-West Territories, and its post office was opened two years later. The North-West Mounted Police built Fort Battleford there in 1876. It was eclipsed in 1905, when the Canadian Northern Railway built its line on the north side of the North Saskatchewan. The Battle River, which rises in Battle and Pigeon lakes southwest of Edmonton, recalls several battles between the Cree and the Blackfoot.

Bawlf AB Southeast of Camrose, the Canadian Pacific Railway station in this village (1906) was named in 1905 after Nicholas Bawlf, of Winnipeg, co-founder and president of the Northern Elevator Company and the N. Bawlf Grain Company. The post office, first called *Molstad*, was renamed Bawlf in 1907.

Bay Bulls NF A town (1986) on the east coast of the Avalon Peninsula, it was named after the bay. The bay was identified in 1592 as *B of Bulls* by map-maker Thomas Hood. It may have been called after the bull bird (*Plautus alle alle*), a common seabird also called the dovekie, which winters along the coasts of Newfoundland.

Bay de Verde NF A town (1950) near the north side of **Bay de Verde Peninsula**, an extension of Avalon Peninsula, its name may relate to the Portuguese *y verde* on the La Cosa map of 1500. However, the first direct reference to it was John Guy's *Greene bay* in 1612, with variants of the present name not occurring until the late 1600s.

Bay du Vin NB At the mouth of **Bay du Vin River**, 26 km east of Chatham (Miramichi), this community's post office was named in 1853. The bay was originally called *Baie des Vents* after an incident about 1760, when a boat of the first French-speaking settlers was driven into the bay by a strong wind. Although Abraham Gesner wrote in 1847 about 'Big Baie des Vents corrupted to big Betty Wind', a map as early as 1786 had *Baye du Vin*. A post office 22 km upriver was called

Wine River from 1908 to 1951, wrongly assuming the name of Bay du Vin River was associated with wine.

Bayfield NB At the New Brunswick end of the Confederation Bridge leading to Prince Edward Island, this place's post office was named in 1866 after Admiralty surveyor Henry W. Bayfield. He surveyed the waters of the New Brunswick coast before 1841.

Bayfield ON At the mouth of **Bayfield River**, south of Goderich, this village in Huron County was named in 1832 for Henry Wolsey Bayfield (1795–1885), who had surveyed the Great Lakes from 1816 to 1826. He recommended the place as a good site for a town. Established as a village in 1875, it became a police village in 1927, and reverted to a village in 1965. The Canada Company, which developed a large part of Southwestern Ontario, named the river in 1827.

Bayham ON A township in Elgin County, it was named in 1810 for Charles Pratt, Earl Camden and Viscount Bayham (1714–94). The ruins of Bayham Abbey are east of Tunbridge Wells in Kent, England.

Bay l'Argent NF A town (1971) on the west side of the Burin Peninsula and near the head of Fortune Bay, its name was derived from the French for 'bay of silver', perhaps in reference to the silvery light reflected from the surrounding cliffs.

Bay Roberts NF A town (1951) on the west side of Conception Bay, west of St. John's, its name may have originated from a Jersey surname. It was shown in its present form on several maps of the 1600s. The reference to *Bay of Robbers* in the 1681 calendar of state papers suggests its name may relate to an incident.

Bay St. Lawrence NS On Cape Breton Island and between Cape North and **Cape St. Lawrence**, this place was first settled in the 1830s. Its post office was opened in 1853. In 1535 Jacques Cartier called the cape *cap de Lorraine*, which was subsequently rendered on maps as *C S Laurent*. Jacques Nicolas Bellin's 1744 map called the bay *Ance S Laurent*.

Bayside NS On the west side of Shoal Bay, southwest of Halifax, this descriptive name was given to its post office in 1891.

Bayside ON On the Bay of Quinte and in Hastings County, east of Trenton, this place was called *Rhinebeck* in 1883, but its name was changed that year to Bayside.

Baysville ON Located in the district municipality of Muskoka and at the southern end of the Lake of Bays, where the lake is drained by the South Branch Muskoka River, this place was named in 1874.

Baytona NF On Birchy Bay, an arm of the Bay of Exploits, this place was called *Gayside* from 1976 to 1985. It was renamed, presumably by joining 'bay' to the last two syllables of Daytona Beach, the Florida city, widely known for its car racing. It had previously been called *Birchy Bay North*.

Beachburg ON A village (1959) in Renfrew County, southeast of Pembroke, it was named for David Beach, a United Empire Loyalist, who received a grant of 405 ha (1000 a) here in 1835. His family of five sons and four daughters ran grist and sawmills, a tannery, a hotel, and the post office. In 1848 the post office was called *South Westmeath*, after Westmeath Township, and was changed to Beachburg 10 years later.

Beachside NF On the southwestern side of Notre Dame Bay, this municipal community was incorporated in 1962 as *Wild Bight*, after its location on a small bight whose waters are usually rough during the prevailing northeasterly winds. It was renamed Beachside in 1980.

Beachville ON In Oxford County, midway between Woodstock and Ingersoll, this place was named in 1832 by William Merigold for Andrew Beach, the first postmaster, who had built a grist mill here.

Beaconsfield QC A city in the Communauté urbaine de Montréal, between Baie-d'Urfé and Pointe-Claire, it was named in 1910, six years after the post office was opened. The post office derived its name from an estate named in 1870 by J.H. Menzies after Benjamin Disraeli, Earl of Beaconsfield (1804–81), prime minister of Great Britain, 1868 and 1874–80.

Beamsville ON In Niagara Region, west of St. Catharines, it was settled in 1788 by Jacob Beam, a United Empire Loyalist, who developed the place. Incorporated as a town in 1963, it was united with the township of Clinton in 1970 to form the town of Lincoln.

Beardmore ON In Thunder Bay District, west of Geraldton, this municipal township (1976) was named about 1920 for Walter William Beardmore, a Toronto industrialist, who was the son-in-law of Sir William Mackenzie (1849–1923), developer of the Canadian Northern Railway. It had been created an improvement district in 1945.

Béarn QC Situated in the regional county of Témiscamingue, near the town of Ville-Marie, its post office was named in 1941 after a regiment which served in the 1750s under the command of the Marquis de Montcalm. The original Béarn is in southwest France.

Bear River NS Southeast of Digby, this community is situated on the steep banks of the **Bear River**. Its post office was established as *Bear River West* in 1870, and changed to its present name in 1901. Portrayed as *Rivière Hébert* on Marc Lescarbot's 1606 map, the river was named after either apothecary Louis Hébert, or after Capt. Simon Imbert-Sandrier, or after both. Presumably, about 1612 Imbert-Sandrier sought refuge during a storm in the river's mouth. Several maps in the 1700s called the river *R Imbert*.

Beatton River BC Rising in the Rocky Mountains, west of the Alaska Highway, this river flows southeast into the Peace River, southeast of Fort St. John. It was named after Frank Wark Beatton (1865–1945), who was in charge of the Hudson's Bay Company post at Fort St. John for several years.

Beauce-Sartigan QC A regional county, south of the city of Québec, it was named in 1982. The original county of Beauce was created in 1829, taking its name from the seigneury of Nouvelle Beauce, established in 1739. Sartigan is a variant of *Mechatigan* and *Msakkikhan*, old Abnaki forms of the Rivière Chaudière, meaning 'shaded river', or 'noisy river'.

Beauceville QC Created as a town in 1904, it is in the regional county of Robert-Cliche, north of the city of Saint-Georges. Renamed *Beauceville-Ouest* in 1930 on the incorporation of the town of *Beauceville-Est*, the two were amalgamated as the town of Beauceville in 1973.

Beaudette, Rivière ON, QC This river rises near Apple Hill in Kenyon Township, and flows east into Quebec, entering the St. Lawrence River 3 km east of the Lancaster Township boundary. The name relates to the finding in the 1600s of a wire-frame bed (*baudet*) in the remains of a burned shack on nearby **Pointe Beaudette**.

Beaufort Sea NT, YT North of the Yukon, and west of Banks and Prince Patrick islands in the Northwest Territories, this part of the Arctic Ocean was named in 1826 by John (later Sir John) Franklin after Capt. (later Adm.) Sir Francis Beaufort (1774–1857), who was then the hydrographer to the British Admiralty. He designed the Beaufort scale to measure the force of wind, with 0 being calm and 12 hurricane force.

Beauharnois-Salaberry QC This regional county was erected in 1982, succeeding the former county of Beauharnois, which was created in 1829. Salaberry recalled the name of the largest municipality in the regional county, Salaberry-de-Valleyfield, which took its name from Charles-Michel d'Irumberry de Salaberry (1788–1829), the hero of the Battle of Châteauguay, 1813.

Beauharnois QC A town (1863) in the regional county of Beauharnois-Salaberry, southwest of Montréal, it was created as a village in 1846, taking its name from the seigneury of Beauharnois. The seigneury was granted in 1729 to Claude de Beauharnois de Beaumont de Villechauve (1674–1738), and to Charles de Beauharnois de La Boische, Marquis de Beauharnois (1671–1749).

Beaulac QC A village in the regional county of L'Amiante, south of the city of Thetford Mines, it was named in 1896 after its location on the western shore of Lac Aylmer.

Beaumont AB A town (1973) south of Edmonton, its post office was named in 1895, possibly by John Royer, because its site revealed a good view of the surrounding landscape.

Beauport QC In the Communauté urbaine de Québec, east of the city of Québec, this city traces its roots back to the municipal parish of *Notre-Dame-de-Beauport*, established in 1855, to the religious parish of Notre-Dame-de-Miséricorde-de-Beauport, created in 1694, and to the reference to *Beau port* on a 1631 map by Jean Guérard.

Beaupré QC A town in the regional county of La Côte-de-Beaupré, northeast of the city of Québec, it was named in 1962, having been established as the municipal parish of *Notre-Dame-de-Rosaire* in 1928. The name has its roots in the early seventeenth century phrase '*le beau pré*', agglutinized into the name **Sainte-Anne-de-Beaupré** in 1657.

Beausejour MB Northeast of Winnipeg, this town (1912) was founded in 1877 by the Canadian Pacific Railway. It may have been named by a French-speaking engineer, who was relieved to encounter an elevated gravel knoll, and pronounced his campsite *beau séjour*, 'a good resting place'. The post office was established in 1881.

Beauval SK Northwest of Prince Albert, this community's post office was first called *Lac la Plonge* in 1910, after a nearby lake. It was changed to *Bauval* in 1912, and then to Beauval in 1915. The site of a Catholic mission in the early 1900s, Beauval means 'beautiful valley' in French.

Beaverbank NS North of Bedford, this community's post office was called *Beaver Bank* from 1863 to 1962. It reopened as Beaverbank in 1966, two years after RCAF *Beaverbank* post office closed. The name may be derived from beavers damming brooks in the area.

Beaver Brook Station NB Located 16 km north of Newcastle (Miramichi), this place's post office was named in 1907. The brook is a tributary of Bartibog River. When Newcastle-raised Max Aitken became a peer in 1917, he intended to take the title Lord Miramichi. Local historian Louise Manny said he might be called 'Lord Merry Mickey' and encouraged him to choose another name. Recalling a Catholic priest at Beaver Brook Station who had once loaned him a surrey for an Orange parade, he took the title Lord Beaverbrook. It also gave him the opportunity to draw attention to Canada's national animal.

Beaver Creek YT At 140°38´W, and near the Alaska border, this is the farthest west community in Canada. It was established on **Beaver Creek**, a tributary of Snag Creek, about 1955, with its post office opening in 1958. The creek was named about 1902, and was prospected between 1909 and 1914.

Beaverdell BC Located on the Kettle River, east of Penticton, this place was established in 1905 by blending the names of two adjacent communities, *Beaverton* and *Rendell.*

Beaverlodge AB On the north bank of the Beaverlodge River and west of Grande Prairie, this town (1956) was first known as *Redwillow* in 1910, since Beaverlodge already named a post office 20 km to the east at Lake Saskatoon. The townsite was developed in 1928 by the Edmonton, Dunvegan and British Columbia Railway, and Beaverlodge post office was transferred here.

Beaverton ON On the east shore of Lake Simcoe and at the mouth of the Beaver River, it was first called *Calder's Mills*, *Mill Town* and *Milton*, before being named Beaverton in 1835. It is located in Brock Township, Durham Region.

Becaguimec Stream NB Rising east of Woodstock, this river flows into the Saint John River at Hartland. The name is from the Maliseet *abekaguimek*, possibly meaning 'going up to the salmon bed'.

Bécancour QC This town in the regional county of **Bécancour** was created in 1965 through the amalgamation of several neighbouring municipalities. The name evolved from the seigneury of Bécancour, established in 1684 by Pierre Robinau, 2nd Baron de Portneuf (1654–1729). The regional county was created in 1982 from the county of Nicolet. **Rivière Bécancour** rises just west of Thetford Mines, and follows a circuitous course of 125 km to empty into the St. Lawrence, 12 km downriver from the mouth of the Rivière Saint-Maurice.

Beckwith ON A township in Lanark County, it was named in 1816 for Sir Thomas Sydney Beckwith (1772–1831), quartermaster general of British North America, 1813–23. He was assigned responsibility for the settlement of the military townships north of the Rideau River.

Bedeque PE Located southwest of Summerside, this place takes its name from **Bedeque Bay**. Noted as early as 1806, Bedeque's post office was opened in 1827, although it was referred to as late as 1880 as *Centreville*. Derived from Mi'kmaq *petekook*, 'place that lies on the backward turn', the bay was identified by Jacques Nicolas Bellin in 1744 as *Bedec*. In 1765 Samuel Holland renamed it *Halifax Bay*, after George Montagu Dunk, Earl of Halifax (1716–71), but it did not supplant the older name. *See also* Baddeck, NS, *and* Central Bedeque, PE.

Bedford NS An urban community in the regional municipality of Halifax since 1 April 1996, it had been incorporated as a town in 1980 at the head of **Bedford Basin**. *Fort Sackville* was located here in 1749, with the place becoming known as *Sackville*. Its post office was opened as *Bedford Basin* in 1856, with the Nova Scotia Railway station being called Bedford the same year. The basin was named after John Russell, 4th Duke of Bedford (1710–71), who was the secretary of state of the southern department, when Halifax was founded in 1749. He had been the first lord of the Admiralty, 1744–8.

Bedford ON A township in Frontenac County, it was named in 1798 for Francis Russell, 5th Duke of Bedford (1765–1802), noted for his experiments in agriculture and sheep breeding in Bedfordshire, England.

Bedford QC A town in the regional county of Brome-Missisquoi, southeast of the town of Cowansville, it was created in 1890. The place was named earlier in the 1800s, possibly after Francis Russell, 5th Duke of Bedford (1765–1802), who was noted for his experiments in agriculture and sheep breeding at Woburn in Bedfordshire, England, and was a close friend of the Prince of Wales (later George IV). It may have been given after Bedford, NH, named after the 4th Duke of Bedford (1710–71), a statesman who negotiated the end of the Seven Years' War, 1757–62.

Beebe Plain QC Part of the town of Stanstead on the Vermont border, and in the regional county of Memphrémagog, south of Magog, it had been an incorporated village from 1873 to 1995. Loyalist settlers Zeeba, David, and Calvin Beebe had come from Vermont in 1789. It is locally known as Beebe.

Beechey Island NT On the south side of Devon Island, this island was named in 1819 by William (later Sir William) Parry after Lt Frederick (later Rear-Adm. Sir Frederick) William Beechey (1796–1856), who had served on the *Hecla* during Parry's 1819–20 expedition. **Cape Beechey**, on the south side of Melville Island, was also named for him in 1820 by Parry.

Beechwood NB Located on the east bank of the Saint John River, north of Woodstock, this place's post office was named *Bumfrau* in 1877, taking its name from the French *bois franc*, 'hardwood'. It was renamed Beechwood in 1893. Bumfrow Brook flows into the river at Beechwood.

Beeton ON In the town of New Tecumseth, Simcoe County, it was first known as *Clarksville*, after Robert Clark, the first settler. It was renamed in 1876 in honour of David Alanson Jones, noted as the 'bee king', and the founder of the Ontario Beekeepers Association. It is also said to have been named after Beeton Castle in Scotland, near the birthplace of Jones's mother-in-law.

Begbie, Mount BC Located southwest of Revelstoke in the Monashee Mountains, this mountain (2732 m) was named in 1886 by Dr Otto Klotz after Sir Matthew Baillie Begbie (1819–94), chief justice of British Columbia, 1869–94. He firmly established British law during turbulent times in the colony after 1858. There are two other mountains in the province named after him, one on the Queen Charlotte Islands, southwest of Masset Inlet, the other in the Cariboo, midway between 70 Mile House and 100 Mile House.

Beiseker AB Northeast of Calgary, this village (1921) was the centre of the 101,000 ha (250,000 a) Rosebud Tract purchased by the Calgary Colonization Company from the Canadian Pacific Railway. The company was owned by two wealthy North Dakota businessman, one of whom was Thomas Beiseker, who donated the townsite in 1910. The CPR station was named that year.

Belcarra BC A village (1979), on the east side of the mouth of Indian Arm, and northwest of Port Moody, it was named in the 1870s by Judge Norman Bole after the Gaelic words *bal*, 'sun', and *carra*, 'cliff' or 'rock'. He stated that the name meant 'fair land of the sun'.

Belcher Islands NT On the southeast side of Hudson Bay, these islands may have been named as early as 1727 by William Coats of the Hudson's Bay Company, after James Belcher, who supplied the HBC posts from 1715 to 1724.

Belfast PE A municipal community (1972), southeast of Charlottetown, it had been named about 1770 by Capt. James Smith of HMS *Mermaid* after Belfast, Ireland. Its post office was opened in 1832. The municipality embraces the areas of Eldon, Point Prim, Pinette, Belle River, and Wood Islands.

Belfountain ON On the Credit River and in the town of Caledon, Peel Region, south of Orangeville, it was named in 1853 by postmaster Thomas J. Bush, possibly after the clear water of the river. It had been earlier called *Tubtown* and *McCurdy's Village*.

Belgrave ON In Huron County, northeast of Goderich, this place was named in 1865 after Belgrave in Leicestershire, England, 4 km north of Leicester. It was first called *Haggerty's Corners* after Dan Haggerty, who managed a tavern here.

Bella Coola BC Situated at the head of North Bentinck Arm and at the mouth of the **Bella Coola River**, this community's post office was established here in 1900, having been located at the site of Hagensborg, five years earlier. The name of the river, derived from the Kwakiutl description of the Bellacoola tribe of the Coast Salish, was noted in 1862 as *Bel-houla* by Cdr Richard Mayne.

Bellechasse QC A regional county, east of the city of Québec, it was created in 1982 by amalgamating the counties of Bellechasse and Dorchester. The name traces its roots to the township of Bellechasse, created in 1871, to the seigneury of Bellechasse, established in 1672, and to *Isle de Bellechasse*, noted by Samuel de Champlain in 1632.

Belledune NB A village (1968) on the south shore of Chaleur Bay, northwest of Bathurst, its post office was named in 1845 after **Belledune Point**. In 1839 the point was referred to as *Grand Belle doune*, suggesting the sand dune there was perceived as being especially pretty (*belle*) by the French.

Belleisle Bay NB A northeasterly extension of the Saint John River, 28 km north of Saint John, this bay was named after Alexandre Le Borgne de Belle-Isle, who lived here after

1736. His father, also Alexandre (*c.* 1643–*c.* 1693), was an acting governor of Acadia in the late 1600s.

Belle Isle, Strait of NF Separating the island of Newfoundland from Labrador, this strait was first named *Grande Baye* before Jacques Cartier's 1534 voyage, when he called it *Baye des Chasteaux*, after the striking castle-like basaltic rocks on present Castle and Henley islands. The island at its entrance was first called *I de la Fortuna* by the Portuguese about 1500, and then *Belle Isle de la Grande Baye* by pilot Jean Fonteneau *dit* Alfonse in his 1544 *Cosmographie*. The strait was subsequently named after the island.

Belleoram NF A town (1946) on the west side of Fortune Bay, it was named after a French adventurer who spent several winters there in the late 1600s. It was occupied by English settlers after 1713, and was initially known as *Belorme's Place*.

Belle River ON In Essex County, east of Windsor, this town (1969) was settled in the early 1800s by pioneers of French extraction, with the nearby stream being called *Belle Rivière*. The post office was called *Rochester*, after the township, in 1854. Twenty years later, when Belle River became a village, the postal name was changed to agree.

Belle River PE At the mouth of Belle River, soouthwest of Montague, this place's post office was named *Belle Creek* in 1867. Exactly a century later it was renamed Belle River. Military engineer Louis Franquet identified the river in 1751 as *Belle Rivière*. The adjacent **Bell Point** was settled about 1800 by a Bell family from Scotland.

Belleterre QC A town in the regional county of Témiscamingue, east of Ville-Marie, it was founded in 1942, and named after the Belleterre Gold Mines, which opened a mine here in 1935.

Belleville ON A city (1877) in Hastings County, it was called *Singleton's Creek* in the 1780s after Capt. George Singleton, who built a trading post in 1784 on the east bank of the Moira River. In 1790, a year after Singleton died, Capt. John Walden Meyers (1745–1821) built the first grist mill above the mouth of the Moira, and later built a sawmill, a distillery, a trading post, and a brick kiln. The place was informally called *Meyers' Creek* until 1816, when Lt-Gov. Francis Gore renamed the new *Singleton's Corners* post office after his wife, Annabella (Bella) Gore.

Bellevue AB Part of the municipality of Crowsnest Pass since 1979, this former village (1957) was named in 1905 by J.J. Fleutôt, founder of the West Canadian Collieries Ltd. He proposed the name after his daughter had exclaimed *Quelle une belle vue*, 'what a beautiful view'.

Bellevue NF Located on Tickle Harbour, near the head of Trinity Bay, this place was originally called *Tickle Harbour*. Because of rocks, tricky currents, and hazardous tides, the harbour is quite ticklish to negotiate. Since there were other places known by the name, it was changed in 1896 to promote the good view of the bay.

Bell Ewart ON In the town of Innisfil, Simcoe County, southeast of Barrie, this place was named in 1855 after James Bell Ewart (1801–53) of the town of Dundas, who had laid out village lots here in 1853.

Bell Island NF In Conception Bay, northwest of St. John's, it may have been named by the French after Belle Isle, off the south coast of Bretagne, France. John Guy referred to it in 1612 as *great Belile*, and it was called *Belle Isle* on several maps between 1669 and 1868. It became known as Bell Island only in the late 1800s. The story that it was named because it appeared to be an inverted bell cannot be traced before 1819.

Belliveaus Cove NS In Digby County, southwest of Digby, this community was founded after 1755 by Charles Belliveau, who returned to Nova Scotia after the expulsion of the Acadians. *Belliveau Cove* post office was opened in 1856.

Bell, Mount BC In the Coast Mountains, northwest of Mount Waddington, this mountain (3252 m) was named in 1928 by W.A.D. (Don) Munday after climber Col Dr F.C. Bell.

Bell, Mount YT West of Carcross, this mountain (1929 m) was named in 1909 by geologist D.D. Cairnes after Walter Andrew Bell (d. 1969), his field assistant. Bell became the director of the Geological Survey of Canada, 1949–53.

Bells Corners ON An urban centre in the city of Nepean, Ottawa-Carleton Region, it was named after Hugh Bell (*c.* 1798–1872), who opened a tavern here in 1834, which he continued to run until 1863.

Belly River AB Rising in Montana, this river flows north to join the Oldman River, upriver from Lethbridge. In 1858 surveyor Thomas Blakiston adapted the Blackfoot *mokowanis*,'big bellies', referring to the Atsina tribe, into Belly River, applying it all the way to its junction with the Bow River, 90 km downriver from Lethbridge. Disgusted with having Lethbridge on a river with an embarrassing name, residents appealed for a new one. A solution was reached in 1915 by making the Belly a tributary of the Oldman.

Belmont MB Southeast of Brandon, this place's post office was first called *Craiglea* in 1884. John Bell had settled here in 1883, and when the Canadian Northern Railway arrived five years later, he requested its station be called *Bellsmount*, which was contracted to Belmont. The post office was changed to Belmont in 1891.

Belmont NS Northwest of Truro, this place's post office was named in 1879, possibly after Belmont, NH, where some of its first settlers came from. It was known as *Vil Nigeganish* by the Acadians in the early 1700s, after the Mi'kmaq name of the Chiganois River, *Nesakunechkik*, 'eel weirs'. The first post office in 1865 was called *Chigonaise River.*

Belmont ON A township in Peterborough County, it was named in 1823, but the source of the name is not known. Belmont House in Ireland's County Waterford and Belmont Castle in Scotland's Perthshire are possible sources. It is part of the municipal township of **Belmont and Methuen**. The village (1961) of **Belmont** in Elgin County, northeast of St. Thomas, was named in 1854, when its post office was opened.

Belœil QC In the regional county of La Vallée-du-Richelieu and on the west bank of the Rivière Richelieu, east of Montréal, this city (1914) was founded as a village in 1903. Its origin may be traced to the religious parish of Saint-Mathieu-de-Belœil, established in 1772, and is believed to be named after the beauty of the view from the summit of Mont Saint-Hilaire.

Belwood ON At the head of **Lake Belwood**, a widening of the Grand River, this place in Wellington County, northeast of Fergus, was first called *Skenesville*, after George Skene, then *Garafraxa*, after West Garafraxa Township, and *Douglas*, after George Douglas Fergusson, who, with John Watt, laid out village lots. In 1884 Dr J.G. Mennie organized a court of the Canadian Order of Foresters in *Douglas.* He called it Court Belwood, likely after Bellwood, near Edinburgh, Scotland. When confusion resulted in mail being sent to Douglas in the Ottawa Valley, the place was renamed Belwood in 1885.

Bengough SK Established in 1911 by the Canadian Northern Railway, this town (1958) is southeast of Moose Jaw. It was named after John Wilson Bengough (1851–1923), founder of the satirical weekly *Grip*, 1873-94, and a brilliant cartoonist and popular lecturer. Its post office was opened in 1912.

Benito MB A village (1940), almost on the Saskatchewan border, southwest of Swan River, its post office was established in 1903. It may have been named by the wife of the first postmaster, A.C. Dykeman, after a character in a novel.

Benmiller ON In Huron County, east of Goderich, this place was named in 1855 after Benjamin Miller (1801–58), an Englishman, who had arrived here in 1832 with his brothers, Daniel and Joseph.

Bennett Lake YT Extending southwest of Carcross to the British Columbia border, this lake was named in 1883 by American army Lt Frederick Schwatka after James Gordon Bennett, editor of the *New York Herald* and a promoter of American exploration.

Benoits Cove NF Part of the town of Humber Arm South since 1991, this place is located west of Corner Brook. It was probably named after descendants of Michael Benoit (1810–80), a native of Cape Breton Island, who settled on the Port au Port Peninsula.

Bentinck ON A township in Grey County, it was named in 1840 for Lord William Henry Cavendish Bentinck (1774–1839), governor general of Bengal, 1828–35.

Bentley AB Northwest of Red Deer, the post office in this village (1915) was named in 1900 after mill worker George Bentley, in preference to *Macpherson*, after merchant Maj. Macpherson.

Berens River MB, ON Rising in Ontario, this river flows west through Family Lake into Lake Winnipeg. The Cree knew it as *Omeemee Sibi*, 'pigeon river'. Pigeon River also flows into Lake Winnipeg, 10 km to the south, from the same source, Family Lake. The river was named after Cree chief Joseph Berens. **Berens River** post office, at the river's mouth, was opened in 1900.

Beresford NB A town (1984) on the northwest side of Bathurst, its post office was named in 1883. It took its name from **Beresford Parish**, named in 1814 after William Carr Beresford, Viscount Beresford (1768–1854). He served with distinction during the Peninsular War in Spain, 1808–14 and was the marshal of the Portuguese Army, 1809–19.

Berkeley ON In Holland Township, Grey County, southeast of Owen Sound, this community was named *Holland* in 1853, but was renamed Berkeley four years later. An early settler may have named it after Berkeley Square, in the Mayfair district of London, England.

Bernières QC A municipality in the regional county of Les Chutes-de-la-Chaudière, southwest of the city of Québec, it was named in 1968 after the French commune of Bernières-sur-Mer, in Normandie, France. It had been established in 1912 as *Saint-Nicolas-Sud*.

Bernierville QC Located in the regional county of L'Érable, west of Thedford Mines, this village was named in 1898 after Julien-Melchoir Bernier (1825–87), the pastor of the religious parish of Saint-Ferdinand-d'Halifax from 1850 to 1886. The local residents prefer to call the place Saint-Ferdinand-d'Halifax.

Berthierville QC A town in the regional county of D'Autray, and on the north side of the St. Lawrence River, opposite Sorel, it was named as the village of *Berthier* in 1852, and changed to Berthierville in 1942. The original county of *Berthier* was possibly named after Isaac (Alexandre) Berthier, Sieur de Villemur (1638–1708), who had a seigneury east of the city of Québec, near Montmagny.

Bertrand NB A village (1968) on the west side of the town of Caraquet, its post office was named in 1893 after a local Acadian family name.

Berwick NB Located 10 km northwest of Sussex, this place was named in 1864 by the Revd Richard Smith after Berwick, NS. Its post office was called *Fenwick* from 1862 to 1928, with F. Fenwick as the first postmaster. Matthew Fenwick had settled here in 1801.

Berwick NS A town (1923) in the Annapolis Valley, west of Kentville, it was named in

1851 after Berwick, ME, on the New Hampshire border. A local resident had passed through there and was impressed with its neatness. In the early 1800s, it was called *Congdon Settlement* for pioneer settlers Benjamin and Enoch Congdon. In the 1820s it was known as *Pleasant Valley*, and was later called *Davisons Corner* after William Davison, who arrived here in 1835.

Berwick ON First known as *Cockburn's Corners*, after early settlers Adam and Peter Cockburn, this place in the united counties of Stormont, Dundas and Glengarry, 35 km northwest of Cornwall, was named *Finch*, after the township, when the post office was opened, 1842. It was renamed in 1858 after the brothers' Scottish birthplace in Berwickshire.

Berwyn AB North of the Peace River, and west of Grimshaw, this village (1936) was named in 1922 after the Berwyn Hills in Denbighshire, Wales.

Bethany ON This place in Victoria County, southwest of Peterborough, was named in 1859 after Bethany in the Holy Land. Bethany (now called Al Ayzariyah) is on the eastern side of Jerusalem and on the flank of the Mount of Olives, from where Jesus made his triumphant return to Jerusalem.

Bethune SK Located 50 km northwest of Regina, this village (1912) was named in 1890 by the Qu'Appelle, Long Lake and Saskatchewan Railway and Steamboat Company, possibly after entomologist Charles James Stewart Bethune (1838–1932).

Betsiamites QC A Montagnais reserve, east of the mouth of Rivière Betsiamites, and southwest of the city of Baie-Comeau, the roots of its name have had varied spellings, beginning with *Bersiamiste* by Champlain in 1632. Its post office was opened in 1863 as *Bersimis*, and was later *Moulin-Bersimis* and *Rivière Bersimis*. Although the reserve was named *Betsiamites* in 1861, Bersimis persisted until 1981, when the band persuaded postal authorities to accept its preference. The name may refer to 'place of leeches, lampreys, or eels'.

Bewdley ON In Northumberland County and at the west end of Rice Lake, north of Port Hope, it was named in 1857 by William Banks, a native of Bewdley, Worcestershire, England, southwest of Birmingham.

Bible Hill NS On the northeast side of Truro, this village (1953) was first settled about 1761. It owes its name to the practice of the Revd Dr William McCulloch, pastor of the First Presbyerian Church in Truro, 1839–85, of giving away copies of the Bible. He lived in the area of the present-day village.

Biddulph ON A township in Middlesex County, it was named in 1830 for Robert Biddulph, a director of the Canada Company, a British company organized in the 1820s to develop the Huron Tract, a large area of what is now Southwestern Ontario.

Bide Arm NF On the east coast of the Northern Peninsula and between Roddickton and Englee, this municipal community (1970) was founded in 1969 as a centre for resettlement.

Biencourt QC A municipality in the regional county of Témiscouata, southeast of Trois-Pistoles, it was created in 1947, taking its name from the township, which was proclaimed in 1920. It was given for Charles de Biencourt de Poutrincourt de Saint-Just (*c.* 1591–*c.* 1623), vice-admiral of the seas of New France, and commander of Port-Royal in Acadia.

Bienfait SK A town (1958) east of Estevan, its post office was opened in 1892, closed in 1895, and reopened in 1903, when the Canadian Pacific Railway was built to Estevan. It was named after Antoine Charles Bienfait, a financier with the Amsterdam financial firm of Adolphe Boissevain and Company. He was born in Amsterdam in 1857.

Bienville, Lac QC A lake with an area of nearly 1000 km^2 in northern Québec, near the head of the Grande rivière de la Baleine, it was named in 1916 after Jean-Baptiste Le Moyne de Bienville (1680–1767), who took part in the capture of English trading posts on Hudson and

James bays at the end of the seventeenth century.

Biggar SK West of Saskatoon, this town (1911) was founded in 1908 by the Grand Trunk Pacific Railway and named after its general counsel, William Hodgins Biggar. Its post office was opened in 1909. The town promotes itself with the slogan 'New York is Big, but this is Biggar'.

Big Presque Isle Stream NB Rising near the city of Presque Isle in Maine (where it is called Prestile Stream), this stream flows southeast to enter the Saint John River south of Florenceville. It was named after a peninsula (*presqu'île* in French) opposite the stream's mouth, not after the Maine city, where there is another Presque Isle Stream, a tributary of the Aroostook River.

Big River SK Situated northwest of Prince Albert, this town (1966) was established in 1909 by the Canadian Northern Railway. It is located at the mouth of **Big River**, which flows into a narrow, 50-km-long lake, called by the Cree *Oklemow Sipi*, 'big river'.

Big Sevogle River NB A tributary of the Northwest Miramichi River, its name is derived from Mi'kmaq *sewokulook*, 'river of many cliffs', a description that may have originally applied to the Little Sevogle River. The Big Sevogle may have been *Elmunakuncheech*, 'little beaver hole', with *Elmunakun,* 'beaver hole', being the Northwest Miramichi.

Big Valley AB South of Stettler, the post office in this village (1942) was named in 1907 after a wide valley extending southwesterly from Ewing Lake to the Red Deer River.

Billings, Mount YT In the Logan Mountains, north of Watson Lake, this mountain (2106 m) was named in 1887 by geologist George Dawson after Elkanah Billings (1820–76), palaeontologist with the Geological and Natural History Survey of Canada.

Binscarth MB A village (1917) northwest of Brandon, it was established as a pure-bred stock farm by the Scottish, Ontario and Manitoba Land Company and managed by William Bain Scarth. The post office, called *Binscarth Farm* in 1893, and renamed Binscarth in 1891, was named after Binscarth, on the Orkney island of Mainland, the location of Robert Scarth's farm.

Birch Hills SK Southeast of Prince Albert, the post office in this town (1960) was named in 1910. It took its name from birch trees lining the hills on both sides of the town.

Birchtown NS West of Shelburne, this place was established by African-Americans in 1783 and named after the birch trees found here. Its post office was opened in 1877.

Birds Hill MB Located northeast of Winnipeg, this community has its roots in the grant of 504 ha (1245 a) of land to James Curtis Bird (*c.* 1773–1856) on the east side of the Red River, after he retired in 1824 as the Hudson's Bay Company's chief factor at Lower Red River. Birds Hill post office was named in 1879 by Bird's son, Dr Curtis James Bird (1838–76), who was born at the Birds Hill estate. The settlement was also known as *Roseneath.*

Birtle MB A town (1884) northwest of Brandon, its post office was named in 1879 by contracting the name Birdtail Creek. The branches of the creek are said to resemble a bird's tail. Hudson's Bay Company maps identified *Fort Birdtail* at the junction of the creek and the Assiniboine River, a stopping place on the way to Fort Edmonton. When the settlers arrived in 1878, they wanted to call the place *St. Clair City*, but it was rejected because of similar names elsewhere in Canada.

Biscotasing ON On the shore of **Biscotasi Lake** and in Sudbury District, northwest of Sudbury, this place was established on the transcontinental Canadian Pacific Railway line in 1882. In Ojibwa, the name means 'water body with long arms'.

Bishop's Cove NF On the north side of Spaniard's Bay and south of Harbour Grace, this municipal community (1969) was first known as *Bread and Cheese Cove*, after the local description of the hawthorne tree's leaves and berries. It was renamed in 1827 or 1832 by the Rt Revd John Inglis (1777–1850), who made two lengthy tours of Newfoundland, when it was part of his Church of England diocese of Nova Scotia.

Bishop's Falls NF This town (1961) is on the Exploits River, northeast of Grand Falls-Windsor. It was named after the Rt Revd John Inglis (1777–1850), who made two lengthy tours of Newfoundland, when it was part of his Church of England diocese of Nova Scotia. He visited the site of the falls in 1827.

Bishopton QC Situated in the regional county of Le Haut-Saint-François and northeast of Sherbrooke, this village was named in 1932. It had earlier been called *Bishop's Crossing* after the family of John Bishop, a Vermont native, who had served in Benedict Arnold's regiment in 1775.

Bittern Lake AB Between Wetaskiwin and Camrose, the post office here was named in 1899 after the nearby lake, whose name was derived from a genus of birds allied to the heron. The village was incorporated as *Rosenroll* in 1904 and changed to Bittern Lake seven years later.

Bizard, Île QC Located on the northwestern side of Île de Montréal, it was named after Jacques Bizard (1642–92), who was appointed in 1678 as the governor of the island (then called *Bonaventure*).

Bjorkdale SK Situated southeast of Tisdale, this village (1968) was first settled by Sid Coppendale and Charlie Bjork. Its post office was named in 1911 by combining parts of their surnames.

Blackburn Hamlet ON In the city of Gloucester, Ottawa-Carleton Region, this neighbourhood was developed in the 1960s, and by the 1970s the name of the former community of *Blackburn* fell out of use. *Blackburn* had been named in 1876 for Robert Blackburn (1828–94), an early mill owner, Gloucester Township reeve, and the Liberal member in the House of Commons for Russell, 1874–8.

Blackcomb Peak BC In the Coast Mountains, southeast of Whistler, this mountain (2438 m) was named by W.A.D. (Don) and Phyllis Munday, who observed its serrated ledge of black rock in 1923.

Black Diamond AB Southwest of Calgary, the post office in this town (1956) was named in 1907 after Addison McPherson's Black Diamond Coal Mine. The Arnold brothers, who were to operate the post office in their general store, had wanted Arnoldville, but an elderly Aboriginal pulled Black Diamond from a hat.

Black Donald Lake ON A reservoir since 1967 on the Madawaska River, and on the Lanark-Renfrew boundary, upriver from Arnprior, it recalls early lumberman Black Donald McDonald, and the Black Donald graphite mine, now inundated. The upper part of the reservoir is called Centennial Lake.

Blackfalds AB Between Red Deer and Lacombe, this town (1980) was first known as *Waghorn*, when a post office was opened here in 1891 and called after the postmaster. It was renamed in 1902 by a Canadian Pacific Railway engineer, after Blackfaulds, possibly an estate in Scotland.

Blackhorn Mountain BC In the Coast Mountains, northeast of Mount Waddington, this mountain (3033 m) was named in 1933 by mountaineer H.S. Hall Jr of Cambridge, MA. A black rock at its summit reminded him of the horn of a saddle, which he deemed appropriate for this cattle ranching country.

Blackie AB Southeast of Calgary, the post office in this village (1912) was named in 1911 after Scottish novelist John Stuart Blackie (1809–95), founder of the publishing firm Blackie and Sons.

Black Lake QC A town in the regional county of L'Amiante, it was named a village in 1906, and a town two years later. The nearby lake was called Lac Noir when the developers of the asbestos mine in the 1890s translated the name for the new urban centre into English, and Lac Noir remains its name.

Black, Mount YT Northeast of Whitehorse, this mountain (2158 m) was named in 1935 by geologist Hugh S. Bostock after George Black (1872–1965), commissioner of the Yukon, 1912–6, and member of the House of Commons, 1921–35 and 1940–9. He was the speaker of the House of Commons, 1930–5.

Black River ON Rising 135 km east of Washago, this river flows into the Severn River downstream from the point where the West, Centre, and East branches of the Severn flow from Lake Couchiching and unite before reaching the mouth of the Black River. The clear waters of the Severn are turned much darker by the Black's waters.

Black River-Matheson ON A municipal township in Cochrane District, it was created in 1969 with the union of the municipal township of Black River and the town of Matheson. Playfair Township and the improvement district of Kingham were added four years later.

Blacks Harbour NB A village (1972) on the Bay of Fundy coast, southwest of Saint John, its post office was named in 1885. It is noted for fish processing and packing. The harbour was identified on a chart by William Fitz William Owen in 1847.

Blackstock ON In Scugog Township, Durham Region, east of Port Perry, this place was first known as *Tooley's Corners* and *Williamsburg* before *Cartwright* post office, after the name of the original township, was opened in 1851. It was changed to Blackstock in 1887 for George Tait Blackstock, a Toronto lawyer and defeated candidate for Durham West in the House of Commons.

Blackville NB On the west bank of the Southwest Miramichi River, southwest of Newcastle (Miramichi), the post office in this village (1966) was named in 1842. It was named after **Blackville Parish**, which was established in 1830 and called after William Black (1771–1866), administrator of the provincial government during the absence of Lt-Gov. Sir Howard Douglas. Earlier it was called *The Forks*, for its location at the mouth of the Bartholomew River.

Blaine Lake SK Founded by the Canadian Northern Railway in 1911, this town (1954) southwest of Prince Albert was called after a nearby post office. The post office was named in 1907 after a surveyor who had drowned in one of the nearby Blaine Lakes, which were then named after him.

Blainville QC Situated in the regional county of Thérèse-De Blainville, this city northwest of Montréal was named in 1968. In 1845 it had been proclaimed as the parish municipality of *Sainte-Thérèse-de-Blainville*, which acquired its name from Louis-Jean-Baptiste Céleron de Blainville (1696–1756), whose wife inherited the seigneury of Mille-Îles in 1730.

Blair ON An urban centre in the city of Cambridge, and on the west side of the Grand River, it was first known as *Shinglebridge*, *Durhamville*, and *New Carlisle*. It was renamed Blair in 1858 in honour of Adam Johnston Fergusson Blair (1815–67), a son of Adam Fergusson, the founder of Fergus. He added the name Blair on inheriting a Scottish estate. He served in the Legislative Assembly of the province of Canada, 1849–57, the Legislative Council, 1860–7, and the Senate in 1867, when he was president of the Privy Council in John (later Sir John) A. Macdonald's first cabinet. **Blair Township** in Parry Sound District was also named for him.

Blairmore AB In the municipality of Crowsnest Pass since 1979, it was named in 1898 after two Canadian Pacific Railway contractors called Blair and More (or Moore). It had been incorporated as a town in 1911.

Blakeney ON On the Mississippi River and in Lanark County, this place was first called *Norway Pine Falls*, *Snedden's Mills*, and *Rosebank* before being named in 1874 after Dominick Edward Blake (1833–1912), premier of Ontario, 1871–2, member of the House of

Commons, 1873–91, and leader of the Liberal Party of Canada, 1880–7.

Blaketown NF A local service district (1993) on the west side of Dildo Pond and north of Whitbourne, it had been called *Dildo Pond* until 1888, when it was renamed after Sir Henry Arthur Blake (1840–1918), governor of Newfoundland, 1887–8.

Blanc-Sablon QC A municipality in the regional county of Côte-Nord-du-Golfe-Saint-Laurent, and just west of Newfoundland's Labrador border, it was noted as early as 1534 in the *Relation* of Jacques Cartier, who had erected a cross at nearby Lourdes-de-Blanc-Sablon. The name means 'little white sand'. The occurrence of a cove called Blancs-Sablons, near Saint-Malo, France, from where Cartier sailed, may only be coincidental. The community of **Lourdes-de-Blanc-Sablon** was attached to the municipality when it was created in 1990.

Blandford ON Part of the municipal township of **Blandford-Blenheim** in Oxford County since 1975, Blandford Township was named in 1798 for one of the titles of John Churchill, the Duke of Marlborough (1650–1722), who had been made the Marquess of Blandford in 1702.

Blanshard ON A township in Perth County, it was named in 1830 for Richard Blanshard, a director of the Canada Company, a British company founded in 1826 to develop the Huron Tract, a large area of Southwestern Ontario.

Blanshard, Mount BC Between Pitt and Alouette lakes, and north of Haney, this mountain (1715 m) was named in 1859 by Capt. George Richards after Richard Blanshard (1817–94), the first governor of Vancouver Island, 1849–51. Its two peaks are called Golden Ears, mentioned by Cdr Richard Mayne in 1862. Golden Ears Provincial Park embraces a large area of this part of the province.

Blenheim ON A town (1885) in Kent County, it was named in 1883 after a battle that took place in 1704, when John Churchill, the Duke of Marlborough, defeated the French and Bavarian forces. Blenheim had been proposed in 1849 for the post office, but because an office was then called that in Oxford County, it was called *Rondeau* after nearby Rondeau Harbour until 1883.

Blenheim ON Part of the municipal township of **Blandford-Blenheim** in Oxford County since 1975, Blenheim Township was named in 1798 after Blenheim Palace, the home of John Churchill, the Duke of Marlborough (1650–1722) at Woodstock in England's Oxfordshire. In 1704 Churchill won a battle at Blenheim, Germany, near Augsburg.

Blewett BC This community, on the south bank of the Kootenay River, west of Nelson, was named in 1923 after first postmaster W.J. Blewett.

Blind River ON At the mouth of the **Blind River**, this town (1906) in Algoma District, east of Sault Ste. Marie, was named in 1877 after the hidden outlet of the river, as perceived from a boat in the North Channel of Lake Huron.

Blockhouse NS West of Mahone Bay, this community's post office was named *Block House* in 1876 after a fortified blockhouse, which had been built on a hill top in 1753.

Blomidon, Cape NS The eastern extremity of North Mountain, north of Wolfville, its name is attributed to a contraction of 'blow me down', a nautical phrase used throughout the Atlantic provinces for places susceptible to high winds. Samuel de Champlain called it *Cap de Poutrincourt* after Jean de Biencourt de Poutrincourt et de Saint-Just, lieutenant-governor of Acadia, 1606–14, who had scaled its cliff. It was also known as *Cap Baptiste* and *Cape Porcupine*.

Bloomfield NF On Goose Bay, an arm of Bonavista Bay, and north of Clarenville, this place was settled after the mid-1850s. Good soils at the site permitted the growing of vegetables, with 60 barrels of potatoes having been produced in 1865.

Bloomfield ON A village (1906) in Prince Edward County, west of Picton, it was named about 1832 after the profusion of wild flowers in the vicinity.

Blue Nose Lake NT Northwest of Coppermine, this lake (391 km^2) was named in 1953 by wildlife biologist John P. Kelsall. He chose the name because both he and his assistant, James Mitchell, were natives of Nova Scotia. He had in mind both the well-known nickname of Nova Scotians and the famous schooner *Bluenose.*

Blue River BC On the west bank of the North Thompson River, this community was named in 1917 after **Blue River**, which rises in the Cariboo Mountains. The Revd George Monro Grant explained in his *Ocean to Ocean* (1873) that the river was named after 'the deep soft blue of the distant hills, which are seen from its mouth well into the gap through which it runs'.

Blue Rocks NS East of the town of Lunenburg, this seaside community was settled about 1760. Its name is derived from the stone along the seashore, which casts a dark blue when it is wet. A post office was opened here in 1882.

Blumenort MB Located west of Emerson and on the North Dakota border, this Mennonite community's post office was named in 1880, closed 10 years later, and reopened in 1962. In German, it means 'flower place', or 'flower nook'.

Blyth ON A village (1876) in Huron County, east of Goderich, its post office was named *Blythe* in 1856 after James Blyth of England, who owned a large part of the site. The postal name was changed to Blyth in 1888.

Bobcaygeon ON A village (1876) in Victoria County and on **Bobcaygeon River**, it was named after the river in 1832. The river's name is from from Mississauga *bob-ca-je-won-unk*, 'shallow rapids'.

Boiestown NB In the valley of the Southwest Miramichi River, north of Fredericton, this place was named in 1842 after Thomas Boies. He came from the United States in the early 1820s, and established sawmills and a grist mill at the mouth of the Taxis River.

Boisbriand QC In the regional county of Thérèse-De Blainville, this city west of the cities of Sainte-Thérèse and Blainville was named in 1974. The name recalls Michel-Sidrac Dugué de Boisbriand (1638–88), who was granted the seigneury of Mille-Îles in 1672.

Boischatel QC A municipality in the regional county of La Côte-de-Beaupré, northeast of the city of Québec, it was called *Saint-Jean-de-Boischatel* in 1920, and shortened to its present form in 1991. It was named for Jean-François de Beauchatel, a military officer who died in 1760.

Bois-des-Filion QC A town (1980) in the regional county of Thérèse-De Blainville, east of the cities of Sainte-Thérèse and Blainville, its post office was named *Bois-de-Filion* in 1855, and modified to its present form in 1932. The name recalls a large maple bush owned by a man called Filion.

Bois, Lac de NT This lake (469 km^2), northwest of Great Bear Lake, was identified on Fr Émile Petitot's map of 1875. He had visited the area three years earlier. The name means 'lake of the woods'.

Boissevain MB Situated directly south of Brandon, this town (1906) was first known as *Cherry Creek*. Its post office was named in 1886 after Athanase Adolphe Boissevain (1843-1921) of the Amsterdam firm of Adolphe Boissevain and Company, which floated the shares of the Canadian Pacific Railway in the European financial markets.

Bolton ON An urban centre in the town of Caledon, Peel Region, it was first called *Bolton Mills* in 1824 after James and George Bolton had built mills on the Humber River. *Albion*, after the township, was the post office from 1832 to 1892. Bolton was a village from 1873 to 1974.

Bompas, Mount YT This mountain (3056 m) in the St. Elias Mountains, west of Kluane Lake, was named in 1918 after the Rt Revd William Carpenter Bompas (1834–1906). He was the first Anglican bishop of Athabasca, (1874–84), of Mackenzie River (1884–91), and of Selkirk (Yukon), 1891–1906.

Bon Accord AB North of Edmonton, this town (1964) received its name in 1896, when Alexander 'Sandy' Florence proposed Bon Accord for the name of a new school district. He had given the name to his farm, deriving it from the motto of his native Aberdeen, Scotland. The post office was opened in 1901.

Bonanza Creek YT Flowing into the Klondike River, 5 km upriver from Dawson, it was known as *Rabbit Creek*, until the discovery of gold in its gravel beds in 1896 by George Carmack, Skookum Jim Mason, and Tagish Charlie touched off the great Klondike Gold Rush. The name was given by Carmack after the Spanish word for rich ore deposits.

Bonaventure QC The regional county is located southwest of Gaspé. The source of its name may be traced to the first discoveries of Jacques Cartier, who noted in 1534 the existence of **Île Bonaventure** at the mouth of Chaleur Bay. The island's name may have related to the excellence of the fishery around the island. The municipality of **Bonaventure** in the regional county was named in 1955, having been called *Saint-Bonaventure-de-Hamilton* in 1884.

Bonavista NF This town (1964) is located at **Cape Bonavista** and on the southeast side of **Bonavista Bay**. The cape may have been named about 1501 by Portuguese explorer Gaspar Corte-Real after Boa Vista, one of the Cape Verde Islands, off west Africa. There is also a claim that John Cabot exclaimed '*O buona vista, O happy sight*', on reaching its shore in 1497, but the claim lacks any specific evidence. In 1534 Jacques Cartier reported sighting *Cap de Bonne Vista*, suggesting it was well known to explorers and fishermen.

Bond Head ON In the town of Bradford West Gwillimbury, Simcoe County, northwest of Newmarket, this place was named in 1832 for Sir Francis Bond Head (1793–1875), lieutenant-governor of Upper Canada, 1835–8.

Bon Echo ON In Frontenac County, north of Kaladar, this place is noted for its Algonkian pictographs on a 115-m cliff, on the east shore of Upper Mazinaw Lake in **Bon Echo Provincial Park**. From the top of the cliff, the acoustics are excellent, with sounds echoing across the narrows between the Upper and Lower Mazinaw lakes.

Bonfield ON A township in Nipissing District, it was named in 1881 after lumberman James Bonfield (1825–83), Liberal member for Renfrew South in the Ontario Legislature, 1875–83. The community of **Bonfield** was an incorporated town from 1905 to 1974.

Bonne Bay NF On the northwestern coast of the island of Newfoundland, this bay has two arms, which deeply penetrate the lofty cliffs of Gros Morne National Park. In the 1500s the Basques called it *Baya Ederra*, 'beautiful bay', which the French subsequently translated. In 1767 Capt. James Cook adopted the French name for his chart of the coast.

Bonnechere River ON A major tributary of the Ottawa River, it rises in Algonquin Park. As early as 1688 *R de la Bonnechere* occurred on maps, but its origins are obscure. With the meaning of 'good flesh', it possibly reflects the eating of a good meal by French voyageurs at the river's mouth.

Bonne-Espérance QC A municipality in the regional county of La Côte-Nord-du-Golfe-Saint-Laurent, it is located west of Blanc-Sablon. It was created in 1990, with the name having been used for a post office in 1878 and for a township in 1907.

Bonnet Plume River YT Rising in the Mackenzie Mountains, this river flows northwest to join the Peel River, just south of the Arctic Circle. It was named after Gwitch'in chief Andrew Flett, nicknamed Bonnet Plume, who lived on the river with his band. He guided many lost gold-seekers during the Klondike

Gold Rush, and was also an interpreter for the Hudson's Bay Company.

Bonney, Mount BC In the Selkirk Mountains, south of Rogers Pass, this mountain (3107 m) was named in 1889 by climber William S. Green after British geologist Thomas George Bonney (1833–1923), who was then president of the Alpine Club in London.

Bonnyville AB Northeast of Edmonton, the post office of this town (1948) was named in 1910 after Fr F.S. Bonny, who built the first Catholic church here that year. It had been earlier called *St. Louis de Moose Lake.*

Bonshaw PE Located southwest of Charlottetown, this municipal community (1977) was named by W.W. Irving, who had come from near Bonshaw Tower, Dumfriesshire, Scotland. Noted in a reference in 1839, its post office was opened 20 years later.

Boothia, Gulf of NT Separating Melville Peninsula on the east, and Boothia and Simpson peninsulas on the west, this large inlet of the Arctic Ocean was named in 1830 by John (later Sir John) Ross after Sir Felix Booth (1775–1850), sheriff of London, England, prominent distiller, and patron of Ross's expedition of 1829–33. Ross wrote that Booth was a 'singularly generous and spirited individual, whose fame and deeds will go down in posterity among the first of those whose character and conduct have conferred honour on the very name of a British merchant'.

Boothia Peninsula NT The most northerly (72°N) point of mainland Canada, this peninsula (32,300 km²) was named *Boothia Felix* in 1830 by John (later Sir John) Ross after Sir Felix Booth, patron of his expedition 1829–33. It was renamed by the Geographic Board before 1910.

Borden ON Located in Simcoe County, west of Barrie, this place was called *Camp Borden* when a military training centre was named in 1916 after Sir Frederick William Borden (1847–1917), minister of militia and defence, 1896–1911. The camp was renamed Canadian Forces Base Borden in 1966. It occupies 7,165 ha (17,704 a).

Borden SK A village (1907) northwest of Saskatoon, it was first called *Baltimore*, but it was renamed by the Canadian Northern Railway in 1905 after Sir Frederick William Borden (1847–1917), minister of militia and defence in Sir Wilfrid Laurier's cabinet, 1896–1911. Its post office also opened in 1905.

Borden-Carleton PE A municipal community (1995) at the end of the Confederation Bridge, joining the province to the mainland at Bayfield, NB, it was created through the amalgamation of the town (1919) of Borden and the municipal community (1977) of Carleton Siding. **Borden** was established in 1917 as the provincial terminal for ferry service to the mainland and was called after Sir Robert Borden (1854–1937), then prime minister of Canada. Its post office opened in 1918 as Port Borden. *See also* Carleton Siding.

Borden Island NT The most northerly of the Parry Islands, and in the Queen Elizabeth Islands, this island (2794 km²) was named in 1915 by Vilhjalmur Stefansson after Sir Robert Borden (1854–1937), prime minister of Canada, 1911–20.

Borden Peninsula NT A broad peninsula at the north end of Baffin Island, this peninsula was named in 1913 by Alfred Tremblay, who explored the area in 1912–13. He named it after the prime minister, Sir Robert Borden.

Bosanquet ON A town in Lambton County, northeast of Sarnia, it had been a township until 1 January 1995. The township had been named in 1830 after Charles Bosanquet, a director of the Canada Company, a British company set up in the 1820s to develop the Huron Tract, a large area of what is now Southwestern Ontario.

Boston Bar BC In the Fraser Canyon, between Hope and Lytton, this place was named in the late 1850s by the local Lillooet, who described Americans panning for gold on a bar in the river 'Boston men'. The site was the scene of the Battle of Boston Bar, where the Lillooet provoked a fight with the Boston men, with seven Lillooet being killed. The post office was on the west side of the river at the site of North Bend from 1884 to 1887. It was reopened at its current site in 1917.

Botha AB A village (1911) east of Stettler, it was named by the Canadian Pacific Railway in 1909 after Gen. Louis Botha (1862–1919), the first prime minister of the Union of South Africa, 1910–19. Its post office was opened the following year.

Bothwell ON A town (1886) in Kent County, northeast of Chatham, it was named in 1854 by the Toronto *Globe*'s George Brown (1818–80) after Bothwell in Lanarkshire, southeast of Glasgow, Scotland.

Botwood NF At the head of the Bay of Exploits and northeast of Grand Falls-Windsor, this town (1960) was first known as *Ship Cove*. In the late 1800s it was named *Botwoodville* after the Revd Edward Botwood (1828–1901), a Church of England archdeacon, 1894–1901. The name was subsequently shortened to Botwood.

Boucherville QC A city (1957) in the regional county of Lajemmerais and on the east side of the St. Lawrence, opposite Montréal, it was created a village in 1857. Pierre Boucher de Grosbois (1622–1717), the most respected Canadian of his time in his many roles of judge, writer, soldier, and governor, was granted the **Îles de Boucherville** in the St. Lawrence in 1672.

Bouchette QC Located in the regional county of La Vallée-de-la-Gatineau, south of Maniwaki, the municipality's name was first used in 1858 for a township, which had been named in 1858 after the noted geographer and surveyor general of Lower Canada, Joseph Bouchette (1774–1841).

Bouctouche NB A town (1985) at the mouth of **Buctouche River**, its post office was named *Buctouche* in 1842. It had been incorporated as the village of *Buctouche* in 1966. Persistent use of Bouctouche by its French-speaking residents resulted in its name being respelled in 1985, but leaving associated names unchanged. The river was identified on Pierre Jumeau's 1685 map as *R. chibouchouch*. The name may be derived from the Mi'kmaq *chebuktoosk*, 'big bay', in reference to **Baie de Buctouche**, at the mouth of the river.

Boularderie Island NS Separated from Cape Breton Island by the channels of the Great Bras d'Or and the Little Bras d'Or, this island was named after Louis-Simon Le Poupet et de la Boularderie (*c.* 1674–1738). He was the commandant of *Port d'Orléans*, at present-day Ingonish, from 1719 to 1738. In 1719 he was granted the island, then known as *Île de Verderonne*, and set out to establish an agricultural settlement here.

Bourget ON Called *The Brook* in 1880, this place in the united counties of Prescott and Russell, east of Ottawa, was renamed in 1910 after Ignace Bourget (1799–1885), bishop of Montréal, 1840–76.

Bowden AB A town (1981) southwest of Red Deer, its post office was named in 1892, when the Canadian Pacific Railway was built from Calgary to Edmonton. It was likely named by a CPR surveyor called Williamson, after the birth name of his wife. It is also said to have been named after either Bowdon, a suburb of Manchester, England, or after Bowden near Galashiels, Roxburghshire, Scotland.

Bowen Island BC In the entrance to Howe Sound, northwest of Vancouver, this island was named in 1860 by Capt. George Richards after Rear-Adm. James Bowen (1751–1835), master of HMS *Queen Charlotte*, Lord Howe's flagship during the remarkable British victory over the French on the Glorious First of June, 1794.

Bow Island AB Located 23 km east of the junction of the Bow and Oldman rivers (where they become the South Saskatchewan River), this town (1912) was named by the Canadian Pacific Railway in 1907 after an island in the Bow River. Presumably the names Grassy Lake, south of the junction, and Bow Island were accidentally switched by the post office department in Ottawa.

Bowmanville ON The largest urban centre in the municipality of Clarington, Durham Region, it was first called *Barber's Creek*, and then *Darlington Mills* and *Darlington*, before being named in 1854 after Montréal merchant, Charles Bowman, who had opened a dry goods store here 30 years earlier, and donated land for a school and a church.

Bow River AB Rising in the Rocky Mountains northwest of Lake Louise, it flows southeast through Calgary to join the Oldman River between Lethbridge and Medicine Hat, where the two together become the South Saskatchewan River. It took its name from the Cree *manachaban*, 'bow', in reference to the bows they made from Douglas fir saplings along its banks. The French called it *Rivière des Arcs*, which remains in the name of Lac des Arcs, an enlargement of the Bow at Exshaw.

Bowsman MB Located north of Swan River, and near the Saskatchewan border, this village (1949) received its name from **Bowsman River**, named by geologist Joseph Burr Tyrrell after Cree trapper and hunter Bowsman Moore, who was then living at Shoal River. First named Bowsman in 1901, the post office was called *Bowsman River* from 1908 to 1952, when Bowsman was restored.

Boyle AB A village (1953) southeast of Athabasca, it was named by the Alberta and Great Waterways Railway in 1914 after John Robert Boyle (1871–1936), then minister of education, and later a judge of the supreme court of Alberta. Its post office was opened in 1916.

Bracebridge ON A town (1889) in the district municipality of Muskoka, it was named in 1864 by postal official William D. LeSueur after Washington Irving's novel, *Bracebridge Hall, or the Humorists* (1822). It was extensively enlarged in 1971 by annexing several adjoining townships. *See also* Gravenhurst.

Brackley PE A municipal community (1983) 8 km north of Charlottetown's city centre, its post office was named *Brackley Point Road* in 1868. It took its name from **Brackley Bay**, 13 km further north, where legislative clerk Arthur Brackley apparently drowned in 1776. Samuel Holland had called the bay *Petersham Cove* in 1765 after William Stanhope, Viscount Petersham (1719–79), but the local name supplanted it.

Bradford West Gwillimbury ON This town was created in Simcoe County in 1991. **Bradford** was named in 1840 by Joel Flesher Robinson after the city of Bradford, in West Yorkshire, England, where he was born. It was an incorporated town from 1960 to 1991. **West Gwillimbury** was created as a township in 1798, and was possibly named after Lt-Col Thomas Gwillim (d. 1762), the father of Elizabeth Posthuma Gwillim Simcoe (1762–1850).

Braeside ON A village (1921) in Renfrew County, northwest of Arnprior, it was named in 1872 by W.J. McDonald, likely for Braeside in Renfrewshire, Scotland, west of Greenock.

Bragg Creek AB Southwest of Calgary, this community was named in 1911 after Albert W. Bragg, who homesteaded here in 1894.

Bralorne BC This place north of Pemberton was named in 1932 after the Bralorne Mines Limited, formed the previous year by the Bralco Development and Investment Company. It had acquired the Lorne Mine, opened here in 1897. The acronym 'Bralco' was created from British, Alberta, and Columbia.

Bramalea ON Developed in 1958 as a multi-use project, it is now a well-known area of the expanded city of Brampton in Peel Region. The name was composed from the initial letters of Brampton and Malton, plus the word 'lea', meaning meadow.

Brampton ON A city (1974) in Peel Region, Brampton was laid out about 1834 by John Elliott and William Lawson, natives of Brampton, Cumberland, England, northeast of Carlisle. Its post office was called *Chinguacousy*, after the township, from 1832 to 1851, when it was renamed Brampton.

Branch NF A municipal community (1966) on the southwest side of St. Mary's Bay, it was established about 1790 as a fishing settlement. It took its name from **Branch River**, which flows into **Branch Bay**.

Brandon MB The province's second largest city, it was incorporated in 1882, a year after its post office was opened. It took its name from Brandon House, erected southeast on the Assiniboine River in 1793 by the Hudson's Bay Company. It had been named after Douglas Hamilton, 8th Duke of Hamilton (1756–99), the head of the Douglas family, who in 1882 became the 4th Duke of Brandon, after a place in Surrey, England, when he entered the House of Lords.

Brant ON The county was named in 1852 after Joseph Brant, or Thayendanegea (1742–1807), the great Mohawk chief, who fought on the British side during the American Revolution. **Brant Township** in Bruce County was named in 1850, also for Joseph Brant.

Brantford ON In 1784, following the American Revolution, Joseph Brant led the Six Nations confederacy to the site of this city (1877) on the Grand River. It became known as *Brant's Ford*. Its post office was named Brantford in 1825 in preference to *Biggar's Town* and *Lewisville*.

Brantville NB Located south of Tracadie, this place's post office was named in 1905 after the migration route of brant geese (*Branta bernicla*) over it in May and October.

Bras d'Or Lake NS Enclosed by Cape Breton and Boularderie islands, this sea-level lake likely derived its name from the Portuguese *lavrador* (small landholder) João Alvares Fagundes. He explored the coasts of Newfoundland and Nova Scotia about 1521. The name may have been given to reinforce Portuguese control of lands near the boundary of Spanish claims, set by the Treaty of Tordesillas in 1494. Subsequently concluding the name was French for 'arm of gold', map-makers called the water body, with many bays and channels, Bras d'Or Lake, and designated the channels, separating Boularderie Island from Cape Breton Island, **Great Bras d'Or** and **Little Bras d'Or**. The lake is informally called the Bras d'Or Lakes.

Bratnober, Mount YT West of Whitehorse, this mountain (1923 m) was named in 1897 by surveyor James J. McArthur after Henry Bratnober, who helped him clear the Dalton Trail that year. Subsequently, Bratnober was a prospector, riverboat captain, and agent in the north for the mining interests of N.M. Rothschild and Sons.

Bray Island NT Located in Foxe Basin, west of Baffin Island, this island (689 km^2) was named in 1946 by Thomas H. Manning after Reynold J.O. Bray, the first non-Aboriginal to explore the area.

Brazeau, Mount AB In the Rocky Mountains, west of Rocky Mountain House, this mountain (3470 m) was named in 1902 by Arthur P. Coleman after Hudson's Bay Company factor Joseph Étienne Brazeau, a native of St. Louis, MO. Brazeau was in charge of the post at Rocky Mountain House, 1858–9, and at Jasper, 1861–2, and was a polyglot, speaking English, French, Spanish, Blackfoot, Cree, Crow, Saulteaux, Siouan, and Stoney.

Breadalbane PE A municipal community (1991) east of Summerside, its post office was named in 1889 after Breadalbane, Perthshire, Scotland. Both its railway station and school names were spelled *Bradalbane*.

Brechin ON In Ramara Township, Simcoe County, southeast of Orillia, this place was named in 1863 by postmaster J.P. Foley after his wife's home town of Brechin in Scotland, northeast of Dundee.

Bredenbury SK Located southeast of Yorktown, this town (1913) was named in 1886, when the Manitoba and North Western Railway was built from Winnipeg to Saskatoon. It was called after A.E. Breden, a railway land inspector. Postal service was established in 1890.

Brent's Cove NF On the north coast of the island of Newfoundland and east of Baie Verte, this municipal community (1966) was founded in the 1850s. The origin of the name is unknown. It may be a transplanted place name from the West Country of England, or a descriptive name relating to its steep cliffs or its burnt shoreline.

Breslau ON In Waterloo Region, east of Kitchener, its post office was named in 1856 after Breslau, then in German Silesia, but now known as Wrocław in Poland.

Breton AB Southwest of Edmonton, this village (1957) was named by the Canadian Pacific Railway in 1926 after Douglas Corey Breton (1883–1953), then a member of the provincial legislature. He migrated to Canada from South Africa in 1903 and moved to England in 1934. The post office, called *Keystone* in 1912, was renamed Breton in 1927.

Brewers Mills NB A community northwest of Fredericton, its post office was named in 1901 after the first postmaster, W.A. Brewer. Abraham Brewer had an earlier land grant on nearby McKiel Brook.

Brew, Mount BC In the Coast Mountains, west of Whistler, this mountain (2286 m) was named in 1911 after Chartres Brew (1815–70), who was appointed the first inspector of police in British Columbia in 1858, chief gold commissioner the following year, and a member of the legislative council, 1864–8. Another **Mount Brew** (2886 m) south of Lillooet was named for him in 1950.

Bridgedale NB In the town of Riverview, opposite the city of Moncton, this place was the site of *Lower Coverdale* post office from 1853 to 1868. The post office was then called Bridgedale; it closed in 1919. Bridgedale had been an incorporated village from 1966 to 1973. Its name may have been taken from a bridge over Mill Creek.

Bridgenorth ON A community in Peterborough County, north of Peterborough, its post office was named in 1854, when a proposal was made to build a bridge across Chemong Lake. The bridge was not opened until 1870.

Bridgetown NS A town (1897) in the Annapolis Valley, it was named in 1824 after a bridge constructed across the Annapolis River. This place was earlier called *Hick's Ferry*. Postal service was provided to Bridgetown in 1825.

Bridgewater NS The site of this town (1899) was first settled in the 1780s. It was named in 1837 after a bridge which had been built over the LaHave River, 18 km above its mouth.

Brigden ON In Lambton County, southeast of Sarnia, this place's post office was named in 1873 after John Brigden, a construction engineer with the Canada Southern Railway.

Brigham QC A municipality in the regional county of Brome-Missisquoi, northwest of Cowansville, it was named in 1980. Its post office opened here in 1860, and was named after brickmaker Erastus Oakley Brigham.

Bright ON Northeast of Woodstock, in Oxford County, its post office was named in 1863 after John Bright (1811–89), a noted British statesman and parliamentary reformer. **Bright** and **Bright Additional** townships in Algoma District were also named for him.

Brighton NF On the north coast of the island of Newfoundland and on the Tickle Islands and Triton Island, this municipal community (1976) was first known as *Dark Tickle*. It was renamed in the early 1900s by the Nomenclature Board, but the reason for choosing it is not in the official records.

Brighton ON A town (1980) in Northumberland County, this place's post office was named in 1831 after the popular English seaside resort of Brighton in Sussex. **Brighton Township**, surrounding the town, was named in 1851.

Brights Grove ON A community on the shore of Lake Huron and in the city of Sarnia, 14 km northeast of the city centre, it was named in 1935 after John Bright developed a summer resort among a grove of trees on his lakeshore farm.

Brigus NF A town (1964) on the western shore of Conception Bay, it was noted as *Brega* on an early map of the 1600s. It occurred on later French maps as *Brigues*, a word that suggests 'intrigue', or 'underhanded'. Its present spelling was established on an English marine pilot of 1689.

Bristol NB A village (1966) on the east bank of the Saint John River, north of Woodstock, its post office was called *Shiktehawk* in 1866 after Shiketehawk Stream. It was renamed in 1878 after Bristol, Gloucestershire, England.

Bristol QC Located in the southern part of the regional county of Pontiac, this place was created in 1845 and named after the township of Bristol, shown on a 1795 map, but not proclaimed until 1834. The name recalls the famous English seaport in Gloucestershire.

Britannia Beach BC On the east shore of Howe Sound, south of Squamish, this community was established in 1907. It took its name from the **Britannia Range** of the Coast Mountains, named in 1859 by Capt. George Richards after HMS *Britannia*, which served during the naval battle off Cabo Trafalgar, southwest Spain, 1805.

British Columbia The roots of the province's name may be traced to the naming of the Columbia River in 1792 by Capt. Robert Gray, after his ship *Columbia*, which had begun its journey in Boston. Subsequently the Hudson's Bay Company named its southern district, west of the continental divide, Columbia. In 1858 Queen Victoria noted Columbia on many maps of the area, but after observing it was the poetical name for the United States as well as a country in South America (actually Colombia), she proposed British Columbia in a letter to Sir Edward Bulwer-Lytton, the colonial secretary.

British Columbia, Mount YT In the Centennial Range of the St. Elias Mountains, this mountain (3109 m) was named in 1967 in honour of the one hundredth anniversary of Confederation.

British Empire Range NT Located at the northwestern side of Ellesmere Island and between Clements Markham Inlet and Yelverton Bay, this range of mountains was named by A. Winston Moore, a member of the Oxford University Ellesmere Island expedition of 1934–5.

British Mountains YT Extending for 175 km from Mount Fitton to Mount Greenough in northern Yukon, and across the border into Alaska, this range of mountains was named *British Chain* in 1826 by John (later Sir John) Franklin in honour of the people of Great Britain.

Britt ON On the north side of Byng Inlet, northwest of Parry Sound, this place's post office was known as *Byng Inlet North* from 1885 to 1913. The Canadian Pacific Railway named its station about 1920 after Thomas Britt, a former head of the CPR's fuelling depot in Montréal. The post office was reopened as Britt in 1927.

Broadback, Rivière QC Rising in west central Québec, north of Chibougamau, this river flows west into Rupert Bay, in the southeastern part of James Bay. The origin of the name is uncertain, with its first appearances being in the 1890s reports by geologist Robert Bell and surveyor Henry O'Sullivan.

Broadview SK A town (1907), east of Regina, its post office and station were named in 1882, when the transcontinental line of the Canadian Pacific Railway was built that far. The name was given, because, as far as the eye could see, there was a 'broad view' of the relatively level prairie.

Brochet MB A place at the north end of Reindeer Lake, it derived its name from the French word for 'pike'. The site of Hudson's Bay Company trading post Fort du Brochet before 1921, its post office was established in 1950.

Brock ON In Durham Region, north of Whitby, the present municipal township was created in 1974 by amalgamating Brock and Thorah townships, and the villages of Beaverton and Cannington. The original Brock Township was named after Maj.-Gen. Sir Isaac Brock (1769–1812), who was killed at the Battle of Queenston Heights.

Brock SK East of Kindersley, this village (1910) was founded by the Canadian Northern Railway in 1909 and named after Maj.-Gen. Sir Isaac Brock, hero of the Battle of Queenston Heights (1812). The post office was opened in 1910.

Brocket AB A community on the Peigan Indian Reserve, southwest of Fort Macleod, it was named by the Canadian Pacific Railway in 1897–8 after Brocket Hall, in Hatfield, Hertfordshire, England. In 1888 it became the residence of George Stephen, 1st Baron Mount Stephen, first president of the CPR, 1880–88.

Brock Island NT Part of the Parry Islands, and in the Queen Elizabeth Islands, this island (764 km^2) was named in 1916 by Vilhjalmur Stefansson after Reginald Walter Brock (1874–1935). He was the director of the Geological Survey of Canada, 1907–14, and briefly deputy minister of mines in Ottawa, before becoming dean of applied science at the University of British Columbia, 1914–34.

Brockville ON First known as *Buell's Bay*, after first settler William Buell (1751–1832), and *Elizabethtown*, after the township, this city (1962) in the united counties of Leeds and Grenville was renamed in 1812 for Maj.-Gen. Sir Isaac Brock, who was killed that year at the Battle of Queenston Heights.

Brodeur Peninsula NT A 290-km peninsula at the northwest side of Baffin Island, and between Prince Regent and Admiralty inlets, it was named in 1907 by explorer Joseph-Elzéar Bernier after Louis-Philippe Brodeur (1862–1924), who was then the minister of marine and fisheries in Ottawa. Later he became a judge of the Supreme Court of Canada, 1911–23, and lieutenant-governor of Québec, 1923–4.

Brodhagen ON In Perth County, northwest of Stratford, it was named in 1863 after the first postmaster, Charles Brodhagen, a native of Hannover, Germany, who settled here in 1860, opened a store and a hotel, and built a sawmill.

Brome-Missisquoi QC A regional county on the Vermont border, its double name, created in 1983, has deep roots in the names of this part of Québec. The township of **Brome**, adjacent to **Lac Brome**, east of Cowansville, was named in 1795 after Brome Hall, the English residence of Charles, 1st Marquess Cornwallis and Viscount Brome (1738–1805), governor general of India, 1786–93. Missisquoi was created a county in 1853, taking its name from the large northeastern bay of Lake Champlain, whose name in Abnaki may mean 'place of gun flint'.

Bromley ON A township in Renfrew County, it was named in 1843 after Bagot's Bromley in Staffordshire, England, the estate of Sir Charles Bagot (1781–1843), governor of the province of Canada, 1841–3.

Bromont QC Located in the regional county of La Haute-Yamaska, south of Granby, this town was named in 1964 after the township of Brome and nearby **Mont Brome**, which has an elevation of 553 m.

Bromptonville QC A town in the regional county of Le Haut-Saint-François, north of Sherbrooke, it was named *Brompton Falls* in 1902, and renamed Bromptonville the following year. The name was derived from the township of Brompton, named in 1795 and proclaimed in 1801, likely after Brompton, north of Northallerton in North Yorkshire, England.

Bronte ON At the mouth of **Bronte Creek** and in the town of Oakville, Halton Region, this urban centre was named for one of the titles of Adm. Lord Horatio Nelson (1758–1805). He received the title of Duke of Brontë in 1800 from King Ferdinand III of Naples, whom Nelson had helped restore to his throne.

Brooke ON A township in Lambton County, it was named in 1834 after Sir James Brooke (1803–68), the rajah of Sarawak, 1838–68. Sarawak is now a state in Malaysia.

Brooke, Mount YT Located in the St. Elias Mountains, near the Alaska border, this mountain (3290 m) was named in 1918 after Pvt. William Brooke (1893–1917), who died in a German prisoner-of-war camp.

Brookfield NS South of Truro, this descriptive name was given in 1784 by pioneer settler William Hamilton. Its post office was opened in 1852.

Brooklin ON In the town of Whitby, Durham Region, this place was first called *Winchester* in 1840. It was renamed Brooklin in 1847, when the post office was opened. Possibly it was named after either Brooklin, ME, or Brooklyn, NY.

Brooklyn NS East of Liverpool and at the mouth of the Mersey River, this municipal community (1995) was first called *Herring Cove*. Its post office was established in 1861. Possibly called after a small stream flowing through the community, its name replicated the noted borough of Brooklyn, part of New York City. A second **Brooklyn**, east of Windsor, has a descriptive name, which was given in the early 1800s. Because of other places called Brooklyn in the province, its post office has been known as Newport since 1831, but it has failed to supplant Brooklyn. There is another **Brooklyn** on the northeast side of Yarmouth, and a **Brooklyn Corner**, on the west side of Kentville.

Brooks AB The largest town (1910) between Calgary and Medicine Hat, its post office was named in 1904 after Noel Edgall Brooks (d. 1926), divisional engineer of the Canadian Pacific Railway at Calgary, 1903–13.

Brossard QC Situated in the regional county of Champlain and across the St. Lawrence from Montréal, this city was named in 1958, and enlarged in 1978. In the mid-1800s the first municipal parish was called *Brosseau* after its mayor Pierre Brosseau. Its modern founder was Georges-Henri Brossard, who was the mayor of the parish and then of the city from 1944 to 1967. In 1958 the local residents wanted to call it Forgetville, after Mgr Anastase Forget, but many, including Premier Maurice Duplessis, opposed a name suggesting forgetful citizens.

Brougham ON In the town of Pickering, Durham Region, this community's post office was named in 1836, probably after Henry Peter Brougham, 1st Baron Brougham and Vaux (1778–1868), a native of Edinburgh, who was lord chancellor of England, 1830–4. His father was raised at Brougham Castle, in Westmorland, England. **Brougham Township** in Renfrew County was named in 1851 for Lord Brougham.

Broughton Island BC Between Vancouver Island and the mainland and on the north side of Queen Charlotte Strait, this island was named in 1792 by Capt. George Vancouver after Lt-Cdr (later Comm.) William Robert Broughton (1762–1821). He commanded the armed tender *Chatham* that year when Vancouver explored the inlets and passages of the west coast.

Broughton Island NT A hamlet (1979) on **Broughton Island** and on the northeast coast of Baffin Island, it was founded in 1956–7 as a Distant Early Warning site. The island was named in 1818 by John (later Sir John) Ross after Comm. William Robert Broughton. In 1811 he had led a British expedition that captured the island of Java, in present-day Indonesia.

Brownsburg QC A town (1935) in the regional county of Argenteuil, west of Lachute, its post office was named in 1854 after postmaster George Brown. A native of England, Brown settled here in 1818, and built grist and sawmills.

Bruce ON The county was named in 1849 for James Bruce, 8th Earl of Elgin (1811–63), governor of the province of Canada, 1846–54. **Bruce Township** in the county was created in 1850. **Bruce Peninsula** is the 90-km-long peninsula that divides Georgian Bay from the main part of Lake Huron.

Bruce Mines ON A town (1903) in Algoma District, southeast of Sault Ste. Marie, it was founded in 1842 and copper mines were operated here from 1847 to 1876. It was named in 1848 after James Bruce, Lord Elgin, who was then governor of the province of Canada.

Brudenell PE A municipal community (1973) north of Montague, and between the Brudenell and Montague rivers, its post office was named in 1898. The river was named in 1765 by Samuel Holland after George Brudenell, 4th Earl of Cardigan (1712–90).

Bruderheim AB A town (1980) northeast of Edmonton, its post office was named in 1895 by Andreas Lilge and a group of Moravians, including his brothers Ludwig and William, who established a colony the year before, after migrating from Volhynia, Russia. Their congregation was called Brethrens Church, with the post office name meaning 'Brethrens home'.

Brule NS On the south side of Amet Sound, and east of Tatamagouche, this community's post office was first called *Point Brule* in 1859, and was changed to Brule in 1876. Derived from French *brûlé*, 'burnt', the name is said to relate to an incident when Aboriginals from Québec encountered some Mi'kmaq at the point, and set the woods on fire.

Bruno SK Located west of Humboldt, this town (1962) was named after Fr Bruno Doerfler, who arranged in 1902 for the German American Land Company to buy 40,470 ha (100,000 a) in the area, with 1000 German Americans applying for homesteads before the end of the year. Postal service was set up here in 1906.

Brussels ON A village in Huron County, east of Goderich, it was called *Ainleyville* in 1855, after developer William Ainley. The next year, the post office was called *Dingle*, but *Ainleyville* continued in use until the village was incorporated in 1872 as Brussels, after the capital of Belgium.

Bryants Cove NF Located southeast of Harbour Grace, this municipal community (1977) was settled in 1675, one of the first settlements in Conception Bay. The origin of the name is not available in the official records.

Bryce, Mount BC South of the Columbia Icefield, this mountain (3507 m) was named in 1899 by climber Norman Collie, after James Bryce, Lord Bryce (1838–1922), British ambassador to the United States, 1907–12, and an alpinist.

Bryson QC In the regional county of Pontiac, and on the Ottawa River, west of Shawville, the village was named in 1873 after prominent lumberman and legislative councillor George Bryson (1813–1900).

Buchanan SK Located west of Canora, this village (1907) was established by the Canadian Northern Railway in 1904 and named after rancher R. Buchanan. The post office was opened here in 1906.

Buchans NF Midway between Corner Brook and Grand Falls-Windsor, this town (1979) was named in 1927 by developers of a copper, zinc, and lead mine after **Buchans Island**, 7 km to the southeast in Red Indian Lake. The island was called after Capt. David Buchans (1780–*c.* 1838), a British naval officer, who had served in Newfoundland. He made contact with the Beothuk at Red Indian Lake in 1811 and 1820.

Buckhorn ON At the north end of **Buckhorn Lake** in Peterborough County, north of Peterborough, this place was first known as *Hall's Bridge* in 1828, after John Hall, an American, who built grist and sawmills, homes for his workers, and a bridge. Having settled here permanently in the 1850s, he became the postmaster of *Hall's Bridge* in 1860. It became Buckhorn in 1941, and was called after the lake, river, and falls, named in the late 1800s after Hall's collection of animal horns.

Buckingham QC A town (1890) in the Communauté urbaine de l'Outaouais, east of the city of Gatineau, it took its name from the township of Buckingham, which had been proclaimed in 1795. Although the first settlers were from England's Buckinghamshire, northwest of London, the name may have been given after George Nugent Temple Grenville, 1st Marquess of Buckingham (1753–1813), twice lord lieutenant of Ireland in the 1780s.

Buena Vista SK Situated at the south end of Last Mountain Lake, north of Regina, this village (1983) had a summer post office from 1930 to 1954. Its name reflects the beautiful view across the lake.

Bugaboos, The BC Located in the Purcell Mountains, northwest of Invermere, these peaks are a popular skiing destination. Bugaboo was noted on an 1893 map by gold commissioner A.P. Cummins. It was reported in 1906 to chief geographer James White that the name referred to the loneliness of the place.

Bulkley River BC Rising northwest of Burns Lake, this river (257 km, to the head of Maxan Creek) flows west, and then north through Houston and Smithers to join the Skeena River at Hazelton. It was named about 1864–6 after Col Charles Bulkley, chief engineer of the Western Union Extension Company, during construction of the Collins Overland Telegraph. The line only reached this part of the province, when it was suspended on learning of the successful laying of the transatlantic cable in 1866.

Bulyea, Mount AB, BC On the continental divide, northwest of Calgary, this mountain was named in 1920 after George Hedley Vicars Bulyea (1859–1926), first lieutenant-governor of Alberta, 1905–15.

Bunbury PE A place in the town of Stratford, across Charlottetown Harbour from Charlottetown, it was possibly named about 1798 by John Bovyer, after his ancestral home in Cheshire, England. Formerly a municipal community (1969), it became part of Stratford on 1 April 1995.

Burdett AB A village (1913) between Lethbridge and Medicine Hat, its station was named in 1907 by the Canadian Pacific Railway after Baroness Angela Georgina Burdett-Coutts (1814–1906), an English philanthropist, and a shareholder in the North Western Coal and Navigation Company of Lethbridge. *See also* Coutts.

Burford ON This township, first in Oxford County, but in Brant County since 1853, was established in 1798 and named for Burford, in England's Oxfordshire, west of the city of Oxford. The community of **Burford**, west of Brantford, was named in 1819.

Burgeo NF This town (1950) is located on an island on the southwestern coast of the island of Newfoundland. It took its name from the **Burgeo Islands**, which were called after an archipelago, in the area of Saint-Pierre and Miquelon, named *Onze myll virgines* by Portuguese explorer João Alvares Fagundes in 1521, and shown on a map of about 1524. That name related to the supposed slaughter by the Huns of St Ursula and her 11,000 virgins, while returning from a pilgrimage to Rome, in either the fourth or fifth centuries. The French apparently transferred the name in the form of *Vierges* to the Burgeo Islands.

Burgess, Mount BC In Yoho National Park, north of Field, this mountain (2599 m) was named in 1886 by Dr Otto Klotz, after Alexander Mackinnon Burgess (1850–1908), deputy minister of the interior, 1883–97. The Burgess Shale Site at the mountain has been declared a World Heritage Site for its unique invertebrate fossil remains. Burgess was also honoured in the naming of **Mount Burgess** in the Yukon's Ogilvie Mountains in 1888 by surveyor William Ogilvie.

Burgessville ON In Norwich Township, Oxford County, south of Woodstock, this place was named in 1853 for Edward N. Burgess, who had been born 5 km east of the community. A blacksmith and carriage maker, he served as postmaster of Burgessville until 1871.

Burin NF A town (1950) on the eastern side of **Burin Peninsula**, its name may have originated in the French word *burin*, 'graver', or 'chisel', and was possibly given by French fishermen. There might be a connection with the Basque word *burua*, meaning 'head', with Basque fishermen having visited the area in the early 1600s. Spanish records of the 1600s called the area *Buria*.

Burk's Falls ON A village (1890) in Parry Sound District, north of Huntsville, it was named in 1879 for its first permanent settler, David Francis Burk. He had come from Oshawa in 1876 and lived here until he died in 1901, aged 49.

Burleigh ON This township was named in 1822, likely after Burghley House, near Stamford, England, 45 km east of Leicester. The house was built by William Cecil, Baron Burghley (1520–98), Queen Elizabeth I's chief minister. For municipal purposes, Burleigh is united with Anstruther Township. The community of **Burleigh Falls**, in adjacent Harvey and Smith townships, is situated where Lovesick Lake and the Trent Canal empty into Stony Lake. Its post office was named in 1864.

Burlington NF A municipal community (1953) on Green Bay, an extension of Notre Dame Bay, it was known as *North West Arm* until 1914, when it was renamed, possibly after Burlington, ON.

Burlington ON A city (1974) in Halton Region, it was first known as *Brant's Block*, after a grant to Joseph Brant for his loyalty to the British Crown. Its post office became *Wellington Square* in 1826 after the Duke of Wellington. In 1873 it was united with Port Nelson to become the village of Burlington, named after *Burlington Bay*. The bay was named in 1792 by Lt-Gov. John Graves Simcoe for the town of Bridlington, in East Yorkshire, England, north of Kingston upon Hull. The bay was renamed Hamilton Harbour in 1919, although Burlington Bay remains in common use, as in Burlington Bay James N. Allan Skyway.

Burnaby BC On the east side of Vancouver, this city (1992) had been created a district municipality in 1892. It was named after **Burnaby Lake**, in the central part of the city. The lake was called after Robert Burnaby (1828–78), the private secretary to Col Richard Moody, commanding officer of the Royal Engineers, 1858–9, during a survey of New Westminster in 1859.

Burns Lake BC Located at the geographical centre of British Columbia, this village (1923) is at the west end of **Burns Lake**. The lake, a 21-km widening of the Endako River, was named in 1866 after surveyor Michael Byrnes, who established a route through the area for the Collins Telegraph Company.

Burnstown ON In Renfrew County, southeast of Renfrew, this community's post office was named at a public meeting in 1853 after the great Scottish poet Robert Burns (1759–96), when it was noted that the whole settlement was Scottish.

Burnt Church NB Located at the mouth of **Burnt Church River**, northeast of Chatham (Miramichi), this place's post office was called *Church Point* from 1873 to 1904. Burnt Church post office was 5 km to the west at New Jersey from 1865 to 1904. The name recalls the burning of French settlements here, including a chapel, by the British in 1758, three years after the expulsion of the Acadians.

Burnt Islands NF A town (1977), east of Channel-Port aux Basques, it was developed as a fishing community in the late 1800s. Presumably the vegetation on the islands was destroyed by a fire.

Burrard Inlet BC Separating Vancouver from the north shore, this arm of the Strait of Georgia was named in 1792 by Capt. George Vancouver, after Sir Harry Burrard (1765–1840), who served with Vancouver in the West Indies in 1785. On marrying in 1795, he became Sir Harry Burrard-Neale. His career in the Royal Navy progressed to his appointment as an admiral in 1830.

Burritts Rapids ON Mostly situated on an island in the Rideau River, northeast of Smiths Falls, this place was settled in 1793 by brothers Stephen and Daniel Burritt. A townsite was laid out in 1830, with its post office being opened in 1839.

Burstall SK Located near the Alberta border, west of Swift Current, this town (1921) was founded in 1920 by the Canadian Pacific Railway. It was named after Gen. E. Henry Burstall, who was killed during the First World War while leading the 2nd Canadian Division. Postal service was established here in 1921.

Burton BC At the north end of Lower Arrow Lake, this place was named in 1896 after first postmaster Reuben Burton. He and a brother established the townsite three years earlier.

Burton NB On the south bank of the Saint John River, and on east side of the town of Oromocto, this place's post office was named *Upper Burton* in 1878. Burton post office was opened in 1856, 6 km downriver at **Lower Burton**. **Burton Parish** was created in 1786, having been established as Burton Township in 1765. The township was named after Brig.-Gen. Ralph Burton (d. 1768), lieutenant-governor of the district of Three Rivers (Trois-Rivières), 1760–2 and of Montreal, 1763–4. He was then appointed military commander-in-chief of the Northern Department (Quebec), retiring to England in 1766.

Burtts Corner NB Located 20 km northwest of Fredericton, this community's post office was named in 1892 after its first postmaster, Elwood Burtt.

Bury QC Situated in the regional county of Le Haut-Saint-François, east of Sherbrooke, this municipality was named in 1845 after the township of Bury, proclaimed in 1803. The township may have been named after the borough of Bury St. Edmunds, in Suffolk, England.

Bute Inlet BC A 65-km sinuous fiord extending into the Coast Mountains, northeast of Campbell River, this inlet was named in 1792 by Capt. George Vancouver after John Stuart, 3rd Earl of Bute (1713–92), described as the 'dearest friend' of George III. He was Britain's prime minister, 1762–3, when the Treaty of Paris was signed.

Byam Martin Island NT In the Parry Islands of the Queen Elizabeth Islands, this island (1150 km^2) was named in 1819 by William (later Sir William) E. Parry after Vice-Adm. Sir Thomas Byam Martin (1773–1854), comptroller of the British Navy. **Byam Martin Channel** separates Melville Island, on the west, from Alexander, Vanier, Massey, and Cameron islands on the east. **Byam Martin Mountains** and **Cape Byam Martin**, both on Bylot Island, were named in 1818 by John (later Sir John) Ross, who served with Sir Thomas Byam Martin, when the latter commanded the British fleet in the Baltic Sea in 1812.

Bylot Island NT Located on the north side of Baffin Island, this island (11,067 km^2) was likely named in 1819 by William (later Sir William) E. Parry after Robert Bylot (d. 1622). With William Baffin, Bylot had explored the waters of Davis Strait and Baffin Bay in 1615–6, but failed to find an eastern entrance leading to the Northwest Passage.

Byng Inlet ON A long narrow inlet in the northeastern part of Georgian Bay, it was named in the 1820s by Admiralty surveyor Henry W. Bayfield after Adm. John Byng (1704–57), who had been executed for cowardice, because he failed to capture Minorca in the Mediterranean Sea. The community of **Byng Inlet**, opposite Britt, was named in 1868.

Byron ON A western suburb of London, its post office was named in 1857 after George Gordon Noel Byron, 6th Baron Byron (1788–1824), the celebrated English poet and satirist.

C

Cabano QC A town (1962) in the regional county of Témiscouata, southeast of Rivière-du-Loup, its post office was named in 1898 after the township of Cabano, which was proclaimed in 1866. The name may have been derived from *cabaneau,* because of the beaver-lodge appearance of the mountains adjacent to Lac Témiscouata.

Cabonga, Réservoir QC Located in the regional county of Vallée-de-l'Or, southeast of Val-d'Or, it was created in 1928–9 from Lac Cabonga as part of a scheme to regulate the headwaters of the Ottawa and Gatineau river systems. Derived from the Algonquin *kakibonga*, 'entirely blocked by sand', the lake's name was first recorded in 1864 as *Kakibonga.*

Cabot Head ON The northeast point of Bruce Peninsula, it was named in the 1790s by Lt-Gov. John Graves Simcoe after John Cabot, who had explored North America's east coast in the 1490s.

Cabot Strait NF, NS Part of the Gulf of St. Lawrence, this strait has a width of 89 km between Cape North on Cape Breton Island, and Cape Ray on the island of Newfoundland. It was named by the British Admiralty in 1885 after John Cabot, who had made a landing in the New World in 1497, but was lost at sea during a second voyage the following year.

Cabri SK A town (1917) northwest of Swift Current, its post office was opened in 1912, and its station was established by the Canadian Pacific Railway the following year. In an Aboriginal language its name may mean 'antelope'.

Cache Bay ON On an enclosed bay of Lake Nipissing, west of North Bay, this town (1903) was established as a Canadian Pacific Railway station in 1883, and its post office was open six years later. The bay was a natural location for fur traders and voyageurs to hide (*cache*) their supplies.

Cache Creek BC A village (1967), west of Kamloops, it is located at the point where Highway 97 leads north from the Trans-Canada Highway to Prince George. It was named after the nearby creek, noted as *Rivière de la Cache* in 1859 by Cdr Richard C. Mayne. Noted even earlier as Cache Creek in 1835, it was likely a point where furs were collected by traders, and hidden, the meaning of *cache* in French. Its first post office was opened in 1868.

Cacouna QC Located in the regional county of Rivière-du-Loup, northeast of the city of Rivière-du-Loup, this Maliseet reserve was created in 1891. In Maliseet, the name may mean 'place of the porcupine', or 'place of the turtle'.

Cadillac QC A town in the regional county of Rouyn-Noranda, northwest of Malartic, it was named in 1948 after the township of Cadillac, proclaimed in 1916. The name recalls Henry Preyssac de Cadillac, an officer who served with the Marquis de Montcalm in the 1750s.

Cadorna, Mount BC In the Rocky Mountains, east of Invermere, this mountain (3145 m) was named in 1918 after Gen. Luigi Cadorna (1850–1928), commander-in-chief of the Italian armies during the First World War.

Caesarea ON A community in Scugog Township, Durham Region, north of Oshawa, its post office was named in 1853 after four Caesar families who had settled east of Lake Scugog in the 1830s. The spelling replicated the Roman name of Jersey, one of the Channel Islands, as well as a number of historic places in the Holy Land.

Cairnes, Mount BC In the Rocky Mountains, northwest of Field, BC, this mountain (3060 m) was named by surveyor Arthur O. Wheeler in 1919 after geologist Delorme Donaldson Cairnes (1879–1917). A second **Mount Cairnes** (2789 m), in the Kluane Range of the St. Elias Mountains, southeast of

Kluane Lake in the Yukon, was named about 1920 after him. There is another **Mount Cairnes** in the Yukon's Ogilvie Mountains, northeast of Dawson, named in 1968 by geologist Dirk Tempelman-Kluit after the same geologist, who explored and mapped the geology of the Yukon from 1905 to 1917.

Caissie Cape NB Located north of Shediac, this community was served by *Cap-Caissie* post office from 1930 to 1955, after the nearby cape, **Cap de Caissie.** The cape was named after an Acadian family with Irish ancestry, with the variant forms Casey, Caissy, Kuessey and Quessy being used in land grant records. After local investigation in 1991, the province endorsed the dual forms of Caissie Cape and **Cap-des-Caissie** for the community.

Calabogie ON On the north shore of **Calabogie Lake**, a widening of the Madawaska River, and in Renfrew County, south of Renfrew, this place's post office was named in 1857. The lake's name may have originated in a rum and molasses drink called 'calabogie', which a Scots Highlander developed at his tavern near the foot of the lake.

Calder SK East of Yorkton, this village (1911) was established in October 1909 by the Canadian Northern Railway, and named in January 1911 after J.A. Calder, provincial minister of education. Its post office was called *Calder Station* in 1911 and was changed to Calder in 1936.

Caledon ON A town in Peel Region, it was created in 1974 when the townships of Caledon and Albion, the north half of Chinguacousy Township, and the villages of Bolton and Caledon East were amalgamated. **Caledon Township** was named in 1819 after a poetic description of Scotland. **Caledon Mountain** is the steep climb up the Niagara Escarpment, in the area of Highway 10. **Caledon Hills** embrace the rugged landscape between Belfountain and Caledon East. **Caledon East**, at the centre of the town of Caledon, Peel Region, was known as *Tarbox Corners*, *Munsie's Corners*, and *Paisley*, before it was named Caledon East in 1851, after its location on the eastern boundary of Caledon Township. The post office at **Caledon Village** was opened as *Charleston* in 1838, but was changed to Caledon in 1839, after the township. However, *Charleston* continued in use as the preferred local name until the end of the 1800s. The postal name was amended in 1975 to distinguish it from the town of Caledon. In its early years the place was also called *Reburn's Corners*, after innkeeper Robert Reburn (*c.* 1785-1870).

Caledonia NS This community, northwest of Liverpool, was settled in 1820 by Scots, and named after the ancient Roman name for northern Scotland. Its post office was called *Caledonia Corner* in 1855, and the present name was adopted in 1904. Earlier the place was called *County Line*, after its location on the border of Kings and Queens counties.

Caledonia ON A place in Haldimand-Norfolk Region, southwest of Hamilton, it was served by *Seneca* post office from 1839 to 1880, named after Seneca Township. It was then renamed Caledonia, a name introduced several years earlier for part of the community, and recalling the poetic name given by the Romans to the northern part of Scotland.

Calgary AB The city (1893) was briefly called *Fort Brisebois*, when it was established by Inspector E.A. Brisebois in 1875. However, Asst Com. A.G. Irvine of the North-West Mounted Police informed Minister of Justice Edward Blake of the proposal by Lt-Col James F. Macleod in 1876 to name it *Fort Calgary*, after the ancestral estate of his Mackenzie cousins on the west side of the Island of Mull, Scotland. Macleod's proposal was shortened to Calgary, which he stated meant 'clear running water' in Gaelic, and it was accepted. Its post office was opened in 1883.

Callander ON Renowned as the nearest rail centre to the birthplace of the Dionne quintuplets in 1934, Callander is located in Parry Sound District, south of North Bay. In 1881 first settler and postmaster George Morrison named it for his native Callander in Scotland's Perthshire, northwest of Stirling.

Calmar AB Southwest of Edmonton, this town (1954) was named in 1900 by the first postmaster, C.J. Blomquist after his hometown of Kalmar, Sweden, southwest of Stockholm.

Calumet QC Situated in the regional county of Argenteuil, west of Lachute, this village was named in 1918 after Calumet post office, opened in 1887, and called after the **Rivière du Calumet**, which flows into the Ottawa River at that point. Excellent stone here was used by the Algonquin for the production of the pipes of peace (*calumets*).

Calvert NF On the the east coast of the Avalon Peninsula, this place was named in 1922 after Sir George Calvert, 1st Baron Baltimore (*c.* 1580–1632). He established a colony at nearby Ferryland in 1621, and lived there with his family in 1628. In 1597 it was called *Caplen Bay*. This name may have been derived from the Portuguese *capelinas*, a small fish called capelin in English.

Cambray ON In Fenelon Township, Victoria County, northwest of Lindsay, it was named in 1853 by postmaster Joseph Wilkinson after Cambrai in France, near the Belgian border, where the noted François de Fenélon was the archbishop, 1695–1715. The township had been named after his half-brother, missionary François de Salignac de la Mothe Fenélon. *See also* Fenelon *and* Frenchman's Bay.

Cambridge ON This city in Waterloo Region, a union of the former city of Galt, the former towns of Hespeler and Preston, and parts of adjacent townships, was incorporated on 1 January 1974. Galt officials suggested calling it Blair, after a place on the west side of the Grand River, while the councils of Preston and Hespeler proposed Cambridge after an early flour mill at Preston. By the narrow vote of 11,728 to 9800 (1674 ballots were spoiled), Cambridge was chosen. *See also* Galt, Hespeler, *and* Preston.

Cambridge NS In the Annapolis Valley, west of Kentville, this community in the village of Cornwallis Square was first known as *Sharpe's Brook*. In 1872 its post office was opened as *Cambridge Station*, with the name having been suggested by a family with relatives in Cambridge, MA.

Cambridge Bay NT This hamlet (1984) on the south coast of Victoria Island was founded in 1947 as a loran (long-range navigational) site, becoming a Distant Early Warning site in 1955. The bay was likely named in 1851 by explorer John Rae after the Duke of Cambridge (1774–1850).

Cambridge-Narrows NB A village (1966) where Highway 695 crosses Washademoak Lake, 45 km east of Fredericton, it was formerly two separate communities. *Cambridge* post office, opened on the west side in 1860, was named after Cambridge Parish, called after Queen Victoria's uncle, Adolphus Frederick, the Duke of Cambridge (1774–1850). *Narrows* post office served the east side from 1855 to 1969.

Camden ON A township in Kent County, it was named *Camden West* in 1794 after Charles Pratt, Earl Camden and Viscount Bayham (1714–94), but it was soon called simply Camden Township. Earl Camden was a British government official, including attorney general and chief justice of common pleas.

Camden East ON A township in Lennox and Addington County, it was named *Camden* in 1787 after Charles Pratt, Earl Camden and Viscount Bayham (1714–94), a British government official. It was renamed Camden East in 1866 to distinguish it from Camden Township in Kent County. The community of **Camden East** was named in 1835.

Cameron Hills AB North of Peace River, these hills were named in 1922 by surveyor Bruce Waugh after his assistant, Maxwell George Cameron. Cameron (d. 1951) later became the chief cartographer of the surveys and mapping branch, department of mines and technical surveys in Ottawa.

Cameron Island NT In the Parry Islands of the Queen Elizabeth Islands, and between Bathurst and Melville islands, this island (1059 km²) was named in 1952 after Maxwell George Cameron (d. 1951), the chief cartographer of the surveys and mapping branch, department of mines and technical surveys.

Campbellford ON Named in 1854, this town (1906) in Northumberland County recalls brothers Lt-Col Robert Campbell and Maj. David Campbell, who were granted 890 ha (2200 a) on the Trent River in 1831. It had been first called *Seymour West*, after Seymour Township.

Campbell's Bay QC The post office of this village in the regional county of Pontiac, west of Shawville, was named in 1888 after early settler Donald Campbell. The village was incorporated in 1904.

Campbell Range YT On the west side of Frances Lake, this range of mountains was named in 1887 by geologist George Dawson after Robert Campbell (1808–94), a Hudson's Bay Company chief factor. He extensively explored the Yukon between 1840 and 1852.

Campbell River BC A district municipality (1964) on the northeast side of Vancouver Island, it had been incorporated as a village in 1947. Its post office was opened in 1907, taking its name from the river, which drains a series of lakes into Discovery Passage. The river may have been named in 1859 by Capt. George Richards after Dr Samuel Campbell, who was the assistant surgeon on Richards's survey ship *Plumper*.

Campbellton NB A city (1958) on the south side of the Restigouche River, it was named in 1833 by Robert Ferguson after Sir Archibald Campbell (1769–1843), lieutenant-governor of the province, 1831–7. In 1700 the site was called *Pointe-des-Sauvages*. It may have been the site of the Acadian settlement *Petite-Rochelle*, established in 1757 but destroyed by the British three years later. In the early 1800s it was variously known as *Cavenick's Point*, *Kavanagh's Point*, *Quinton's Point*, and *Martin's Point*.

Campbellton NF On the Indian Arm of the Bay of Exploits, this town (1972) was first known as *Indian Arm*, after a Beothuk camp located here in the early 1800s. It was renamed in the early 1900s, possibly after John Campbell, the manager of a local sawmill.

Campbellville ON In the town of Milton, Halton Region, west of the town centre, it was named after John Campbell (1784–1854), who settled here in 1832 and built a sawmill in 1838. Malcolm Campbell, the first postmaster in 1849, kept the office in William Campbell's house.

Camperville MB Located on the west side of Lake Winnipegosis and north of Dauphin, this community's post office was named in 1905 after Fr Joseph Charles Camper (1842–1916), who was a missionary here for more than 30 years.

Campobello Island NB In Passamaquoddy Bay, and joined to Lubec, ME, by a bridge, this island was named in 1770 by its grantee William Owen after the governor of Nova Scotia, Lord William Campbell (*c.* 1730–78). Impressed by its fine soil and appearance, Owen modified Campbell's name into the Spanish and Italian forms of 'fair field'. *Campo Bello* post office was located at Welshpool from 1837 to 1924.

Camrose AB A city (1955) southeast of Edmonton, its post office was named in 1905, by selecting the name of Camrose, Pembrokeshire, Wales from the British Postal Guide. Sparling had been proposed, but it was turned down because it was too similar to other names.

Canal Flats BC A community on the Kootenay River, only 1 km from Columbia Lake, the headwaters of the Columbia River, its post office was named in 1913. In 1889 British/Austrian capitalist William Adolph Baillie-Grohman constructed a canal uniting the two river systems. The scheme essentially failed because of fears of flooding downriver on the Columbia.

Canborough ON Part of the town of Dunnville, Haldimand-Norfolk Region since

1974, this township was named in 1825 after Benjamin Canby of Niagara, who had bought 7690 ha (19,000 a) of land here in 1799. Canby laid out the community of **Canborough** in 1800, with the post office being opened in 1836.

Candiac QC A town in the regional county of Roussillon, south of Montréal, and on the south shore of the St. Lawrence, it was named in 1957 after an estate near Nîmes, in Languedoc, France, where the Marquis de Montcalm was born in 1712.

Canfield ON In the town of Haldimand, Haldimand-Norfolk Region, since 1974, this place was named in 1859, when the Buffalo, Brantford and Goderich Railway station was named after local contractor Albert Canfield.

Caniapiscau QC A regional county in northeastern Quebec, its area of nearly 82,000 km^2 is centred on the **Rivière Caniapiscau** and the **Réservoir de Caniapiscau**. The 500-km-long river flows north to join the Rivière Koksoak, near Kuujjuaq. The reservoir was created in the 1980s as part of the vast Hydro-Québec project in northwestern Québec. In both the Cree and Montagnais languages, *caniapiscau* means 'rocky point'. In 1895 geologist Albert P. Low had observed a prominent rock projecting above the surface of the lake, and which has remained above the reservoir.

Canmore AB On the Bow River, west of Calgary, this town (1955) was named in 1884 by the Canadian Pacific Railway, possibly on the suggestion of Lord Strathcona, in honour of Malcolm III Canmore (*c.* 1031–93), King of the Scots, 1057–93.

Cannifton ON In Hastings County, north of Belleville, this community was founded in 1808 by miller John Canniff (1757–1843). *Bridgewater* post office was opened here in 1853, but was renamed Cannifton the next year, with Jonas Canniff, John's nephew, as the postmaster.

Canning NS Northwest of Wolfville, this village (1968) was first known as *Apple Tree Landing* and *Habitant Corner*. It was renamed about 1830 after George Canning (1770–1827), who was briefly prime minister of Great Britain in 1827, after having been the secretary of foreign affairs for five years. Postal service was established here in 1846.

Cannington ON In Durham Region, west of Lindsay, Cannington was named in 1849 after George Canning (1770–1827), British foreign secretary, 1822–7, and briefly prime minister in 1827.

Canoe Lake ON This lake in Algonquin Park was named in 1853 by geologist Alexander Murray, after his party took several days to build a canoe here. The lake became famous through the many sketches and paintings of Tom Thomson, who was accidentally drowned in it in 1917.

Canora SK This town (1910) was established by the Canadian Northern Railway in 1903–4, and its name was formed from the first two letters of each of the words of the company's name. Postal service was established in 1904.

Canso NS This town (1901), at the eastern extremity of mainland Nova Scotia, was identified as *Canseau* on Jean-Baptiste-Louis Franquelin's 1686 map of Acadia. Surveyor General Charles Morris introduced the present spelling in 1749. The name is derived from Mi'kmaq *kamsook*, 'opposite the lofty cliff', in reference to either the steep cliffs of Chedabucto Bay, or the mirage of an impressive height on Isle Madame, 13 km to the north.

Canso, Strait of NS This narrow passage between the mainland and Cape Breton Island was called *Passage de Fronsac* and *Détroit de Fronsac* by the French between the 1660s and the 1750s, after a title of an uncle and a son of Acadian Gov. Nicolas Denys, and of Cardinal Richelieu. In 1715 map-maker Herman Moll described it as *Straits of Canceaux*, with chart-maker J.F.W. DesBarres introducing the present name in 1776, replacing *Gut of Canso*.

Canterbury NB A village (1966) west of Fredericton, it was called *Howard Settlement* in the 1820s after Lt-Gov. Sir Howard Douglas

(1776–1861). After the arrival of the St. Andrews and Quebec Railway in 1859, its post office was called *Canterbury Station*, after Canterbury Parish, established in 1855. The parish was named after Thomas Manners-Sutton (1814–77), lieutenant-governor of New Brunswick, 1854–61. He succeeded his father as Lord Canterbury in 1869.

Cantley QC Created in 1989 when this municipality was severed from the city of Gatineau, its post office had been named in 1857, possibly after a place in Norfolk, England, midway between Norwich and Great Yarmouth, birthplace of an early pioneer.

Canwood SK Northwest of Prince Albert, this village (1916) was established by the Canadian Northern Railway in 1909 to exploit the timber resources of the area. Its name is a contraction of Canadian and woods. Its post office was called *Mcoun* and *Mcowan* in 1911, changed to *Forgaard* in 1912, and finally called Canwood in 1913.

Cap-à-l'Aigle QC A village (1916) in the regional county of Charlevoix-Est, east of La Malbaie, its name had been noted as early as 1812. It was named after the nearby **Cap à l'Aigle**, meaning 'eagle cape'.

Cap-aux-Meules QC Located in the regional county of Les Îles-de-la-Madeleine, this village was named in 1950. It was named after **Île du Cap aux Meules**, noted as *Île aux Meules* by Joseph Bouchette in 1815. Either grindstones (*meules*) had been quarried here, or some hills resembled haystacks (*meules*).

Cap-Chat QC In the regional county of Denis-Riverin, northeast of Matane, this town was incorporated in 1968. The township of Cap-Chat was proclaimed in 1842, taking its name from a striking rock profile of a cat, which was noted as early as 1660 by mapmaker Fr François Du Creux.

Cap-de-la-Madeleine QC This city (1918) at the mouth of the Rivière Saint-Maurice and adjacent to Trois-Rivières, took its name from a grant to the Jesuits in 1651 by Jacques de La Ferté de La Madeleine, abbot of Sainte-Marie-Madeleine de Châteaudun, southwest of Paris, France. The religious parish of Sainte-Marie-Madeleine-du-Cap-de-la-Madeleine was established in 1651. The parish municipality was called *Sainte-Marie-Madeleine* in 1855.

Cap-d'Espoir QC A place in the town of Percé and in the regional county of Pabok, southwest of the town centre, it was named in 1935 after the cape, which was first identified on a 1685 map by Emmanuel Jumeau. Often called 'Cape Despair' in English, it was identified as that on Thomas Jefferys's 1775 map. Although *espoir* means 'hope', there is reliable linguistic evidence that it was pronounced like 'despair' in seventeenth and eighteenth century French.

Cap-des-Rosiers QC A community in the town of Gaspé and in the regional county of La Côte-de-Gaspé, northeast of the town centre, its harbour was visited by fishermen in the middle of the 1600s. The cape's name, which refers to wild roses, was noted by Samuel de Champlain in 1632.

Cape Breton Island NS This island is called after the most southeasterly cape of the island. The cape's name, one of the oldest on the Atlantic coast, may have been named in the early 1500s after English fishermen. However, there are claims it was called after fishermen from Bretagne, France or after Capbreton, in southwestern France, near ports where Basque whalers and fishermen set out to exploit the resources of the New World. The first cartographic reference to it is *c. dos bretoes* on a map produced about 1521. Before 1550 the island was commonly called *tiera de bretones* and *Terre des Bretons*. During the French regime it was called *Isle Royale*. Cape Breton was a separate crown colony from 1784 to 1820, when it became part of Nova Scotia. Cape Breton County was established in 1820 and became the regional municipality of **Cape Breton** in 1995. **Cape Breton Highlands National Park** (951 km^2) was created in 1936.

Cape Broyle NF A locality at the head of **Cape Broyle Harbour** and on the east coast of the Avalon Peninsula, it was named after the cape at the southeast side of the harbour. Identified as *Cape Broile* by John Guy in 1612, the cape may have been named after the turbulence of the sea crashing on the rocks. It is possible that the name may have been derived from Portuguese *albrolho*, 'pointed rock', but there is no cartographic evidence for this suggestion.

Cape Croker ON On the east side of Bruce Peninsula, this community on **Cape Croker Indian Reserve** was named in the 1850s after the nearby cape. The cape was named in 1822 by Admiralty surveyor Henry W. Bayfield for John Wilson Croker (1780–1853), secretary at the British Admiralty, 1809–30.

Cape Dorset NT On **Dorset Island** and on the southwest coast of Baffin Island, this hamlet (1982) was established in 1913 by the Hudson's Bay Company. The cape was named in 1631 by Luke Foxe after Edward Sackville, 7th Earl of Dorset (1590–1652). The island was named after the cape in 1930 by naturalist J. Dewey Soper.

Cape Sable Island NS At the southwestern end of the province, this island takes its name from the cape, identified as *Cap de Sable*, 'sandy cape', by Samuel de Champlain in 1607. After 1761 the English adapted the name from the French *Isle du Cap de Sable.*

Cape Tormentine NB A community at the most easterly point of the province, its post office was named in 1847 after the cape. The cape was called *Cap de Tourmentin* on Nicolas Denys's map of 1672, with the possible meaning of 'cape of storms'. Thomas Jefferys's 1755 map had *Stormy Point* and John Mitchell's 1755 map had *C. Storm*. Frederick Holland's 1791 map introduced *C. Tormentine.*

Cape Traverse PE East of Borden-Carleton, this place's post office was named about 1841. In 1765 Samuel Holland named the traditional terminus for the water crossing to the mainland *Cape Traverse*. This remained its official name until 1966, when the local name, **Bells Point**, was endorsed. In 1880 George Bell was living on the point.

Caplan QC In the regional county of Bonaventure, southeast of New Richmond, this municipality's name has its roots in the name **Rivière Caplan**. The river's name was likely derived from the Portuguese *capelinas*, 'capelin', a small fish of the smelt family.

Cap-Pelé NB A village (1969), 21 km east of Shediac, its post office was called *Tedish* in 1861, and was renamed *Cape Bald* from 1879 to 1903 and 1906 to 1947. The post office was *Cap Pelé* from 1903 to 1906, and has been Cap-Pelé from 1947. In the 1600s and 1700s the cape was variously called *C au Haran, C. Harang, C. Herring, C. Heron*, and *C. Scott*. A land petition in 1807 called it *Cap a Lee.*

Capreol ON A town (1918) and Canadian National Railways divisional point in Sudbury Region, north of Sudbury, it was named in 1911 after Frederick Chase Capreol (1803–86), a Toronto railway promoter in the mid-1800s.

Cap-Rouge QC In the Communauté urbaine de Québec, southwest of the city of Québec, it was established as the parish municipality of *Saint-Félix-du-Cap-Rouge* in 1872, and the town of Cap-Rouge in 1983. The cape of red schist rocks was noted as **Cap Rouge** as early as 1637. The place's name was deformed into Carouge from the seventeenth to the nineteenth centuries.

Cap-Saint-Ignace QC In the regional county of Montmagny and on the south shore of the St. Lawrence, northeast of the city of Québec, this municipality was incorporated in 1981, having previously been the parish municipality of *Saint-Ignace* since 1855. The cape's name, given after Jesuit founder St Ignatius of Loyola (*c.* 1491–1556), was shown on Jean Bourdon's 1641 map.

Cap-Santé QC A municipality (1979) in the regional county of Portneuf, southwest of the city of Québec, its parish of La Sainte-Famille-du-Cap-Santé was founded in 1679, with the cape being noted on Vincenzo Coronelli's 1689 map. The meaning of its name is uncertain.

Caradoc ON A township in Middlesex County, it was named in 1820 after a revered king of Wales, who was killed during a battle in 795 AD between the Welsh and the Saxons.

Caraquet NB A town (1961) on **Baie de Caraquet**, an extension of Chaleur Bay, it was settled about 1760 by shipwrecked sailors, and later by Acadians and French-speaking settlers from Quebec. **Rivière Caraquet**, which flows into the bay, was identified as Caraquet on Nicolas Denys's map of 1672. The name may have been derived from Mi'kmaq *pkalge*, possibly 'junction of two rivers', in reference to the point where the Rivière du Nord joins the main river.

Carberry MB This town (1905) is located east of Brandon. Its Canadian Pacific Railway station was supposed to be located at De Winton in 1881, 3 km to the east. However, the next year, when the landowner proposed an excessive price for his land, the CPR moved its station to its present site. CPR director James J. Hill named it after Carberry Tower in Musselburgh, Scotland, east of Edinburgh. The tower was the seat of Lord Elphinstone, an early director of the CPR, who accompanied Hill in 1882.

Carbon AB West of Drumheller, this village (1912) was named in 1904 by settler John M. Bogart, after the local coal mines. It was also called *Kneehill*, after the Kneehill Coal Company, which mined the coal deposits on the banks of Kneehills Creek, a tributary of the Red Deer River.

Carbonear NF A town (1948) on **Carbonear Bay** and on the western shore of Conception Bay, it was first identified as *Carboneare* in the calendar of estate papers of 1674, with the present spelling occurring in 1705. John Guy called the bay *Carbonera Bay* in 1612, with this form persisting into the 1700s alongside *Carbonere* and *Carboniere*. The name, of either Spanish or French origin, probably relates to the production of carbon there.

Carcross YT Located where Bennett and Tagish lakes meet, this place was first known as *Caribou Crossing*. To avoid confusion with other places with similar names, the Rt Revd William Bompas appealed to the Canadian government in 1901 to rename the place Carcross. The post office department complied with the request the next year, but the White Pass and Yukon Railway did not change its station name until 1916.

Cardiff ON A township in Haliburton County, it was named in 1862 after the city in Wales. The community of **Cardiff** was named in 1957. It is in the municipal township of Bicroft, which is surrounded by the municipal township of Cardiff.

Cardigan PE North of Montague, this municipal community (1983) had been incorporated as a village in 1954. Its post office was opened in 1863, and called after **Cardigan River**. The river was named in 1765 by Samuel Holland after George Brudenell, 4th Earl of Cardigan (1712–90). He also named **Cardigan Bay**, into which the Montague, Brudenell, and Cardigan rivers flow. Together they were called the *Trois Rivières* during the French regime.

Cardinal ON On the St. Lawrence River and in the united counties of Leeds and Grenville, northeast of Brockville, the post office in this village (1878) was called *Edwardsburg* in 1837, and was changed to Cardinal in 1880. **Point Cardinal** was named by the French, possibly after Cardinal Richelieu, but a tradition claims that there was a person called Cardinal living there during the French regime.

Cardston AB Located on Lee Creek, a tributary of the St. Mary River, southwest of Lethbridge, this town (1901) was founded by Charles Ora Card (1839–1906), a son-in-law of Mormon leader Brigham Young. He had settled here in 1886, having left Utah to avoid being tried on charges of polygamy. First called *Lees Creek*, it became Cardston when a post office was opened in 1892.

Cargill ON In Bruce County, northwest of Walkerton, this place was named in 1880 after miller Henry Cargill (1838–1903), Conservative member for Bruce East in the House of Commons, 1887–1903.

Cariboo BC First given to the Quesnel and Barkerville area during the gold rush of the early 1860s, the Cariboo eventually extended to a huge area from Cache Creek in the south to Prince George in the north. In adopting the word in 1861, Gov. James Douglas observed it should more properly be spelled *cariboeuf*. Actually its roots are the French word *caribou*, adapted from the Mi'kmaq *xalibu*, in reference to the way this member of the deer family shovelled snow with its wide feet to get at lichen, its favourite food. The **Cariboo Mountains**, part of the Columbia Mountains, extend from southeast of Purden Lake (east of Prince George) to Valemount and Blue River.

Caribou NS North of Pictou, this is the site of a ferry terminal connecting with Wood Islands, PE. The adjacent harbour and island were noted on an 1838 Arrowsmith map as *Carriboo Harbour* and *Carriboo Island*, with these spellings persisting into the latter part of the twentieth century. In 1609 French historian Marc Lescarbot used the spelling Caribou for the Mi'kmaq *xalibu*, meaning 'shoveller', referring to the habit of these members of the deer family scraping away snow to uncover lichens, a favoured food.

Carievale SK A village (1903) near the southeastern corner of the province, its post office was named in 1891 by John Young, who had settled here in 1886. Of Scottish origin, the name is said to mean 'lovely valley'.

Carignan QC Established as a town (1965), it is located in the regional county of La Vallée-du-Richelieu, and beside the city of Chambly. Called the municipal parish of *Saint-Joseph-de-Chambly* in 1855, it was renamed after the famous Carignan-Salières Regiment, which settled a large part of the province, especially the Chambly area. The name originates in the town of Carignano in Italy's province of Turino.

Carleton NB On the west side of the province and centred on the town of Woodstock, this county was named in 1831 after Thomas Carleton (1736–1817), the first lieutenant-governor of the province, 1784–1817. In 1899, naturalist William F. Ganong named the highest point (820 m) in the province, **Mount Carleton**, after him.

Carleton NS Located northeast of Yarmouth, this place was first known as *Temperance*. Its post office was named in 1857, after nearby *Lake Carleton* (now Raynards Lake), given earlier after Sir Guy Carleton, Lord Dorchester. **Carleton Village**, south of Shelburne, was named in 1900 after his brother, Thomas Carleton, the first lieutenant-governor of New Brunswick, 1784–1817.

Carleton ON The county was named in 1798 after Sir Guy Carleton, Baron Dorchester (1724–1808), governor of Quebec, 1768–78, commander-in-chief of the British forces in North America 1782–6, and governor-in-chief of British North America, 1786–96. In 1969 the municipalities within it, plus Cumberland Township in the united counties of Prescott and Russell, became the regional municipality of **Ottawa-Carleton**.

Carleton QC A town in the regional county of Avignon, southwest of Gaspé, it was established in 1972 by merging the township municipalities of Carleton and *Carleton-sur-Mer*. The original township of Carleton was named in 1786 after Sir Guy Carleton, Lord Dorchester, governor of Quebec, 1768–78, commander-in-chief of the British forces in North America, 1782–6, and governor-in-chief of British North America, 1786–96. The mountain rising to 609 m to the rear of the town centre was first known as *Mont Tracadigache*, from the Mi'kmaq for 'many herons', with **Mont Carleton** replacing it during the early years of the twentieth century.

Carleton Place ON In Lanark County, this town (1890) was called *Morphy's Falls* in 1819 for Edmond Morphy, a native of County Tipperary, Ireland, and his three sons, William, John, and James. It was renamed Carlton Place in 1829 by innkeeper Alexander Morris, after a street in central Glasgow. However, postal officials spelled it Carleton Place in 1830, following the spelling of nearby Carleton County.

Carleton Siding PE In the municipal community of Borden-Carleton since 1995, it had been created as a separate municipal community in 1977. It has been the postal designation in this place since 1925. Its post office had been called *Carleton* in 1870, taking its name from **Carleton Cove**, given by Samuel Holland in 1765 after Sir Guy Carleton, who subsequently became the governor-in-chief of British North America.

Carlsbad Springs ON In the city of Gloucester, southeast of Ottawa, it was named in 1906 after a famous health spa in the Czech Republic, now known as Karlovy Vary. In the 1860s the site had been known as *Cathartic*, after the healing qualities of the waters. The post office had been called *Eastman's Springs* from 1867 to 1906, after Danny Eastman's inn, which operated here from 1867 to 1870.

Carlyle SK Established with a post office in 1883, this town (1906) may have been named by member of Parliament J.G. Turriff after a couple who had operated a store there, and were buried in West Carlyle Cemetery. It has also been claimed that it was called after Scottish writer Thomas Carlyle (1795–1881).

Carmacks YT On the west bank of the Yukon River and at the mouth of the Nordenskiold River, the post office at this village was named in 1908 after George Washington Carmack (1860–1922), a native of California, who developed the coal deposits and operated a fur trading post here after 1893. Three years later Carmack, Skookum Jim, and Tagish Charlie discovered gold in the gravelly bottom of Bonanza Creek.

Carman MB Situated southwest of Winnipeg, the post office in this town (1905) was named in 1880 by Premier Rodmond Roblin, after the Revd Dr Albert Carman (1833–1917), the bishop of the Methodist Episcopal Church in Ontario until 1884, and general superintendent of the Methodist Church of Canada until 1915.

Carmangay AB On the Little Bow River, northwest of Lethbridge, the post office in this village (1910) was named in 1907 after Charles Whitney Carman, his wife, the former Gertrude Gay, and their son, Gay. Two years later the Canadian Pacific Railway adopted the name for its station.

Carmanville NF On the north shore of the island of Newfoundland, this town (1955) was named in 1906, after the Revd Albert Carman, general superintendent of the Methodist Church of Canada. It had been earlier called *Rocky Bay*.

Carnduff SK A town (1905) in the southeastern part of the province, east of Estevan, its post office was named in 1884 after first postmaster John Carnduff. A native of Belfast, Ireland, he settled first in southern Manitoba, and then moved to the area of Carnduff the year the post office was opened.

Carnes Peak BC Located in the Selkirk Mountains, north of Revelstoke, this mountain (3051 m) was named in 1956 by climber William Putnam after prospector Henry Carnes, who had explored for gold in the creeks flowing into the Columbia in the 1860s. In 1912 alpinist Howard Palmer had called it *Serenity Mountain*.

Caroline AB Southwest of Red Deer, the post office in this village (1951) was named in 1908 after Caroline Langley, the daughter of postmaster Harvey Langley.

Caronport SK A village (1988) west of Moose Jaw, it was founded in 1946 as the site of a Bible school, having been located in nearby **Caron** in 1934. It took over an abandoned airport, which had been used to train British Commonwealth airmen during the Second World War. Caronport post office was opened in 1947. Caron had been named in 1883 by the Canadian Pacific Railway after Sir Adolphe Caron (1843–1908), minister of militia and defence 1880–92, who organized the forces to put down the North-West Rebellion.

Carp ON In West Carleton Township, Ottawa-Carleton Region, it was named in 1854 after the **Carp River**, itself called after the suckers (*carpes à cochon*) found there by French voyageurs. It is commonly called 'The Carp' and 'Carp Village'.

Carrick ON A township in Bruce County, it was named in 1850, likely after the Carrick district of Ayrshire, where Robert the Bruce, the King of Scotland, 1306–29, may have been born in 1274. He succeeded his father as the Earl of Carrick in 1292.

Carrot River SK A town (1948) northwest of Tisdale, it was established in 1921 as a terminus of the Canadian Northern Railway. It is located near **Carrot River**, a tributary of the Saskatchewan River, which it joins just upriver from The Pas, MB. The first Carrot River post office was situated closer to the river in 1907, but was changed the next year to *Silver Stream*. The second Carrot River post office was opened in 1921 at the station, changed to *Carrot River Station* in 1931, with Carrot River being restored in 1936. The name of the river is a translation of Cree *Oskatask Sipi*.

Carrying Place ON On the isthmus joining Northumberland and Prince Edward counties, south of Trenton, this place recalls the Aboriginal practice of carrying their canoes from the Bay of Quinte on the east to Wellers Bay on the west. *Murray* post office, after the township, operated in the area from 1820 to 1893, when it became *Wellers Bay*, and was changed again to Carrying Place in 1913.

Carstairs AB A town (1966) north of Calgary, it was named in 1893 by the Calgary and Edmonton Railway Company, after Carstairs, Lanarkshire, Scotland, southeast of Glasgow. Its post office was opened in 1900.

Cartier ON In **Cartier Township**, Sudbury District and on the Canadian Pacific Railway main line, northwest of Sudbury, this place's post office was named in 1888. The township had been named three years earlier by surveyor H.B. Proudfoot for Sir George-Étienne Cartier (1814–73), a Father of Confederation, and co-premier of the province of Canada with Sir John A. Macdonald, 1857–62.

Cartier, Mount BC In the Selkirk Mountains, southeast of Revelstoke, this mountain (2610 m) was named in 1886 by Dr Otto Klotz after Sir George-Étienne Cartier, a Father of Confederation.

Cartwright MB A village (1947) near the North Dakota border, southeast of Brandon, its post office was named in 1882 after Sir Richard Cartwright (1835–1912), Canadian minister of finance, 1873-8. A native of Kingston, ON, he was a member of the House of Commons from 1867 to 1904, first as a Conservative, and then as a Liberal from 1869.

Cartwright NF On the Atlantic coast of Labrador, this municipal community (1956) was named after Capt. George Cartwright (1739/40–1819), who traded on the coast from 1770 to 1786.

Cascade Mountain AB Overlooking the town of Banff, this mountain (2998 m) was named in 1858 by Dr James (later Sir James) Hector, translating an Aboriginal designation meaning 'mountain where the water falls'.

Cascumpec Bay PE South of the town of Alberton, this bay's name is derived from Mi'kmaq *kaskampek*, 'bold sandy shore', in reference to the **Cascumpec Sand Hills**, which enclose the bay. On a 1631 map by Jean Guérard the bay is called *Caisiqupet*, and in 1744 Jacques Nicolas Bellin referred to it as *Casquembec*. In 1765 Samuel Holland identified it as *Holland Bay by the French Kascumpeck*. He likely intended to name it after Henry Fox, 1st Baron Holland of Foxley. The Geographic Board spelled it *Cascumpeque Bay* in 1901, but the local spelling was restored in 1966. **Cascumpec** was a postal designation in Alberton South from 1839 to 1868, and is now a locality on the southwest side of the bay.

Casselman ON A village (1888) in the united counties of Prescott and Russell, and on the east side of the South Nation River, some 50 km east of Ottawa, it was settled in 1843 by farmer and lumberman Martin Casselman. Casselman post office was opened in 1857 on the west side of the river. The post office was moved to the east side in 1898, replacing the *South Casselman* office, opened in 1886.

Cassiar Mountains BC, YT Located in the northwestern part of the province, with the Coast Mountains on the west and the Rocky Mountains on the east, their name was derived from the Kaska, an Aboriginal division of the Nahani. *Kaska* may mean 'creek' or 'small river', their name for McDame Creek, where they gathered in summer to fish and trade, or 'old moccasins', a term of derision given by the neighbouring Tahltan. In 1952 the community of **Cassiar** was established north of Dease Lake by the Cassiar Asbestos Corporation and attained a population of over 800 by 1976. In 1992 the corporation ceased operations and all of the place's buildings were sold and moved away.

Castleford ON On the Ottawa River and in Renfrew County, east of Renfrew, it was named in the 1820s by Lt Christopher James Bell after his birthplace in West Yorkshire, England, 15 km southeast of Leeds.

Castlegar BC A city (1974) at the confluence of the Columbia and Kootenay rivers, it was founded in 1902 by the Canadian Pacific Railway. Its newly constructed station reminded someone of the Castle Garden immigration centre in New York. Following the practice of shortening the names of stations, the CPR called it Castlegar. A fortification in New York's Battery Park was called Castle Clinton in 1807, and after some remodelling to create a concert hall in 1845, it was called Castle Garden. It served as the immigrant landing depot from 1859 to 1890.

Castle Mountain AB Northwest of Banff, this 11-km-long mountain (2862 m) was named in 1858 by Dr James (later Sir James) Hector, because he was reminded of a great crenellated castle in Britain. In 1946 Prime Minister Mackenzie King instructed the Geographic Board of Canada to rename it after Gen. Dwight Eisenhower, then supreme commander of the allied forces in Europe. Persistent public pressure resulted in the original name being restored in 1979, with **Eisenhower Peak** being assigned to its prominent eastern elevation.

Castleton ON In Northumberland County, northeast of Cobourg, its post office was named in 1852, likely after Castleton in Vermont, west of Rutland, where some of the first settlers came from.

Castor AB A town (1909) east of Stettler, it was named by the Canadian Pacific Railway in 1909 after nearby **Castor Creek**, a tributary of Battle River. The French word *castor* means 'beaver'.

Catalina NF South of Bonavista, this town (1958) was probably named after *p de S Catarina*, a Portuguese name portrayed on maps of the east coast of the island of Newfoundland in the 1500s. In 1534 Jacques Cartier spent ten days in *H Saincte Katherine*, where he prepared for further exploration. It was superseded by the Spanish name *Cataluña*.

Catalone NS Southeast of Sydney and on the west side of Catalone Lake, its post office was opened in 1842. The lake was earlier called *Barachois de Catalogne*, after Gédéon de Catalogne (1662–1729), who commanded a company at the Fortress of Louisbourg after 1723, and undertook the growing of crops in the area of the lake and on the Mira River.

Cataract ON In the town of Caledon, Peel Region, south of Orangeville, and at a point where the Credit River plunges through a deep gorge of the Niagara Escarpment, this place's post office was named in 1865.

Cataract Peak AB In the Rocky Mountains, north of Lake Louise, this peak (3333 m) was named in 1908 after several waterfalls on the adjacent Pipestone River.

Cataraqui ON On the **Little Cataraqui River** and in Frontenac County, west of Kingston, this place was known as *Sandville* and *Waterloo* before it was renamed Cataraqui in 1868. The **Cataraqui River** rises north of Kingston, and, from Seeleys Bay, coincides with the Rideau Canal, draining into Lake Ontario on the east side of Kingston. *Cataraqui* is an Iroquois word meaning 'rocks (or clay bank) rising out of the water'.

Cathedral Mountain BC In the Rocky Mountains, east of Field, this mountain (3189 m) was named in 1900 by Sir James Outram for its imposing outline.

Caubvik, Mount NF The highest point (1650 m) in the Torngat Mountains, it was named Mont d'Iberville in 1971 by the Commission de toponymie du Québec. Ten years later the government of Newfoundland and Labrador named it after Caubvik, one of the Inuit who went to London with George Cartwright in 1772. She was the only one to survive smallpox, and later returned to the Labrador coast. *See also* d'Iberville, Mont, QC.

Causapscal QC Located in the regional county of La Matapédia, south of Matane, this town was named in 1965. It had previously been the village of Causapscal (1928), and the parish municipality of *Saint-Jacques-le-Majeur-de-Causapscal* (1897). The original township, *Casupscull*, was proclaimed in 1864. In Mi'kmaq, the name means 'stony and glittering bottom'.

Cavan ON A township in Peterborough County, it was named in 1816 after the north-central county in Ireland. The post office in the community of **Cavan** was named in 1830.

Cavendish PE Northwest of Charlottetown and adjacent to Prince Edward Island National Park, this place was named about 1772 by William Winter, probably after Field Marshal Lord Frederick Cavendish (1729–1803). Its post office, opened in 1833, is now called Green Gables.

Cawston BC A community in the Similkameen River valley, southeast of Keremeos, its post office was named in 1917 after pioneer rancher and magistrate Richard Lowe Cawston (1849–1923).

Cayley AB Between High River and Nanton, south of Calgary, this place was named in 1893 by the Canadian Pacific Railway after Hugh St Quentin Cayley (1857–1934), then the member for Calgary in the Assembly of the North-West Territories. He moved in 1897 to Vancouver and was appointed a judge in 1917. Cayley was an incorporated village from 1904 to 1996.

Cayuga ON In the town of Haldimand, Haldimand-Norfolk Region, this place's post office was named in 1851 after **North Cayuga** and **South Cayuga** townships, themselves named in 1835 after the Cayuga nation, one of the Six Nations. There are several interpretations of Cayuga, including 'place where the canoes are drawn out', and 'advising nation'.

Cedar Lake MB Located northwest of Lake Winnipeg, this enlargement of the Saskatchewan River, east of The Pas, is located at the most northerly limits of the cedar tree. Discoverer Pierre Gaultier de La Vérendrye called it *Lac Bourbon*, after the reigning house in France, and built *Fort Bourbon* on its shore. Alexander Henry the younger called it *Lac Bourbon or Cedar Lake* in 1808.

Centennial Range YT This range of 13 peaks in the St. Elias Mountains was named in 1967 by the Yukon Alpine Centenial Expedition in honour of the one hundredth aniversary of Canada's Confederation. The expedition's organizers named the peaks after Canada's provinces and territories, and its climbers ascended nine of them in a five-week period during that summer.

Central Butte SK Established with a post office in 1907, this town (1967) is northwest of

Moose Jaw. It was named after the central of three buttes, the first being 27 km west near Demaine, and the third near Eyebrow, 14 km to the east.

Central Bedeque PE On the Dunk River and adjacent to the municipal community of Bedeque, the post office of this municipal community (1966) was named in 1884. It was earlier called *Weatherbie's Corner* and *Strong's Corner.*

Centralia ON In Huron County, south of Exeter, this community was named *Devon* in 1852, after the many early settlers who came from Devon, England. It was changed to Centralia in 1873, probably after its location in the middle of a rich agricultural area.

Central Kings PE A municipal community (1975) in the central part of Kings County, it embraces the localities of Albion Cross, Bridgetown, Dundas, and Poplar Point.

Central Saanich BC A district municipality (1950) on the Saanich Peninsula, north of Victoria, it takes the second part of its name from the Saanich tribe. In Straits Salish it means 'elevated', in reference to Mount Newton (307 m), the profile of which looks like a raised rump. The community of **Saanichton**, named in 1922, is in the northern part of the municipality. *See also* Saanich *and* North Saanich.

Centreville NB A village (1966), northwest of Woodstock, its post office was named in 1862 after its location among a cluster of surrounding rural communities.

Centreville NS Located midway between the Cornwallis River and Minas Channel, this community's post office may have been opened in 1859. A second **Centreville**, in the middle of Digby Neck, and halfway between the town of Digby and the southwest end of the neck, had a post office from about 1875 to 1970. A third **Centreville**, on Cape Sable Island, is halfway between the town of Clark's Harbour and the ferry terminal at North East Point.

Centreville ON In Lennox and Addington County, north of Napanee, its post office was named in 1849, for its location in the middle of Camden East Township.

Centreville-Wareham-Trinity NF This town on the northwest side of Bonavista Bay was established in 1992. **Centreville**, a neighbourhood midway between Trinity (Bonavista) and Indian Bay (Parson's Point), was created in 1959–60 as a resettlement community for the families on Fair Island in the bay. It was the only completely new settlement established in the province during that period, when remote and isolated communities were encouraged to move to areas served by land-based communication and transportation links. **Wareham**, on the south side of Indian Bay, and on the northwest side of Bonavista Bay, was named in 1931 by the Firmage family, natives of Wareham, Dorset, England. The family had settled here in 1918. **Trinity**, at the head of Trinity Bay, a small bay on the northwestern side of Bonavista Bay, first appeared in the census of 1901.

Cerberus Mountain BC In the Coast Mountains, southeast of Bella Coola, this mountain (3155 m) was named in 1953 by a member of the Alpine Club of Canada because its three summits reminded him of the mythical three-headed dog guarding the gates of hell.

Cereal AB East of Hanna, the post office in this village (1914) was named in 1911, on the suggestion of the Revd R.J. McMillan, encouraged by the prospects of the region becoming the 'bread basket of the west'.

Ceylon ON First known as *Virginia*, *Walterville*, and *Flesherton Station*, it was named Ceylon in 1899, when a postal inspector was inspired after finding a package of Ceylon tea in Tristam Chislett's store.

Ceylon SK Situated southwest of Weyburn, on the Radville-Willow Bunch line of the Canadian Northern Railway, both the village and the station were established in 1911. Earlier its post office had been called *Aldred* after pioneer settler John Aldred, but he recommended Ceylon for the station. *Ceylon Station* post office was opened in 1911 and was changed to Ceylon in 1957. It may have been called after a yacht used in races by Sir Thomas Lipton. There is a Ceylon Street in Lipton, SK.

Chaffeys Lock ON In the united counties of Leeds and Grenville and on the Rideau Canal, it was settled after the War of 1812 by Samuel Chaffey, who built several mills. He died in 1827, before the lock was constructed.

Chaleur Bay NB; **Baie des Chaleurs** QC This wide bay, extending from the mouth of the Restigouche River to Miscou Island, was named 9 July 1534 by Jacques Cartier, who observed that the great heat (*chaleur*) reminded him of Spain. English speakers in the Gaspé and northern New Brunswick usually call it Bay Chaleur.

Chalk River ON In Renfrew County, northwest of Pembroke, the post office in this village (1954) was named in 1875, after the nearby river. The river's name may have been derived from chalk used to mark timber at its outlet in Rafting Bay.

Chambly QC A city on the Richelieu, east of Montréal, it was created in 1965 when the towns of Chambly and *Fort-Chambly* were merged. In 1672 Jacques de Chambly (*c.* 1640–87) was granted the seigneury of Chambly at the rapids on the Richelieu, where he had constructed Fort Saint-Louis seven years earlier, with the post being subsequently named Fort Chambly.

Chambord QC This municipality (1916) is located on the south shore of Lac Saint-Jean and in the regional county of Le Domaine-du-Roy. Its post office was named in 1872 after Henri V de Bourbon, comte de Chambord (1820–83), a pretender to the French throne.

Champion AB Northwest of Lethbridge, this village (1911) was named by the Canadian Pacific Railway in 1909 after H.T. Champion of the Winnipeg banking firm of Alloway and Champion. *Clerverville* post office, opened nearby in 1907 and named after settler Martin G. Clever, was moved into Champion, and renamed in 1910.

Champlain QC Established in 1983, this regional county is on the southeast side of the St. Lawrence, and embraces the cities of Longueuil, Saint-Hubert, Saint-Lambert, and Brossard. It took its name from the founder of Québec, Samuel de Champlain. Also named after him were the huge lake separating the states of New York and Vermont, and an electoral district east of Trois-Rivières, where the municipality of **Champlain** is located at the mouth of a river also named after him.

Champlain, Mount NB Northwest of Saint John and on the perimeter of the Canadian Forces Base Gagetown, this mountain (446 m) was named in 1901 by naturalist William F. Ganong after Samuel de Champlain. Before that it was called *Bald Mountain*, a common name in the province.

Chance Cove NF On the Isthmus of Avalon and on the southeast shore of Trinity Bay, this town (1972) is located on **Big Chance Cove** and **Little Chance Cove**. The coves may have been named because they were difficult to enter in seeking refuge during storms. *See also* Come By Chance.

Chancellor Peak BC This mountain (3280 m) in Yoho National Park, south of Field, was named in 1898 after Sir John Alexander Boyd (1837–1916), chancellor of Ontario, who arbitrated a dispute between the Canadian Pacific Railway and the Dominion of Canada over some mineral rights.

Chandler QC A town since 1958 in the regional county of Pabok, southwest of Gaspé, it was named a village in 1916 after Percy Milton Chandler, of Philadelphia, who was the first president of the St Lawrence Pulp and Paper Corp., which built a mill here in 1912.

Change Islands NF This town (1951) is located on a narrow tickle separating two of the **Change Islands**, which are between Fogo and New World islands. It was established in the late 1700s as a centre of the English fishery on the Labrador coast.

Channel-Port aux Basques NF This town (1945) on the southwestern coast of the island of Newfoundland is comprised of the former communities of Channel and Port aux Basques, and adjoining settlements. The harbour called **Port aux Basques** was likely named by French explorers after Basque whalers, encountered there in the late 1500s. Samuel de Champlain recorded it as *po[t] aux basque*. In the 1700s it was called *Swift Harbour* and *Swints Harbour*, with Capt. James Cook identifying it as *Port aux Basque* in 1764. Since 1893 it has been the main terminus linking the island of Newfoundland with Nova Scotia. The original settlement of Channel was named after a narrow channel separating a small island from the peninsula on which it is located.

Chantrey Inlet NT An inlet of the Arctic Ocean, and at the mouth of Back River, it was named in 1921 after the *Chantrey Hills*, shown on maps on the west side of the inlet. The hills were named in 1834 by Capt. George (later Sir George) Back after sculptor Sir Francis Legatt Chantrey (1781–1842). However, when air photography did not reveal hills in that location in 1952, this name was rescinded.

Chapais QC A mining town northwest of Lac Saint-Jean and southwest of Chibougamau, it was established in 1955, and named after Senator Thomas Chapais (1858–1946), a prominent journalist, historian, and politician.

Chapeau QC A village since 1874 in the regional county of Pontiac, north of Pembroke, ON, its name is attributed to a hat-shaped rock in the Rapides de la Culbute of the Ottawa River.

Chapel Arm NF This town (1970) is located on the Isthmus of Avalon, and on **Chapel Arm**, at the south end of Trinity Bay. The arm was possibly named after some steep hills, whose sharp elevations resembled the spires of chapels.

Chapleau ON In **Chapleau Township**, Sudbury District, this community was named in 1886, after Joseph-Adolphe (later Sir Joseph-Adolphe) Chapleau (1840–98), premier of Québec, 1879–82, secretary of state in the government of Canada, 1882–92, and lieutenant-governor of Québec, 1892–8. The township was incorporated in 1906.

Chaplin SK Situated on the Canadian Pacific Railway, midway between Moose Jaw and Swift Current, the station of this village (1912) was named in 1883 after **Chaplin Lake**, which extends some 30 km to the south of the village. The lake had been named in 1861 by Sir John Rae, after a hunting companion, Viscount Chaplin. The post office was opened in 1907.

Chapman, Mount BC Located in the Selkirk Mountains, north of Revelstoke, this mountain (3075 m) was named in 1917 by climber Howard Palmer after geologist Robert Hollister Chapman (1868–1920), who had accompanied Palmer in a survey of the Selkirks.

Charing Cross ON In Kent County, southeast of Chatham, this place was named in 1860 after Charing Cross, adjacent to Trafalgar Square, in central London, England.

Charlemagne QC Located in the regional county of L'Assomption, northeast of Montréal, it was named the municipality of *Laurier* in 1907, and was changed the next year to Charlemagne, which became a town in 1969. It was given in honour of Romuald-Charlemagne Laurier (1852–1906), half-brother of Sir Wilfrid Laurier, and member for L'Assomption in the House of Commons, 1900–6.

Charlesbourg QC In the Communauté urbaine de Québec, north of the city of Québec, it was created a village in 1914, became the municipality of *Charlesbourg-Est* in 1928, and the city of Charlesbourg in 1949. The name was mentioned in the civil registers of 1666, taking its name from the first chapel built at *Bourg-Royal*, which was dedicated to St Charles Borromée (1538–84). The municipal parish of Saint-Charles-Borromée was created in 1845 in the area of the present city.

Charleston ON On **Charleston Lake**, west of Brockville, this community was settled in 1836, with its post office being named in 1853, possibly after Brockville developer Charles Jones (1781–1840). With a moderate climate and an untamed landscape, and with numerous islands, coves, and narrow passages, Charleston Lake became one of Canada's first vacation resorts.

Charleswood MB A neighbourhood in the southwestern part of Winnipeg, its post office was named in 1910, after the parish of St. Charles, and after the attractive woods along the south bank of the Assiniboine River. The post office was earlier called *Kelheau*, with P.H. Kelly as its first postmaster.

Charlevoix QC A regional county northeast of the city of Québec and centred on the town of Baie-Saint-Paul, it was named in 1982, succeeding the former county of Charlevoix-Ouest. The regional county of **Charlevoix-Est**, proclaimed the same year in the area of La Malbaie, was formerly a county with the same name. The names originate from the noted historian, Jesuit Fr Pierre-François-Xavier de Charlevoix (1682–1761), who wrote the *Histoire et description générale de la Nouvelle-France* (1744).

Charlo NB A village (1969), southeast of Dalhousie, it had been incorporated in 1966 as *Colborne*, after the local territorial parish. Its post office was called *Charlo Station* in 1881 and was renamed Charlo in 1968. It took its name from the **Charlo River**, which may have been called after an early settler, such as Charles 'Charlo' Doucet, the founder of the village of Petit-Rocher, or after a trapper from Québec.

Charlotte NB In the southwestern part of the province, this county was named in 1785 after Queen Charlotte (1744–1818), the consort of George III.

Charlottenburgh ON A township in the united counties of Stormont, Dundas and Glengarry, it was named in 1787 after Queen Charlotte Sophia (1744–1818), the wife of George III.

Charlottetown NF First known as *Brown's Cove*, this place was renamed at the end of the nineteenth century, possibly after Charlotte (Hussey) Spracklin, the first woman settler, or possibly after Charlottetown, PE. There is a second **Charlottetown** on the coast of Labrador, north of Port Hope Simpson. First known as *Old Cove*, it was renamed about 1949 by lumberman Ben Powell after the capital of Prince Edward Island.

Charlottetown PE This city (1855) was established in Charlottetown Royalty in 1765 by Surveyor General Samuel Holland. He named both of them after Charlotte Sophia (1744–1818), the consort of George III. The royalty, one of the original 70 divisions laid out by Holland, is now entirely within the limits of the city.

Charlton ON In Timiskaming District, northwest of New Liskeard, and Ontario's smallest town (1996 population, 275), it was named in 1904 after brothers John Charlton (1829–1910), Liberal member for Norfolk North in the House of Commons, 1872–1904, and William Andrew Charlton (1841–1930), Liberal member for Norfolk South in the Ontario Legislature, 1890–1904, speaker of the Ontario Legislature, 1903–4, and Liberal member for Norfolk in the House of Commons, 1911–21. It became a town in 1914.

Charlton, Mount AB On the west side of Maligne Lake and southeast of Jasper, this mountain (3217 m) was named in 1911 by Mary Schaffer, after Henry Ready Charlton (1866–1919), general advertising agent of the Grand Trunk Pacific Railway, 1898–1919.

Charny QC A town in the regional county of Les Chutes-de-la-Chaudière, south of the city of Québec, it was created in 1965, having been incorporated as a village in 1924. The name was given in honour of Jean de Lauson de Charny (*c.* 1620–1661), a son of New France Gov. Jean de Lauson. Appointed by the father as the grand seneschal (judge) of New France, he was killed by the Iroquois on Île d'Orléans.

Chartersville NB East of Moncton, this place was named after cattle dealer Samuel Charters, who had come from England about 1800. Its post office was opened about 1893, with Ethel B. Charters as the first postmistress. It was incorporated as a village in 1966, and was annexed by the town of Dieppe in 1973.

Chase BC On Little Shuswap Lake, northwest of Salmon Arm, this village (1969) was named in 1908 by James A. Magee, secretary of the Adams River Lumber Company, after Whitfield Chase (1820–96), the first settler, who established a ranch here in 1865. An American, Chase joined the Cariboo Gold Rush in the early 1860s. He was brought to the site of Chase by Aboriginals, after having been rescued in the Adams River country, where he had become lost.

Château-Richer QC A town (1968) in the regional county of La Côte-de-Beaupré, northeast of the city of Québec, its post office was named in 1832. Of an uncertain origin, the name occurred on a 1641 map by Jean Bourdon.

Châteauguay QC Situated at the mouth of the **Rivière Châteauguay** and in the regional county of Roussillon, southwest of Montréal, this city was created in 1975 through the merger of the towns of Châteauguay and *Châteauguay-Centre*. The name has its roots in the seigneury of Charles Le Moyne de Longueuil et de Châteauguay (1626–85), granted by Governor Frontenac in 1673.

Chatham NB Part of the city of Miramichi since 1 January 1995, it was named about 1800 by Francis Peabody after William Pitt, 1st Earl of Chatham (1708–78), the father of William Pitt (1759–1806), who was then the prime minister of Great Britain. Its post office was opened in 1825, and it became an incorporated town in 1896.

Chatham ON In 1793 Lt-Gov. John Graves Simcoe proposed naming a town Chatham on the Thames, downriver from London, as they are in England. **Chatham Township** in Kent County was named in 1794, and remains a separate municipality from the city. Incorporated as a village in 1851, Chatham became a town four years later, and finally a city in 1895.

Chatham QC A township municipality (1845) in the regional county of Argenteuil, west of Lachute, it was named after the township of Chatham, proclaimed in 1799, and called after William Pitt, 1st Earl of Chatham (1708–78), the father of noted British prime minister William Pitt.

Chats, Lac des ON, QC A 40-km enlargement of the Ottawa River at Arnprior, its name reflects the description of *Chats Falls* (dammed in the 1930s for hydroelectric development), which either had the appearance of a wildcat's paw, or its plunging waters sounded like hissing wildcats.

Chatsworth ON A village (1904) in Grey County, south of Owen Sound, it was named in 1857 by postmaster Henry Caldwell, after his birthplace near Chatsworth House in Derbyshire, southwest of Sheffield, England.

Chaudière Falls ON, **Chaudières, Chute des** QC On the Ottawa River and between the cities of Ottawa and Hull, they were first described in 1613 by Samuel de Champlain. The Algonquin called the large deep basin, shaped by the swift eddies below the falls, *Asticou*, which Champlain interpreted as *chaudière* (kettle).

Chaudière, Rivière QC Rising in Lac Mégantic, near the border of Maine, it flows almost 200 km to the north to empty into the St. Lawrence, upriver from the city of Québec. Champlain called it *Rivière des Etechemins* on his 1612 and 1632 maps, but after 1650, it usually had a variation of Rivière Chaudière, referring to a deep circular hole in relatively soft rock surrounded by more resistant rock in the

Chutes de la Chaudière, 3 km above the mouth of the river.

Chauvin AB Near the Saskatchewan border, southeast of Wainwright, this village (1912) was named by the Grand Trunk Pacific Railway in 1908, after George von Chauvin, a director of the GTP's parent, the Grand Trunk Railway. The post office was opened the following year.

Cheadle, Mount BC This mountain (2660 m) in the Monashee Mountains, south of Valemount, was named in 1863 by Viscount Milton after Dr Walter Butler Cheadle (1835–1910), who undertook an arduous journey through the western mountains in 1862–3.

Chedabucto Bay NS Separating the mainland from Isle Madame, this bay extends for 60 km from Cape Canso to the head of the bay, at Guysborough. In Mi'kmaq it was called *Sedabooktook*, 'a bay running far back'.

Chelmsford ON An urban centre in the town of Rayside-Balfour, Sudbury Region, northwest of Sudbury, it was named in 1886 after Chelmsford in Essex, England, northeast of London. It was an incorporated town from 1910 to 1969.

Chelsea QC A municipality in the Communauté urbaine de l'Outaouais, northwest of Hull, it replaced *Hull-Partie-Ouest* in 1990. The community of Chelsea developed in 1819 at the present site of Old Chelsea, and was likely named by Thomas Brigham, and his nephew Thomas Brigham Prentiss, after Chelsea, VT, south of Barre.

Cheltenham ON In the town of Caledon, Peel Region, northwest of Brampton, this place was named in 1852 after millwright Charles Haines's home town, Cheltenham, in Gloucestershire, England.

Chemainus BC In the district municipality of North Cowichan, southeast of Ladysmith, this community was named in 1871 after the Tsiminnis, a Salish tribe. In Island Halkomelem it means 'bitten breast', in reference to the crescent shape of **Chemainus Bay**, which legend claims was a bite taken from a bystander by an excited shaman during a tribal ceremony.

Chemong Lake ON One of the Kawartha Lakes, north of Peterborough, its name is derived from the Ojibwa *Tchiman*, 'canoe', in reference to its long curving shape. Curve Lake Indian Reserve is at its north end.

Chénéville QC Located in the regional county of Papineau, northeast of Buckingham, this village's post office was called *Hartwell* from 1876 to 1884, when it was renamed after postmaster Hercule Chéné.

Chephren, Mount AB Named *Black Pyramid* by British climber Norman Collie in 1898, this mountain northeast of Howse Pass was renamed in 1918 by the Alberta-British Columbia Boundary Commission. It was named after the son of King Khufu (Cheops), who was the builder of the Great Pyramid in 2565 BC.

Chepstow ON In Bruce County, west of Walkerton, this place was named *Emmett* in 1865 by miller John Phelan after the Irish revolutionary hero Robert Emmet, who was hanged for treason in 1803. But the postal authorities named it Chepstow, well known by Irish nationalists as the home of the 2nd Earl of Pembroke (1130–76), known as Strongbow, the first English invader of Ireland. Chepstow is northeast of Cardiff, Wales.

Cherry Valley ON In Prince Edward County, south of Picton, it was named in 1812 by Alva Stephens, possibly after Cherry Valley, NY, southeast of Utica, or after cherry trees he found in the valley. The post office was named in 1848.

Chertsey QC Situated in the regional county of Matawinie, west of Joliette, this municipality was named in 1991 after the parish municipality, created in 1856. It had taken its name from the township, proclaimed in 1795, and named after Chertsey, in Surrey, England, southwest of London.

Chesley ON A town (1906) in Bruce County, north of Walkerton, it was named in 1868 by the post office department after Solomon Yeomans Chesley (1795–1880), an official of the department in Ottawa. From 1820 to 1845 he was the Indian agent at the St. Regis Indian Reserve near Cornwall.

Chester NS On Mahone Bay, this village (1963) was first called *Shoreham*, after Shoreham Township, which had been created in 1759, and named after Shoreham, Sussex, England. Both were changed to Chester within a year, possibly after Chester, PA. Postal service was provided at Chester in 1832 and at nearby **Chester Basin** in 1855.

Chesterfield Inlet NT A hamlet (1980) on the south side of the mouth of **Chesterfield Inlet**, it was named in 1946. The inlet, which extends for 160 km to the mouth of the Thelon River, was named in 1747 by William Moor and Francis Smith after Philip Dormer Stanhope, 4th Earl of Chesterfield (1694–1773). Moor and Smith were sponsored by the North West Committee to seek a western outlet of the Northwest Passage.

Chestermere AB A town (1995) east of Calgary, it had been developed since 1959 as a summer community surrounding **Chestermere Lake**. The lake was created in 1906 by the Canadian Pacific Railway as a reservoir and may have been named after a director.

Chesterville ON A village (1889) in Winchester Township and the united counties of Stormont, Dundas and Glengarry, it was first known as *Armstrong's Mills*, after miller Thomas Armstrong. In 1845 its post office was called *Winchester*, but in 1876 it was renamed for merchant and telegraph operator Chester Casselman. *See also* Winchester.

Chéticamp NS On the west side of Cape Breton Island, this community's name is of French origin, and may mean 'poor' or 'miserable camping'. Its first post office was opened in 1837. **Chéticamp Island** is connected to Cape Breton Island by a narrow isthmus.

Chetwynd BC West of Dawson Creek, this district municipality (1983) was incorporated as a village in 1962. Its post office was called *Little Prairie* in 1949, but it was renamed in 1959 after Ralph L.T. Chetwynd, minister of railways, 1952–4. He had the Pacific Great Eastern Railway (now the British Columbia Railway) extended from Prince George to Dawson Creek and Fort St. John.

Chezzetcook Inlet NS East of Dartmouth, this inlet's name was derived from Mi'kmaq *sesetcook*, 'flowing rapidly in many channels'. Among places surrounding the inlet, with postal opening dates, are **West Chezzetcook** (1864, as *Chezzetcook*), **Head of Chezzetcook** (1880), **East Chezzetcook** (1880), and **Lower East Chezzetcook** (1897).

Chibougamau QC A town located 200 km northwest of Lac Saint-Jean, it was established as a mining centre in 1952, taking its name from **Lac Chibougamau**, which in Cree may mean 'a lake crossed from side to side by a river', 'water blocked by a narrow outlet', or 'meeting place'.

Chic-Chocs, Monts QC This mountain chain in the Gaspé has several high summits, including Mont Jacques-Cartier, which rises to 1268 m. The Mi'kmaq word *sigsôg* signifies 'steep rocks', or 'rocky mountains'.

Chicoutimi QC The premier city of the Saguenay, it is in the regional county of Le Fjord-du-Saguenay. It had its beginnings in 1845 on the creation of the township of Chicoutimi on the south bank of the river. The name in Montagnais means 'the end of the deep water'.

Chidley, Cape NF, NT The most northerly point of the province of Newfoundland and the most southeasterly point of the Northwest Territories, it was named in 1587 by explorer John Davis, probably after John Chidley, a friend and neighbour, near Exeter, in Devon, England.

Chignecto Bay NB, NS The northeastern head of the Bay of Fundy, this bay's name may be derived from Mi'kmaq *sigunikt*, 'footcloth'. It may have applied originally to **Cape Chignecto** in Nova Scotia, which separates the bay from Minas Channel, or from *saukanicktook*, 'forks at the mouth', referring to the LaPlanche and Missaguash rivers flowing together into the head of Cumberland Basin. Fr Pierre Biard called it *Chinictou* in 1611. The 30-km-wide **Chignecto Isthmus** joins New Brunswick to Nova Scotia.

Chilcoot Pass BC This pass (1140 m) on the Alaska-British Columbia border, just west of White Pass, was named in the Alaska Coast Pilot of 1883 *Chilkoot Portage* and *Shasheki Pass*. The Chilcoot, a Tlingit tribe, live in the areas of Skagway and Haines AK.

Chilcotin River BC Rising 160 km west of Quesnel, this 235-km river flows southeast to join the Fraser River, 45 km southwest of Williams Lake. The area is occupied by the Chilcotin, an Athapaskan tribe. Their name means 'ochre river people', in reference to red and yellow mineralized substances valued as a base for dyes and paints. The area west of Williams Lake is commonly called 'the Chilcotin' and 'the Chilcotin country'.

Chilkat Pass BC North of the Alaska-British Columbia border, and midway between Haines, AK and Haines Junction, YT, this pass was named in 1957 after the **Chilkat River**. The river, which rises east of the pass, was named in 1898 after a Tlingit tribe, whose name means 'salmon storehouse'.

Chilko Lake BC In the Coast Mountains, this lake is drained by the **Chilko River** northeast into the Chilcotin River. The river was identified on Joseph Trutch's 1871 map as *Chilcote*, but the Geographic Board of Canada authorized the present name in 1911. *Chilko* is a variant of Chilcotin, which means 'ochre river' in the Chilcotin language.

Chilliwack BC A district municipality since 1980, its urban centre, where Chilliwack post office was opened in 1872, had been created as a village in 1873 and a city in 1908. The incorporated township (1873), which surrounded the city before 1980, was called *Chilliwhack*. The **Chilliwack River** rises to the east in **Chilliwack Lake**, and flows west to join the Vedder River, in the southern part of the district municipality. The name of the Chilliwack, a Salish tribe, means either 'quieter water on the head', or 'travel by way of a backwater or slough'.

Chinguacousy ON This township was named in 1819 by Lt-Gov. Sir Peregrine Maitland, after the Mississauga name for the Credit River, which means 'young pine'. In 1974 it was divided equally between the city of Brampton and the town of Caledon, in Peel Region.

Chipman AB Northeast of Edmonton, the village (1913) was founded in 1905 by the Canadian Northern Railway, and named after Clarence Campbell Chipman, private secretary to Sir Charles Tupper, when he was minister of railways and canals in 1882. He was appointed chief commissioner of the Hudson's Bay Company in 1891.

Chipman NB A village (1966) on the Salmon River, northeast of Fredericton, its post office was named in 1865 after **Chipman Parish**. In 1835 the parish was called after Ward Chipman Jr (1787–1851), chief justice of the province, 1834–50. About 1870 W.C. King tried to rename the place *Lillooet*, after a schooner built by his brother George, which he had named after the place in British Columbia, but by 1903 that name had almost been forgotten.

Chippawa ON In the city of Niagara Falls, this former village (1849–1970) was first settled in 1783 at the mouth of *Chippawa Creek* (renamed the Welland River in 1792 by Lt-Gov. John Graves Simcoe). Its post office was named *Chippewa* in 1801, and was changed to Chippawa in 1844.

Chiputneticook Lakes NB On the New Brunswick-Maine border and near the head of the St. Croix River, the name of the lakes was taken from the Passamaquoddy word meaning 'great fork river', in reference to that part of the river above Grand Falls Flowage, a lake northwest of St. Stephen.

Chisasibi QC This Cree village at the mouth of La Grande Rivière, where it flows into James Bay, was created in 1978 by Hydro-Québec to provide a new community for the residents of Fort George. In Cree it is called *Tschishasipi*, meaning 'big river'.

Choiceland SK A town (1979) located west of Nipawin, its post office was opened in 1927. It was named by local resident Pete Rotz in recognition of the high quality of soils in the area.

Chomedey QC Created as a town in 1961 by amalgamating Saint-Martin, Renaud, and L'Abord-à-Plouffe, it was named after Paul de Chomedey (1612–76), first governor of the Île de Montréal, 1641–65. In 1965 Chomedey was united with all municipalities on the Île Jésus to form the city of Laval.

Chortitz MB A Mennonite community southeast of Morden, it was named about 1875 after Chortitza, a village north of Zaporizhzhya in present-day Ukraine, the site of the first Mennonite settlement in the late 1700s in the area of the Dnieper River. It was the name of the post office from 1884 to 1967, when it was changed to Randolph.

Chown, Mount AB In the northwestern part of Jasper National Park, this mountain (3381 m) was named in 1912 by climber H.A. Stevens after the Revd Samuel Dwight Chown (1853–1933), a Winnipeg Methodist clergyman, and a founder of the United Church of Canada in 1925.

Christina Lake BC Located between Trail and Grand Forks, this community's post office was named in 1912 after the adjacent lake. The lake was named in 1870 after Christina MacDonald McKenzie Williams (1847–1925), a daughter of Hudson's Bay Company trader Angus MacDonald. That year she had saved valuable HBC records during a rafting accident on the lake, and the following year the lake was identified on Joseph Trutch's map.

Christopher Lake SK Located north of Prince Albert, the post office of this village (1985) was named in 1925. The nearby lake was named about 10 years earlier after Christopher Monier by his brother, a surveyor.

Churchbridge SK This town (1964) southeast of Yorkton was named in 1886, when the Manitoba and North Western Railway was built from Winnipeg to Saskatoon. The place was developed by the Church Colonization Land Society of the Anglican Church. The name was formed from the society's title and from the name of the Revd John Bridges, a society director. Its post office was opened in 1889.

Churchill MB Manitoba's seaport, and the Hudson Bay Railway's terminus, this place is located at the mouth of the **Churchill River** on Hudson Bay. Its post office was named in 1932, after Fort Churchill, constructed here in 1686, and named after John Churchill, the Duke of Marlborough (1650–1722). He had been made a governor of the Hudson's Bay Company the previous year. The river was subsequently named after the fort, having been called *Danish River* and *Manoteusibi* (Cree for 'stranger's river', in reference to a visit in 1619–20 by the Danish naval officer Jens Munk), *Missinipi* in 1714 (Cree for 'big river',) and *English River*, given by fur trader Joseph Frobisher in 1775.

Churchill PE Located west of Charlottetown, this place's post office was named in 1900 after Winston (later Sir Winston) Spencer Churchill (1874–1965), who had then received considerable fame as a reporter of the South African War. Although the post office existed for only 13 years, its name was reinforced by a church built here on a hillside.

Churchill Falls NF Located on the **Churchill River** in Labrador, this set of falls (305 m) was called *Grand Falls* in 1839 by Hudson's Bay Company trader John MacLean. It was renamed on 4 February 1965 by Premier Joseph R. Smallwood after the British wartime leader, Sir Winston Churchill, honouring a pledge made to Churchill to name a major feature after him. An unincorporated community called **Churchill Falls** was established at the falls. The river (335 km), also called after Sir Winston in 1965, had been named *Hamilton River* after Hamilton Inlet. This 215-km arm of the Atlantic Ocean was named after Sir Charles Hamilton (1767–1849), governor of Newfoundland, 1818–24.

Churchill Peak BC Located in the Battle of Britain Range of the Muskwa Ranges of the Rocky Mountains, this mountain (2819 m) was named in 1944 by provincial minister of lands E.T. Kenny after British prime minister Winston (later Sir Winston) Churchill. It was one of the names given in the area to honour Allied leaders and victories of the Second World War.

Church Point NS On the southern shore of St. Marys Bay and southwest of Digby, this place was named after a Catholic church first built in 1808, rebuilt in 1829, and replaced by the present church in 1905. The highest and largest wooden church in North America, its spire rises 56.3 m above the ground, and its dimensions are 58 m long and 41 m wide. Church Point post office was opened in 1857, but was changed to *Port Acadie* six years later. In 1893 the name Church Point was restored. The place is known to its Acadian residents as Pointe-de-l'Église.

Chute-à-Blondeau ON A locality in the united counties of Prescott and Russell, east of Hawkesbury, it was named after Adolphe Blondeau, who drowned in 1876 in the Carillon Rapids of the Ottawa River. The rapids disappeared when the Carillon Dam was built, 1959–64.

Chutes-aux-Outardes QC A village beside rapids on the **Rivière aux Outardes**, southwest of Baie-Comeau, it was founded in 1951. In 1535 Jacques Cartier identified the wild goose as an *outarde*, which is really the French word for a bird known in Europe as the bustard. The Rivière aux Outardes is a major river on the north shore, flowing south parallel to Rivière Manicouagan for almost 400 km, to enter the St. Lawrence just west of the latter. The village of **Pointe-aux-Outardes** was erected in 1964, taking its name from its post office, opened in 1901.

Clair NB A village (1966) southwest of Edmundston, its first post office was called *Middle St. Francis* in 1867. It was renamed in 1892 after Peter Clair (1817–1902), a native of County Clare, Ireland.

Claireville ON At the point where the cities of Brampton, Etobicoke, and Vaughan meet, its post office was named in 1853 by Upper Canada College French master Jean du Petit Pont de la Haye, for his daughter Claire. Called John de la Haye by his neighbours, he had a farm here called Les Ormes (meaning the elms).

Clandeboye ON In Middlesex County, northwest of London, this place was named Clandeboye in 1877, after the home in County Down, northeast of Belfast, of the Marquess of Dufferin and Ava, governor general of Canada, 1872–8.

Clara ON This township in Renfrew County was named in 1863. As Herbert Gardiner wrote in *Nothing But Names* (1899), 'the most diligent inquiry has failed to discover a single person who knows Clara's other name'. For municipal purposes Clara Township is united with Head and Maria townships.

Claremont ON In the town of Pickering, Durham Region, this place was named in 1851 by William Henry Michell, after the nearby cottage of his uncle, Mr Watkins. Watkins named it for an estate in Surrey, near London, England, itself named by the Earl of Clare in 1714.

Clarence ON A township in the united counties of Prescott and Russell, it was named in 1798 after Prince William Henry, Duke of Clarence (1765–1837), who was crowned William IV in 1830, and reigned for seven years. The community of **Clarence** in the township was known as *New England* until 1848, when Clarence post office was opened here.

Clarence Creek ON In the united counties of Prescott and Russell, the name of this community was chosen in 1867, because it was in the middle of Clarence Township and near the head of Fox Creek. To reflect the community's French character, it was renamed *Lafontaine* in 1935, but, because of confusion with Lafontaine in Simcoe County, it was changed back to Clarence Creek the next year. Assuming the proper name of the creek was **Clarence Creek**, the Geographic Board of Canada officially approved it in 1931.

Clarendon QC A municipal township in the regional county of Pontiac and adjacent to the town of Shawville, it was created in 1855, taking its name from the township, proclaimed in 1833. The township was named after Clarendon Park, an estate near Salisbury, Wiltshire, England.

Clarenville NF On the Northwest Arm of Random Sound, an extension of Trinity Bay, this town (1951) was named *Clarenceville* about 1892, possibly after the Duke of Clarence, the eldest son of the Prince of Wales (later Edward VII), who had died that year. It is also said to have been named after a son of Newfoundland prime minister Sir William Whiteway (1829–1908), but he did not have a son by that name. The town had amalgamated the coastal settlements of Brook Cove, Dark Hole, Lower Shoal Harbour, Broad Cove, and Red Beach, and shortly after the name was changed to Clarenville.

Claresholm AB Northwest of Lethbridge, this town was named in 1891 by John Niblock, a Canadian Pacific Railway superintendent, after his home in Medicine Hat. His house was called 'Clare's Home', and was named after his wife. Its incorporation on 31 August 1905 was the last official act of the North-West Territories administration, based in Regina; the following day Alberta became a province.

Clarington ON In 1993 the town of Newcastle became the municipality of Clarington, a name devised from the township names Clarke and Darlington. In 1974, on the creation of the regional municipality of Durham, the town of Bowmanville, the village of Newcastle, and the two townships had been united as Newcastle, named after Newcastle District, which existed from 1798 to 1850. Confusion between the new town and the former village, and displeasure by the residents of Bowmanville, resulted in a referendum in favour of a change, and a consultative process by a committee which recommended Clarington.

Clark's Harbour NS This town (1919) on Cape Sable Island was named after a New England captain who had his Nova Scotia headquarters here while on fishing expeditions. Its post office was opened in 1858.

Clarke City QC A sector of the city of Sept-Îles since 1970, it was founded at the beginning of the twentiethth century by four Clarke brothers, George, James, John, and William. They were New York publishers who wanted the best quality paper for the production of such books as the *Century Dictionary* and the *Encyclopædia Britannica.*

Clarke's Beach NF At the head of Bay de Grave and on the west side of Conception Bay, this town (1965) may have been named before 1800, although it was not noted in a census until 1857. John Clarke was a fisherman at Port de Grave in 1772.

Clark, Mount NT In the Franklin Mountains and on the east side of the Mackenzie River, this mountain (1462 m) was named before 1910 after a chief factor of the Hudson's Bay Company. After he climbed it, he ate breakfast and deposited a flagon of brandy for future climbers.

Clarksburg ON On Beaver River and in Grey County, east of Owen Sound, this community was named in 1862 for woollen miller W.A. Clarke. William Marsh had proposed Marshville, but as the chairman of the public meeting, he broke a tie in favour of Clarksburg.

Clarkson ON Warren Clarkson (d. 1882) built a house in 1819 at this place, now in the city of Mississauga, and a store in 1835. In 1853 he sold land to the Great Western Railway for the right of way. Clarkson post office was opened in 1875, with Warren's grandson, William, as the postmaster.

Clavet SK A village (1978), southeast of Saskatoon, its post office was called *French* in 1904. When the Grand Trunk Pacific arrived in 1908, its name was changed to Clavet, to maintain the alphabetical order of station names between Portage la Prairie and Saskatoon and beyond.

Clayoquot Sound BC An intricate group of bays, channels, and inlets on the west coast of Vancouver Island, west of Port Alberni, it took its name from the Clayoquot, a Nootka tribe. The tribe's name may mean either 'different people', because they had become quarrelsome after having been once quiet and peaceful, or 'people of a place where it becomes the same even when disturbed'.

Clayton ON In Lanark County, west of Almonte, this community was known as *Bellamy's Mills*, after miller Edward Bellamy, until 1858, when it was changed briefly to *Clifton*, and then called Clayton.

Clearview ON Created in 1994 in the reorganized Simcoe County, Clearview was the name chosen for the new municipal township when the town of Stayner, the village of Creemore, and the townships of Nottawasaga and Sunnidale were merged into a single municipality.

Clearwater BC At a bend of the North Thompson River, where the **Clearwater River** flows in from the north, this community's post office was called *Clearwater Station* in 1925. It was changed to Clearwater in 1961.

Clemenceau, Mount BC This mountain (3658 m), the fourth highest in the Canadian Rockies, was named in 1919 by the Alberta-British Columbia Boundary Commission, after Georges Clemenceau (1841–1929), premier of France, 1906–9 and 1917–20. The **Clemenceau Icefield** extends east toward the Columbia Icefield.

Clementsport NS On the south side of Annapolis Basin, this place was first called *Moose River*, after a river identified on Marc Lescarbot's 1609 map as *Rivière de l'Orignal*, which translates as 'moose river'. *Clements* post office was opened in 1840, taking its name from Clements Township. It was renamed Clementsport in 1852. The post office at nearby **Clementsvale** was opened in 1863. Clements was the family name of the mother of John Parr, governor of Nova Scotia, 1782–91.

Clermont QC Located in the regional county of Charlevoix-Est, it was created as a village in 1949, and a town in 1967. It took its name from the municipal parish of *Saint-Philippe-de-Clermont*, established in the early 1930s by writer Félix-Antoine Savard, who may have had in mind Clermont-Ferrand, the French birthplace of celebrated writer Blaise Pascal (1623–62).

Cleveland QC A municipal township in the regional county of Le Val-Saint-François, adjacent to Richmond, it was named in 1855 after prominent landowner George Nelson Cleveland.

Clifford ON A village (1873) in Wellington County, south of Hanover, it was named in 1856 by miller and innkeeper Francis Brown, after his native village in West Yorkshire, England, northeast of Leeds.

Climax SK A post office was opened in 1913 near the Montana border, southwest of Swift Current, and named after Climax Township, ND. When the railway arrived in 1923, the office was moved 5 km to the present site of the village (1923).

Cline, Mount AB West of Rocky Mountain House, this mountain (3361 m) was named in 1902 by British alpinist Norman Collie. He named it after Michael Cline (Klyne, Clyne, Klein, Kline), who was in charge of the Hudson's Bay Company's Jasper House, 1824–5 and 1829–34.

Clinton BC A village (1963) north of Cache Creek, it was originally known as *Cut-Off Valley* and *47 Mile House*. On the completion of the Cariboo Road to Barkerville in 1863, it was named after Henry Pelham Clinton, 5th Duke of Newcastle, who was the British colonial secretary, 1859–64.

Clinton ON A town (1874) in Huron County, southeast of Goderich, it was named in 1853 by tavern keeper William Rattenbury after Sir Henry Clinton (1771–1829), who had served with Wellington in the Peninsular War in Spain. Rattenbury's father had been a tenant farmer on Lord Clinton's estate in Hampshire, England.

Clinton-Colden Lake NT Located northeast of Great Slave Lake, this lake (596 km^2) was named in 1834 by explorer Capt. George (later Sir George) Back 'as a mark of respect to the memory of those distinguished' American lawyers and statesmen De Witt Clinton (1769–1828) and Cadwaller David Colden (1769–1834).

Clive AB A village (1912) east of Lacombe, it was named in 1909 after Robert Clive, Baron Clive (1725–74), who took a large part in establishing British power in India.

Cloverdale BC On the east side of the district municipality of Surrey, this place's post office was named in 1872. It was called after **Clover Valley**, named by pioneer settler William Shannon, who had observed the luxuriant wild clover growing throughout the valley of the Nikomekl River.

Cloyne ON Divided between the municipal township of Kaladar, Anglesea and Effingham, Lennox and Addington County, and the municipal township of Barrie, Frontenac County, north of Kaladar, this community was named in 1859 after a village in County Cork, Ireland, east of the city of Cork.

Clutterbuck, Mount BC In the Purcell Mountains, northeast of Kaslo, this mountain (3063 m) was named in 1930 by climber Dr James Monroe Thorington after British author Walter John Clutterbuck (1853–1937), who had visited the area and wrote several travel books, including *B C 1887*.

Clyde AB A village (1914) north of Edmonton, its post office was named in 1905 after homesteader George Clyde, who settled here in 1900.

Clyde River NS Situated on the **Clyde River** southwest of Shelburne, this place was founded in 1850. The river was called *Cape Negro River* in the latter half of the 1700s, but by 1785 Clyde River, after Scotland's River Clyde, was being used.

Clyde River NT Near the mouth of **Clyde Inlet** and on the northeastern coast of Baffin Island, this hamlet was established in 1978. The site was named *Clyde* in 1950 by Patrick D. Baird, and the post office was called *Clyde* in 1956, but preferred usage became Clyde River, after a small river flowing into Patricia Bay, on the north side of the inlet. The inlet was named in 1818 by John (later Sir John) Ross.

Clyde River PE A municipal community (1974) west of Charlottetown and on the **Clyde River**, a tributary of West River, it was created as a school district about 1864. Its post office was opened in 1886. The river was formerly called *Dock River*, sometimes rendered as *Dog River*. In 1765 Samuel Holland called it *Edward River*, after Edward Eliot, 1st Baron Eliot (1727–1804).

Coachman's Cove NF A municipal community (1970) on the north coast of the island of Newfoundland, it was first known as *Pot d'étain* by the migratory French fishermen. It may have been named in English because the French hired English guards (or coaches) to maintain their fishing stages.

Coaldale AB East of Lethbridge, this town (1952) was named in 1909 by the Canadian

Pacific Railway, after Elliott Galt's first residence in Lethbridge. Galt was general manager of the Alberta Railways and Irrigation Company.

Coalhurst AB A town (1995) northwest of Lethbridge, it took its name from the official designation of Lethbridge before 1885. Coal mining was begun in the area in 1872, with the first settlers describing it as *The Coal Banks*.

Coast Mountains BC Extending from Vancouver north to the Yukon, this coastal chain of the Cordillera was named in 1902. Its highest peak is Mount Waddington.

Coaticook QC A town (1883) in the regional county of **Coaticook**, south of Sherbrooke, it had been incorporated as a village in 1864, taking its name from the **Rivière Coaticook**, a tributary of the Saint-François. The regional county was named in 1982. In Abenaki the name means 'river of the pine land'.

Coats Island NT At the north end of Hudson Bay, south of Southampton Island, this island (5498 km^2) was named after Capt. William Coats (d. 1752), who compiled valuable notes on the geography of Hudson Bay. He made several trips between 1727 and 1751 to supply the Hudson's Bay Company forts in the bay.

Cobalt ON A town (1906) in Timiskaming District, Cobalt was named in 1904 by provincial geologist Willet Green Miller. He had observed traces of that element in the silver ore, discovered the previous year, when the route of the Temiskaming and Northern Ontario Railway (now Ontario Northland) was being laid out.

Cobden ON A village (1900) in Renfrew County, it was founded in 1849 by Jason Gould, and named two years later after Richard Cobden (1804–65), a British pacifist and free trader admired by Gould.

Cobequid Bay NS One of the two heads of the Bay of Fundy (the other is Cumberland Basin), it extends for 50 km from Lower Truro to Economy Point. The name was noted as *Cocobequy* in a 1689 grant and as *Gobetick* in a 1738 report. It was possibly derived from Mi'kmaq *wakobetquick*, 'end of flowing water', referring to the extent that the tidal bore flows up the rivers on each side of the bay. The rugged terrain, rising to 373 m north of the bay and Minas Basin, is known as the **Cobequid Mountains**.

Coboconk ON In Victoria County, north of Lindsay, it was named in 1859, after Ojibwa *kakapikang*, 'waterfalls', or 'swift water', in reference to the Gull River. In 1873 the Toronto and Nipissing Railway named the station after its president John Shedden, who was accidentally killed that year at Cannington. However, eight years later local residents had it renamed Coboconk.

Cobourg ON A town (1850) in Northumberland County, it was first named *Buckville*, for settler Elijah Buck, *Amherst*, after Jeffery Amherst, commander-in-chief of the British forces in North America, 1760–3, and *Hamilton*, after the township. In 1819 it was named Cobourg in honour of the 1816 marriage of the future George IV's daughter, Charlotte, to Prince Leopold of Saxe-Coburg, Germany. She had died in 1817, while Leopold became the King of the Belgians from 1831 to 1865. Coburg is in northern Bavaria, north of Nürnberg. The extra 'o' was apparently a clerical error.

Cocagne NB Situated on the **Cocagne River**, 16 km northwest of Shediac, this place's post office was named in 1837 after the river. The river was named *R de Cocanne* before 1672 by Acadian Gov. Nicolas Denys, after a pleasant sojourn he had here during an eight-day period of inclement weather. A thirteenth century English satire called *Land of Cockayne* described a medieval land made of cakes.

Cochenour ON In Kenora District, north of Red Lake, it was named in 1939 after brothers Bill and Ed Cochenour, who had staked mining claims here in 1924.

Cochrane AB Northwest of Calgary, this town (1971) was named in 1884 by the Canadian Pacific Railway after Senator Matthew Henry Cochrane. He established the Cochrane Ranche Company Limited nearby in 1881.

Cochrane ON When the Temiskaming and Northern Ontario Railway (now Ontario Northland) reached the site of this town (1910) in 1908, it was named after Frank Cochrane (1852–1919), then Ontario minister of lands and forests, and an avid promoter of the railway's route to James Bay. In 1911 he was appointed the minister of railways and canals in Sir Robert Borden's government. **Cochrane District** was created in 1922.

Codette SK South of Nipawin, this village (1929) was named by the Canadian Pacific Railway in 1924, after the **Codette Rapids**, in the Saskatchewan River, 6 km to the west. The rapids were named after Jean-Baptiste Cadot (*c.* 1723–*c.* 1801), a partner in the late 1700s of Alexander Henry the elder. The post office was called *Codette Station* in 1925 and was changed to Codette in 1936.

Colborne ON A village (1858) in Northumberland County, Colborne was named in 1829 for Sir John Colborne, 1st Baron Seaton (1778–1863), lieutenant-governor of Upper Canada, 1828–36. **Colborne Township**, in Huron County, was named in 1830 after him.

Colchester NS This county was separated from Halifax County in 1835. It was named after the district of Colchester, used in the area of Truro as early as 1780. The district was likely named after Colchester, a borough in Essex, England.

Colchester ON The post office in the community of Colchester, west of Leamington, was named after Colchester Township in 1831. The township was established in Essex County in 1792 and called after the city of Colchester, in Essex, England. It was divided into the municipal townships of **Colchester North** and **Colchester South** in 1880.

Coldbrook NS West of Kentville, this place's post office was called *Coldbrook Station* in 1870. It may have been named after Coldbrook Park, Monmouthshire, Wales.

Cold Lake AB A town (1955) on the southwestern shore of **Cold Lake**, its post office was named in 1910. In 1996 it annexed the town of Grand Centre. The lake, on the Saskatchewan border, was referred to as *Coldwater Lake* on a 1790 map by trader Philip Turnor. It is quite cold during every season of the year.

Coldstream BC A district municipality (1906) in the Okanagan Valley, south of Vernon, it was named after the Coldstream Ranch. The ranch was developed in 1863 by Col Charles F. Houghton and named after cold springs that rise on **Coldstream Creek**. Its main urban centre is Lumby.

Coldwater ON On the **Coldwater River** and in Simcoe County, northwest of Orillia, this place was named in 1835. The river's name in Ojibwa is *Kassina Nibish*, 'cold water'. Incorporated as a village in 1908, it was united in 1994 with the townships of Matchedash and Orillia to form the municipal township of Severn.

Colebrook ON In Lennox and Addington County, northeast of Napanee, its post office was named in 1851 after Cole Warner, in whose store the post office was located.

Cole Harbour NS East of Dartmouth, and on the west side of the shallow **Cole Harbour**, this place was identified in a 1765 grant. It was possibly named after an early settler. Its post office was opened in 1899.

Coleman AB Located in the municipality of Crowsnest Pass, this urban centre was an incorporated town from 1910 to 1979. Its post office was named in 1904 after Florence Coleman Flumerfelt, the youngest daughter of A.C. Flumerfelt, president of the International Coal and Coke Company.

Coles Island NB At the mouth of the Canaan River, northwest of Sussex, this place's post office was named in 1858. The island, in the mouth of the river, was named after David Cole, a Loyalist, who settled here in 1809.

Coleville SK A post office was opened north of Kindersley in 1908 and named after its first postmaster, Malcolm Cole. It was relocated to the present site of this village (1953) in 1913, when the Grand Trunk Pacific Railway was built southwest from Biggar.

Colinet NF This municipal community (1974) was named after **Great Colinet Island.** Recorded in the late 1600s on maps as *Collinett*, *Colonet Isle*, and *Collemot*, the island's name may have been given for a French or Channel Islands fisherman, or after a French word meaning 'a place of little hills'.

College Bridge NB On the west side of Memramcook River, southeast of Moncton, this community's post office was named in 1885 after the nearby Collège Saint-Joseph (now Memramcook Institute), in the village of Saint-Joseph.

Colliers NF Located at the head of **Colliers Bay**, on the southwest side of Conception Bay, this town (1972) may have been named because it was a site where wood was gathered to produce charcoal. Collier was a surname in Newfoundland as early as the mid-1700s, although there is no evidence the town might have been named after a person.

Collingwood ON The township in Grey County was named in 1840 for Cuthbert Collingwood, Baron Collingwood (1748–1810), a British admiral who won a sea battle in 1797 off Cabo São Vincente, Spain, and served as Lord Nelson's second-in-command off Cabo Trafalgar in 1804. The town (1858) in Simcoe County was named in 1853, when Frederic Cumberland, general manager of the Toronto, Simcoe and Lake Huron Railway, chose it as the terminus of its line on Georgian Bay. He named it after the nearby township.

Collins Bay ON A community and bay in Kingston Township, Frontenac County, they were named for Deputy Surveyor General John Collins (d. 1795), who laid out the township in 1783 and was granted land on the bay. The post office was opened in 1854.

Colonsay SK Located southeast of Saskatoon, this town (1977) was named by the Canadian Pacific Railway in 1906 after one of the Inner Hebrides islands, off the west coast of Scotland. Its streets and avenues recall other islands in the group, including Jura, Islay, Oronsay, Bute, Kintyre, and Tiree. Postal service was provided in 1908.

Colpoy's Bay ON The bay on the east side of the Bruce Peninsula was named in 1826 by Admiralty surveyor Henry W. Bayfield after Rear-Adm. Sir Edward Griffith Colpoys (d. 1832). **Colpoy's Bay** post office was open from 1863 to 1917. The community's name was changed in 1946 to *Colpoy Bay*, properly spelled *Colpoys Bay* in 1950, and then changed to the locally preferred Colpoy's Bay in 1978. The bay's name was also changed to Colpoy's Bay the same year.

Columbia, Cape NT Canada's most northerly point of land (83° 06′41.35″N), this cape was named in 1876 by George (later Sir George) S. Nares of a British naval exploring expedition after the poetic name of the United States. Pelham Aldrich determined on 1 May of that year that it was the most northerly point of land. In 1987 the Mapping and Charting Establishment of the department of defence ascertained its precise position.

Columbia, Mount AB, BC Located at the point on the continental divide where waters drain to three oceans — the Atlantic, the Arctic, and the Pacific — this mountain (3747 m) was named in 1898 by climbers Norman Collie and Herman Woolley after the Columbia River. The river drains the area of the **Columbia Icefield** into the Pacific Ocean.

Columbia River BC Rising in **Columbia Lake** in the Rocky Mountain Trench, this river (2044 km, 917 km in Canada) flows north then south to cross the British Columbia-Washington boundary at Waneta. It was named on 11

May 1792 by Capt. Robert Gray of Boston, MA after his ship *Columbia Rediviva*.

Columbus ON In the city of Oshawa, Durham Region, it was named in 1847, likely after the famous fifteenth century explorer, after whom many places in the United States were named.

Colville Lake NT Northwest of Great Bear Lake, this lake (452 km^2) was named in 1857 by Hudson's Bay Company trader Roderick MacFarlane, after Andrew Colvile, deputy governor of the HBC, 1839–52, and governor, 1852–6.

Colwood BC This city (1985) west of Victoria was named after one of the four farms of the Puget Sound Agricultural Company, which belonged to the Hudson's Bay Company. Capt. E.E. Langford arrived here in 1851 to manage it, naming it after a property of his in Sussex, England. The farm became better known as Esquimalt, with the Langford residence retaining the designation of Colwood. Its post office was opened in 1881.

Combatant Mountain BC This mountain (3756 m) in the Coast Mountains, between Mount Waddington and Mount Tiedemann, was named in 1933 by climber W.A.D. (Don) Munday, possibly after the difficulty in climbing it.

Comber ON In Essex County, north of Leamington, this place was named in 1851 by John Gracey, a native of Comber, County Down, Ireland, east of Belfast.

Combermere ON In Renfrew County, south of Barry's Bay, this community was first known as *Dennison's Bridge*, after Capt. John Dennison. He renamed it in 1865 after Sir Stapleton Cotton, Viscount Combermere (1773–1865), with whom he may have served. Lord Combermere commanded cavalry in the Peninsular War in Spain, 1808–14, was commander-in-chief in India, 1825–30, and became a field marshal in 1855. His home was Combermere Abbey in Cheshire, England.

Comeauville NS Located on the French Shore, southwest of Digby, this place's post office was first called *Clare* in 1841, after the municipality of Clare, which comprises the western part of Digby County. It was renamed in 1877 after its first postmaster, Augustin F. Comeau.

Come-By-Chance NF A town (1969) at the north end of the Isthmus of Avalon, it was named after an extension of Placentia Bay called **Come by Chance**. There is a reference to this bay as *Comby Chance* as early as 1706. It may have been named because it was a chancy proposition for fishermen to bring their boats to shore here during rough weather. In 1612 John Guy identified it as *Passage Harbour*.

Comfort Cove-Newstead NF This municipal community (1967), which extends into the Bay of Exploits, originally comprised the separate centres of Comfort Cove (at **Comfort Head**), *New Harbour*, and *Turtle Cove*. *New Harbour*, the largest of the three, became Newstead in 1921.

Committee Punch Bowl AB, BC On the continental divide at the head of Athabasca Pass, it was named in 1824 by Sir George Simpson after the Hudson's Bay Company's governing committee. The small basin, measuring 18 m across, impressed Simpson because it drained toward both the Pacific and the Arctic oceans.

Comox BC On the east coast of Vancouver Island and east of Courtenay, this town (1967) was named in 1868, after the Kamuckway, a Salish tribe. Its name means 'place of plenty', in reference to the excellent game and berries in the Puntledge River valley and around **Comox Lake**.

Compton QC A village in the regional county of Coaticook, midway between Coaticook and Lennoxville, it was named in 1893 after the township of Compton, proclaimed in 1802. Its name was possibly taken from a village in Surrey, England, southwest of London. Compton post office was opened in 1829.

Compton Station, an adjacent municipality, had a post office from 1890 to 1964.

Compton Mountain BC In the Coast Mountains, east of Bute Inlet, this mountain (2873 m) was named in 1933 by George G. Aitken, provincial member of the Geographic Board of Canada, after the birthplace in England of Sir Humphrey Gilbert. *See also* Mount Gilbert.

Conception Bay NF This bay was named *baia da conceição*, possibly by the Portuguese in the early 1500s, in honour of the feast day (8 December) of Our Lady of the Immaculate Conception. As early as 1588 it was called Conception Bay, but only four years later it was identified as *Consumption Bay* on a map by Thomas Hood, presumably after local usage. Possibly as early as 1497 it had been named *Assumption Bay*, after the feast day (15 August) of Our Lady of the Assumption, with the two names becoming confused. Conception Bay and *Consumption Bay* were used interchangeably during the 1600s, with the present name prevailing after 1700. Nevertheless some people were still calling it *Consumption Bay* after the turn of the twentieth century.

Conception Bay South NF A town (1971) on the eastern shore of Conception Bay and southwest of St. John's, it amalgamated the settlements of Topsail, Chamberlains, Codner, Long Pond, Manuels, Kelligrews, Upper Gullies, and Seal Cove. It took its name from the provincial electoral district of Conception Bay South.

Conception Harbour NF On **Conception Harbour** and at the head of Conception Bay, this town had been called *Cats Cove* (1830s) and *Avondale North* (1906), but was renamed in 1972 after the harbour.

Conche NF On the northeast side of the Northern Peninsula, this municipal community (1960) may have been named in the 1700s by fishermen, after the shellfish or after the shape of the peninsula, which is shaped like a conch shell. It may also have been named after a place on Guernsey, in the Channel Islands.

Concord ON In the city of Vaughan, York Region, this place was named in 1854, possibly after Concord, VT, or perhaps as an antidote when people here could not agree on a name.

Conestogo ON At the junction of the Grand and Conestogo rivers and in Waterloo Region, this place was named in 1856 after the river, itself named in 1806 by Benjamin and George Eby, who had observed its likeness to Conestoga Creek in Lancaster County, PA. The four-wheeled Conestoga covered wagon helped Americans open the West.

Coniston ON Incorporated as a town in 1934, it became part of the town of Nickel Centre in Sudbury Region in 1973. It had been named in 1903 by first postmaster Dennis O'Brien, on the suggestion of Canadian Pacific Railway engineer T.R. Johnson, who had been reading an English novel set in the Coniston Lake district of northwestern England.

Conne River NF On the south bank of the **Conne River**, where it flows into the Bay d'Espoir, this town (1972) was settled by the Mi'kmaq in the mid-1800s, after they had lived a nomadic existence for several years in the southern and central parts of the island of Newfoundland. The river was recorded as *rivière le con* by Capt. James Cook in the 1760s, and was subsequently anglicized to its present form. The meaning of the original name is not known.

Connolly, Mount YT Northeast of Faro, this mountain (2139 m) was named in 1976 after Thomas Osborne Connolly (1918–75), an outfitter and guide at Ross River, 1945–64.

Conquerall Mills NS On Fancy Lake, south of Bridgewater, this place was possibly named about 1805, when miller George Fancy was struck by the beauty of the lake, and exclaimed 'this conquers all'. However, he may have used the word on his lumber to promote the quality of his sawmill. Its post office was established in 1882. Earlier, in 1855, **Conquerall Bank** post office was opened on the south bank of the LaHave River, east of Bridgewater.

Conquest SK Southwest of Saskatoon, this village (1911) was founded in 1904 by pioneer settlers, who performed a 'conquest' of the South Saskatchewan by ferrying their families and possessions across it. The Canadian Pacific Railway was built through it in 1908, with the post office opening in 1911.

Conrad, Mount BC This mountain (3252 m) in the Purcell Mountains, northwest of Invermere, was named in 1935 by British alpinist Igor A. Richards after climber and guide Conrad Kain (1833–1934). Kain, a native of Austria, had immigrated to Canada in 1909 and made the first ascent of Mount Robson.

Consecon ON In Prince Edward County, south of Trenton, its post office was named in 1836 after **Consecon Lake**, a name derived from the Mississauga word *ogans*, 'pickerel'.

Consort AB Southeast of Stettler, this village (1912) was named in 1911 by the Canadian Pacific Railway in honour of Queen Mary, the consort of recently crowned George V. Its post office was previously called *Sanderville.*

Constance Bay ON Adjacent to the Ottawa River and in West Carleton Township, Ottawa-Carleton Region, it was named after **Constance Lake** and **Constance Creek**, both called after Constance Pinhey (1819–98), the wife of John Hamnett Pinhey (1827–1901) and the daughter of Hamnett Kirkes Pinhey (1784–1857). The latter settled nearby on the Ottawa River in 1820. It was developed as a summer-cottage community. Most of Constance Bay's residents now live here year round.

Constantine, Mount YT This mountain in the St. Elias Mountains was named in 1900 by surveyor James J. McArthur after Charles Constantine (1849–1912), an inspector with the North-West Mounted Police. He established a NWMP post at Fortymile River in 1895 and represented Canadian law during the Klondike Gold Rush.

Contrecœur QC Located in the regional county of Lajemmerais, southwest of Sorel, this municipality's post office was opened in 1849 as *Sainte-Trinité-de-Contrecœur,* and shortened to Contrecœur in 1955. In 1672 Antoine Pécaudy de Contrecœur (1596–1688), an officer with the Carignan-Salières Regiment, was granted a seigneury here by Jean Talon.

Contwoyto Lake NT Located north of Great Slave Lake, this lake (933 km^2) was named in 1821 by John (later Sir John) Franklin, after a Chipewyan word for 'rum'. Samuel Hearne provided rum to his guide Mattonabee and his companions here in 1771.

Conway NS On the southwest side of Digby, this place's post office was opened in 1927. It took its name from the first designation given to Digby, likely given for Henry Seymour Conway, British secretary of state for the southern department in the 1760s.

Cook's Harbour NF On the southeastern side of the Strait of Belle Isle, the inlet at this town (1956) was named in 1764 by Capt James Cook (1728–79), who undertook a hydrographic survey of the coasts of the island of Newfoundland from 1763 to 1768. After extensively exploring the Pacific Ocean during the following decade, he was killed in the Sandwich Islands (Hawaiian Islands).

Cook, Cape BC The northwestern point of Brooks Peninsula, and on the west coast of Vancouver Island, it was named in 1860 by Capt. George Richards after Capt. James Cook (1728–79), who in 1778 was the first British navigator to explore the West Coast. Cook had named it *Woody Point.*

Cook, Mount YT On the Yukon-Alaska boundary and in the St. Elias Mountains, this mountain (4194 m) was named in 1874 by William H. Dall, of the United States Coast Survey, after Capt. James Cook (1728–79), who explored the northwestern coast in 1778.

Cook's Bay ON The southern arm of Lake Simcoe, it was named in 1793 by Lt-Gov. John Graves Simcoe after Capt. James Cook (1728–

79). Cook undertook hydrographic surveys on the east coast of North America, 1758–68, and served with Simcoe's father, after whom Lake Simcoe was named.

Cookshire QC A town (1892) in the regional county of Le Haut-Saint-François, east of Sherbrooke, it was founded in 1799 by Loyalists from Vermont and New Hampshire, and named after first settler John Cook (1770–1819). Its post office was opened in 1851.

Cookstown ON In the town of Innisfil, Simcoe County, since 1991, its post office was named in 1847 after Maj. James Cooke, who laid out village lots that year.

Cooksville ON In the city of Mississauga, Peel Region, this urban centre was named in 1836 after Jacob Cook, who bought land in 1814 at the corner of Dundas and Hurontario streets and settled here in 1819. The post office from 1826 to 1836 was called *Toronto*, after the township.

Coombs BC South of Qualicum Beach and northwest of Nanaimo, this settlement was named in 1911 after Capt. Coombs of the Salvation Army, which helped labourers and their families from Leeds, England, to relocate to Vancouver Island.

Copenhagen ON In Elgin County, southeast of St. Thomas, this community was settled by Charles Kuntze, a native of Denmark, who built the Copenhagen Inn at what became *Kuntze's Corners*. The post office was called Copenhagen in 1870.

Copper Cliff ON An urban centre on the west side of the city of Sudbury, it was founded in 1883. A mine was opened in 1886 and a smelter was built two years later. Its post office was named in 1890. The place was incorporated as a town in 1901 and was annexed by Sudbury in 1973.

Coppermine River NT Rising in Lac de Gras, north of Great Slave Lake, this river (845 km) flows northwest into Coronation Gulf. It was named in 1771 by Samuel Hearne, who observed outcroppings of copper on its banks. In Inuktitut the river is called *Kugluktuk*, 'big river'. The name of the hamlet of *Coppermine* was changed to Kugluktuk 1 January 1996.

Coquitlam BC A district municipality (1891) east of Burnaby, and south and east of Port Moody, it was named after the Coquitlam, a Salish tribe. Their name means 'stinking with fish slime', in reference to their butchering salmon for their masters, the Kwantlen, a Cowichan tribe. Their name may also mean 'small red salmon'. *See also* Port Coquitlam.

Coral Harbour NT This hamlet (1972), located on the south side of Southampton Island, was established in 1924 by the Hudson's Bay Company. The harbour was named in 1907–9 by Capt. George Comer after the unusual red rock he retrieved while taking soundings in the harbour.

Corbeil ON In East Ferris Township, Nipissing District, southeast of North Bay, its post office was named in 1906, for lumber millers Jean-Baptiste and Joseph Corbeil. The Dionne quints were born on the west side of the community in 1934.

Corbetton ON In Dufferin County, northwest of Shelburne, its post office was named in 1881 after James Corbett (d. 1902), who had built a tavern on the nearby road joining Shelburne to Owen Sound.

Corbyville ON In 1855 Henry Corby (1806–81) bought a mill north of Belleville and added a distillery in 1859. In 1882 the new post office was named in his honour.

Cordova Bay BC In the district municipality of Saanich, north of Victoria, this community was named in 1920 after the adjacent bay. The bay was named about 1842 by the Hudson's Bay Company. It transferred the name *Puerto de Cordova*, given in 1790 by Spanish Sub-Lt Manuel Quimper to Esquimalt Harbour, west of Victoria. Quimper named it after the forty-sixth viceroy of Mexico, Don Antonio Maria Bucareli y Ursua Henestrosa Lasso de la Vega Villacis y Cordova.

Corinth ON In Elgin County, between Tillsonburg and Aylmer, its railway station and post office were both named in 1871 after the Greek city.

Cormack NF Located north of Deer Lake, this municipal community (1964) was named in 1948, after William Epps Cormack (1796–1868), in 1822 the first non-Aboriginal to travel by foot across the island of Newfoundland.

Corner Brook NF A city (1956) on the south side of Humber Arm, it was created through the amalgamation of the towns of *Corner Brook, Corner Brook East, Corner Brook West,* and *Curling*, and named after the brook draining **Corner Brook Lake** into the Humber Arm of the Bay of Islands. The town of *Humbermouth* was annexed two years later. Capt. James Cook named the brook in 1767, presumably because the shoreline took a sharp turn at its mouth.

Cornwall ON The township in the united counties of Stormont, Dundas and Glengarry was named in 1787, after a title of George III's eldest son, the Duke of Cornwall (1762–1830), who reigned as George IV, 1820–30. The city (1944) was named *New Johnstown* in 1789, after Johnstown, NY, which had been given in honour of Sir William Johnson, whose son, Sir John Johnson, brought many Loyalists to Upper Canada. It was renamed **Cornwall** in 1797.

Cornwall PE A town (1995) west of Charlottetown, it had been created as a village in 1966, and a municipal community in (1983). Established as a school district in 1849 and a post office in 1859, it may have been named after the title of the Duke of Cornwall, received at birth by Albert Edward (1841–1910), the eldest son of Queen Victoria, and the future Edward VII. On 1 April 1995 Cornwall annexed the adjoining municipal communities of North River and Eliot River.

Cornwallis NS On the Annapolis Basin, east of Digby, this place was established during the Second World War as a Canadian military base. It took its name from Edward Cornwallis (1712/13–76), the first governor of Nova Scotia, 1749–52.

Cornwallis Square NS This village (1977) west of Kentville, and on the Cornwallis River, is an amalgamation of the former villages of Cambridge, Grafton, and Waterville. It took its name from the **Cornwallis River**, which rises near Berwick, and flows east into Minas Basin.

Cornwallis Island NT The most southeastern of the Parry Islands of the Queen Elizabeth Islands, this island (6995 km^2) was named in 1819 by William (later Sir William) Parry after Adm. Sir William Cornwallis (1744–1819). Parry served with Cornwallis in the Channel Fleet in 1803.

Cornwall Island NT The most southeastern of the Sverdrup Islands of the Queen Elizabeth Islands, this island (2258 km^2) was named in 1852 by Sir Edward Belcher after Prince Edward, the Prince of Wales and Duke of Cornwall.

Coronach SK This town (1978) near the Montana border, southeast of Assiniboia, was founded by the Canadian Pacific Railway in 1926, and named by its English settlers after a horse that had won the Epsom Derby. Its post office was opened in 1927.

Coronation AB Located southeast of Stettler, this town (1912) was named by the Canadian Pacific Railway in honour of the crowning of George V on 22 June 1911. Its post office was opened the same year.

Coronation Gulf NT Between the mainland and Victoria Island, this 290-km gulf was named *George IV's Coronation Gulf* in 1821 by John (later Sir John) Franklin in honour of the king's coronation. The Geographic Board authorized Coronation Gulf in 1920.

Coronation Mountain AB, BC On the continental divide near the head of the North Saskatchewan River, this peak (3176 m) was named on 9 August 1902 by British climber Norman Collie on the day of the coronation of Edward VII and Queen Alexandra.

Corunna ON In Moore Township, Lambton County, this place was named in 1823 by Visc. William Carr Beresford, after the 1809 Peninsular War battle in La Coruña, Spain, where he had been an officer in Sir John Moore's forces.

Coteau-Landing QC Located in the regional county of Vaudreuil-Soulanges and on the north shore of Lake St. Francis, this village's post office was named in 1847, six years before the village was incorporated. The municipality of **Coteau-Station**, established as *Coteau* in 1855 on the Grand Trunk Railway line between Montréal and Toronto, was named *La Station-du-Coteau* in 1887 and changed to its present form in 1984. The municipality of **Coteau-du-Lac**, downriver, had a post office by that name in 1789. Each of the names recalls a hillside on the portage around rapids at the foot of the lake.

Côte-Saint-Luc QC One of two remaining '*cités*' in the province (the other is Dorval), it is located in the Communauté urbaine de Montréal, southwest of Montréal. The name was given for an ancient shoreline terrace, and Luc de La Corne Saint-Luc (1711–84), military officer, merchant, interpreter, and legislative councillor.

Cottam ON In Essex County, northwest of Leamington, this place's post office was named in 1877 after Cottam in East Yorkshire, England, north of Kingston upon Hull.

Couchiching, Lake ON The name of this pale blue lake, which drains Lake Simcoe into the Severn River, means in Ojibwa either 'the lake source of a river', or 'water running out'.

Coulonge, Lac ON, QC A widening of the Ottawa River, east of Pembroke, it was named after **Rivière Coulonge**, which enters the lake from the north. These features recall a noted family of the French regime, especially trader Nicolas d'Ailleboust de Manthet (1663–1709), whose grandfather was Nicolas d'Ailleboust de Coulonge-la-Madeleine. *See also* Fort-Coulonge.

Courcelette, Mount BC In the Rocky Mountains, east of Invermere, this mountain (3044 m) was named in 1918 after a village in northeastern France, where Canadian forces fought between 1916 and 1918.

Courtenay BC A city (1915) on the east coast of Vancouver Island, it was named in 1891 after the nearby **Courtenay River**. The river had been named about 1860 after Capt. George William Conway Courtenay, commander of HMS *Constance* on the Pacific station, 1846–9.

Courtenay Bay NB An inlet on the east side of Saint John Harbour, it was named *Courtney Bay* about 1775 by chartmaker J.F.W. DesBarres after John Courtenay (1741–1816), master surveyor of the British ordnance.

Courtice ON Halfway between Bowmanville and Oshawa, this place's post office was named in 1882 after carpenter James Courtice. His parents, Thomas and Mary Courtice, natives of Devon, England, came to Canada about 1831 and settled nearby in Darlington Township.

Courtland ON In Norfolk Township, Haldimand-Norfolk Region, between Tillsonburg and Delhi, this community's post office was named in 1864 after the divisional court, which had been located here in 1861.

Courtright ON A community in Lambton County, south of Sarnia, its station was named in 1869 after Canada Southern Railway president Milton Courtright.

Coutts AB On the Montana border, southeast of Lethbridge, this village (1960) was named in 1890 by the Alberta Railways and Irrigation Company, after either Sir William Lehman Ashmead Bartlett Burdett-Coutts (b. 1851), a British banker and director of the company, or his wife, Baroness Angela Georgina Burdett-Coutts (1814–1906), a stockholder of the company. *See also* Burdett.

Covehead Road PE Located north of Charlottetown, this place's post office was named in 1856 after the road leading north to **Covehead**

Bay. Called *Stanhope Cove* in 1765 by Samuel Holland, the bay likely acquired its present name after the locality of **Covehead**, which developed at the head of the bay.

Cowansville QC Located in the regional county of Brome-Missisquoi, it was created a village in 1876, and a town in 1931. It was first called *Nelsonville* and *Churchville*, before the Cowansville post office was opened in 1841, and named after postmaster and merchant Peter Cowan.

Cow Head NF On the northwestern coast of the island of Newfoundland, and surrounded on the land side by Gros Morne National Park, this town (1964) may have been named after a boulder shaped like the profile of a cow's head.

Cowley AB East of Crowsnest Pass, the post office in this village (1906) was named in 1900 by rancher F. Godsal, possibly inspired by the line 'The lowing herd winds slowly o'er the lea' in Thomas Gray's *Elegy* (1750).

Coxheath NS Southwest of Sydney, this community was named about 1791, after Capt. William Cox, who received a grant here the year before. Its post office was established in 1861.

Cox's Cove NF On the Middle Arm of the Bay of Islands, and on the west side of the island of Newfoundland, this town (1969) was settled about 1840 by fishermen George, John, and William Cox.

Crabtree QC A municipality (1945) in the regional county of Joliette, it was the site of a saw and pulp mill established in 1905 by Edwin Crabtree. **Crabtree Mills** post office was opened a year later.

Craigellachie BC A railway siding in Eagle Pass, between Revelstoke and Salmon Arm, it was at this point that the Last Spike of the Canadian Pacific Railway was driven on 7 November 1885 by Donald Smith, Lord Strathcona. It was named after a prominent crag in Morayshire, Scotland. A post office operated here from 1912 to 1970.

Craighurst ON In Simcoe County, north of Barrie, this place was called *Flos*, after the township, in 1837. In the 1850s it was renamed after pioneer settler Thomas Craig.

Craigleith ON In Grey County, west of Collingwood, it was named in 1857, possibly after the island of Craigleith in the Firth of Forth, northeast of Edinburgh, Scotland.

Craig, Mount YT This mountain (4039 m) was named in 1916 after John Davidson Craig (1875–1936), who surveyed the Yukon-Alaska boundary from Mount St. Elias north to the Beaufort Sea, 1905–10. From 1924 to 1931 he was director of surveys in the department of the interior.

Craik SK On a Canadian National Railways line, but built for the Canadian Pacific in 1889–90, this town (1907) northwest of Regina was named in 1902 after an engineer with the CPR's construction party. Its post office was opened the following year.

Cramahe ON A township in Northumberland County, it was named in 1792 after Hector Theophilus Cramahé (1720–1788), a native of Dublin, Ireland. He administered the government of British North America in the absence of Sir Guy Carleton, 1770–74, and was lieutenant-governor of Quebec, 1771–82.

Cranberry Portage MB Southeast of Flin Flon, this place is located on a narrow portage separating the basins of the Saskatchewan and Nelson rivers. Its post office was established in 1928. It was noted as *Cranberry Carrying Place* on a 1796 Arrowsmith map, and *Cranberry Lake* on David Thompson's map of 1812–3.

Cranbrook BC The largest city (1905) in the southeastern part of the province, it was laid out in lots in 1897. Twelve years before, Col James Baker had settled here and named his farm after Cranbrook, Kent, England, south of Maidstone. In 1898 it became a divisional point on the Canadian Pacific Railway's Crowsnest line.

Crapaud PE A municipal community (1983) on Westmoreland River, east of Borden-Carleton, it had been incorporated in 1950 as a village. Its name was first noted in 1832, and its post office was opened in 1857. It took its name from the French designation of the river, *Rivière aux Crapauds*, 'river of toads', in reference to the *crapaud de mer*, meaning 'eel'.

Craven SK A village (1905) in the Qu'Appelle River valley, north of Regina, its first settlers wanted to call their post office Sussex. Postal authorities rejected it because that name was in use in New Brunswick, and sent a list from which the settlers chose Craven in 1884, after William Craven, Earl of Craven (1606–97), noted for his devotion to Elizabeth I.

Crediton ON In Huron County, southwest of Exeter, this place was named in 1861 after Crediton in Devon, England, which is north of the city of Exeter.

Credit River ON During the 1700s French traders met with the Mississauga at the mouth of the *Rivière au Crédit* each spring, and extended credit for supplies until the following spring. *See also* Port Credit.

Creemore ON In Simcoe County, southeast of Collingwood, this urban centre was named in 1849 by Barrie judge James (later Sir James) Gowan from the Gaelic words *creagh mhor*, 'big heart'. Incorporated as a village in 1889, it was merged with the municipal township of Clearview in 1994.

Creighton SK A town on the Manitoba border, adjacent to the city of Flin Flon, its post office was named in 1958, after Tom Creighton (d. 1952), a prospector who staked claims in the area in 1915.

Creignish NS On the west coast of Cape Breton Island and northwest of Port Hawkesbury, this place was named after Loch Craignish, Argyllshire, Scotland. *Creignish Station* post office opened in 1913 at Craigmore, 5 km to the north.

Cremona AB Northwest of Calgary, this village (1955) was named in 1906 by the post office department, possibly after the city on the Po in Italy, northwest of Bologna.

Creston BC A town (1924) south of Kootenay Lake, and near the Idaho border, it was named in 1898 by the Canadian Pacific Railway, after the Iowa home town of settler Fred Little. Little agreed to sell land to the CPR after it consented to name its station Creston. The post office opened a year later. Creston, IA is southwest of Des Moines.

Creston NF This place was proclaimed in 1914 by the government. Its name was coined, and had no particular local connection. Known informally as Creston North and Creston South, these two communities were annexed by the town of Marystown in 1967.

Crofton BC In the district municipality of North Cowichan, northeast of Duncan, it was named in 1902 after Henry Croft. A brother-in-law of coal magnate James Dunsmuir, he was appointed the manager of the Mount Sicker Copper Company in 1900. He developed the townsite as a location for a copper smelter.

Crossfield AB Between Airdrie and Carstairs, north of Calgary, this town (1980) was named in 1893 by the Calgary and Edmonton Railway Company, after one of its engineers. The post office was opened in 1902.

Cross Roads PE Part of the town of Stratford since 1 April 1995, it had been created as a municipal community in 1972. It took its name from the crossing of the Old Georgetown Road (now the Trans-Canada Highway), by another road leading south over Tea Hill.

Crotch Lake ON On the Mississippi River and in Frontenac County, this cross-shaped reservoir was known as Crotch Lake in the mid-1800s, when surveyor John Harper gave it the euphemism *Cross Lake*. Crotch Lake was still well known in the 1960s, so the name was restored.

Crown Prince Frederik Island NT In the Gulf of Boothia and west of Fury and Hecla Strait, this island (401 km^2) was named in 1922 by the Fifth Thule Expedition after a son of Christian X of Denmark. The name was submitted to the Geographic Board in 1925 by Knud Rasmussen and endorsed the following year. In 1911 explorer J.T.E. Lavoie had named it *Hall Island*, after Charles Francis Hall, who had explored the area in 1868.

Crowsnest Pass AB This municipality was named in 1979 through the amalgamation of the towns of Blairmore and Coleman, the villages of Bellevue, Frank, and Hillcrest Mines, and adjacent communities. The pass (1357 m) was explored in 1858 by Thomas Blakiston. **Crowsnest Mountain** (2785 m) may have been named after a hill 30 km to the east, where the raven's nest of Blackfoot and Cree legends was located, with their designations being translated as 'crow's nest', and being applied to the river, the pass, and subsequently to the mountain above the pass.

Crysler ON In the united counties of Stormont, Dundas and Glengarry, northwest of Cornwall, it was named in 1860 for miller John Crysler (1770–1852), who settled here on the South Nation River in 1843. He served in the House of Assembly of Upper Canada, 1808–26. In 1784, his father, John Crysler, had settled on the St. Lawrence River east of present-day Morrisburg. During the War of 1812 his land was the site of the Battle of Crysler's Farm, where the British forces defeated the much larger American forces on 11 November 1813.

Crystal Beach ON In the town of Fort Erie, Niagara Region, this urban community was established in the 1890s as a summer resort, and a religious meeting ground for the residents of Buffalo, seeking peace and spiritual renewal.

Crystal City MB A village (1947) near the North Dakota border and southeast of Brandon, its post office was named in 1879 by Thomas Greenway, premier of Manitoba, 1888–90, who had settled here that year.

Cudworth SK Founded in 1911 by the Grand Trunk Pacific, the post office of this town (1961) was also named in 1911. Located south of Prince Albert, it was named after Ralph Cudworth (1617–88), a noted English moralist and freethinker.

Culloden ON A community in Oxford County, northwest of Tillsonburg, it was named in 1854 after the famous 1746 Battle of Culloden Moor in Inverness-shire, Scotland.

Culross ON A township in Bruce County, it was named in 1850 after the burgh of Culross in Scotland's Fifeshire, northwest of Edinburgh. Sir George Bruce built a palace in Culross at the beginning of the seventeenth century, and developed an export trade in salt and coal.

Cultus Lake BC On the north shore of **Cultus Lake** and south of Chilliwack, this community was named in 1913. The lake's name in Chinook jargon means 'worthless', 'foul', 'anything bad'.

Cumberland BC A village (1958) south of Courtenay, it was first known as *Union* after the Union Coal Company. It was renamed in 1898 by coal mine owner James Dunsmuir, after Cumberland, England, from where many of the miners had come.

Cumberland ON A municipal township in Ottawa-Carleton Region, it was named after the original township, established in 1799 and named for Prince Ernest Augustus (1771–1851), who was created that year the Duke of Cumberland and Teviotdale. With a 1996 population of 44,630, Cumberland is now Ontario's most populous municipal township. An application for city status was submitted to the minister of municipal affairs in 1996. The community of **Cumberland** in the municipal township was settled in 1801, with its post office opening in 1839. It was changed to *Osborne* in 1856, but Cumberland was restored seven years later.

Cumberland Basin NB, NS At the head of Chignecto Bay, this basin was called *Beau Basin* by the French. It was renamed in 1755 after Fort Cumberland, the name given that year by the captors of the French *Fort Beauséjour*. It was named after the Duke of Cumberland (1721–65), who won the decisive victory over the supporters of Bonnie Prince Charlie at Culloden, 16 April 1746. **Cumberland County**, NS was established in 1759.

Cumberland House SK On the south shore of **Cumberland Lake**, northeast of Nipawin, it was founded in 1774 by Samuel Hearne as the Hudson's Bay Company's first inland post, and named after Prince Rupert, Duke of Cumberland (1619–82), first governor of the company. Postal service was established here in 1891.

Cumberland Sound NT On the southeast side of Baffin Island, this 260-km sound was named in 1587 by explorer John Davis after George Clifford, 3rd Earl of Cumberland (1558–1605), a mathematician and navigator, who became an admiral in 1598.

Cupar SK A town (1968) on the Canadian Pacific Railway, northeast of Regina, its post office was called *Dalrymple* in 1903. It was renamed two years later after a burgh in Fifeshire, Scotland.

Cupids NF This town (1965), on the western side of Conception Bay, has the distinction of being the first year-round British colony in what is now Canadian territory. It was founded in 1610 by Bristol merchant John Guy, who arrived in 'Cuperres coue' with 39 carpenters, shipwrights, and stonemasons. *Cupers Cove*, as he later wrote it, likely took its name from the practice of fish packers sending coopers to the adjacent well-treed slopes to cut wood for staves to make barrels for the pickling of fish.

Curling NF A neighbourhood in the west side of the city of Corner Brook, it was formerly called *Birchy Cove*. It was named in 1904 by the Most Revd Michael Howley, after the Revd James Joseph Curling (1844–1906), who had been appointed the pastor of the Catholic St. Mary's Parish in 1873. Incorporated as a town in 1947, it was annexed to Corner Brook in 1956.

Curran ON In the united counties of Prescott and Russell, southwest of Hawkesbury, this place was named in 1862, likely after John P. Curran (1750–1817), an Irish politician, who had worked for Irish independence and Catholic emancipation.

Cut Knife SK West of Battleford, the post office in this town (1968) was established in 1905. It was named after a Sarcee chief, who was killed by a Cree party, possibly at the Battle of Cut Knife Hill in 1885. The Cree were so proud of eliminating the famed Sarcee fighter that they thereafter called the hill after him.

Cypress Hills AB, SK These hills were called by the French *Montagnes des Cyprès*, mistaking the lodgepole pines for the cypress trees of Eastern Canada. Attaining a height of 1460 m, the hills have the highest elevation between the Torngats, on the Québec-Newfoundland boundary, and the Rocky Mountains.

Cypress River MB Situated on a tributary of the Assiniboine, southeast of Brandon, this place was named in 1886, when the Canadian Pacific Railway was built west from Winnipeg toward Souris. *Littleton* post office was located here from 1881 to 1887, when it was changed to Cypress River.

Czar AB A village (1917) south of Wainwright, it was named in 1909 by the Canadian Pacific Railway, possibly as a recognition of the Anglo-Russian Treaty (1907), which resulted in the creation of the Allied forces during the First World War. It may have been given for a boss of the railway's construction workers, or after the nickname of Arthur Sifton, premier of Alberta, 1910–17.

D

Dacre ON A community in Renfrew County, southwest of the town of Renfrew, its post office was named in 1865, possibly after Dacre in Cumberland, England, southwest of Penrith.

Dagon, Mount BC In the Coast Mountains, southeast of Bella Coola, this mountain (2880 m) was named in 1953 by a member of the Alpine Club of Canada after the mythical chief lord of Satan.

Dalhousie NB A town (1905) at the mouth of the Restigouche River, it was named in 1826, probably by Lt-Gov. Sir Howard Douglas after George Ramsay, 9th Earl of Dalhousie (1770–1838), governor-in-chief of British North America, 1820–8.

Dalhousie ON A township in Lanark County, it was named in 1823 for George Ramsay, 9th Earl of Dalhousie, governor-in-chief of British North America, 1820–8. For municipal purposes it is united with Lavant and North Sherbrooke townships. *See also* Ramsay.

Dalkeith ON In the united counties of Stormont, Dundas and Glengarry, north of Alexandria, this place's post office was named in 1867 after the title of the Duke of Buccleuch's eldest son. Dalkeith, the former seat of the dukes of Buccleuch, is southeast of Edinburgh, Scotland.

Dalmeny SK North of Saskatoon, this town (1883) was founded by the Canadian Northern Railway in 1905. It was likely named after Dalmeny, West Lothian, west of Edinburgh. Its post office was opened the following year.

Dalvay Beach PE At the east end of Prince Edward Island National Park, this beach was named after Dalvay-By-The-Sea, a summer home built in 1896 by Alexander MacDonald of Cincinnati, OH, vice-pres. of the Standard Oil Company. His home in Ohio was also named Dalvay, in commemoration of his original home in Scotland. The national park's headquarters is at **Dalvay by the Sea**.

Daniels Harbour NF A municipal community (1965), on the northwest coast of the Northern Peninsula, it may have been named after Daniel Regan, an Englishman, who sought refuge in its harbour in the 1820s.

Danville QC A town in the regional county of Asbestos, north of Sherbrooke, it was settled in 1801 by Loyalists from Danville, VT. That place northeast of Barre had been named in 1784 after French map-maker Jean-Baptiste Bourgignon d'Anville (1697–1782).

Darlington PE A municipal community (1983) northwest of Charlottetown, its post office was named in 1870, possibly after Darlington, County Durham, England.

Dartmouth NS A metropolitan area in the regional municipality of Halifax since 1 April 1996, it had been created as a town in 1873 and a city in 1961. Founded in 1750 by 355 settlers, who came on the ship *Alderney*, it was named after William Legge, 1st Earl of Dartmouth, who died that year, although he had not been involved in the affairs of state after 1714.

Dartmouth, Mount BC In the Coast Mountains, southeast of Mount Waddington, this mountain (2972 m) was named in 1958 by Capt. Dr G.V.B. Cochran of the United States Air Force after Dartmouth College in Hanover, NH. He and English climbers A. Morrison and J. Rucklidge climbed it in 1957.

Dashwood ON In Huron County, west of Exeter, this community was named *Friedsburg* in 1860 after brothers and millers Absolom and Noah Fried. In 1871 the post office was changed to Dashwood, possibly after Dashwood House in London, England, the headquarters in Britain of Canada's Grand Trunk Railway.

Dauphin MB A town (1910), north of Riding Mountain National Park, and west of **Dauphin Lake**, its post office was first called

Lake Dauphin in 1890, with the present name being adopted in 1897, a year after the site was laid out in lots. The lake took its name from Fort Dauphin, founded by explorer Pierre Gaultier de la Vérendrye in 1741 at present Winnipegosis, 55 km to the north. He named it after the Dauphin, the eldest son of Louis XV, who did not survive his father. The **Dauphin River** unites Lake Manitoba with Lake Winnipeg, via Lake St. Martin.

D'Autray QC A regional county on the north shore of the St. Lawrence, between the regional counties of L'Assomption and Maskinongé, it was named in 1982 after military officer and explorer Jacques Bourdon d'Autray (1652–88).

Daveluyville QC In the regional county of Arthabaska, northwest of Victoriaville, it was named in 1901 after merchant Adolphe Daveluy (1841–1915), who was the first mayor of *Sainte-Anne-du-Sault*, the place's earlier name.

Davidson SK A town (1906), midway between Regina and Saskatoon, its station was named in 1902 after Col A.D. Davidson of Minnesota, who organized the Saskatchewan Valley Land Company. He sold over 2,023,500 ha (5,000,000 a) to American settlers. Its post office was opened in 1903.

Davis Inlet NF This place, now located on Iluikoyak Island on the Labrador coast, was formerly established as a Hudson's Bay Company trading post in 1831 on Ukasiksalik Island. It was named after the inlet, given after explorer John Davis (*c.* 1550–1605), who entered it while exploring the coast in 1585. It has been the site of the Innu (Naskapi) settlement of Utshimassit (meaning 'storekeepers place') since 1967, but it is being relocated to Natuashish at Little Sango Pond on the mainland, 18 km to the south, in the late 1990s.

Davis Strait NT Separating southern Baffin Island from Kalaallit Nunaat (Greenland), this strait was possibly named in 1616 by William Baffin and Robert Bylot after John Davis, who had explored the area in 1585, 1586, and 1587.

Dawn ON A township in Lambton County, it was named in 1829. Its name may have been given because surveyors encountered sombreness in Sombra Township, but found Dawn more pleasant. Perhaps fugitive blacks, who arrived here in 1828, expressed thanks for the 'dawn of liberty'.

Dawson YT After the discovery of gold on Bonanza Creek in early August 1896, Joseph Ladue started laying out a townsite at the junction of the Klondike and Yukon rivers. The following January William Ogilvie agreed to survey it into lots, on the condition that the place was named for the finest person he knew, George Mercer Dawson (1849–1901), who was then the director of the Geological Survey of Canada. Despite physical handicaps, this brilliant and versatile scientist undertook a number of incredible exploratory journeys. His most arduous trip took him in 1879 up the Stikine and Liard rivers, down the Pelly River to the Yukon River, and then up that river, and over the Chilcoot Pass to the Pacific, 2200 km in all. Dawson was declared the capital of the Yukon in 1897, but in 1951 the territory's government was moved to the larger centre of Whitehorse, and two years later Dawson ceased to be the capital. Commonly called *Dawson City*, it is incorporated in the statutes as the 'town of the City of Dawson'. It had been an incorporated city from 1902 to 1904.

Dawson Creek BC A city (1958) in the Peace River country, its post office was named in 1921 after a small tributary of Pouce Coupé River. The creek was likely named in 1879 by Henry A.F. MacLeod, a Canadian Pacific Railway exploration engineer, who met geologist George M. Dawson that year. Dawson had undertaken an arduous reconnaisance from the mouth of the Skeena River to Edmonton. On his maps he called the creek *Dawson's Brook*.

Dawson, Mount BC In the Selkirk Mountains, east of Revelstoke, this mountain (3390 m) was named in 1888 by British alpinist William Green, after geologist George M. Dawson.

Dawson, Mount NB Southwest of Nictau Lake, it was named in 1899 by naturalist William F. Ganong after geologist George M. Dawson. Although he was better known for his geological work in British Columbia and Yukon, he also undertook some studies in New Brunswick.

Daysland AB A town (1907) southeast of Camrose, it was named in 1905 by the Canadian Pacific Railway after Egerton W. Day, who first settled here in 1888. He returned to Ontario soon after, coming back to central Alberta in 1904. Its post office was opened in 1906.

Dayspring NS On the southeast side of Bridgewater, this place was first known as *Summerside*. Its post office was called *Gibbon* in April 1884, but was renamed Dayspring two months later. The name not only reflects early dawn, it conveys the idea of the advent of Christ.

Dease Lake BC In northwestern British Columbia, this place was established in 1834 by Hudson's Bay Company chief trader John McLeod, and named after fur trader Peter Warren Dease (1788–1863). Dease served with the XY Company, the North West Company, and the Hudson's Bay Company. He also took part in Arctic expeditions in the 1820s and '30s.

Dease Strait NT Uniting Coronation and Queen Maud gulfs and between the mainland and Victoria Island, this strait was named in 1826 by John (later Sir John) Franklin, after Hudson's Bay Company chief factor Peter Warren Dease. Dease joined Franklin's British Arctic land expedition, 1825–7.

Deauville QC A village on the north shore of Lac Magog, southwest of Sherbrooke, it was known as *Petit-Lac-Magog* until 1952, when it was renamed by Mayor Edgar Genest after the prominent French resort community in Normandie, France.

Debden SK Northwest of Prince Albert, this village (1922) was founded by the Canadian Northern Railway in 1909. It was named after Debden in Essex, England, southeast of Cambridge. Postal service was established in 1912.

Debec NB Southwest of Woodstock, this community's post office was named *Debeck Station* in 1870 after miller George Debec, who had settled here in 1835. It became *Debeck* in 1875, was changed to *DeBec* before 1885, and to its present form about 1896. The place was also known as *Blairs Mills*, after miller Andrew Blair, father of Andrew G. Blair, premier of the province, 1883–96.

Debert NS Northwest of Truro and on the **Debert River**, this place's post office was called *River Debert* in 1855, became *Debert Station* in 1875, and was changed to the present name in 1963. The river's name, noted as early as 1820, was adapted from *Rivière de Bourq*, its designation during the French period before 1755.

DeCew Falls ON On the southwest side of the city of St. Catharines, this mill site was settled by John DeCew (1766–1855) in 1788. It was to his home that heroine Laura Secord walked in 1813 from Queenston to warn the British of a possible American attack. After the first Welland Canal destroyed his mill site in 1825, DeCew moved in 1834 to Haldimand County, where he founded **DeCewsville**.

Deep River ON A town (1959) in Renfrew County, it was created in 1944 as the residential community for the families of scientists and other staff of the Chalk River Nuclear Laboratories. In 1686 Chevalier de Troyes named the 70-km stretch of the Ottawa River *rivière creuse*, which Alexander Henry the younger translated as 'Deep-river' at the end of the next century. The river is more than 75 m deep at its deepest point, 10 km downriver from the town, near the site of the laboratories.

Deer Island NB In Passamaquoddy Bay, its name may have been derived from Passamaquoddy *edokwemeneek*, 'deer island'. There is a legend that wolves were chasing a moose and a deer, but they turned into rock, resulting in Moose Island in Maine, Deer Island, and The Wolves in the Bay of Fundy.

Deer Lake NF Located at the north end of **Deer Lake** and northeast of Corner Brook, this town (1950) was first known as *Nicholsville*. It had been named after George Nichols, a native of Cape Breton Island, who established a timber-cutting operation here in 1867.

Dégilis QC Located in the regional county of Témiscouata, southeast of Rivière-du-Loup, it was known as *Sainte-Rose-du-Dégelé* from 1879 to 1968, when it was modified to *Sainte-Rose-du-Dégilis*, and abbreviated the following year to Dégilis. It took its name from the unusual phenomenon of the adjacent Rivière Madawaska, that does not freeze at the outlet of Lac Témiscouata during the winter. In French such a phenomenon is called a *dégilis*.

De Grau NF This place, on the south shore of the Port au Port Peninsula, was presumably named after a French phrase meaning 'we are finished', in reference to having completed a season of summer fishing. It has been part of the municipal community of Cape St. George-Petit Jardin-Grand Jardin-De Grau-Marches Point-Loretto since 1969.

Delaware ON A township in Middlesex County, it was named in 1798 after an Algonquian confederacy of tribes, which settled on the Thames River. The name is traced to the Delaware River, a major river in the eastern United States. The river and the state of Delaware are named after Thomas De La Warr (1577–1618), governor of Virginia, 1610–18. The community of **Delaware**, southwest of London, was named in 1820.

Delburne AB East of Red Deer, this village (1913) was named in 1911 by the Grand Trunk Pacific Railway. It may have been named after Della Mewburn, sister of pioneer medical doctor F.H. Mewburn. On the other hand, it may possibly have been chosen by landowners M.J. Manning and W.C. Clendening from a GTP list.

Delhi ON An urban centre in Haldimand-Norfolk Region, it was known *as Sovereen's Corners*, and *Fredericksburg*, both after innkeeper Frederick Sovereen, and *Middleton*, after the township, before it was named in 1853 in honour of the capital city of India. In 1974 it was united with parts of the townships of Charlotteville, Middleton, South Walsingham, and Windham to form the municipal township of **Delhi**.

Delia AB A village (1914) west of Hanna, its post office was named in 1909 after the wife of merchant A.L. Davis, who earlier operated Delia's Stopping House here. In 1913 the Canadian Northern Railway called its station *Highland*, as it was its highest point between Kindersley and Drumheller, with Delia post office being moved to the line. Subsequently, it was agreed to change the station name to Delia.

Délįne NT Located near the point where Great Bear River flows west from Great Bear Lake, this charter community (1993) was the site of Fort Franklin, established by the Hudson's Bay Company in 1825 to serve an expedition led by John (later Sir John) Franklin. Its post office was opened in 1965, with *Fort Franklin* becoming a hamlet in 1972. In 1993 it was renamed after the Slavey word for 'bear river'.

Delisle QC A municipality in the regional county of Lac-Saint-Jean-Est, north of Alma, it had been created as the township municipality of Delisle in 1890. The latter had been named after the township of *De l'Île*, itself named in 1857 after Île d'Alma, with the township's name being respelled Delisle in 1861.

Delisle SK Southwest of Saskatoon, the post office in this town (1913) was named in 1905 after four brothers, Amos, Eddie, Eugene, and Fred Delisle. It was moved to the line of the Canadian Northern Railway, when it was laid in 1907–8.

Della Falls BC The highest falls (440 m) in Canada, it is located on Vancouver Island, northwest of Port Alberni. It was named in 1948 after Della Drinkwater, whose husband, Jo, had been a prospector in the area. Jo Creek drains **Della Lake** into Drinkwater Creek.

Deloraine MB Located southwest of Brandon, near the North Dakota border, this town (1907) was founded in 1882 by Scots settlers and named that year by the first postmaster, after Deloraine, Roxburghshire, Scotland. Four years later, when the Canadian Pacific Railway arrived, the townsite was shifted south to the tracks.

Deloro ON A village (1919) in Hastings County, east of Marmora, it was named in 1881 after an early discovery of gold in Ontario. Refining of gold, as well as silver and cobalt after 1903, continued here until 1961.

Delphine, Mount BC In the Purcell Mountains and west of Windermere, this mountain (3399 m) was named in 1911 by climber Edward Harnden, after a mine opened by George Starke. The mine was named after his wife, Delphine Francour, who died in 1921 at Wilmer, where Starke had opened a hotel.

Delson QC A town (1957) in the regional county of Roussillon, south of Montréal, its post office was named in 1912 by devising a name from Delaware and Hudson, the title of a railway company.

Delta BC A district municipality south of Vancouver and between the Fraser River and Boundary Bay, it was established as the corporation of Delta in 1879, taking its name from the flat deltaic land at the Fraser's mouth. Among its urban areas are **North Delta** and Ladner.

Delta ON In the united counties of Leeds and Grenville, west of Brockville, it was first known as *Stevenstown* after first settler Abel Stevens, *Stone Mills*, after Stevens's grist mill, and *Beverley*, for Chief Justice Sir John Beverley Robinson. It was called Delta because Upper and Lower Beverley lakes, and the place's site between the lakes, all have the triangular shape of the Greek capital letter 'delta'.

Deltaform Mountain AB, BC On the continental divide, south of Lake Louise, this mountain (3424 m) was named in 1898 by American writer and climber Walter Wilcox after its distinctive shape reminded him of the Greek capital letter 'delta'.

Demarcation Point YT The most northwesterly point of Canada, it was noted in 1828 by John (later Sir John) Franklin 'from its being situated in longitude 141°W, the boundary between the British and Russian dominions'.

Demorestville ON In Prince Edward County, northwest of Picton, it was named for miller Guillaume Demorest (1769–1848), a native of Orange County, NY, who settled here in 1793. His son Thomas was the first postmaster in 1828.

Dempster Highway NT, YT Extending northeast from the Klondike Highway, 42 km east of Dawson, to Fort McPherson, NT, this highway was created after 1959. It was named in 1963 after William John Duncan 'Jack' Dempster (1876–1964), a Royal North-West Mounted Police officer. He served in the Yukon from 1897 to 1934.

Denbigh ON A township in Lennox and Addington County, it was named in 1859 after Denbighshire in Wales. The post office in the community of **Denbigh** was named in 1864.

Denman Island BC On the east coast of Vancouver Island, and southeast of Courtenay, this island was named about 1864 by Capt. George Richards, after Rear-Adm. Joseph Denman (1810–74), who served on the Pacific station, 1864–6.

Denmark NS A place located east of Tatamagouche, and on property that once belonged to the Rt Revd Charles Inglis (1734–1816), bishop of Nova Scotia, 1787–1816. The perimeter of the property was identified with the dean's mark. The post office opened as *Denmark Road* in 1888, and became Denmark 10 years later.

Denys, Mount NB On the south side of the Nepisiguit River, this mountain was named in 1899 by naturalist William F. Ganong after Nicolas Denys (1598–1688), who was governor of Acadia in the late 1600s, and author of a book on the geography and natural history of North America in 1672.

Denzil SK Located southwest of Battleford, this village (1911) was founded by the Canadian Pacific Railway in 1910. It was named after Ansel Olson, in appreciation of the friendly service of his father, A.F. Olson. However, through some confusion (possibly Olson's accent) led to the name Denzil. The post office was opened in 1911.

Derby ON A township in Grey County, it was named in 1842 after Edward George Geoffrey Smith Stanley, 14th Earl of Derby (1799–1869), who was prime minister of Great Britain for three terms.

Deroche BC On the north side of the Fraser River, west of Chilliwack, this place's post office was named in 1891 after Joseph Deroche (d. 1922), who settled here after working as a teamster in the Cariboo during the 1860s. His name may have been DesRochers, as the first postmaster was E. DesRochers.

Derwent AB Northwest of Lloydminster, this village (1930) was named in 1928 by the Canadian Pacific Railway after the River Derwent in Cumberland, England. Its post office opened the same year.

Desbarats ON Montréal businessman and printer George-Paschal Desbarats (1808–64) worked the mines in 1840 at this place in Algoma District, southeast of Sault Ste. Marie.

DesBarres, Mount NB This mountain, on the south side of the Nepisiguit River, was named in 1899 by naturalist William F. Ganong after Joseph Frederic Wallet DesBarres (1722–1824), producer of the hydrographic charts in the *Atlantic Neptune*, lieutenant-governor of Cape Breton, 1784–7, and of Prince Edward Island, 1804–12.

Desbiens QC Located in the regional county of Lac-Saint-Jean-Est and on the south shore of Lac Saint-Jean, its post office was called *Moulin-Desbiens* in 1922, after early settler and miller Louis Desbiens, and was shortened to Desbiens the following year. It was the village of *Saint-Émilien* from 1926 to 1960, when it became the town of Desbiens.

Desboro ON In Grey County, south of Owen Sound, this community's post office was named in 1869 after Desborough in Northamptonshire, England, north of the city of Northampton.

Deschaillons-sur-Saint-Laurent QC On the south shore of the St. Lawrence, east of Trois-Rivières, this municipality traces its origins to the seigneury of *Rivière-Duchesne* (or *Deschaillons*), granted in 1674 to Pierre de Saint-Just de Saint-Louis (1640–1724), a captain in the Carignan-Salières Regiment. Échaillon was a family estate near Grenoble, France.

Deschambault QC Located in the regional county of Portneuf and on the north shore of the St. Lawrence, southwest of the city of Québec, this municipality was created in 1989 from the village of Deschambault and the parish municipality of *Saint-Joseph-de-Deschambault*. In 1683 Jacques-Alexis de Fleury Deschambault (*c.* 1642–1715) became the seigneur of Chavigny, with his name superseding it in 1690.

Deschênes, Lac ON, QC A widening of the Ottawa River, extending for 40 km northwest from Ottawa to Fitzroy Harbour, its name was taken from *Portage des Chênes* on the north side of the river, used during the French regime to bypass the **Deschênes Rapids**. The name of the portage means 'carrying place at the oak trees'.

D'Escousse NS On the north shore of Isle Madame, this place may have been named after Martin Descouts, a surgeon who was engaged in the fishery off the Atlantic coast from 1714 to 1725. He lost property at Canso in 1718, and spent the years from 1726 to 1745 at Port la Joie, on the southwest side of Charlottetown Harbour.

Deseronto ON On the north side of the Bay of Quinte, east of Belleville, this town (1889) was known as *Mill Point* from 1851 to 1881. In 1855 John Culbertson laid out village lots and named the site Deseronto after his grandfather, John Deserontyon, also known as Odeserundiye (*c.* 1740–1811), who led some Mohawk here in 1784. The postal and village names were changed to Deseronto in 1881.

Desjardins QC A regional county centred on the city of Lévis, opposite the city of Québec, it was named in 1982 after Alphonse Roy *dit* Desjardins (1854–1920), the founder in 1900 of the co-operative movement and the first *caisse populaire* (people's bank) in Québec.

Des Ruisseaux QC A municipality northwest of Mont-Laurier, it was created in 1974, and named after the numerous streams in the area. In 1897 it had been called the united townships of *Robertson-et-Pope*, named after Quebec Central Railway founder Joseph Gibb Robertson (1820–99) and former minister of railways and canals Joseph Henry Pope (1819–89).

Destruction Bay YT On the western shore of Kluane Lake, this place was named in 1942 after a construction camp, built by United States Army engineers, had been destroyed by high winds.

Detroit River ON Antoine de la Mothe, Sieur de Cadillac built Fort Pontchartrain du Détroit, commonly called Fort Détroit, on the Michigan side of the river in 1701. By 1744, *Rivière du Détroit* was commonly used on maps for the drainage connecting Lakes Huron and Erie, but the name Detroit River was restricted to the present river in English-language records after 1759. Although the name suggests the French perceived the feature to be a strait (*étroit*) there is in fact a drop of one m from Lake Huron to Lake St. Clair, and 1.7 m from Lake St. Clair to Lake Erie.

Deux-Rivières ON In Renfrew County, east of Mattawa, its post office was named in 1866, reflecting its location where the Rivière Maganasipi, from the north in Québec, joins the Ottawa River.

Deux-Montagnes QC A town at the eastern end of **Lac des Deux Montagnes** and in the regional county of **Deux-Montagnes**, it was named in 1963. The regional county was created in 1983, taking its name from the former county. The lake's name first occurred on a 1684 map by Jean-Baptiste-Louis Franquelin. The two mountains in question may be Mont Rigaud on the south side, and Mont Oka on the north, although claims for the two highest summits at Oka are often advanced.

Devon AB On the south bank of the North Saskatchewan, southwest of Edmonton, this town (1950) was established by the Imperial Oil Company in 1947. Its name was proposed by a Mr Johnson of Toronto after the Devonian formation, from which the Leduc No. 1 gushed forth. The post office was opened in 1948.

Devon Island NT The second largest island (55,247 km^2) of the Queen Elizabeth Islands, it was named in 1819 by William (later Sir William) Parry after Devon, England, the home county of his second-in-command, Lt Matthew Liddon.

Dewberry AB A village (1957) northwest of Lloydminster, its post office was named in 1907. Difficulty in choosing a name was resolved when a woman brought a basket of dewberries into the store where the post office was to be located. The dewberry, *Rubus pubescens*, is a trailing raspberry common to the damp woods of the parkland country of central Alberta.

Dewdney, Mount BC In the Cascade Mountains, southeast of Hope, this mountain (2255 m) was named in 1936 after Edgar Dewdney (1835–1916), who built the Dewdney Trail in the mid-1860s from Fort Hope to present-day Trail. **Dewdney Creek**, which rises 14 km to the north and flows northwest into the Coquihalla River, was named when the trail was laid out.

Diadem Peak AB In the Winston Churchill Range, north of the Columbia Icefields, this peak (3371 m) was named in 1898 by British climber Norman Collie after the band of snow around its crest.

Diadem Peak BC In the Coast Mountains, southeast of Mount Waddington, this mountain (2868 m) was named in 1982 by climber Bruce Fairley of the Alpine Club of Canada's Vancouver section. It was named because its profile reminded him of a crown.

Diamant, Cap QC On the heights at the city of Québec, this name was given by Jacques Cartier in 1541 after several stones he found upriver at Cap Rouge, which he concluded were diamonds. By the beginning of the 1800s the furthest east point of the heights was identified as Cap Diamant. A comment about 1575 by French writer André Thevet provided the oft-repeated proverb 'false as a Canadian diamond'.

Diamond Jenness Peninsula NT A mid-western extension of Victoria Island, this 260-km peninsula was named in 1971 by William E. Taylor of the National Museum of Man after Diamond Jenness (1886–1969). Jenness joined the Geological Survey of Canada in 1913 as an anthopologist, and was the chief anthropologist at the museum from 1926 to 1948. He undertook extensive anthropological studies in Northern Canada and wrote a number of books on the Aboriginals of Canada.

D'Iberville, Mont QC The highest point (1622 m) of a massif in the Torngats on the Quebec-Newfoundland boundary was named in 1971 after Pierre Le Moyne d'Iberville (1661–1706), noted for expeditions to Newfoundland and Hudson Bay in the 1690s. The same mountain was named Mount Caubvick in 1981 by the Newfoundland Geographical Names Board.

Didsbury AB North of Calgary, this town (1906) was named by the Canadian Pacific Railway in 1892–3, after a southside district of Manchester, England.

Diefenbaker, Lake SK This enlargement of the South Saskatchewan River extends some 200 km upriver from the Gardiner and Qu'Appelle dams. It was created in 1958 and named in 1967 after John George Diefenbaker (1895–1979), prime minister of Canada, 1957–63.

Dieppe NB On the east side of Moncton, this town (1952) was named in 1946 after Dieppe, France, where many Canadian troops were killed on 19 August 1942, while assaulting the German defences. The post office was called *Legere Corner* in 1896, and renamed *Leger Corner* in 1930.

Digby NS A town (1890) on the Annapolis Basin, it was settled by New Englanders in 1766 and named *Conway*, likely after Henry Seymour Conway, British secretary of state for the southern department. It was renamed in 1783 after Rear-Adm. Robert Digby (1732–1815), commander of HMS *Atlanta*, the flagship of the convoy, which brought Loyalist settlers to the site in 1783. In 1782 he had been named a joint commissioner with Sir Guy Carleton to reconcile the differences between Great Britain and its North American colonies. **Digby County** was separated from Annapolis County in 1837. **Digby Gut** is the narrow entrance into Annapolis Basin, and **Digby Neck** is a long peninsula extending southwest, between St. Marys Bay and the Bay of Fundy.

Digdeguash River NB Rising near McAdam, this river flows southeast into Passamaquoddy Bay. Of unknown meaning in Passamaquoddy, its current spelling appeared on various plans and maps in the late 1700s, but it is pronounced more like *Diggedeguash River*, shown on a 1799 map by Surveyor General Thomas Sproule.

Dildo NF The colourful name of this place on the east side of Trinity Bay has an obscure origin. Adjacent **Dildo Island** was recorded as *Dildoe* in 1711. The harbour (**Dildo Arm**) may have reminded someone of a cylinder, an old English meaning of the word. Some vegetation may have recalled a low shrub called the Dildo-tree. It is also said that a Spanish mariner may have been reminded of a similar name back home. **Dildo Run** is a narrow channel separating the north side of the island of Newfoundland from New World Island.

Dingwall NS On Aspy Bay and south of Cape Breton Island's Cape North, this place was first called *Young's Cove*. Its post office, called *Aspy Bay* in 1855, was renamed in 1883 after its postmaster, Robert Dingwall.

Dinorwic ON At the north end of Dinorwic Lake and in Kenora District, southeast of Dryden, it was named in 1897 by Canadian Pacific Railway superintendent William Whyte after Dinorwic, in northwestern Wales.

Dinsmore SK In 1907 a post office was opened at this village (1913) southwest of Saskatoon, and named after a local farmer. When the Canadian Northern Railway built a line through the area in 1913, it retained the name for its station.

Disraeli QC In the regional county of L'Amiante, south of Thetford Mines, this place was incorporated as the village of *D'Israëli* (or *d'Israëli*) in 1904, with the spelling modified to Disraëli in 1953, and incorporated as the town of Disraeli in 1969. The original parish municipality was created in 1882 and named after Benjamin Disraeli, 1st Earl of Beaconsfield (1804–81), prime minister of Great Britain, 1867–8 and 1874–80.

Dixon Entrance BC Separating the Queen Charlotte Islands from Alaska, it was named in 1788 by Sir Joseph Banks, president of the Royal Society of London, after Capt. George Dixon (d. 1791). Dixon explored the west coast with Capt. James Cook in 1778, and was the commander of the *Queen Charlotte*, 1786–8.

Dixville QC A village (1874) in the regional county of Coaticook, south of the town of Coaticook, its post office was called *Drew's Mills* from 1835 to 1874. It was then renamed after pioneer miller and postmaster Richard (Dick) Baldwin, replicating a name 45 km to the southeast in New Hampshire.

Doaktown NB A village (1966) on the Southwest Miramichi River, its post office was named in 1854 after Robert Doak (d. 1857), who arrived from Scotland about 1812.

Dodsland SK This village (1913) northeast of Kindersley, was established in August 1913, when the Grand Trunk Pacific Railway built a line southwest from Biggar. It was named after Dr Dodds, who had come from Nebraska and acquired 810 ha (2000 a) of land in the area. The post office was opened in 1914.

Dolbeau QC On the Rivière Mistassini, north of Lac Saint-Jean, and in the regional county of Maria-Chapdelaine, this town was named in 1927 after the township of Dolbeau, created in 1904 and named after Récollet missionary Jean Dolbeau (1586–1652).

Dollard-des-Ormeaux QC A city (1960) in the Communauté urbaine de Montréal, west of the city of Montréal, it was established as a municipality in 1924. The name honours one of the most celebrated figures in early Canadian history, Adam Dollard des Ormeaux (1635–60). In May 1660, with 16 French companions and several Huron, he defended *Ville-Marie* (Montréal) by turning back 800 Onondaga, Mohawk, and Oneida at Long Sault, on the Ottawa River, near present-day Hawkesbury ON.

Dolphin and Union Strait NT A shallow passage connecting Coronation Gulf with Amundsen Gulf, this strait was named in 1826 by John (later Sir John) Richardson, after the two little boats that transported Richardson and Edward Kendall along the coast from the Mackenzie Delta to the Coppermine River.

Dominion NS An urban community in the regional municipality of Cape Breton since 1 August 1995, it had been incorporated as a town in 1906. It was named after the Dominion No. 1 coal shaft, sunk into a seam in 1894, having been incorporated earlier as the village of *Old Bridgeport Mines*. The post office was called *Bridgeport Mines* in 1886, *Old Bridgeport Mines* in 1887, and *Old Bridgeport* in 1906. That was changed to Dominion in 1940.

Dominion City MB North of Emerson, this small community of some 385 people was first called *Roseau Crossing*, after a tributary of the Red River. It was renamed Dominion City by the Canadian Pacific Railway in 1878, taking its name from the Dominion of Canada and from the hope it would become a large urban centre. Its post office was opened two years later.

Domremy SK A village (1921) south of Prince Albert, it was founded in 1895 by settlers from France, and named after their home town Domrémy-la-Pucelle, the birthplace of St Joan of Arc. Postal service was set up the following year. It was moved 3 km to the south, when the Canadian Northern Railway was built in 1918.

Domville ON In the united counties of Leeds and Grenville, north of Prescott, this place was named in 1880 by Augusta Township reeve John Dumbrille, who proposed the original French Huguenot form of his family name. Earlier, he had suggested Frogmore, after one of Queen Victoria's homes, but the local residents thought he had the noisy frogs in nearby Long Swamp in mind.

Donalda AB Located southeast of Camrose, this village (1912) was named in 1911 by the Canadian Northern Railway, after Donalda Crossway of Cobourg, ON, a niece of Sir Donald Mann, vice-president of the company. The post office was called *Harker* from 1907 to May 1911, when it was called *Eidswold* for just four months, before adopting Donalda.

Donjek River YT A tributary of the White River, this river rises in Kluane Lake. It was likely named by Dr Charles Willard Hayes during a trip in the area in 1891. In Gwich'in *donyak* or *donchek* means 'pea-vine', a good source of winter feed for game and domestic animals.

Donkin NS East of Glace Bay, this place's post office was called *Dominion No. 6* from 1906 to 1940. It was then renamed after Hiram F. Donkin, the general manager of the Dominion Coal Company, when Dominion No. 6 mine was opened in 1904.

Don Mills ON An urban centre in the city of North York, it was farmland until 1954, when Canada's first large planned residential community was built. In the early 1800s, mills were operated on the **Middle Don River** by James, William, and Alexander Gray.

Donnacona QC Located in the regional county of Portneuf, southwest of the city of Québec, this place became a village in 1915 and a town five years later. The name recalls the Iroquois chief of Stadaconé, at the present site of the city of Québec, where Jacques Cartier met him in 1535 and took him back to France the following year.

Donnelly AB In the Peace River country, between Falher and McLennan, this village (1956) was named in 1915 after an official of the Edmonton, Dunvegan and British Columbia Railway. The post office was opened in 1917.

Don River ON Flowing into the east side of Toronto Harbour, this river was named by Elizabeth Simcoe, wife of Lt-Gov. John Graves Simcoe, because it reminded her of the River Don in East Yorkshire, England.

Doon ON A neighbourhood in southwestern Kitchener, it was founded as *Doon Mills* in 1834 by Adam Ferrie (1813–49). He named the mills after the River Doon, which flows into the Firth of Clyde at Ayr, Scotland.

Dorchester NB West of Sackville, this town (1911) was founded in 1801 and its post office was opened in 1825. It was named after Dorchester Parish, established in 1787 and named after Sir Guy Carleton, 1st Baron Dorchester (1724–1808), who was then governor-in-chief of British North America. He had been created Lord Dorchester in 1786.

Dorchester ON In **North Dorchester Township**, Middlesex County, east of London, it was first called *Edwardsburg* in the 1820s. The post office was called *Dorchester Station* in 1855, and was shortened to Dorchester six years later. However, *Edwardsburg* remained the name of the police village until 1915, when it was changed to Dorchester.

Dorion ON A township in Thunder Bay District, northeast of Thunder Bay, it was named in 1874 after Sir Antoine-Aimé Dorion (1818–91), co-premier of the Province of Canada during two terms between 1858 and 1864, and chief justice of the Court of Queen's Bench of Québec from 1874. The post office was called *Dorion Station* in 1912, and shortened to Dorion in 1954.

Dorion QC A town (1916) in the regional county of Vaudreuil-Soulanges, west of Montréal, it was named in honour of Sir Antoine-Aimé Dorion (1818–91), member of the Legislative Assembly of Canada, 1854–67 and of the House of Commons, 1872–4. He was co-premier of the Province of Canada with George Brown for a few days in 1858, and with John (later Sir John) A. Macdonald in 1863–4. He was the chief justice of the Court of Queen's Bench of Québec from 1874.

Dornoch ON In Glenelg Township, Grey County, south of Owen Sound, this place was named *Glenelg* in 1852. Philip McIntosh renamed it Dornoch in 1880, after his father's birthplace at the mouth of Dornoch Firth, 50 km northeast of Inverness, Scotland.

Dorset ON In the district municipality of Muskoka, southeast of Huntsville, this community was first called *Colebridge* after the Coles family. It was named Dorset in 1880, possibly after Dorset, England. The adjacent townships of Ridout and Sherborne were named after Thomas Ridout (1754–1829), surveyor general of Upper Canada, 1810–29, and his birthplace in Dorset.

Dorval QC One of two municipalities still designated '*cité*' in the province (the other is Côte-Saint-Luc), it was created a village in 1892, a town in 1903, and a city in 1956. The post office was named in 1878. In 1691 Jean-Baptiste Bouchard d'Orval (*c.* 1658–1724) bought property in the area of the present-day city.

Douglas MB Located east of Brandon, this place was first known as *Currie's Landing* after homesteader and ferry operator William Currie. It was moved north to the Canadian Pacific Railway main line in 1882, and renamed Douglas after James Douglas Sutherland Campbell, Marquess of Lorne (1845–1914), governor general of Canada, 1878–83, who travelled on the newly laid track in 1881. Its post office was called *Douglas Station* in 1883.

Douglas ON On the Bonnechere River in Renfrew County, west of Renfrew, this community was named in 1853 by Judge John Glass Malloch (1806–73) of Perth, who surveyed lots, and named it for Douglas in Scotland's Lanarkshire, southeast of Glasgow. His wife's father and Perth pioneer, the Revd William Bell, was from Douglas, Scotland.

Douglas Channel BC Extending southwest from Kitimat Arm, this channel was named by Hudson's Bay Company officers after Sir James Douglas (1803–77), governor of Vancouver Island, 1851–63, and of British Columbia, 1858–64.

Douglas, Mount AB Northwest of Banff, this mountain (3235 m) was named in 1884 by geologist George M. Dawson after Scottish botanist David Douglas (1798–1834). He studied trees and plants during several trips through North America, and the Douglas fir, an evergreen closely related to the larch, was named after him. It is considered the most valuable timber tree in North America.

Douglastown NB In the city of Miramichi and between Newcastle and Chatham, this place's post office was named in 1835. It had been named about 10 years earlier by lumbermen Allan Gilmour and John Rankin, after Sir Howard Douglas (1776–1861), lieutenant-governor of the province, 1823–32. It may have been named when Sir Howard was in Newcastle to lay a stone for a Presbyterian church, just before a great fire swept the area. It was also known in the early years as *Gretna Green*.

Douglastown was an incorporated village from 1966 to 1995.

Douglastown QC A community in the town of Gaspé, southeast of the town centre, it was named after Scottish surveyor John Douglas, who was commissioned by the government in 1775 to lay out a model village for Loyalists.

Douro ON A township in Peterborough County, it was named in 1821 after one of the titles of the Duke of Wellington, Baron Douro. The duke was given the title after a victory over French forces, after crossing the Rio Douro at Oporto, and driving them from Portugal.

Dover NF On the northwest shore of Bonavista Bay, this town was first called *Shoal Bay* in the 1890s. In 1949 Dover had been selected by the local residents from a list submitted by the post office department of Canada, after *Wellington* had been rejected. The town was incorporated as *Wellington (Dover)* in 1971, but after a local petition, it was shortened to Dover two years later.

Dover NS Southwest of Canso, this village (1969) was probably named in the 1820s after Dover, England. Its post office was called *Little Dover* in 1890.

Dover ON A township in Kent County, it was named in 1794 after the seaport in Kent, England.

Dowling ON A township in Sudbury Region, northwest of the city of Sudbury, it was named in 1885 after John F. Dowling, member for Renfrew South in the Ontario Legislature, 1884–6 and 1891–4. In 1973 most of the township became part of the new town of Onaping Falls. The community of **Dowling** is one of the main urban centres in the town.

Downie ON A township in Perth County, it was named in 1830 for Robert Downie, a director of the Canada Company, a British company established in 1826 to develop the Huron Tract, a large area of what is now Southwestern Ontario.

Downie Peak BC Located in the Selkirk Mountains, north of Revelstoke, this mountain (3015 m) was named in 1866 by surveyor Walter Moberly after William Downie (1819–93), who prospected in the Goldstream River valley in 1865. Mountaineer Howard Palmer called it *Mount Arrowhead* in 1912 and *Eldorado Mountain* in a report in 1914. Earlier in 1849 Downie had taken part in the gold rush to California, where Downieville is named for him.

Downsview ON An urban neighbourhood in the city of North York, its post office was named in 1869 after the Downs View Farm of John Perkins Bull (1822–1902).

Drake SK Located southeast of Saskatoon, between Lanigan and Nokomis, this village (1910) was founded by the Canadian Pacific Railway in 1907. It was named after Sir Francis Drake (1540–96), the famous English sea captain, who may have been the first European explorer to cruise the west coast of Canada. The post office was opened in 1908.

Drayton ON A village (1873) in Wellington County, northwest of Fergus, it was named in 1851 after Drayton Manor, the English home of Sir Robert Peel in Staffordshire, northeast of Birmingham.

Drayton Valley AB A town (1957) on the northwest side of the North Saskatchewan River, southwest of Edmonton, its post office was named in 1920 after the home town of the wife of postmaster W.J. Drake in Hampshire, England, a northern suburb of Portsmouth. Previously its post office was called *Powerhouse*.

Dresden ON A town (1882) on the Sydenham River in Kent County, its post office was named in 1854 by postal authorities after the city in Germany. Fairport had been suggested for part of the town, but the name soon disappeared. Dresden was the end of the Underground Railroad for Josiah Henson and other fugitive slaves. It is the site of Josiah Henson's cabin, which inspired Harriet Beecher Stowe to write *Uncle Tom's Cabin*.

Drumbo ON In Oxford County, northeast of Woodstock, this community was named in 1854 after Drumbo in County Down, Ireland, south of Belfast.

Drumheller AB A city (1930) on the Red Deer River, northeast of Calgary, its post office was named in 1911 after Samuel Drumheller (1864–1925). He arrived here in 1910 and developed the townsite in the centre of coal fields. The Canadian Northern Railway arrived in 1912.

Drummond NB A village (1967) located southeast of Grand Falls, its post office was named in 1891 after **Drummond Parish**, which had been established in 1872. It may have been named after Drummond Castle, near Perth, Scotland, where various Barons Drummond were also called Lord Perth. Perth Parish is immediately south of Drummond Parish.

Drummond ON A township northeast of Perth in Lanark County, it was named in 1816 after Gen. Sir Gordon Drummond (1772–1854), administrator of Lower and Upper Canada, 1814–6.

Drummondville QC A city on the west bank of the Rivière Saint-François, and in the regional county of **Drummond** (1982), it was named by Maj.-Gen. Frederick Heriot in 1815 after Gen. Sir Gordon Drummond, a native of the city of Québec. The administrator of Lower and Upper Canada, 1814–16, he had directed Heriot to organize a military agricultural establishment along the river.

Dryden ON A town (1910) in Kenora District, it was founded in 1895 by provincial agriculture minister John Dryden (1840–1909). He opened an experimental farm called New Prospect at nearby Wabigoon Lake.

Dubawnt Lake NT Midway between Great Slave Lake and Hudson Bay, this lake (3629 km^2) was named in 1893 by geologist Joseph Burr Tyrrell after the Chipewyan word *tobutua*, 'water shore lake', possibly because the lake remains ice-covered in the summer, except for a narrow fringe along the shore. The **Dubawnt River** rises north of Lake Athabasca and flows northeast to join the Thelon River, upstream from Aberdeen Lake.

Dublin ON In Perth County, northwest of Stratford, this place's post office was named *Carronbrook* in 1854, after a tributary of the Bayfield River. It was renamed Dublin in 1878, after the birthplace in Ireland of merchant and miller Joseph Kidd.

Dublin Shore NS On the Atlantic coast, southwest of the mouth of the LaHave River, it was named in 1857 after New Dublin Township. The township was established here in the 1760s and called after the capital of Ireland. The post office in **West Dublin**, west of Dublin Shore, was opened in 1864.

Dubreuilville ON In Algoma District, northeast of Wawa, it was founded in 1951 by brothers Napoléon, Joachin, Augustin, and Marcel Dubreuil, natives of Taschereau, Abitibi, QC. Its post office was called *Magpie Mine* in 1954, but it was renamed in 1961 after the founders of the lumbering community. It became a municipal township in 1989.

Duchess AB A village (1921) north of Brooks, it was named in 1911 by the Canadian Pacific Railway after Duchess Louise Marguerite (d. 1917), wife of the Duke of Connaught and Strathearn, son of Queen Victoria, and governor general of Canada, 1911–16.

Duck Lake SK A town (1911) southwest of Prince Albert, it was established in 1890 by the Canadian Northern Railway, and named after the lake to its west, made famous in 1885 when the first shots of the North-West Rebellion were fired. *Stobart* post office, after merchant William Stobart, was opened in the area in 1879, and was changed to Duck Lake in 1894. In 1907 the postmaster reported that 'The ducks are so plentiful that when generally disturbed by sportsmen the vibration of their wings sounds like thunder.'

Dufferin ON Provisionally created as a county in 1874 and legally established seven years later, its name was proposed in 1872 for

Frederick Temple Blackwood, 1st Marquess of Dufferin and Ava (1826–1902), governor general of Canada, 1872–8 and viceroy of India, 1884–8. The title of Dufferin came from a barony in County Down, Ireland.

Dugald MB East of Winnipeg, this community's post office was first called *Sunnyside* in 1879, but was renamed in 1890 after Dugald Gillespie, who had arrived here from Ontario in 1880.

Dummer ON A township in Peterborough County, it was named in 1821 after William Dummer Powell (1755–1834), chief justice of Upper Canada, and speaker of the Legislative Council, 1816–25.

Duncan BC A city (1912) on Vancouver Island, northwest of Victoria, it was first named *Duncan's* in 1887 by the Esquimalt and Nanaimo Railway, after William Chalmers Duncan. He arrived in Victoria in 1862 from Ontario, and settled at the site of the city shortly after. He named his farm and the new community *Alderlea*. *Duncan's Station* post office was opened in 1889 and changed to Duncan in 1926.

Duncan, Mount BC North of Kootenay Lake, this mountain (3228 m) was named in 1890 by British alpinist Harold Topham after prospector John Duncan. **Duncan Lake** and **Duncan River**, both north of Kootenay Lake, were also named for him.

Dunchurch ON In Parry Sound District, northeast of Parry Sound, it was named in 1877 by postmaster George Kelcey, after his birthplace in Warwickshire, England, southwest of Rugby.

Dundalk ON A village (1886) in Grey County, northwest of Shelburne, it was named in 1874 by Elias Brabazon Grey, after his home town in County Louth, Ireland.

Dundas ON A county, part of the united counties of **Stormont, Dundas and Glengarry** since 1850, it was named in 1792 after Henry Dundas, 1st Viscount Melville (1742–1811), who served in several British ministries, and communicated on questions bearing on British North America.

Dundas ON A town (1848) in Hamilton-Wentworth Region, it was named *Dundas Mills* in 1800, after its location at the east end of **Dundas Street**, and Dundas post office was opened in 1814. In 1793 Lt-Gov. John Graves Simcoe ordered the building of Dundas Street to the Forks of the Thames (London), naming it after Henry Dundas, Viscount Melville, then secretary of state for the home department, 1791–1801. Dundas Street was subsequently extended to York (Toronto).

Dundas Peninsula NT A southwestern extension of Melville Island, this 140-km peninsula was named by William (later Sir William) Parry, after Robert Saunders Dundas, 2nd Viscount Melville (1771–1851), who was then the first lord of the Admiralty.

Dundela ON In the united counties of Stormont, Dundas and Glengarry, northwest of Morrisburg, its post office was named in 1866 after local resident Delia Dillabough. It was earlier called *McIntosh's Corners*, where pioneer settler John McIntosh discovered some wild apple trees and his son Allen propagated the famous McIntosh apple.

Dundurn SK A town (1980) south of Saskatoon, it was named in 1894 by postmaster James Leslie, after Dundurn Castle in Hamilton, ON, built by Sir Allan Napier MacNab in the 1830s.

Dunedin ON In Simcoe County, south of Collingwood, this place was named in 1866 by first postmaster John Carruthers, after the New Zealand city of Dunedin, which he had visited.

Dungannon ON A community in Huron County, northeast of Goderich, it was laid out in village lots in 1855 by William Mallough and named the next year after his home town in County Tyrone, Ireland.

Dunham QC A town in the regional county of Brome-Missisquoi, south of Cowansville, it was created a township municipality in 1845, taking its name from the township, proclaimed in 1796. It became a village in 1867 and a town in 1971.

Dunk River PE This river rises near the community of Hunter River, and flows west into Bedeque Bay, south of Summerside. It was named in 1765 by Samuel Holland, after George Montagu Dunk, 2nd Earl of Halifax (1716–71), who was then British secretary of the southern department.

Dunlop ON Just north of Goderich, it was named *Millburn* in 1878. Four years later it was renamed after Capt. Robert Graham Dunlop (1790–1841), and Dr William 'Tiger' Dunlop (1792–1848). Robert served in the House of Assembly of Upper Canada, 1836–41, and was followed in the Legislative Assembly of the Province of Canada by his brother William, who became the superintendent of the Lachine Canal, at Montréal, in 1846.

Dunmore NS Southeast of Antigonish, this place's post office was named in 1879 after Dunmore Hill, Perthshire, Scotland, 13 km south of Amulree, and west of Perth.

Dunnottar MB Situated on Lake Winnipeg, south of Gimli, this village (1947) was named by Alex Melville, who united the resort developments of Whytewold, Ponemah, and Matlock. He chose the name after Dunnottar Castle, at Stonehaven, south of Aberdeen, Scotland.

Dunnville ON In Haldimand-Norfolk Region, this town (1900) was named in 1830 for John Henry Dunn (1792–1854), receiver general of Upper Canada 1820–43, and president of the Welland Canal Company, 1825–33. In 1974 the town was considerably enlarged by annexing the townships of Dunn, Canborough, Moulton, and Sherbooke.

Dunrobin ON In Ottawa-Carleton Region, northwest of Kanata, this community was first called *Torbolton*, after the township, in 1864. Four years later it was named after Dunrobin Castle, the home of the Duke of Sutherland, in Sutherlandshire, northeast of Inverness, Scotland.

Duntroon ON A community in Simcoe County, south of Collingwood, it was known as *McNab's Corners*, *Scotch Corners*, *Nottawasaga* (after the township), and *Bomore*, before Presbyterian minister John Campbell renamed it in 1864 after Duntroon Castle in Argyllshire, northwest of Glasgow.

Dunvegan NS Situated northeast of Inverness, this place's post office was called *Broad Cove Marsh* in 1866. It was renamed 20 years later after Dunvegan Castle, on the Isle of Skye, Scotland.

Dunvegan ON North of Cornwall, in the united counties of Stormont, Dundas and Glengarry, this place's post office was first called *Kenyon*, after the township, in 1848. It was renamed in 1862 after Dunvegan on Scotland's Isle of Skye, the Clan Macleod home for over 700 years.

Dunville NF Located on the north bank of the Northeast Arm of Placentia Harbour, this place is a neighbourhood in the town of Placentia. It had been incorporated as a town in 1963, but was annexed in 1994. It was first known as *North East Arm*. It was renamed Dunville about 1888, when a branch of the Newfoundland Railway was built to Placentia.

Dunwich ON A township in Elgin County, it was named in 1792 after the village of Dunwich on the North Sea coast of Suffolk, England.

Duparquet QC Located in the regional county of Abitibi-Ouest, northwest of Rouyn-Noranda, its post office was opened in 1933, receiving its name from the township, created in 1916. The township's name honoured a captain in the La Sarre Regiment, who was killed in a battle at the city of Québec in 1760.

Durham ON Durham County was proclaimed in 1792 after England's County Durham. It was united with Northumberland County for municipal purposes from 1850 to

1974, when it annexed most of Ontario County (while losing three townships and the town of Port Hope to adjacent counties) to form the regional municipality of Durham.

Durham ON A town (1872) in Grey County, it was named in 1848 by George Jackson, when he moved the Crown lands office here from *Sydenham* (Owen Sound). Jackson was born in the city of Durham, County Durham, England. Its post office was *Bentinck*, after the township, from 1847 to 1857, when it was called Durham.

Durham-Sud QC In the regional county of Drummond, south of Drummondville, this municipality and its post office were first named *South Durham* in 1854. The present municipality merged the municipality of Durham-Sud (1865) and the village of Durham-Sud (1918) in 1975. The original township of Durham was named in 1795 after County Durham in England.

Dutton ON A village (1890) in Elgin County, southwest of St. Thomas, it was named in 1872, after a civil engineer of the Canadian Southern Railway.

Dwight ON In the district municipality of Muskoka, east of Huntsville, it was named in 1871 for Harvey Prentice Dwight (1828–1912). He was then with the Montreal Telegraph Company in Toronto and was later president of the Great Northwestern Telegraph Company, also in Toronto.

Dyment ON In Kenora District, southeast of Dryden, its post office was named in 1899, probably after Thessalon lumberman Albert Edward Dyment (1869–1944), a Liberal member of Parliament, 1896–1908.

Dysart ON A township surrounding the community of Haliburton, in Haliburton County, it was named in 1860 after the town of Dysart, in Fifeshire, Scotland, north of Edinburgh. Since 1874 it has been the senior member of the municipal township of Dysart, Bruton, Clyde, Dudley, Eyre, Guilford, Harburn, Harcourt and Havelock, which is frequently referred to as 'Dysart et al.'.

Dysart SK Located northeast of Regina, the post office serving this village (1909) was named in 1906, after one of the surveyors of the Canadian Pacific Railway. The father of First World War flying ace Billy Bishop owned the land through which the line ran. He agreed to name the place after Dysart, if Dysart agreed to run the line through his land.

Eagle Pass BC This pass (561 m) through the Monashee Mountains, and between Revelstoke and Salmon Arm, was named in 1865 by Canadian Pacific Railway explorer Walter Moberly, who was guided by two eagles up the valley of **Eagle River**. Twenty years later Donald Smith, Lord Strathcona, drove the Last Spike of the CPR at Craigellachie in the pass.

Eaglesham AB This community, northeast of Grande Prairie, was named in 1916 by the Edmonton, Dunvegan and British Columbia Railway after Eaglesham, Renfrewshire, Scotland, south of Glasgow. Its post office was opened in 1929, and it was an incorporated village from 1967 to 1996.

Ear Falls ON A municipal township in Kenora District, southeast of Red Lake, its post office was named in 1948. **Ear Falls** in Ojibwa is called *Otak Powitik*, after a rock ledge at the lip of the falls, shaped like a huge human ear. French fur traders called it *Portage de l'Oreille*.

Earl Grey SK North of Regina, the post office at this village (1906) was opened in 1905 and called after Albert Henry George Grey, 4th Earl Grey (1851–1917), governor general of Canada, 1904–11. In 1909 he donated the Grey Cup for the Canadian football champions of each year. The Canadian Pacific Railway arrived here in 1907.

Earlton ON In Timiskaming District, north of New Liskeard, its post office was named in 1905 for George Earl Brasher, the eldest son of postmaster Albert Edward Brasher.

East Angus QC A town (1912) northeast of Sherbrooke, its post office was named in 1882 after Montréal industrialist William Angus, who built a sawmill and a pulp and paper mill here.

East Broughton QC A municipality in the regional county of L'Amiante, northeast of Thedford Mines, it was first known as Sacré-Cœur-de-Jésus. It was renamed East Broughton in 1931 after Broughton Township, named in 1795, and proclaimed in 1800. The village of **East Broughton Station**, one km to the northwest, was created in 1954, with its post office having been named in 1899.

East Coulee AB Southeast of Drumheller, this community was named in 1929, when the post office opened adjacent to a steep-sided coulee (valley), on the east side of the Red Deer River.

Eastend SK Located on the east side of the Cypress Hills, southwest of Swift Current, the post office of this town (1920) took its name in 1900 from a North-West Mounted Police post established here in 1877. The post office was first called *East End*, because it was the furthest east of the NWMP's Division A, whose headquarters was located in Maple Creek.

Eastern Kings PE A municipal community (1974) on the east side of Kings County, it embraces several localities, including Baltic, Bayfield, East Point, Elmira, Kingsboro, North Lake, and South Lake.

Easterville MB In 1964 a Manitoba Hydro dam at Grand Rapids raised the level of Cedar Lake, and inundated the Aboriginal community of *Chemahawin*. It was resettled on the south shore of the lake and named for Donald Easter, a former chief of the community. The post office opened the following year.

East Farnham QC A village in the regional county of Brome-Missisquoi, north of Cowansville, it was established in 1914, with its post office having been named in 1837. It is situated in the eastern part of the township of Farnham, proclaimed in 1798, and named after a village in Surrey, England, southwest of London.

East Ferris ON A municipal township in Nipissing District, southeast of North Bay, it was created in 1921 when Ferris Township was divided into East and West Ferris municipal townships, with West Ferris being annexed by North Bay in 1968. Ferris Township had been named in 1880 after James M. Ferris, Liberal member for Northumberland East in the Ontario Legislature, 1875–83.

East Garafraxa ON A municipal township in Dufferin County, it was separated from West Garafraxa Township in 1869, and added to Dufferin in 1874. There are many guesses as to the meaning of Garafraxa, with none of them more convincing than any other.

East Gwillimbury ON A town (1977) in York Region, it was created as a township in 1798. It was possibly named after Lt-Col Thomas Gwillim (d. 1762), the father of Elizabeth Posthuma Simcoe (1762–1850).

East Hawkesbury ON A municipal township in the united counties of Prescott and Russell, it was separated in 1884 from Hawkesbury Township. The latter had been named in 1798, either for Charles Jenkinson (1727–1808), Baron Hawkesbury and Lord Liverpool, or for his son, Robert Banks Jenkinson (1770–1828), created Baron Hawkesbury in 1796. The son, who became Lord Liverpool in 1808, was prime minister of Great Britain, 1812–27.

East Kildonan MB Annexed by the city of Winnipeg in 1972, this neighbourhood, on the east side of the Red River, had been incorporated as a city in 1957. Kildonan, on both sides of the river, had been named in 1817 by Lord Selkirk after the parish of Kildonan in Sutherlandshire, Scotland.

East Luther ON Luther Township, named in 1831 after Protestant reformer Martin Luther (1483–1546), was divided in 1879, with East Luther joining Dufferin County in 1883. In 1995 it was united with the village of Grand Valley to form the new municipal township of **East Luther Grand Valley**.

Eastmain QC A Cree village at the mouth of **Rivière Eastmain**, which flows for 700 km from central Québec into the southeastern part of James Bay, it is located at the site of the Hudson's Bay Company post of Eastmain, established in 1730. Taking its name from its location on the east side of James Bay's 'main' (water body), *East Main* was attached to the river in the late 1800s, and shortened to Eastmain in the early 1900s.

Eastman QC A village in the regional county of Memphrémagog, west of Magog, its post office was called *Bolton Forest*, after the township of Bolton, in 1865. It was renamed in 1881 after pioneer settler John Eastman.

Eastport NF On **Eastport Bay**, an extension of Bonavista Bay, this town (1959) was established in the 1860s by families from nearby Salvage. Its post office was called *Salvage Bay* before 1900, when it was changed to *Brighton*. Because of duplication, *Salvage Bay* was restored in 1904. Eleven years later the residents asked for Eastview, but the Newfoundland Nomenclature Board imposed Eastport, although the place did not have harbour facilities.

East Redonda Island BC Between Vancouver Island and the mainland, northeast of Campbell River, this island was considered by Spanish explorers Dionisio Alcalá Galiano and Cayetano Váldes y Bazan in 1792 to be part of *Isla Redonda*, which included present West Redonda Island. The two names were endorsed in 1950. In Spanish *redonda* means round.

East Riverside-Kingshurst NB A village (1966), 13 km northeast of the centre of Saint John, its first post office from 1898 to 1901 was called *Riverside Beach*. It was reopened as East Riverside in 1911, and closed in 1969. Kingshurst was a former Canadian National Railways station.

East Royalty PE Within the eastern limits of the city of Charlottetown, this neighbourhood, part of the original territorial division of Charlottetown Royalty, was a municipal community from 1983 to 1995.

East Wawanosh ON Wawanosh Township was named in 1840 for Chief Joshua Wawanosh of the Chippewas on the Sarnia Reserve, who was one of the chiefs who transferred a large area of Southwestern Ontario to the British Crown in 1827. It was divided in 1867 into East Wawanosh and West Wawanosh municipal townships, with both remaining in Huron County.

East Williams ON Williams Township, created in Middlesex County in 1830, was named after William Williams, a deputy govenor of the Canada Company, a British company established in 1826 to develop the Huron Tract, a large area of what is now Southwestern Ontario. It was divided into East Williams and West Williams municipal townships in 1860.

Eastwood ON In Oxford County, east of Woodstock, this place was named in 1853 after Eastwood Park, the home of Rear-Adm. Henry Vansittart (1777–1843). Built here in the 1830s, the home was named after his sister, a widow, who lived with him, and whose married name was East.

East York ON The only borough in Ontario, it is one of the six municipalities in Metropolitan Toronto. York Township was named in 1793, with East York Township being severed from it in 1924. It and the town of Leaside were joined in 1967 to form the borough of East York.

East Zorra ON The original Zorra Township was named in 1819 by Lt-Gov. Sir Peregrine Maitland after the Spanish word *zorro*, 'female fox'. It was divided into East Zorra and West Zorra municipal townships in 1845. In 1975 East Zorra was joined with the village of Tavistock to form the municipal township of **East Zorra-Tavistock**.

Eaton QC Located in the regional county of Le Haut-Saint-François, this township municipality east of Sherbrooke was named in 1855 after the township, named in 1795, and proclaimed in 1800. It may have been named after one of the places called Eaton in England. Eaton post office operated from 1825 to 1969.

Eatonia SK Southwest of Kindersley, this town (1954) was founded in 1917 by the Canadian Northern Railway, and named *Eaton* after Timothy Eaton (1834–1907), founder of the T. Eaton Company department store in Toronto in 1869. He established the popular Eaton's catalogue, which allowed frontier communities access to a variety of merchandise. The first post office in the area was called *Cornfeld* in 1913 and was renamed Eatonia in 1921, amending the station's name to avoid confusion with nearby Eston.

Ebenezer SK North of Yorkton, the post office at this village (1948) was named *Ebenezer Valley* in 1891, and renamed Ebenezer in 1911. When the Grand Trunk Pacific Railway constructed a line though the area in 1909, it named its station *Anoka*. The German Baptist settlers disliked the name, so the company agreed to call its station Ebenezer.

Echo Bay NT On the east shore of Great Bear Lake, this was the site of *Cameron Bay* post office in 1933. It was changed to *Port Radium* in 1937. Radium had been discovered in 1930 on Echo Bay by Gilbert Labine, and mining was undertaken from 1932 to 1940, and 1942 to 1960. The mine was reopened by Echo Bay Mines in 1963, and the place was renamed Echo Bay two years later, although *Port Radium* post office continued until 1982.

Eckville AB West of Red Deer, the post office in this town (1966) was named in 1905 after pioneer settler Andy E. Eckford. Presumably a naming contest was held, with Hattie Mitzner (later Stephens) winning with Eckville.

Eclipse Sound NT Located between Baffin and Bylot islands, this sound was named in 1904, possibly by geologist Albert P. Low. He encountered the Scottish whaler *Eclipse*, under the command of Capt. W.F. Milne, working the waters west of Pond Inlet.

Economy NS On the west side of **Economy Point**, and west of Truro, its post office was opened in 1836. It was an Acadian settlement called *Vil Conomie* in the mid-1700s. Its name was taken from the point, called by the Mi'k-maq *Kenomee*, 'large point jutting out into the sea'.

Ecum Secum NS On the Eastern Shore, between Sheet Harbour and Sherbrooke, this place's post office was opened in 1879. Its name may be derived from Mi'kmaq *megwasaagunk*, 'red bank'.

Edam SK Settled in 1907, the post office at this village (1911), northwest of North Battle-ford, was named in 1908 after Edam, in the Netherlands, famous for its cheese. In 1911 the Canadian Northern Railway established a station here, when it built its line through the Turtlelake River valley.

Edberg AB This village (1930) was named in 1902 by postmaster Johan Edstrom (1850–1910), who arrived here from Sweden not long before. He chose the name by joining the first syllable of his name with Swedish *berg*, 'hill'.

Eden ON In Elgin County, south of Tillson-burg, it was first settled in 1817 and named in 1854, likely after the biblical paradise.

Eden Mills ON East of Guelph and in Wellington County, this place was first called *Kribbs Mills* after David and Aaron Kribbs. Adam Argo bought the mills in 1846 and renamed the site Eden Mills, after the garden where Adam and Eve first lived.

Edenwold SK Northeast of Regina, the first post office at this village (1912) was opened in 1890 and moved around among a number of farm residences, before it was fixed at its present location in 1912, when the Grand Trunk Pacific Railway was built from Melville to Regina. The name was originally spelled *Edenwald*, from the Garden of 'Eden' and the German for 'forest', but the postal officials misspelled it Edenwold, perhaps misreading the handwritten application.

Edgerton AB Southeast of Wainwright, this village (1917) was named in 1908 by the Grand Trunk Pacific Railway, after H.H. Edgerton, a GTP engineer. The post office was also opened that year.

Edgetts Landing NB Adjacent to the village of Hillsborough, and southeast of Moncton, this place was named after Loyalist Joel Edgett (1760–1841), who came from New York State. Its post office was opened in 1854, with Ward Edgett as the first postmaster.

Edgewater BC In the Columbia River valley, northwest of Invermere, this place was named in 1913 after its location on the east bank of the river. It was earlier called *Marlborough*.

Edith Cavell, Mount AB, BC After English nurse Edith Cavell (1865–1915) was executed in Belgium by the Germans during the First World War, British Columbia Premier Sir Richard McBride and Prime Minister Sir Robert Borden urged the naming of a mountain after her in Canada. In 1916 the Geographic Board chose this mountain (3363 m) south of Jasper, overlooking the valley of the Athabasca River.

Edmonton AB Edmonton House was established by the Hudson's Bay Company in 1795, 32 km downstream, near present-day Fort Saskatchewan, probably taking its name from clerk John Peter Pruden's birth-place on the northeast side of London, England. It was abandoned in 1807 and relocated near the present site of the city. In 1821 it was called Fort Edmonton, and was rebuilt in 1830 near the present legislative building. The city was incorporated in 1904. The summer village (1959) of **Edmonton Beach**, west of Edmonton, was created in 1926.

Edmundston NB At the confluence of the Madawaska and Saint John rivers, this city (1952) was named in 1851 after Sir Edmund Walker Head (1805–68), lieutenant-governor of the province, 1848–54. It was originally called *Petit-Sault*, after falls at the mouth of the Madawaska. The first post office, from 1847 to 1851, was *Little Falls*. The city's name is commonly misspelled Edmunston.

Edson AB A town (1911) on the north side of the McLeod River, west of Edmonton, it was named in 1910 by the Grand Trunk Pacific Railway after Edson J. Chamberlain (1852–1924), then vice-president and general manager of the company. He served as the company's president from 1912 to 1917. Earlier its post office was called *Heatherwood.*

Edwards ON In Ottawa-Carleton Region, southeast of Ottawa, this community's post office was named in 1901, probably after Rockland lumberman William Cameron Edwards (1844–1921), Liberal member for Russell in the House of Commons, 1887–1903, and then a senator.

Edwardsburgh ON A township in the united counties of Leeds and Grenville, it was named in 1787 after Prince Edward Augustus, Duke of Kent (1767–1820), the fourth son of George III, and father of Queen Victoria.

Edwardsville NS Across Sydney Harbour from Sydney, this place's post office was named in 1894, after nearby Point Edward. The point was called after Prince Edward, the Duke of Kent and Strathearn (1767–1820), the father of Queen Victoria.

Edziza, Mount BC East of the Stikine River, this mountain (2480 m) was first called *Edziza Peak* on a Geological Survey map of 1926, and was shown as Mount Edziza in a GSC report the following year. It was named after Johnny Edzerza, who was killed by an avalanche while crossing the mountain. **Mount Edziza Provincial Park** embraces the area of the mountain, where craters, reddish volcanic ash and pumiceous sand are characteristics of its area.

Eel Ground NB Located southwest of Newcastle (Miramichi) and on **Eel Ground Indian Reserve**, this place in Mi'kmaq was called *Nendookun*, 'where eels are speared in the mud'.

Eel River Crossing NB This village (1966), southwest of Dalhousie, was named after the river, which flows northeast into Chaleur Bay. Its post office was *Eel River* from 1861 to 1886, when it was changed to its present name. In 1783 the river was referred to as *Longuil River*, a reference to the French word *anguille*, 'eel'.

Eganville ON A village (1890) in Renfrew County, west of the town of Renfrew, its post office was named in 1852 after lumberman John Egan (1811–57), who bought a lumbering depot in 1837 at the Fifth Chute, the village's site on the Bonnechere River.

Eglinton Island NT In the Parry Islands of the Queen Elizabeth Islands, and between Melville and Prince Patrick islands, this island (1541 km^2) was named in 1853 by George F. Mecham after Archibald William Montgomerie, 13th Earl of Eglinton (1812–61), who was then lord lieutenant of Ireland.

Egmondville ON A community in Huron County, south of Seaforth, its post office was named in 1855 after merchant Constant Van Egmond. He and his brothers Leopold, William, and August developed several businesses here. Their father, Anthony Van Egmond (1775–1838) built the road from Guelph to Goderich in 1832, but after a dispute with the Canada Company, he joined William Lyon Mackenzie's rebel forces in 1837, and was put in jail, where he died.

Egmont, Cape PE Extending into Northumberland Strait, this cape was named in 1765 by Samuel Holland after John Perceval, 2nd Earl of Egmont (1711–70), then first lord of the Admiralty. Holland also named the bay northwest of the cape after the earl.

Egremont ON A township in Grey County, it was named in 1840 after George O'Brien Wyndham, 3rd Earl of Egremont (1751–1837), a brother-in-law of Gov. Gen. Lord Durham's mother.

Ekfrid ON This township in Middlesex County was named in 1821 after the Northumbrian king Egfrith, killed in AD 685 in a battle with the Picts, near present-day Forfar, Scotland.

Elbow SK Located due south of Saskatoon, this village's site is at a sharp turn of Lake Diefenbaker, described in 1804 as an 'elbow' in a journal by John McDonald of Garth, when ascending the South Saskatchewan River. The post office was called *River View* in 1906, but it was renamed Elbow three years later, when the village was incorporated.

Elderslie ON A township in Bruce County, it was named in 1850 after the village of Elderslie, west of Glasgow, where the Scottish national hero Sir William Wallace (1270–1305) was born. *See also* Wallaceburg *and* Wallacetown.

Eldon ON A township in Victoria County, it was named in 1823 after John Scott, 1st Earl of Eldon (1751–1838), who was then lord chancellor of Great Britain.

Eldorado ON In Hastings County, north of Madoc, its post office was named in 1867, following a gold rush. Eldorado is the legendary city of gold in South America.

Eldorado Creek YT A tributary of Bonanza Creek, it was named in September 1896 by Knut Halstead, who had earlier prospected for gold on Fortymile River. It turned out to be one of the richest gold streams in the world, producing some $20 million of gold in a 5-km stretch.

Elfros SK This village on the Canadian Pacific Railway, northwest of Yorkton, and its post office were both established in 1909, and named after the Icelandic for 'garden of roses'.

Elgin MB A community southwest of Brandon, it was laid out in 1898 and named by the Canadian Northern Railway after James Bruce, 8th Earl of Elgin (1811–63), governor of the province of Canada, 1847–54. Its post office opened the following year.

Elgin ON Elgin County was named in 1851 for James Bruce, Earl of Elgin, governor of the province of Canada, 1847–54. The title was taken from the city of Elgin, in Morayshire, Scotland.

Elgin ON In the united counties of Leeds and Grenville, southwest of Smiths Falls, this community was named in 1850 after James Bruce, Earl of Elgin, governor of the province of Canada, 1847–54.

Elginburg ON North of Kingston, in Frontenac County, it was named in 1853 after James Bruce, Earl of Elgin, governor of the province of Canada, 1847–54.

Elginfield ON In Middlesex County, north of London, its post office was named in 1849 after James Bruce, Earl of Elgin, governor of the province of Canada, 1847–54.

Elie MB Situated midway between Winnipeg and Portage la Prairie, this place was founded by the Canadian Northern Railway in 1888–9, and named after merchant Elie Dufresne, described on his tombstone as 'The Founder of Elie'. Its post office was opened in 1898.

Eliot River PE Created as a municipal community in 1975, it was annexed by the town of Cornwall in 1995. Its name was taken from the original one given to the West River by Samuel Holland in 1765, after Edward Eliot, 1st Baron Eliot (1727–1804), then lord commissioner of trade and plantations. The river's name West River superseded *Eliot River* in 1970.

Elizabeth, Mount NB Southeast of Mount Carleton, this mountain (655 m) was named in 1903 by naturalist William F. Ganong after Louise-Élisabeth de Joybert de Soulanges et de Marson (1673–1740), who was born near present-day Jemseg, NB. She was the mother of Pierre de Rigaud de Vaudreuil de Cavagnial, Marquis de Vaudreuil, the last French governor of New France.

Elizabethtown ON A township in the united counties of Leeds and Grenville, it was named in 1787 for Princess Elizabeth (1770–1840), third daughter of George III.

Elkford BC A district municipality (1981) in the **Elk River** valley, north of Sparwood, it is, at an elevation of 1400 m, the highest populated place in Canada. It was established as a village in 1971, the same year that the post office was opened. The name was composed from Elk River and its tributary, Fording River. Elkford was developed in 1971 by the Fording Coal Company.

Elkhorn MB Located northwest of Virden, near the Saskatchewan border, this village (1905) was established on the main Canadian Pacific Railway line in 1882. Surveyors had observed a large pair of elk horns set up on a nearby hill. Postal service was set up the next year.

Elk Island National Park AB Located 45 km northeast of Edmonton, this park (194 km^2) was created in 1913 as a fenced 'island' for elk, bison, mule deer, and other wildlife.

Elk Lake ON In Timiskaming District, northwest of New Liskeard, this place was named in 1907 after a widening of the Montreal River. It was also known as *Elk City*, *Bear Creek*, *Cookville*, and *Smythe*.

Elko BC A community south of Fernie and in the Elk River valley, its post office was named in 1899 after the river.

Elk Point AB This town (1962) on the north side of the North Saskatchewan River, northwest of Lloydminster, owes its name to a Mr Higby, originally from Elk Point, SD, who had moved first to Vegreville. To escape the temptation to drink, he went down the river in 1905, and lived in a hole in the river bank near the site of the future town. Deeming the site to be promising, he persuaded others from Vegreville to homestead here the next spring. In South Dakota, Elk Point is in the very southeastern corner, near Sioux City, IA.

Ellef Ringnes Island NT In the Sverdrup Islands of the Queen Elizabeth Islands, this island (11,295 km^2) was named *Ellef Ringnes's Land* in 1900 by Otto Sverdrup, during his 1898–1902 Norwegian Arctic expedition, after Ellef Ringnes, a Norwegian patron.

Ellershouse NS Southeast of Windsor, this place was named after Baron Franz von Ellershausen (1820–1914). A German native, he settled here in 1864 and built homes for German immigrants. He subsequently built a pulp and paper mill and a sugar refinery. The post office opened as *Ellerhausen* in 1866 and was changed to Ellershouse ten years later.

Ellerslie-Bideford PE North of Tyne Valley, this municipal community (1977) embraces the localities of **Ellerslie** and **Bideford**. Ellerslie was named before 1845 by a man called Wallace, after Elderslie, Scotland, west of Glasgow, where the Scottish hero Sir William Wallace (*c.* 1270–1305) was born. Bideford was named by William Ellis and Thomas Chanter, who had come from Bideford, Devon, England in 1818. First known as *New Bideford*, its post office opened as Bideford in 1912. In 1765 Samuel Holland named the adjacent stream *Goodwood River*, after a residence near Chichester, Sussex, England of the 3rd Duke of Richmond. By the late 1800s it had become better known as **Bideford River**.

Ellesmere Island NT Canada's third largest island (196,236 km^2), after Baffin and Victoria islands, this island was named in 1852 by Edward (later Sir Edward) Inglefield, after Francis Leveson-Gower, 1st Earl of Ellesmere (1800–57), who was then president of the Royal Geographical Society. **Ellesmere Island National Park Reserve** (37,775 km^2) was established in 1988.

Ellice ON A township in Perth County, it was named in 1830 for Edward Ellice (1783–1863), a director of the Canada Company, a British company established in 1826 to develop the Huron Tract, a large area of what is now Southwestern Ontario.

Elliot Lake ON A city (1991) in Algoma District, it is the site of an uranium ore discovery in 1948. Laid out as a planned community in 1955, its population gyrated between 7000 and 25,000 over the next 30 years. Anticipating the closing of the mines, Elliot Lake has strongly promoted its potential as a retirement and recreation centre since 1983, resulting in a stable population of some 12,500. The lake may have been named by a geologist in the late 1800s.

Elliston NF A town (1965) southeast of Bonavista, it was first known as *Bird Island Cove*. It was renamed in 1892 after the Revd William Ellis (1780–1837), a Methodist missionary. He is believed to have held the first Protestant service here in 1812.

Elma ON A township in Perth County, it was named in 1849 after Lady Elma Bruce, daughter of James Bruce, Lord Elgin, governor of the Province of Canada, 1847–54.

Elmira ON An urban centre in Woolwich Township, Waterloo Region, its post office was called *West Woolwich* in 1849. It was renamed in 1867 after Elmira, NY.

Elmsdale NS Located northeast of Halifax, this place's post office was named in 1856. It may have been called after a farm established before 1818 by John Archibald.

Elmvale ON In Simcoe County, south of Midland, this former village was first called *Elm Flats* in the early 1850s, with Elmvale post office opening in 1859. It was united with Flos and Vespra townships in 1994 to form the municipal township of Springwater.

Elmwood ON A community on the boundary of Grey and Bruce counties, north of Hanover, its post office was named in 1864 after a giant elm tree in its main intersection, which had shattered into many pieces when it had been cut down 10 years earlier.

Elnora AB Southeast of Red Deer, this village (1929) requested *Stewartville*, after a student minister, as a post office name in 1908, but it was turned down. A new name was then created from the first names of Elinor Hogg and Nora Edwards, whose husbands, Alex Hogg and William Edwards, were the first postmasters.

Elora ON A village (1858) in Wellington County and on the Grand River, northwest of Guelph, it was founded in 1832 by retired British Capt. William Gilkison (d. 1837), and named after his brother John's ship *Ellora*. The ship had been called after the famous cave temples and sculptures at Ellora, India, northeast of Mumbai (formerly Bombay).

Elphinstone MB Situated south of Riding Mountain National Park and northwest of Neepawa, this community was named in 1879 by Hudson's Bay Company factor John A. Lauder after the 15th Earl of Elphinstone, who owned 13 sections south of the park, and actively promoted the settlement of Manitoba. The post office was opened in 1887.

Elrose SK South of Rosetown, this town (1951) was founded about 1909, and named *Laberge*, after a pioneer settler. When the Canadian Northern Railway arrived in 1913 it was renamed by contractor Louis Chirhart, possibly after a place in the United States, such as Elrosa, a village in central Minnesota. Its post office was also opened that year.

Elsa YT East of Dawson, this incorporated hamlet was founded in 1929 as a residential community for miners at the Calumet silver-lead mine. Production did not get underway until 1937, and its post office was not opened until 1949. The reason for choosing the name Elsa is not available in official records.

Embree NF On the Bay of Exploits, northeast of Lewisporte, this town (1871) was first known as *Salt Pond*. It was renamed in 1931, possibly after the Revd Jerimiah Embree (1844–90), a Methodist missionary, who served on the north coast of the island of Newfoundland from 1871 to 1888.

Embro ON In Oxford County, west of Woodstock, this place was first called *Edinborough* and *Palmerston Depot*, before it was named Embro in 1836 after Embo, in Sutherlandshire,

Scotland, an area where many Embro settlers came from. Embo is among many British Isles names that were misspelled when transferred across the Atlantic.

Embrun ON In the united counties of Prescott and Russell, southeast of Ottawa, this urban centre was named in 1858, by either Fr François Coopman or Fr François-Joseph Michel, after Embrun in France's Hautes-Alpes, southeast of Grenoble.

Emerald Isle NT In the Parry Islands of the Queen Elizabeth Islands, this island (549 km^2) was named in 1853 by Francis (later Sir Francis) McClintock, after a poetic name for Ireland, and because of its proximity to Prince Patrick Island.

Emerald Junction PE In the central part of the island, this place became the point where the Prince Edward Island Railway, originating in Borden, was built east to Charlottetown and west to Summerside. Because of its location on the Prince-Queens county boundary, its post office was called *County Line* from 1875 to 1887. It was renamed in 1887 after the 'Emerald Isle', a poetic name for Ireland.

Emerson MB This town (1887), at the point where the Red River flows north into Manitoba, was founded in 1879 by Thomas Carney and W.H. Fairbanks, and named after American essayist Ralph Waldo Emerson (1803–83). *West Lynne* post office was opened here in 1873, and was changed to Emerson in 1879, with West Lynne remaining a neighbourhood in the town.

Emily ON A township in Victoria County, it was named in 1820 after Lady Emily Charlotte Berkeley, a sister of Charles Lennox, the Duke of Richmond, governor-in-chief of British North America, 1818–9. Her niece, Lady Sarah Maitland, was the wife of Sir Peregrine Maitland, lieutenant-governor of Upper Canada, 1818–28.

Emo ON A place in the municipal township of **Emo**, Rainy River District, west of Fort Frances, it was first called *Emo River* in 1887 and shortened to Emo 10 years later. Alex Luttrell named it after Emo in Ireland's County Laois, southwest of Dublin, and near his birthplace in County Kildare.

Empress AB On the Saskatchewan border and near the junction of the South Saskatchewan and Red Deer rivers, this village (1914) was named in 1913 by the Canadian Pacific Railway after Queen Victoria's title of Empress of India, proclaimed in 1876.

Endeavour SK Established in 1928 by the Canadian Northern Railway, this village (1953) was named after the first passenger aircraft to fly from England to the United States. Its post office had been called *Annette* in 1915, after a Ukrainian settler, but was renamed Endeavour in 1929.

Enderby BC A city (1905) on the Shuswap River and at the north end of the Okanagan Valley, it first had a variety of names, including *Spallumcheen*, *Steamboat Landing*, *Fortune's Landing*, *Lambly's Landing*, and *Belvidere*. On a day in 1887, when the river was in flood, a lady at an afternoon tea party was inspired to recite Jean Inglelow's *The high tide on the coast of Lincolnshire.* The line 'Play uppe "the Brides of Enderby"' so caught the imagination of the others that Enderby was proposed as the name for the post office.

Enfield NS Northeast of Halifax and 4 km southwest of Elmsdale, this place was named in 1863 by Thomas B. Donaldson, a native of Enfield, CT, north of Hartford.

Engelfeld SK Situated east of Humboldt, the post office at this village (1916) was named in 1907 in honour of Abbot Peter Engel of St John's Abbey, Collegeville, MN. He helped the Benedictines establish the parish of the Holy Guardian Angels of Engelfeld.

Englee NF On the north side of Canada Bay and on the east coast of the Northern Peninsula, the name of this town (1948) was possibly derived from *Éguillettes,* a French fishing station located here during the seventeenth and eighteenth centuries.

Englehart ON When the Temiskaming and Northern Ontario Railway was extended from New Liskeard in 1906, its station 40 km to the north was named after Jacob Lewis Englehart, the popular chairman of the railway, 1906–21. It became a town in 1908. In 1880 Englehart had founded the Imperial Oil Company in Petrolia, selling it to Standard Oil (Esso) in 1898.

English Bay BC On the northwest side of Vancouver and at the entrance of Burrard Inlet, this bay (and nearby Spanish Bank) commemorate the friendly meeting on 22 June 1792 of Capt. George Vancouver and Spanish explorers Dionisio Alcalá Galiano and Cayetano Valdés.

English Harbour East NF A municipal community (1974) on the northeastern shore of Fortune Bay, its port was recorded as English Harbour by Capt. James Cook in 1764. By the 1860s it was given its present name to distinguish it from another fishing port on the western side of Fortune Bay called **English Harbour West**, which was settled by English fishermen in the early 1800s.

English River ON Flowing west from Lac Seul to join the Winnipeg River at the Manitoba border, this river was part of the traditional British fur trade route from Lake Winnipeg to Hudson Bay.

Enilda AB West of Lesser Slave Lake, the post office in this community was named in 1913, by reversing the first name of Adline Tompkins, the wife of first postmaster J. Tompkins.

Ennadai Lake NT On the Kazan River and east of Great Slave Lake, the name of this lake (669 km^2) was derived from the Chipewyan word for 'slaughter'. Presumably the Cree disguised themselves as stags in an attempt to ambush the Chipewyan, but the ruse was discovered, and the Cree were slaughtered. Or, as legends often evolve, it may have been the Cree who slaughtered the Chipweyan.

Enniskillen ON A township in Lambton County, it was named in 1834 by Lt-Gov. Sir John Colborne in honour of Gen. Sir Galbraith Cole (1772–1842). A native of Enniskillen in Ireland's County Fermanagh, he was an associate of Colborne in the Peninsular War in Spain.

Ennismore ON A township in Peterborough County, it was named in 1829 after William Hare, Viscount Ennismore, 2nd Earl of Listowel (1802–1856). A native of the townland of Ennismore near Listowel in County Kerry, Ireland, he was a member of Parliament, 1826–30.

Enterprise ON In Lennox and Addington County, north of Napanee, it was first called *Thompson's Corners* after merchant Robert Thompson. When a visitor observed that it was an enterprising place, Thompson had its post office named Enterprise in 1854.

Entwistle AB On the Pembina River, west of Edmonton, this village (1955) was named in 1908 after first postmaster James G. Entwistle. He had formerly been a locomotive engineer with the Grand Trunk Railway in Ontario and had lived in Winnipeg before moving to this area of Alberta about 1904–5.

Eramosa ON A township in Wellington County, it was named in 1822, possibly after *un-ne-mo-sah,* the Mississauga word for 'dog'. One of the first surveyors may have found a dead dog.

Erasmus, Mount AB Near the head of the North Saskatchewan River, this mountain (3265 m) was named in 1859 by Dr James (later Sir James) Hector after Peter Erasmus (1834–1931), a fur trader, interpreter, and noted guide, who assisted Hector in 1858–9 during the Palliser expedition's investigation of what is now Western Canada.

Erickson BC A community in the Kootenay River valley and adjacent to the town of Creston, its post office was named in 1908 after E.G. Erickson (1857–1927), the Canadian Pacific Railway superintendent at Cranbrook, 1904–8.

Erickson MB Located south of Riding Mountain National Park, northwest of Neepawa, the post office in this village (1953) was first called *Avesta* in March 1908, but was renamed in August after postmaster E. Albert Erickson, who owned the site of the village.

Erieau ON A village (1917) at the entrance of Rondeau Harbour in Kent County, southeast of Chatham, it was named in 1886 by joining 'Erie' to the last syllable of 'Rondeau'.

Erie Beach ON A village on the shore of Lake Erie and in Kent County, southeast of Chatham, it was incorporated in 1916, but never had a post office. There was, however, an Erie Beach post office from 1913 to 1959, on the south side of Fort Erie.

Erie, Lake ON The Erihon ('wildcat') clan of the Iroquois lived on the south shore of the lake. As early as 1650, the lake was shown on maps as *Lac du Chat*, but 20 years later René de Bréhant de Galinée identified it on a map as *Lac Érié*.

Eriksdale MB East of Lake Manitoba, and northwest of Winnipeg, this place was founded by the Canadian Northern Railway in 1911 and named after Swedish settler Jonas Erikson. Postal service was also established here that year.

Erin ON A township in Wellington County, it was named in 1820 after the poetic name for Ireland. The village (1879) of **Erin** was first called *MacMillan's Mill*, but was renamed Erin in 1832.

Erindale ON On the Credit River and in the city of Mississauga, it was called *Credit* and *Springfield-on-the-Credit*, before being named in 1900 after Erindale Farm, established by James Magrath in 1827.

Erinsville ON In Lennox and Addington County, north of Napanee, this community was named in 1853 by settler John Murphy, after the poetic title of Ireland, with 'ville' being added, because he believed it had the prospect of becoming a thriving village.

Ernestown ON A township in Lennox and Addington County, it was named in 1784 for Prince Ernest Augustus, Duke of Cumberland and Teviotdale (1771–1851), fifth son of George III. It was locally spelled *Ernesttown* until the 1950s, when the surveyor general of Ontario urged the municipality to abide by the original spelling.

Errington BC On Vancouver Island, northwest of Nanaimo, this place was named in 1891, possibly by early settler Duncan McMillan after a reference to Errington in Sir Walter Scott's poem *Jock of Hazeldean*. The Errington of Scott's poem is west of Newcastle upon Tyne, Northumberland, England.

Errol ON Laid out as a planned town in 1833 in Lambton County, northeast of Sarnia, it was named after Errol in Scotland's Perthshire, east of Perth. A post office was opened in 1837. With no harbour and poor land communications, it became a ghost town by the 1880s. New building has taken place in recent years, because of its desirable location on the Lake Huron shore.

Erskine AB West of Stettler, this community's post office was named in 1905, possibly after Thomas Erskine (1750–1823), a British jurist. It may have been named after Millicent Fanny St Clair-Erskine (1867–1955), wife of the 4th Duke of Sutherland (1851–1913), and eldest daughter of the 4th Duke of Rosslyn. Prior to 1905 its post office was called *Liberal*.

Escuminac QC A municipality (1912) in the regional county of Avignon and on the north side of the estuary of the Restigouche River, it took its name from the **Rivière Escuminac**, which flows into **Baie Escuminac**, and in front of **Pointe d'Escuminac**. The sharp turn of the point replicates Point Escuminac, on the southeast side of Miramichi Bay, NB, whose name was derived from Mik'maq *eskumunaak*, 'lookout place'.

Escuminac, Point NB At the northwestern end of Northumberland Strait, this point's name was derived from Mik'maq *eskumunaak*, 'lookout place'. In 1685 map-maker Pierre Jumeau referred to it as *Pte echkoumenak*. The present spelling was established in 1791 on a map by Frederick Holland. The community of **Escuminac**, west of the point, had a post office from 1858 to 1968.

Eskasoni NS On the north shore of Bras d'Or Lake's East Bay and on the **Eskasoni Indian Reserve**, this place's name was derived from the Mi'kmaq *eskasoonig*, 'green boughs'.

Eskimo Lakes NT Situated southeast of Tuktoyakyuk, this chain of lakes was named in 1928 after the English-language designation of the local native people, the Inuvialuit. The form *Eskimo (Husky) Lakes* was authorized in 1957, but the original form was restored in 1968. The word 'Eskimo' is the English rendering of 'Esquimaux', acquired by the French from the Algonquian-speaking Montagnais or Naskapi on the north shore of the Gulf of St. Lawrence, with the meaning 'they who eat their fish raw'.

Espanola ON A town (1958) in Sudbury District, it was established in 1899 by the Spanish River Pulp and Paper Company. The feminine form of 'Spanish', it was given to the planned community in 1902.

Espoir, Bay d' NF Penetrating the south coast of the island of Newfoundland for some 45 km, this bay may be the same as Jacques Cartier's *hable du Sainct Esperit*, which he entered on 4 June 1536, the holy day of Whitsuntide. The French spelling, and the former French pronunciation, which sounded like 'bay despair', have been maintained.

Esquesing ON Now divided between the towns of Halton Hills and Milton in Halton Region, this township was named in 1819 after the Mississauga word *eshkwessing*, 'the last in a row or a range', presumably referring to present Sixteen Mile Creek.

Esquimalt BC A district municipality on the west side of Victoria, it was incorporated in 1912 as a township. Its post office had been named in 1864 after **Esquimalt Harbour**. The harbour's name is derived from Straits Salish 'place of gradually shoaling', in reference to the flats at the mouth of Mill Stream at the head of the harbour. The bay was named *Puerto de Cordova* in 1790 by Spanish Sub-Lt Manuel Quimper. About 1842 the Hudson's Bay Company restored the Aboriginal designation, and relocated Cordova Bay north of Victoria.

Essa ON A township in Simcoe County, it was named in 1822, but its origin is obscure. One tradition claims Shawnee chief Tecumseh had a wife called Essa, but it is believed he had only one wife, Mamate, who died after her only child was born. Other traditions point to the name of either an Aboriginal chief or to the daughter of a chief. Ojibwa chief Wasson (*c.* 1730–*c.* 1790) was the best known Aboriginal in the period 1750–1820 with a name similar to Essa. He signed a peace treaty with the British at Detroit in 1764–5.

Essex ON The county was named in 1792 for Essex, in England. *Essex Centre* became a railway station in 1872, but when it was incorporated as a town in 1890, the name was shortened to **Essex**.

Estcourt QC A community in the town of Pohénégamook, south of Rivière-du-Loup, and on the border of the most northerly point of Maine, it was called after the township of Estcourt, proclaimed in 1842. The township was named after British boundary commissioner Lt-Col James B.B. Estcourt (1802–55).

Esterhazy SK Southeast of Melville, this town (1957) was named after Count Paul Esterhazy, who helped Hungarian coal miners in 1886 to move from Pennsylvania to this area north of the Qu'Appelle River. Hungarian migrants came directly to the settlement from Hungary two years later. *Esterhaz* post office was opened in 1886, with both *Kolin* and Esterhazy replacing it in 1903.

Estevan SK A city (1957), just north of the North Dakota border, it was founded in 1892, when the Soo Line was extended by the Canadian Pacific Railway from North Portal, on the border, to Moose Jaw. Its post office was composed from the name of Esther Van Horne, a daughter of Sir William Van Horne (1843–1915), the second president of the CPR, 1888–99. Claims that it was composed from the names of George Stephen, Baron Mount Stephen, the first president of the CPR, 1880–8, and of Sir William Van Horne, have no validity.

Estevan Point BC On the south side of Nootka Sound and on the west coast of Vancouver Island, this point was the first feature to be named by a European explorer on Canada's west coast. It was named in 1774 by Spanish Lt-Cdr Juan Perez, after his second lieutenant, Estevan Jose Martinez. Capt. James Cook named it *Breakers Point* four years later, but the earlier name was restored in 1849 by the British Admiralty.

Eston SK The post office in this town (1928), southeast of Kindersley, was named in 1912 after the hometown of A.J. Harris in West Yorkshire, England, east of Middlesbrough, and near the North Sea coast.

Estrie QC An administrative region of southeastern Québec since 1981, the name was devised in 1946 by Sherbrooke historian Maurice O'Bready as a French equivalent of 'Eastern Townships', which had been usually rendered as *Cantons de l'Est*. The region, embracing 10,698 km^2, is centred on Sherbrooke, with Stanstead on the south, Asbestos on the north, Magog on the west, and Thetford Mines and Lac-Mégantic on the east.

Ethelbert MB A village (1905), northwest of Dauphin, it was named in 1898 by K.J. Mackenzie, construction supervisor of the Canadian Northern Railway, after his niece, Ethel Bertha, a daughter of the company's president, Sir William Mackenzie. Its post office was opened the next year.

Etobicoke ON The city of Etobicoke was laid out as a township in 1792 and was named after **Etobicoke Creek**. The creek's name is a variant of the Mississauga designation, *Wah-do-be-kaung*, 'place where the black alders grow'. The municipal township was united in 1967 with the towns of Mimico and New Toronto and the village of Long Branch, to become the borough of Etobicoke, which was elevated to city status in 1983.

Euphemia ON A township in Lambton County, it was named in 1848 by Malcolm Cameron, the member for Kent in the Legislative Assembly of the province of Canada, after his mother, Euphemia McGregor Cameron.

Euphrasia ON A township in Grey County, it was named in 1823 after the Greek word meaning 'delight' and a genus of flowers known as eyebright.

Eureka NS South of Stellarton, this community's post office was named in 1887 after the Eureka Milling Company, established here on the East River of Pictou five years earlier.

Eureka NT Located on the Fosheim Peninsula of Ellesmere Island, this site of a weather station was established in 1948, near Eureka Sound, which separates Ellesmere and Axel Heiberg islands. The sound was named *Heureka Sund* in 1901 by Ivar Fosheim, a member of the Sverdrup expedition, who discovered the 280-km passage.

Eutsuk Lake BC Located in Tweedsmuir Provincial Park, southeast of Kitimat, this name was adapted in 1910 by Frank Swannell, a provincial surveyor, from the Carrier word *teootsabungut*, which may have the meaning of 'far from Ootsa Lake'. Fr Gabriel Morice claimed in 1904 that a Carrier companion had named it after him, but his proposal had not been submitted to the provincial names authority for consideration.

Evansburg AB A village (1953) on the Pembina River, west of Edmonton, its post office was named in 1914 after Henry Marshall Erskine Evans (1876–1973), who prospected for coal on the river in 1904. Subsequently he settled in Edmonton, where he became president of the board of trade and mayor of the city.

Evans, Mount AB Near the head of the North Saskatchewan River, this mountain (3210 m) was named in 1935 after Capt. Sir Edward Ratcliffe Garth Russell Evans (1880–1957), who had performed heroic exploits while commanding HMS *Broke* during the First World War.

Everett ON In Simcoe County, northwest of Alliston, this place's post office was named in 1878. Local stories state it was named by Thomas Gordon for his father's birthplace in England, or called after the birthplace of Henry Baycroft in England, or given for a Mr Fisher's farm. No place is identified as Everett on the largest scale of British ordnance survey maps.

Exeter ON A town (1951) in Huron County, northwest of London, it was first called *Francistown* and *Carling's Corners*, before merchant James Pickard may have named it in 1851 after the city of Exeter in Devon, England.

Exploits River NF Rising in the central part of the island of Newfoundland, this river (246 km) flows northeast into the **Bay of Exploits**, on the north coast of the island. Capt. James Cook recorded the name of the river on his 1774 chart. The origin and meaning of the name are not provided in various standard references.

Exshaw AB A community east of Canmore, it was named in 1905 by Sir Sandford Fleming after his son-in-law, William Exshaw (1866–1927). Both of them were directors of the Western Canada Cement and Coal Company. Exshaw managed the company's cement factory here.

Eyebrow SK A village (1909) northwest of Moose Jaw, it was named in 1907 when the Canadian Pacific Railway was built north to Outlook. It received its name from the outline of a hill above Eyebrow Lake, on the south side of the village. Its post office was first called *Eyebrow Station* in 1908, but 'Station' was dropped later that year.

Eyebrow Peak BC In the Purcell Mountains, the name of this peak (3353 m) was proposed in 1910 by surveyor Arthur O. Wheeler after the distinctive character of the peak that he observed. Subsequent investigation by geologist Peter Robinson revealed that Wheeler was really looking at Mount Farnham, 14 km to the east.

Faber Lake NT Situated between Great Bear and Great Slave lakes, this lake was named in 1864 by Fr Émile Petitot after Fr Frederick William Faber (1814–63), an English Catholic priest and orator.

Fair Haven NF On the east side of Placentia Bay and on the southwest side of the Isthmus of Avalon, this place was called *Famish Gut* and *Famish Cove* until June 1940, when the Newfoundland Nomenclature Board authorized the change to this more positive designation. In 1971 Edgar R. Seary suggested *Famishgut*, and its adjacent settlement of *Pinchgut*, may have been named because hydrographer Michael Lane's survey crew had become short of rations in the 1770s.

Fairmont Hot Springs BC In the Columbia River valley, southeast of Invermere, this resort community was named *Fairmont Springs* in 1888, after the Fairmont Hotel, established here by Mr and Mrs Brewer. The hotel closed in 1909, but reopened under its present name two years later.

Fairvale NB A village (1966) 15 km northeast of Saint John, its first post office, called *Fairleigh*, opened in 1885 and closed in 1890. *Fair Vale* post office served the area of the village from 1908 to 1946, and *Fairvale Station* post office was open from 1925 to 1970.

Fairview AB On the north side of the Peace River, north of Grande Prairie, this town (1949) was founded by the Edmonton, Dunvegan and British Columbia Railway in 1928, and named by E.J. Martin after the local municipal district. In 1910 H.L. Propst had established a farm that he called Fairview. Four years later the municipal district of **Fairview** was created.

Fairweather Mountain BC On the British Columbia-Alaska border, southwest of Whitehorse, YT, this mountain (4663 m) was named *Mount Fair Weather* in 1778 by Capt. James Cook. It is the highest point in the province.

Falconbridge ON In the town of Nickel Centre, northeast of Sudbury, the former township was named about 1885 after Sir William Glenholme Falconbridge (1846–1920), chief justice of the Supreme Court of Ontario, 1900–20. Falconbridge post office was named in 1930.

Falher AB West of Lesser Slave Lake, this town (1955) was named in 1915 by the Edmonton, Dunvegan and British Columbia Railway, after Oblate missionary Fr Constant Falher. A native of Bretagne, France, he had arrived at nearby Grouard in 1889, with the mission of Saint-Jean-Baptiste-de-Falher being established 5 km from the present town's site. Falher post office was opened in 1923.

Falkland BC In the Salmon River valley, southwest of Salmon Arm, this community was established in 1893 by Col Falkland G.E. Warren, who had retired from the Royal Horse Artillery after serving in India. Its post office opened in 1898. After closing in 1912, it succeeded the post office called *Slahaltkan* (1906) the next year.

Fallowfield ON In the city of Nepean, this community was named as early as 1861, when Bishop Bruno Guigues inspected the ongoing building of the present handsome stone St Patrick's Church. The name was likely inspired by a nearby field left fallow during a growing season. Its post office was opened in 1872.

Fall River NS Located at falls between Miller Lake and Lake Thomas, north of Dartmouth, this place became a station on the Intercolonial Railway in 1857. Fall River post office was opened in 1878, but it was renamed Goffs the following year, after postmaster William Goff. The place called Goffs developed 12 km to the northeast.

Falmouth NS On the west side of the Avon River, this place was named after *Falmouth Township*, established in 1761. The township was named after Adm. Edward Boscawen, 2nd Viscount Falmouth (1711–61), who led the successful British naval attack on Louisbourg in 1758.

False Creek BC This small inlet in the city of Vancouver was named in the late 1850s by Capt. George Richards, who expected to find a stream flowing into it at the head of the inlet. However, he only found mud flats there.

Farnham QC In the regional county of Brome-Missisquoi, southeast of Granby, this place was first created as the village of *West Farnham (Farnham-Ouest)* in 1862 and became the town of Farnham in 1876. It took its name from the township of Farnham, proclaimed in 1798, and named after the town of Farnham, in Surrey, England, southeast of London.

Farnham, Mount BC In the Purcell Mountains, west of Invermere, this mountain (3468 m) was named in 1911 by climber Edward Harnden after mine developer Paulding Farnham. A native of New York State, Farnham spent a considerable fortune prospecting and developing mines in southeastern British Columbia.

Faro YT A town in the Pelly River valley, northeast of Whitehorse, it was founded in 1969 to provide homes for the miners of the Anvil lead, zinc, and silver mine. The name was probably first given by Dr Aro Aho to the first claim staked there, choosing a word of a gambling card game which was popular in mining camps.

Farquhar, Mount AB, BC On the continental divide, north of Crowsnest Pass, this mountain (2905 m) was named in 1919 after Maj. Francis Douglas Farquhar (1874–1915), who was the commander of the Princess Patricia's Canadian Light Infantry in Europe, and was killed during the First World War.

Farrellton QC A community in the municipality of La Pêche, 13 km north of Wakefield, it was named *Farrelltown* in 1872 after first postmaster Patrick Farrell. The spelling was amended to Farrellton in 1887.

Fascination Mountain BC In the Coast Mountains, southeast of Mount Waddington, this mountain (3048 m) was named about 1927 by climber W.A.D. (Don) Munday because he found it a challenging and shapely mountain.

Fassett QC Situated in the regional county of Papineau, east of Montebello, the post office of this municipality (1951) was named in 1906 after S. Jonathan Fassett, who established a lumber company here the previous year. It had been first called *Petite-Nation* in 1845, after the Rivière de la Petite-Nation, and was renamed the parish municipality of *Notre-Dame-de-Bon-Secours* 10 years later.

Fathom Five National Marine Park ON Located in Georgian Bay and on the northeast side of the Bruce Peninsula, this park (113 km^2) was created in 1987 to protect valuable underwater sites and flowerpot formations such as Flowerpot Island. Its depth is five fathoms (about 9 m).

Fatima QC Located in the regional county of Les Îles-de-la-Madeleine, this municipality was named in 1959, 10 years after its post office was called after the place in Portugal where three young children saw a vision of the Virgin Mary in 1917.

Fauquier BC On the east shore of Lower Arrow Lake, this community's post office was named in 1913 after F.G. Fauquier, an early rancher and fruit grower here. He had also worked in Nakusp in the 1890s as a customs officer, mining recorder, government agent, and police officer.

Fauquier ON In the municipal township of **Fauquier-Strickland**, Cochrane District, east of Kapuskasing, this place was named in 1910 after G.E. Fauquier, a National Transcontinental Railway contractor. *Fauquier Township*, west of the community of Fauquier, was incorporated in 1921, and became the municipal township of Moonbeam in 1982.

Faust AB On the south shore of Lesser Slave Lake, this place was named in 1914 after Edmonton, Dunvegan and British Columbia locomotive engineer E.T. Faust. Its post office was opened in 1920.

Fay, Mount AB, BC On the continental divide and south of Lake Louise, this mountain (3234 m) was named in 1902 after American alpinist and editor Charles Ernest Fay (1846–1931), who had climbed several mountains in the Canadian Rockies.

Fenelon ON A township in Victoria County, it was named in 1823 after Fr François de Salignac de la Mothe Fénelon (1641–79). In 1668 he founded a mission on the Bay of Quinte, and spent the next year at Frenchman's Bay, in the present-day town of Pickering. The village (1874) of **Fenelon Falls** was named in 1838 after the township.

Fenwick NS South of Amherst, this place's post office was named in 1866 after Sir William Fenwick Williams (1800–83), a native of the province. He was the hero of the Russian siege of Kars, in present-day Turkey, in 1855, and lieutenant-governor of Nova Scotia, 1865–7.

Fenwick ON In the town of Pelham, Niagara Region, this place was named in 1853, possibly by Pelham Township reeve Dr John Fraser, after his birthplace, northeast of Kilmarnock, Ayrshire, Scotland.

Fergus ON A town (1952) in Wellington County, it was named in 1834 after Adam Fergusson (1783–1862), a native of Perthshire, Scotland, and a member of the Legislative Council of Upper Canada, 1839–62.

Ferguson Lake NT Southwest of Baker Lake, this lake was named in 1894 by Joseph Burr Tyrrell after Robert Munro Ferguson, an aide-de-camp to Gov. Gen. Lord Aberdeen. The lake is drained by **Ferguson River** into Hudson Bay, northeast of Arviat. At his own expense, Ferguson accompanied Tyrrell on his geological reconnaissance through northern Manitoba and Keewatin District.

Ferme-Neuve QC A village north of Mont-Laurier, it was named in 1917 after a model farm that eight Montréal journalists established in 1902. The parish municipality of Ferme-Neuve was named in 1930. From 1902 to 1930 it was called the united township municipality of *Würtele-Moreau-et-Gravel*. The three separate townships were named in 1899 after legislator Jonathan S.C. Würtele, and priests Louis-Zépherin Moreau, and Elphège Gravel.

Fermeuse NF On the east coast of the Avalon Peninsula, the name of the municipal community (1967) was taken from **Fermeuse Harbour**. A name of Portuguese origin, the latter was recorded on a 1519 map as *R Fermoso*, meaning 'beautiful river'.

Fermont QC Located some 300 km north of Sept-Îles, this town was founded in 1972 by the mining company Québec Cartier on the Newfoundland border, near Labrador City. Its name means 'mountain of iron'. Iron ore, mined at nearby Mont Wright, is shipped south by rail to the seaport of Port-Cartier.

Fernie BC A city (1904) in the Elk River valley, east of Cranbrook, its post office was named in 1898 after William Fernie (1837–1921), who discovered coal in 1887 on a tributary of Michel Creek. He had been gold commissioner in the Kootenay district, 1873–82, and had become a director of the Crows Nest Pass Coal Company. After achieving considerable wealth in the coal fields, he retired in 1906 to Oak Bay, beside Victoria, and became a prominent philanthropist.

Ferryland NF A municipal community (1971) on the east coast of the Avalon Peninsula, it was named after the adjacent **Ferryland Head**. The head was named in Portuguese in the early 1500s as *Farilham*, after *farelhão*, 'steep rock', in reference to the high cliffs on both the head and adjacent Bois Island, between which is the narrow passage leading to Ferryland's harbour. By the mid-1500s, the French had called it *Forillon*, and early in the following century it became Ferryland in English.

Feuilles, Rivière aux QC A major river in northwestern Québec, it flows 480 km northeast from Lac Minto to Ungava Bay. Known by the Inuit as *Kuugaaluk*, 'grand river', it was well known in the late 1800s and early 1900s as Leaf River. In 1914 it was rendered as *Rivière des Feuilles*, and changed 11 years later to the present form. At the northern limit of trees, the name is probably derived from dwarf willow and birch.

Field BC The first Canadian Pacific Railway station west of the continental divide, and in Yoho National Park, it was named in 1884 after Cyrus West Field (1819–92), the founder of the American Telegraph Company. The ATC won the race in 1866 to connect North America with Europe, by laying the first successful transatlantic telegraph cable. **Mount Field** (2643 m) overlooks the station.

Field ON A township in Nipissing District, it was named in the 1880s after Corelli Collard Field, Liberal member for Northumberland West in the Ontario Legislature, 1886–90. The post office in the community of **Field**, in the township, was named in 1899.

Fillmore SK Northeast of Weyburn, this village (1905) was named for Charles Wesley Fillmore, president of the Winnipeg Oil Company. He owned the site of the village at the time of the survey of the route of the Canadian Pacific Railway, which was built through its site in 1900. The post office was named the following year.

Finch ON A township in the united counties of Stormont, Dundas and Glengarry, it was named in 1798, either after George Finch-Hatton, a son-in-law of the 7th Viscount Stormont, or after Lady Elizabeth Finch-Hatton, the viscount's aunt by marriage. The village (1906) of **Finch** was known as *South Finch* from 1856 to 1899, when it was changed to Finch.

Findlay, Mount BC In the Purcell Mountains, northeast of Kaslo, this mountain (3162 m) was named in 1954 after **Findlay Creek**. The latter was named in 1915 after George Finlay, who prospected the area in 1863. He was a son of John (or Jacques) Raphael 'Jaco' Finlay, a partner in the North West Company, who was in charge of the post on Lake Athabasca in the early 1800s, and ran a trail through Howse Pass in 1806.

Fingal ON A place in Elgin County, southwest of St. Thomas, it was named in 1830 by Col Thomas Talbot after Fingall in County Dublin, Ireland, north of the River Liffey.

Finlay River BC Rising in the Omineca Mountains in the north central part of the province, this river was once one of the two major tributaries (the Parsnip is the other) of the Peace River, before the construction of the W.A.C. Bennett Dam in 1968–9 and the subsequent creation of Williston Lake. The river was named after John Finlay of the North West Company, who had travelled up it in 1797.

Firth River YT Rising in the Davidson Mountains in Alaska, this river flows northeast through the British Mountains into the Beaufort Sea, opposite Herschel Island. It was named *Mountain Indian River* in 1826 by John (later Sir John) Franklin, after some interior Aboriginals, who used to come to the coast to trade with the Inuit. It was renamed in 1890 by John H. Turner after John Firth, Hudson's Bay Company agent at Rampart House (on the Porcupine River where it flows west across the Alaska-Yukon border), who joined Turner in his trip along the coast.

Fisher Branch MB Located on the **Fisher River**, north of Winnipeg, this place's post office was opened in 1911. The French called the river *Rivière aux Pécans*, 'river of the fisher', after a fur-bearing animal of the marten family, *Martes pennanti*.

Fisher, Mount BC In the Rocky Mountains, northeast of Cranbrook, this mountain (2846 m) was named about 1900 after Jack Fisher, who had found gold on nearby Wild Horse Creek in 1863. The mining camp that developed here was called Fisherville.

Fisher Ridge NB South of the Nepisiguit River, this ridge was named in 1964 by Arthur F. Wightman, then the provincial names authority, after Charles Fisher (1808–80), premier of

New Brunswick, 1854–61 and a Father of Confederation.

Fitzroy ON A ward in the township of West Carleton in Ottawa-Carleton Region since 1974, it was named as a township in 1823 after Sir Charles Augustus Fitzroy (1796–1858), a son-in-law of the Duke of Richmond, governor-in-chief of British North America, 1818–9. The community of **Fitzroy Harbour** was named in 1832.

Fitzwilliam, Mount BC On the south side of Yellowhead Pass, west of Jasper, this mountain (3385 m) was named in 1863 by Dr Walter Butler Cheadle after his companion, William Wentworth Fitzwilliam, Viscount Milton (1839–77).

Five Houses PE Located east of St. Peters Bay, this place was probably named after five French dwellings that had been destroyed. In 1765 Surveyor General Samuel Holland wrote 'Ruined Village' on his map, and 'Ruined Village of 5 Houses' in his notes.

Flaherty Island NT The largest of the Belcher Islands, on the southeast side of Hudson Bay, it was named in 1919 by geologist G.A. Young after Robert Joseph Flaherty, who investigated the minerals of the islands in 1914. In 1922 Flaherty produced the documentary film *Nanook of the North.*

Flamborough ON A town (1985) in Hamilton-Wentworth Region, it had been formed as a municipal township in 1974, through the amalgamation of the village of Waterdown, the municipal township of Beverley, and parts of the municipal townships of East Flamborough and West Flamborough. East Flamborough and West Flamborough were named in 1792, after Flamborough Head and the town of Flamborough, East Yorkshire, England.

Flat Rock NF North of St. John's, this town (1975) was named after 'the thick red sandstone, [which] dips at a slight angle towards the sea, forming a long, smooth, sloping pavement', according to Joseph B. Jukes in 1840. References to the place were made on maps as early as 1630.

Fleming SK Just west of the Manitoba border, this town (1907) is located on the main Canadian Pacific Railway line, between Winnipeg and Regina. Its station was named in 1882, after Sir Sandford Fleming (1827–1915), Canada's most distinguished railway engineer in the nineteenth century, and the initiator of the system of time zones. The post office was opened in 1884.

Fleming Peak BC A peak (3164 m) of Mount Rogers in the Selkirk Mountains, northeast of Revelstoke, it was named in 1901 by surveyor Arthur O. Wheeler after Sir Sandford Fleming, first engineer-in-chief of the Canadian Pacific Railway. *See also* Sir Sandford, Mount.

Flesherton ON A village (1912) in Grey County, southeast of Owen Sound, it was founded in 1853 by miller William Kingston Flesher (1827–1907). The post office was called *Artemesia*, after the township, from 1851 to 1867, when it was changed to Flesherton.

Fletchers Lake NS The name of this place, located north of Dartmouth, has its origins in pioneer settler William Fletcher, and his son Robert. William bought land at the head of the lake in the 1780s, and Robert received a grant of 101 ha (250 a) in 1812.

Fleur de Lys NF On the western side of Notre Dame Bay, and north of Baie-Verte, this town (1967) was a French fishing harbour from the 1500s to the 1700s, with its name being noted in 1640. It was given by French fishermen after a striking rock formation, rising to 249 m, said to resemble the national symbol of France.

Fleurimont QC On the north side of Sherbrooke, it was known as the municipality of *Ascot-Nord* from 1937 to 1971, when it became the town of Fleurimont. It took its name from Nicolas-Joseph de Noyelles de Fleurimont (1695–1761), who arrived in Canada at the beginning of the 1700s.

Flin Flon MB A city (1970) on the Saskatchewan border, its fascinating name has its roots in Joyce Emerson Preston Muddock's dime novel *The Sunless City,* in which Professor Josiah Flintabbatey Flonatin found a strange world under the Rocky Mountains paved with gold and ruled by strong women. The professor escaped through a hole of an inactive volcano. In the winter of 1915 several prospectors led by Tom Creighton were searching for gold, when Creighton presumably fell through the ice. While drying off beside a fire, Creighton observed that the heat revealed gold at the bottom of a hole, prompting him to exclaim: 'That must be the hole where old Flin Flon came up and shook his whiskers, so what do you say we call the discovery Flin Flon?' The post office was opened as *Flinfon* in 1929, and was changed to Flin Flon in 1937.

Flinton ON In Lennox and Addington County, northeast of Tweed, it was named in 1858 after Belleville merchant and lumberman Billa Flint (1805–94). He was a member of the Legislative Assembly of the province of Canada, 1847–51, of the Legislative Council, 1863–7, and of the Senate until his death.

Florence NS Northwest of Sydney Mines, this place's post office was called *Stubbert* in October 1904 and *Cantley* in December 1904. The next year it was named after Florence McKenzie, the wife of Daniel D. McKenzie, Liberal member for Cape Breton North-Victoria in the House of Commons, 1904–6 and 1908–23, and solicitor general of Canada, 1921–3.

Florence ON In Euphemia Township, Lambton County, it was named *Zone Mills* in 1840, when it was in Zone Township. Postmaster George Pigeon Kirby wanted to call it *Victoria* in 1856, after the Queen, but he was persuaded by postal authorities to rename it after the city in Italy. It was likely merely a coincidence that Florence Nightingale became famous then as the heroine of the Crimean War.

Florenceville NB A village (1966) north of Woodstock, it was named in 1853 by Lt-Gov. Lemuel A. Wilmot after Florence Nightingale, the heroine of the Crimea. Settled in 1832, it was first called *Buttermilk Creek.*

Flowers Cove NF A town (1961) on the northwestern coast of the island of Newfoundland, it includes the nearby localities of Nameless Cove and Mistaken Cove. It was earlier called *French Island Harbour,* as the French continued to fish here well into the late 1800s. The cove was named in 1764 by Capt. James Cook, who named it *Flour Cove* after the foam created by breakers. The word 'flour' is an old English word for froth formed by breaking waves on a shoreline.

Foam Lake SK This town (1924) northwest of Yorkton acquired its name from the nearby shallow lake, identified by early settler Joshua Milligan, because its shore always had a ring of foam around it. Its post office was opened in 1900.

Foch, Mount AB Southwest of Calgary, this mountain (3180 m) was named in 1918 after Marshal Ferdinand Foch (1851–1929), supreme commander of the Allied forces at the end of the First World War.

Fogo NF On **Fogo Island**, this town (1948) derived its name from the Portuguese name of the island, *y do Fogo,* 'fire island', possibly in reference to forest fires, or to fogs that appeared to be smoke. It may also have been named after the island of Fogo in the Cape Verde islands, west of Dakar, Senegal.

Foleyet ON A station on the main line of the Canadian National Railways, and in Sudbury District, west of Timmins, it was named accidentally in 1912, when Canadian Northern Railway president Sir Donald Mann insisted on naming a station after contractor Timothy Foley. Having been told by superintendent A.J. Hills there already was a station called Foley, Mann blurted out 'I'll name that place Foley yet!' 'Foleyet will be fine, Sir,' replied Hills, and the post office was opened in 1916.

Foley Island NT In Foxe Basin and west of Baffin Island, this island was named in 1949 after John Hilliard Foley, the navigator of a special Royal Canadian Air Force flight, which had investigated the recently discovered islands in the basin the previous year. Foley was killed in 1949 in a flying accident, near Maxville, ON.

Fond-du-Lac SK West of the mouth of the **Fond du Lac River** and on the north shore of the east end of Lake Athabasca, this place was the site of a Hudson's Bay Company post built in 1851. There had been both HBC and North West Company posts nearby in the earlier part of the nineteenth century. A post office was opened here in 1934. The French phrase *fond du Lac* means 'far end of the lake'.

Fonthill ON In the town of Pelham, Niagara Region, it was first known as *Osborne's* and *Temperanceville*. When *Pelham* post office was located here in 1856, it was called Fonthill, either after Fonthill Abbey in Wiltshire, England, or after a public reservoir and fountain on the road leading from this place to Canborough, and beside the Short Hills.

Forbes, Mount AB Northwest of Lake Louise, this mountain (3612 m) was named in 1859 by Dr James (later Sir James) Hector after Edward Forbes (1815–54), professor of natural history at the University of Edinburgh, who was one of Hector's teachers.

Fordwich ON In Howick Township, Huron County, north of Listowel, this community was first called *Howick* and *Lisadel*. It was named Fordwich in 1873 after a village in Kent, England, northeast of Canterbury.

Foremost AB Located southwest of Medicine Hat, this village (1950) was established in 1913 by the Canadian Pacific Railway. It was the 'foremost' point reached that year, and the railway was not extended further east until two years later. Foremost post office was opened in 1914, replacing *Webber*, opened four years earlier, 3 km to the south.

Forest ON A town (1889) in Lambton County, northeast of Sarnia, it was named in 1859, when the Grand Trunk Railway was built through a heavily wooded area. Forest post office was opened in 1862. To qualify as a town with 2000 people, the village delayed a Grand Trunk train for an hour in 1888 and registered the passengers and crew.

Forestburg AB Southeast of Camrose, this village (1919) was established in 1915 by the Canadian Northern Railway. It was likely named by a pioneer settler after Forestburg, SD, northwest of Sioux Falls. A claim that it may have been named after Forestburg in Ontario is doubtful, as this name has never been used for a place in that province.

Foresters Falls ON In Renfrew County, north of Renfrew, it was named in 1854 after first postmaster Oliver Forester.

Forest Grove BC In the Cariboo, northeast of 100 Mile House, this place's post office was named in 1917 by the sister of storekeeper E.C. Phillips after a grove of trees here.

Forest Hill ON Part of the city of Toronto since 1967, the former village was named in 1923 after John Wickson's residence, built in the mid-1860s. It is noted for its handsome residences, sedate ambiance, and Upper Canada College.

Forestville QC On the north shore of the St. Lawrence, midway between Tadoussac and Baie-Comeau, it was first called *Sault-au-Cochon*, after the river that flows into the St. Lawrence. From 1870 to 1890 Price Brothers operated a sawmill here, and the place was called *Forrest-Ville*, after the company's superintendent, Grant William Forrest. The town of Forestville was incorporated in 1944, seven years after the Anglo Pulp Company opened a mill here.

Forillon National Park QC Created in 1970 on the eastern end of the **Presqu'île de Forillon**, it became the first national park (240 km^2) in Québec. In 1626 Samuel de Champlain noted that there was a *forillon* (rocky point extending into the sea), which looked like a mill wheel with a rock shaped like a hat on top.

Formosa ON In Bruce County, southwest of Walkerton, it was named about 1856 by Fr Gaspar Matoga, of Guelph, after the Asian island now known as Taiwan, but called Formosa ('beautiful') by the Portuguese in 1590. Its post office was opened in 1861.

Forres, Mount BC West of Williston Lake and on the north side of Ingenika River, this mountain (2072 m) was named in 1933 by surveyor James R. Mackenzie after his birthplace in Morayshire, Scotland, northeast of Inverness.

Fort Albany ON Near the mouth of Albany River, northwest of Moosonee, it was established in 1679 by the Hudson's Bay Company, and named about 1683 after the Duke of York and Albany, who became King James II in 1685.

Fort Alexander MB Located 5 km above the mouth of the Winnipeg River, this was the site of the French Fort Maurepas, established in 1733–4 on the north side of the river, but abandoned in 1745. In 1792 the North West Company built Fort Bas de la Rivière (also called Winnipeg House) on the south side, which the Hudson's Bay Company called Fort Alexander in 1821. The post office was called *Pine Falls* in 1879, and was renamed Fort Alexander two years later.

Fort Assiniboine AB On the north bank of the Athabasca River, northwest of Edmonton, this community was the site of a Hudson's Bay Company post in 1823. Its post office was opened in 1913, and it was an incorporated village from 1958 to 1991.

Fort Augustus PE Northeast of Charlottetown, this place was named about 1853 by Fr John Macdonald, probably after Fort Augustus, Inverness-shire, Scotland.

Fort Beauséjour NB Located on the boundary with Nova Scotia, it was declared a national historic site in 1926. Built by the French in the early 1750s, it was captured by Col Robert Monckton in 1755 and renamed Fort Cumberland. The fort had been named after a nearby point, part of the property of Laurent Chatillon, Sieur de Beauséjour.

Fort Chipewyan AB On the north side of the west end of Lake Athasbasca and at the point where the lake is drained by the Slave River, this place was the site of a North West Company post established in 1804. The post had been created on the south shore of the lake by explorer Alexander Mackenzie and fur trader Roderick Mackenzie, naming it after the Chipewyan nation, whose name is Cree for 'pointed skins'. Its post office was opened in 1912.

Fort-Coulonge QC Situated at the mouth of the **Rivière Coulonge**, where it flows into **Lac Coulonge**, a widening of the Ottawa River, and in the regional county of Pontiac, the village's post office took its name in 1853 from the trading post established here in 1784 by the North West Company. The river and the post, and subsequently the lake, were named after Nicolas d'Ailleboust de Mathet (1663–1709), who was also called Sieur de Coulonge, after a title of his grandfather, Nicolas d'Ailleboust de Coulonge-la-Madeleine.

Forteau NF Located on **Forteau Bay**, and on the Labrador side of the Strait of Belle Isle, this municipal community (1971) was named after the bay. The bay's name is said to derive from either the strong (*fort*) southerly winds producing heavy swells, or the strong tide rips. In 1535 Jacques Cartier called the bay *H de la Balaine*, meaning 'whale harbour'.

Fort Erie ON The fort was established in 1764, and the post office was called Fort Erie in 1820. However, the town was only named in 1931, after the amalgamation of the village of *Waterloo*, named in 1857, and the village of *Bridgeburg*, named in 1895. In 1970 Fort Erie annexed Bertie Township and the village of Crystal Beach.

Fort Fraser BC On the Nechako River, just east of Fraser Lake, and west of Prince George, this place has its roots in the establishment of a North West Company trading post here in 1806 by Simon Fraser. Its post office was called *Fraser Lake* in 1907, but was renamed Fort Fraser in 1913.

Fort Frances ON A town (1903) in Rainy River District, it was named in 1830 during a visit by Sir George Simpson and his wife Lady Frances Simpson, when the fort was christened in her honour. Lady Frances (*c.* 1812–53) was also the first cousin of Sir George, governor-in-chief of the Hudson's Bay Company, 1820–60. The post office was opened in 1876.

Fort Good Hope NT A settlement on the east bank of the Mackenzie River, just below the Arctic Circle, this place was established in 1804 as a trading post on the west bank by the North West Company. After a number of moves, the fort was re-established at the present site by the Hudson's Bay Company in 1836. It was approved as *Good Hope* in 1915 by the Geographic Board, but the older name was restored in 1939.

Fortierville QC A village in the regional county of Bécancour, northeast of the city of Trois-Rivières, it was named in 1913 after a pioneer family.

Fort Lawrence NS West of Amherst and on the New Brunswick border, this place had postal service as early as 1825. The fort was built in 1750 by Lt-Col Charles Lawrence (1709–60) to protect the province from invasion. Appointed governor of Nova Scotia in 1753, he directed the expulsion of the Acadians from the province two years later.

Fort Liard NT This hamlet (1987) on the Liard River, where the Petitot River flows in from the east, was the site of the Hudson's Bay Company post of Fort Rivière-au-Liard, established about 1800. It was familiarly called Fort Liard. The name of the place was approved by the Geographic Board as *Liard* in 1921, but Fort Liard was endorsed in 1944.

Fort MacKay AB On the west bank of the Athabasca River, north of Fort McMurray, this place was named about 1870 after Dr William Morrison MacKay (1836–1917), a Hudson's Bay Company surgeon and fur trader. He settled in Edmonton in 1898, and was elected first president of the Northern Alberta Academy of Medicine in 1902.

Fort Macleod AB A town (1892) on the Oldman River, west of Lethbridge, it originated with the establishment in 1874 of the first North-West Mounted Police fort in the area of present-day Alberta. It was named after Lt-Col James Farquharson Macleod (1836–94), who became commissioner of the NWMP from 1877 to 1880, and was made a judge of the supreme court of the North-West Territories in 1887.

Fort McMurray AB At the junction of the Athabasca and Clearwater rivers, this urban service centre was established by the Hudson's Bay Company in 1870 at the site of the North West Company's Fort of the Forks (1788). It was named by factor H.J. Moberly after William McMurray (1822–77), chief factor at Île-à-la Crosse, located southeast on the Churchill River, in present-day Saskatchewan. Incorporated as a city in 1980, it was joined with Improvement District No. 143 on 1 April 1995 to form the regional municipality of Wood Buffalo.

Fort McPherson NT On the Peel River, southwest of Inuvik, this hamlet (1986) was established in 1840 as a Hudson's Bay Company trading post. It was named in 1848 after Murdoch McPherson, a HBC chief trader.

Fort Nelson BC A town (1987) in the northeastern part of the province, it was founded as a trading post in 1805 by George Keith of the North West Company. It was likely named after Adm. Viscount Horatio Nelson, who died that year. It had been incorporated as a village in 1971.

Fort Providence NT On the north bank of the Mackenzie River, and 75 km downriver from the point where the river drains Great Slave Lake, this hamlet (1987) is the site of a fort built by the Hudson's Bay Company about 1850. In 1915 the Geographic Board authorized the name *Providence*, with the historic name being restored in 1945. In 1790, Alexander Mackenzie had built another Fort Providence at the outlet of Yellowknife River, 210 km to the northeast.

Fort Qu'Appelle SK Northeast of Regina, this town (1951) is located in the valley of the Qu'Appelle River. It was the site of the Hudson's Bay Company's fort, established by Peter Hourie in 1864. Its post office was called *Qu'Appelle* in 1880 and was renamed Fort Qu'Appelle in 1911. *See also* Qu'Appelle.

Fort Resolution NT On Great Slave Lake, south of the mouth of the Slave River, this settlement (1988) is the site of a Hudson's Bay Company trading post, established in 1815. The name is suggestive of the hardships endured at the most northerly HBC post, before the union with the North West Company in 1821. The Geographic Board called it *Resolution* in 1915, but the older name was restored in 1940.

Fort St. James BC At the east end of Stuart Lake, northwest of Prince George, this village (1952), the oldest continuously occupied community in the province, was founded in 1806 by Simon Fraser as the North West Company trading post of *Stuart Lake*. It was renamed Fort St. James in 1822, but the reason for the name change is not recorded in the official names records. Its post office was opened in 1899.

Fort St. John BC On the Alaska Highway, north of the Peace River, this city (1975) is located near the original site of a trading post called Rocky Mountain House in 1799, and later called Fort St. John. The city's present site was fixed in 1942 at *North Fort St. John*, when the Alaska Highway was built northwest from Dawson Creek.

Fort Saskatchewan AB A city (1985) northeast of Edmonton, its site was selected in 1875 for the first North-West Mounted Police post north of Calgary. It was anticipated that the transcontinental Canadian Pacific Railway line would cross the North Saskatchewan River at this point, but it was subsequently decided to build the railway though the Kicking Horse Pass, west of Calgary, and not the Yellowhead Pass.

Fort Simpson NT A village (1973) on the Mackenzie River and on the west side of the mouth of the Liard River, it was established about 1820 by the North West Company as the site of *Fort of the Forks*. It was renamed in 1821 after George (later Sir George) Simpson, the governor-in-chief of the Hudson's Bay Company 1820-1860, who was instrumental in uniting the HBC with the NWC in 1821. The Geographic Board approved the name *Simpson* in 1915, but Fort Simpson was restored in 1944.

Fort Smith NT A town (1966) on the Slave River just north of the Alberta border, it was established as a Hudson's Bay Company trading post in 1870. It was named after Donald A. Smith, 1st Baron Strathcona and Mount Royal (1820–1914). In 1871 he was appointed chief commissioner of the HBC.

Fortune NF A town (1946) on the west side of the Burin Peninsula, it took its name from **Fortune Bay**. The bay's name came from the Portuguese *fortuna*, 'luck', perhaps referring to an incident involving something fortunate in the early 1500s. It is identified on the maps of Maggiolo (1527), Verrazzano (1528), and Ribiero (1529).

Fort Vermilion AB On the south bank of the Peace River, west of Lake Athabasca, this place was established about 1828 by the Hudson's Bay Company as a post at the mouth of the **Vermilion River**. The fort had been built by the North West Company in 1788 farther downriver. Its post office was opened in 1905.

Fort William ON A large sector of the city of Thunder Bay, it was established in 1803 by the North West Company and named after William McGillivray (1764–1825), who was appointed the principal director of the company in 1804. Fort William, Port Arthur, and parts of two neighbouring townships were amalgamated in 1970 as the city of Thunder Bay.

Fortymile River YT Rising in Alaska, this river flows 100 km northeast across the Yukon border to enter the Yukon River, about 66 km (or 40 mi) below the site of the Hudson's Bay Company's Fort Reliance. The name was given about 1886 by prospectors who found coarse gold on the river.

Fosheim Peninsula NT On the west side of Ellesmere Island, this 160-km peninsula was named before 1910 after Ivar Fosheim, a member of the Sverdrup expedition, 1898–1902. He and Oluf Raanes explored the west side of the island in 1901.

Foulon, Anse au QC Located at the foot of the cliff below Sillery, west of the city of Québec, it was the site of two fulling mills in the early 1700s. Fulling (*foulon*) describes a process of cleaning and thickening cloth. The earliest written evidence of the cove is a reference to Gen. James Wolfe's ascent with his troops to the Plains of Abraham in 1759. Thereafter, the site became widely known in English as *Wolfe's Cove*.

Fournier ON This place in the united counties of Prescott and Russell, 28 km southwest of Hawkesbury, was named in 1856 for Cajetan Fournier, who had opened a store here the previous year, and became the first postmaster. It was also known as *Fournierville* in the late 1800s.

Foxboro ON In Hastings County, northwest of Belleville, its post office was called *Smithville* in 1851, after either pioneer settler Richard Smith, or blacksmith Smith Demorest. Because of duplication, its name was changed in 1861 to Foxboro. It may have been chosen by first postmaster William Ashley, possibly after Foxboro Hall in Suffolk, England, or after Foxborough Hall in County Roscommon, Ireland.

Fox Creek AB A town (1983) northwest of Whitecourt, it was established in 1967, and named after a stream flowing north into Iosegun Lake. The creek was noted in a surveyor's notes in 1901, but was portrayed on a 1928 map as *Fox Brook*.

Foxe Basin NT North of Hudson Bay, with Baffin Island on its east and Melville Peninsula on its west, this basin was named in 1921 after Capt. Luke Foxe (1586–1635). In 1631 Foxe explored Hudson Bay and the west coast of Baffin Island. **Foxe Peninsula**, which extends for 240 km from southwestern Baffin Island, was named *Fox Land* in 1905, and the present name was substituted in 1920. **Foxe Channel**, which unites Foxe Basin with Hudson Strait, was called *Fox Channel* in 1905. It was corrected to Foxe Channel in 1920.

Fox Harbour NF A municipal community (1964) on the east side of Placentia Bay and north of Placentia, it was first called *Little Glocester* after the fishing town of Gloucester, MA. It was known as Fox Harbour as early as 1836, possibly because foxes were observed eating fish drying on flakes.

Fox Valley SK Located near the Alberta border, west of Swift Current, this village was established in 1908. It received rail service in 1924–5 when the Canadian Pacific Railway extended a line from Pivot in Alberta. The name probably came from the occurrence of numerous red foxes in the area. Its post office was opened in 1911.

Foxwarren MB A community northwest of Brandon, near the Saskatchewan border, it was developed in 1879–80 by a Mr Dawson. He set out to replicate W. Barneby's English estate of Fox Warren, near Cobham in Surrey, southwest of London. The post office was opened as *Fox Warren* in 1889, and given its present name in 1909.

Foymount ON At 523 m the province's most elevated community, Foymount is in Renfrew County, southeast of Barry's Bay. It was named in 1873 after tavern keeper and first postmaster John Foy.

Frances River YT Draining **Frances Lake**, this river flows south to join the Liard River, upriver from Watson Lake. It and the lake were named in 1840 by Hudson's Bay Company chief factor Robert Campbell after Lady Frances Simpson, the wife of HBC governor-in-chief Sir George Simpson.

Francheville QC A regional county on the north shore of the St. Lawrence, centred on the city of Trois-Rivières, and extending between Portneuf and Maskinongé, it replaced the county of Champlain in 1982. It was named after Montréal merchant and Saint-Maurice seigneur François Poulin de Francheville (1692–1733). He developed the iron at the Forges du Saint-Maurice, a national historic site.

Francis SK Southeast of Regina, this town was named in 1900 after J. Francis, who donated land for the town site. Earlier, the area was called *Wascana*, after Wascana Creek, before the railway arrived.

François Lake BC South of Burns Lake and west of Prince George, this name resulted from a misinterpretation by French speakers of the Carrier name *Nitapoen*, 'lip lake', named because its two shores provide the outline of a smiling mouth. Believing the name was *Netapoen*, 'white man's lake', French traders translated it in the 1800s as *Lac des Français*, which was subsequently rendered in English as François Lake.

Frank AB In the municipality of Crowsnest Pass since 1979, this coal mining community was named in 1901 after A.L. Frank, a partner in the Canadian American Company. Nineteen months after it was founded, it was nearly destroyed by a massive rock slide on Turtle Mountain, killing 70.

Frankford ON In Hastings County, northeast of Trenton, this village (1921) was named in 1836, when a bridge was constructed across the Trent River. The name was suggested by Lt-Gov. Sir Francis Bond Head, who had descended the river by canoe that year. The post office was opened in 1838.

Franklin Mountains NT Extending over 400 km on the east side of the Mackenzie River, this range of mountains was named in 1944 after Sir John Franklin, who passed this way in 1825, during his second Arctic expedition.

Franktown ON Col Francis Cockburn built a storehouse in 1818 at this place in Lanark County, midway between Perth and Richmond. Its post office, opened in 1832, was called after his forename.

Frankville ON In the united counties of Leeds and Grenville, northwest of Brockville, it was first known as *Wilson's Corners*, *Brennan's Corners*, and *Brennanville*. *Brandenburgh* was suggested as a possible postal name, but Frankville was adopted in 1852. Claims that it was named after a Frank Wilson cannot be verified.

Fraser Lake BC A village (1966) on the south shore of **Fraser Lake** and west of Prince George, its post office was named in 1920 after the lake. In 1806, Simon Fraser (1776–1862) established a North West Company trading post on the east end of the lake. *Fraser Lake* post office had been opened here in 1907, but was renamed Fort Fraser in 1913.

Fraser, Mount AB, BC This mountain (3338 m) on the continental divide and near the head of the North Saskatchewan River was named in 1917 after Simon Fraser, a North West Company fur trader and discoverer of the Fraser River. **Simon Peak** (3322 m), just to the north on the divide, was also named after him.

Fraser River BC Rising in the Yellowhead Pass, west of Jasper, this river (1370 km) drains into the Strait of Georgia at Vancouver. In 1808, believing he was travelling down the Columbia, Simon Fraser was so disappointed on reaching its mouth that he failed to name it. Five years later geographer David Thompson named it after him. It had been previously called *Rio Floridablanca* by the Spanish, and *New Caledonia River* and *Jackanet River* by the fur traders.

Frayn, Mount BC In the Rocky Mountains, west of Sparwood, this mountain (2914 m) was named in 1964 after F/O Richard P. Frayn of Fernie, who was killed in action in Europe on 8 January 1943.

Fredericton NB The capital of New Brunswick, this city (1848) was named in 1785 by Lt-Gov. Thomas Carleton after Prince Frederick (1763–1827), Bishop of Osnabrück and second son of George III. In a letter that year to Lord Sydney, Carleton called it *Fredericstown*, indicating an early elision of the 'k'. The French had named the site *Sainte-Anne*, with the Pre-

Loyalists calling it *St. Anne's Point*. Briefly in the 1780s it was also called *Osnaburg*.

Fredericton Junction NB Located south of Fredericton, this village (1966) was first known as *Hartt's Mills* after Thomas Hartt, who established a lumber business here in 1804. Its post office was opened in 1871, having been served by the Blissville way office from 1852.

Freels, Cape NF On the northwestern side of Bonavista Bay, it was named after a nearby island, which the Portuguese had called in the early 1500s *Ilha de frey luis*, 'island of Brother Louis'. Perhaps a brother accompanied one of the early expeditions, possibly even John Cabot's of 1497.

Freeport NS At the southwestern end of Long Island, and southwest of Digby, this village (1929) was first called *Lower Cove*. It was renamed Freeport in 1865, although its post office, opened as *Long Island* ten years before, was not changed to Freeport until 1883. It may have been named after Freeport, ME, northeast of Portland.

Freetown PE Southeast of Summerside, this place was named after the extensive freehold land available there in the 1800s, in contrast to the mostly rental property elsewhere. Its post office was opened in 1854. Its area was earlier called *Burns Settlement*.

Frelighsburg QC A municipality (1985) in the regional county of Brome-Missisquoi, south of Cowansville, it was settled in 1800 by Abram (Abraham) Freligh, a native of Clinton, NY, who managed a sawmill here for many years. The municipality annexed the village (1867) and the parish municipality (1967) of Frelighsburg.

Frenchman's Cove NF A municipal community (1974) on the west side of the Burin Peninsula, its harbour was known by this name as early as the 1760s, when Capt. James Cook undertook hydrographic surveys here. It may have been named by English fishermen after one or more French fishermen, who continued to fish along the shore after the close of the French regime.

French, Mount AB At the head of the Kananaskis River, southwest of Calgary, this mountain (3234 m) was named in 1918 by surveyor M.P. Bridgland after Sir John Denton Pinkstone French (1852–1925), commander-in-chief of the British Expeditionary Force in France during the First World War.

French River ON An important navigation route joining Lake Nipissing with Georgian Bay and Lake Huron, it was well known as *Rivière des Français* during the French regime. After 1759 it was first called *Frenchman's River* on English maps, with the current form becoming the only English name used on maps after 1847.

French Village NS On St. Margarets Bay, and west of Halifax, this place took its name from descendants of French families, who had been settled by Gov. Charles Lawrence in Lunenburg in 1753. They were encouraged in 1783 to resettle on the east side of the bay by Gov. John Parr.

Freshfield, Mount AB, BC On the continental divide and southwest of Howse Pass, this mountain (3336 m) was named in 1897 by British climber Norman Collie after alpinist Douglas William Freshfield (1845–1934), president of the Alpine Club in England in 1893.

Freshwater NF A neighbourhood in the town of Placentia and on the north side of the Northwest River, it had been incorporated as a town in 1950. Its name likely refers to the finding of a good supply of fresh water by fishermen. Locally, it is believed the site was named *Fontaine* by the French, in reference to the fresh water obtained from a brook draining from Larkins Pond.

Fresnoy Mountain AB, BC This mountain (3240 m) on the continental divide, southeast of the Columbia Icefield, was named in 1919 after a village in northeastern France, which had been captured by Canadian troops two years earlier.

Frobisher SK East of Estevan, this village (1904) was established by the Canadian Pacific Railway in 1884. Its post office was called *Frobyshire* on 1 February 1902, with the name being respelled Frobisher on 1 June 1902. In 1905 chief geographer James White stated that it was named after Benjamin Frobisher (*c.* 1742–87), a fur trader in the West, 1764–87, and a shareholder in the North West Company after 1779. It may have been named after the Elizabethan explorer Sir Martin Frobisher (1535–94).

Frobisher Bay NT On the southeast side of Baffin Island, this 240-km inlet was named in 1900 after Sir Martin Frobisher (1535–94), who, when exploring the area in 1576, assumed the bay was a strait. In 1587 John Davis called it *Lumley Inlet*, after Lord Lumley, one of his patrons. *See also* Iqaluit.

Frontenac ON The county was named in 1792 after Fort Frontenac. It had been built as Fort Cataraqui in 1673, but was renamed soon after by René-Robert Cavelier de La Salle after Louis de Buade, comte de Frontenac et de Palluau (1622–98), governor of New France, 1672–82 and 1689–98. Col John Bradstreet destroyed the fort in 1758.

Frontenac QC Located in the regional county of Le Granit, east of Lac Mégantic, this municipality replaced the united townships of *Spalding-et-Ditchfield* in 1959. It was named after Louis de Buade de Frontenac et de Palluau (1622–98), governor of New France, 1672–82 and 1689–98.

Frontier SK Near the Montana border, southwest of Swift Current, this village (1930) took its name from the rural municipality of **Frontier** (1913). Its post office was opened in 1917, with a Canadian Pacific Railway branch line arriving in 1923.

Front of Escott ON A municipal township in the united counties of Leeds and Grenville, it was named in 1905. Escott Township was established in 1840, and was named after the family residence of Sir George Yonge at Escott, near Honiton, Devon. In 1850 it became part of Front of Yonge and Escott Township. *See also* Rear of Yonge and Escott.

Front of Leeds and Lansdowne ON This municipal township was created in 1850 from the front portions of the townships of Leeds and Lansdowne. They were named in 1788, the one for Francis Osborne, 5th Duke of Leeds (1751–99), secretary of state for foreign affairs, 1783–91; and the other for William Petty Fitzmaurice, 2nd Earl of Shelburne and Marquess of Lansdowne (1737–1805), prime minister of Great Britain, 1782–3. *See also* Leeds.

Front of Yonge ON Yonge Township in Leeds County was named in 1786 after Sir George Yonge (1732–1812), secretary of war, 1782–94. In 1850 it was divided between Front of Yonge and Escott, and Rear of Yonge and Escott townships. Front of Yonge was made a separate municipal township in 1905.

Fruitvale BC In the Beaver Creek valley, east of Trail, this village (1952) was named in 1907, when the post office was opened.

Fryatt, Mount AB Northwest of the Columbia Icefield, this mountain (3361 m) was named in 1920 by the Alberta-British Columbia Boundary Commission after Capt. Charles Algernon Fryatt (1872–1916). He had rammed a German submarine with his unarmed merchant ship during the First World War and was subsequently shot by his captors.

Fullarton ON A township in Perth County, it was named in 1830 for John Fullarton, a director of the Canada Company, a British company established in 1826 to develop the Huron Tract, a large area of what is now Southwestern Ontario.

Fundy, Bay of NB, NS Separating the provinces of New Brunswick and Nova Scotia, this 50 to 70-km-wide inlet was called *La Baye Françoise* in 1604 by explorer Pierre du Gua de

Monts. In 1624 Sir William Alexander referred to it as *Argall's Bay*, after a raider who was here 11 years before. On a 1679 map it was identified as *Bay of Funda*, with the present form appearing on a map in 1711. It would appear the name may have been derived from *Cap Fendu*, the French name of Cape Split, in reference to the prominent cape separating Minas Channel from Minas Basin. **Fundy National Park** (206 km^2), in Albert County, was created in 1948.

Fury and Hecla Strait NT Uniting Foxe Basin and the Gulf of Boothia, this strait was discovered and named in 1822 by William (later Sir William) Parry, after the two ships under his command, the *Fury* and the *Hecla*. Because the strait was ice-bound, he was unable to navigate through it, but a journey overland confirmed that a sea lay beyond.

Fusilier Peak BC In the Tower of London Range of the Muskwa Ranges of the Rocky Mountains, this mountain (2819 m) was named in 1960 by Capt. M.F.R. Jones after the Royal Fusiliers in London, England, and the Third Battalion Royal Canadian Regiment (London and Oxford Fusiliers). A party of British and Canadian fusiliers climbed it.

G

Gabarus NS On the south side of **Gabarus Bay** and southwest of Louisbourg, its post office was called *Gabarouse* in 1842, taking its spelling from the British Admiralty charts. The bay was identified as *Gabor* on Jean Guérard's map of 1631, and *Baye de Gabari* on Jacques-Nicolas Bellin's map of 1764. Jurist and writer Thomas C. Haliburton used the present spelling in 1829. Although there has been much speculation on the linguistic origin of the name, its source and meaning remain uncertain.

Gabriola Island BC One of the Gulf Islands, offshore from Nanaimo, it took its name from its eastern point, named *Punta de Gaviota*, 'seagull point', in 1791 by Spanish explorer Jose Maria Narvaez. It was subsequently rendered as *Punta de Gaviola*, and was further changed to Gabriola when given to the island, possibly in 1859 by Capt. George Richards.

Gads Hill ON In Perth County and northeast of Stratford, it has a name of English literary origin, but there is no certainty as to its source. It may be from Shakespeare's *Henry IV* (Part I), where it occurs in the form Gadshill, in reference to Gads Hill, northwest of Rochester, Kent, England, where Charles Dickens lived from 1857 to 1870.

Gagetown NB A village (1966), 45 km southeast of Fredericton and on the west bank of the Saint John River, it was named in 1825 after **Gagetown Parish**, established in 1786. The parish was called after *Gage Township*, named in 1765 after Thomas Gage (1721–87), commander-in-chief of British forces in North America, 1764–75 and a grantee of land in the township. **Canadian Forces Base Gagetown** was established in 1952. It embraces 111,000 ha in Sunbury and Queens counties, between Fredericton and Saint John.

Gainsborough SK The province's oldest village (1894), it is located near both the Manitoba and North Dakota borders. It was named by J. Sadler, after his hometown in Lincolnshire, England, northwest of the city of Lincoln. Postal service was provided here the next year.

Galahad AB Southeast of Camrose, this village (1918) was named after Galahad post office, established in 1907 at present Alliance, 14 km to the southeast. After Alliance was named in 1916 by the Canadian Northern Railway, Galahad was reassigned to its present site. The name may have been given after the thirteenth century legendary knight of the Round Table.

Galiano Island BC One of the Gulf Islands, northeast of Saltspring Island, it was named in 1859 by Capt. George Richards after Dionisio Alcalá Galiano (d. 1805), who, as commander of the Spanish vessel *Sutil,* explored the waters of the Strait of Georgia in 1792.

Galt ON A city of 38,000 until 1973, it was united with the towns of Preston and Hespeler, and part of the township of North Dumfries to form the new city of Cambridge. First called *Schade's Mill*, after miller Absolom Schade, it was renamed in 1825 by legislative councillor and land developer William Dickson for John Galt (1779–1839), the founder of Guelph. Galt was the commissioner of the Canada Company, which developed the Huron Tract, a large part of what is now Southwestern Ontario.

Gambo NF A town (1980) at the head of Freshwater Bay, the most westerly extension of Bonavista Bay, its name, first given to **Gambo Ponds** and **Gambo Brook**, may have been derived from the word *gambo*, a type of sledge used to haul timber in the woods. It may have its origin in a Mi'kmaq word. A suggestion in the early 1900s by the Most Revd Michael Howley that the name might have originated in the sixteenth century Spanish or Portuguese name *Baie de las Gamas*, 'bay of does', cannot be verified. The town also comprises the adjacent communities of Dark Cove and Middle Brook.

Gananoque ON Joel Stone founded the town of Gananoque at the mouth of the **Gananoque River** in 1791. Although Lord Dorchester used the present spelling as early as 1788, it was not standardized until the 1820s. French surveyor Jean Deshayes was told in 1685 that the river in Onondaga was called *Ganonocouy*, 'flint at the mountain', still the preferred meaning. Incorporated as a town in 1890, it became a separated town in 1922.

Gander NF On the north shore of **Gander Lake**, this town (1959) took its name from the lake and the **Gander River** (156 km), which may have been named after wild geese. It had its beginnings in 1937–8, when a large airport was constructed to strengthen the transatlantic links between North America and Europe. The modern townsite was developed in the mid-1950s.

Ganges BC The main service centre of Saltspring Island, its post office was named in 1906. It is located at the head of **Ganges Harbour**, which was named in 1859 by Capt. George Richards, after HMS *Ganges*, which served on the Pacific station from 1857 to 1860. The ship was built at Bombay (Mumbai), India in 1821.

Ganong, Mount NB Northeast of Mount Carleton, this mountain (655 m) was named in 1901 by American naturalist Mauran Furbish after William Francis Ganong (1864–1941). Ganong was a professor of botany at Smith College, Northampton, MA, who spent several summers exploring the geography, geology, and natural history of his native province. He wrote two monographs on New Brunswick's place names and several articles on Aboriginal names.

Gargantua, Cape ON Located on the eastern shore of Lake Superior, this cape was named during the French regime after the legendary giant in the tales of *Gargantua* and *Pantagruel* by François Rabelais (1494–1553). **Pantagruel Bay**, after Gargantua's son, and **Grangousier Hill**, after Gargantua's father, Grandgousier, are located nearby.

Garibaldi, Mount BC Northeast of Squamish and in **Garibaldi Provincial Park** (1927), this mountain (2678 m) was named in 1860 by Capt. George Richards after Giuseppe Garibaldi (1807–82), an Italian patriot who had a major role in achieving Italian unity.

Garnish NF Located on the west side of the Burin Peninsula, west of Marystown, this town (1971) is located on **Garnish Barasway**, a small harbour protected by a bar. Identified in 1775 as *Little Garnish* by Capt. James Cook, neither the meaning of this name, nor that of **Great Garnish Barasway**, 3 km to the southwest, is known.

Garry Lake NT On the Back River, northwest of Baker Lake, this lake was named in 1834 by Capt. George (later Sir George) Back after Nicholas Garry, who strongly supported research in the Arctic. After the Army Survey Establishment of the department of national defence found in 1959 that there were three different levels, it was divided into Garry, Upper Garry, and Lower Garry lakes. The present Garry Lake has an area of 916 km^2.

Garson MB Midway between Selkirk and Beausejour and northeast of Winnipeg, this village was incorporated in 1915 as *Lyall*, but was changed in 1927 to Garson, after the Canadian Pacific Railway station established in 1902. William Garson opened a quarry here about 1900 and **Garson Quarry** post office was opened in 1908.

Garson ON An urban centre in the town of Nickel Centre, Sudbury Region, it was named in 1907 after Garson Township. In 1887 the township was called after William Garson, who represented Lincoln in the Ontario Legislature, 1887–90.

Gaskiers-Point La Haye NF This municipal community (1970) is located on the east shore of St. Mary's Bay. Gaskiers may have been derived from French *casse-cœur*, a variant of *crève-cœur*, meaning 'heart-break'. Or it may be from a French family name such as Gasquié or Gascoigne. Or it could be a variant spelling of the surname of the Marquis de Castries (1727–1801), who served as the French minister of marine in the 1760s. Both *Point La Haye* and *La Haye Point* are shown on Capt. James

Cook's 1762 map. La Haye is both a French family name and place name. The 1869 census listed *Gaskin Point, La Haye.*

Gaspé A town (1959) near the inner part of the **Baie de Gaspé**, and not far from **Cap Gaspé**, the most eastern extremity of the **Péninsule de Gaspé**, it is near the site of Jacques Cartier's famous planting of his first cross in the New World in 1534. The name, most often spelled *Gaspay* or *Gaspei* until the late 1600s, first appears in records about 1590. Its exact origin remains uncertain. It seems likely that its form is associated with the Mi'kmaq root *kesp(i)-/gesp(i)-*, which conveys the meaning of 'end' and occurs in the names of other promontories within Mi'kmaq territory in the present-day Maritime provinces. Assuming that this is correct it could be concluded that it originally designated the headland of Cap Gaspé. However, one or more European languages almost certainly influenced the name through re-interpretation or outright creation, before the prevailing pronunciation, spelling, and the use of the name Gaspé were arrived at. It is quite probable that Basque mariners, who were active in the Gulf of St. Lawrence in the late 1500s, adapted the final syllable to conform to the suffix *-pe*, 'below' in their language, and, using it to refer to the bay and harbour, took the complete name to be a compound Basque word such as *gerizpe/kerizpe*, 'harbour'. Documentary evidence is too scanty either to prove or disprove this theory.

Gaspereau NS On the **Gaspereau River**, south of Wolfville, this place's post office was called *Gaspereaux* from 1842 to 1962. The river was named during the French regime after a common salt-water fish of the herring family known in English as 'alewife'.

Gaspereaux PE Situated southeast of Montague, this place's name means 'alewives', a variety of herring that spawns in May in the ponds beside Northumberland Strait. Its post office was opened about 1873.

Gatineau QC A city in the Communauté urbaine de l'Outaouais and at the mouth of the **Rivière Gatineau**, it was created in 1975 by merging the towns of Gatineau and *Pointe-Gatineau*, the villages of *Touraine*, *Templeton*, and *Templeton-Est*, and the territory of the present municipality of Cantley, separated from Gatineau in 1989. The river's name recalls the seventeenth century fur trader Nicolas Gatineau (or Gastineau) *dit* Duplessis, who traded on the river between the Ottawa and the Saint-Maurice, and his sons, Louis and Jean-Baptiste, who had a trading post at the mouth of the river. **Gatineau Park**, just north of Canada's capital, is a popular all-season recreation area.

Gaultois NF On Hermitage Bay and on the southwestern coast of the island of Newfoundland, this town (1962) took its name from *Havre le Galtois*. In 1764 Capt. James Cook referred to it as Gaultois. It was possibly named by French or Jersey fishermen in the 1600s, after the many rocky pinnacles found here, deriving it from the Norman French word *galtas*, 'pinnacle'.

Geary NB A community south of Oromocto, it was founded about 1810 by families from Rhode Island, who had lived briefly in the Niagara Peninsula. They called their settlement *New Niagara*, but the land petition of 1810 had *New Gary*, and eight years later it was spelled Geary on a map.

Geddes, Mount BC In the Coast Mountains, northwest of Mount Waddington, this mountain (3223 m) was named in 1928 by W.A.D. (Don) Munday after climber M.D. Geddes of Calgary, who had been killed the previous year while ascending Mount Lefroy in the Rockies.

George, Rivière QC This 563-km river rises north of Smallwood Reservoir and flows north to Ungava Bay. It was called *George River* in 1811 by Moravian missionaries Gottlieb Kohlmeister and George Kmoch after George III. They noted its local name was *Kangertlualuksoak*, which in modern Inuktitut is *Kangirsualujjuap Kuunga*, 'river of the very great bay'.

Georgetown ON An urban centre in the town of Halton Hills, Halton Region, it was settled in 1821 by George Kennedy (1799–1870), who built several mills on Silver Creek. It was named in 1837 after Kennedy, who had been named for George III, the king when he was born.

Georgetown PE Northeast of Montague, this town (1912) was created in 1765 by Samuel Holland and named after George III. Its port was called *Three Rivers* in the early 1800s, after the French *Trois Rivières*, describing the Brudenell, Cardigan, and Montague rivers. Georgetown post office opened in 1827. It is located in Georgetown Royalty, one of the 70 territorial divisions laid out by Holland.

Georgian Bay ON A large arm of Lake Huron, it was called *Mer Douce*, 'freshwater sea', by Samuel de Champlain in 1615. Admiralty surveyor Henry W. Bayfield charted its shores in the 1820s and named it after George IV. The municipal township of **Georgian Bay** in the district municipality of Muskoka was created in 1971. **Georgian Bay Islands National Park** (14 km^2) was created in 1929.

Georgia, Strait of BC Separating southern Vancouver Island from the mainland, this water body was named on 4 June 1792 the *Gulph of Georgia* by Capt. George Vancouver on the occasion of the birthday of George III. On 27 November 1858 Capt. George Richards informed the hydrographer of the British Navy that he had changed *Gulf of Georgia* to Strait of Georgia, as 'it seemed a more appropriate name and I thought it better to alter it at once.' The Gulf Islands have remained as evidence of Vancouver's earlier name. In 1791 Spanish Lt Francisco Eliza had named the gulf *Gran Canal de Nuestra Señora del Rosario la Marinera*, which Vancouver shifted to the channel (now Malaspina Strait) between Texada Island and the mainland.

Georgina ON This original township was named in 1815 after George III. Georgina was proposed in 1819 as a name for the new royal princess, the future Queen Victoria. In 1971 the municipal township of Georgina was united with the village of Sutton and the municipal township of North Gwillimbury to form the new municipal township of Georgina, which has been a town since 1986.

Gerald SK Located east of Esterhazy, this village (1953) was established by the Grand Trunk Pacific Railway in 1907 and named after one of its officials. Its post office was opened two years later.

Geraldton ON This town (1937) in Thunder Bay District was the site of a major gold strike in the early 1930s. In 1933 the Canadian National Railways proposed naming its new station Fitzgerald, after Stanley J. Fitzgerald, vice-president of the Little Long Lac Gold Mines Ltd. Informed it was already in use in Western Canada, the CNR proposed Geraldton, composed from syllables from Fitzgerald and Joseph Errington, the company's president.

Gesner Ridge NB On the east side of Little Tobique River and west of Mount Carleton, it was called *Gesner Mountains* by naturalist William F. Ganong in 1902 after geologist Abraham Gesner (1797–1864), the inventor of kerosene. The name was amended to Gesner Ridge in 1975.

Gibbons AB A town (1977) northeast of Edmonton, it was first known as *Astleyville* and *Battenburg*. Its post office was named in 1920 after William Reynolds Gibbons, who moved to the area in 1892 from Orillia, ON.

Gibsons BC A town (1982) on the west side of Howe Sound, where it joins the Strait of Georgia, its post office was called *Howe Sound* in 1892, and was renamed *Gibson's Landing* in 1907 after George William Gibson (1829–1913). He and his sons settled here in 1866. In 1948 local businessmen persuaded the post office department to rename the office Gibsons.

Gilbert, Mount BC In the Coast Mountains, east of Bute Inlet, this mountain (3117 m) was named in 1933 by George G. Aitken, provincial member of the Geographic Board of Canada, after Sir Humphrey Gilbert (*c.* 1537–83). He had applied for a charter in 1566 to trade on the west coast of North America, and received it 12 years later. After taking possession of the island of Newfoundland in 1583, he drowned in the Atlantic Ocean before attaining his goal to reach the Pacific.

Gilbert Plains MB Located west of Dauphin, the post office at this village (1906) was named in 1892, after the adjacent plains, identified by residents of Dauphin with Métis scout Gilbert Ross.

Gillam MB A community on the south side of Stephens Lake, an enlargement of the Nelson River, it was founded in 1927 by the Hudson Bay Railway, and named after Hudson's Bay Company employee Capt. Zachariah Gillam (1636–82), who lost his life near the mouth of the river. His son, Capt. Benjamin Gillam (1662/3–1706) arrived earlier in 1682, sailing on behalf of the governor of Massachusetts. Subsequently, the HBC based its claim over the Nelson River's basin on Benjamin Gillam's arrival there ahead of Pierre Radisson. Gillam post office was opened in 1928.

Gillams NF On the north side of Humber Arm and northwest of Corner Brook, this municipal community (1971) was named after Gillams Cove. The cove, noted in 1849, may have been named after a French settler, possibly Blanchard, whose forename was Guillaume, meaning 'William'.

Gimli MB On the southwestern shore of Lake Winnipeg, this town (1947) was founded in 1875 by the first Icelandic settlers in Western Canada. It was named after the Great Hall of Heaven, where all brave men gather with the Norse god Odin. Gimli post office was opened at present Riverton in 1877 and was transferred to its current site in 1886.

Girardville QC A municipality northwest of Lac Saint-Jean, it was named in 1921 after Joseph Girard (1853–1933), a member of the Legislative Assembly for Lac Saint-Jean, 1892–1900, and of the House of Commons for Chicoutimi, 1900–08.

Girouxville AB A village (1951) west of Falher in the Peace River country, it was founded in 1915 by the pioneering Giroux family, with the post office opening two years later. Earlier it was called *Fowler* by the Edmonton, Dunvegan and British Columbia Railway.

Gjoa Haven NT A hamlet (1981) on the southeast shore of King William Island, it was established in 1947. The haven was named in 1904 after the vessel *Gjoa*, under the command of Roald Amundsen. Amundsen wintered here from the fall of 1903 to the spring of 1905. In 1906 the *Gjoa* became the first vessel to complete the navigation of the Northwest Passage. The ship was anchored at San Francisco until 1972, when it was returned to Norway.

Glace Bay NS An urban community in the regional municipality of Cape Breton since 1 August 1995, it had been incorporated as a town in 1901. Postal service was established at *Little Glace Bay* in 1864 and the name was changed to Glace Bay in 1902. During the French regime, the water body now known as **Big Glace Bay** was called *Baie de Glace*, 'icy bay'. Samuel Holland named it *Dyson Bay* in 1769, and J.F.W. DesBarres referred to it as *Windham River* and *Ice Bay* in 1780, and *Glass Bay* in 1786.

Glacier National Park BC Located in the Selkirk Mountains and on both sides of the Rogers Pass, this park (1350 km^2) was established in 1886. Among its mountains are several glaciers and its valleys are often the scenes of avalanches.

Gladstone MB Located halfway between Lake Manitoba and Neepawa, this town (1882) was first called *Third Crossing*, because the old pioneer Carlton Trail to Edmonton had already crossed the Whitemud River twice. Its post office had been called *Palestine* in 1872, because the bounty from the land recalled the biblical 'promised land'. It was renamed Gladstone in 1879 for William Ewart Gladstone (1809–98), prime minister of Great Britain, 1880–5, 1886, and 1892–4.

Glanbrook ON A municipal township in Hamilton-Wentworth Region, it was created in 1974 through the amalgamation of Binbrook and Glanford townships. **Glanford Township** had been named in 1798 after the parish of Glanford Brigg in England's Lincolnshire, north of the city of Lincoln. **Binbrook Township** had been named the same year after a town in England's Lincolnshire, northeast of the city of Lincoln.

Glanworth ON In the city of London, south of the city centre, its post office was named in 1857 after the birthplace of pioneer Minchin Jackson in County Cork, Ireland, north of the city of Cork.

Glaslyn SK North of North Battleford, this village (1929) was settled in 1908 by Welsh pioneers and chose a name meaning 'clear water' in Welsh. Its post office was established in 1911.

Gleichen AB A town (1910) southeast of Calgary, it was named in 1883 by the Canadian Pacific Railway after Count Albert Edward Wilfred Gleichen, a financial supporter of the CPR, who had travelled on the line.

Glenavon SK Southeast of Regina, this village (1910) was first called *Kendal.* Its post office was named in 1908 by R.T. Young, possibly after Glendevon, in Perthshire, Scotland, southwest of Perth.

Glenboro MB Situated southeast of Brandon, this town (1949) was settled in 1879 and its post office was opened in 1883. It was named by Scottish settler James Duncan after the glens of his homeland, with 'boro' being added to reflect the word 'town'.

Glenburnie-Birchy Head-Shoal Brook NF This municipal community (1978) on the South Arm of Bonne Bay is surrounded on the land side by Gros Morne National Park. **Glenburnie** was named by the first Scottish fishermen who settled here about 1800, likely after a community in Fifeshire, Scotland. The neighbourhoods of **Birchy Head** and **Shoal Brook** are located adjacent to physical features with these names.

Glencoe ON In Middlesex County, southwest of London, this village (1873) was named in 1856 by surveyor A.P. Macdonald, after the vale of Glencoe in Argyllshire, Scotland, where the terrible massacre of Macdonalds took place in 1692.

Glendon AB A village (1956) southwest of Cold Lake, its post office was named in 1912 after the birth name of the mother of postmaster J.P. Spencer.

Glenelg ON A township in Grey County, it was named in 1840 after Charles Grant, who had become Lord Glenelg in 1836, choosing his title after Glenelg, on the Scottish mainland opposite the Isle of Skye. He was British secretary of state for war and the colonies, 1835–41.

Glenella MB Northeast of Neepawa, this community's station was named in 1897 after Ella Williams, a sister-in-law of Sir Donald Mann, vice-president of the Canadian Northern Railway. Its post office was opened the following year.

Glen Ewen SK East of Estevan and between the towns of Oxbow and Carnduff, this village (1904) was named in 1884 by the Canadian Pacific Railway after Thomas Ewen, who had homesteaded here two years earlier. Ewen was its first postmaster in 1890.

Glengarry ON A county in the united counties of **Stormont, Dundas and Glengarry** since 1850, it was named in 1792 after the well-known glen in Inverness-shire, Scotland. The name was given in honour of Scottish Highlanders, who supported George III during the American Revolution, and received land grants in Glengarry through the intercession of Bishop Alexander McDonell.

Glen Morris ON Located in Brant County, between Brantford and Cambridge, its post office was named in 1851 after James Morris (1798–1865), appointed Canada's first postmaster general that year.

Glen Robertson ON In the united counties of Stormont, Dundas and Glengarry, east of Alexandria, it was named in 1874 when the

Canada Atlantic Railway was built across Alexander Robertson's farm.

Glen Williams ON In the town of Halton Hills, Halton Region, north of Georgetown, this place was first called *Williamsburg*, after Benajah Williams (1765–1851), who settled here in 1825. Because of duplication, it was renamed *Glen William* in 1852, when his son Charles became the first postmaster. It was changed to Glen Williams about 1870.

Glenwood AB A village (1961) southwest of Lethbridge, its post office was named *Glenwoodville* in 1908, after Glen Wood, the son of Edward J. Wood, president of the Alberta stake of the Church of Jesus Christ of Latter-day Saints. In 1906 the stake acquired the Cochrane Ranche, whose headquarters was at the site of the village. The post office was changed to Glenwood in 1979.

Glenwood NF On the west bank of the Gander River, this town (1962) was founded in the 1890s by G.L. Phillips, of Point Leamington, NF, and W.J. Sterritt, one of the owners of the Glenwood Lumber Company of Yarmouth, NS. The name was used for the sawmilling centre by 1901.

Gloucester NB In the northeastern part of the province, this county was named in 1826 after Princess Mary, Duchess of Gloucester and Edinburgh (1776–1857), fourth daughter of George III, and wife of her first cousin, William Frederick, 2nd Duke of Gloucester (1776–1834).

Gloucester ON A city in Ottawa-Carleton Region since 1981, it had been created a township in Russell County in 1798, but was transferred that year to Carleton County. It was named after Prince William Frederick, Duke of Gloucester (1776–1834), a nephew of George III, and husband of George III's daughter, Princess Mary.

Glovertown NF A town (1954) on the Middle Arm of Alexander Bay, an extension of Bonavista Bay, it was first known as *Bloody Bay*, after an earlier name of Alexander Bay. It was renamed after Sir John Hawley Glover (1829–85), governor of Newfoundland, 1876–81 and 1883–5.

Godbout QC Located at the mouth of **Rivière Godbout** in the regional county of Manicouagan, east of Baie-Comeau, this municipality was created in 1933. The river, noted by Fr Charles Albanel in 1670, and referred to by map-maker Jean-Baptiste-Louis Franquelin in 1684 as *Godebou*, probably owes its name to Nicolas Godbout (1634–74), a navigator and pilot on the St. Lawrence.

Goderich ON A town (1850) and a municipal township in Huron County, they were named in 1828 by John Galt and Dr William 'Tiger' Dunlop after Frederick John Robinson, Viscount Goderich and Earl of Ripon (1782–1859), British chancellor of the exchequer, 1823–7, secretary of state for war and the colonies, 1827, and prime minister of Great Britain, August 1827 to January 1828.

Godmanchester QC A township municipality on the south side of Lake St. Francis, southwest of Salaberry-de-Valleyfield, it was created in 1845 and named after the township of Godmanchester, proclaimed in 1811, and named after a town in Huntingdonshire, England, northwest of Cambridge.

Gods Lake MB With an area of 933 km^2, this lake, in the east central part of the province, took its name from a translation of the Cree name honouring the great spirit Manitou. The Hudson's Bay Company established a trading post at **Gods Lake Narrows** in 1880. *God's Lake* post office was opened in 1935 on an island in the lake. It was renamed God's Lake Narrows, when it was moved here in 1964.

Gogama ON On the intercontinental line of the Canadian National Railways, and in Sudbury District, northwest of Sudbury, this place was named in 1919 after the Ojibwa words *gogaam gigo*, 'the fish leaps up over the surface of the water'.

Gold Bridge BC This place north of Pemberton was named in 1934 after the bridge over the Bridge River, which provided access to the gold mining country around Bralorne.

Golden BC In the Columbia River valley and at the mouth of the Kicking Horse River, this town (1967) was first known as *The Cache* and *Kicking Horse Flats*. In 1883, on learning about Silver City being developed west of Banff, surveying assistant Frederick W. Aylmer was inspired to call this place *Golden City*. Subsequently 'City' was dropped from the name when the Canadian Pacific Railway arrived in 1885. The post office was opened two years later.

Golden ON A municipal township north of Red Lake, and in Kenora District, it was established in 1985, when the improvement district of Balmertown was renamed. The name reflects the gold-mining area of Red Lake, Balmertown, and Cochenour.

Golden Hinde BC Southwest of Campbell River, this highest mountain (2200 m) on Vancouver Island was named in 1939 by R.P. Bishop after Sir Francis Drake's ship, the *Golden Hind*. It is possible Drake may have seen it during his voyage through the North Pacific in 1578, although accounts of his voyage suggest he did not get this far north.

Golden Lake ON A community and a lake in Renfrew County, southwest of Pembroke, they were possibly named about 1830 by Hudson's Bay Company trader Charles Thomas, because of the beauty of the sunlight's golden reflection upon the waters and sandy shore. However, the French may have earlier translated an Algonquin description as *Lac Doré*. Coincidentally, Lake Doré, 10 km to the east, may mean 'golden lake', but may also mean 'yellow pike lake'.

Gold River BC A village (1981) on **Gold River**, southwest of Campbell River, it had been created a district municipality in 1965. Its post office was named *Muchalat* in 1952 and changed to Gold River in 1961. The river flows south from **Gold Lake** into Muchalat Inlet, a southern arm of Nootka Sound. The name may have come from the fact that some Chinese had panned the river for gold in the 1860s.

Gondola Point NB A village (1966) northeast of Saint John, its post office was named in 1883. The point, which extends into Kennebecasis River, was identified on a 1787 map. In the 1700s 'gondolas' were dugout canoes used for transporting people on rivers.

Good Hope Mountain BC In the Coast Mountains, southeast of Mount Waddington, this mountain (3240 m) was named *Mount Good Hope* before 1930 after HMS *Good Hope*, sunk in action 1 November 1914 off Coronel, Chile. The current form of the name was adopted in 1950.

Good Neighbour Peak YT In the Centennial Range of the St. Elias Mountains, 3 km north of Mount Vancouver, this mountain (4785 m) was named in 1967 by the Yukon Alpine Centennial Expedition, comprising four American and four Canadian climbers.

Goodsir, Mount BC In Yoho National Park, south of Field, this mountain (3581 m) was named in 1858 by Dr James (later Sir James) Hector after Dr John Goodsir (1814–67), professor of anatomy at the University of Edinburgh, where Hector studied medicine.

Goodsoil SK This village (1960) northwest of Meadow Lake was named in 1929, because the good soil of the surrounding farmland appealed to those suggesting it as a postal name.

Goodwood ON In Durham Region, east of Aurora, it was named in 1852 after tavern keeper and first postmaster Michael Chapman's birthplace at Goodwood House, in West Sussex, England, north of Chichester.

Goose Cove East NF This municipal community (1971) is located on the eastern coast of the Northern Peninsula, just south of St. Anthony. The place was originally called *Les Petites Oies* (meaning 'little geese') by French migratory fishermen in the late 1600s.

Gordon, Mount NB Northeast of Mount Carleton, this mountain was named in 1899 by naturalist William F. Ganong after Arthur Charles Hamilton-Gordon, 1st Baron Stanmore (1829–1912), lieutenant-governor of the province, 1861–6. He visited the area in 1863.

Gore Bay ON A town (1890) on **Gore Bay** and in Manitoulin District, it was named in 1860 after the bay, called after the steamship *Gore,* which was frozen in the North Channel in the 1850s. The steamboat, which had run between Collingwood and Sault Ste. Marie until 1870, was named after Sir Charles S. Gore, who helped put down the Rebellion of 1837–8.

Gores Landing ON On the south shore of Rice Lake and in Northumberland County, north of Cobourg, it was named in 1848 after British naval captain Thomas Sinclair Gore (1820–58). He was a native of County Antrim, Ireland, who came to Canada in 1841, and settled at the lake in 1844.

Gorrie ON In Huron County, east of Wingham, it was named in 1857 by James and Nathaniel Leech after their home town of Gorey in County Wexford, Ireland.

Gosfield North; **Gosfield South** ON These municipal townships in Essex County were separated in 1887 from Gosfield Township, which had been named in 1792 after the village of Gosfield, in Essex, England.

Gouin, Réservoir QC A huge reservoir covering an area of over 1300 km^2 at the head of Rivière Saint-Maurice, it was named in 1918 by the Shawinigan Water and Power Company after Sir Lomer Gouin (1861–1929), premier of Québec, 1905–20.

Goulbourn ON A municipal township in Ottawa-Carleton Region, the original township was named in 1816 after Henry Goulburn (1784–1856), under-secretary of state for war and the colonies, 1812–26.

Goulds NF A neighbourhood in the city of St. John's since 1992, south of the city centre, it had been incorporated as a town in 1971. It is said to have been named after a farm, which Joseph B. Jukes visited about 1840. He stated that the farm was called after a bright gold-coloured flower, common along the brooks of the Avalon Peninsula.

Govan SK North of Regina, the post office in this town (1911) was named after homesteader Walter Govan. The Canadian Pacific Railway arrived here the following year.

Gowganda ON On the east shore of Gowganda Lake and in Timiskaming District, west of New Liskeard, it was founded in 1907, after the discovery of silver here. The lake's name in Ojibwa is *Mitchi-oga-gan*, 'place of big pickerel'.

Gracefield QC A village (1905) in the regional county of La Vallée-de-la-Gatineau, south of Maniwaki, its post office was named in 1883 after Patrick Grace, the first merchant here, and the mayor of the township municipality of Wright, 1885–90.

Grafton NB On the east side of the Saint John River and opposite Woodstock, this place's post office was named in 1875 by Sperry Shea and George Stickney after their extensive grafting and nursery operations. The form of the name may have been chosen to honour Crimean War Gen. Augustus Fitzroy, 7th Duke of Grafton (1821–1918).

Grafton NS West of Kentville, this community has been in the village of Cornwallis Square since 1977. Settled about 1821, it may have been named after Grafton, MA.

Grafton ON In Haldimand Township, Northumberland County, east of Cobourg, this place was first known in 1814 as *Newcastle*, after the district. It was called *Haldimand* in 1832, and then renamed in 1858 after Grafton, MA.

Graham Island BC The largest of the Queen Charlotte Islands, it was named in 1853 by Cdr James C. Prevost after Sir James Robert Graham George (1792–1861), who was then the first lord of the Admiralty.

Graham Island NT In Norwegian Bay and between Ellesmere and Cornwall islands, this island (1378 km^2) was named in 1853 by Sir Edward Belcher after Sir James Robert

George Graham, then the first lord of the Admiralty.

Granby QC A city (1916) in the regional county of La Haute-Yamaska, east of Montréal, it had been created a village in 1859. The latter was named after the township of Granby, proclaimed in 1803, and named after John Manners, 4th Duke of Rutland and Marquess of Granby (1721–70), who fought bravely during the Seven Years' War in Germany, and returned to England in 1763 to a hero's welcome. The township municipality of **Granby** totally surrounds the city.

Grand Bank NF A town (1943) on the west side of the Burin Peninsula, its site was a French fishing port in the 1600s, which they described as *Grand Banc* or *Grand Banq*. It took its name from a prominent bank on the west side, which extends from Grand Bank Harbour to L'Anse au Loup Point.

Grand Bay NB A village (1972) northwest of Saint John, its post office was named in 1869, after the large bay formed by the Saint John River at that point, and named during the French regime. The bay was identified as *Le Grand Baye* by Robert Monckton in 1758, and as Grand Bay by Joseph Peach in 1762.

Grand Bend ON A village (1951) in Lambton County, west of Exeter, it was named in 1872 after a sharp bend in the Ausable River. Twenty years later the bend was altered by diverting the river into Lake Huron at the village, rather than following its original course to enter the lake, 16 km to the south at Port Franks.

Grand-Calumet QC This township municipality in the regional county of Pontiac, and surrounded by two branches of the Ottawa River, was created in 1847, a year after the township of Grand-Calumet was proclaimed. The name of the island of Grand Calumet recalls the practice by the Algonquin of smoking the pipe of peace, called *calumet* in French.

Grand Centre AB This place was incorporated as a town in 1958, but was annexed by the town of Cold Lake in 1996. Its post office was named in 1937, when it was perceived it would become an important centre in a prosperous agricultural area.

Grand Coulee SK The first Canadian Pacific Railway station west of Regina, this place was an incorporated village from 1907 to 1919, and was reincorporated in 1984. It was named after Cotton Wood Coulee, 4 km to the west, which, as the largest coulee (valley) crossed by the line in 1883, was given the designation 'grand'. It received postal service in 1903.

Grande-Anse NB A village (1968) west of Caraquet, its post office was named *Grand Ance* in 1851, after a cove of Chaleur Bay. The post office name was respelled Grande-Anse by 1883.

Grande Cache AB This town (1983) northwest of Jasper was created as a coal mining 'new town' in 1966, with its post office opening three years later. Its name recalls the hiding of furs about 1820 by trader Ignace Giasson, who decided a load of furs he was carrying, from the west of the Rocky Mountains down the Smoky River, was too difficult to carry through deep snow, prompting him to leave them in a large hiding place, a '*grande cache*'.

Grande Prairie AB A city (1958) in the Peace River country, it was founded as the site of a Hudson's Bay Company fort in 1881 and named by Mgr Émile-Jean-Marie Grouard, who identified the vast undulating landscape as a '*grande prairie*'. Promoted as a 'city' as early as 1910, it did not develop as a regional centre until the Edmonton, Dunvegan and British Columbia Railway arrived in 1916. Its post office was opened in 1911.

Grande-Rivière QC A town in the regional county of Pabok, southwest of Gaspé, it was created in 1967 through the merger of the village and the municipality of Grande-Rivière. The post office was called *Grand River* from 1853 to 1933, when it was changed to the French form. In 1697 the seigneury of Grande-Rivière was granted to Jacques Cochu. The river is rather less than grand, rising only 40 km to the northwest among many larger rivers.

Grande Rivière, La QC Rising in central Québec, this river flows west for nearly 900 km before emptying into James Bay. It has three reservoirs, created by Hydro-Québec in the 1970s and 1980s to harness its magnificent waterpower potential. It was known as *Big River* until the beginning of the twentieth century. Its Cree name is *Tschishasipi* (meaning 'big river'), a name reflected in the Cree village of Chisasibi at its mouth.

Grandes-Bergeronnes QC A village in the regional county of La Haute-Côte-Nord, northeast of Tadoussac, it was developed after 1845 when two sawmills were built at the mouth of the Rivière des Grandes-Bergeronnes, and was incorporated in 1929. The river's name originated with Samuel de Champlain in 1626, who likely mistook some plover for a type of sparrow, known in French as *bergeronnette*.

Grandes-Piles QC Located in the regional county of Mékinac, north of Shawinigan, this village's post office was called *Saint-Jacques-des-Piles* from 1882 to 1966, when it was changed to Grandes-Piles. The parish municipality of *Saint-Jacques-des-Piles*, created in 1885, became the village of Grandes-Piles in 1988. The name is derived from the sedimentary beds along the banks of the Saint-Maurice, described locally as strata being piled one on top of another.

Grand-Étang NS South of Chéticamp, this place was settled in the 1790s by Acadians from Prince Edward Island. The name means 'big pond'. Its post office was opened in 1871.

Grande-Vallée QC Located northwest of Gaspé, this municipality's name was derived from a seigneury, granted in 1691 to François Hazeur. Its post office was opened in 1872. The valley's name, noted by explorer Louis Jolliet in 1699, reflects its larger river in contrast to the nearby Rivière de la Petite Vallée.

Grand Falls NB A town (1890) on the Saint John River, its post office was opened in 1837. It was laid out as *Colebrooke* that year, after Sir William MacBean George Colebrooke (1787–1870), lieutenant-governor of New Brunswick, 1841–7. The post office was called *Colebrooke* in 1848, but Grand Falls was restored soon after. The French name, **Grand-Sault** is also official for the town's name. In 1688 Mgr Jean-Baptiste Saint-Vallier had called the falls *Grand Sault Saint Jean-Baptiste*. John Gyles, in his 1689 memoir of his life among the Maliseet, called it *Checanekepeag*, which means 'destroyer place' in Maliseet, probably in reference to a legend where Malabeam, a Maliseet woman, led Mohawks to their death.

Grand Falls-Windsor NF The present town at the **Grand Falls** on the Exploits River was incorporated in 1991. **Grand Falls** was laid out as a planned town in 1909, when the Anglo-Newfoundland Development Company constructed a pulp and paper mill here, and was incorporated as a town in 1961. **Windsor** was founded in the early 1900s, because only the company's mill workers could live in Grand Falls. It had first been named *Grand Falls Station*, but was renamed in 1938 after the British royal family.

Grand Forks BC A city (1897) at the junction of the Kettle and Granby rivers, near the Washington State border, its post office was named in 1895. Earlier the place was called *Grande Prairie*.

Grand Harbour NB In the village of Grand Manan since 8 May 1995, it had been incorporated as a village in 1968. Its post office was called *Grand Manan* in 1837 and was renamed Grand Harbour in 1868. The harbour was identified on maps in 1770.

Grand Lake NB The largest lake in the province, its name appeared on plans and maps in the 1770s. It was earlier known as *Grand Lac* in French. On a 1632 map it was named *Lac Freneuse* after Mathieu Damours, Sieur de Freneuse, who was then living in the area of Jemseg.

Grand Le Pierre NF A municipal community (1969) on the northeastern shore of Fortune Bay, its name may have been derived from *Grande Île à Pierre*, 'great stone island', possibly in reference to the steep cliff here.

Grand Manan Island NB The largest island in the Bay of Fundy, it was known in Passamaquoddy, Maliseet, and Penobscot as *Munanook*, 'island'. Samuel de Champlain referred to it as *Menane* on his 1607–8 and 1632 maps. The French called it Grand Manan after 1686, to distinguish it from Petit Manan, on the Maine coast, east of Bar Harbor. The present name was established on English maps by 1755. On 8 May 1995 the whole island was incorporated as the village of **Grand Manan**, with North Head, Grand Harbour, and Seal Cove becoming urban communities.

Grand Marais MB Situated on the southeastern shore of Lake Winnipeg, this community's name in French means 'big marsh'. Alexander Henry the younger referred to the place in his account of 1800. Its post office was established in 1923.

Grand-Mère QC A town on the right bank of the Saint-Maurice, north of Shawinigan, it was made a village in 1898, elevated to city status in 1920, and became a town in 1970. Its post office was opened in 1888 and named after the Algonquin *kokomis*, 'grandmother', in reference to a split rock that resembled the profile of an aged woman. It has often been misspelled *Grand'Mère*.

Grand Pré NS Settled as early as the 1680s by Acadians from Port-Royal, this place was resettled by New Englanders in the 1760s. In French the name means 'big meadow', referring to the extensive dyked farmland created by the Acadians. The post office was first called *Lower Horton* in 1842, and was given its present name in 1879.

Grand Rapids MB Located at the point where the Saskatchewan River enters Lake Winnipeg, this place takes its name from the former fast rapids, silenced by a hydroelectric development in the 1960s. In 1874 a tramway had been constructed around the rapids to facilitate the transfer of goods and people between the river and the lake. Postal service was set up here in 1900.

Grand-Remous QC On the Rivière Gatineau, west of Mont-Laurier, this township municipality was known from 1937 to 1973 as *Sicotte*, after Louis-Victor Sicotte, member of the Legislative Assembly, 1857–63. Its post office was also called *Sicotte* from 1927 to 1933, when it was changed to Grand-Remous, which describes some fast water and swirling eddies in the river.

Grand River ON The Grand rises in Grey and Dufferin counties, and drains into Lake Erie below the town centre of Dunnville, after passing through Fergus, Kitchener, Cambridge, Paris, and Brantford. It was called *Grande Rivière* during the French regime. It was also called *Rivière d'Urfé*, probably after missionary François-Saturnin Lascaris d'Urfé (1641–1701). Lt-Gov. John Graves Simcoe tried in 1793 to impose Ouse River, after a tributary of England's River Humber, but Grand River prevailed.

Grand Tracadie PE The name of this municipal community (1984), northeast of Charlottetown, was noted in the *Journal of the Assembly* in 1854. It took its name from **Tracadie Bay**, on its east side, and its post office was opened in 1871. The bay's name may have been derived from Mi'kmaq *tulakadik*, 'camping ground'. Samuel Holland called it *Bedford Bay*, with *Tracadi Harbour* as a secondary name. He named it after John Russell, 4th Duke of Bedford (1710–71), who was first lord of the Admiralty, 1744–8.

Grand Valley ON On the Grand River and in Dufferin County, west of Orangeville, it was called *Joice's Corners*, *Little Toronto*, *Manasseh*, and *Luther*, before its post office was called Grand Valley in 1883. Incorporated as a village in 1898, it was merged with East Luther Township in 1995 to form the municipal township of **East Luther Grand Valley**.

Grandview MB Situated west of Dauphin and midway between Riding Mountain National Park and Duck Mountain Provincial Park, this town and its post office were both named in 1901, presumably because a passenger on a Canadian Northern Railway train exclaimed that the place had such a 'grand view' of the Gilbert Plains to the south and west.

Granger, Mount YT Southwest of Whitehorse, this mountain (2035 m) was named after William P. Granger, who came to the Yukon to prospect in 1895. He died in the Copper King Mine in May 1907.

Granisle BC A village (1971) on the west shore of Babine Lake, northwest of Burns Lake, it was founded in 1965 by the Zapata Granby Mining Company to exploit copper on Sterret Island in the lake. Its name was created by combining 'Gran', from Granby, with 'isle'.

Grantham QC Located in the regional county of Drummond, west of the city of Drummondville, this municipality was created in 1936, taking its name from the township, proclaimed in 1800. The township was named after pioneer settler William Grant, and after the town of Grantham, in Lincolnshire, England.

Granton ON In Middlesex County, north of London, it was named by the Grand Trunk Railway in 1864 after Granton, Scotland, on the northwest side of Edinburgh, and after farmers Alexander, William, and James Grant. Two other farmers, William Levitt and George Foreman, wanted to call it *Awmik*, from the Ojibwa word for 'beaver', and it was selected at a public meeting. However, both the railway and postal officials rejected it.

Granum AB A town (1910) between Claresholm and Fort Macleod, its post office was named in 1907 by Malcolm McKenzie, a member of the House of Commons, taking the Latin for 'grain', because it was located in good grain-growing land. It was formerly called *Leavings*, because the trail from Fort Benton, MT to Calgary left the banks of Willow Creek at that point.

Granville Ferry NS Located across the Annapolis River from Annapolis Royal, this place's post office was established in 1836. It took its name from Granville Township, created in 1764, and named after John Carteret, Earl Granville (1690–1763), who had been president of the council in the British government, 1751–63.

Gras, Lac de NT Located 300 km northeast of Yellowknife, this lake (577 km^2) was recorded in 1890 by British explorer and hunter Warburton Pike, who wrote in his *Barren Ground* (1892) that it was known as Lac de Gras by his companions, presumably his Métis assistants. The name was submitted for approval in 1924 by surveyor G.H. Blanchet. The reason a name meaning 'fat lake' was given to the lake is not explained in the official records.

Grassy Lake AB Midway between Lethbridge and Medicine Hat, this place, incorporated as a village from 1950 to 1996, was named in 1893 by the Canadian Pacific Railway after a nearby lake, long since having gone dry. The lake's name was known to the Blackfoot as *Moyi-kimi*, 'grassy waters'.

Grassy Narrows ON In Kenora District, northeast of Kenora, it was named in 1960 after a thick growth of grass on the shore of **Grassy Narrows Lake**, an enlargement of the English River.

Grattan ON A township in Renfrew County, it was named in 1851 after Henry Grattan (1746–1820), who promoted the independence of Ireland and encouraged the emancipation of Catholics.

Gravelbourg SK Southwest of Moose Jaw, this town (1916) was founded in 1906 by Fr Louis-Joseph-Pierre Gravel (1868–1926) and five brothers, including Dr Henri Gravel and Émile Gravel, and a sister. They promoted a colonization scheme to resettle Québec and New England francophones in the West. Its post office was opened in 1907.

Gravenhurst ON A town (1887) on the southern end of Lake Muskoka, and in the district municipality of Muskoka, it was named in

1862 by William Dawson LeSueur. A secretary of the post office department, he called it after a reference to Gravenhurst in Washington Irving's novel, *Bracebridge Hall, or the Humorists* (1822).

Gray, Mount NB South of the Nepisiguit River, this mountain was named in 1964 by Arthur F. Wightman, then the provincial names authority, after John Hamilton Gray (1814–89), a Father of Confederation and chief justice of the British Columbia supreme court, 1872–89.

Grayson SK Southeast of Melville, this village (1906) was first called *Nieven*. It was renamed in 1903 by the Canadian Pacific Railway after construction contractor Harry Grayson. Its post office was opened the next year.

Great Bear Lake NT The largest lake (33,764 km^2) entirely within Canada, and the fourth largest in North America, it was noted in 1783–4 by fur trader Peter Pond. It took its name from the river, which flows west for 90 km into the Mackenzie River. The river was then known to the Dene as *Sacschohetha*, in reference to the bears in the area.

Great Harbour Deep NF Located on the eastern coast of the Northern Peninsula, this municipal community (1971) was known to French fishermen as *Baie l'Orange*. A 1693 British marine chart labelled it *Harver Deep*. The harbour is surrounded on three sides with steep cliffs leading to the plateau of the Long Range Mountains.

Great Slave Lake NT The second largest lake (27,048 km^2) in Canada, it was known to Samuel Hearne as *Arabasca Lake*, when he travelled to the mouth of the Coppermine River, 1770–2. The present name was noted on a map of the 1819–22 expedition of John (later Sir John) Franklin, and was named for the Slavey, an Athapascan tribe.

Great Village NS On the north side of Cobequid Bay and west of Truro, this place was the site of an Acadian settlement before 1755. Its post office was opened in 1855. It may have been named because it was judged to be a great site to establish a village.

Greely ON In Ottawa-Carleton Region, south of Ottawa, this community was named *Greeley* in 1885, after American explorer Adolphus Washington Greely (1844–1935), who sailed into the Canadian Arctic in the 1880s. It was correctly spelled Greely in 1903.

Greely Fiord NT On the western side of Ellesmere Island, this fiord was named in 1883 by Lt James B. Lockwood after Adolphus Washington Greely (1844–1935), who led the United States Polar Year expedition to Ellesmere Island, 1881–4.

Greenfield NS Situated on the Medway River, north of Liverpool, its post office was opened in 1857. Its name was likely descriptive of a field at the outlet of Ponhook Lake.

Greenfield Park QC Created in 1911, this city in the regional county of Champlain, east of Montréal, was possibly named after its green and park-like surroundings, although it has also been suggested its name was a translation of Parc des Prés-Verts.

Green Gables PE This name was given in 1953 to the post office in Cavendish. It was taken from the book *Anne of Green Gables*, published in 1908 by internationally acclaimed author Lucy Maud Montgomery. Its title relates to the green-gabled house now in Prince Edward Island National Park.

Green Lake SK At the north end of **Green Lake** and northwest of Prince Albert, this Métis community received a post office in 1901. The lake was named after a water plant, whose leaves gave the lake a very green colour, especially in summer.

Green, Mount YT In the St. Cyr Range, south of Ross River, this mountain (2033 m) was named in 1951 after Lt A.D. Green, who was killed in action during the Second World War.

Greenmount-Montrose PE Located between Alberton and Tignish, this municipal community (1977) comprises the localities of **Greenmount** and **Montrose**. The former had a post office from 1895 to 1914. The latter was named in 1864 by Donald McIntyre after Montrose, Angusshire, Scotland, with its post office being open from 1871 to 1913.

Greenock ON A township in Bruce County, it was named in 1850, possibly after Charles Cathcart (1783–1859), 2nd Earl Cathcart and 2nd Earl Greenock, who commanded the British forces in North America in the 1840s, and administered the government of the province of Canada for 14 months. His title was taken from the great Scottish seaport on the Firth of Clyde, northwest of Glasgow.

Green's Harbour NF On the eastern shore of Trinity Bay and southwest of Harbour Grace, its harbour was noted on maps as early as 1775. It has been speculated locally that the harbour may have been called after trees found there, providing lumber for boat building. It may, however, have been named after an English family.

Greenspond NF On **Greenspond Island**, and on the northwest side of Bonavista Bay, this town (1951) was settled by the Green and Pond families soon after 1697, with Thomas Green and Edward Pond being listed here in 1800. The place is identified on Herman Moll's map of 1801.

Green Valley ON In the united counties of Stormont, Dundas and Glengarry, south of Alexandria, it was named in 1878 after the Green Valley Hotel, established in the 1840s by Mary McDonald.

Greenwich NS Between Kentville and Wolfville, this place's post office was named in 1857, possibly after Greenwich, CT. It was called *Port Williams Station* from 1870 to 1970, when the original name was restored.

Greenwood BC Canada's least populous city (725 in 1991), it was established by American prospectors as a mining community in 1895, and named after the Greenwood mining camp in Colorado, west of Pueblo. It was incorporated in 1897.

Greenwood NS In the Annapolis Valley, southwest of Kentville, this village (1961) was first known as *Greenwood Square*. Its post office was opened in 1874 and closed in 1916. It reopened in 1966, after the closing of the *RCAF Station Greenwood* office.

Greenwood ON In the town of Pickering, Durham Region, it was first called *Norwood*, but was renamed in 1852 after miller Frederick Green.

Grenfell SK On the main Canadian Pacific Railway line east of Regina, the station and post office in this town (1911) were named in 1883 after CPR director Pascoe du Prés Grenfell.

Grenfell Heights NF On the Exploits River, east of Grand Falls-Windsor, this place was developed in the 1960s. It was named after Sir Wilfred Thomason Grenfell (1865–1940), a Methodist medical missionary, who served on the Labrador coast from 1892–6 and 1899–1939. Sir Wilfred Grenfell College in Corner Brook was named after him in 1975.

Grenville ON The county was named in 1792 after Lord William Wyndham Grenville (1759–1834), secretary of state for foreign affairs, 1791–1801. The municipality of the united counties of **Leeds and Grenville** was formed in 1852.

Grenville QC The township municipality of Grenville was created in 1845, abolished in 1847, and reinstated in 1855, taking its name from the township. The township was named in 1795, and proclaimed in 1808, after Lord George Grenville (1712–70), British prime minister 1763–5. The village of **Grenville** was separated from the township municipality in 1876.

Grenville, Mount BC In the Coast Mountains, east of Bute Inlet, this mountain (3079 m) was named in 1933 by George G. Aitken, provincial member of the Geographic Board of Canada, after Sir Richard Grenville (1542–91). He had been granted the first English charter to create a settlement in the Pacific. His submitted a proposal to search for the Northwest Passage, a mission later undertaken by Sir Francis Drake.

Gretna MB A village (1896) on the North Dakota border, west of Emerson, its Canadian Pacific Railway station was named in 1882 after Gretna Green, the place on the Scottish border noted for its many marriages of eloping couples. Its post office was opened in 1884.

Grey ON The county was named in 1852, after Henry George Grey, 3rd Earl Grey (1802–94), secretary of state for war and the colonies, 1846–52. **Grey Township** in Huron County was named in 1848, either for Charles Grey, 2nd Earl Grey (1764–1845), prime minister of Great Britain, 1830–4, or for Henry George Grey, 3rd Earl Grey.

Grey, Point BC West of Vancouver and extending into the Strait of Georgia, it was named in 1792 by Capt. George Vancouver after Capt. George (later Sir George) Grey (1767–1828) of the Royal Navy. He commanded HMS *Victory*, Sir John Jervis's flagship at the Battle of St.Vincent in 1797. In 1791 Spanish Lt Francisco Eliza named the point *Punta de Langara*, after Spanish Adm. Don Juan de Langara (d. 1806).

Grimsby ON The town (1921) in Niagara Region was named in 1816 after Grimsby Township. The township, originally in Lincoln County, was named in 1792 after the town of Grimsby, in England's Lincolnshire. It was divided into the municipal townships of North Grimsby and South Grimsby in 1882. When the region was formed in 1970, North Grimsby was merged with the town of Grimsby and South Grimsby was annexed by the municipal township of West Lincoln.

Grimshaw AB A town (1953) north of the Peace River and west of the town of Peace River, it was founded in 1921 by the Edmonton, Dunvegan and British Columbia Railway. It was named in honour of Dr M.E. Grimshaw, who established a medical practice in 1914 at Peace River and served as the mayor of that town in 1922.

Grindrod BC In the Shuswap River valley, southeast of Salmon Arm, this place's post office was named in 1911 after Edmund Holden Grindrod, the Canadian Pacific Railway's first inspector of telegraphs in the province in 1886. He was later a farmer near Kamloops.

Grinnell Peninsula NT The northwestern extension of Devon Island, this peninsula was named in 1851 by William Penny after Henry G. Grinnell (1799–1874), an American philanthropist. He fitted out two expeditions in 1850 and 1853 to search for Sir John Franklin.

Grise Fiord NT Located at the mouth of Grise Fiord and at the south end of Ellesmere Island, this hamlet (1987) is the farthest north of any permanent communities in Canada. It was founded in 1953 by the government of Canada as a suitable location for the resettlement of impoverished Inuit on the eastern shore of Hudson Bay. It was named *Grisefjorden* by explorer Otto Sverdrup in 1899, with *grise* meaning 'pig' in Norwegian.

Gros Morne NF One of the highest points (806 m) in the Long Range Mountains, and on the west coast of the island of Newfoundland, its name in French means 'great height'. The word *morne* may be derived from Spanish *morro*, 'mound'. **Gros Morne National Park** (1943 km^2) was established in 1970.

Grosse Île, La QC Located in the St. Lawrence River, southeast of Île d'Orléans, this small 3-km-long island gained notoriety from 1832 to 1937 as a quarantine station, where more than 10,000 immigrants, mostly Irish, died from cholera, typhus, and malnutrition. It was called *Île de Grâce* when it was ceded to Gov. Charles Huault de Montmagny in 1646. The name was transformed in 1662 into *Grosse-Île* in a reference to the seigneury of Noël Jérémie *dit* Lamontagne. In 1984 the island became a national historic site.

Grosses Coques NS On the French Shore, southwest of Digby, this place was founded by Acadians soon after their expulsion in 1755. Its name means 'large shells'. Its post office was opened in 1874.

Grouard AB On the northwest side of Lesser Slave Lake, this place's post office was first called *Lesser Slave Lake* in 1903. In 1909 it was renamed after Mgr Émile-Jean-Marie Grouard (1840–1931), vicar apostolic of the diocese of Athabasca, 1890–1929. The adjacent community on the Freeman Indian Reserve is called **Grouard Mission**.

Grouse Mountain BC In the district municipality of North Vancouver, this mountain (1221 m) was named by a climbing party which shot a blue grouse here in 1894.

Grunthal MB Situated southwest of Steinbach, this place's post office was named in 1898. In German the name means 'green valley'.

Guelph ON The city (1879) and the township were founded in 1827 by Canada Company general agent John Galt, choosing Guelph in honour of the royal family. The directors of the company wanted it called *Goderich*, after Lord Goderich, then the British prime minister. The conflict between Galt and the directors led to the erroneous story that the plans for Guelph and Goderich, on Lake Huron, had been switched.

Guernsey SK East of Saskatoon, near Lanigan, the post office at this village (1908) was named in 1907 after Guernsey, in the Channel Islands. Some of the street names recall places on the island of Guernsey. The Canadian Pacific Railway was built through it in 1908.

Guillaume-Delisle, Lac QC Named in 1744 *Sir Atwell's Lake* after Hudson's Bay Company governor Sir Atwell Lake, this saltwater body (712 km^2) was also known as *Winipeq*, *Artiwinipeck*, *Baie Winipeke*, *Golfe de Hazard*, *Golfe de Richmond*, and *Baie de Richmond* before it was officially named *Richmond Gulf* in 1905, a name of uncertain origin. In 1962 it was renamed after Guillaume Delisle (1675–1726), the great French map-maker, who was the first geographer to the King in 1718.

Gull Lake SK Southwest of Swift Current, this town (1911) was named in 1883 by the Canadian Pacific Railway. The nearby lake was named in 1879 by John Macoun, who provided agricultural advice to the CPR to justify the route chosen to cross the prairies. The name was a translation of the Cree *kiaskus*. The post office was opened in 1889.

Gunningsville NB In the town of Riverview, across the Peticodiac River from Moncton, Gunningsville was the name of a post office from 1910 to 1919 and 1939 to 1958. Its was an incorporated village from 1966 to 1973. The first settler here was Hazen Gunning.

Guysborough NS At the head of Chedabucto Bay, this place was founded by disbanded troops and their families in 1784, and named after Sir Guy Carleton, who was then commander-in-chief of British forces in North America. The previous year they had settled at Port Mouton, in the southwestern part of the province, and called the place *Guysborough*, but moved to the eastern part of the province after the settlement was wiped out by a fire. Postal service was established here in 1825. **Guysborough County** was created in 1836.

Gwaii Haanas National Park Reserve BC Located at the south end of Moresby Island of the Queen Charlotte Islands, this park (1470 km^2) was established as the South Moresby National Park Reserve in 1988 to preserve its wilderness and ancient Haida cedar villages. It was called South Moresby/Gwaii Haanas National Park Reserve the following year, and the present name was confirmed in 1993. In Haida the name means 'islands of wonder and beauty'.

Gypsumville MB This place, between Lake Manitoba and Lake Winnipeg, was the site of a gypsum mine in 1890, but it did not develop as a community until the post office opened in 1905 and the Canadian Northern Railway was built to it in 1912.

Hafford SK East of North Battleford, this town (1981) was established in 1912–13 by the Canadian Northern Railway, and may have been named after a survey engineer, who later lost his life in an accident in British Columbia. *Redberry* post office had been opened on 1 September 1906, but was changed to *Luxemburg* on 1 December 1906. It was renamed Hafford in 1913.

Hagar ON Located in Sudbury District, east of Sudbury, this place's post office was named in 1927 after **Hagar Township**, given in 1882 for Albert Hagar, Liberal member for Prescott in the Ontario Legislature, 1880–6.

Hagensborg BC On the Bella Coola River, upstream from Bella Coola, this place was named in 1900 after first postmaster Hagen B. Christenson. He added the Norwegian word for 'fortified place where chiefs used to live' to his own first name.

Hagersville ON An urban centre in the town of Haldimand, Haldimand-Norfolk Region, it was settled in the 1840s by David and Charles Hager and its post office named in 1852.

Hague SK Established in 1890 by the Canadian Pacific Railway, this village (1903) was named after bridge and railway engineer Jenkins Harry Hague (d. 1962 in Vancouver) during a meeting of CPR railway engineers in Prince Albert. The line became part of the Canadian Northern Railway in 1906 and of the Canadian National Railways in 1923. Its post office was opened in 1896

Ha! Ha!, Baie des QC Appearing as though it is an extension of the impressive Saguenay fiord, this bay, east of Chicoutimi, comes to an abrupt end only 10 km above the point where the Saguenay appears to divide in two rivers of equal size. *Haha* is an old French word describing an unexpected barrier blocking progress or penetration. At the head of this bay are two coves that give the illusion of leading to other waterways, thus explaining its plural form. The exclamation marks came about when surveyors, historians, and others presumed in the nineteenth century that the phrase represented a shriek of surprise by travellers. *See also* Saint-Louis-du-Ha! Ha!

Haileybury ON A town (1904) in Timiskaming District, it was named in 1889 by developer Charles C. Farr, after the exclusive Haileybury public school in Hertfordshire, England.

Haines Junction YT At the junction of the Haines and Alaska highways, this village was created in 1942, when the Alaska Highway was connected with Haines, AK, on Lynn Canal, an extension of the Pacific Ocean. The Alaska village was named about 1881 by Presbyterian missionary S. Hall Young, after Francine E. Haines, secretary of the committee of home missions.

Haldimand ON The county was named in 1783 after Sir Frederick Haldimand (1718–91), a native of Switzerland and a British Army officer, who was governor of Quebec from 1778 to 1786. The county was united in 1974 with Norfolk County to form the regional municipality of **Haldimand-Norfolk**. The town of **Haldimand** was also formed in 1974 by merging the town of Caledonia, the villages of Cayuga and Hagersville, the municipal townships of North Cayuga, Oneida, Seneca, and South Cayuga, and parts of the municipal townships of Rainham and Walpole. **Haldimand Township**, in Northumberland County, was named in 1792 after Sir Frederick Haldimand.

Haliburton ON The community in **Haliburton County** was named in 1864 after Nova Scotia humorist Thomas Chandler Haliburton (1796–1865), who had been a judge of the Nova Scotia Supreme Court. He had moved to England in 1856 and in 1859 he had been elected to the British House of Commons. He was also appointed chairman of the

London-based Canadian Land and Emigration Company, which in 1861 bought 402,979 acres in central Canada West, embracing ten townships. The county was created as a provisional county in 1874 and did not achieve full county status until 1983.

Halifax NS Located on the west side of **Halifax Harbour** this metropolitan area, in the regional municipality of **Halifax** since 1 April 1996, had been incorporated as a city in 1841. It was named in June 1749 by Gov. Edward Cornwallis after George Montagu Dunk, 2nd Earl of Halifax (1716–71), then British secretary of the southern department. In July of that year the seat of Nova Scotia government was moved here from Annapolis Royal. During the French regime the harbour was called *Chebucto*, derived from Mi'kmaq *chebookt*, 'great long harbour'. The new regional municipality embraces all the former municipal county of Halifax, extending east to Ecum Secum Bridge and west to Hubbards.

Halkirk AB East of Stettler, this village (1912) was named in 1910 by pioneer settlers from a list provided by the Canadian Pacific Railway. It was named after Halkirk, Caithness, Scotland, south of Thurso.

Hall Beach NT A hamlet (1978) on the eastern side of Melville Peninsula, it was founded in 1970. It took its name from nearby **Hall Lake**. The lake (474 km^2) was named by the Fifth Thule expedition in 1921–4 after Charles Francis Hall, who explored the area, 1864–9.

Hallebourg ON In Cochrane District, southeast of Hearst, it was called *Hallewood* in 1922 after Mgr Joseph Hallé (1874–1938), appointed the first Catholic apostolic vicar of Northern Ontario in 1919. The name was changed to Hallebourg in 1935.

Hallowell ON A township in Prince Edward County, it was named in 1797 after Benjamin Hallowell (1724–99), a Loyalist who had been granted 486 ha (1200 a) at the head of Picton Bay. He lived in England until his son-in-law, John Elmsley, became chief justice of Upper Canada in 1796, when he joined the Elmsley family in York (Toronto) in 1797 and died there two years later.

Hall Peak BC Located in the Purcell Mountains, northeast of Kaslo, this mountain (3040 m) was named in 1961 after Canadian Army private John H. Hall, of Marysville, BC, who was killed in action on 24 May 1945.

Hall Peninsula NT On the east side of Baffin Island and northeast of Iqaluit, this 280-km peninsula was named by the Geographic Board in 1901 after Christopher Hall, the master of the vessel *Gabriel* during the 1576 voyage by Martin (later Sir Martin) Frobisher. Frobisher had named *Hall Island* in Frobisher Bay in 1576 and this was endorsed by the board in 1928. It was changed to **Christopher Hall Island** in 1961.

Hallville ON In the united counties of Stormont, Dundas and Glengarry, west of Winchester, this place was named in 1879 after the hall of the local lodge of the Orange Order.

Halton ON The county was named in 1816 after Maj. William Halton (d. 1821), secretary to Lt-Gov. Francis Gore, 1806–11 and 1815–7. In 1974 it became the regional municipality of **Halton**. The town of **Halton Hills** in Halton Region was created in 1974 by amalgamating the towns of Georgetown and Acton, most of the municipal township of Esquesing, and a small part of the town of Oakville.

Halvorson, Mount BC In the Cariboo Mountains, west of McBride, this mountain (2782 m) was named in 1965 after Tpr Frank Halvorson of McBride, who was killed in action on 29 October 1944.

Hamill, Mount BC In the Purcell Mountains, northeast of Kaslo, this mountain (3243 m) was named in 1911 after prospector Thomas Hamil. He was murdered in 1885 by Robert Evan Sproule, a fellow prospector, who accused Hamil of jumping his claim. Sproule was hanged the following year in Victoria.

Hamilton ON The city (1846) was founded and named by George Hamilton, a native of

Queenston, and the son of Robert Hamilton, the founder of St. Catharines. He represented Wentworth in the House of Assembly of Upper Canada from 1820 to 1830. **Hamilton Harbour** was proclaimed *Burlington Bay* in 1792 by Lt-Gov. John Graves Simcoe, but by provincial order in council in 1919, it was renamed Hamilton Harbour. The high level bridge carrying the Queen Elizabeth Way over the harbour entrance is called the Burlington Bay James N. Allan Skyway, retaining its original designation, and adding the name of a former provincial minister of highways.

Hamilton ON A township in Northumberland County, it was named in 1792 after Henry Hamilton (*c.* 1734–96), lieutenant-governor of Quebec, 1781–5, and administrator of British North America, 1784–5. Subsequently, he was governor of Bermuda, where its capital bears his name. Fort Henry, at Kingston, was also named for him.

Hamilton Inlet NF An extended entrance (212 km to the head of Lake Melville) into the coast of Labrador, it was named after Sir Charles Hamilton (1767–1849), governor of Newfoundland, 1818–24. Earlier, the inlet was called in French *Baie des Esquimaux*, *Baie des Sauvages*, and *Baie Saint-Louis*. Churchill River, which flows into the inlet, was officially called *Hamilton River* until 1965. **Hamilton Sound**, between Fogo Island and the island of Newfoundland, was also named after Sir Charles.

Hamilton-Wentworth ON The regional municipality was created in 1974, and comprises the cities of Hamilton, Burlington, and Stoney Creek, the towns of Flamborough, Dundas, and Ancaster, and the municipal township of Glanbrook.

Hamiota MB Situated northwest of Brandon, the post office at this village (1906) was first called *Hamilton* in 1882 after first postmaster Thomas Hamilton, but confusion with the Ontario city prompted the selection of another name. In devising a new postal name in 1884, Hamilton may have been inspired by the rural municipality of Miniota, 30 km to the west, which was named in 1883 from the Sioux words for 'plenty' and 'water'.

Hammond ON In the united counties of Prescott and Russell and east of Ottawa, this place was named in 1895 by its first postmaster W.F. Empey after George Henry Hammond (1838–86), an American meat packer, who had slaughterhouses in Omaha, NE and Hammond, IN. Empey reported in 1906 that he did not know Hammond, but liked the appearance of the name.

Hammond, Mount BC In the Purcell Mountains, southwest of Invermere, this mountain (3368 m) was named in 1910 by mine developer Charles Ellis after Toronto financier Herbert Carlyle Hammond.

Hammonds Plains NS Northwest of Halifax, this place was likely named in the 1780s after Sir Andrew Snape Hamond (1738–1828), lieutenant-governor of Nova Scotia, 1780–2.

Hampden NF At the head of White Bay, northeast of Deer Lake, this municipal community (1959) was known as *Riverhead* before 1911. It may have been named after Hampden House, in Buckinghamshire, England, built by member of Parliament John Hampden, and later occupied by the Earl of Buckingham.

Hampshire PE A municipal community (1974) northwest of Charlottetown, its post office was named in 1897, probably after the English county.

Hampstead QC A town in the Communauté urbaine de Montréal, it was developed at the beginning of the First World War, although by 1932 it only had a population of less than 800. The name was inspired by the wealthy district northwest of the centre of London, England.

Hampton NB A village (1966) on the Kennebecasis River, northeast of Saint John, its first post office was opened in 1831. It was named after **Hampton Parish**, created in 1795, and possibly called after Hampton, NY, Hampton, NJ, or Hampton, southwest of London, England. In 1859 its European and North American Railway station was named *Ossekeag*, after the Ossekeag River. It was renamed *Hampton Station* in 1916.

Hampton ON In the municipality of Clarington, Durham Region, this community was first called *Elliott's Mills*, after a mill built in 1840 by Henry Elliott. When an application was made for a post office in 1851, *Millville* was proposed, but was rejected because of duplication. Elliott's suggestion of Hampton, after Kilkhampton, his birthplace in north Cornwall, England, was accepted.

Haney BC In the district municipality of Maple Ridge and on the north bank of the Fraser River, its post office was named in 1936 after Thomas Haney (1841–1916), who had settled here in 1876. In its early years it was known as *Haney's Landing*. *Port Haney* post office was also opened in 1936 and was renamed Maple Ridge in 1970.

Hanley SK Located south of Saskatoon, this town (1906) was established in 1890 by the Qu'Appelle, Long Lake and Saskatchewan Railway and Steamboat Company, and became part of the Canadian Northern Railway in 1906. It may have been named after Hanley Falls, MN, west of Minneapolis, on the Yellow Medicine River. Postal service was provided here in 1903.

Hanmer ON Gilbert Hanmer moved north from Brant County about 1900 and settled in what is now the town of Valley East, north of Sudbury. Hanmer post office was opened in 1904.

Hanna AB A town (1914) northeast of Calgary, it was named in 1913 by the Canadian Northern Railway after David Blythe Hanna (1858–1938), then third vice-president of the railway. After the Canadian government bought the railway in 1918, he became the first president of the Canadian National Railways, 1918–22.

Hanover ON A town (1904) in Grey County, it was first known as *Buck's Crossing* and *Adamstown*. Most residents disliked *Adamstown*, so in 1856 first postmaster Abraham Gottwals proposed Hanover, because many of the first settlers came from the area of Germany now known as Hannover.

Hansen ON A township in Sudbury District, southwest of Sudbury, it was named in 1986 after Rick Hansen, whose Man in Motion world tour focused on the need for research into spinal cord injuries. It had been named Stalin Township during the Second World War after Joseph Stalin, whose perceived villainy led to a unanimous vote in the Ontario Legislature to change it.

Hants NS A county in central Nova Scotia, it was named in 1781 after the colloquial name of England's Hampshire. For municipal purposes, it is divided into the districts of **East Hants** and **West Hants**.

Hants Harbour NF On the east shore of Trinity Bay and north of Carbonear, this town (1970) was named after the harbour, which was identified in the 1697 report by Abbé Jean Baudoin as *Lance arbe* and *Anse Arbre*, meaning 'tree cove'. By 1775 it had become known by its present name.

Hantsport NS A town (1895) on the west side of the Avon River, north of Windsor, it was first called *Halfway River*, as it was midway between *Pisiquid* (Windsor) and Grand Pré. It was named in 1849 by Ezra Churchill, because it was the chief seaport of Hants County. Postal service was established here that year.

Happy Adventure NF This municipal community (1960) on **Happy Adventure Bay** may have been named after an incident when a fishing ship eluded pirates in the adjacent maze of channels and coves. The place was first recorded in the 1869 census of the province.

Happy Valley-Goose Bay NF A town at the mouth of the Churchill River, it was incorporated in 1974 by amalgamating the town of Happy Valley (1961) and the local improvement district of Goose Bay (1970). In 1941 an airport had been established at Goose Bay, at the river's mouth, to provide a shorter transatlantic link with the British Isles. Happy Valley, a name suggesting pleasure and contentment, was created as the residential community of *Refugee Cove* in 1942, and was renamed in 1955.

Harbour Breton NF On the south coast of the island of Newfoundland and on the north side of Fortune Bay, this town (1952) was likely named after fishermen from Bretagne, France, in the 1600s, although it was also noted as *Havre Bertrand* during those years.

Harbour Grace NF A town (1945) on the west side of Conception Bay, it was named after its harbour, called *Havre de Grâce* by the French in the 1600s. It was likely named after the earliest name of Le Havre, at the mouth of the Seine, in France, which François I created as the port of *Havre de Grâce* in 1517. The English form of the name came into use as early as 1681.

Harbour Main-Chapel Cove-Lakeview NF Near the head of Conception Bay, this town (1964) comprises three separate built-up centres. **Harbour Main**, which occurred on a 1630s map as *Harbour Maine*, may have been called after a French family surname. **Chapel Cove**, which was first noted in the 1830s, may have been given for an English surname, either Chappel or Chapple. **Lakeview** was developed inland in the 1930s on the west side of First Pond, Second Pond, and Third Pond.

Harcourt NB This place was named in 1869 when the Intercolonial Railway was built northwest of Moncton. It was called after **Harcourt Parish**, named in 1826 after William Harcourt, 3rd Earl of Harcourt (1743–1830), field marshal of British forces in 1820, and a friend of Lt-Gov. Sir Howard Douglas. Its post office was named *Weldford* in 1871, after Weldford Parish, and was called Harcourt in 1894.

Hardisty AB Southwest of Wainwright, this town (1911) was named in 1906 by the Canadian Pacific Railway after Richard Charles Hardisty (*c.* 1832–89), the last chief factor of the Hudson's Bay Company at Fort Edmonton, 1872–88. Appointed to the Senate in 1888 by Sir John A. Macdonald, he died the next year in Winnipeg, after being accidentally thrown from his buggy at Broadview, SK.

Hardwicke Island BC Between Vancouver Island and the mainland and north of Campbell River, this island was named in 1792 by Capt. George Vancouver after Philip Yorke, 3rd Earl of Hardewicke (1757–1834), lord lieutenant of Ireland, 1801–04 and promoter of Catholic emancipation. He encouraged the naval career of Spelman Swaine, Vancouver's master's mate on the *Discovery*.

Hare Bay NF A town (1964) on the northwestern side of Bonavista Bay, it was possibly named in the 1860s, when a trapper caught hares in the rabbit snare he had set. There is also a **Hare Bay** on the northeastern side of the Northern Peninsula. It was identified on seventeenth-century French maps as *B. aux Lievres*, 'bay of hares', with Hare Bay occurring on several eighteenth century maps.

Harkin, Mount BC In the Rocky Mountains, northeast of Invermere, this mountain (2938 m) was named in 1923 by surveyor Morrison Bridgland after James Bernard Harkin (1875–1955), Dominion commissioner of parks, 1911–36.

Haro Strait BC The main channel uniting the Strait of Georgia to Juan de Fuca Strait, it was named *Canal de Lopez de Aro* in 1790 by Sub-Lt Manuel Quimper after his first mate on the captured British vessel *Princess Royal*, Gonzalez Lopez de Haro. The present form was adopted by the British Admiralty, possibly on the recommendation of Capt. George Richards in 1859.

Harper, Mount YT In the Ogilvie Mountains, northwest of Dawson, this mountain (1874 m) was named in 1887–8 by surveyor William Ogilvie after Arthur 'Cariboo' Harper (1835–98), one of the earliest (1873–4) prospectors in the Yukon.

Harricana, Rivière QC Rising just west of the city of Val-d'Or, this river flows north for 480 km to James Bay, crossing the Ontario border 30 km above its mouth, where it becomes the **Harricanaw River**. In Algonquin the name means 'river of biscuits', in reference to the hardtack carried by exploration parties and troops.

Harrietsfield NS South of Halifax, this place was likely named in the late 1700s after Harriet Thompson, the wife of Col William Thompson. He had a grant of 40.5 ha (100 a) surveyed in 1787. The place had its own post office from 1947 to 1958.

Harrington PE Located north of Charlottetown, this place's school district was named in 1855, followed by its post office in 1888. It was named after *Harrington Bay*, the original name given in 1765 to Brackley and Covehead bays by Samuel Holland. He had given that name in honour of William Stanhope, Viscount Petersham and 2nd Earl of Harrington (1719–79).

Harrington Harbour QC A fishing community situated on four islands on the north shore of the Gulf of St. Lawrence, its post office was named *Harrington* in 1890, and changed to Harrington Harbour seven years later. The islands may have been named about 1831 by Gov. Gen. Lord Aylmer after his friend Charles Stanhope, 3rd Earl of Harrington, who had died in 1829.

Harris SK Southwest of Saskatoon, the post office at this village (1909) was originally established in 1906, 3 km to the north, in the home of Richard Harris, who had arrived here from Huron County, ON, two years earlier. When the Canadian Northern Railway passed through in 1908, the post office was moved to the rail line.

Harris, Mount AB North of Lake Louise, this mountain (3299 m) was named in 1957 after Ley Edward Harris, a surveyor with the Topographical Survey of Canada, who had climbed it in 1919.

Harris, Mount BC On the British Columbia-Alaska boundary and in the Coast Mountains, this mountain (1948 m) was named by the United States Geographic Board in 1923, and by the Geographic Board of Canada in 1924, after Dennis R. Harris (1851–1932), who participated in the survey of the boundary in 1904. He was a son-in-law of Sir James Douglas, a chief trader of the Hudson's Bay Company, the second governor of Vancouver Island, and the first governor of British Columbia.

Harrison Hot Springs BC A village (1949) at the south end of Harrison Lake, northeast of Chilliwack, its post office was named in 1889. In 1828 Hudson's Bay Company governor Sir George Simpson named the lake after Benjamin Harrison, then a director of the company, and later its deputy governor, 1835–9.

Harrison, Mount BC In the southern Rocky Mountains, this mountain (3359 m) was named in 1964 after PO Francis A. Harrison of Cranbrook, who was killed 14 October 1944, while serving overseas.

Harriston ON Incorporated as a town in 1878, it is in Wellington County, northeast of Listowel. Its post office had been named in 1856 after miller Archibald Harrison and first postmaster Joshua Harrison.

Harrow ON In Essex County, south of Windsor, this town (1930) was named in 1857 by John O'Connor, a member of Parliament, 1867–82, and postmaster-general, 1873–82. It was named after the exclusive public school in Harrow, within present-day Greater London, England, northwest of the city of London.

Harrowsmith ON In Frontenac County, northwest of Kingston, its post office was named in 1857 in honour of Sir Henry Smith (1812–68), member for Frontenac in the Legislative Assembly of the province of Canada, 1841–61, and speaker, 1858–61. His friends called him Harry Smith.

Hartland NB A town (1918) on the east bank of the Saint John River, north of Woodstock, its post office was first called *Beckaguimeck* in 1852, and changed the following year to *Becaguimic*, both after the adjacent Becaguimec Stream. It was renamed Hartland in 1868, possibly for its central location in Carleton County, although claims have been made for James R. Hartley, a member of the Legislative Assembly, and the Revd Samuel Hartt.

Hartney MB A town (1905) in the Souris River valley, southwest of Brandon, its post office was named in 1885 after postmaster and farmer James H. Hartney. He was later an immigration agent in Toronto.

Harvey NB A village (1966) southwest of Fredericton, it was named in 1837 by Andrew Inches after Sir John Harvey (1778–1852), lieutenant-governor of New Brunswick, 1837–41. Its post office was named in 1842, but was closed in 1873, because of confusion with **Harvey** in Albert County. **Harvey Station** post office in the village had been opened in 1870.

Harvey ON A township in Peterborough County, it was named in 1821 for Sir John Harvey (1778–1852), who had a distinguished military career, including service in Upper Canada during the War of 1812. Between 1836 and 1852, he served as lieutenant-governor of Prince Edward Island, New Brunswick, and Nova Scotia, as well as governor of Newfoundland.

Harwich ON A township in Kent County, it was named in 1794 after the town of Harwich, in Essex, England, on the North Sea coast.

Harwood ON On the south shore of Rice Lake and in Northumberland County, north of Cobourg, this place was named in 1854 after first postmaster Euphrasia Vivian Harwood, an English gentleman, who moved to Canada East (present Québec) soon after his appointment.

Hastings ON The county was named in 1792 after Francis Rawdon-Hastings, Baron Rawdon, Earl of Moira (1754–1826), who became the Marquess of Hastings in 1817. The title was inherited through his mother's side of the family.

Hastings ON A village (1874) on the Trent River and in Northumberland County, it was named in 1851 by miller Henry Fowlds after Flora Muir Campbell, the Countess of Loudoun (1780–1840), wife of the Marquess of Hastings.

Hatfield Point NB On Belleisle Bay, and north of Hampton, this place's post office was named *Sprague's Point* in 1855. It was renamed Hatfield Point in 1880. Thomas Spragg (d. 1812) was the first settler here. Among settlers in 1866 were Daniel, Edmund, Uriah, and Weeden Hatfield. Variant spellings of surnames, such as Sprague and Spragg, were common in Canada until the early 1900s.

Havelock NB West of Moncton, this community's post office was called *Butternut Ridge* in 1848. It was renamed in 1964 after **Havelock Parish**, established in 1858, and named after Sir Henry Havelock (1795–1857), who was besieged at Lucknow, India, in 1857, and died a hero seven days after he was relieved by Sir Colin Campbell.

Havelock ON A village (1892) in Peterborough County and east of the city of Peterborough, it was named in 1859 after Sir Henry Havelock, who fought with distinction in various wars in Asia. During the Sepoy Mutiny in India in 1857, Havelock was besieged at Lucknow. A township in Haliburton County was also named for him in 1872.

Havre-aux-Maisons QC A municipality since 1875 in the regional county of Les Îles-de-la-Madeleine, its post office was called *House Harbour* from 1870 to 1964, when it was changed to the present name, taken from the name of the island. In the eighteenth century the island had been called *Allright* and *Alright*, and in the nineteenth century it was known as *Saunders*, after Adm. Charles Saunders, who accompanied Gen. James Wolfe to the capture of the city of Québec in 1759.

Havre Boucher NS East of Antigonish, the site of this village (1960) was occupied in the winter of 1759 by Capt. François Boucher, according to an 1812 report by Mgr Joseph-Octave Plessis. In 1785 John and Paul Bushee were reported as being here and in 1811 Paul Boucher was recorded as living at *Havre au Bouchee*. The post office was called *Harbour Au*

Bouche in 1854, and was renamed Havre Boucher in 1909.

Havre-Saint-Pierre QC Located 200 km east of Sept-Îles, this municipality was settled in the 1850s by Acadian families from the Îles de la Madeleine. Incorporated as *Pointe-aux-Esquimaux* in 1873, it was renamed Havre-Saint-Pierre in 1927, three years after *Esquimaux Point* post office had been changed to reflect the importance of the place's harbour and the name of its religious parish.

Hawke's Bay NF At the head of Ingornachoix Bay and on the northwestern coast of the Northern Peninsula, this town (1956) was called after the adjacent bay. The bay was named in 1767 by Capt. James Cook after Adm. Edward Hawke, 1st Baron Hay (1705–81). Two years later Cook assigned the same name on the east coast of New Zealand's North Island.

Hawkesbury ON A town (1896) in the united counties of Prescott and Russell, it was first known as *Hawkesbury Mills* and *Hamilton's Mills* before Hawkesbury post office was named in 1819. The original Hawkesbury Township was created in 1798, and named either for Charles Jenkinson, Baron Hawkesbury and Lord Liverpool (1727–1808), or for his son, Robert Banks Jenkinson (1770–1828), who, as Lord Liverpool, was prime minister of Great Britain. 1812–27. *See also* East Hawkesbury *and* West Hawkesbury.

Hawkestone ON In Simcoe County, south of Orillia, this place was named in 1846 by James Patton of Barrie after Anthony Bewden Hawke (d. 1867), chief immigrant agent for Upper Canada, 1835–64.

Hawkesville ON A community in Waterloo Region, northwest of Waterloo, it was named in 1852 for Gabriel Hawke, the first postmaster. His father and three of his brothers settled here about 1846.

Hay ON A township in Huron County, it was named in 1835 after R.W. Hay, then a joint secretary of state for the colonies. The other was Lord Stanley, after whom the adjacent Stanley Township was named.

Hayes River MB Rising in Molson Lake, northeast of Norway House, this river flows 420 km northeast into Hudson Bay, just east of the mouth of the Nelson. Its lowest part was named by Pierre Radisson in 1684 after Sir James Hayes, Prince Rupert's secretary and a charter member of the Hudson's Bay Company. In 1901 the name was extended to its head.

Hayes Peak YT On the west side of Teslin Lake, southwest of Whitehorse, this peak (1849 m) was named in 1897 by surveyor Arthur St Cyr, after Dr Charles Willard Hayes. A geologist with the United States Geological Survey, Hayes explored southwestern Yukon in 1891 with US Army Lt Frederick Schwatka.

Hay Lakes AB A village (1928) southeast of Edmonton, its post office was named in 1913 after nearby lakes, known in Cree as *Apichikoo Obiwas*, 'little swamp'. The lush grass of such lakes was called 'prairie wool' by the pioneer settlers.

Hay River NT Originally established on Vale Island at the mouth of **Hay River**, the centre of the town (1980) was moved 6 km upriver to its present site, after severe flooding in 1963. The river (702 km), identified on an Arrowsmith map of 1854, was called after grass along its banks.

Haysville ON Located in Wilmot Township, Waterloo Region, southwest of Kitchener, this place's post office was named *Wilmot* in 1837. It was renamed in 1854 after miller and first postmaster Robert John Hays.

Hazelbrook PE East of Charlottetown, this place's school district was named in 1893. Its post office was called *Hazel Brook* from 1906 to 1912. Possibly the name was given after hazelnut trees in the valley of Fullertons Creek.

Hazelton BC A village (1956) at the north side of the junction of the Skeena and Bulkley rivers, it was developed in 1871 by Edgar Dewdney, and named by Thomas Hankin, when he found a large amount of hazelnuts ripening there. Its post office was opened in 1899. **South Hazelton**, on the south side of the junction, received its own postal service in 1923. The district municipality of **New Hazelton** (1980) is located east of Hazelton, in the Bulkley River valley. New Hazelton post office was opened here in 1912.

Hazen, Lake NT In northern Ellesmere Island, this lake (537 km^2) was named in 1882 by explorer Adolphus Greely, after Gen. W.B. Hazen, chief signals officer of the United States Army, 1880–7.

Hazen Strait NT Between Melville and Mackenzie King islands, this strait was named in 1917 by Vilhjalmur Stefansson after Sir John Douglas Hazen (1860–1937), who was then the minister of both the marine and fisheries, and the naval service.

Headingley MB A rural municipality west of Winnipeg, it had been united with Winnipeg from 1972 to 1992, when it was separated from it. It was named about 1853 by the Revd Griffith Owen Corbett after his former parish in Leeds, West Yorkshire, England, and *Headingly* post office was opened in 1871. In 1951 the name was respelled Headingley.

Head-Smashed-In Buffalo Jump AB West of Fort Macleod, this provincial historic site and interpretative centre is designated a UNESCO World Heritage Site. Buffalo jumps were part of the ancient hunting culture of Plains Aboriginals. This particular one recalls the story of a curious brave, who was trapped on a ledge as the buffalo plunged over the jump, and was subsequently found with his skull crushed.

Hearst ON In Cochrane District and at the point where the main line of the Canadian National Railways meets the Algoma Central Railway, this town (1922) was founded shortly after 1900. First named *Grant*, it was renamed in 1912 after Sir William Howard Hearst (1864–1941), minister of lands, forests and mines, 1911–4, and premier of Ontario, 1914–9.

Heart's Content NF Located on the eastern shore of Trinity Bay and northwest of Carbonear, this town (1967) was identified by John Guy in 1612 as *Hartes content*, and by Abbé Jean Baudoin in 1697 as *Havre-Content*. Its name may have been given by English fishermen either after a ship or as a place of relaxation and protection from stormy seas.

Heart's Delight-Islington NF A town (1972) on the eastern shore of Trinity Bay and west of Carbonear, it comprises the separate communities of **Heart's Delight** and **Islington**. Heart's Delight may have been named in the early 1800s, because the first settlers to arrive were delighted with its beauty. First known as *Island Cove*, Islington was named in 1911, possibly after the district in London, England.

Heart's Desire NF A local improvement district on the eastern shore of Trinity Bay and west of Carbonear, it was named in the 1700s, possibly because its pleasant site fulfilled the 'desires' of its first settlers.

Hebbville NS Southwest of Bridgewater, this village (1975) may have been named after pioneer settler John George Hebb. George, Adam, and Nicholas Hebb bought land here in 1805. Its post office was opened in 1913. **Hebbs Cross**, 7 km to the southwest, had a post office in 1873.

Hébertville QC A municipality east of Lac Saint-Jean, it was created in 1859 and named after Nicolas-Tolentin Hébert (1810– 8), who led settlers here from Kamouraska in 1849 and founded the parish of Notre-Dame-d'Hébertville in 1857. The village of **Hébertville-Station**, 3 km to the north, was incorporated in 1903, eight years after its post office had been opened.

Hebron NS Located north of Yarmouth, this place was developed around a house built by pioneer settler and shoemaker Anthony Landers, and named by him after the biblical Hebron. Postal service was established in 1838.

Hecate Strait BC Separating the Queen Charlotte Islands from the mainland, this water body was named about 1862 by Capt. George Richards after his Royal Navy survey ship *Hecate*, which he commanded in 1861 and 1862.

Hectanooga NS Northeast of Yarmouth, this place was named about 1874, when the Western Counties Railroad was built from Yarmouth to Digby. It had a post office from 1887 to 1963. The name was devised from the Mi'kmaq language by a daughter of a lieutenant-governor, possibly Adams George Archibald.

Hector, Mount AB North of Lake Louise, this mountain (3394 m) was named in 1884 by geologist George Dawson, after Dr (later Sir) James Hector (1834–1907), the surgeon and geologist with the Palliser Expedition, 1857–60. Hector named many geographical features in Western Canada, including Castle Mountain and Kicking Horse Pass. He headed the Geological Survey of New Zealand from 1865 to 1903.

Hedley BC In the Similakameen River valley, east of Princeton, this place was named in 1898 by prospector Peter Scott after Robert R. Hedley, manager of the Hall Mines smelter at Nelson. Hedley had financially helped Scott stake some of the earliest claims in the area of the community. Its post office was opened in 1903.

Heidelberg ON In Waterloo Region, northwest of Waterloo, this community's post office was named in 1854 after the city in present-day Baden-Württemberg, Germany, noted for its castle and university.

Heisler AB A village (1961) southeast of Camrose, its post office was named in 1915 after pioneer settler Martin Heisler.

Hemmingford QC Located south of Montréal, near the New York State border, the municipal township of Hemmingford was established in 1845. It took its name from the township, proclaimed in 1799, and named after Hemingford in Huntingdonshire, England, northwest of Cambridge. The village of **Hemmingford** was separated from the township municipality in 1878.

Henrietta Maria, Cape ON The point where the west side of James Bay meets Hudson Bay was named in 1631 by explorer Thomas James after his ship the *Henrietta Maria,* which was called after the wife of Charles I of England.

Henryville QC Situated east of the Richelieu, and south of Saint-Jean-sur-Richelieu, this village's post office was called *Saint-Georges* in 1827, taking its name from the Anglican parish of St George, established in 1794, and named after George III. The post office was renamed Henryville in 1927. The parish municipality of *Saint-Georges*, established in the 1840s, was called **Henryville** in 1957. In 1815 Napier Christie Burton, the seigneur of Noyan, appointed Edmund Henry (1760–1841) to develop his lands on both sides of the Richelieu.

Hensall ON A village (1896) in Huron County, southeast of Goderich, its post office was named in 1876 after Hensall in West Yorkshire, England. Brothers George and James Petty had migrated from that place, southeast of Leeds, in 1851.

Hepburn SK North of Saskatoon, this village (1919) was named about 1908, when the Canadian Northern Railway was built from Dalmeny to Carlton. Gordon Hepburn gave 16 ha (40 a) for the village centre. Its post office was opened in 1909.

Hepworth ON A village (1907) in Bruce County, northwest of Owen Sound, it was named in 1866 after Epworth, the birthplace of John Wesley, the founder of Methodism, in Lincolnshire, England, 40 km southwest of Kingston upon Hull. The Revd Josias Greene proposed the name, and prominent resident William Plows's mispronunciation resulted in Hepworth.

Herbert SK East of Swift Current, this town (1912) was founded by the Canadian Pacific Railway in 1883 and named after English diplomat Sir Michael Henry Herbert. Its post office was established in 1904.

Hermitage-Sandyville NF On the south coast of the island of Newfoundland and northwest of Harbour Breton, this municipal community (1963) comprises **Hermitage**, a settlement on **Hermitage Bay**, and **Sandyville**, a second settlement on Connaigre Bay. Hermitage Bay was possibly named by Channel Islands fishermen, who perceived that a small island in the bay resembled the Hermitage, an island outside the port of St. Helier, Jersey. Sandyville was first called *Dawson's Cove*, noted by Capt. James Cook in 1765 as *Dawsson's Cove*. There is a wide sandy beach at the head of Sandyville's harbour.

Herring Cove NS South of Halifax, this place's post office was named in 1875 after the cove. The cove was identified as *Ance du Hareng* on a 1779 French marine chart. In 1779 Charles Morris showed it on a map as *Dunk alias Herring Cove*. He may have named it after George Montagu Dunk, 2nd Earl of Halifax (1716–71), who had been British secretary of the southern department when Halifax was founded.

Herschel Island YT Off the north coast of the Yukon, this island was named in 1826 by John (later Sir John) Franklin after Sir William Herschel (1738–1822), the noted British astronomer.

Hespeler ON An urban centre in the city of Cambridge, it was settled in 1845 by miller Jacob Hespeler, who was appointed the first postmaster of *New Hope* in 1851. When it was incorporated as a village seven years later, the name was changed to honour the place's leading citizen.

Hewitt Bostock, Mount BC In the Cascade Mountains, southwest of Merritt, this mountain (2179 m) was named in 1932 by Charles Taggart after Hewitt Bostock (1864–1930), a Cambridge graduate, who had settled on a ranch east of Kamloops in 1893 and was elected to the House of Commons three years later. Appointed to the Senate in 1904, he served as both its Liberal leader and speaker.

Hiawatha ON On the north shore of Rice Lake and in Peterborough County, southeast of Peterborough, it is located in **Hiawatha Indian Reserve 36**. It was named in 1860 after Henry Wadsworth Longfellow's epic poem, *The Song of Hiawatha* (1855). The hero of the poem is a legendary co-creator of the League of Five Nations, also known as the Iroquois.

Hibbert ON A township in Perth County, it was named in 1830 after William T. Hibbert, Jr, a director of the Canada Company, a British company established in 1826 to develop the Huron Tract, a large area of what is now Southwestern Ontario.

Hibbs Cove NF This small fishing centre on Bay de Grave, northeast of Bay Roberts, was founded in the mid-1700s as *Hibbs Hole*. It may have been named a century earlier for Thomas Hibbs, one of the earliest settlers in Conception Bay. Although the word 'hole' has been long established in the English language as a small cove, it has acquired an indelicate connotation in the twentieth century. The name was changed to Hibbs Cove in 1969.

Hickman's Harbour NF On the southern side of Random Island and on Random Sound, an inlet on the western side of Trinity Bay, this place's harbour may have been named in 1770 by Capt. James Cook, after Jonathan Hickman (1747–1847), the pilot of Cook's survey vessel in Trinity Bay.

Hickson ON Located in Oxford County, north of Woodstock, this place's post office was named in 1883 for Sir Joseph Hickson (1830–97), general manager of the Grand Trunk Railway, 1874–90.

Hickson, Mount BC In the Coast Mountains, north of Mount Waddington, this mountain (3170 m) was named in 1928 by W.A.D. (Don) Munday after climber J.W.A. Hickson.

Highgate ON A village (1917) in Kent County, northeast of Chatham, it was named in 1865, after Highgate in north central London, England.

High Level AB Located in northwestern Alberta and at the height of land separating the Peace and Hay rivers, the post office in this town (1965) was named in 1958. Its site was known to the Slavey as *Tloc-Moi*, 'hay meadow'.

High Prairie AB A town (1950) west of Lesser Slave Lake, it was first known as *Prairie River*. Its post office was named in 1910, five years before the Edmonton, Dunvegan and British Columbia Railway arrived.

High River AB Located on the Highwood River, south of Calgary, this town (1906) was first called *The Crossing*. It was named High River by the Canadian Pacific Railway in 1892. The river's name, identified as *High Woods River* by Thomas Blakiston in 1858, was a translation of Blackfoot *ispitsi*, 'tall timbers', in reference to the trees along its banks rising higher above the surrounding lands than usually found along rivers flowing from the Rocky Mountain Foothills.

Hilden NS Located south of Truro, this place was known as *Halifax Road*, *Clarksville*, *Johnstons Crossing*, and *Slabtown*, before it was named Hilden by an act of the Legislative Assembly in 1895. It may have been named after Hillden, near Lisburn, south of Belfast, Ireland.

Hillier ON A township in Prince Edward County, it was named in 1824 after Maj. George Hillier (d. 1840), aide-de-camp and secretary to Lt-Gov. Sir Peregrine Maitland.

Hillsborough NB A village (1966) on the west bank of the Petitcodiac River, southeast of Moncton, its post office was opened in 1842, and named after **Hillsborough Parish**. The parish, named in 1786, was created as a township the year before, and called after Wills Hill, Earl of Hillsborough (1718–93), then lord commissioner of trade and plantations.

Hillsborough River PE Rising at the community of **Head of Hillsborough**, southwest of Morell, this river flows into **Hillsborough Bay**, on the south side of Charlottetown. The names of the river and the bay were given in 1765 by Samuel Holland after Wills Hill, Earl of Hillsborough (1718–93), lord commissioner of trade and plantations, 1763–6. The French knew the river as *Rivière du Nord-Est*. It has been commonly called the *East River* during the twentieth century, as a counterpoint to the North and West rivers.

Hillsburgh ON A community in Wellington County, northeast of Guelph, its post office was named in 1851 after Hiram Hill, and his son Nazareth, who had built the first sawmill here the previous year.

Hillsdale NF Located between Witless Bay and Cape Broyle on the Avalon Peninsula, this place was originally called *Brigus South*. The roots of that name may be traced to a 1636 map, where its harbour is identified as *Abra de Brigas*, likely a Portuguese description of its rough waters. Because of confusion with Brigus in Conception Bay, the place was renamed Hillsdale in the 1960s.

Hillsdale ON A post office was opened in 1867 at this place in Simcoe County, 22 km north of Barrie, and named after tavern keeper Alexander Hill.

Hill Spring AB Located southwest of Lethbridge, the post office at this village (1961) was named in 1911, after a large spring on the north end of a hill, which provided water for the village. It had been traditionally called *Spring Hill*, but as that name had already been taken for a post office elsewhere in Canada, the two words were reversed.

Hilton ON This township on St. Joseph Island, and in Algoma District, was named in 1887. *Marksville* post office was opened in 1879 and changed to *Hiltonbeach* in 1928. **Hilton Beach** became a village in 1923 and the postal name was amended in 1956.

Himsworth North; Himsworth South ON Himsworth Township in Parry Sound District was named in 1876 after William Alfred Himsworth (1820–80), clerk of the Queen's Privy Council, 1872–80. The township was divided into the municipal townships of Himsworth North and Himsworth South in 1886.

Hinchinbrooke ON A township in Frontenac County, it was named in 1798 after Viscount Hinchinbrooke, 5th Earl of Sandwich, who, as a member of Britain's Parliament, voted for the Canada Bill in 1791.

Hinchinbrooke QC A municipal township south of Salaberry-de-Valleyfield, it was first created in 1845, abolished in 1847, and re-established in 1855. It took its name from the township, named in 1795, and proclaimed in 1799, which was called after an estate in Huntingdonshire, England.

Hines Creek AB North of the Peace River and west of the town of Peace River, the post office at this village (1951) was named in 1928 after the nearby creek. It was relocated 5 km when the Edmonton, Dunvegan and British Columbia Railway arrived later that year. The creek may have been named after an Anglican missionary, who served in the area from 1875 to 1888.

Hinton AB A town (1958) in the Athabasca River valley and between Edson and Jasper, its post office was named in 1910 after the Hinton Trail, which led from Jasper to the Yukon. The trail was called after William Pitman Hinton, who became president and general manager of the Grand Trunk Pacific Railway in 1917.

Hinton, Mount YT Northeast of Mayo, this mountain (2059 m) was named in 1904 after Thomas Hinton, a mining recorder in the early 1900s at both Dawson and Mayo.

Hixon BC Midway between Prince George and Quesnel, this place's post office was named in 1923 after Joseph Foster Hixon, who prospected for gold here in 1866.

Hobbema AB An urban community serving several adjacent Indian reserves south of Wetaskiwin, it was named in 1891 by the Canadian Pacific Railway after the Dutch landscape painter Meyndert Hobbema (1638–1709). CPR president Sir William Van Horne collected his paintings.

Hochfeld MB A Mennonite community south of Morden, it was founded in 1886. In German, the name means 'high field'. It had a post office from 1909 to 1918.

Hodgeville SK Southeast of Swift Current, the post office at this village (1921) was named in 1908. It was named after the first postmaster, a Mr Hodges. The Canadian Northern Railway arrived here in 1921.

Hodnett, Mount YT South of Whitehorse, this mountain (1993 m) was named in 1906 after David Hodnett, who discovered lode gold near it that year.

Holberg BC At the head of **Holberg Inlet** and near the north end of Vancouver Island, it was named in 1895 by Danish settlers, after Danish historian and dramatist Baron Ludvig Holberg (1684–1754). Its post office was opened in 1909.

Holden AB Southeast of Edmonton, the post office at this village (1909) was named in 1907 after James Bismark Holden (1876–1956), who settled at Vegreville in 1900. He was a member of the Alberta Legislature, 1906–13, and served as mayor of Vegreville from 1921 to 1945. The post office here had been earlier called *Vermilion Valley*.

Holdfast SK Located northwest of Regina, this village (1911) was named in 1910, when farmer John A. Fahlman declined to let the Canadian Pacific Railway go through his land, forcing it to go farther west, before turning north. Because he was determined to 'hold fast', the CPR named the station after his obstinacy. *Frohlich* post office had been opened in 1907, with G. Frohlich as the postmaster. It was changed to Holdfast in 1912.

Holland MB Southwest of Portage la Prairie, this place's post office was named in 1880 after first postmaster A.C. Holland. When the Canadian Pacific Railway built a line from Winnipeg west to Souris in 1886, the post office and school were moved 3 km east to the present location.

Holland ON A township in Grey County, it was named in 1840 after Henry Richard Vassall Fox, 3rd Baron Holland (1773–1840), a minister in the 1830s in the governments of Lord Grey and Lord Melbourne. *Williamsford Station* post office was opened in 1874, but was changed to **Holland Centre** in 1886.

Holland Landing ON On the **Holland River East Branch** and in the town of East Gwillimbury, York Region, this place was an important transfer point before the availability of reliable land and rail transportation. Lt-Gov. John Graves Simcoe named it *Gwillimbury*, after his wife's father, Lt-Col. Thomas Gwillim (d. 1762). It was also known as *St. Albans* and *Beverley*, before Holland Landing post office was opened in 1831.

Holland River ON Flowing north into Lake Simcoe, this river was named in 1793 by Lt-Gov. John Graves Simcoe after Samuel Holland (1728–1801), surveyor general of Quebec from 1764, and of the northern district of North America, 1764–78. Its west branch (officially *Schomberg River*, 1927–65), now the head of the river, is the location of the **Holland Marsh** market gardens, first developed in 1927.

Holman NT On the southwestern shore of Diamond Jenness Peninsula, on Victoria Island, this hamlet (1984) was established at this site in 1960 by the Hudson's Bay Company. It was previously called *Holman Island* after the island, 8 km to the south. The island may have been named after John R. Holman, who was the surgeon general in 1853 and 1854 on the voyages of Edward Inglefield.

Holstein ON Located in Grey County, north of Mount Forest, this community's post office was named in 1864, when Prussia won the uneven contest with Denmark over control of Holstein and Schleswig.

Holway, Mount BC Located in the Selkirk Mountains, north of Revelstoke, this mountain (3047 m) was named in 1912 by climber Howard Palmer after Ewart Willett Dorland Holway (1853–1923), a botany professor at the University of Minnesota, who led the first ascent of this mountain.

Holyrood NF At the head of Conception Bay, this town (1962) was named after **Holyrood Bay**. The bay may have been named after Holyrood, in the parish of Crewkerne, Somerset, England. It was noted as *Hollyrude* as early as the 1630s.

Homathko River BC Rising northeast of Mount Waddington, this river flows southwest into Bute Inlet. It was named in 1911 after the Mainland Comox word for 'swift'. **Homathko Icefield**, east of the river, rises to an elevation of 2400 m. In the 1860s Alfred Waddington pursued financing to build a railway through the river's valley in order to connect Victoria and Vancouver Island with Eastern Canada.

Honey Harbour ON In the district municipality of Muskoka and northeast of Midland, this resort community was developed with tourist facilities after 1900. with its post office being named *Royal Honey Harbour* in 1910. 'Royal' was dropped from the name in 1912, when the post office was moved to the mainland.

Honeywood ON In Dufferin County, north of Shelburne, this community was settled in the 1840s by young men from Yorkshire, England, who bought the land from a Mr Wood of Toronto. *Rosewood*, composed from Mr Wood's surname and his wife's first name, was not accepted as a postal name, but Honeywood was approved in 1865.

Honguedo, Détroit d' QC The large channel at the mouth of the St. Lawrence River, between Île d'Anticosti and Péninsule de Gaspé, was first noted by Jacques Cartier in his

Relation of 1535–6. However, it was not portrayed on a map until the twentieth century. It may mean 'gathering place' in Mi'kmaq, or it may be connected to *hehonguesto*, 'his own nose', a word from an Iroquoian tribe's language listed in Cartier's vocabulary.

Hooker, Mount AB, BC This mountain (3286 m) east of Athabasca Pass was named in 1827 by Scottish botanist David Douglas after one of his patrons, Sir William Jackson Hooker (1785–1865), English botanist and director of Kew Museum and the Royal Botanical Gardens, southwest of London.

Hope BC This district municipality (1992) is located at the confluence of the Fraser and Coquihalla rivers. It was the site of the Hudson's Bay Company's Fort Hope, established in 1848–9 after an all-British route from Fort Kamloops to Fort Langley was found north of the forty-ninth parallel, answering a 'hope' of the company. O.J. Travaillot and Cpl William Fisher laid out the townsite in 1858 and the post office opened the same year. It had been incorporated as a village in 1929 and a town in 1965.

Hope ON A township in Northumberland County since 1974, it was in Durham County in 1792, when it was named for Henry Hope (*c.* 1746–89), lieutenant-governor of Quebec, 1785–9. *See also* Port Hope.

Hopedale NF On the coast of Labrador and due north of Happy Valley-Goose Bay, this municipal community (1969) was founded in 1774 by Moravian missionaries, and named after their ship *Hope*. Settlement began here in 1781, with 25 missionaries being assigned to the station the following year.

Hopes Advance, Cap QC One of the most northerly points of Québec, at the northwestern side of Ungava Bay, it had been named *Prince Henries Foreland* by Henry Hudson in 1610. Two years later Thomas Button, while searching for the abandoned Hudson, made references to *Hopes Checked* and *Hopes Advance*. While on board the *Advance* in 1850–1 in search for Sir John Franklin, Capt. Henry Grinnell may have assigned Hopes Advance to this particular cape.

Hopewell NS South of Stellarton, this place was founded in 1761 by Ulster Scots, and was presumably named after the ship which brought them across the ocean. Its post office was opened in 1838.

Hopewell Cape NB South of Hillsborough, this place's post office was named *Hopewell 'The Cape'* in 1846, after the cape extending into Shepody Bay. In the mid-1700s the cape was called *Cap des Demoiselles*, after the shape of the Cape Rocks, whose curvaceous outlines probably reminded those who named them of the figures of young women (*demoiselles*). English speakers call similar rocks 'flowerpots'.

Hornby Island BC One of the Gulf Islands, southeast of Courtenay, it was named about 1850 by the Hudson's Bay Company after Rear-Adm. Phipps Hornby (1785–1867), commander-in-chief of the Pacific station, 1847–51, and a lord of the Admiralty, 1851–2.

Hornepayne ON In Algoma District, southwest of Hearst, the station in this municipal township (1986) was first named *Fitzback* in 1916. It was renamed in 1920 after British financier Robert M. Horne-Payne, an advisor to Canadian Northern Railway builder Sir William Mackenzie.

Horning's Mills ON A community in Dufferin County, north of Shelburne, it was founded far from other settlements in 1830 by Lewis Horning. He built mills on the Pine River, but returned to his native Hamilton in 1838, a few years after his son Lewis and three children of other families were lost in the woods.

Horsefly BC A community northeast of Williams Lake and near **Horsefly Lake**, south of Quesnel Lake, its post office was called Harper's Camp in 1897 after pioneer rancher Thaddeus Harper. When it closed in 1921 Horsefly post office, which had opened in 1895 nearer to Williams Lake, was moved here. The name recalls the pesky fly that causes misery among both man and beast in the spring and summer.

Horton ON A township in Renfrew County, it was named in 1826 for Sir Robert John Wilmot Horton (1784–1841), under-secretary of state for war and the colonies, who corresponded about the union of Upper and Lower Canada. Born Robert John Wilmot, he married Anna Beatrix Horton in 1823, and added her surname to his.

Horton River NT Rising north of Great Bear Lake, this river flows northwest to Harrowby Bay, part of Beaufort Sea. It was named in 1826 by John (later Sir John) Richardson, who was a member of the British Arctic Land expedition, which explored the lands north of the lake. He named it after Sir Robert Wilmot Horton, under-secretary of state for war and the colonies.

Hortonville NS This place, east of Wolfville, was named in 1897, after its post office had been called *Horton Landing* in 1871. The latter was named in 1760, when New England settlers came ashore here in Horton Township, established a year earlier. The township may have been named after Horton Hall, in Northamptonshire, England, the residence of George Montagu Dunk, 2nd Earl of Halifax. He was then the president of the board of trade in the British government.

Horwood NF On Dog Bay of Hamilton Sound and on the north coast of the island of Newfoundland, this place was formerly called *Dog Bay*. It was renamed about 1914 after William Frederick Horwood (1856–1927), president of the Horwood Lumber Company in St. John's, 1902–27.

Hosmer BC In the Elk River valley, northeast of Fernie, this place was named in 1906 after Charles R. Hosmer (1851–1927), a manager of the Canadian Pacific Railway's telegaph system, and later a CPR director.

Hottah Lake NT South of Great Bear Lake, this lake (840 km^2) was named before 1910 after the Slavey word for 'two-year-old moose'.

Houston BC A district municipality (1957) in the Bulkley River valley, it was first called *Pleasant Valley*. A newspaper contest was held in 1910 to chose a new name, and John Houston (1850–1910), Prince Rupert's first newspaperman and a railway surveyor, won the contest. Previously he had lived in Golden and Nelson, where he had edited newspapers, had served as the mayor of Nelson, and had been elected to the provincial legislature. A creek, pass, and glacier, west of the head of Duncan River and north of Kootenay Lake, were also named for him.

Howard ON A township in Kent County, it was named in 1794 after Thomas Howard, 2nd Earl of Effingham and 8th Baron Howard (1714–91). His daughter, Lady Mary Howard, was the wife of Sir Guy Carleton, Lord Dorchester, governor-in-chief of British North America, 1786–96.

Howe Island ON One of the Thousand Islands, it became a township in Frontenac County in 1792. It was named for Gen. William Howe, 5th Visc. Howe (1729–1814), who was commander-in-chief of the British forces in North America, 1776–8.

Howe Sound BC Extending 40 km north from the Strait of Georgia, this water body was named in 1792 by Capt. George Vancouver, after Adm. Richard Howe, 4th Visc. Howe (1726–99), who won a decisive victory over the French in 1794. The victory was thereafter called 'The Glorious First of June'.

Howick ON A township in Huron County, it was named in 1850 after Henry George Grey, 3rd Earl Grey and Viscount Howick (1802–94), under-secretary of state for the colonies when his father, the 2nd Earl Grey was prime minister of Britain, 1830–3, and secretary of state for the colonies, 1846–52. He was born and died at Howick, Northumberland, north of Newcastle upon Tyne.

Howick QC A village in the Châteauguay valley, southeast of Salaberry-de-Valleyfield, it was called *Howick Village* by English settlers about 1833 after Henry George Grey, 3rd Earl

Grey and Viscount Howick (1802–94), British under-secretary of state for the colonies, 1830–3, and secretary of war and the colonies, 1846–52. Its post office was named in 1851 and the village was incorporated in 1915.

Howlan PE Located north of O'Leary, this place was called after *Howlan Road*, named about 1870 after George William Howlan (1835–1901), then a member of the House of Assembly, who was appointed lieutenant-governor of the province, 1894–9. The post office was called *Mill River* in 1883, and renamed Howlan in 1912.

Howley NF At the foot of Grand Lake and east of the town of Deer Lake, this town (1958) was named after geologist James Patrick Howley (1847–1918), author of the *Geography of Newfoundland* (1877) and *The Beothucks or Red Indians* (1915). He was a brother of the Most Rev Michael Francis Howley (1843–1914), a writer about Newfoundland's history and place names, and the Catholic archbishop of St. John's, 1904–14.

Howse Pass AB, BC Hudson's Bay Company explorer and fur trader Joseph Howse (1773–1852) crossed this pass on the continental divide, northwest of Kicking Horse Pass, in 1810. **Howse Peak** (3290 m) is on the east side of the pass.

Howser Spires BC North of Kootenay Lake, these spires (3339 m) are spectacular granite walls. They were named after **Howser Creek**, which in turn was called after the locality of **Howser** on the west side of Duncan Lake. Howser replaced *Duncan* in 1900, when mail was frequently misdirected to the city of Duncan on Vancouver Island. Howser was named after a prospector whose surname was Hauser.

Hubbard, Mount YT On the Yukon-Alaska border, and in the St. Elias Mountains, this mountain (4577 m) was named in 1890 by American explorer and geologist Israel C. Russell after Gardiner Green Hubbard (1822–97), founder and president of the National Geographic Society. The Society was a joint sponsor with the United States Geological Survey of Russell's 1890–1 expeditions.

Hubbards NS West of Halifax and at the northwestern side of St. Margarets Bay, this place's post office was called *Hubbards Cove* in 1855, and was shortened to Hubbards in 1905. It may have been named after a pioneer family.

Huber, Mount BC A mountain (3368 m) in Yoho National Park, east of Field, it was named in 1903 by American writer Walter Wilcox after Emil Huber (1865–1939), a noted Swiss alpinist, who also climbed in the Canadian mountains.

Hudson ON A transportation and distribution centre in Kenora District, northeast of Dryden, it was first known as *Rolling Portage*. Its post office was named in 1932 after a surveyor who worked for the Canadian National Railways.

Hudson QC In the regional county of Vaudreuil-Soulanges and on the south shore of Lac des Deux Montagnes, it became a town in 1969 through the merger of the villages of Hudson and **Hudson Heights**. *Pointe-à-Cavagnol* post office was renamed Hudson in 1865, given in honour of Elisa Hudson, the wife of George Matthews, who established a glassworks here 20 years earlier.

Hudson Bay SK The post office in this town (1947) was named *Hudson Bay Junction* in 1909, after plans were made to connect the Canadian Northern Railway with a seaport on Hudson Bay. The line was built as far as The Pas before 1915, but it was not until 1931 that it reached Churchill. 'Junction' was dropped from the postal designation in 1947.

Hudson Bay NT A massive interior sea (637,000 km^2), it is united to the Atlantic Ocean by **Hudson Strait**. Henry Hudson explored the Hudson River, in present-day New York State, as far as Albany in 1609. The following year he and his crew sailed into the strait and the bay, and wintered at the bottom of James Bay. Most of his crew abandoned him with seven others on present Charlton Island on 24 June 1611, with eight of the mutineers being able to return to London. The fate of Hudson and companions is not known. The Hudson's Bay Company was created in London in 1670, possibly making The Bay the oldest merchandising company in the world at the present time. In 1900, following a standard then used throughout the English-speaking world, the Geographic Board modified the names of the bay and the strait by lopping off the apostrophe and the 's'. Attempts to rename the bay Canada Sea during the twentieth century have not been widely supported.

Hudson's Hope BC A district municipality (1965) on the Peace River, west of Fort St. John, it was named after the Hudson's Bay Company's Hudson's Hope post, established in 1873. The name may have been given as an ironic comment on the poor prospects of success at this seasonal trading post. Since 1913 its post office has been called **Hudson Hope**.

Hughenden AB South of Wainwright, this village (1917) was named by Charles E. Stockdill of the Canadian Pacific Railway after Hughenden, north of High Wycombe, Buckinghamshire, England. It was the estate of Benjamin Disraeli, Earl of Beaconsfield, prime minister of Great Britain, 1867–8 and 1874–80.

Hull QC Located in the Communauté urbaine de l'Outaouais, opposite the city of Ottawa, it became a city in 1875. Its post office was named in 1819, taking its name from the township of Hull, noted on a map as early as 1795, and proclaimed in 1806. It was named after Kingston upon Hull, Yorkshire, England as part of a series of Yorkshire names in what was then York County.

Hullett ON A township in Huron County, it was named in 1830 after John Hullett, a director of the Canada Company, a British company established in 1826 to develop the Huron Tract, a large area of what is now Southwestern Ontario. Hullett was a hops supplier based at Sillery, a suburb of the city of Québec.

Humber River NF Draining Grand and Deer lakes, this river (153 km) flows into the **Humber Arm** of the Bay of Islands. The river was named after England's River Humber, which flows into the North Sea. It was named by Capt. James Cook, who charted Humber Arm in 1767.

Humber River ON Rising in Dufferin County, this river flows southeast to enter Lake Ontario on the western boundary of the city of Toronto. Known as *Rivière Taronto*, after the French name of Lake Simcoe (*Lac de Taronto*), and *St. John's River*, after trader Jean-Baptiste Rousseau (1758–1812), it was named about 1793 by Lt-Gov. John Graves Simcoe after the River Humber, which separates the English counties of Yorkshire and Lincolnshire.

Humboldt SK East of Saskatoon, this town (1907) was named as early as 1870, when the Dominion Telegraph established an office nearby. It was named after Baron Alexander von Humboldt (1769–1859), a German natural scientist who had investigated the geography of North and South America, 1799–1804. Postal service was established in 1905, after the Canadian Northern Railway was built from Winnipeg to Edmonton.

Hungabee Mountain AB, BC On the continental divide and west of Lake Louise, this mountain (3492 m) was named in 1894 by climber Samuel Allen after the Stoney word for 'chieftain'.

Hungerford ON A township in Hastings County, it was named in 1798 after one of the titles of Sir Francis Rawdon-Hastings (1754–1826). The title of Baron Hungerford was granted to Edward Hastings in 1482 and was derived from the town of Hungerford in Berkshire, England, west of Reading.

Hunker Creek YT A tributary of the Klondike River, this creek was named after prospector Andrew 'Old Man' Hunker, who discovered gold on it in September 1896. A native of Germany, he prospected for gold in the Cariboo, British Columbia, and on the Fortymile River, near the Alaska border.

Hunter River PE In the central part of the province and at the head of Hunter River, the post office in this municipal community (1983) was named in 1875. The community had been incorporated as a village in 1974. The post office was called *Hunter's River* from 1901 to 1967. The river was named in 1765 by Samuel Holland after Thomas Orby Hunter, lord of the Admiralty in 1761. During the nineteenth century the river became known as both *Clyde River* and *New Glasgow River*, but neither designation has survived.

Huntingdon ON A township in Hastings County, it was named in 1798 after the Earl of Huntingdon, the father-in-law of Sir Francis Rawdon-Hastings. The title was taken from Huntingdonshire, in east central England.

Huntingdon QC Situated on the Rivière Châteauguay, south of Salaberry-de-Valleyfield, this town's post office was named in 1830 after the county of Huntingdon, created in 1792, and named after Huntingdonshire, in England. It became a village in 1848 and a town in 1921.

Huntley ON A ward of the municipal township of West Carleton in Ottawa-Carleton Region since 1974, it was named as a township in 1823 after the Huntly estate in Scotland. The estate was inherited by the wife of the Duke of Richmond (1764–1819), governor-in-chief of British North America, 1818–9. Huntly Castle is a grand ruin, northwest of the city of Aberdeen.

Huntsville ON A town (1901) in the district municipality of Muskoka, its post office was named in 1870 after Capt. George Hunt (1830–82), a British military officer, who had settled at Huntsville the previous year, and was appointed the first postmaster.

Huron ON The county was named in 1841 after the Huron Tract, which was established by the Canada Company in 1826 on the eastern shore of **Lake Huron**. The lake was named by the French after a confederacy of five Iroquoian tribes, with the name signifying 'boar's head' or 'ruffian', because their hair stood up in bristles. **Huron Township** in Bruce County was named in 1849. After the closing in 1966 of Royal Canadian Air Force Station Centralia, south of Exeter, the civilian community was renamed **Huron Park**, after the county.

Hythe AB A village (1929) northwest of Grande Prairie, its post office was named in 1914 by Mr and Mrs Harry Hartley after their hometown of Hythe, Kent, England, southwest of Dover.

I

Iberville QC A town opposite the city of Saint-Jean-sur-Richelieu, it was called the village of *Christieville* in 1846 in honour of seigneur Napier Christie Burton, who had donated land for a church and presbytery. When it became a town in 1859, it was renamed after Pierre Le Moyne d'Iberville (1661–1706), described in the *Dictionary of Canadian Biography* as 'the most renowned son of New France'.

Ibex Valley YT A hamlet on the Alaska Highway, west of Whitehorse, it was named in 1990 after the valley of the **Ibex River**. The river was named in 1947 after wild goats found in its valley.

Iconoclast Mountain BC In the Selkirk Mountains, northwest of Rogers Pass, this mountain (3240 m) was named in 1902 by surveyor Arthur O. Wheeler after its steep black face. Five years later surveyor Percy Carson described it as a very prominent majestic mountain whose name was quite suitable.

Igloolik NT On **Igloolik Island**, at the east end of Fury and Hecla Strait and offshore of Melville Peninsula, this hamlet (1976) was named in 1951 after the Inuktitut for 'there are houses'. Archaelogical studies reveal the site has been continuously occupied for 4000 years.

Ignace ON Located in Kenora District, southeast of Dryden, it was named in 1879 by Sir Sandford Fleming after Ignace Mentour, a Mohawk from the *Caughnawaga* (now Kahnawake) Reserve, south of Montréal, who guided him across Canada in 1872. After it became a municipal township in 1908, an effort was made to rename it *Elmsdale*, but the place's historical significance of linking east to west prevented such disrespect.

Ilderton ON In Middlesex County, northwest of London, this place was named in 1864 by postmaster George Ord after his birthplace in Northumberland, England, south of Berwick upon Tweed.

Île-à-la-Crosse SK This place is situated on the west side of the cross-shaped **Lac Île-à-la-Crosse**, an enlargement of the Churchill River, northwest of Prince Albert. North West Company trader Roderick McKenzie (d. 1859) based his headquarters here in 1830. The first mission was founded in 1846 at this northern village by Fr (later Bishop) Alexandre Taché and Fr (later Bishop) Louis-François Laflèche. Its post office was opened in 1910.

Ilford MB A station on the Hudson Bay Railway, southwest of Gillam, its post office was named in 1928 after the constituency of Ilford in the British House of Commons. Its member, Sir Frederick Wise, took a strong interest in the construction of the railway. Ilford is in Greater London, northeast of the London city centre.

Illecillewaet River BC Rising in the Selkirk Mountains, this river (77 km) flows southwest into the Columbia River at Revelstoke. It was named in 1865 by explorer Walter Moberly, adapting its Okanagan name, meaning 'swift water'.

Imperial SK Northwest of Regina, this town (1962) was founded by the Canadian Pacific Railway in 1910, and was likely named by an admirer of the monarchy. Among its streets are Royal, King, Queen, Prince, and Princess. Its post office was first called *Harkness* in 1907 and was renamed Imperial in 1911.

Incomappleux River BC Rising in the Selkirk Mountains, south of Rogers Pass, this river flows southwest into the Northeast Arm of Upper Arrow Lake. The name was derived from the Okanagan word for 'head of lake' or 'end of the water'.

Indian Bay (Parson's Point) NF Located on the northwestern side of Bonavista Bay and northeast of Gambo, this municipal community (1971) was probably settled soon after 1800 by William Parsons. Although Parson's Point is considered a former name of the place, the province authorized the double name for the municipality.

Indian Head SK A town (1902) east of Regina, it was founded by the Canadian Pacific Railway in 1882, and named after an Aboriginal's skull found near the site of the railway station by the CPR survey crew. The skull was still in the possession of a farmer in 1905. In 1854, a missionary referred to a hill called Indian Head, noting its Aboriginal name was *Ustiquanuci*.

Ingenika River BC Rising west of Williston Lake and on the west side of the **Ingenika Range**, this river was named after the Sekani word for 'bearberry'. The low bearberry bush is also known as kinnikinnick.

Ingersoll ON In 1817 Charles Ingersoll (1791–1832) settled at *Oxford-upon-the-Thames* and became the first postmaster of *Oxford Centre* in 1821. It was incorporated as the village of Ingersoll in 1852, in honour of his father Thomas, and the postal name was also changed the same year. In 1795 Thomas Ingersoll (1749–1812) was granted 486 ha (1200 a) here, where he lived from 1797 to 1799. Established as a town in 1860, it became a separated town in 1914 and reverted to a town in 1975.

Ingleside ON In the united counties of Stormont, Dundas and Glengarry, west of Cornwall, this planned community was created in 1955 for the residents of *Aultsville*, *Farrans Point*, *Dickinson Landing* and *Wales*, flooded in 1957 by the St. Lawrence Seaway project. Osnabruck Township reeve Thorold Lake observed the name Ingleside on a house and the township council endorsed it for the new place.

Inglewood ON In the town of Caledon, northwest of Brampton, this place was named *Riverdale* in 1860. Because of confusion with Riversdale in Bruce County, residents asked Thomas White, their member of Parliament, to suggest a new name. He proposed Inglewood in 1885, possibly after the old tale *The Farmer of Inglewood Forest*. Two years later the name was transferred from Ontario to a new community in California, southwest of the centre of Los Angeles. It is now a city of 125,000.

Ingonish NS On the north side of **North Bay Ingonish** and on the northeast coast of Cape Breton Island, this place was noted by Samuel de Champlain in 1632 as *Niganis*, and by Nicolas Denys in 1672 as *Niganiche*. The present form of the name may have been a misprint in Thomas Chandler Haliburton's *Nova-Scotia* (1829). Ingonish post office was opened in 1846. Likely of Mi'kmaq origin, the name's meaning is not noted in official names records. **Ingonish Centre** is at the head of North Bay Ingonish, and **Ingonish Beach** is on the north side of **South Bay Ingonish**.

Ingornachoix Bay NF South of Port au Choix, this bay's name was derived from the Basque *aingura charra*, 'bad anchorage'. In the late 1500s many of the bays on the west coast of the island of Newfoundland were either unprotected, or were too deep for the small Basque fishing craft to anchor.

Ingram, Mount YT West of Whitehorse, this mountain (2158 m) was named in 1897 by surveyor James J. McArthur, after he had discovered the gravesite headboard of a prospector, inscribed 'Ingram'. The prospector had apparently drowned in the Takhini River and was buried beside the Dalton Trail.

Inkerman NB Southeast of Caraquet, this community's post office was named in 1882 after **Inkerman Parish**. The parish was created in 1855 and called after the British-French victory in 1854 in the Crimea.

Innerkip ON In Oxford County, northeast of Woodstock, this community was named in 1853 by Susan Barwick, the wife of Woodstock postmaster Hugh Barwick, after Inverkip, in Renfrewshire, Scotland, west of Glasgow. Innerkip is an older variant of the Scottish Inverkip.

Innisfail AB A town (1903) south of Red Deer, its post office was named in 1892 by

Estella (Wildman) Scarlett, after the poetic name for Ireland, *Inisfail*, 'isle of destiny'. It was previously called *Poplar Grove*.

Innisfil ON A town in Simcoe County, it was created in 1991 by uniting the municipal township of Innisfil, the village of Cookstown and parts of the municipal townships of West Gwillimbury and Tecumseth. Innisfil Township had been named in 1822 after *Inisfail*, 'isle of destiny', a poetic name of Ireland.

Innisfree AB Between Vegreville and Vermilion, the post office at this village (1911) was first called *Delnorte*, but was renamed in 1909 on the suggestion of Sir Edmund Walker, president of the Canadian Bank of Commerce. The area of Birch Lake reminded him of the view from his summer residence of Innisfree on Lake Simcoe, ON. Having been offered a bank in their community, the residents agreed to the change.

Inukjuak QC Located on the east coast of Hudson Bay and at the mouth of **Rivière Innuksuac**, it was first named *Port Harrison* in 1901 by geologist Albert Peter Low in honour of a shareholder of a mining company. *Port Harrison* post office was opened in 1935 and was renamed *Inoucdjouac* 30 years later. The spelling was modified in 1980, after the residents were consulted on their preference for the newly incorporated northern village.

Inuvik NT On the East Channel of the Mackenzie River, the site of this town (1967) was selected in 1955 as an alternate community for the residents of Aklavik, and was first called *Aklavik East Three* and *New Aklavik*. Inuvik, from the Inuvialuktun for 'place of man', was proposed 1 May 1958 by Graham Rowley, who was then the secretary-coordinator of the Advisory Committee on Northern Development in the northern affairs and natural resources department. It was adopted by the Northwest Territories Council on 18 July 1958.

Inverary ON A community in Storrington Township, Frontenac County, north of Kingston, it was named *Storrington* in 1841. It was renamed in 1856, after Inveraray, in Argyllshire, Scotland, northwest of Glasgow.

Inverhuron ON In Bruce County, north of Kincardine, its post office was named in 1854 from Gaelic *inver*, 'confluence', and Huron, because of its location where the Little Sauble River flows into Lake Huron.

Invermay SK Located west of Canora, this village (1908) was named in 1904, when the Canadian Northern Railway arrived here. *Tullock* post office had been opened the year before, 5 km to the southeast, and was named after Walter Tulloch. It was moved to the proposed station, and was first called *Inverness*. As there was already another Inverness post office in Canada, Invermay was put forward and accepted. Invermay is in Perthshire, Scotland, southwest of Perth.

Invermere BC At the point where the Columbia River flows northwest from Windermere Lake, this district municipality (1984) was first called *Copper City* by prospector Edmund T. Johnson in 1890. It was renamed *Canterbury* in 1900 by the Canterbury Townsite Company after the famous cathedral city in Kent, England. The name was changed again in 1912 to Invermere by R. Randolph Bruce, after the Irrigated Fruit Lands Company bought the site. The name means 'at the mouth of a lake'.

Inverness NS In **Inverness County**, and on the west side of Cape Breton Island, this town (1904) was first known as *Broad Cove Marsh* in 1838. It was renamed after the county in 1903. The latter was called after Inverness-shire, Scotland, in 1837 by Sir William Young (1799–1887), its first member in the Legislative Assembly, 1837–59. It had been called *Juste au Corps County* two years earlier.

Inverness QC Northwest of Thetford Mines, the village's post office was named in 1832 after the township, proclaimed in 1802, and named after the Scottish royal burgh, where many of the first settlers came from. The township municipality was established in 1845, abolished two years later, and re-established in 1855. The village was incorporated in 1900.

Inwood MB Located between Lake Manitoba and Lake Winnipeg, southwest of Gimli, this place's post office was first called *Cossette* in July 1906, but was renamed Inwood in December of that year. Its name reflected its setting in the woods.

Inwood ON This place in Lambton County, southeast of Sarnia was named in 1872 after Inwood, a neighbourhood in New York City at the north end of Manhattan Island.

Iona ON In Elgin County, southwest of St. Thomas, this community's name was proposed in 1848 by Duncan McCormick after the holy isle of Iona in Scotland. The post office was opened four years later. In 1872 a railway bypassed Iona 3 km to the northwest and the new community here was named **Iona Station** three years later.

Ipperwash Beach ON A resort centre in Lambton County, northeast of Sarnia, it was developed in the 1920s around the Ipperwash Hotel. The hotel's name may have been derived from Norwegian *yppare vass*, 'better water'.

Iqaluit NT Located at the head of Frobisher Bay, this town (1980) was established in 1949 as *Frobisher Bay*, when the Hudson's Bay Company moved its post here from a site 70 km southeast. It became a municipal hamlet in 1971 and a village three years later. In December 1984 its residents voted 310 to 213 to rename the place Iqaluit, meaning 'place of fish' in Inuktitut.

Irishtown-Summerside NF On the north side of Humber Arm, opposite Corner Brook, this town was incorporated in 1991. **Irishtown** was settled by Irish in the early 1800s. **Summerside** was named because it was on the north, or sunny, side of Humber Arm.

Irlande QC A municipality in the regional county of L'Amiante, west of Thetford Mines, it was established as *Ireland* in 1845 by merging the townships of Ireland and Wolfestown. It was abolished two years later, and then re-established in 1855. It took its name from the township of *Ireland*, proclaimed in 1802, with both the township and the municipality being renamed Irlande in 1985.

Irma AB A village (1912) northwest of Wainwright, its post office was named in 1909 after a daughter of William Wainwright, second vice-president of the Grand Trunk Pacific Railway.

Iron Bridge ON A village (1960) in Algoma District, west of Blind River, it was named in 1886, two years after a steel bridge over the Missisagi River replaced a wooden one. In 1949 a concrete structure replaced the old iron bridge.

Iroquois ON A village (1857) in the united counties of Stormont, Dundas and Glengarry, southwest of Morrisburg, it was named in 1857 after nearby **Point Iroquois**, a name dating from the French regime, and said to be a site where the Iroquois held councils. In 1958 the entire village was relocated, when the St. Lawrence Seaway project inundated the old site.

Iroquois Falls ON A town (1915) in Cochrane District, it was founded in 1912 by Montreal businessman Frank Anson, who developed the Abitibi Pulp and Paper Company at a falls on the Abitibi River. There is a legend of Huron women cutting loose the canoes of Iroquois men and sending them to their death over the falls, as a retaliation for killing all the Huron men.

Irricana AB A village (1911) northeast of Calgary, its post office was established in 1909 and took its name from the words 'irrigation' and 'canal'. In 1894 the Dominion government granted the Canadian Pacific Railway a block of land on condition that an irrigation system, drawing on the waters of the Bow River, was developed. The system was abandoned before reaching the village's site, but the name Irricana remains as a memory of it.

Irvine AB Located east of Medicine Hat, this place's post office was named in 1883 by the Canadian Pacific Railway after Col Acheson Gosford Irvine (1837–1916), commissioner of the North-West Mounted Police, 1880–6, and a member of the council of the North-West Territories. It was an incorporated town from 1909 to 1996.

Isachsen, Cape NT At the northwestern point of Ellef Ringnes Island, this cape was named *Kap Isachsen* in 1901 after cartographer Gunerius Ingvald Isachsen, a member of the Sverdrup expedition, 1898–1902. **Isachsen** weather station, on the south side of **Isachsen Peninsula**, was established in 1949. The peninsula was named in 1959.

Ishpatina Ridge ON Ontario's highest elevation, with summits of 670 m to 685 m (2200 ft to 2250 ft) in Timiskaming and Sudbury districts, it is located north of Sudbury and west of New Liskeard. It was named in 1972 from the Ojibwa word *ishpadina*, 'it is high', and used in reference to a hill or mountain.

Island Lake MB Located near the Ontario border and due east from the north end of Lake Winnipeg, this lake (1425 km^2) was crossed by Samuel Hearne in 1772. He reported that it was entirely full of islands, making it appear to be a jumble of separate rivers and creeks. **Island Lake** post office was opened in 1905.

Island View NB West of Fredericton, this place's post office was named in 1908 after its view of Savage, Sugar, and Keswick islands, at the head of tide in the Saint John River.

Islands, Bay of NF Located on the west coast of the island of Newfoundland and north of Corner Brook, this bay was named *Baie St-Jullian* by Jacques Cartier in 1534. He found it full of islands, calling them *Les Coulombiers*, from *colombier*, 'little dovecote'. Capt. James Cook recorded the present name in 1767, and repeated it two years later on the northeast coast of New Zealand's North Island.

Isle aux Morts NF Located on the southwestern coast of the island of Newfoundland and east of Channel-Port aux Basques, this town (1956) was likely named after an island, where, after a shipwreck, some bodies had been discovered near the mouth of **Isle aux Morts River**.

Islington ON An urban neighbourhood in the city of Etobicoke, it was named at a public meeting in 1858, when Elizabeth Wilson Smith, the wife of hotel keeper Thomas Smith, proposed Islington after her birthplace in east central London, England.

Italy Cross NS South of Bridgewater, this place was first known as *New* Italy, after the European country. Its post office was opened under its present name in 1889.

Ituna SK Located northwest of Melville, this town (1961) was named by the Grand Trunk Pacific Railway in 1908. It may have been chosen by a railway official familiar with the Celtic name for the Solway Firth, used by Rudyard Kipling in one of his books. The firth is the wide bay that separates northwestern England from southwestern Scotland. Ituna post office was opened in 1909.

Ivujuvik QC The most northerly of Québec's northern villages, and at the most northeasterly point of Hudson Bay, it was the site of the Hudson's Bay Company post *Wolstenholme*, established in 1909 and named after the nearby cape. It was renamed Ivujuvik in 1981, from the Inukititut word for 'place where the ice is carried by strong currents'.

Ivvavik National Park YT This park (10,168 km^2), on the shore of the Beaufort Sea, was called the Northern Yukon National Park in 1984, and 10 years later was given its present name. It was created to protect the migratory route of the Porcupine herd of the barren-ground caribou. In Inuvialuktun the name means 'place of giving birth and raising young'.

J

Jackson's Arm NF Located on a small arm of White Bay, this municipal community (1982) was settled in the mid-1800s and possibly named after one of the first settlers.

Jacksons Point ON In the town of Georgina, York Region, this resort community's post office was named in 1902 after John Mills Jackson, who had settled here in 1812.

Jacksonville NB Located north of Woodstock, this community's post office was named in 1859 after John Jackson, who settled in the area about 1810. It was then considered part of **Jacksontown**, whose post office had opened 8 km to the north, seven years earlier.

Jacobsen, Mount BC In the Coast Mountains, southeast of Bella Coola, this mountain (3027 m) was named in 1948 by surveyor Maj. F.V. Longstaff after Thorwald Jacobsen. He and the Revd Mr Saugstad brought Norwegian settlers to the area about 1892.

Jacques-Cartier, Détroit de QC Located at the mouth of the St. Lawrence River, and between the north shore and Île d'Anticosti, it was named in 1934 to celebrate the epic voyage by Jacques Cartier 400 years before. Cartier himself had called it *Détroit Saint-Pierre*, because he discovered the passage on 1 August, the feast day of St Peter.

Jacques-Cartier, Mont QC The highest point (1268 m) in the Monts Chic-Chocs, west of the town of Gaspé, it was named in 1934 in memory of the four hundredth anniversary of Cartier's discovery of Canada.

Jacquet River NB A village (1966), southeast of Dalhousie, its post office was named *Jacquet River Station* in 1892 after the adjacent river. It was renamed *Durhamville* in 1892, after the parish and became Jacquet River in 1898. The river was identified on Thomas Sproule's 1799 map as *River Jaques*, suggesting it was named after a local French resident. It was called *Jacket* in 1803, with the present form occurring in Robert Cooney's 1832 history. The river may have been called after pioneer James 'Jock' Doyle, who settled here about 1790.

Jaffray BC On the Canadian Pacific Railway, southwest of Fernie, this place was named in 1901 after Toronto businessman and senator Robert Jaffray (1832–1914), who was the president of the Crows Nest Pass Coal Company.

Jaffray Melick ON A town on the northeastern side of the town of Kenora, it had been created as the municipal township of *Jaffray and Melick* in 1908. It became a town in 1988 and the name was amended to Jaffray Melick in 1991. Jaffray Township was named after businessman and senator Robert Jaffray (1832–1914). Melick Township was named after the parish of Meelick in County Mayo, Ireland.

Jakes Corner YT On the Alaska Highway, southeast of Whitehorse, this community was named in the 1940s, either after a Teslin called Jake Jackson, who had camped near here, or after Capt. 'Jake' Jacoby (or Jacobson) of the United States Army engineers. The engineers were responsible for the contruction of the highway through this area in 1942.

James Bay NT This 460-km extension of Hudson Bay was named before 1662 after explorer Thomas James, who had been seeking the Northwest Passage on behalf of Bristol merchants. He wintered in the bay in 1631–2, returning to Bristol in July 1632.

Jansen SK Located on the Canadian Pacific Railway, between Lanigan and Wynyard, both the village and the post office were named in 1908 after pioneer settler John Jansen.

Jarvis ON An urban centre in the city of Nanticoke, Haldimand-Norfolk Region, east of Simcoe, its post office was named in 1851 after William Jarvis (1756–1817), provincial secretary and registrar of Upper Canada, 1792–1817.

Jasper AB At the junction of Athabasca and Miette rivers, this resort centre owes its name to the North West Company's Jasper House, located on the west side of Brûlé Lake in 1817, with Jasper Hawes in charge of it. It was rebuilt in 1829 by the Hudson's Bay Company, 20 km northeast of the present community, but was abandoned in 1884. Named *Fitzhugh* in 1911 by the Grand Trunk Pacific Railway, the present place was renamed Jasper the next year. **Jasper National Park** (10,878 km^2) was established as a park reserve in 1907.

Jasper ON In the united counties of Leeds and Grenville, southeast of Smiths Falls, it was known for most of the 1800s as *Irish Creek*, after a tributary of the Rideau River. When a railway was built from Brockville to Smiths Falls in 1859, the railway officials refused to use *Irish Creek*, and urged the adoption of a new name. Jasper was selected in 1864 from a list provided by a postal inspector.

Jean Marie River NT On the west bank of the Mackenzie River, 60 km southeast of Fort Simpson, this settlement was established after 1884 and named for Bro. Jean-Marie Beaudet (1866–1949), a native of Bretagne, France. He became a missionary among the Dene in 1884 and was based at Fort Resolution, 380 km to the east.

Jeckell, Mount YT Northeast of Dawson, this mountain (1951 m) was named about 1957 after George Allan Jeckell, comptroller of the Yukon from 1932 to 1947. He performed the duties of commissioner during those years.

Jeddore Harbour NS East of Dartmouth, this harbour was described by Nicolas Denys in 1672 as *Riviere de Theodore.* Some maps in the 1700s called it *R St Theodore* and *Riviere Peliodore.* By 1769 it was known as *Jedore Harbour,* when John Perceval, the 2nd Earl of Egmont advertised for settlers, with its identity becoming *Egmont Harbour, Jedore.* Among the post offices there were **Jeddore** (1862), **East Jeddore** (1865), **Head of Jeddore** (1875), **West Jeddore** (1878), and **Jeddore Oyster Ponds** (1884).

Jeddore Lake NF North of Bay d'Espoir, this reservoir may have been named after Noel Jeddore (1865–1944), the chief of the Mi'kmaq at Conne River, 1918–25. A defender of the traditional life of the Mi'kmaq, he left Conne River in 1925 after a dispute with Fr Stanley St Croix.

Jellicoe ON In Thunder Bay District, west of Geraldton, its post office was named in 1914 after Adm. John Rushworth Jellicoe, 1st Earl Jellicoe (1859–1935), commander-in-chief of the British Navy during the First World War.

Jellicoe, Mount AB This mountain (3246 m), southwest of Calgary, was named in 1919 after Adm. John Rushforth, 1st Earl Jellicoe (1839–1935), commander of the British grand fleet during the First World War and victor over the Germans at the Battle of Jutland in 1916.

Jemseg NB Located on **Jemseg River**, the drainage of Grand Lake, and 7 km northeast of Gagetown, this was the site of a trading post in the mid-1600s and the French capital of Acadia in the late 1600s. Its post office was open from 1850 to 1969. The river's name was derived from the Maliseet *kadjimusek*, 'jumping across place', referring to the narrowness of Grand Lake at the river's head.

Jenny Lind Island NT In Queen Maud Gulf and on the southeast side of Victoria Island, this island (420 km^2) was named *Lind Island* in 1851 by explorer John Rae after the Swedish singer Johanna Maria Lind (1820–87), popularly known as Jenny Lind, 'whose sweetness of voice and noble generosity have been the theme of every tongue'. It was amended to Jenny Lind Island in 1946.

Jenpeg MB Situated on the south end of Cross Lake, north of Norway House, this place was named in 1973 after Jennie (Kotoski) Kaniuga and Peggy (Johnston) Stitchbury, employees of the Winnipeg office of the water resources branch, department of mines and natural resources.

Jens Munk Island NT At the northwest side of Foxe Basin, this island (920 km^2) was named in 1921–4 by the Fifth Thule expedition after Danish explorer Jens Munk. Commissioned to find the Northwest Passage in 1619 by Christian IV of Denmark, Munk spent the following winter at the mouth of Churchill River, on the west side of Hudson Bay, returning across the ocean the following summer.

Jerseyside NF On the north side of Placentia Harbour, this has been a neighbourhood in the town of Placentia since 1994. A former town (1950), it was named in the mid-1800s after fishing families from Jersey, in the Channel Islands, who had settled here after 1713.

Jerseyville ON In the town of Ancaster, Hamilton-Wentworth Region, it was first called *Jersey Settlement* by Loyalist settlers from New Jersey. Jerseyville post office was named in 1852.

Jervis Inlet BC Extending 90 km inland from Malaspina Strait, east of Powell River, this fiord-like inlet was named in 1792 by Capt. George Vancouver after Rear-Adm. Sir John Jervis, Earl of St.Vincent (1735–1823). He won considerable fame in 1797 for his victory over the stronger Spanish fleet off Portugal's Cabo São Vincente.

Jésus, Île QC Located north of Montréal and embracing the city of Laval, it was named in 1636 when it was granted to the Jesuits by the Company of New France. Although it was only considered a temporary description, and the Jesuits themselves called it *Isle de Montmagny*, after the governor, the original designation prevailed.

Jetté, Mount YT Located at the point where the Yukon, British Columbia, and Alaska meet, this mountain (2579 m) was named in 1905 after Sir Louis-Amable Jetté (1836–1920), lieutenant-governor of Québec, 1898–1908, and Canadian member of the Canada-Alaska Boundary Tribunal in 1903.

Joe Batt's Arm-Barr'd Islands-Shoal Bay NF A town (1989) on the north side of Fogo Island, the first two had been joined as a rural district in 1972 and became a town in 1980. **Joe Batt's Arm** took its name from a small bay, on the west side of **Joe Batt's Point**. The arm and the point may have been named after the same Joe Batt 'who was sentenced to receive fifteen lashes for stealing a pair of shoes and buckles valued at 7s 6p about 1754 at Bonavista'. He may have worked with Capt. James Cook, during his hydrographic survey of the north coast, before settling on Fogo Island in 1765, and was possibly a native of Ringwood, Hampshire, England. In 1901, the town was renamed *Queenstown*, but the local people resented the change. **Barr'd Islands**, which was probably settled before 1800, may have been named after the narrow necks of land which formed its harbour. **Shoal Bay** is on a long but shallow indentation of the island.

Joffre, Mount AB This mountain (3449 m) southeast of Banff, was named in 1818 by the Alberta-British Columbia Boundary Commission after Joseph-Jacques-Césaire Joffre (1852–1931), commander-in-chief of the French armies in the First World War, 1914–16.

Joggins NS On Chignecto Bay and southwest of Amherst, this place's name was derived from Mi'kmaq *chegoggin*, 'fish weir place', and it appeared on maps as early as 1750. Its post office opened as *Joggins Mines* in 1856 and became Joggins in 1937. Coal was mined here commercially from 1847 to 1961.

John Oliver, Mount BC In the Premier Range, west of Valemount, this mountain (3120 m) was named in 1928 after John Oliver (1856–1927), premier of British Columbia, 1918–27.

Johnson, Mount NB South of Nepisiguit River, this mountain was named in 1964 by Arthur F. Wightman, then the provincial names authority, after John Mercer Johnson (1818–68), a Father of Confederation.

Johnstone Strait BC This passage, linking the Strait of Georgia with Queen Charlotte Sound and the open Pacific, was named in

1792 by Capt. George Vancouver after James Johnstone, master of the tender *Chatham*, who discovered that year that Vancouver Island was separated from the mainland.

Johnstown ON On the St. Lawrence River, and in the united counties of Leeds and Grenville, northeast of Prescott, it was named about 1790 after Sir William Johnson (*c.* 1715–74), superintendent of the northern Indians in the American colonies, who had died at Johnstown, NY, northwest of Albany. His son, Sir John Johnson, acquired land at Johnstown, which was the administrative centre of Upper Canada's Eastern District, 1793–1808.

Joliette QC A city in the regional county of Joliette, northeast of Montréal, it was founded in 1823 by Barthélemy Joliette (1789–1850), who built saw and gristmills, and a foundry there, and called the place *L'Industrie.* It was renamed Joliette in 1864 after the county, the judicial district and the post office had been called Joliette. The regional county was named in 1982, embracing the area of the former county, established in 1853.

Jones Sound NT Separating Ellesmere and Devon islands, this 275-km body of water was named *Alderman Jones Sound* in 1616 by Robert Bylot and William Baffin. They named it after Sir Thomas Jones, a London merchant and lord mayor, who was a generous patron of their expedition. The name was subsequently shortened to Jones Sound.

Jonquière QC A city west of Chicoutimi, it was named in 1866 after the township of Jonquière, proclaimed in 1850, and named after Jacques-Pierre de Taffanel, Marquis de Jonquière (1685–1752), governor of New France, 1746–52. In 1975 it annexed the adjoining towns of Kénogami and Arvida and the parish municipality of *Saint-Dominique-de-Jonquière.*

Jordan ON In the town of Lincoln, Niagara Region and west of St. Catharines, it was named in 1840, possibly by first postmaster William Bradt after the biblical River Jordan.

Jordan Falls NS On the Jordan River and northeast of Shelburne, this place's post office was named *Head of Jordan River* in 1857. Its present name was adopted in 1913. Portuguese maps published in the mid-1500s identified *Ribeira de jardines* in this area, and it evolved into *R des Jardins* during the French regime and **Jordan River** after 1783.

Joseph, Lake ON This lake, one of the three largest of the Muskoka Lakes, was named in 1860–1 after Joseph Dennis, the father of surveyor John Stoughton Dennis. Joseph Dennis, who was then living in Weston, ON, was captured during the War of 1812, while a captain of a boat on Lake Ontario.

Jourimain, Cape NB Extending into Northumberland Strait, north of Cape Tormentine, this cape may have been named after an Acadian settler or after pioneer Germain Allen. Referred to as *Jeauriman* in an 1809 land petition, its present form was established on John Wilkinson's 1859 map.

Joyceville ON In Frontenac County, northeast of Kingston, this place's post office was named *Bermingham* in 1852, and changed in 1855 to *Birmingham*, after postmaster and hotel keeper James Birmingham. In 1901 it was named Joyceville after postmaster Luke J. Joyce and the many Joyces living in the area.

Juan de Fuca Strait BC Separating Vancouver Island from the state of Washington, this 23-km-wide entrance to Puget Sound and the Strait of Georgia was named in 1787 by fur trader Charles W. Barkley after a Greek sea captain known as Juan de Fuca (1525–1602). He had told a tale in 1596 of having discovered a wide inlet, four years before, between the forty-seventh and forty-eighth parallels, while sailing under the instructions of the Viceroy of Mexico. Although the navigator's Greek name has been rendered as Apostolos Valerianos in various accounts, his real Greek name was Ioannis Phokas. He came from Valerianos on the island of Kefallinía (Cephalonia) in the Ionian Sea, west of Athens.

Judique NS Located on the southwestern coast of Cape Breton Island, this place's post office was named in 1832 after the **Judique River**. The latter's name may have evolved from *jou-jou-dique*, derived from the French words *jouer* and *dique*, in reference to the river playfully cutting through the sandbar at different places in front of **Judique Harbour**, after storms closed the entrance.

Jules-Léger, Lac QC Situated in north central Québec, near Mont Otish, it was named in 1982 after Jules Léger (1913–80), governor general of Canada, 1974–9, and brother of Paul-Émile Cardinal Léger.

Jumbo Mountain BC In the Purcell Mountains, west of Invermere, this mountain (3399 m) was named in 1911 by climber Edward Harnden after the Jumbo Mine, staked in 1890 and named after the famous elephant of P.T. Barnum's circus. At that time Jumbo's body was on display in the museum of Boston's Tufts University.

Juniper NB A community at the head of the Southwest Miramichi River, its post office was named in 1917 to serve the lumbering operations of James K. Flemming and Alexander Gibson. **Juniper Brook**, a tributary of the Southwest Miramichi, was named after the evergreen tree, which is part of the cypress family.

K

Kahnawake QC A Mohawk community and reserve, south of Montréal, the name was spelled *Caughnawaga* from 1716 to 1980, when the preferred Mohawk form was acknowledged. The name means 'at the rapids', in reference to the Rapides de Lachine.

Kakabeka Falls ON Located in Thunder Bay District and on the Kaministiquia River, west of Thunder Bay, it plunges down a 47-m cliff. The name in Ojibwa is *Kakabika*, 'steep rock with a waterfall'. The post office in the community of **Kakabeka Falls** was named in 1906.

Kaladar ON A township in Lennox and Addington County, it was named in 1820. Its origin is uncertain, but it may be named after John Campbell, Baron Cawdor of Castlemartin (1753–1821), a Conservative member of Britain's Parliament, who took his title from the Scottish parish of Cawdor in Nairn, Scotland. A post office called *Cawdor* in the township from 1865 to 1869 suggests a connection between Cawdor and Kaladar. The post office at the present community of **Kaladar** was called *Kaladar Station* from 1889 to 1950.

Kaleden BC In the Okanagan Valley, south of Penticton, this community was named in 1909 as a result of a contest. The Revd Walter Russell devised it from the Greek word *kalos*, 'beautiful', and *Eden*, the Biblical garden of paradise. Its post office opened a year later.

Kaministiquia River ON Rising 120 km west of Thunder Bay, this river flows east over Kakabeka Falls and enters Lake Superior in the Fort William sector of Thunder Bay. The name in Ojibwa means 'island river'. The post office at **Kaministiquia**, northwest of Thunder Bay, was established in 1875.

Kamloops BC A city (1893) at the forks of the South and North Thompson rivers, it is at the site where fur trader David Stuart established Fort Kamloops for the Pacific Fur Company in 1812. Its post office was opened in 1870. The Shuswap called the place *Kahmoloops*, 'meeting of the waters'.

Kamouraska QC A regional county fronting on the St. Lawrence, upriver from Rivière-de-Loup, it was named in 1982 after the former county of Kamouraska. The name goes back to 1674 when Olivier Morel de la Durantaye (1640–1716) was granted the seigneury of Kamouraska. **Kamouraska** post office was opened in 1816. Created as a village in 1858, it was united in 1987 with the parish municipality of *Saint-Louis-de-Kamouraska* to form the municipality of Kamouraska. Derived from Algonquin, the name means 'where there are reeds along the edge of the water'.

Kamsack SK This site of this town (1911) on the Assiniboine River and near the Manitoba border was created in 1904 by the Canadian Northern Railway. Its post office had been established nearby in 1888 and named after an Aboriginal chief. He was a short, deformed person, who was derisively called Kamsack, 'big man'. He had helped the settlers when they arrived, and died about 1896.

Kananaskis River AB In 1858 explorer John Palliser named this river, and the **Kananaskis Pass** over the continental divide, southeast of Banff, after a Cree called Kineahkis, who was said to have recovered from an axe blow to the head. Subsequently the name was given to a group of lakes, a set of falls, a range of mountains, a hamlet near Exshaw, and an alpine village in Peter Lougheed Provincial Park.

Kanata ON A city (1978) in the regional municipality of Ottawa-Carleton, it was named in the 1960s because its clusters of neighbourhoods replicated the concept of *Kanata*, meaning 'a cluster of dwellings' in an Iroquoian language. It was a word encountered by Jacques Cartier in the 1530s, which eventually evolved into the name for the country.

Kane Basin NT On the east side of Ellesmere Island and extending for 220 km from Cape Sabine to Cape Lawrence, this basin was named after Dr Elisha Kent Kane (1820–57), who commanded the second Grinnell expedition in search of Sir John Franklin. He explored the Kalaalluit Nunaat (Greenland) and Ellesmere Island coasts north of Baffin Bay, 1853–5.

Kanesatake QC Formerly called *Oka*, the Mohawk community on the north shore of Lac des Deux Montagnes was officially named Kanesatake in 1986. In Mohawk the name means 'at the bottom of the hill'. *See also* Oka.

Kangiqsualujjuaq QC A northern village at the mouth of Rivière George in northeastern Québec, it is near the site of the Hudson's Bay Company's George River trading post, which operated from 1869 to 1952. The place was called *Port-Nouveau-Québec* in 1961, an unpopular choice among the local Inuit. Given its present name in 1980, it means 'the very large bay' in Inuktitut.

Kangiqsujuaq QC Located in northern Québec, this northern village was the site of the Stupart meteorological station, established in 1884 and named after meteorologist Frederick Stupart. In 1897 geologist Albert Peter Low named the inlet here *Wakeham Bay*, after Capt. William Wakeham, the commander of his vessel. Fur traders Revillon Frères opened their post at *Wakeham Bay* in 1910. In 1965 it was renamed *Maricourt* after Paul Le Moyne de Maricourt (1663–1704), who took part in an expedition to Hudson Bay in 1686, but it was not accepted by the local Inuit. It was officially renamed Kangiqsujuaq in 1980, after the Inuktitut for 'the big bay'.

Kangirsuk QC Located at the mouth of Rivière Arnaud, on the west side of Ungava Bay, this northern village was called *Payne River*, where the Hudson's Bay Company established a post in 1920. In 1961, the place was named *François-Babel*, after a missionary, and was renamed *Bellin* the following year, after the famous French hydrographer and map-maker Jacques-Nicolas Bellin (1703–72). The place was then called *Bellin (Payne)* until 1980, when it was renamed Kangirsuk, which is Inuktitut for 'the bay'.

Kapuskasing ON A town (1921) in Cochrane District, its post office was named in 1917 after the **Kapuskasing River**, which rises in **Kapuskasing Lake**, and flows north to join the Mattagami River, north of the town. The name was derived from the Cree *kepuskaskikwa*, 'it branches from a river'.

Kars ON On the Rideau River and in Ottawa-Carleton Region, south of Ottawa, it was named *Wellington* in 1856 after the Duke of Wellington. Because of duplication, Kars was selected to recognize the courage of Gen. William Fenwick Williams, who defended the city of Kars in Turkey in 1855 against a Russian siege and arranged for an honourable surrender.

Kasba Lake NT North of the point on the sixtieth parallel where the provinces of Manitoba and Saskatchewan meet, the name of this lake (1341 km^2) was ascertained in 1894 by geologist Joseph Burr Tyrrell. In Chipewyan it means 'small ptarmigan'.

Kaslo BC A village (1893) on the west bank of Kootenay Lake, northeast of Nelson, it was first called *Kane's Landing*, after George and David Kane, who settled here in 1889. Its post office was named in 1892 after **Kaslo Creek**. David Kane, who was the village's second mayor in 1894, claimed the creek was named after Johnny Kasleau, who had prospected for lead for the Hudson's Bay Company many years before. According to William Cockle, the postmaster in 1905, the creek's name was derived from the Kootenay word *cassoloe*, 'blackberry'. However, it has been stated that blackberries have never grown on the creek's banks.

Kativik QC Embracing 500,164 km^2 of northern Québec, this regional territory was created in 1978 to supervise the northern villages and to manage the health, welfare, economic development, and environment of the province north of the fifty-fifth parallel. In

Inuktitut the name means 'the place where they go to assemble'.

Kaumajet Mountains NF On the coast of northern Labrador, these mountains rise some 1300 m straight from the sea. In Inuktitut, the name means 'shining top'.

Kawartha Lakes ON These lakes, north of Peterborough and on the Trent Canal system, were named *Kawatha* in 1895 by Martha Whetung, a resident of the Curve Lake Indian Reserve, who said it meant 'land of reflections'. Resort promoters respelled it Kawartha, and claimed it meant 'bright waters and happy lands'.

Kazabazua QC A municipality in the regional county of La Vallée-de-la-Gatineau, midway between Hull and Maniwaki, it was named in 1976 after the well-known community of Kazabazua, a name in use before 1873. The place was called after **Rivière Kazabazua**, whose name in Algonquin means 'water flowing underground', in reference to the river flowing under a natural stone bridge near its mouth.

Kearney ON A town (1907) in Parry Sound District, north of Huntsville, it was settled in 1879 by William Patrick Kearney and the post office adopted his name the next year.

Kedgwick NB A village (1966) northeast of Edmundston, its post office was called *Richards Station* in 1912. It was renamed three years later after the **Kedgwick River**, which rises in Québec, and flows 70 km southeast into the Restigouche River, west of the village. Its name was derived from Mi'kmaq *madawamkedjwik*, meaning 'large branch', or 'flowing underground in many places'. It was spelled in many ways until the Geographic Board standardized it in 1901.

Keele Peak YT North of Macmillan Pass, this peak (2975 m) in the Hess Mountains of the Selwyn Mountains was named in 1909 by geologist R.G. McConnell after Joseph Keele (1863–1923), who extensively surveyed the Yukon and the western Northwest Territories. In 1911 geologist D.D. Cairnes named **Keele Range**, on the Yukon-Alaska boundary, south of Porcupine River, for him.

Keele River NT A tributary of the Mackenzie River, it rises in the Mackenzie Mountains. It was named by the Geographic Board in 1924 after geologist Joseph Keele, who had explored this river and the Ross River, in the Yukon. It had been approved in 1909 as *Gravel River*, shown on an 1887–8 map and recommended by Keele. In Slavey it was known as *Mbekonityeh*, 'dried meat river'.

Keene ON On the north shore of Rice Lake and in Peterborough County, southeast of the city of Peterborough, this place was first called *Gilchrist's Mills* in 1825, after its founder, Dr John Gilchrist. Gilchrist renamed it Keene in 1858 after his home town in New Hampshire, west of Manchester.

Keewatin ON A town in Kenora District, west of the town of Kenora, it was called *Keewatin Mills* in 1880, after John Mather's sawmill and the district of Keewatin, established in 1876. The name is from the Ojibwa for 'north wind', in reference to any cold winds. The name was soon shortened to Keewatin, and the place became an incorporated town in 1908.

Kejimkujik Lake NS In south central Nova Scotia, this lake is midway between Digby and Liverpool. In Mi'kmaq, its name may mean 'swelled private parts', in reference to the exertion required to row across the lake, or 'attempting to escape', or 'fairy lake', as it was formerly called *Fairy Lake*. **Kejimkujik National Park** (382 km^2) was created in 1968.

Kelligrews NF In the town of Conception Bay South, southeast of St. John's, this built-up centre was named after nearby **Kelligrew's Point**. The point was likely called after a Kelligrew family, which established a store here in the early 1800s. The family may have come from Port de Grave, across Conception Bay.

Kelliher SK Established in 1908, this village (1909) northwest of Melville, was named after the Grand Trunk Pacific Railway chief engineer B.B. Kelliher. The name preserves the alphabetical arrangement of stations from Portage la Prairie to Edmonton, with Jasmin to the east, and Leross to the west. Its post office was opened in 1908.

Kelowna BC On the eastern shore of Okanagan Lake, this city (1905) was named in 1892 after the Okanagan word for 'female grizzly bear'. With the largest beach on the lake, it was early known as *L'Anse au Sable* by French-speaking trappers and fur traders.

Kelvington SK A town (1944) northwest of Canora, its post office was opened in 1905 and named after Baron William Thomson Kelvin (1824–1907), a famed physicist and inventor, who lived at Largs on the Firth of Clyde, west of Glasgow. Mrs John McQuarrie, who was raised near Largs, proposed the name.

Kelwood MB Situated east of Riding Mountain National Park and north of Neepawa, this place's post office was named in 1905, by blending the names 'Callie' and 'Wood'. Callie Barber was the first white child born here and Angus Wood was the first postmaster. The post office had been called *Glensmith Station* in 1904.

Kempenfelt Bay ON A western arm of Lake Simcoe, it was named by Lt-Gov. John Graves Simcoe in 1793, after Rear-Adm. Richard Kempenfelt (1718–82). He commanded the Blue Squadron of His Majesty's Fleet, and died when his ship *Royal George* went down off Spithead and Portsmouth, England.

Kemptville ON A town (1963) in the united counties of Leeds and Grenville, it was named in 1828 after Sir James Kempt (*c.* 1765–1854), lieutenant-governor of Nova Scotia, 1820–8, and administrator of the government of Canada, 1828–30. It was earlier called *Clothier's Mills*, after Lyman Clothier and his four sons, who settled here in 1812.

Kenaston SK This village (1910) was founded as *Bonnington* in 1890 by the Qu'Appelle, Long Lake and Saskatchewan Railway and Steamboat Company, which became part of Canadian Northern Railway in 1906. *Bonnington* post office was opened in 1904, but, because of confusion with Bonnyville, AB, it was renamed Kenaston in 1906. It was named after F.E. Kenaston of Minneapolis, vice-president of the Saskatchewan Land Valley Company, which developed the area between Davidson and Hanley.

Kenmore ON In Ottawa-Carleton Region, southeast of Ottawa, this community's post office was named in 1857 by Peter McLaren, after his birthplace in Perthshire's Strath Tay, northwest of Perth in Scotland.

Kennebec ON A township in Frontenac County, it was named in 1823, possibly after the Ojibwa word *kenibig*, meaning 'snake', or after the Kennebec River in Maine, where in Penobscot it means 'long bay place'.

Kennebecasis River NB Rising east of Sussex, this river flows southwest to empty into **Kennebecasis Bay**, an arm of the Saint John River. In Mi'kmaq the name means 'little bay place', suggesting the lower Saint John River may have been once called Kennebec, 'big bay place', as far as the head of tide west of Fredericton. Many spellings were used after 1686, with the present one being adopted by surveyor Joseph Bouchette in 1815.

Kennedy SK The post office at this village (1907) was named in 1904 after Findlay Kennedy, who previously carried the mail from Whitewood on the main Canadian Pacific Railway line, 35 km to the north. The CPR built a line through Kennedy in 1907.

Kennedy Channel NT Between Ellesmere Island and Kalaalluit Nunaat (Greenland), this channel unites Kane Basin with Hall Basin. It was named in 1854 by explorer Dr Elisha Kent Kane, after John Pendleton Kennedy, secretary of the United States Navy.

Kennedy, Mount YT The proposal to name this mountain in the St. Elias Mountains, east of Mount Alverstone, was made in December 1964 by Walter Wood, president of the American Geographical Society, after John Fitzgerald

Kennedy (1917–63), the thirty-fifth president of the United States. A few weeks earlier the Canadian government had named another peak after him, but Wood pointed out that it was only a lower part of the uplift of Mount Logan.

Kennetcook NS Northeast of Windsor and near the head of the **Kennetcook River**, this place's post office was established as *Kennectcook Corner* in 1867. It was changed to the present name in 1934. In Mi'kmaq the name is said to mean 'place nearby or further on'.

Kénogami QC A sector of Jonquière, it was an incorporated town from 1958 to 1975. It was named after **Rivière Kénogami**, which drains **Lac Kénogami** north to the Saguenay. Fr Charles Albanel made reference to the lake in 1672. The name in Montagnais means 'long lake'.

Kenora ON A town in Kenora District, it became the Manitoba town of *Rat Portage* in 1882, when the British Privy Council awarded it to that province, reversing Canada's decision of a year earlier. But in 1884 the Privy Council put it back into Ontario, where it became an incorporated town in 1892. The place took its name from the abundance of muskrat in the Winnipeg River basin. It was renamed in 1905, taking the first two letters of each of Keewatin, *Norman*, and Rat Portage. *Norman*, a locality midway between Kenora and Keewatin, had been named after the son of a lumber company manager. *See also* Keewatin.

Kenosee Lake SK A village (1987) within Moose Mountain Provincial Park and north of Carlyle, its post office was called *Kenosee Park* in 1923. It was renamed Kenosee Lake in 1974. In Ojibwa *kenosee* means 'long creek'.

Kensington PE A town (1914) east of Summerside, its name was chosen about 1863 for a school district after Kensington in London, England. First called *Glover's*, its post office was *Barretts Cross* from 1851 to 1886, when it was changed to its present name. *Barretts Cross* may have been named after a Miss Barrett, who operated a tavern here.

Kent BC A district municipality (1895) on the north side of the Fraser River, northeast of Chilliwack, it was likely named after Kent in southeastern England.

Kent NB Established in 1826, this county was named after Prince William, the Duke of Kent (1765–1837), who became William IV, 1830–7.

Kent ON The county was named in 1792 after the southeastern county in England. Its territory then went all the way to Hudson Bay, but was confined to Southwestern Ontario in 1800 and to its present area in 1850.

Kenton MB Northwest of Brandon, this place was called *Ralphtown* in 1884 after the son of the postmaster, W.J. Helliwell. It was renamed in 1904 after postmaster A.W. Kent, who had settled in the area in 1881.

Kent Peninsula NT Part of the mainland south of Victoria Island and between Coronation and Queen Maud gulfs, this peninsula was named in 1838 by explorer Thomas Simpson, after the Duchess of Kent. The mother of Queen Victoria, she was Princess Victoria of Saxe-Coburg-Saalfeld, and the widow of Emich, the Prince of Leinigen. She married the Duke of Kent in 1818. After he died in 1820, she lived as a widow again for another 41 years.

Kentville NS This town (1886) was named in 1826 after Prince Edward, the Duke of Kent and Strathearn (1767–1820), the father of Queen Victoria. He visited the place in 1794, while he was the commander of the British forces in Halifax. Earlier it was called in Mi'kmaq *Penooek*, 'crossing place', and *Horton Corner*, after the township.

Kenyon ON A township in the united counties of Stormont, Dundas and Glengarry, it was named in 1798 after Lloyd, Lord Kenyon (1732–1802), a chief justice of England's Court of King's Bench.

Keppel ON A township in Grey County, it was named in 1855 by William Coutts Keppel, Viscount Bury, superintendent of Indian affairs in Canada, after his uncle Henry Keppel, the captain of the ship that took James Brooke, the future rajah, to Sarawak in 1841. Keppel wrote an account of the voyage.

Keppoch PE Formerly part of the municipal community of *Keppoch-Kinlock*, south of Charottetown, it became part of the town of Stratford in 1995. It had been named by Alexander Macdonald after Keppoch, Arisaig, Inverness-shire, Scotland.

Keremeos BC In the Similkameen River valley, southwest of Penticton, this village (1956) served as the site of the Hudson's Bay Company's Fort Okanogan from 1860 to 1872. Its post office was opened in 1887 and named after **Keremeos Creek**, whose name in Okanagan means either '(land) cut across in the middle', or 'flat cut through by water'.

Kerrobert SK Located west of Biggar, this town (1911) was founded in 1911 by the Canadian Pacific Railway and named after Robert Kerr, the CPR's traffic manager. The post office was called *Kerr-Robert* in 1910 and was changed to Kerrobert in 1924.

Keswick NB West of Fredericton, this community is located at the mouth of the **Keswick River**. Its post office was called *Mouth of Keswick* in 1849. The river derived its name from Maliseet *nookamkeechwak*, 'gravelly river'. Its present name occurred in a land petition of 1811, but for many years between 1783 and 1820 it was called *Madamkeswick River*. Nearby **Keswick Ridge** was named after Maliseet *quesawednek*, 'the end hill', and its spelling was adjusted to agree with the name of the river.

Keswick ON In the town of Georgina, York Region, and north of Newmarket, its post office was located in 1835 at present Roches Point, 3 km to the northwest, and was moved to its present site about 1870. It was possibly named after the town of Keswick in Cumberland, England.

Ketch Harbour NS South of Halifax and on the Atlantic coast, this place had an important fishing harbour in the 1750s. It was likely named after a ketch, a sailing vessel with two masts, although it may have been named because a good 'catch' of fish could be assured offshore for fishermen based in the harbour. Its post office was opened in 1855.

Ketepec NB A neighbourhood in the city of Saint John and on the west side of the Saint John River, its name was created in 1902 by naturalist William F. Ganong for a Canadian Pacific Railway station. He adapted it from Maliseet *pekwitepekek*, 'grand bay', which may have been a Maliseet translation of the French name.

Kicking Horse Pass AB, BC In 1858 Dr James (later Sir James) Hector, after ascending a tumultuous tributary of the Columbia River, was kicked by a horse and knocked senseless near the summit of the pass (1627 m). By association, the river was named **Kicking Horse River**.

Kidd, Mount AB This mountain (2958 m) in the Kananaskis Range, southeast of Banff, was named in 1907 by geologist Donaldson B. Dowling after trader Stuart Kidd (1883–1956). Kidd managed posts at Morley and Nordegg, and became fluent in the language of the Stoney, who made him Honorary Chief Tah-Osa ('moose killer') in 1927.

Kilbride NF On the south side of the city of St. John's, this urban centre may have been named about 1860. It may have been named after Kilbride Falls on the Waterford River, which possibly inspired the naming of St. Bridget's Church in 1863. The Irish for Kilbride, *Cill-Bhrigde*, means 'Brigid's church'.

Killaloe ON The original Killaloe in Renfrew County was named in 1850 by lumberman James Bonfield after his birthplace in County Clare, Ireland. In 1894 the railway passed 2 km to the north, with the post office (1896) and village (1911) becoming *Killaloe Station*. The post office was changed to Killaloe in 1961, and the village became Killaloe in 1989, while the original place became **Old Killaloe** in 1990.

Killam AB Southeast of Camrose, the station in this town (1965) was named in 1906 by the Canadian Pacific Railway after Albert Clements Killam (1849–1908), Western Canada's first appointment to the Supreme Court of Canada, 1903–4, and chief commissioner of the Board of Railway Commissioners, 1904–8. The post office was opened in 1907.

Killarney MB Southeast of Brandon and near the North Dakota border, the post office in this town (1907) was named in 1883 by John O'Brien after the town and lakes in County Kerry, Ireland. It was previously called *Oak Lake* after the adjacent lake, but O'Brien thought it was as pity that such a beautiful gem of water should go by such a plain name.

Killarney ON In Manitoulin District and on the north shore of Georgian Bay, this community's post office was named in 1854 after the town in County Kerry, Ireland. **Killarney Provincial Park** was created in 1964.

Kimberley BC A city (1944) northwest of Cranbrook, it was named in 1897 by Col William Ridpath of Spokane, WA. He likely chose the name in the hope that the mines, which he had bought the previous year for his syndicate, would produce riches similar to those of South Africa's Kimberley. The South African diamond centre was named in 1873 after John, 1st Earl of Kimberley (1826–1902), then British secretary of state for the colonies, who placed the mines under British protection.

Kimmirut NT On the south coast of Baffin Island and southwest of Iqaluit, this hamlet (1982) was first called *Lake Harbour*. In Inuktitut, *kimmirut* means 'looks like a heel', because a nearby large rock resembles an upturned heel. *Lake Harbour* was the site of an Anglican mission in 1910 and a Hudson's Bay Company store in 1911. A Royal Canadian Mounted Police post and a post office were established there in 1927.

Kinbasket Lake BC Impounded by the Mica Dam on the Columbia River, this reservoir was created in 1973 and first named after Gen. A.G.L. McNaughton, commander of the Canadian forces in Britain, 1939–43, and chairman of the Canadian section of the International Joint Commission, 1950–62. In 1980 the name Kinbasket Lake, a small lake submerged by the reservoir, which had been named officially in 1912, was restored. In 1866 surveyor Walter Moberly had named it *Kinbaskit Lake* after Shuswap chief Paul Ignatius Kinbaskit, who provided much appreciated assistance to Moberly.

Kinburn ON In Ottawa-Carleton Region, southeast of Arnprior, its post office was named in 1855 after the fortress of Kinburn at the mouth of the Dneiper River, in present-day Ukraine. It was attacked by the British/French forces during the Crimean War.

Kincaid SK West of Assiniboia, this place was reached by the Canadian Pacific Railway in early July 1913 and the village was promptly incorporated that month. It and the post office were named after pioneer settler Charlie Kincaid.

Kincardine ON The township in Bruce County was named in 1849 after a title of James Bruce, 8th Earl of Elgin and 12th Earl of Kincardine (1811–63), governor of the province of Canada, 1846–54. The town (1875) of **Kincardine**, first called *Penetangore* in 1851, after the Penetangore River, was renamed after the township in 1857.

Kindersley SK In 1909 the Canadian Northern Railway ran a line from Rosetown to Alsask on the Alberta border. The post office in the town (1910) was named in 1909 after Sir Robert Kindersley, a prominent stockholder in the Canadian Northern Railway. Kindersley has grown into the largest urban centre between Saskatoon and Medicine Hat.

King ON A township in York Region, it was named in 1798, probably after John King, then a British under-secretary of state for the colonies. The post office at **King City** was called *King* from 1853 to 1953, when local residents persuaded the postal authorities to accept King City, then the name of the police village.

King Christian Island NT In the Sverdrup Islands of the Queen Elizabeth Islands and south of Ellef Ringnes Island, this island (645 km^2) was named *King Christian Land* in 1901 by Otto Sverdrup after Christian IX (1818–1906), who became king of the Danes in 1863.

King Edward, Mount AB, BC On the contintental divide and near the head of the Athabasca River, this mountain (3490 m) was named in 1906 by climber Mary Schaffer, after Edward VII (1841–1910), who succeeded his mother, Queen Victoria, in 1901.

King George IV Lake NF In the southwestern part of the island of Newfoundland, this lake was named in 1822 by William E. Cormack, after the reigning monarch.

King George, Mount BC This mountain (3422 m) in the Royal Group of the Rocky Mountains, east of Invermere, was named in 1916 by the Alberta-British Columbia Boundary Commission after George V (1865–1936), the first sovereign of the House of Windsor (1917), who ascended the throne in 1910. Other peaks in the group were named at the same time after Queen Mary, Princess Mary, and the princes Albert, Edward, George, Henry, and John.

King George, Mount YT Located north of Mount Vancouver in the St. Elias Mountains, this mountain (3734 m) was named in 1935 by the National Geographic Society Yukon Expedition after George V, on the occasion of the twenty-fifth anniversary of his accession to the throne.

King, Mount YT Near the Alaska border and west of Mount Logan, this mountain (5173 m) was named about 1918 after William Frederick King (1854–1916), chief astronomer of Canada (1890–1905), director of the Dominion Observatory (1905–16), and member of the International Boundary Commission, 1892–1916.

King Peak BC In the Muskwa Ranges of the Rocky Mountains, this mountain (2972 m) was named in 1967 by W.R. Young, provincial member of the Canadian Permanent Committee on Geographical Names, after William Lyon Mackenzie King (1874–1950), prime minister of Canada during the Second World War. It and several other names were given that year to honour world leaders, places, and battles associated with the war.

Kings NB Established in 1785, north of Saint John, this county was named to express loyalty to the monarch, George III.

Kings NS A county in the eastern part of the Annapolis Valley, it was named in 1786 after George III.

Kings PE The eastern county of the province, it was named in 1765 by Samuel Holland after George III, who had become the monarch in 1760.

Kingsclear NB Located west of Fredericton, this community was named in 1849 after **Kingsclear Parish**. The parish was established in 1786, and was possibly named by Edward Winslow, the surveyor of the king's woods, devising the name from the clearings made in the woods to provide masts and timber for the navy.

King's Cove NF On the south side of Bonavista Bay and southwest of the town of Bonavista, the name of this municipal community (1966) is likely a variant of *Canning's Cove*, an eighteenth century name. It was settled in the early 1800s by English and Irish immigrants, who pursued the local inshore and Labrador fishery and sealing.

Kingsey Falls QC Located in the regional county of Arthabaska, south of Victoriaville, the municipality was erected in 1865, taking its name from the township. The latter was named in 1792, proclaimed in 1803, and likely called after a village in Oxfordshire, England. The post office in the village (1922) of **Kingsey Falls** had been opened in 1849. The western part of the township was incorporated as the township municipality of **Kingsey** in 1845, after its post office had been opened in 1836.

Kingsmere QC This community in Gatineau Park, north of Aylmer, was named in 1879 after a small lake, which was earlier named after nearby **Mont King**. It is reported that the mountain was named after an early settler in the area of the present park. Dr William Frederick King, who was appointed the Dominion Astronomer in 1905, established the first geodetic triangulation station that year on the mountain, and subsequently it was incorrectly assumed the mountain had been named after him.

King Solomon Dome YT The highest point (1234 m) in the Klondike area, it was initially called *The Dome*, because of its shape. Recognizing some of the richest goldfields in the world radiated from it, the miners named it after the biblical king's lost gold mines.

Kingsport NS On the shore of Minas Basin, north of Wolfville, this place's post office was named in 1874 after its situation as a port in Kings County. It was earlier known as *Indian Point* and *Oak Point*.

Kingston NS In the Annapolis Valley, west of Kentville, the first post office of this village (1957) was called *Kingston Station* in 1869, when the Dominion Atlantic Railway was built from Digby to Halifax, and it was changed to Kingston in 1922. **Kingston Village** post office was established in 1842, 3 km to the south. Both had been named after Kings County.

Kingston PE West of Charlottetown, this place's school district was called *South Wiltshire* in 1864. Its post office was opened as *South Wiltshire* in 1870, which became Kingston in 1898. Created a municipal community in 1974, it embraces the area of Emyvale.

Kingston ON The township and townsite of the city (1846) of Kingston were surveyed in 1783 and were first named *King's Town*, after George III, but were soon shortened to Kingston. The French had built Fort Cataraqui in 1673 at the mouth of the Cataraqui River. It was renamed Fort Frontenac that year by Sieur de La Salle in honour of Louis de Buade, (1622–98), comte de Frontenac et de Palluau, governor of New France 1672–82, and 1689–98. Kingston was the capital of the province of Canada from 1841 to 1843, when the provincial capital was moved to Montréal.

Kingsville ON A town (1901) in Essex County, west of Leamington, it was named in 1843 by developer Andrew Stewart after Col James King, who took up Stewart's offer to have the place named after the first person to build a house here. Subsequently, King was a customs collector and court clerk here.

King William Island NT Between Boothia Peninsula and Victoria Island, this island (13,111 km^2) was named in 1830 by explorer John (later Sir John) Ross after William IV (1765–1837), who became the king of Great Britain and Ireland that year.

Kinistino SK The area of this town (1952) west of Melfort was known as the *Carrot River Settlement* in the 1870s. Its post office was named *Kinisteno* in 1883, taken from the Ojibwa name for the Cree. Four years later it was respelled Kinistino.

Kinkora PE A municipal community (1983) northeast of Borden-Carleton, it had been incorporated as a village in 1955. Its school district was called *Somerset* about 1841. Its post office opened as *Somerset* in 1867 and was changed to Kinkora 20 years later, after the ruins of the palace of Irish king Brian Boru (926–1014), near Killaloe, County Clare, Ireland. The area of Kinkora was settled by Irish immigrants about 1841.

Kinlock PE Formerly part of the municipal community of *Keppoch-Kinlock*, this place, south of Charlottetown, became part of the town of Stratford in 1995. It was named either by Alexander Macdonald, or another Macdonald, after Kinloch, Isle of Skye, Scotland.

Kinloss ON A township in Bruce County, it was named in 1850, after one of the titles of

James Bruce, 8th Earl of Elgin (1811–63), governor of the province of Canada, 1846–54. Kinloss in Scotland is northeast of Inverness.

Kinmount ON In Victoria County, north of Lindsay, this place was named in 1859 by Mrs Malcolm Bell after Kinmount in Dumfriesshire, Scotland, northwest of Annan.

Kinuso AB A village (1949) south of Lesser Slave Lake, it was first known as *Swan River*, but its post office was called Kinuso in 1915, after the Cree word for 'fish'.

Kipawa, Lac QC A lake with a vast labyrinth of arms and bays, near the town of Témiscaming in western Québec, its name in Algonquin means 'it is closed' or 'it is blocked', possibly in reference to the lake's outlet at Laniel, where **Rivière Kipawa** carries its waters to Lake Timiskaming.

Kipling SK This town (1954) southwest of Grenfell was founded in 1909 by the Canadian Northern Railway, and named after popular writer Rudyard Kipling (1865–1936). The post office was named *Kipling Station* that year, with the name being shortened to Kipling in 1955.

Kippens NF Located on St. George's Bay and on the southwestern coast of the island of Newfoundland, this town (1968) may have been named after a Capt. Kippen, who was shipwrecked here in the 1840s, or after a local Keeping family.

Kirkfield ON A community in Victoria County, northwest of Lindsay, its name was selected by residents in 1864 from a list of eight names provided by the post office department. As most of the area's settlers were Highland Scots, it appropriately reflected the Scottish kirk (church) and the farming community.

Kirkland QC A city in the Communauté urbaine de Montréal, between Pointe-Claire and Pierrefonds, it was named in 1961 after Dr Charles-Aimé Kirkland (1896–1961), member of the Legislative Assembly, 1939–61.

Kirkland Lake ON In 1907 surveyor Lewis Rorke named a small lake in Timiskaming District after Winifred Kirkland, a secretary in the provincial department of mines. Gold was discovered here four years later, and a mining community quickly developed. In 1972 the municipal township of Teck, which included this urban centre as well as the communities of Swastika and Chaput Hughes, was incorporated as the town of Kirkland Lake.

Kirkton ON On the Huron-Perth boundary, west of Stratford, this place was named in 1856 after five Kirk brothers, who settled here in the 1840s, as well as after one of many places called Kirkton in Scotland.

Kisbey SK Situated northeast of Estevan, this village (1907) was established by the Canadian Pacific Railway in 1905 and named after R. Claude Kisbey, then the agent of the Crown lands department based in nearby Alameda.

Kitchener ON A city (1912) in Waterloo Region, it was named *Berlin* in 1833 after the capital of Prussia. Because patriotism of local residents was doubted during the First World War and because of fear of business losses from products labelled 'made in Berlin', the city was renamed in 1916 after British military hero Lord Horatio Herbert Kitchener, who was drowned that year off the Orkney Islands. In the early years, it had been known as *Sandhills*, and also *Ben Eby's* and *Ebytown* after Mennonite bishop Benjamin Eby.

Kitchener, Mount AB This mountain (3480 m) on the north side of the Columbia Icefield, was named in 1919 after Horatio Herbert Kitchener, 1st Earl Kitchener (1850–1916), a prominent British military leader, who was lost at sea when the HMS *Hampshire* struck a mine north of Scotland. There are three mountains named after him in British Columbia, one south of Kootenay Lake, a second in the Coast Mountains, near Knight Inlet, and a third in the northwest, near the Stikine River.

Kitimat BC A district municipality (1953) southeast of Prince Rupert and at the head of Kitimat Arm, it was founded in 1951 by the

Aluminum Company of Canada. **Kitimat Arm** was named in 1837 by the Hudson's Bay Company, after the Kitamaat, a Kwakiutl tribe, whose name in Coast Tsimshian means 'people of the falling snow'.

Kitley ON A township in the united counties of Leeds and Grenville, it was named in 1798 after the Bastard family home in Devon, England. The Bastard family acquired the Kitley estate in 1710. John Pollexfen Bastard was a British member of Parliament when Kitley Township was named. *See also* Bastard.

Kitscoty AB A village (1911) west of Lloydminster, its post office was named in 1907 after Kit's Coty House in Kent, England, north of Maidstone. The name may have been given by Canadian Pacific Railway contractor George Still. The ancient English 'house' comprises three large upright stones with a fourth balanced over them.

Kitwanga BC At the confluence of the Kitwanga and Skeena rivers, this place's post office was named *Gitwangak* in 1910. It was changed to Kitwanga in 1917. The river's name in Gitksan means 'people of the place of rabbits'.

Klahowya Mountain BC In the Purcell Mountains, northeast of Kaslo, this mountain (2922 m) was named in 1931 after **Klahowya Creek**. The creek was called after a riverboat built in 1910 by Capt. Francis P. Armstrong, who had operated boats on the upper Columbia River between 1885 and 1914. In Kootenay the name means 'greetings friend'.

Kleefeld MB West of Steinbach, this Mennonite community was established in 1896, and its post office was named that year after a village in Ukraine, where they had previously lived. In German the name means 'clover field'. It had been briefly called *Gruenfeld*, meaning 'green field'.

Kleinburg ON In the city of Vaughan, northwest of Toronto, it was named in the 1830s after John Nicholas Klein Sr, who built a sawmill, a grist mill, and a cooperage on the West Humber River. The post office was called *Klineburg* in 1852, and corrected to Kleinburg five years later.

Kliniklini River BC Rising northeast of Mount Waddington, this river flows west and then south into the head of Knight Inlet. In Kwakwala the name means 'eulachon grease', a valuable component in the diet of both coastal and interior Aboriginal tribes. The oily eulachon is also called 'candle fish'. The river provided a 'grease-trail' route into the interior, where the Kwakwala traded with the Chilcotin and other tribes. The post office at **Kleena Kleene**, where the Chilcotin-Bella Coola Highway crosses the river, was named in 1927.

Klondike River YT Rising in the Ogilvie Mountains, this river flows west into the Yukon River at Dawson. Derived from the Gwich'in *Thronduik*, 'hammer water', the name recalls their practice of driving stakes into the river's bottom to trap migrating salmon. On learning of the discovery of gold on Bonanza Creek, a tributary of the Klondike, North-West Mounted Police Insp. Charles Constantine, the mining recorder at Fortymile River, rendered the name in 1896 as *Klondyke*. Two years later the Geographic Board of Canada made the present spelling official.

Kluane Lake YT Northwest of Whitehorse, this lake (405 km^2) was named in 1882 by Aurel Krause, who undertook an expedition on behalf of the Bremen Geographical Society. In Tlingit the name means 'whitefish place'. **Kluane National Park Reserve** (22,015 km^2) was established in 1972 in the area of Mount Logan and the St. Elias Mountains.

Knight Inlet BC A deep 95-km fiord extending inland from Queen Charlotte Strait, it was named in 1792 by Lt-Cdr William Broughton after Capt. (later Adm. Sir) John Knight (*c.* 1748–1831). Broughton and Knight had served together in 1775–6.

Knowlton QC Established as a village in 1888, it became part of the town of Lac-Brome in 1971. It had been named after Paul Holland Knowlton (1787–1863), who settled on a farm on the south shore of Lac Brome in 1815, built a store and a distillery, and exercised considerable influence in the area. The post office was opened in 1851.

Knutsford PE West of O'Leary, this place was established as a school district in 1869. It may have been named after Knutsford, on the southwest side of Manchester, England. Its post office from 1871 to 1913 was called *O'Leary Road*.

Koch Island NT In the north part of Foxe Basin and southwest of Baffin Island, this island (458 km^2) was named in 1921–4 by the Fifth Thule expedition after Danish polar explorer J.P. Koch.

Koksoak, Rivière QC Flowing into the south side of Ungava Bay, this wide river is the main outlet for several rivers draining north from central Québec. The name, derived from the Inuktitut for 'big river', was known in the early 1800s, but was not made official until 1916, after having been called *Big River* or *South River* for many years.

Komoka ON A community in Lobo Township, Middlesex County, west of London, it was first called *Lobo Junction* in 1853. It was renamed Komoka in 1855, when a railway was built from London to Sarnia. The name was derived from Ojibwa *kokomiss*, meaning either 'young grandmother', or 'quiet resting place of the dead'.

Kootenay River BC Rising in the Rocky Mountains, southeast of Field, this river (780 km) flows south, crossing the border into Montana, northeast of Libby, and recrossing the border from Idaho, 100 km to the west. After flowing into **Kootenay Lake** (407 km^2), it turns west to join the Columbia at Castlegar. The river and the lake were named after the Blackfoot pronunciation of the name of the Kootenay, a distinct linguistic group, whose name means 'water people'. In 1808 geographer David Thompson named the river *McGillivray's River* after brothers Duncan and William McGillivray, North West Company traders. **Kootenay National Park** (1378 km^2) was created in 1920.

Kouchibouguac River NB This river flows northeast into Northumberland Strait. Its name was derived from Mi'kmaq *pijeboogwek*, 'river of the long tideway'. Nearby is the **Kouchibouguacis River**, whose Mi'kmaq name means 'little river of the long tideway'. Extending along the shore of the strait is **Kouchibouguac National Park** (225 km^2), established in 1969.

Krestova BC North of Castlegar, this community was named by Peter Veregin, who led Doukhobors from Saskatchewan in 1908 to the Slocan River valley. The name in Russian means 'place of the cross'.

Kugluktuk NT On Coronation Gulf and at the mouth of the Coppermine River, this hamlet (1981) was officially given this name 1 January 1996. First known as *Coronation*, it was established as a trading post by Charles Klengenberg in 1916. It was renamed *Coppermine* in 1930. The name Kugluktuk means 'place of rapids' in Inuktitut, probably in reference to Bloody Fall, where Chipewyan, accompanying Samuel Hearne, massacred a group of Inuit on 17 July 1771.

Kuujjuaq QC Located on the west bank of Rivière Koksoak, this largest of Québec's northern villages became the site in 1830 of the Hudson's Bay Company's Fort Chimo. The word *chimo* in Inuktitut means 'greeting', in the sense of 'are you friendly?' The community of *Fort-Chimo* was renamed Kuujjuaq in 1979, using the contemporary spelling of Koksoak, meaning 'big river'.

Kuujjuarapik QC Situated at the mouth of the Grande rivière à la Baleine, in northwestern Québec, it was first known during the Second World War as *Great Whale River*. By 1960 the Inuit residents started calling it *Kuujjuaq*, but they were persuaded by the residents of Fort-Chimo (Kuujjuaq) on the larger 'grand river' to call their place Kuujjuarapik, 'the little grand river'. In 1961 the Québec government changed the name to *Grande-Baleine*, and the following year it became Poste-de-la-Baleine, which remains the postal name. Kuujjuarapik was recognized as the name of the northern village in 1980, and the Cree name Whapmagoostui was also authorized. In English the place is commonly called Great Whale.

Kyle SK Located north of Swift Current and Lake Diefenbaker, the post office in this town (1959) was called *Kyleville* in 1909 after early settler Jerry Kyle. When the Canadian Pacific Railway built its line south from Rosetown in 1923, it shortened the name to Kyle, and the postal name was shortened the following year. There had been another *Kyle* post office (1905–1918) northwest of Melfort, which had been named after Kyle, in Ayrshire, Scotland, near the birthplace of poet Robert Burns.

L

La Baie QC A city in the regional county of Le Fjord-du-Saguenay and at the head of the Baie des Ha! Ha!, it was created in 1975 through the merger of the towns of Bagotville, Grande-Baie, and Port-Alfred, and the parish municipality of Bagotville. Grande-Baie had been incorporated as a town in 1859. The town of Port-Alfred, created in 1919, was named after Julien-Édouard-Alfred Dubuc (1871–1947), who founded a pulp company here at the beginning of the century. The town of Bagotville was named after the township of Bagot, proclaimed in 1848, and named after Sir Charles Bagot, governor of the province of Canada, 1841–3.

Labelle QC A municipality in the regional county of Les Laurentides, west of Mont Tremblant, it was created in 1973 through the amalgamation of the village of Labelle and the township municipality of *Joly*. The post office was called *Chute-aux-Iroquois* from 1881 to 1902, when it was changed to Labelle, the year that the village was incorporated. It took its name from the township, proclaimed in 1894, and named in honour of Mgr Antoine Labelle (1833–91), who promoted settlement in the area north of Montréal.

Laberge, Lake YT A widening of the Yukon River, directly north of Whitehorse, this lake was named *Lake Lebarge* in 1870 by American naturalist William H. Dall, who was working on behalf of the Western Union Telegraph Company in choosing a route for a line from the United States to Russia. The United States Board on Geographic Names adopted it in 1890, noting it was named after 'Michael Lebarge'. This and other decisions made for names on Canadian territory prompted Canada in 1897 to set up its own names board in 1897. The Canadian board consulted church officials in Chateauguay, QC, and learned that Michel Laberge was still living there. The name was corrected to Lake Laberge, but its pronunciation rhymes with 'le marge'.

Labrador NF The mainland district of the province of Newfoundland, which also appears in the title of the 'government of Newfoundland and Labrador', its name may be traced to the Portuguese *lavrador* (meaning small landholder) João Fernandes. Fernandes explored the coasts of North America at the end of the 1400s and beginning of the 1500s. In 1500 he identified Kalaallit Nunaat (Greenland) as *Tiera del Lavrador*, and the present Labrador was called *Tiera del Corte Real*, given for another *lavrador*, Gaspar Corte-Real. About 1560, after the name 'Greenland' was found to be in general use, map-makers shifted Labrador to the mainland.

Labrador City NF In western Labrador, this town (1961) was founded as *Carol Lake* in 1959 by the Iron Ore Company of Canada. It has the distinction of being the most populous incorporated municipality in Canada with the word 'City' as the terminal part of its name.

La Broquerie MB East of Steinbach, this place's post office was named in 1882 by Mgr Alexandre Taché after his uncle, Joseph B. La Broquerie (1759–1830). The rural municipality of **La Broquerie** was erected a year earlier.

L'Acadie QC Situated west of Saint-Jean-sur-Richelieu and on the **Rivière L'Acadie**, this municipality's post office was called L'Acadie in 1835. The municipality was incorporated as *Lacadie* in 1926 and was changed to L'Acadie in 1976. Acadians settled here in 1768, a few years after their expulsion from Acadia.

Lac-à-la-Tortue QC A municipality east of Grand-Mère, it was created in 1981, taking its name from the post office, which had been opened in 1882. The name likely reflected the presence of many turtles (*tortues*) here.

Lac-au-Saumon QC Situated in the regional county of La Matapédia and midway between the towns of Causapscal and Amqui, this village was named in 1905 after Lac au Saumon, an enlargement of the Rivière Matapédia. Its post office had been opened eight years earlier.

Lac-Baker NB A village (1967) west of Edmundston, its post office was named *Lake Baker* in 1893, after the 8-km-long lake, which extends into the province of Québec. It was renamed Lac-Baker in 1938. The lake was named after John Baker (1796–1868), who strenuously resisted having his land being considered British territory in the 1820s, inflaming a boundary dispute between New Brunswick and Maine that was not resolved until the Webster-Ashburton Treaty of 1842. *See also* Baker Brook.

Lac-Beauport QC A municipality north of the city of Québec, and inland from the city of Beauport, it was named in 1989, taking its name from the parish municipality of *Saint-Dunstan-du-Lac-Beauport*, created in 1855. The area is a popular all-season resort centre.

Lac-Bouchette QC Located south of Lac Saint-Jean, the municipality was established in 1930, taking its name from the post office, opened in 1902. The lake was named in 1828, after survey party leader Joseph Bouchette (1774–1841), who was surveyor general of Lower Canada, 1801–40.

Lac-Brome QC A town in the regional county of Brome-Missisquoi, it was created in 1971 through the merger of the villages of Knowlton and Foster, and the township municipality of *Brome*. The township municipality had been created in 1845, abolished in 1847, and re-established in 1855, taking its name from the township, proclaimed in 1797.

Lac-Carré QC A village south of Mont Tremblant, it was named in 1947 after a small square-shaped (*carré*) lake. It had been established as the municipality of *Saint-Faustin-Station* in 1922. The latter had been the postal designation from 1900 to 1946, when it was changed to Lac-Carré.

Lac-des-Écorces QC Located on the east side of Mont-Laurier, this village was created in 1955, with its post office having been opened in 1903. The name of the nearby lake relates to bark (*écorce*) stripped from birch trees.

Lac-Drolet QC A municipality north of Lac-Mégantic, it was named in 1968 after the nearby lake, called after pioneer settlers. In 1885 it had been created as the township municipality of *Saint-Samuel-de-Gayhurst*, which was also the postal name from 1879 to 1969.

Lac du Bonnet MB A village (1947) on the Winnipeg River, northeast of Winnipeg, its post office was named in 1900, after the **Lac du Bonnet**, a widening of the river. The lake's name recalls the practice of Aboriginals laying stones in a circle and crowning them with a wreath of plants and branches.

Lac-Etchemin QC In the regional county of Les Etchemins, southeast of the city of Québec, and at the head of Rivière Etchemin, its post office was called *Lake Etchemin* in 1866, and was changed to Lac-Etchemin in 1920. Lac-Etchemin was established as a village in 1960 and became a town six years later. In Abenaki, the name means 'where there is leather for snowshoes'.

Lachenaie QC A town (1972) fronting on Rivière des Mille-Îles, northeast of Montréal, it was created as a municipality in 1845, abolished two years later, and re-established in 1855 as the parish municipality of *Saint-Charles-de-Lachenaie*. The name was derived from fur trader Charles Aubert de La Chesnaye (1632-1702).

Lachine QC A city in the Communauté urbaine de Montréal and at the **Rapides de Lachine**, its name may be traced to René-Robert Cavelier de La Salle, who had planned an expedition to China in 1669. After returning that year far short of his objective, his land, which had been granted to him in 1667, derisively became known as La Chine. The religious parish of Saints-Anges-de-la-Chine was created in 1678. Lachine post office was opened in 1829, the place became a village in 1848, and a city in 1909.

Lachute QC A town (1885) in the regional county of Argenteuil, west of Montréal, it is located at falls (*chute*) on the Rivière du Nord. Known as *La Chute Settlement* in the late 1700s, its post office was opened in 1835, although between 1880 and 1957 it was called *Lachute Mills*.

Lac-Kénogami QC Located southwest of Jonquière, it became a township municipality in 1897, taking its name from the township of Kénogami, proclaimed in 1865, with its name meaning 'long lake' in Montagnais. Its post office was opened in 1899 and it became a municipality in 1986.

Lac La Biche AB A town (1950) on the south shore of **Lac la Biche**, northeast of Edmonton, it was established in 1798 as a North West Company trading post by geographer David Thompson. The lake's name, which means 'red doe lake' in English, was recorded in 1790 by Hudson's Bay Company surveyor Philip Turnor.

Lac la Hache BC On the east shore of **Lac la Hache**, northwest of 100 Mile House, this place's post office was named *Lake La Hache* in 1864 and renamed *Lac La Hache* 10 years later. The present form was confirmed in 1955. The lake was named after an incident in the nineteenth century, when a Hudson's Bay Company fur brigade lost a load of axes (*haches*) or hatchets, after a horse carrying them broke through the ice and drowned.

Lac-Mégantic QC In the regional county of Le Granit, east of Sherbrooke, it became the town of *Mégantic* in 1907, through the merger of the villages of *Mégantic* (1885) and *Agnès* (1895), and the present name was adopted in 1958. Located at the north end of **Lac Mégantic**, the source of Rivière Chaudière, its name has its roots in the Abenaki *namagôtegw*, 'at the camp of the salmon trout'.

Lac-Nominingue QC A municipality east of Mont-Laurier, it was created in 1971 through the merger of the municipalities of *Loranger* (1896), *Loranger-Partie-Sud-Est* (1920), and *Nominingue* (1904). The post office, called *Nominingue* in 1887, and changed to Lac-Nominingue in 1986, has its roots in the Algonquin *Onamani Sakaigan*, 'vermilion lake'.

Lacolle QC A village south of Saint-Jean-sur-Richelieu and near the New York State border, it was named in 1920. It was severed from the municipality of *Saint-Bernard-de-Lacolle*, which had been created in 1855, having been called Lacolle ten years earlier, but abolished in 1847. The name may be traced to the seigneury of La Colle, granted in 1743 to Daniel-Hyacinthe-Marie Liénard de Beaujeu (1711–55). The name may have been derived from a hill (*colline*) 7 km to the southwest, where Rivière Lacolle was first identified as *Rivière à la Colle*.

Lacombe AB A town (1902) north of Red Deer, its post office was named in 1891 after Fr Albert Lacombe (1827–1916), an Oblate missionary to the Cree, Blackfoot, and Métis from 1849. The post office was earlier called *Barnett*. *See also* St. Albert.

La Côte-de-Beaupré QC A regional county northeast of the city of Québec, it was named in 1982. The name has its roots in the early seventeenth century phrase *le beau pré*, 'the pretty meadow', which was blended into Beaupré by 1657.

La Côte-de-Gaspé QC A regional county embracing an area of nearly 4000 km^2 centred on the town of Gaspé, it takes it name from the town, the peninsula, and the bay. It is usually called simply La Côte.

Lac-Saint-Charles QC Located in the Communauté urbaine de Québec, 16 km northwest of the city of Québec, this municipality was named in 1947 after the lake. It and the **Rivière Saint-Charles** were named in the early 1700s by Récollet missionaries, after Charles Desboues (or de Boves), the vicar of

Pontoise, 30 km northwest of Paris. The post office was called *Lake St. Charles* from 1905 to 1936, when it was changed to its current form.

Lac-Saint-Jean-Est QC Located on the east side of Lac Saint-Jean, this regional county (1982) is centred on the city of Alma, and embraces 2709 km^2.

Lac-Supérieur QC A municipality east of Mont Tremblant, it was first named *Wolfe* in 1881, after the township, which was named in honour of Gen. James Wolfe, the British commander at the Battle of the Plains of Abraham in September 1759. In 1944 it was renamed *Saint-Faustin*, after the parish, founded in 1878, and was changed to its present name in 1957, which had been its postal name since 1913.

Ladner BC In the district municipality of Delta, south of Vancouver, and on the south bank of the Fraser River, this urban centre took its name from brothers William H. and Thomas E. Ladner. After arriving in 1858 from Cornwall, England, to go prospecting in the Cariboo, they settled here 10 years later. The post office was called *Ladner's Landing* in 1876, which was shortened to Ladner in 1895.

La Doré QC Located northwest of Lac Saint-Jean, this parish municipality was called *Saint-Félicien-Partie-Nord-Ouest* in 1906, renamed *Notre-Dame-de-la-Doré* in 1915, and given its present name in 1983 after the Rivière au Doré. *Doré* is the French for 'yellow pike'.

Lady Slipper PE A municipal community (1983) northwest of Summerside, it was named after the Lady Slipper Drive, which extends along the southern and western coasts of the province from Summerside to North Cape. The drive was called after the provincial flower. Among localities in the municipal community are Victoria West, Enmore, Springhill, Inverness, and Portage.

Ladysmith BC A town (1904) on the east coast of Vancouver Island and on the forty-ninth parallel, it was originally called *Oyster Harbour*. When coal-mine owner James Dunsmuir received word on 1 March 1900 that Ladysmith in South Africa had been relieved the previous day, he decided to call the place Ladysmith. The post office was renamed the following January. The original Ladysmith had been named in 1850 after Lady Juana (1798–1872), the wife of Sir Harry Smith, governor of Cape Colony and high commissioner to South Africa, 1847–52. He had rescued her in 1812, when she was 13, after the British had overrun Badajos in Spain, and married her the same year.

Ladysmith QC A community north of Shawville in the regional county of Pontiac, it was named 28 February 1900 in honour of the lifting of the siege of Ladysmith in Natal, South Africa. That town had been named after the wife of Sir Harry Smith, governor of the Cape Colony, 1847–52.

Lafleche SK A town (1953) west of Assiniboia, its post office was named in 1909 after Mgr Louis-François Laflèche (1818–98), a missionary in the West, 1844–54, and bishop of Trois-Rivières, QC, from 1869.

Lafontaine ON In Simcoe County, west of Midland, this French-speaking community was named in 1856 after Sir Louis-Hippolyte Lafontaine (1807–64), who formed union governments with Robert Baldwin in 1842–3 and 1848–51.

Lafontaine QC This village (1958), on the north side of the city of Saint-Jérôme, was named after Sir Louis-Hippolyte Lafontaine (1807–64), a prominent Reformer after the union of Canada in 1841, who formed union governments with Robert Baldwin, 1842–3, and 1848–51.

La Guadaloupe QC This village in the regional county of Beauce-Sartigan, southwest of the city of Saint-Georges, was known as the village of *Saint-Évariste-Station* from 1929 to 1949, when it was changed to La Guadaloupe. It was named after the religious parish of Notre-Dame-de-la-Guadaloupe, created in 1945, and named in commemoration of the four hundredth anniversary of three apparitions of the Virgin Mary to Mexican Juan Diego. The place in Mexico had been named in honour of Our Lady of Guadaloupe, whose image had been found by a shepherd in Spain's Extramadura.

La Haute-Côte-Nord QC Located between Tadoussac and Betsiamites, this regional county, embracing over 12,500 km^2, was named in 1982 after its location on the upper north shore of the St. Lawrence River.

La Haute-Yamaska QC In the upper watershed of the Rivière Yamaska, and centred on the city of Granby, this regional county was named in 1982.

LaHave River NS Rising within 20 km of the Bay of Fundy, this river (75 km) flows southeast to the Atlantic Ocean. It was named after **Cape LaHave**, at the outer side of **Cape LaHave Island** and 12 km south of the river's mouth. The cape was named *Cap de la Hève* in 1604 by Pierre du Gua de Monts, after Cap de la Hève, at the entrance to the port of Le Havre, France, where his voyage had originated.

Laird SK North of Saskatoon, the post office at this village (1911) was named *Tiefengrund* in 1900. It was renamed nine years later after David Laird (1833–1914), first lieutenant-governor of the North-West Territories, 1876–81, and Indian commissioner for the West, 1898–1909.

La Jacques-Cartier QC Extending north of the Communauté urbaine de Québec, this regional county was named in 1981 after its central feature, the **Rivière Jacques-Cartier**, named in homage to the discoverer of Canada.

Lajemmerais QC A regional county fronting on the St. Lawrence from Boucherville to Contrecœur, it was named in 1982 after Marie-Marguerite Dufrost de Lajemmerais (1701–71), also known as Marguerite d'Youville, the founder of the Grey Nuns in 1753.

Lake ON A township in Hastings County, it was named in 1822 for Viscount Gerard Lake (1744–1808), who conquered Delhi in India in 1803. **Marmora and Lake** townships comprise a single township municipality.

Lake Cowichan BC A village (1944) west of Duncan, it is located at the point where the **Cowichan River** flows east from **Cowichan Lake**. The Cowichan, a Salish tribe that once occupied a large area on both sides of the Strait of Georgia, took their name from an Island Halkomelem word meaning 'land warmed by the sun', referring to a rock formation on the side of Mount Tzuhalem, which resembled a frog basking in the sun. The Cowichan called the lake *Kaatza,* 'big lake'. *See also* North Cowichan.

Lake Echo NS This community is on the north side of Lake Echo, east of Dartmouth. The lake had been named for its acoustic qualities.

Lakefield ON North of the city of Peterborough, this village (1874) was variously known as *Nelson's Falls*, *Thompson's Rapids*, *Herriot's Falls*, and *Selby*. By 1853 it had been laid out and named Lakefield, and the *North Douro* post office, named in 1856, became Lakefield in 1875.

Lake Lenore SK Located between Humboldt and Melfort, the post office at this village (1921) was named *Lenora Lake* in 1906 after the lake, 5 km to the north. The lake had been named after a surveyor's daughter, Lenora, but it became **Lake Lenore** after the village was incorporated. The postal name was changed to Lake Lenore in 1939.

Lake Louise AB A resort community northwest of Banff, it was named in 1916 after the lake, which had been named in 1884 after Princess Louise Caroline Alberta (1848–1939), fourth daughter of Queen Victoria, and wife of the Marquess of Lorne, governor general of Canada, 1878–83. First known in 1883 as *Holt City*, the station was renamed *Laggan* that year, by Lord Strathcona, after Laggan, Inverness-shire, Scotland.

Lakelse Lake BC Located south of Terrace and on the eastern shore of **Lakelse Lake**, this place's post office was called *Lakelse* from 1913 to 1935. The lake's name in Tsimshian means 'fresh water mussel'.

Lake of Bays ON A municipal township created in 1971 in the district municipality of Muskoka, it was named after its main feature, itself named in 1853 by geologist Alexander Murray after its numerous coves and bays. The lake had been earlier identified as *Lake Baptiste*, *Trading Lake*, and *Forked Lake*.

Lakeside NS West of Halifax, this place is located on the shore of Governors Lake. Its post office was opened in 1952.

Lakeville NS Northwest of Kentville, this place was named in 1860 after a nearby lake. Its post office was opened the following year.

La Loche SK On the east shore of **Lac La Loche** and east of Fort McMurray, AB, this settlement was established by the Chipewyan in the early 1800s. The French word *loche* identifies a fish called 'mariah', which resembles a tadpole. When Peter Pond travelled in 1778 over the 20-km Methy Portage, separating the waters of the Mackenzie and Churchill basins, the lake was called *Methy Lake*. The North West Company had a post here in 1808, and the Hudson's Bay Company opened the Portage La Loche post here in 1853. La Loche post office was opened in 1926.

La Macaza QC Erected in 1930 as a municipality northwest of Mont Tremblant, it took its name from **Lac Macaza**, which may have been named after an Algonquin chief, or after an Aboriginal who camped on its shore.

La Malbaie QC In 1608 Samuel de Champlain named the **Rivière Malbaie**, because the bay at its mouth had insufficient depth to anchor vessels. In 1762 John Nairne and Malcolm Fraser, retired officers from Gen. James Wolfe's army, received grants at the mouth of the river, and named it, the bay, and the settlement *Murray Bay*, after Gen. James Murray (*c.* 1721–84), Wolfe's successor. *Murray Bay* was the post office from 1832 to 1914, when it was changed to La Malbaie, although *Murray Bay* continued to have wide international usage well into the latter half of the twentieth century. Incorporated as a village in 1896, La Malbaie became a town in 1958.

Lamaline NF On the south side of the Burin Peninsula, this town (1963) may have been named after the dangerous shoals offshore, described by the French as *La Maligne*, 'the evil one'. A map of 1750 identified *cap de la Meline*, with islands offshore called *Isles de la Meline*.

La Matapédia QC Embracing the central valley of **Rivière Matapédia**, this regional county was named in 1982. It embraces 20 municipalities, including the towns of Amqui and Causapscal.

La Mauricie National Park QC Established in 1970 on the west side of the Rivière Saint-Maurice and in the regional county of Le Centre-de-la-Mauricie, this national park (544 km^2) took its name from the regional designation, Mauricie, proposed in 1931 by Mgr Albert Tessier.

Lambart, Mount YT Situated in the St. Elias Mountains, this mountain (3269 m) was named in 1918 after geodesist Howard Frederick John Lambart (1880–1946). From 1906 to 1917 he worked on the Yukon-Alaska boundary, and determined the position and height of this mountain. He led a United States-Canada expedition to climb it in 1925.

Lambeth ON A suburban community of London, southwest of the city centre, it was named *Westminster*, after the township, in 1840. In 1857 it was called after the district of Lambeth in Greater London, England, on the south side of the River Thames.

Lambton ON This county was named in 1849 by Malcolm Cameron, assistant commissioner of public works in the Province of Canada, for Sir John George Lambton, Earl of Durham (1792–1840), governor-in-chief of British North America, 1838.

Lambton QC A municipality on the east side of Lac Saint-François and north of Lac-Mégantic, it was created a township municipality in 1845, and a municipality in 1913, taking its name from the township. The township was proclaimed in 1848 and named after Lord Durham, governor-in-chief of British North America, 1838. The author of the famous Durham Report, he recommended the union of the two Canadas under a single government.

Lamèque NB A town (1982) on **Île Lamèque**, east of Caraquet, its post office was opened in 1867. In 1672 Acadia Gov. Nicolas Denys referred to the island as *Grande Ile de Miscou*. It was identified in 1777 as *Shippegan Island* by chart-maker J.F.W. DesBarres. After a local petition in 1974, the present name of the island was officially acknowledged.

L'Amiante QC A regional county centred on the city of Thetford Mines, it was named in 1982 after the principal industrial activity of the area, the extraction of asbestos (*amiante*).

La Mitis QC Situated on the south shore of the St. Lawrence and centred on the **Rivière Mitis**, east of Rimouski, this regional county was established in 1982. The name of the river has its roots in the Mi'kmaq *miti sipo*, 'poplar river'.

Lamont AB A town (1968) northeast of Edmonton, its post office was named in 1906 after John Henderson Lamont (1865–1936), who moved from Toronto to Prince Albert in 1899 to practise law. He served in both the House of Commons and the Saskatchewan Legislature before being appointed to the Supreme Court of Saskatchewan in 1907, and to the Supreme Court of Canada in 1927.

Lampman SK A town (1963) northeast of Estevan, it was established by the Canadian Northern Railway in 1909 as a station on its Poet's Line. It was named after poet Archibald Lampman (1861–99), whose lyrical poetry was rooted in nature. Postal service was set up here in 1910.

Lanark ON The village (1862) of Lanark was founded by Scottish settlers in 1820, who named it *New Lanark*, for the county and town east of Glasgow. That year **Lanark Township**, which surrounds the village, was also named. Lanark post office was established in 1824 at the village. The following year, **Lanark County** was separated from Carleton County.

Lancaster ON Ontario's most easterly township, this village (1887) was named in 1787 after the Duke of Lancaster, one of the titles of George III. Lancaster post office was first located in 1789 at the mouth of the Raisin River, now the site of **South Lancaster**. When the Grand Trunk Railway was built from Montreal to Brockville in 1855, its station 2 km to the north was called *New Lancaster*. It soon became simply Lancaster.

Lancaster Sound NT Between Baffin and Bylot islands on the south, and Devon Island on the north, this 350-km-long entrance to Parry Channel was named *Sir James Lancaster Sound* by explorers Robert Bylot and William Baffin in 1618. Sir James died that year. The name was subsequently shortened to Lancaster Sound.

L'Ancienne-Lorette QC Located in the Communauté urbaine de Québec, west of the city of Québec, this city (1967) had been incorporated as the village of *Notre-Dame-de-Lorette* in 1948. Its post office was called *Ancienne-Lorette* in 1854, in contrast to *Nouvelle-Lorette*, the present town of Loretteville, to the north. In 1673 the Huron settlement here was placed under the protection of Notre-Dame de l'Annonciation, whose order was modelled on the Holy House of the Virgin, established in 1295 in Loreto on Italy's east coast.

Landis SK Located northwest of Biggar, this village (1909) was founded by the Grand Trunk Pacific Railway in 1908, and named after the American judge Kenesaw Mountain Landis (1866–1944). He was well known as the commissioner of baseball, 1920–44. The post office had been opened as *Daneville* in 1907 and was changed to Landis in 1909.

Lang SK Situated northwest of Weyburn, this village (1906) was established by the Canadian Pacific Railway's extension of the Soo Line in 1893 and named after its resident engineer living in Moose Jaw. Its post office was opened in 1904.

L'Ange-Gardien QC Located northeast of the city of Québec, this parish municipality was established in 1845, abolished two years later, and re-established in 1855, and *Ange-Gardien* post office was opened in 1861. Its area was the site of the religious parish of Saints-Anges-Gardiens, erected here in 1664. The village of **L'Ange-Gardien**, southwest of Granby, was long known as *Ange-Gardien-de-Rouville*, *Saint-Ange-Gardien*, and *Canrobert*. The municipality of **L'Ange-Gardien** was created in 1980, having been part of the town of Buckingham since 1975. Its roots may be traced to the religious parish, created in 1861, and the parish municipality, incorporated in 1881.

Langenburg SK This town (1959), southwest of Yorkton and near the Manitoba border, was established by the Manitoba and North Western Railway in 1886. It was named by German settlers after Prince Hohenlohe Langenburg, a former chancellor of Germany. The post office was first called *Hohenlohe*, but it was renamed Langenburg in 1888.

Langford BC A district municipality (1992) west of Victoria, its post office was named *Langford Station* in 1911 after **Langford Lake**. The lake received its name from Capt. Edward E. Langford, who came to Victoria in 1851 to manage the Esquimalt farm of the Puget Sound Agricultural Company, which was owned by the Hudson's Bay Company. He returned to England 10 years later.

Langham SK Located near the east bank of the North Saskatchewan, northwest of Saskatoon, this town (1907) was founded in 1905 by the Canadian Northern Railway and named after E. Langham, then general passenger agent for the line.

Langley BC The district municipality (1936) and the city (1955) of Langley took their names from Fort Langley. The fort was established in 1827 on the Fraser River, opposite the site of Haney, by the Hudson's Bay Company, and named after HBC director Thomas Langley. Twelve years later, the fort was relocated 4 km upriver to better agricultural land. Abandoned in 1896, the fort was restored in the 1950s as a national historic site.

Langruth MB Located near Lake Manitoba, northwest of Portage la Prairie, this place's post office was named in 1912 after two pioneer landowners, Langdon and Ruth.

Langton ON In Haldimand-Norfolk Region, southwest of Simcoe, this place was named in 1862, possibly after a place near Great Walsingham, in England's county of Norfolk. Its post office was named *North Walsingham*, after the township of that name, in 1854.

Lanigan SK Southeast of Saskatoon, the site of this town (1908) was established by the Canadian Pacific Railway in 1905 and named after the CPR's freight traffic agent, W.B. Lanigan. Its post office was also opened that year.

L'Annonciation QC A village northwest of Mont Tremblant, it was first known as *Ferme-du-Milieu*, because it furnished provisions for the timber shanties in the surrounding woods. L'Annonciation became a village in 1908, its name having been chosen by Fr Antonio Labelle in 1880 for the religious parish in honour of the annunciation by the Angel Gabriel that the Virgin Mary would be the mother of Christ.

Lanoraie-d'Autray QC A municipality on the St. Lawrence, southeast of Joliette, it was established in 1948. The double name honours two seigneurs, Louis de Niort de La Noraye (1639–1708) and Jacques Bourdon d'Autray (1652–88).

La Nouvelle-Beauce QC This regional county (1982) is centred on the town of Sainte-Marie, south of the city of Québec. The county of Beauce had been established in 1829 in the valley of the Rivière Chaudière, taking its name from the valley's regional designation, La Beauce, in use as early as 1739.

Lansdowne ON A community in **Front of Leeds and Lansdowne Township**, united counties of Leeds and Grenville, northeast of Gananoque, its post office was named in 1852 after Lansdowne Township, itself given for William Petty-Fitzmaurice, Earl of Shelburne (1780–1863), who was made the Marquis of Lansdowne in 1784. When the Grand Trunk Railway was built in 1856 from Montréal to Kingston, the place was relocated 2 km to the north.

L'Anse-Amour NF On the north side of the Strait of Belle Isle, this place was named after the adjacent Pointe aux Morts, 'dead man's point', because of the many shipwrecks here. The first reference to 'Amour', meaning 'love', as part of the name, was recorded in 1820. A burial mound dating from 5500–5000 BC is located here.

L'Anse-au-Clair NF On the north side of the Strait of Belle Isle and just east of the Québec-Newfoundland boundary, it is believed the cove was first known as *L'Anse à l'Eau Claire*, 'clear water cove'. It was used by the French as a fishing and sealing station in the 1700s.

L'Anse-au-Loup NF On the north side of the Strait of Belle Isle, this place's name means 'wolf cove'. Augustin Le Gardeur de Courtemanche built a fort here in the early 1700s, and a French migratory fishery was established here later that century. Permanent settlement took place after the 1830s.

L'Anse aux Meadows NF At the northern end of the Northern Peninsula and on the south side of the Strait of Belle Isle, this place is the site of a Norse settlement, established about 1000 AD. The name is derived not from 'grassy meadows', but from *Anse aux Meduses*, 'jellyfish cove'.

L'Anse-Saint-Jean QC A municipality on the south side of the Saguenay and east of Chicoutimi, it was named in 1981, having been created the township municipality of *Saint-Jean* in 1859, with the religious parish having been called Saint-Jean-Baptiste-de-l'Anse-Saint-Jean in 1839.

Lantz NS On the Shubenacadie River, northeast of Halifax, this place was named *Lantz Siding* about 1898, when Harvey and Croft Lantz built a brick factory here. Its post office was opened as *Lantz Siding* in 1913 and was shortened to Lantz in 1967.

Lantzville BC Northwest of Nanaimo, this community was named in 1923 after Harry Lantz, an American investor, who financially supported the mining of coal at nearby Nanoose.

La Patrie QC A village east of Sherbrooke, it was named in 1941 after the newspaper *La Patrie*, founded about 1870 by Jérôme-Adolphe Chicoyne, and after the religious parish, named Saint-Pierre-de-Ditton ou La Patrie in 1878, given in honour of Québécois repatriating from the United States. Miller Pierre-V. Vaillant was one of the first to return from New England, with the first post office here being called *Vaillantbourg* in 1875.

La Pêche QC This municipality, north of Hull, was created in 1975 through the merger of the township municipalities of Wakefield and Aldfield, the village of Wakefield, and the municipality of Sainte-Cécile-de-Masham. The name was taken from the **Rivière la Pêche**, which drains **Lac la Pêche**, noted for its good quality of fish (*pêche*).

La Plaine QC Located east of Saint-Jérôme, this parish municipality was first named *Sainte-Anne-des-Plaines* in 1922, and shortened to La Plaine in 1969, the name given to its post office in 1879. The flat land here was described as 'la plaine' in the 1700s by the first seigneur of Terrebonne, Louis Lepage de Sainte-Claire, with an addition to his seigneury being called Belle Plaine in 1731.

La Pocatière QC Established as a town in 1961 in the regional county of Kamouraska, it had been erected as the village of *Sainte-*

Anne-de-la-Pocatière the previous year. The name was derived from the first seigneur, François Pollet de La Combe-Pocatière, who was granted it in 1672. His widow called her property Sainte-Anne-de-la-Pocatière, which became the name of the religious and civil parishes.

La Prairie QC On the southeast side of the St. Lawrence, opposite the city of Montréal, the area of this town was known as a 'prairie' in the mid-1600s, and the parish of Notre-Dame de Laprairie de la Magdelaine was established in 1692. The village of *Laprairie de la Magdelaine* was incorporated in 1846, became the town of *Laprairie* in 1909, and adopted the present spelling in 1972.

La Présentation QC Located northwest of Saint-Hyacinthe, this parish municipality was named in 1845, abolished two years later, and re-established in 1855, taking its name from the religious parish of La Présentation-de-la-Vierge, named in 1806 in honour of the presentation of the Virgin Mary in the temple. The post office of La Présentation was opened in 1857.

Lardeau BC On the west shore of Kootenay Lake and just south of the mouth of **Lardeau River**, this place's post office was named *Lardo* in 1899 after prospector Johnny Lardeau. It was renamed Lardeau in 1947.

Larder Lake ON A municipal township in Timiskaming District, east of Kirkland Lake, it was laid out in 1907, after a nearby gold strike, and called *Larder City* after the nearby lake. Larder Lake was a town from 1938 to 1946, when it became a municipal township. The lake was known to the Ojibwa as *Asandjikaning*, 'hanging storehouse', a container where fish and game are hung to dry.

L'Ardoise NS On the south coast of Cape Breton Island, this place was named during the French regime after the slate (*ardoise*) cliffs. Postal service was established here in 1838, followed by **Lower L'Ardoise** (1865), **L'Ardoise Highlands** (1904), and **L'Ardoise West** (1912).

La Rivière-du-Nord QC Created in 1983, this regional county embracing the area of Saint-Jérôme was named after the river, which flows southwest to the Ottawa River, south of Lachute.

La Rivière MB Situated on the Pembina River, west of Morden, this community's post office was named in 1887 after Alphonse Alfred Clément La Rivière, who served in both the provincial legislature and the House of Commons, before being appointed to the Senate in 1911. In 1964 the Winnipeg radio station CKY offered to advertise the area's ski resort for free, if it would change its name to Seekayewye. It was vigorously opposed by the Manitoba government, especially when it was learned that, by sheer coincidence, the station's call letters also meant 'he who urinates' in Cree.

Lark Harbour NF On the southwest side of the Bay of Islands and northwest of Corner Brook, this municipal community (1974) was named after the harbour. The harbour was named in 1767 by Capt. James Cook, after HMS *Lark*, which he used during his hydrographic surveys of the coasts of Newfoundland.

La Ronge SK Located on the west shore of **Lac la Ronge**, 205 km north of Prince Albert, this town (1976) was named after the lake. It may have named either because the jagged shoreline had the appearance of having been gnawed by *Kitchi Amik*, the Great Beaver, or because early voyageurs discovered trees along its shoreline gnawed by beavers. Its post office was called *Lac la Ronge* in 1911, and was changed to La Ronge in 1949.

Larrys River NS On Tor Bay and south of Guysborough, this place's post office was named in 1869. The small river had been named after Larry Keating, of Halifax, who had earlier hunted moose here.

Larsen Sound NT Surrounded by Boothia Peninsula, and Victoria, King William, and Prince of Wales islands, this body of water was named in 1965 after Henry Asbjorn Larsen (1899–1964). A member of the Royal Canadian Mounted Police, he commanded the patrol vessel *St. Roch* from 1928 to 1948. He made the first east-to-west traverse in 1940–2 through the Northwest Passage, and in a single season (1944) made the return west-to-east voyage.

La Salle MB On the **La Salle River**, south of Winnipeg, this place may have been named about 1882 after René-Robert Cavelier de La Salle (1643–87), the noted explorer of the Mississippi. Its post office was named in 1891. The place's name was inspired by its original designation of the river, *Rivière Salé*, 'foul river', because of its smell and alkaline taste. The river was officially renamed after the place in 1975.

LaSalle ON A town in Essex County, on the west side of Windsor, it was originally named in 1924 to honour explorer René-Robert Cavelier, Sieur de La Salle (1643–87). It was annexed in 1959 by the municipal township of *Sandwich West*, which became the town of LaSalle in 1992.

LaSalle QC A city in the Communauté urbaine de Montréal, southwest of the city of Montréal, it was named in 1912 after René-Robert Cavelier de La Salle (1643–87), the noted explorer of the Mississippi, 1681–2, and 1684–7. It was previously called the parish municipality of *Saint-Michel-de-Lachine* (1845) and *Saints-Anges-de-Lachine* (1855).

La Sarre QC In the regional county of Abitibi-Ouest, it became a village in 1937, and a town in 1949, taking its name from the township of La Sarre, proclaimed in 1916. The township was named after the La Sarre Regiment, which came to New France in 1756. The township municipality of La Sarre was established in 1917, and united with the town in 1980.

L'Ascension-de-Notre-Seigneur QC A parish municipality in the regional county of Saint-Jean-Est, north of Alma, it was named in 1919 after the religious parish, which was established on 8 June 1916, the day of the ascension of Christ. It is usually shortened to L'Ascension.

La Scie NF On the north coast of the island of Newfoundland and east of Baie-Verte, this town (1955) was named in the 1700s by the French, for the serrated landscape surrounding the harbour, which reminded them of a saw.

Lashburn SK Southeast of Lloydminster, the first post office in this town (1979) was called *Wirral* in 1905, after Wirral Peninsula, in Cheshire, England, opposite Liverpool. The Canadian Northern Railway arrived that year and the following year renamed it after Z.A. Lash, the company's solicitor, and after the Scottish word *burn*, meaning a 'small stream'.

Lasqueti Island BC In the Strait of Georgia and south of Texada Island, it was named in 1791 by Spanish sailing master Jose Maria Narvaez after Juan Maria Lasqueti, a prominent naval officer.

L'Assomption QC The names of the town and the regional county of **L'Assomption** may be traced to the **Rivière de l'Assomption** and the seigneury of L'Assomption, established in 1647, and granted to Pierre Legardeur de Repentigny. The present area of the town was created in 1992 through the merger of the parish municipality of L'Assomption (1855) and the town (1888). The regional county was incorporated in 1983.

La Tabatière QC A fishing community on the lower north shore of the Gulf of St. Lawrence, it was founded in 1820 and was first known as *Spark Point*. In 1885 the mission of Saint-Joseph-de-la-Tabatière was established here, with the La Tabatière post office opening in 1907. The word *tabatière* has its roots in a Montagnais word meaning 'sorcerer', with Montagnais who visited the mission calling upon their *tabatière* before returning to their camps.

Latchford ON A town (1907) with the lowest population in Ontario (1996 population, 328), it is located in Timiskaming District, southwest of New Liskeard. It was named in 1903 after Francis R. Latchford, then Ontario minister of public works, who turned the first sod of the Temiskaming and Northern Ontario Railway in North Bay that year, and drove the last spike at Moosonee in 1932.

Laterrière QC A town south of Chicoutimi, it was erected in 1989, having merged the municipal parish of Laterrière (1883) and the village of Laterrière (1921) in 1983. The original township of Laterrière was proclaimed in 1850, after Marc-Pascal de Sales Laterrière (1792–1872), the member for Saguenay in the Legislative Assembly of the province of Canada, 1845–51.

LaTour, Mount NB North of the Nepisiguit River, this mountain was named in 1899 by naturalist William F. Ganong after Charles de Saint-Étienne de la Tour (1593–1666), governor of Acadia, 1651–66.

La Tuque QC A town on the Rivière Saint-Maurice, midway between the St. Lawrence and Lac Saint-Jean, it was created in 1911 through the merger of the villages of La Tuque (1909) and *La Tuque Falls* (1910). As early as 1806 a portage at La Tuque was mentioned by Jean-Baptiste Perrault. Explorer François Verreault explained in 1823–4 that the name had its roots in a nearby high mountain, shaped like the knitted hat called a tuque.

Laurentides QC Located in the regional county of Montcalm, northeast of Saint-Jérôme, this town, formerly called the village (1856) of Saint-Lin, and its post office were named in 1883. It took its name from the vast mountain complex north of Montréal, which was named in 1845 by historian François-Xavier Garneau. In English the popular all-season tourist destination is widely known as the Laurentians. North of the city of Québec is the 8706-km^2 Laurentides fish and game reserve, established in 1981, replacing a park created in 1895. The regional county of **Les Laurentides** was created in 1983.

Laurier MB Located east of Riding Mountain National Park, southwest of Dauphin, this place's post office was first called *Fosbery* in 1896. It was renamed the next year by settlers from Québec after Sir Wilfrid Laurier (1841–1919), prime minister of Canada, 1896–1911. He donated instruments to the local band.

Laurier, Mount YT Located east of Lake Laberge, this mountain (1779 m) was first named in 1888 by surveyor William Ogilvie after geologist George M. Dawson. However, Dawson asked Ogilvie to rename it after Wilfrid (later Sir Wilfrid) Laurier, who was then the leader of the Liberal Party of Canada.

Laurier-Station QC This village southwest of the city of Québec was incorporated in 1951. Its post office was named in 1900 in honour of Sir Wilfrid Laurier, prime minister of Canada, 1896–1911.

Laurierville QC A village (1902) 20 km south of Laurier-Station, and 35 km northeast of Victoriaville, it was named after Prime Minister Sir Wilfrid Laurier. Its post office was opened in 1903.

Laussedat, Mount BC In the Rocky Mountains, north of Golden, this mountain (3059 m) was named in 1911 by surveyor Arthur O. Wheeler after Col Aimé Laussedat (1819–1907), a French officer who developed photographic surveying techniques during the latter half of the 1800s.

Lauzon QC A sector of the city of Lévis, the name recalls the first grantee of the seigneury of Lauzon (Lauson, Loson), Jean de Lauson (1584–1666), who acquired it in 1636. Erected as a village in 1867, it became a town in 1910, and a city in 1957. It was united with Lévis to become the city of *Lévis-Lauzon* in 1989, but two years later Lauzon was dropped from the city's designation.

Laval QC Île Jésus was granted in 1675 to Mgr François Laval (1623–1708), apostolic vicar of New France, 1658–74, and the first bishop of Québec, 1674–88. The island became the county of Laval in 1853, and the regional county of Laval in 1979. In 1965 all the municipalities in the county were united as the city of Laval, making it the province's second most populous city, after Montréal.

La Vallée-de-la-Gatineau QC A regional county in the valley of the Rivière Gatineau, it extends north from Farrelton to Réservoir

Cabonga. It replaced the former county of *Gatineau* in 1983.

La Vallée-du-Richelieu QC Centred on the valley of the Richelieu, and extending north from Chambly, this regional county was created in 1982.

Lavaltrie QC Located on the north side of the St. Lawrence, downriver from Montréal, and opposite Contrecœur, this village was erected in 1926. The seigneury of Lavaltrie was granted in 1672 to Séraphin Margane, Sieur de Lavaltrie (1644–99), a lieutenant in the Carignan-Salières Regiment.

L'Avenir QC A municipality southeast of Drummondville, it was created in 1862. It was named after *L'Avenirville* post office, itself named in 1853 after the newspaper *L'Avenir*, founded in Montréal in 1847 by Jean-Baptiste-Éric Dorion (1826–66). Dorion was also a member for Drummond in the Legislative Assembly of the province of Canada, 1854–7, and 1861–6.

Lawn NF A town (1952) at the south end of the Burin Peninsula, it was named after **Lawn Bay**, which was identified by the French in the 1600s as *L'Anse Sauvage*, 'wild cove'. The bay was called *Great Lawn* by William Tavener in 1714 and Capt. James Cook adapted it in the 1760s for the present bay's name. Suggestions that a French fisherman named it *l'ane*, 'the donkey', after a doe caribou seen here, or that Cook was inspired by the lushness of grass to call it Lawn are probably just speculative.

Lawrence Station NB Located northeast of St. Stephen, this place was named in 1853, when the St. Andrews and Quebec Railway was being built to the northwest. It was called after settler Wheeler Lawrence.

Lawrencetown NS In the Annapolis Valley, between Bridgetown and Middleton, this village (1953) was named after Lt-Col Charles Lawrence (1709–60), governor of Nova Scotia, 1753–60. Its post office was opened in 1836. A second post office was named **Lawrencetown** after him, east of Dartmouth. It was opened in 1862, but was changed to *East Lawrencetown* in 1924. **Upper Lawencetown**, between Lawencetown and Dartmouth, had a post office from 1907 to 1960.

Lawrenceville QC A village (1905) in the regional county of Le Val-Saint-François, midway between Sherbrooke and Granby, its post office was named in 1859 after Isaac Lawrence, a native of Connecticut, who settled here in 1794.

Lax Kw'alaams BC On the Tsimpsean Peninsula, north of Prince Rupert, this Tsimpsian (Tsimshian) name, which means 'place of the wild rose with small hips', was adopted in 1986 in place of *Port Simpson*. Fort Simpson had been established further up the Nass River in 1831 by Hudson's Bay Company trader Capt. Æmilius Simpson. He died there later that year. The fort was moved in 1834 to the site of *Port Simpson*. It was closed in 1913, and was burned the next year. *Port Simpson* post office was named in 1900.

Lazo BC A community north of Comox, its post office was named in 1910 after **Cape Lazo**. Spanish sailing master Jose Maria Narvaez named the cape *Punta de Lazo de la Vega*, likely after a Spanish official he wished to honour. In Spanish *lazo* means 'snare'.

Leacock, Mount YT In the St. Elias Mountains, this peak (3109 m) was named in 1970 by the Stephen Leacock Centennial Committee after Stephen Leacock (1869–1944), the noted Canadian humorist. He had been a professor of economics at McGill University, in Montréal.

Leader SK Situated south of the South Saskatchewan River and near the Alberta border, the post office at this town (1917) was first called *Happyland* in 1908. It was renamed *Prussia* in 1912 by the first German settlers, but it was changed to Leader in 1917, as the connection with Germany was deemed offensive.

Leading Tickles West NF On an island off the north coast of the island of Newfoundland, and encompassing **Leading Tickles South** on the mainland, this municipal community

(1961) took its name from a rock called The Ladle, and from the very narrow tickles (channels) between the islands and the mainland, which were difficult for fishing boats to pass through. The community was first called *Ladle Tickles* (1836) and *Lading Tickles* (1845).

Leaf Rapids MB Situated at **Leaf Rapids** and on the south bank of the Churchill River, northwest of Thompson, this town (1974) was established in 1970 as a mining community, and its post office was opened in 1972.

Leamington ON A town (1890) in Essex County, it was named in 1854 by miller William Gaines after his home town, Royal Leamington Spa, a borough on the east side of the borough of Warwick, in Warwickshire, England.

Leaside ON In the borough of East York, it was named *Leaside Junction* by postmaster J.H. Lea in 1893, and it was shortened to Leaside in 1915. His father, William Lea, built a brick house, 1851–4, and called it Leaside.

Leask SK A village southwest of Prince Albert, it was founded by the Canadian Northern Railway in 1912 and incorporated that year. Its post office was also opened then, and named after homesteader Robert Leask, who had settled here in 1904.

Leaskdale ON In Uxbridge Township, Durham Region, northeast of Newmarket, it was settled in 1847 by Peter Leask of Banffshire, Scotland, who built a sawmill which his son, James, managed. Another son, George, built a grist mill in 1854 and was appointed the first postmaster three years later.

Le Bas-Richelieu QC A regional county in the area of the mouth of the Richelieu and centred on the city of Sorel, it was named in 1982.

Lebel-sur-Quévillon QC Located on the western shore of Lac **Quévillon**, and northeast of Val-d'Or, this town (1965) was founded in 1948. It owes its name to pioneer Jean-Baptiste Lebel (1887–1966), who built a sawmill on nearby Rivière Bell. The lake was named after noted architect, sculpter, and master cabinetmaker Louis-Amable Quévillon (1749–1823).

Le Bic QC A municipality located southwest of Rimouski, the present form of its name was established in 1987. The municipality of *Bic* had been created in 1972 through the merger of the village of *Bic* and the parish municipality of *Sainte-Cécile-du-Bic*. In 1603 Champlain named a nearby mountain *Pic*, which became *Bic* when the seigneury was created in 1675, and continued in the postal name, established in 1832.

Lebret SK East of Fort Qu'Appelle, this village (1912) was named in 1886 by Senator Girard after its first postmaster, Fr Lebret (d. 1903), who had established a residential school here in 1884 for Métis.

Le Centre-de-la-Mauricie QC Centred on the city of Shawinigan and the Rivière Saint-Maurice, this regional county was named in 1982.

LeClercq, Mount NB North of the Nepisiguit River, this mountain was named in 1899 by naturalist William Francis Ganong after Fr Chrestien LeClercq (1641–*c.* 1700), a Recollet missionary to the Mi'kmaq.

Le Domaine-du-Roy QC A regional county northwest of Lac Saint-Jean and centred on the town of Roberval, it was named in 1983. The name recalls a vast domain, including the whole area of Lac Saint-Jean, which was established in 1674 for the sole pleasure of the King. The special lease was only suspended in 1842, when the lands were opened for settlement.

Leduc AB South of Edmonton, this city (1983) was named about 1890 by Edgar Dewdney, when he established a Dominion Telegraph station here, and named it after Fr Hippolyte Leduc (1842–1918), an Oblate missionary who served in Western Canada after 1865. The famous Imperial Leduc No 1 oil well came into production in 1947 at Devon, in **Leduc County**, 18 km to the northwest.

Leeds ON The county was named in 1792 by Lt-Gov. John Graves Simcoe after Francis Godolphin Osborne, 5th Duke of Leeds (1751–99), secretary of state for foreign affairs, 1783–91. The municipality of the united counties of **Leeds and Grenville** was formed in 1852. *See also* Front of Leeds and Lansdowne, *and* Rear of Leeds and Lansdowne.

Lefaivre ON Located on the Ottawa River and in the united counties of Prescott and Russell, west of Hawkesbury, this place was named in 1877 after postmaster Hercule Lefaivre, whose father, Pierre, was the first settler here.

Le Fjord-de-Saguenay QC A regional county centred on the cities of Chicoutimi and Jonquière, it was established in 1983, taking its name from the submerged glacial valley between Chicoutimi and Tadoussac. The valley with its vertical escarpment walls is the only true fiord on the southern edge of the Canadian Shield.

Lefroy ON In the town of Innisfil, Simcoe County, southeast of Barrie, this community was named by Sir John Beverley Robinson in 1852 after his son-in-law, Sir John Henry Lefroy (1817–90), superintendent of the magnetic laboratory in Toronto, 1842–53.

Lefroy, Mount AB This mountain (3423 m), near Lake Louise, was named in 1858 by Dr James (later Sir James) Hector after Sir John Henry Lefroy, an astronomer noted for his studies of magnetism. Lefroy performed measurements at Fort Edmonton in 1844.

Legal AB This village (1914), north of Edmonton, was named in 1900 after Mgr Émile-Joseph Legal (1849–1920), who arrived in Western Canada in 1881 as a missionary, and subsequently became bishop of St. Albert (1902) and archbishop of Edmonton (1912).

Le Gardeur QC Located in the regional county of L'Assomption, this town was named in 1978, replacing the town of *Saint-Paul-l'Ermite*, erected five years earlier, which had been established as a parish municipality in 1857. The post office had been called Le Gardeur in 1957, to reduce the confusion among the many places called Saint-Paul in the province. In 1647 Pierre Legardeur de Repentigny was granted the vast seigneury of La Chesnaye, but he died the following year, and was succeeded by his son Jean-Baptiste.

Le Goulet NB A village (1986) southeast of Caraquet, its post office was called *Shippigan Gully* from 1904 to 1955, and *Shippegan Gully* until 1961, when it was renamed Le Goulet. The name refers to the narrow entrance to the harbour at Shippagan, between the mainland and Île Lamèque, which is officially called Shippegan Gully.

Le Granit QC A regional county in the southeastern part of the province, surrounding the town of Lac-Mégantic, it was named in 1982 after the granitic mountainous relief, rising to 1105 m at Mont Mégantic.

Le Haut-Richelieu QC Based on the upper reaches of the Richelieu, with Saint-Jean-sur-Richelieu as its main urban centre, this regional county was named in 1982.

Le Haut-Saint-François QC Embracing the upper part of the Rivière Saint-François, with several small towns, including East Angus, Cookshire, Scotstown, and Ascot Corner, this regional county was created in 1982.

Le Haut-Saint-Laurent QC Located south and west of Salaberry-de-Valleyfield, with Huntingdon, Ormstown, and Saint-Chrysostome as its main centres, this regional county was named in 1982, after its location at the upper part of the St. Lawrence that is within the territory of Québec.

Le Haut-Saint-Maurice QC Located in the upper reaches of the Rivière Saint-Maurice, and centred on the town of La Tuque, this regional county was named in 1982.

Leith ON A community in Grey County, northeast of Owen Sound, it was named in 1853 after the Scottish port and Edinburgh suburb. It was expected it would become a shipping centre, but its shallow water ended such a plan.

Leitrim ON In the city of Gloucester, Ottawa-Carleton Region, this place was named in 1883 after County Leitrim, Ireland, because several early settlers came from there.

Lemberg SK Southwest of Melville, this town (1907) was established by the Canadian Pacific Railway in 1903 and its post office was opened the following year. It was named after Lemberg, then the chief centre of Galicia, but now L'viv in Ukraine. As Lwow, it was part of Poland from 1340 to 1772, when it became Lemberg in the Austrian Empire. It was returned to Poland in 1918, and then, as Lvov, it became part of the USSR in 1939. After the breakup of the Soviet Union in 1991, it adopted its Ukrainian spelling. Meanwhile the Saskatchewan town has continued to be called Lemberg.

LeMoyne QC A town in the regional county of Champlain and adjacent to the cities of Longueuil and Saint-Lambert, it was created in 1949. LeMoyne post office existed from 1910 to 1921. The name recalls Charles Le Moyne (1626–85), the first seigneur of Longueuil in 1657. Among his 12 sons was the noted Pierre Le Moyne d'Iberville (1661–1706), the founder of Louisiana, 1700–2.

Lempriere, Mount BC This mountain (3208 m) in the Monashee Mountains was named in 1863, after Lt (later Maj.-Gen.) Arthur Reid Lempriere, an English military engineer, who helped lay out wagon roads in the province's interior between 1858 and 1863.

Lennox and Addington ON Created as separate counties in 1792, each then had only three townships near the Lake Ontario shore. They were joined temporarily in 1798, but in 1860 their union as a single county was affirmed, with several townships to the rear added to it. Lennox was named after Charles Lennox, 3rd Duke of Richmond (1734–1806), the uncle of the 4th Duke of Richmond, governor-in-chief of British North America, 1818–9. Addington was named after Henry Addington, Viscount Sidmouth (1755–1844), who was then speaker of the British House of Commons, and was later chancellor of the exchequer and prime minister.

Lennox Island PE In the northern end of Malpeque Bay, this island was named *Lenox Island* in 1765 by Samuel Holland after Charles Lennox, 3rd Duke of Richmond (1734–1806), who had served as the British ambassador to France. It was established as a Mi'kmaq reserve in 1840, although the Mi'kmaq had long been living here.

Lennoxville QC Situated south of the city of Sherbrooke, this town was first known as *Petites-Fourches* and *Little Forks* for its location at the junction of the Saint-François and Coaticook rivers. It was named about 1820 after Charles Lennox, 4th Duke of Richmond and Lennox (1764–1819), governor-in-chief of British North America, 1818–9. It became a village in 1871 and a town in 1920.

Leoville SK Located northwest of Prince Albert, the post office at this village (1944) was established in the 1930, taking its name from first homesteader Leo Carpenter.

L'Épiphanie QC A parish municipality in the regional county of L'Assomption, south of Joliette, it was named in 1853 in honour of the visit to the Christ child by the three wise men. The village of **L'Épiphanie** was separated from it in 1921 and became a town in 1967.

Lepreau, Point NB Extending into the Bay of Fundy, southwest of Saint John, this point is identified on Jean-Baptiste-Louis Franquelin's map of 1686 as *Pointe aux Napraux*. The origin and meaning of the name is unknown. Variations of 'Little Pro' appeared on several English-language maps between 1711 and 1772, suggesting the name may relate to the point resembling a little boat, the French word *proe* being a sailing boat. The post office in the community of **Lepreau**, at the mouth of the **Lepreau River**, was named *Lepreaux* from 1850 to 1854, when it was changed to Lepreau.

Lequille NS Southeast of Annapolis Royal, this place's post office was named in 1860 after the *Rivière à l'Anguille*, a former name of Allains River. The river had been named after the French word *anguille*, 'eel', possibly by Samuel de Champlain. Allains River was called after Louis Allain, a miller here before 1755.

L'Érable QC Located in the middle of the province's important maple syrup and sugar area, northeast of Victoriaville, this regional county was named in 1982. The word *érable* means 'maple'.

Leroy SK A town (1963) northeast of Lanigan, its post office was first named *Bog End* in 1909. When the Canadian Pacific Railway was built from Lanigan to Watson, it was called *Onwa*, a name disliked by the local people. They suggested commemorating local soldier Jack Leroy, who was killed during the First World War. Leroy post office was opened in 1921.

Léry QC A town on the south shore of Lac Saint-Louis, southwest of Montréal, it was created before the First World War. Its roots can be traced to the religious parish of Saint-Cyprien-de-De Léry, established in 1831. The town was named by one of the town's leading citizens after an ancestor of his wife, perhaps Gaspard-Joseph Chaussegros de Léry (1716–56), who designed several public buildings during the eighteenth century.

Les Basques QC Located between Rivière-du-Loup and Rimouski and centred on the town of Trois-Pistoles, this regional county was founded in 1981, taking its name from Île aux Basques, where Basques dried fish and extracted whale oil during the first half of the seventeenth century.

Les Cèdres QC On the north shore of the St. Lawrence, northeast of Salaberry-de-Valleyfield, the abundant cedars, after which this municipality owes its name, were noted as early as 1695. The municipality was created in 1985, having been incorporated as the village of *Cèdres* in 1967, which had been created in 1852 from the village of *Soulanges* and the parish municipality of *Saint-Joseph-de-Soulanges*. It has been widely identified in English as Cedars.

Les Chutes-de-la-Chaudière QC Located at the mouth of the Rivière Chaudière, southwest of the city of Québec, this regional county was created in 1982.

Les Collines-de-l'Outaouais QC North of the cities of Hull and Gatineau, this regional county was created in 1991 by severing seven municipalities from the Communauté urbaine de l'Outaouais. West of Aylmer, it fronts on the Ottawa River, officially called Rivière de l'Outaouais in French.

Les Éboulements QC A municipality in the regional county of Charlevoix, it is located east of Baie-Saint-Paul. Having received its present name in 1956, its roots can be traced to the seigneury of *Éboulements* granted to Pierre Lessard in 1683. *Éboulements* post office was opened in 1832. The name relates to a massive landslide (*éboulement*) as the result of an earthquake in February 1663.

Les Escoumins QC References were made to the site of this municipality, on the north shore of the St. Lawrence and northeast of Tadoussac, as early as 1603 by Samuel de Champlain. It was called *Les Escoumains* in 1863, and amended to the present form in 1957. The name is derived from Montagnais *iskomin*, 'place of wild fruit'.

Les Etchemins QC A regional county to the rear of Bellechasse, and centred on the town of Lac-Etchemin and the upper reaches of Rivière Etchemin, it was named in 1982. The name in Abenaki means 'where there is leather for snowshoes'.

Les Îles-de-la-Madeleine QC Embracing the Îles de la Madeleine in the Gulf of St. Lawrence, this regional county was created in 1981. *See also* Madeleine, Îles de la.

Les Jardins-de-Napierville QC Located south of Montréal and on the border of New

York State, this regional county was named in 1982 after the former county of *Napierville* and the rich agricultural lands of the area.

Les Maskoutins QC This regional county surrounding the city of Saint-Hyacinthe, east of Montréal, was named in 1982. The name describes the people living along the Rivière Yamaska, which flows north to the St. Lawrence, some 20 km east of the Richelieu.

Les Méchins QC A municipality in the regional county of Matane, northeast of the town of Matane, its name is derived from a legendary evil monster called *Matsi* by the Mi'kmaq, which was rendered by the French as *méchant*, and later deformed to *méchin*.

Les Moulins QC Located north of Laval and centred on the towns of Mascouche and Terrebonne, this regional county was created in 1982. It recalls the ancient flour, saw, carding, and fulling mills (*moulins*) that were built here in the early 1700s.

Les Pays-d'en-Haut QC Centred on the all-season recreational centres of Sainte-Adèle and Saint-Sauveur, this regional county was created in 1983. The name essentially means 'the highlands', an apt description of the rugged hills and valleys between Saint-Jérôme and Mont Tremblant.

Lesser Slave Lake AB This lake (1168 km^2), in the north central part of the province, was named *Slave Lake* in 1792 by Alexander (later Sir Alexander) Mackenzie after the Cree word *Awonak*, their name for any strangers, but which translated literally as 'slaves'. It was not the Slavey, a major group of the Dene people living farther north, that they had in mind, but another people, perhaps either the Blackfoot or the Beaver. 'Lesser' was subsequently added to distinguish it from Great Slave Lake.

Lester Pearson, Mount BC In the Premier Range, west of Valemount, this mountain (3060 m) was named in 1973 after Lester Bowles Pearson (1897–1972), prime minister of Canada, 1963–8.

Lestock SK Northwest of Melville, this village (1912) was established by the Grand Trunk Pacific Railway in 1908, taking its name from land surveyor John Lestock Reid. As part of the alphabetical list of names given by the GTP, it was first called *Mostyn*, as it was the station following Leross. Mostyn is a place in Wales, west of Liverpool. The post office was opened as *Lestock Station* in 1911 and was changed to Lestock in 1947.

Letang NB Located on **Letang Harbour**, south of St. George, its post office was named in 1892. The original French name was *Havre à L'Étang*, 'harbour at the pond'. The name is still commonly spelled *L'Etang*.

L'Étang-du-Nord QC Located in the regional county of Les Îles-de-la-Madeleine, this municipality was named in 1875. Located on the north side of a small harbour, it took its name from the religious parish of Saint-Pierre-de-l'Étang-du-Nord, established in 1848.

Letellier MB On the Red River, north of Emerson, this community was first known as *Catherine*, after Catherine Wright, a local landowner. It was named in 1880 after Luc Letellier de Saint-Just (1820–81), lieutenant-governor of Québec, 1876–9.

Letete NB Situated on **Letete Passage**, southwest of St. George, this place's post office was called *L'Etete* from 1856 to 1941. The name may be derived from French *la tête*, 'the head', possibly in reference to a high hill on Macs Island, or to a white ledge at Greens Point.

Lethbridge AB A city (1906) on the Oldman River, it was first known as *The Coal Banks* in 1872, after a coal mine that had been developed here. It was renamed Lethbridge about 1882 after William Lethbridge (1824–1901), first president of the North Western Coal and Navigation Company, which was created that year. As the name had already been alotted to a rural office in Ontario, the post office department called the new office *Coalhurst*. However, three years later, it consented to Lethbridge for the office in Alberta. William Lethbridge was a partner in the London-based booksellers W.H. Smith and Son.

Lethbridge NF On Goose Bay, which is united to Bonavista Bay by Chandler Reach, this place was first known as *Southeast Arm* and *Hopevale*. It was renamed in 1912 after James Lethbridge, the oldest resident here.

Levack ON In the town of Onaping Falls, Sudbury Region, this place was named in 1914 after **Levack Township**. The township was created in 1885, and named in honour of Helen Levack, the mother of Sir Oliver Mowat, then premier of Ontario.

Le Val-Saint-François QC Located between Sherbrooke and Drummondville, this regional county in the valley of the Rivière Saint-François was named in 1982.

Lévis QC A city across the St. Lawrence from the city of Québec, it has an indirect association with Pointe de Lévy, noted as early as 1629 by Samuel de Champlain. That name was given in honour of Henri de Lévis (or Lévy), Duc de Ventadour, who was the viceroy of New France, 1625–7. The former county of Lévis and the post office were named in 1854. The city itself was known as the *Ville-d'Aubigny* from 1849 to 1861, after a Duc de Richmond et d'Aubigny. In 1860, in honour of the one hundredth anniversary of the heroism of François-Gaston de Lévis, Duc de Lévis (1719–87) at the Battle of Sainte-Foy in 1760, it was renamed after him. In 1989 Lévis and Lauzon became the city of *Lévis-Lauzon*, but two years a later, when the town of Saint-David-de-l'Auberivière was annexed, Lévis became the name of the new city.

Lewin's Cove NF Located at the north end of Burin Inlet and southwest of Marystown, this municipal community (1973) was identified as *Lunes Cove* on an 1860 Admiralty chart and as *Loon's Cove* in the 1891 census. The present form of the name occurred for the first time in a 1945 provincial census. The adjacent cove may have called after the common loons that nest around nearby ponds.

Lewisporte NF On Burnt Bay, an extension of the Bay of Exploits, this town (1946) was originally named *Burnt Bay* (1857) and *Marshallville* (1891). It was renamed Lewisporte in 1900 after the Scottish firm of Lewis Miller and Company, which was involved in lumbering in the area until 1903. *See also* Millertown.

Liard River BC, NT, YT Rising in the Yukon, northwest of Watson Lake, this river flows generally southeast to its confluence with the Fort Nelson River, where it runs generally northeast to enter the Mackenzie River at Fort Simpson. French voyageurs, who accompanied John McLeod of the Hudson's Bay Company in 1834, called it *Rivière aux Liards*, *liard* being the French word for the cottonwood (poplar) trees lining its banks. Four years later Robert Campbell of the HBC called it *Bell River*, after chief trader John Bell, but Liard River eventually prevailed.

Likely BC At the west end of Quesnel Lake, southeast of Quesnel, this place's post office was named in 1923 after gold miner John Likely (1842–1929), who was known as 'Plato John' for his lectures on the Greek philosophers.

L'Île-d'Orléans QC Embracing the entire island, downriver from the city of Québec, this regional county was named in 1982. *See also* Orléans, Île d'.

L'Île-du-Havre-Aubert QC A municipality in the county of L'Île-de-la-Madeleine, it was created in 1875 as *Havre-Aubert*, and was changed to *Bassin* in 1959. The municipality of *Havre-Aubert-Est* (1951) became *Havre-Aubert* in 1964, and it and *Bassin* merged in 1971 under the present name. Its origin is uncertain, with some pointing to a sixteenth century sailor, and others to a seventeenth century pilot or ship's master.

L'Île-Perrot QC Located on Île Perrot, at the point where the Ottawa River drains Lac des Deux-Montagnes into Lac Saint-Louis, this town was named in 1955. The island was named in 1672 after its seigneur, François-Marie Perrot (1644–91), a captain in the

Picardie Regiment, and governor of Montréal in 1670.

Lillooet BC This village (1946) on the Fraser River, west of Kamloops, was named after the Lillooet, a Salish tribe, which occupied the valleys to the west and the south. It received its name because it was the end of the trail leading from **Lillooet Lake**, 65 km to the southwest, and near Mount Currie. The meaning of the name is unknown, with the usual interpretation, 'wild onion', being considered incorrect. The site of the village was earlier called *Cayoosh Flat*, after a dead pony (cayuse) found in Cayoosh River. The **Lillooet River** (165 km) rises in the Coast Mountains, and flows southeast through Lillooet Lake to Harrison Lake. **Alouette Lake**, to the south near Mission, was officially called *Lillooet Lake* until 1914, when it was renamed after the French word for 'lark'.

Limerick SK West of Assiniboia, the post office at this village (1913) was named in 1908 by first postmaster Edward Loosing, a native of Limerick, Ireland.

Limoges ON A community in the united counties of Prescott and Russell, east of Ottawa, it was first called *South Indian* in 1883, after a nearby creek. It was renamed in 1926 for Fr Honoré Limoges (1878–1961), its parish priest, 1913–21, who was a native of Sainte-Scholastique, north of Montréal.

Linacy NS East of New Glasgow, this place was founded in the 1800s. Among its early settlers was Edward Linacy.

Lincoln NB In the southeast side of Fredericton and on the Saint John River, its post office was named in 1867. It took its name from **Lincoln Parish**, established in 1786, and probably named by the Loyalist Glasier family, which came from Lincoln, MA.

Lincoln ON The county was named by Lt-Gov. John Graves Simcoe in 1792 after Lincolnshire, England. In 1970 it was merged into the regional municipality of Niagara. The town of **Lincoln** in Niagara Region was created in 1970, by uniting the town of Beamsville with the municipal township of Clinton and part of the municipal township of Louth.

Lincolnville NS North of Guysborough, this place was founded in the late 1800s by descendents of African-Americans, who had come to Guysborough as Loyalists in 1784. The place was named after Abraham Lincoln, the American president, who freed his nation's slaves.

Linden AB Northeast of Calgary, this village (1964) was named after a school, established in 1904. The name may have been given by a strict sect of Mennonites called Holdmanites after the well-known linden tree, which is widely used in the northern hemisphere for shade and its fragrant flowers.

Lindsay ON Chosen as a townsite in 1825, this town (1857) was first called *Purdy's Mills* and *Purdy's Rapids* after millers William, Jesse, and Hassard Purdy. When John Huston surveyed the townsite in 1834, a man named Lindsay was presumably shot by accident, subsequently died of infection, and was buried in the river bank. Huston made no mention of the incident in his diary of July 1834, although he inscribed the name Lindsay on the cover page of his diary. Lt-Gov. Francis Bond Head authorized it as the postal name in 1836.

Linière QC A village in the regional county of Beauce-Sartigan, south of Saint-Georges, its post office was named in 1875 after the township of Linière, created in 1852, and named after a family of seigneurs, who went by the surname of Linière.

Linkletter PE A municipal community (1972) west of Summerside, it was settled in 1786 by George Linkletter, a Loyalist from Connecticut, and his sons George, John, and James. Its post office was opened in 1896.

Lintlaw SK Southeast of Kelvington, this village (1921) was named when the first school opened here in 1909. The post office opened the following year, and the Canadian Northern Railway arrived in 1919. Its was proposed by John McChesney, likely after Lintlaw in

Berwickshire, Scotland, west of Berwick upon Tweed.

Linwood ON In Waterloo Region, northwest of Waterloo, it was named in 1858 by a local schoolteacher, who stated 'linn' implied a pool, and recalled several natural springs forming the source of Speed Creek, with majestic elm trees lining the creek.

Lion's Head ON A village (1917) in Bruce County, north of Wiarton, it was named in 1875 after a rock formation on a nearby headland that had the profile of a lion's head. Much of the headland collapsed into Isthmus Bay in the early 1900s.

Lions Bay BC On the east shore of Howe Sound, northeast of Vancouver, this village (1971) and the bay took their names from two mountains directly to the east, which had been earlier called *The Sisters* and *Sheba's Paps*. From Vancouver they look like a pair of crouching lions, and about 1890 **The Lions** were named by Judge John Hamilton Gray, who suggested the narrow entrance into Burrard Inlet be called **Lions Gate**. The Lions Gate Bridge, uniting the north shore with downtown Vancouver, through Stanley Park, was built in 1938.

Lipton SK North of Fort Qu'Appelle, this village (1905) was first called *Miles* in 1904 by the Canadian Pacific Railway after surveyor Miles Patrick Cotton. The pioneer settlers, having been here since 1883, objected, proposing Lipton after Sir Thomas Lipton (1850–1931). A famous Scottish provisions merchant, Lipton made five unsuccessful attempts to win the America's Cup, emblematic of the international yachting championship. Lipton post office was named in 1905, replacing *Hayward*, given in 1884 after Henry Hawksworth Hayward (d. 1904).

Lisle ON In Simcoe County, southwest of Barrie, this place was named in 1879 after either the popular song *Annie Lisle*, a favourite of lumberman Thomas H. Wilmott's daughter, or after a village in the southern United States, where Wilmott had lived as a boy.

L'Islet QC The seigneury of L'Islet was granted in 1677 to Geneviève Couillard, and the seigneury of L'Islet-de-Bonsecours was granted the same year to François Bellanger, taking their names from a 244 m by 45 m rock offshore in the St. Lawrence, downriver from Montmagny. The municipality of **L'Islet-sur-Mer** was created in 1989 through the merging of the village of L'Islet and the parish municipality of *Notre-Dame-de-Bon-Secours-de-L'Islet*. The town of **L'Islet**, created in 1966, was earlier called *L'Islet-Station* (1950) and *L'Isletville* (1954). The regional county of **L'Islet** was established in 1982.

L'Isle-Verte QC A village (1955) downriver from Rivière-du-Loup, it had been named as a municipality in 1845, after an offshore island noted in the Jesuit *Relations* of 1664. Subsequently it had become *Saint-Jean-Baptiste-de-l'Isle-Verte* in 1855.

Listowel ON A town (1874) in Perth County, north of Stratford, it was named *Listowell* by the post office department in 1856, when the residents could not agree on either Mapleton or Windham. Ten years later, it was respelled Listowel, when the place was incorporated as a village. Listowel is a town in County Kerry, Ireland.

Little Bay Islands NF In the southwest side of Notre Dame Bay, this municipal community (1955) on Little Bay Island and on adjacent Mack's Island was first settled in the early 1800s.

Little Britain ON A community in Victoria County, southwest of Lindsay, it was named in 1855 by merchant Robert Ferguson Whiteside, after his birthplace in the township of Little Britain, Lancaster County, PA.

Little Burnt Bay NF A town (1975) on Southern Head, where it extends into the Bay of Exploits, north of Lewisporte, its name was first recorded in 1901, although settlement had taken place here as early as the 1830s.

Little Catalina NF This town (1965) is located 5 km northeast of Catalina. Efforts in the 1930s to get out from under the shadow of the town of Catalina by renaming it *Orangeville* or *Dayton* were rejected by the Newfoundland Nomenclature Board.

Little Current ON A town (1890) in Manitoulin District, it was known to the Aboriginals as *Waibejewung*, 'where the waters flow back and forth', and to the French as *Le Petit Courant*. Shaftesbury was proposed in 1865, but the translation of the French name was adopted.

Little Heart's Ease NF On the Southwest Arm of Random Sound, a western extension of Trinity Bay, it was named either after the small secluded cove where fishermen could find comfort from stormy waters, or after a fishing vessel.

Lively ON An urban centre in the town of Walden in Sudbury Region, it was created in 1950 by the International Nickel Company for its workers at nearby Creighton Mine. It was named in 1951 after Charles Lively, a long-time INCO employee. Incorporated as a town in 1953, it was merged into Walden in 1973.

Liverpool NS An urban community (1996) in the regional municipality of Queens and at the mouth of the Mersey River, it had been incorporated as a town in 1897. It had been founded in 1761 and named after *Liverpool Township*, which had been named a year earlier by the governor and council in Halifax, presumably after the British port city. In 1604 Pierre du Gua de Monts called the site *Port Rossignol*, for a French captain who was undertaking a fur trade in the territory granted to de Monts. Postal service was established at Liverpool in 1802. *See also* Rossignol, Lake.

Livingstone, Mount BC In the Clemenceau Icefield of the Rocky Mountains, southeast of Athabasca Pass, this mountain (3094 m) was named in 1927 by alpinist James Monroe Thorington after explorer David Livingstone (1813–73), who was found in Africa by Henry Stanley in 1871. **Mount Livingstone** (3090 m) and **Livingstone Range** in Alberta, northwest of Claresholm, were named in 1858 by Capt. Thomas Blakiston of the Palliser expedition.

Lloyd George, Mount BC In the Muskwa Ranges of the Rocky Mountains, this mountain (2972 m) was named in 1916 by American explorer Paul L. Haworth after David Lloyd George (1863–1945), prime minister of Great Britain, 1915–22.

Lloydminster AB, SK A city (1958) on the Saskatchewan-Alberta border, it was conceived in 1902 by Anglican cleric Isaac Barr to develop a British presence in the area of the North Saskatchewan River, northwest of Saskatoon. The following year 2000 settlers arrived in Saint John, NB, and came by train to Saskatoon, where their chaplain, the Revd George Exton Lloyd took over the project, and led it to the *Britannia Settlement*. In 1930 the Saskatchewan town and the Alberta village were united as the town of Lloydminster by orders in council passed by both provinces.

Lloydtown ON In York Region, west of Aurora, this community was settled in 1812 by Pennsylvania Quakers Jesse and Phoebe Lloyd, and named in 1831. Jesse Lloyd (1786–1838), a supporter of William Lyon Mackenzie, fled to the United States in 1837, after the defeat of the rebels, and died the next year in Tippecanoe, IN. His wife remained at Lloydtown.

Lobo ON A township in Middlesex County, it was named in 1821 by Lt-Gov. Sir Peregrine Maitland after the Spanish word for 'wolf'.

Lochaber NS South of Antigonish, this place on Lochaber Lake was founded in 1808 by Scots from Lochaber, Inverness-shire, Scotland. Lochaber post office was opened in 1852, with subsequent offices being located at **Lochaber West Side** (1861), **South Lochaber** (1877), and **North Lochaber** (1885).

Lochiel ON A township in the united counties of Stormont, Dundas and Glengarry, it was named in 1816 after the Scottish home of the Clan Cameron, Lochiel Castle, which overlooks Loch Lochy in Inverness-shire.

Lockeport NS On the South Shore and east of Shelburne, this town (1907) was named after Jonathan Locke, a native of Rhode Island, who acquired land in *Liverpool Township* in 1764. Shortly after, he and other New England planters and fishermen founded the settlement of *Locke's Island*, which was given its present name in 1870.

Lockport MB On the Red River, southwest of Selkirk, it was first called *Little Britain* by the first settlers from England and Scotland. Its post office was then named *St. Andrews North* in 1894 after the St. Andrews Rapids, given by the Selkirk settlers after the patron saint of Scotland. When a lock was constructed in 1902 to allow boat traffic to pass the rapids, the place and post office were renamed Lockport.

Logan ON A township in Perth County, it was named in 1830 after Hart Logan, a London, England merchant and a director of the Canada Company, a British company established in 1826 to develop the Huron Tract, a large area of what is now Southwestern Ontario. William (later Sir William) Edmond Logan worked in Hart Logan's business in London for some 20 years, before returning in 1842 to his birthplace, Montréal, and to found the Geological Survey of Canada.

Logan Lake BC Southwest of Kamloops, this district municipality (1984) was previously incorporated as a village in 1970, which was named after a small lake here. The lake was named after an Aboriginal called Tslakan, who raised horses and traded furs, but his name was badly rendered as 'Logan'.

Logan, Mont QC Forming one of the highest points (1135 m) of the Monts Chic-Chocs, east of Matane, this mountain was named after an exploratory mineral survey of the mountains in 1844 by geologists William (later Sir William) Edmond Logan (1798–1875) and Alexander Murray.

Logan, Mount YT The highest mountain (5959 m) in Canada, it was named in 1890 by American explorer and geologist Israel C. Russell after Sir William Edmond Logan (1798–1875), the founder of the Geological Survey of Canada in 1842. He served as its director until 1869, undertaking several geological expeditions in present Ontario and Québec. The **Logan Mountains**, in southeastern Yukon and southwestern Northwest Territories, and on the eastern side of Frances Lake, were named in 1887 after Sir William by geologist George M. Dawson.

Loggieville NB Located east of Chatham (Miramichi), this place was settled about 1790 by Robert Logie (later Loggie). Its first post office was *Black Brook*, 1861–95, with Robert Blake as the first postmaster. It was renamed Loggieville in 1895. Incorporated as a village in 1966, it was annexed by the city of Miramichi in 1995.

Loki, Mount BC In the Purcell Mountains, northeast of Nelson this mountain (2771 m) was named before 1890 after a poker game played by miners. Loki is the Norse god of fire, strife, and humour.

Lombardy ON A community in the township of South Elmsley, united counties of Leeds and Grenville, southwest of Smiths Falls, it was settled in the 1820s by Francis Lombard (d. 1864), who operated an inn here. It was called *Lombard Corners* until 1853, when the *South Elmsley* post office was opened. Twenty years later local residents requested Lombardy, after the first resident, and after the area's resemblance to the plain of Lombardy in Italy.

Lomond AB A village (1916) southeast of Calgary, it was named in 1914 by the Canadian Pacific Railway after Loch Lomond, northwest of Glasgow, Scotland.

Londesborough ON In Huron County, east of Goderich, this place was named by Thomas Hagyard in 1861 after the East Yorkshire estate of Lord Londesborough, midway between the English cities of York and Kingston upon Hull, where Hagyard had lived.

London ON In March 1793 Lt-Gov. John Graves Simcoe considered the forks of the Thames as a possible location for the capital of

Upper Canada, which he had called *Georgina*, before arriving in Canada. After looking at the site, he renamed it *New London*. Lord Dorchester, the governor-in-chief, disallowed Simcoe's choice, but consented to his proposal to move the capital to Toronto (renamed York later that year). London post office had been established 5 km west of the forks in 1825, but it was moved here the next year, when London became the administrative capital of London District. It became a city in 1855. **London Township** was named in 1798.

Londonderry NS Located northwest of Truro, this place's postal service was established in 1838. It took its name from *Londonderry Township*, created in 1775, and named after County Londonderry, Ireland. The first settlers had emigrated from there in 1762 and some New Englanders had come from Londonderry, NH, at the same time.

Long Branch ON A neighbourhood in the city of Etobicoke, it was named about 1884 by Thomas J. Wilkie, possibly after the seaside resort of Long Branch, NJ. It was an incorporated village from 1931 to 1967.

Longlac ON A town in Thunder Bay District, east of Geraldton, it is situated at the north end of **Long Lake**. Founded as a Hudson's Bay Company post in 1800, *Long Lake* was a busy settlement and trading centre when the Canadian Northern Railway arrived in 1914, and proposed Longlac for its station. The post office was opened in 1919. Set up as an improvement district in 1952, it became a municipal township in 1964, and a town in 1982.

Long Point ON Stretching for 35 km into Lake Erie, this peninsula was crossed by explorers François Dollier de Casson and René de Bréhant de Galinée in 1670, who named it *La Longue Pointe*. After 1759 it was called Long Point.

Long Range Mountains NF Extending for 500 km from Cape Ray to the northern end of the Northern Peninsula, these mountains achieve heights of 559 m in the Anguille Mountains, 814 m in the Lewis Hills, 806 m in Gros Morne National Park, and 606 m northeast of Port au Choix.

Long Sault ON In the united counties of Stormont, Dundas and Glengarry, west of Cornwall, it was created in 1956 to accommodate the residents of *Mille Roches* and *Moulinette*, inundated during the construction of the St. Lawrence Seaway. Taken from a historic stretch of the St. Lawrence River, the name was spelled Longue Sault by English-speaking settlers. However, the names authorities and French-language purists insisted on spelling it Long Sault.

Longstaff, Mount BC Northwest of Mount Robson, this mountain (3180 m) was named in 1911 by surveyor Arthur O. Wheeler after Dr Thomas George Longstaff (1875–1964), who made several climbs around the world, including an early attempt on Mount Everest.

Longueuil ON A township in the united counties of Prescott and Russell, it was known as the seigniory of Pointe-à-l'Orignal until 1798, when it was renamed after Baron Joseph-Dominique-Emmanuel Le Moyne de Longueuil (1738–1807), who had become its seigneur in 1778.

Longueuil QC The largest city in the regional county of Champlain, across the St. Lawrence from Montréal, it was created a parish municipality in 1845, abolished in 1847, and recreated as a village in 1848. It became a town in 1874, and a city in 1920, when it annexed the town of *Montréal-Sud*. In 1974 it further annexed the city of *Jacques-Cartier* (1947) and the parish municipality of *Saint-Antoine-de-Longueuil* (1855). Charles Le Moyne (1626–85) became the seigneur of Longueuil in 1657, which was elevated to a barony in 1700. He was born in Longueuil, near Dieppe, France, and came to New France in 1641.

Longview AB South of Calgary on the Highwood River, the post office at this village (1964) was named in 1905, after the good view of the mountains and after first postmaster

Thomas Long, who had settled in the area of Okotoks in 1895.

Loon Lake SK West of Meadow Lake, this village (1950) is 1 km east of Loon Lake, one of the seven Makwa Lakes. The Cree word *makwa* means 'loon'.

Lord's Cove NF On the south end of the Burin Peninsula, this municipal community (1966) was established in the latter half of the 1800s and named after the cove. The origin of the cove's name is unknown, although there may be a connection with harlequin ducks, called lords and ladies in Newfoundland.

Loreburn SK On the Canadian Pacific Railway line uniting Moose Jaw with Outlook, the post office at this village (1909) was established in 1908, and named after Sir Robert Reid, 1st Earl of Loreburn (1846–1923). A British politician, he was created Baron Loreburn in 1905, when he was appointed lord chancellor.

Lorette MB Between Winnipeg and Steinbach, its post office was named *Loretto* in 1875, after Loretteville, northwest of the city of Québec. It was renamed Lorette in 1939.

Loretteville QC In the Communauté urbaine de Québec, northwest of the city of Québec, it was named in 1913, and became a town in 1947. Its territory was formerly part of the parish municipality of *Saint-Ambroise-de-la-Jeune-Loretteville*, incorporated in 1845. The original religious parish was created in 1676 to serve the Huron, who had gathered here in the middle of the 1600s. They were placed under the protection of Notre-Dame de l'Annonciation, modelled on the Holy House of the Virgin, founded in 1295 in Loreto on the east coast of Italy.

L'Orignal ON A village (1876) in the united counties of Prescott and Russell, it derived its name from the seigniory of Pointe-à-l'Orignal, granted to François Prévost in 1674. The seigniory was named after *Pointe à l'Orignal* (now Grants Point), where moose (*orignal*) historically crossed the Ottawa River.

Loring ON In Parry Sound District, north of the town of Parry Sound, it was named in 1884 by member of Parliament William Edward O'Brien after his wife, Elizabeth (Loring) Harris, whom he married in 1864. The nearby community of **Port Loring** was named in 1922. She was the daughter of Col R.R. Loring, of Toronto, and the widow of J.F. Harris of London, ON.

Lorne NB Southeast of Dalhousie, this place's post office was named in 1893 after Sir John Douglas Sutherland Campbell, 9th Duke of Argyll and Marquess of Lorne (1845–1914), governor general of Canada, 1878–83.

Lorne, Mount YT On the west side of Marsh Lake, and south of Whitehorse, this mountain (1951 m) was named in 1887 by surveyor William Ogilvie after the Marquess of Lorne, governor general of Canada, 1878–83. The hamlet of **Mount Lorne**, west of the mountain, was named in 1990.

Lorne Valley PE A municipal community (1978) north of Montague, its post office was named in 1878 after a property in Argyllshire, Scotland, owned by the Marquess of Lorne, governor general of Canada, 1878–83.

Lorneville NB In the southwest part of the city of Saint John, its post office was first called *Pisarinco* in 1854. Because that name was perceived to be indelicate, it was renamed in 1902 after the Marquess of Lorne.

Lorneville ON In Victoria County, northwest of Lindsay, this community was named in 1874 after the Marquess of Lorne (1845–1914), governor general of Canada, 1878–83, who had married Princess Louise in 1871.

Lorraine QC A town in the regional county of Thérèse-De Blainville and on the north shore of the Rivière des Mille-Îles, it was created in 1960. It took its name from the historic region of eastern France.

Lorrainville QC Located in the regional county of Témiscamingue, east of Ville-Marie, this village's post office was named in 1889 after Mgr Narcisse-Zéphirin Lorrain (1842–1915), apostolic vicar of the Pontiac 1882–98, and bishop of Pembroke, 1898–1915. The village was incorporated in 1930.

Lotbinière QC A regional county on the south shore of the St. Lawrence, southwest of the city of Québec, it was named in 1982 after the former county. The municipality of **Lotbinière** was created in 1978 through the merger of the village of Lotbinière and the parish municipality of *Saint-Louis-de-Lotbinière*. Its post office was opened in 1831, taking its name from the seigneury of René-Louis Chartier de Lotbinière (1641–1709), who was granted the seigneury in 1672. The name is derived from Chartier's ancestral home of Binière in western France.

Loudon, Mount AB South of Abraham Lake and the North Saskatchewan River, this mountain (3220 m) was named in 1956 after William James Loudon (1860–1951), a professor of mechanics at the University of Toronto.

Loughborough ON A township in Frontenac County, it was named in 1798 after Alexander Wedderburn, Baron Loughborough and Earl of Rosslyn (1733–1805). The title was taken from the borough of Loughborough in Leicestershire, England.

Lougheed AB Southeast of Camrose, this village (1911) was first named *Holmstown*. It was renamed in 1906 by the Canadian Pacific Railway after Sir James Alexander Lougheed (1854–1925), who had set up a law practice with Richard B. (later Viscount Richard) Bennett in Calgary in 1882. He was appointed in 1889 to the Senate, where he served as its Conservative leader from 1906 to 1921. The next station to the east was named after his father-in-law, William Hardisty. **Mount Lougheed** (3105 m), between Calgary and Banff, was named after him in 1928. It had been named *Windy Mountain* in 1858 by botanist Eugène Bourgeau and called *Wind Mountain* by geologist George Dawson in 1886.

Lougheed Island NT In the Parry Islands of the Queen Elizabeth Islands, this island (1308 km^2) was named in 1917 by Vilhjalmur Stefansson after Sir James Alexander Lougheed (1854–1925). He was called to the Senate in 1889, and from 1911 to 1918 served as a minister without portfolio in Sir Robert Borden's cabinet.

Louisbourg NS An urban community in the regional municipality of Cape Breton since 1 August 1995, it is located on the north side of **Louisbourg Harbour**. It was founded by the French on the south side of the harbour on 2 September 1713, following the awarding of Cape Breton Island to France by the Treaty of Utrecht. The harbour was then known as *Havre à l'Anglois*, after English fishermen. The place was first called *Port St-Louis*, after Louis XIV. The site of the French fortress, captured and destroyed by the British in 1758, was declared a national historic park in 1931. Since 1961 the area of the fortress has been reconstructed. The post office was opened as *Louisburg* in 1835, with the current spelling being adopted for it in 1955. The town, incorporated as *Louisburg* in 1901, was renamed Louisbourg in 1966.

Louisdale NS On Cape Breton Island and east of Port Hawkesbury, its post office was called *Barachois St. Louis* in 1893, after a local cove. It was renamed Louisdale in 1905 by postmaster J. Nelson Scott, retaining part of the original name.

Louiseville QC A town in the regional county of Maskinongé, southwest of Trois-Rivières, it was created as a village in 1878, and named the following year after Princess Louise Caroline Alberta (1848–1939), the wife of the Marquess of Lorne, governor general of Canada, 1878–83. Its post office was known as *Rivière-du-Loup-en-Haut* from 1816 to 1880, when it was renamed after the village.

Louis St-Laurent, Mount BC In the Premier Range, west of Valemount, this mountain (3018 m) was named in 1964 after Louis Stephen St-Laurent (1882–1973), prime minister of Canada, 1948–57.

Lourdes NF On the northwest side of the Port au Port Peninsula, this municipal community (1979) was formerly called *Clam Bank Cove*. It was renamed in the 1930s after Lourdes in southwestern France, where the Virgin Mary revealed herself in 1858 as the Immaculate Conception to Marie Bernarde Soubirous (St Bernadette).

Lowe Farm MB Located west of the Red River and southwest of Winnipeg, this community's post office was named in 1900 after John Lowe, deputy minister of agriculture, 1888–95. He had bought 19 sections (4921 ha, 12,160 a) in the area, and sold the land in 1894.

Lower Arrow Lake BC The lower of the two Arrow lakes, this narrow 95-km-long widening of the Columbia River was named in the early 1800s after *Arrow Rock*. The rock was, before the flooding of the lake behind the Keenleyside Dam, a steep cliff on the east side of the lake where the arrows of Aboriginal braves were lodged. Arrows that remained lodged in the rock were believed to bring good luck to the brave.

Lower Mazinaw Lake ON On the Mississippi River system, north of Kaladar, this body of water was considered part of *Mazinaw Lake* until 1981, when it was divided at The Narrows into Lower and Upper Mazinaw lakes. *Mazinaw* is an Algonkian word meaning 'picture', recalling pictographs on the face of **Mazinaw Rock**, on the east shore of Upper Mazinaw Lake.

Lower Montague PE A municipal community (1974) east of the town of Montague and on the south side of the Montague River, its post office served the area from 1871 to 1913 and from 1916 to 1946.

Lucan ON In Middlesex County, northwest of London, this village (1871) was named in 1857 by 'Dublin' Tom Hodgins after Lord Lucan's estate at Lucan in County Dublin, west of the centre of the city of Dublin.

Lucania, Mount YT In the St. Elias Mountains, this mountain (5227 m) was named in 1897 by Italian explorer and mountaineer Luigi Amedeo Abruzzi, the Duke of Abruzzi, after the Cunard liner on which he and his expedition members crossed the Atlantic.

Luceville QC A village in the regional county of La Mitis, northeast of Rimouski, it was created in 1918. Its post office was called *Sainte-Luce-Station* in 1883, and renamed Luceville in 1917. It may have its roots in the name of an eighteenth century grantee of a seigneury, Gertrude-Luce (or Luce-Gertrude) Drapeau.

Lucifer, Mount AB The most northwesterly point of Jasper National Park, this mountain (3060 m) was named in 1956 after one of the titles of the devil.

Lucknow ON A village (1873) in Bruce County, southwest of Walkerton, it was planned in 1858 and named by James Somerville in honour of a fellow Scot, Sir Colin Campbell, who the previous year had led the relief of Lucknow in India.

Lucky Lake SK A village (1920) north of Lake Diefenbaker and northwest of Moose Jaw, its post office was named in 1908. It took its name from a lake 10 km to the north, which had been named by settler Jock Swansen. He considered it lucky to retrieve his team of oxen, which had fallen into what had been previously called *Devil's Lake*. The lake is officially called **Luck Lake**.

Lulu Island BC Part of the city of Richmond and comprising a large part of the delta of the Fraser River, this island was named in 1862 by Col Richard C. Mayne after Lulu Sweet, an actress with the first theatrical company to visit the province.

Lumby BC A village (1955) east of Vernon, its post office was called *White Valley* in 1889. It was named in 1894 after Moses Lumby (1842–93), who had been appointed a government agent in Vernon in 1891. After migrating from England in 1862, he became a miner in the Cariboo, a farmer at Monte Creek, near Kamloops, a mail carrier on the upper Columbia, a farmer again in the north end of the Okanagan Valley, and a vice-president of the Shuswap and Okanagan Railway.

Lumsden NF A town (1968) on the northeastern coast of the island of Newfoundland, it was first known as *Cat Harbour*. It was renamed in 1917, after the Revd James Lumsden (1854–1915), a Methodist minister in Newfoundland, 1881–3 and 1885–92, who was affectionately called the 'skipper parson'. Subsequently he was sheriff of Queens County, NS, and the member for Queens in the provincial legislative assembly.

Lumsden SK Located in the Qu'Appelle River valley, north of Regina, this town (1905) was named in 1890, when the Qu'Appelle, Long Lake and Saskatchewan Railway and Steamboat Company line reached that point. It was named after Hugh D. Lumsden, the railway's supervising engineer. Its post office was opened in 1892.

Lundar MB Situated near the eastern shore of Lake Manitoba and due west of Gimli, this place's post office was named in 1891 by postmaster Henrik Jonson after his wife's family farm in Iceland. The farm was called *Lundi*, 'meadow', but an error occurred when the post office department prepared the cancelling hammer.

Lunenburg NS This town (1888) was founded in 1753 by Gov. Charles Lawrence for the settlement of Protestants from Germany, France, and Switzerland. It took its name from the British royal house of Brunswick-Lüneburg, and its post office was opened in 1819. The site had been occupied by Acadians until 1749, which they called *Merligueshe*, a Mi'kmaq word, which is said to mean 'milky bay'. **Lunenburg County** was created in 1759.

Lunenburg ON A community in the united counties of Stormont, Dundas and Glengarry, northwest of Cornwall, it was named in 1858 after Lunenburg District, given in 1788 out of respect for George III's German roots, but it was renamed Eastern District in 1792 by Lt-Gov. John Graves Simcoe.

Luseland SK Established in 1908 by the Luse Land Development Company of St. Paul, MN, northwest of Kerrobert, this town (1954) was relocated in 1910 to the new Canadian Pacific Railway line built northwest from Rosetown to Macklin. Its post office was also opened that year.

Lushes Bight - Beaumont - Beaumont North NF A municipal community (1968) on Long Island in Notre Dame Bay, it comprises three separate built-up areas. **Lushes Bight** may have been named after William Lush, of nearby Burlington, who fished there seasonally. **Beaumont**, first called *Ward's Harbour*, was named in 1918 after Beaumont Hamel, France, where many soldiers of the Newfoundland Regiment lost their lives in April, 1916, during the First World War.

Luskville QC A community in the municipality of Pontiac, northwest of Aylmer, it was named in 1884 after pioneer Joseph Lusk, who settled there in 1832.

Łutselk'e NT On the southeast shore of Great Slave Lake, this settlement was named *Snowdrift* in 1936. It was renamed Lutselk'e in 1992, after the Chipewyan word for 'place of small fish'.

Lyall, Mount AB, BC On the continental divide, north of Crowsnest Pass, this mountain (2952 m) was named in 1917 after Dr David Lyall (1817–95), a surgeon and naturalist who had served on the International Boundary Commission between the Rockies and the Pacific Ocean, 1858–62.

Lyell, Mount AB, BC On the continental divide and near the head of the North Saskatchewan River, this mountain (3504 m) was named by Dr James (later Sir James) Hector in 1858 after British geologist Sir Charles Lyell (1795–1875).

Lyn ON In the united counties of Leeds and Grenville, west of Brockville, this place was named *Lowell* in 1839, after the Massachusetts city north of Boston, from where some settlers had come. Fearing confusion with similar names in Upper Canada, it was renamed after Lynn, MA, a city northeast of Boston.

Lynden ON In the town of Flamborough, west of Hamilton, this place was named in 1851 by Jeremiah Bishop after Lyndon, VT, north of St. Johnsbury.

Lyndhurst ON A community in the united counties of Leeds and Grenville, north of Gananoque, it was called *Furnace Falls* in 1836 after a local iron foundry. In 1851 it was renamed after John Singleton Copley, Baron Lyndhurst (1772–1863), lord chancellor of England in several ministries.

Lynn Lake MB Located north of Flin Flon, this community has its roots in the discovery in 1941 of copper and nickel deposits, which were developed in 1947. Its post office was named in 1951 after Lynn Smith, chief engineer of the Sherritt Gordon Mining Company.

Lyons Brook NS West of Pictou, this place's first post office was called *Logans Tannery* in 1883, with Dougal Logan as the first postmaster. It was renamed Lyons Brook in 1903 after the Revd James Lyons, a Presbyterian minister, who had brought some settlers from Pennsylvania in 1767.

Lyster QC A municipality in the regional county of L'Érable, northeast of Victoriaville, it was created as a village in 1912. In 1976 it became a municipality by merging with the municipality of *Sainte-Anastasie-de-Nelson*. Lyster post office was named in 1862, and renamed *Sainte-Anastasie* in 1898, but Lyster was restored in 1981. The name was given by lumber merchant Charles King after his birthplace of Lister in Essex, England.

Lytton BC At the confluence of the Fraser and Thompson rivers, this village (1945) was named in 1858 by Gov. James Douglas after Sir Edward Bulwer-Lytton (1803–73), the British secretary of state for the colonies. Before the first detachment of Royal Engineers sailed in 1858 from Cowes, he joined them on board, and wished them well in the new colony.

M

Maberly ON In Lanark County, southwest of Perth, it was named *Maberley* in 1865 by post office department secretary William D. LeSueur after Col Maberly of England's general post office in London. To conform with local preference, it was respelled Maberly in 1976.

Mabou NS At the head of **Mabou Harbour** and on the west side of Cape Breton Island, this place was settled in the 1770s. Its name, derived from Mi'kmaq *molabokek*, of unknown meaning, was assigned to its post office in 1834. **Cape Mabou**, to the north, is a 363-m mountain.

MacAlpine Lake NT On the Arctic Circle, northwest of Baker Lake, this lake (421 km^2) was named in 1929 by geologist G.H. Blanchet after Col C.D.H. MacAlpine, president of Canadian Explorers Ltd. Learning in 1930 that North-West Mounted Police inspector F.H. French had named it in 1918 after NWMP (RCMP from 1919) Commissioner Aylesworth Bowen Perry (1860–1956), the Geographic Board changed it to *Perry Lake*. In 1932 the board reversed its decision, restored MacAlpine Lake, and named the 140-km river draining the lake north to Queen Maud Gulf after Perry.

Macamic QC Located on the south shore of **Lac Macamic** in the regional county of Abitibi-Ouest, north of Rouyn-Noranda, it became a village in 1919 and a town in 1955. The name of its post office was spelled *Makamik* from 1915 to 1949. In Algonquin *makamik* means 'crippled beaver'.

Macaulay, Mount YT In the St. Elias Mountains, this mountain (4663 m) was named in 1958 after territorial judge Charles Daniel Macaulay.

Maccan NS On the **Maccan River**, south of Amherst, this place's name was derived from Mi'kmaq *maakan*, 'fishing place'. Postal service was established here in 1838.

MacCarthy, Mount BC In The Bugaboos of the Purcell Mountains, west of Invermere, this mountain (3062 m) was named in 1954 by climber Peter Robinson after alpinist Albert Henry MacCarthy (1876–1955), who had climbed in the Purcells and had led the first ascent of Mount Logan in the Yukon in 1925.

Macdiarmid ON In Thunder Bay District, northeast of Thunder Bay, this place may have been named after Clyde McDiarmid, a Carleton Place native and a Canadian Northern engineer when the railway was built in 1916. It may also have been called after F.G. MacDiarmid, then provincial minister of public works, who also had responsibility for fish and game.

Macdonald, Mount BC In the Selkirk Mountains, directly south of Rogers Pass, this mountain (2893 m) was named in 1897 by order in council after Sir John Alexander Macdonald (1815–91), prime minister of Canada, 1867–73 and 1878–91.

Maces Bay NB Adjacent to the Bay of Fundy, southwest of Saint John, this bay may have been named by map-maker J.F.W. DesBarres after Benjamin Mace, surgeon with the 22nd Regiment, or derived from Passamaquoody *musquasikik*, 'place for Indian beans', a reference to Salkeld Island. The community of **Maces Bay**, north of Point Lepreau, had a post office from 1855 to 1918.

MacGregor MB West of Portage la Prairie, this village (1947) was named in 1881 by the Marquess of Lorne after his chaplain, the Revd James MacGregor of St Cuthbert's Church, Edinburgh, who accompanied him west. Its post office was called *MacGregor Station* in 1883, with Station being dropped in 1900.

Machias Seal Island NB In the Bay of Fundy and southeast of Machias Bay, ME, this island was known to the Passamaquoddy as *Menascook*, 'at the grassy island'. Map-maker

Herman Moll called it *Seal I* in 1715. The present name was noted in 1876 by historian J.G. Lorimer.

MacKay Lake NT Northeast of Great Slave Lake, this lake (977 km^2) was named *Lake Mackay* in 1892 by explorer Warburton Pike. In 1924 geologist G.H. Blanchet recommended MacKay Lake, as it was named after Dr William Morrison MacKay (1836–1917), a Hudson's Bay Company surgeon and fur trader.

Mackenzie BC A district municipality (1966) on the eastern shore of Williston Lake, it was created as a centre for the exploitation of the rich forest and mineral wealth of the area. It was named after explorer Sir Alexander Mackenzie (1764–1820), who passed by the future site of the municipality in 1793. **Mount Mackenzie**, northeast of Bella Coola, was named in 1953 by the department of national defence after the explorer. *See also* Sir Alexander, Mount.

Mackenzie King Island NT In the Parry Islands of the Queen Elizabeth Islands, this island (5048 km^2) was named in 1949 after William Lyon Mackenzie King (1874–1950), prime minister of Canada, 1921–26, 1926–30, and 1935–48. Until a Royal Canadian Air Force aerial survey in 1947, it had been considered part of Borden Island.

Mackenzie King, Mount BC In the Premier Range, west of Valemount, this mountain (3280 m) was named in 1962 after William Lyon Mackenzie King, prime minister of Canada for three periods between 1921 and 1948. *See also* King, Mount.

Mackenzie, Mount BC In the Selkirk Mountains, southeast of Revelstoke, this mountain (2456 m) was named in 1887 after Alexander Mackenzie (1822–92), prime minister of Canada, 1873–8.

Mackenzie Mountains NT On the border of the Yukon and Northwest Territories, this system of several ranges of mountains was named in 1944 by the Geographic Board after Alexander Mackenzie, prime minister of Canada, 1873–8. The name had occurred as early as 1927 on a map of northwestern Canada.

Mackenzie River NT The longest river system (4241 km, to the head of the Finlay) entirely within Canada, this river was named by John (later Sir John) Franklin in 1825 after Alexander (later Sir Alexander) Mackenzie. He had explored the river to its mouth in 1789, but, discouraged that he had not arrived at the Pacific Ocean, called it *River Disappointment*.

Macklin SK Southwest of Battleford, near the Alberta border, this town (1912) was established in 1906 by the Canadian Pacific Railway and named after Harry Macklin. A Winnipeg *Free Press* executive, he was following the progress of the railway's construction. Its post office was opened in 1908.

Macmillan River YT Rising in the **Macmillan Pass**, and in the Mackenzie Mountains, the **South Macmillan River** joins the **North Macmillan River** north of Faro. The main river then flows west to join the Yukon River at the site of Fort Selkirk, west of Pelly Crossing. Hudson's Bay Company trader Robert Campbell named it in 1843 after HBC chief factor James McMillan. The name's official spelling was made by the Geographic Board in 1898.

Macoun SK Northwest of Estevan, this village (1903) was named after John Macoun (1831–1920), a field naturalist, who had persuaded Canadian Pacific Railway officials in the early 1880s that the southern prairies were ideally suited for agriculture. Its post office was also opened in 1903.

Macoun, Mount BC In the Selkirk Mountains, southeast of Rogers Pass, this mountain (3049 m) was named in 1889 by climber William S. Green after naturalist John Macoun (1831–1920). Earlier, alpinist Harold Topham had called it *North Sentinel*.

Mactaquac Lake NB Extending from a dam west of Fredericton up the Saint John River valley to Woodstock, this reservoir was created by the New Brunswick Electric Power Commission in 1967. It was named after **Mactaquac Stream**, which rises south of Millville. This name may be Mi'kmaq for 'big branch', possibly in reference to the Saint John River itself, and subsequently applied to the first stream above the head of tide.

MacTier ON In the district municipality of Muskoka, southeast of Parry Sound, this place's post office was first called *Muskoka Station*, when the Canadian Pacific Railway built a line in 1908 from Toronto to Sudbury. To avoid confusion with similar names, it was renamed in 1915 after A.D. MacTier, general superintendant of the CPR's eastern division.

Madame, Isle NS On the south side of Cape Breton Island, this island was known to Acadian Gov. Nicolas Denys in the 1660s as *Isle Ste-Marie*, which was rendered by map-makers as early as 1684 as *Isles Madame*. It was called *I. de Maurepas* in the early 1700s after Comte Jean Frédéric Phélypeaux Maurepas (1701–81), a French secretary of state and minister of the marine and colonies. Samuel Holland introduced *Richmond Island* in 1767 after Charles Lennox, 3rd Duke of Richmond (1734–1806), a British official who served as ambassador to France. However, the name Isle Madame was not supplanted.

Madawaska River NB, QC This river drains Lac Témiscouata, in Québec, southeast into the Saint John River, at Edmundston. The name is derived from Maliseet *medaweskak*, 'porcupine place', or 'murmuring at the mouth'. It was noted in a seigneury grant of 1683 as *Madouesca*, with the present spelling being established in 1778 by map-maker James Peachey. **Madawaska County** was created in 1873.

Madawaska River ON Rising in Algonquin Park, this river flows southeast to the Ottawa River at Arnprior. The name may have been derived from an Algonquin band that Samuel de Champlain met in 1613, and called by him *Matouoüescarini*, 'people of the shallows'. The post office in the community of **Madawaska**, on the river, and in Nipissing District, west of Barry's Bay, was named in 1900.

Madeira Park BC On the Sechelt Peninsula, northwest of Vancouver, this place's post office was named in 1946. The name was proposed by Joseph Gonsalos after his native Arquipélago da Madeira, a Portuguese possession in the Atlantic Ocean, west of Casablanca, Morocco.

Madeleine, Îles de la QC Located in the southern part of the Gulf of St. Lawrence, and 250 km southeast of Gaspé, this archipelago was visited in 1534 by Jacques Cartier. Two years later he called them *Les Araynes*, signifying 'the sandy ones'. During the 1600s the group was called *Ramea*, *Ramée*, and *Ramées*, names suggesting branches united by strips of sand. Samuel de Champlain called the largest island (Havre Aubert) *La Magdelene* in 1632. In 1663 their second proprietor, François Doublet, was authorized to rename the group Îles de la Madeleine after his wife, Madeleine Fontaine. Until about 1980 they were commonly called the *Magdalen Islands* in English, but the print media have almost entirely accepted the official designation.

Madoc ON The township in Hastings County was named in 1820 after the Welsh Prince Madoc, who reputedly discovered America in 1170. The post office in the village was named after the township in 1836. In 1855, the police village was called *Hastings*, in the hope it would become the county seat of Hastings County. Ten years later, it was renamed **Madoc**, and incorporated as a village in 1877.

Madsen ON In Kenora District, southwest of Red Lake, it was named in 1938 after Marius Kristian (Matty) Madsen (d. 1967), a Danish scientist and explorer, who prospected for 10 years before striking gold at the site of present Madsen Red Lake Gold Mine.

Mafeking MB Located west of Lake Winnipegosis, and between Swan River and The Pas, this place's station and post office were named in 1904 in honour of the brilliant defence in 1900 of Mafeking (now Mafikeng), South Africa, by Col Robert Baden-Powell. The next two Canadian Northern Railway stations were called Baden and Powell.

Magaguadavic River NB This river drains **Magaguadavic Lake** southeast into the Bay of Fundy, at St. George. Its name was derived from Mi'kmaq, through Passamaquoddy, *Mageecaatawik*, 'river of the big eels'. Spelled many ways in the late 1600s and throughout the 1700s, its present spelling was set by 1859, although some earlier references used *Macadavic*, reflecting its pronunciation.

Magnetawan ON In 1871 the post office was named *Maganetawan*, but the spelling of Magnetawan had been fixed in a 1859 Crown lands report. The postal name was amended in 1918 when the village was incorporated. The **Magnetawan River** rises in Algonquin Park and flows almost straight west to Byng Inlet. The name in Ojibwa means 'long open channel', a description of Byng Inlet.

Magnetic Hill NB In Moncton, 10 km northwest of the city centre, this site of an illusory phenomenon of vehicles seeming to roll up a hill was discovered in the 1930s by Muriel Lutes Sikorski on the side of a hill called Lutes Mountain, and promoted by several writers.

Magog QC Located at the north end of Lac Memphrémagog, which is drained by Rivière Magog northeast to the Saint-François at Sherbrooke, its present form can be traced to the township of Magog, proclaimed in 1849. The township took its name from the lake, which in Abenaki had the significance of 'place where there are salmon trout'. Magog was incorporated as a village in 1888, and was upgraded to a town less than two years later. *See also* Memphrémagog.

Magrath AB A town (1907) south of Lethbridge, it was named in 1899 after Charles A. Magrath (1860–1949), who had been a Dominion land surveyor from 1878 to 1885. Settling in Lethbridge, he managed the Canadian North Western Coal and Irrigation Company, 1885–1906, served as the city's first mayor, 1891, represented Lethbridge in the North-West Territories Assembly, 1891–1902, and was a member of Parliament for Medicine Hat, 1908–11.

Mahone Bay NS Situated on the west side of the 20 km^2 **Mahone Bay**, this place was founded in 1754, with its post office being established in 1848. An attempt in 1857 to rename it Kinburn, after the fortress at the mouth of the Dneiper, in present-day Ukraine, attacked by the British and the French during the Crimean War, was not accepted. The bay was called *Baie de la Mahonne* in French, after low-lying vessels used by pirates.

Maidstone ON A township in Essex County, it was named in 1792 after the county town of Kent in England, southeast of London. The post office at **Maidstone**, southeast of Windsor, was named in 1837, but its postmaster was not appointed until 1851.

Maidstone SK Located southeast of Lloydminster, this town (1955) was named in 1905 by the Canadian Northern Railway, likely after Maidstone, in Kent, England, southeast of London. The post office had been called *Sayers* that year and was renamed Maidstone in 1906.

Main-à-Dieu NS On the east coast of Cape Breton Island, this place's name was derived from Mi'kmaq *Menadou*, 'the spirit of evil'. In 1786 J.F.W. DesBarres interpreted it as *Main a Dieu*, 'hand of God', and this became its postal name in 1837. The Geographic Board amended it to *Mainadieu* in 1909. Assuming it was a French name, the Canadian Board on Geographical Names changed it to its present form in 1954.

Main Brook NF On the south side of Hare Bay, southwest of St. Anthony, this town (1948) was founded in the 1920s, becoming one of the first 11 towns incorporated in the province before 1949.

Mainland NF Located on the west side of Port au Port Peninsula, this place was called in French *Grand'Terre*, 'mainland', as seen by French-speaking fishermen from Red Island (*Île Rouge)*, offshore. Basque fishermen may have earlier called it *certan*, their word for mainland.

Main Point-Davidsville NF This local service district (1984) is located on the east side of Gander Bay and on the north coast of the island of Newfoundland. **Main Point** was named after a nearby point. **Davidsville**, known as *Mann Point* until 1956, was named after the St David's Anglican Church.

Maisonnette NB A village (1986) northwest of Caraquet, it was noted on a survey map of 1755 as *Maisonette*, meaning 'little house'. Its post office was opened as *Mizonette* in 1885, became *Ste Jeanne d'Arc* in 1919, and changed to its present name in 1936.

Maitland NS On the west side of the mouth of the Shubenacadie River and west of Truro, it was named in 1832 by Judge Charles R. Fairbanks after Sir Peregrine Maitland (1777–1854), lieutenant-governor of Nova Scotia, 1828–34. It had been earlier called *Beaver River Settlement* and *Port Shubenacadie*. A second **Maitland**, between Mahone Bay and Bridgewater, was likely named for the lieutenant-governor as well. *See also* Port Maitland.

Maitland ON In the united counties of Leeds and Grenville, northeast of Brockville, this community was named in 1824, possibly by settler Ziba Marcus Phillips, after Sir Peregrine Maitland, lieutenant-governor of Upper Canada, 1818–28, who may have stopped here that year. The post office was opened in 1828.

Maitland River ON Rising in Wellington County, this river flows west to Goderich on Lake Huron. Earlier called *Red River*, and *Menesetung*, Ojibwa for 'healing waters' or 'windings', it was named by the Canada Company about 1822 after Sir Peregrine Maitland, lieutenant-governor of Upper Canada, 1818–28.

Majestic Peak BC In the Coast Mountains, southeast of Mount Waddington, this mountain (2911 m) was named in 1958 by Capt. Dr G.V.B. Cochran of the United States Air Force because he and his English climbing companions A. Morrison and J. Rucklidge were reminded of a characteristic of the Elizabethan era in the late 1500s.

Makinsons NF West of Conception Bay and inland from Brigus, this place began as a railway station in 1898. The land in the area had been purchased about 1863 by George Makinson of Harbour Grace. Previously it was known as *Goulds* and *Cochranedale*.

Makkovik NF On the Labrador coast, northeast of Happy Valley-Goose Bay, this place is located on **Makkovik Bay**. The origin of the name is uncertain, with the possibility that it was named after an Inuit family, or after a French trader, possibly Pierre Marcoux. It was founded in the mid-1800s, and was known among fishermen from the island of Newfoundland as *Flounders Bight*. It became a Moravian mission in 1896.

Malagash NS Located northwest of Tatamagouche, this place's name was derived from Mi'kmaq, either *malegawate*, possibly meaning 'mocking place', in reference to games played in the area, or *muligech*, 'milk', referring to the appearance of the water when it is stirred up. Its post office was established in 1855. Other places in the area with former post offices include **Malagash Point** (1877), **Upper Malagash** (1877), **Malagash Centre** (1901), **Malagash Station** (1937), and **Malagash Mine** (1938).

Malahide ON A township in Elgin County, it was named in 1810 after Malahide Castle, in the village of Malahide, County Dublin, Ireland, where Thomas Talbot was born. Col Talbot built his Castle Malahide at Port Talbot, in Dunwich Township.

Malakwa BC East of Shuswap Lake and northeast of Salmon Arm, the post office in this place was named in 1902 after the Chinook jargon word for 'mosquito'.

Malarctic QC A town (1939) in the regional county of Vallée-de-l'Or, west of the city of Val-d'Or, it was took its name from the township of Malarctic, proclaimed in 1916. The township was named after Anne-Joseph-Hippolyte de Maurès de Malarctic (1730–1800), who fought in the battles of both the Plains of Abraham (1759) and Sainte-Foy (1760).

Malaspina, Mount YT In the St. Elias Mountains, this mountain (3886 m) was named in 1890 by explorer and geologist Israel C. Russell after navigator Alexandro Malaspina (1754–1809). Italian-born Malaspina explored the northwestern coast of North America in 1791 on behalf of Spain.

Malaspina Strait BC Separating Texada Island from Powell River on the mainland, this channel was named in 1859 by Capt. George Richards after Capt. Alexandro Malaspina, an Italian, who explored the west coast of Vancouver Island in 1791 for Spain. **Malaspina Inlet** and **Malaspina Peninsula** are northwest of Powell River.

Malcolm Island BC Between Vancouver Island and the mainland, and at the head of Queen Charlotte Strait, this island was named in 1846 by Cdr George T. Gordon of HMS *Cormorant* after Adm. Sir Pulteney Malcolm (1758–1838), commander-in-chief of the St. Helena station in 1816–7 where Napoléon Bonaparte spent his last days.

Malden ON A township in Essex County, it was named in 1792 after the town of Maldon, in Essex, England, east of London. Fort Malden, named after the township, was constructed from 1797 to 1799 at Amherstburg.

Maligne Lake AB In Jasper National Park, southeast of Jasper, this lake took its name from the **Maligne River**, a tributary of the Athabasca. It had been called *Sore-foot Lake* by railway surveyor Henry MacLeod in 1875. The name of the river, given in 1846, means 'bad' in French. **Maligne Mountain**, east of the lake, was named in 1911.

Maliotenam QC Situated on the north shore of the St. Lawrence River, east of Sept-Îles, this Indian reserve was established in 1949. In Montagnais the name means 'village of Mary'.

Mallery Lake NT Located west of Baker Lake, this lake (467 km^2) was named in 1945 by surveyor Walter Stilwell after Mr and Mrs H.A. Mallery of Calgary, AB.

Mallorytown ON A community in the united counties of Leeds and Grenville, southwest of Brockville, it was founded in 1790 by Vermont native and Loyalist Daniel Mallory, and its post office was opened in 1852.

Malpeque Bay PE On the north side of the province, this bay was referred to as *Magpec* by Jacques Nicolas Bellin in 1744. Its name is derived from Mi'kmaq *makpaak*, 'big bay'. In 1765 Samuel Holland named it *Richmond Bay*, with *Malpeck Bay* as a secondary name, after Charles Lennox, 3rd Duke of Richmond (1734–1806), who served as the British ambassador to France. *Richmond Bay* continued to be used into the 20th century for the west side of the bay, and provided the name Richmond for a municipal community west of it. **Malpeque** is a place in the municipal community (1973) of **Malpeque Bay**, on the east side of the bay. It was originally called *Princetown*, after the original *Princetown Royalty,* which served as its post office from 1827 to 1945, but Malpeque was used for the place as early as 1775, and became the postal name in 1945. The municipal community of Malpeque Bay embraces several localities, including Indian River, Hamilton, Spring Valley, and Darnley.

Malton ON On the east side of the city of Mississauga, this place was named in the 1830s by Richard Halliday, after his birthplace in North Yorkshire, England, northeast of the city of York. Toronto-Lester B. Pearson International Airport (1983) was long known as Malton Airport, after it was opened in 1937.

Mamquam Mountain BC In the Coast Mountains, northeast of Squamish, this mountain (2595 m) was named by a group of climbers of the British Mountaineering Club in 1911 after the **Mamquam River**. The river was named by the Squamish after its smooth and murmuring characteristics in lower stretches.

Manicouagan QC The regional county, on the north shore of the St. Lawrence and centred on the city of Baie-Comeau, was named in 1981. It took its name from the **Rivière Manicouagan**, which drains the **Réservoir Manicouagan**, 210 km north of Baie-Comeau. In 1664 missionary Henri-Nouvel referred to a *grande rivière de Manicouaganistikou*, and the following year called it *rivière Manicoüagan*. In Montagnais the name means either 'where birch bark is gathered to repair canoes', or 'place where they get a drink'.

Manilla ON On the border of Victoria County and Durham Region, west of Lindsay, its post office was called *Mariposa*, after the township, in 1836. In 1851 it was renamed Manilla, after a village in Rush County, IN, which was probably named for the city of Manila in the Philippines.

Manitoba The name of the province was derived from Ojibwa *Manito-Bah* or Cree *Manito-Wapow*, 'the strait of the spirit', in reference to noisy sounds made by pebbles on a beach of **Manitoba Island**, in The Narrows of **Lake Manitoba**. It is less likely from Assiniboine *Mini-Tobow*, 'lake of the prairie', even though Pierre Gaultier de La Vérendrye called it *Lac des Prairies* in 1730.

Manitoba, Mount YT In the Centennial Range of the St. Elias Mountains, this mountain (3399 m) was named in 1967 after the province.

Manitou MB This village (1897) west of Morden was first called *Manitoba City*, but was renamed Manitou by the Canadian Pacific Railway in 1883, selecting a word meaning 'great spirit' in Ojibwa and other Algonquian languages. Its post office was called *Archibald* in 1879, and was changed to Manitou 10 years later.

Manitou, Lake ON A lake on Manitoulin Island, it was named after the great spirit of the Ojibwa and the Ottawa. It may be the world's largest lake on an island in a lake.

Manitoulin Island ON The largest island in a freshwater lake in the world, its name originated in the Ojibwa, Ottawa, and Algonquin concept of the great spirit Manitou, the master of life, the ruler of all things. The island is a sacred place for the First Nations.

Manitouwadge ON A township in Thunder Bay District, southeast of Geraldton, it was developed in the 1950s as a mining centre on the shores of **Manitouwadge Lake**. In Ojibwa the name of the lake means 'cave of the great spirit'.

Manitowaning ON On Manitoulin Island, south of Little Current, this place was founded in 1836 as a centre of the island's Aboriginal education. In Ojibwa and Ottawa, the name means 'home of the great spirit'.

Maniwaki QC A town (1957) in the regional county of La Vallée-de-Gatineau, it had been incorporated as a municipal township in 1904 and a village in 1930. Its post office was called *River Desert* from 1854 to 1875, when it was renamed Maniwaki. In Algonquin the name means 'place of Mary', in reference to the mission established among the Algonquin in 1849 by the Oblate fathers. The Maniwaki Indian Reserve was established in 1851 and its limits were set two years later.

Mankota SK This village (1941) southwest of Assiniboia, was established by early settlers from Mankato, MN, southwest of Minneapolis, and its post office was named in 1911. The Canadian Pacific Railway arrived here in 1928.

Mannheim ON In Waterloo Region, southwest of Kitchener, its post office was named in 1863 after Mannheim on the Rhine in Germany.

Manning AB Originally called *Aurora*, this town (1957) was named after Ernest Charles Manning (1908–96), premier of Alberta, 1943–68, and a member of the Senate, 1970–96.

Mannville AB This village (1906), between Vegreville and Vermilion, was named in 1905 by Winnipeg land agents Davidson and McRae after Donald D. (later Sir Donald) Mann (1853–1934), who was then vice-president of the Canadian Northern Railway.

Manor SK Northeast of Estevan, this village (1902) was named in 1900, when the Canadian Pacific Railway was built from Souris, MB, to Arcola. It took its name from *Cannington*

Manor, 16 km to the northwest, established in the 1880s by English capitalists, who dreamed of a gracious community on the frontier. Its post office was opened in 1901.

Manotick ON Partly on Long Island, in the Rideau River, and in Rideau Township, south of Ottawa, this community was named in 1864 by miller Moss Kent Dickinson, who devised the name from the Ojibwa words for 'long island'.

Manseau QC A village in the regional county of Bécancour, north of Victoriaville, it was named in 1922 after Fr Martial Manseau, who was the parish priest here from 1899 to 1907.

Mansel Island NT This island (3180 km^2) in the northeastern part of Hudson Bay was named *Mancel Island* in 1613 by explorer Thomas Button after Vice-Adm. Sir Robert Mansell (1573–1656), treasurer of the Navy. Button's wife was Sir Robert's niece. It was corrected to Mansel Island by the Geographic Board in 1908, after it had been shown as *Mansfield Island* on some British Admiralty charts.

Mansfield ON A community in Dufferin County, northeast of Shelburne, its post office was named in 1858 after the local Presbyterian church's glebe (field) and residence (manse).

Mansons Landing BC On Cortes Island, east of Campbell River, this place was named in 1941 after Michael Manson, a Shetlander who, with his brother John, operated a trading post here, 1887–95. In 1893 he was the first postmaster of *Cortez Island* at the same location.

Mansonville QC A community west of Lac Memphrémagog, near the Vermont border, its post office was named in 1837 after miller Robert Manson, a Scot, who settled here in 1811.

Manvers ON A township in Durham County when it was named in 1816, it was transferred to Victoria County in 1974. It was named for Charles Pierrepoint, Baron Pierrepoint (1737–1816), who was created Earl Manvers in 1806.

Maple ON In the city of Vaughan, York Region, this urban centre's post office was called *Rupert* in 1852, but it was changed in 1855 to Maple. From 1851 to 1904, its railway station was known as *Richmond Hill*, after the place on Yonge Street, 7 km to the east, now a town of 95,000 (1996).

Maple Creek SK A town (1903) southwest of Swift Current, it originated in 1882–3, when construction of the Canadian Pacific Railway east from Calgary was halted near here, with the crew remaining for the winter. Its post office was opened in 1883. **Maple Creek** rises in the Cypress Hills and flows north to Bigstick Lake. It was named after the fringe of box elder (Manitoba maple) trees along its banks.

Maple Grove QC Located in the regional county of Beauharnois-Salaberry, southwest of Montréal, this town was named in 1918 after the estate of Robert Howden Norval, who purchased the land in 1828 and named the estate about 1840. Maple Grove post office operated here from 1854 to 1954.

Maple Ridge BC A district municipality (1974) on the north side of the Fraser River, and east of Vancouver, its post office was named in 1876, but was closed 10 years later. It was restored in 1959 and replaced *Port Haney* in 1970.

Maquapit Lake NB On the west side of Grand Lake and east of Fredericton, this lake's name was derived from Maliseet *maquahpak*, 'red lake place'. Variously called *Quacopeck*, *Oquapo*, and *Maquako*, its present name was established by 1859 on John Wilkinson's map of the province.

Marathon ON When the Canadian Pacific Railway was built in 1883–4 along the shore of Lake Superior, the site of this town (1988) was called *Peninsula*, after a prominent point extending into Peninsula Harbour. In 1917 *Peninsula* post office was opened here. General

Timber Company of Port Arthur (Thunder Bay), a subsidiary of Marathon Paper Mills of Rothschild, WI, built a paper mill here in 1936. A request in 1944 to rename the post office Everest (after the company's president) was rejected by the post office department, but the parent company's name was accepted.

Marbleton QC In the regional county of Le Haut-Saint-François, northeast of Sherbrooke, the post office in this village (1895) was named in 1855 after a local marble quarry. A request in 1979 to rename it Saint-Adolphe-de-Marbleton was rejected because it would be confused with the nearby locality of Saint-Adolphe-de-Dudswell.

Marcelin SK Southwest of Prince Albert, this village (1911) was settled in 1902 by Antoine Marcelin. Five years later the first school was opened and named after him. When the Canadian Northern Railway arrived in 1912, he proposed St. Albert for the station, but was persuaded to accept Marcellin.

Marchand QC A municipality west of Mont-Tremblant, it was named in 1886 after the township of Marchand, which was named after 1880 and proclaimed in 1892. The township was called after Félix-Gabriel Marchand (1832–1900), who became the premier of Québec, 1897–1900.

Marconi, Mount BC In the Rocky Mountains, southeast of Invermere, this mountain (3105 m) was named in 1919 after Guglielmo Marconi (1874–1919), noted for his experiments with wireless telegraphy.

Margaree River NS On the west coast of Cape Breton Island, this river's name was shown on a 1685 French name as *R de Magre*, possibly derived from the name *Marguerite*, noted on later maps. It was also called *R au Saumon* and *Salmon R*. A post office was opened at Margaree in 1824, and other offices were opened later at **Margaree Forks** (1859), **Margaree Harbour** (1876), **South West Margaree** (1876), **East Margaree** (1877), **Margaree Island** (1894), **Margaree Ford** (1910), **North East Margaree** (1901), **Margaree Brook** (1930), **Margaree Valley** (1961), and **Margaree Centre** (1961).

Margaretsville NS On the Bay of Fundy shore and west of Kentville, this place was named in 1855 after Lady Margaret (Inglis) Halliburton, the wife of Chief Justice Sir Brenton Halliburton, who had a summer residence here. Lady Margaret was the eldest daughter of the Rt Revd Charles Inglis, bishop of the Church of England diocese of Nova Scotia, 1824–50. The place had earlier been called *Reagh's Cove* and *Pete's Point*.

Margate PE Northeast of Kensington, this place was named about 1790 by a man called Smith, after Margate, Kent, England. Its post office was open from 1869 to 1948.

Margo SK A village (1911) northwest of Canora, its post office was opened in 1904 and took its name from a nearby lake. The lake had been named after a farmer's daughter called Margot, who had had a tragic death. The Canadian Northern Railway was constructed through Margo the following year.

Maria QC Situated on the north shore of Chaleur Bay, west of New Richmond, the name of this municipality (1977) was noted as early as 1815. It was given after Lady Maria Howard, the wife of Sir Guy Carleton, Lord Dorchester, governor-in-chief of Quebec, 1768–78, and 1786–96, and after whom nearby Carleton was named. Maria was a township municipality from 1845 to 1977.

Maria-Chapdelaine QC A regional county north of Lac Saint-Jean, it was named in 1983 after the celebrated 1913 novel by Louis Hémon (1880–1913), who wrote about the difficult pioneer life in the area.

Mariapolis MB Located southeast of Brandon, this place was named in 1891 by French contractors building the Canadian Northern Railway line west from Winnipeg. They had first proposed Sainte-Marie, but substituted a name meaning 'St Mary's city'. Its post office also opened that year.

Marieville QC A town in the regional county of Rouville, east of Chambly, it traces it roots to the French seigneury of *Saint-Nom de Marie de Monnoir*, granted to Gov. Claude de Ramezay of Montréal. He was also granted the seigneury of Monnoir in the area of the town in 1708, and the religious parish of Sainte-Marie-de-Monnoir was established in 1801. The same name was given to the parish municipality in 1845. The town was incorporated in 1905, seven years after Marieville post office was opened.

Maringouin, Cape NB Extending into Chignecto Bay, southwest of Sackville, this cape's name was derived from the French for 'mosquito'. Jean-Baptiste-Louis Franquelin called it *C des maringouins* on his map of 1686, and *Mosquito Point* occurred in a 1704 report. However, most English-language maps and reports used variations on the French name, such as *Cape Marangouin*.

Marion Bridge NS Situated on the Mira River, south of Sydney, this place was given postal service in 1858. It may have been named after the builder of a bridge across the Mira.

Mariposa ON A township in Victoria County, it was named by Lt-Gov. Sir Peregrine Maitland after the Spanish word for 'butterfly'. Stephen Leacock's fictional town of Mariposa, in *Sunshine Sketches of a Little Town* (1912), is believed to have been modelled on Orillia.

Markdale ON A village (1888) in Grey County, southeast of Owen Sound, it was called *East Glenelg*, after Glenelg Township, in 1851. In 1864 it was renamed *Cornabuss*, after Cornabus, on Scotland's Isle of Islay. It was renamed again in 1873, when Mark Armstrong sold land for a railway, stipulating its station be named after him.

Markham ON Laid out as a township in 1792, it was named after the Rt Revd William Markham (1720–1806), archbishop of York, 1777–1806. Its largest urban centre was called Markham in 1828. When regional government was established in 1969, Markham Township became a town. With a population of 165,000 (1996) it is the province's most populous town.

Markstay ON In Sudbury District, east of Sudbury, this place was named in 1892 after Marks Tey, in Sussex, England, 18 km southeast of Sudbury, in Suffolk.

Marmora ON The township in Hastings County was named in 1820 after a huge marble rock on Crowe Lake, taking the Latin for 'marble'. It is part of the township municipality of **Marmora and Lake**. The post office at the village (1901) of **Marmora** was first called *Marmora Iron Works* in 1821 and was shortened to Marmora in 1823.

Marmot Towers BC In the Coast Mountains, northwest of Pemberton, this mountain (3124 m) was named in the 1950s by climbers of the Alpine Club of Canada. Earlier they had ascended its Desperation, Deviation, and Inspiration peaks, and at first used the name Desperation Towers for the mountain. They renamed it after the marmots encountered during an ascent.

Marsden SK West of Battleford, the area of this village (1931) was first known as *Manitou*, after a lake to the south. In 1907 Marsden post office was opened and named after the birthplace in England of postmaster Alex Wright's wife. Marsden is in West Yorkshire, southwest of Bradford. The Canadian Pacific Railway adopted the name in 1923.

Marshall SK Southeast of Lloydminster, the post office at this village (1914) was called *Stringer* in 1904, after homesteader Lewis Stringer. When the Canadian Northern Railway arrived in 1905, its station was named Marshall, and the postal name was changed the following year. The reason for choosing this name is not known.

Marshalltown NS This place southwest of Digby was named after pioneer settler Anthony Marshall. The post office, called *Marshall Town* in 1856, was closed in 1951.

Marshfield PE A community located northeast of Charlottetown, its school district was named about 1861 after the farm of Robert Poore Haythorne (1815–91). Haythorne had been a magistrate in Marshfield, Somerset, England, before immigrating to the province in 1842, and buying 4047 ha (10,000 a) on the north side of the Hillsborough River. He was the premier of Prince Edward Island, 1869–70 and 1872–3, and a senator, 1873–91.

Marsh Lake YT A widening of the Yukon River, southeast of Whitehorse, it was first called *Mud Lake* by the early miners. In 1883 American Army officer Lt Frederick Schwatka named it after paleontologist Othniel Charles Marsh (1831–99). He was a professor of vertebrate paleontology at Yale University, 1866–99, and honorary curator of vertebrate paleontology at the United States National Museum, 1887–99.

Marsoui QC This village (1950) is located on the south shore of the St. Lawrence, northwest of Gaspé. Noted on maps as early as 1755, the name was derived from Mi'kmaq *malseoui*, meaning 'flint', which they used to produce tools and ignitors.

Martensville SK Located just north of Saskatoon, this town (1969) had its beginnings when Isaac Martens and his son Dave subdivided their farm into building lots in the 1960s. Its post office was established in 1963.

Martha Black, Mount YT This mountain (2496 m) in the St. Elias Mountains, south of Haines Junction, was named in 1945 by geologist Hugh Bostock after Martha Munger Black. The wife of Yukon commissioner George Black, she replaced him as the Yukon member of Parliament, 1935–40, while he was ill.

Martinon NB In the city of Saint John, west of the city centre, this neighbourhood was named in 1902 for a Canadian Pacific Railway station by naturalist William F. Ganong after Martin d'Aprendestiguy, Sieur de Martignon (*c.* 1616–*c.* 1689). A son-in-law of Gov. Charles de Saint-Étienne de la Tour, he had a seigneury here about 1660–2.

Martins Point NS On the western shore of Mahone Bay, this place's post office was opened in 1876. The adjacent Birch Point had been settled by Martin Westhaver in the late 1700s.

Martintown ON In the united counties of Stormont, Dundas and Glengarry, northeast of Cornwall, this community was first called *MacMartin Mills*, after miller Malcolm MacMartin. It was renamed Martintown in 1824.

Martre, Lac la NT This lake (1687 km^2) was described as both *Martin Lake* and *Marten Lake* by Alexander (later Sir Alexander) Mackenzie in 1801, and *Great Marten Lake* by John (later Sir John) Franklin in the journal of his trip to the polar seas, 1819–22. It appeared on various maps before 1900 as Lac la Martre, the French form of 'marten lake', and may have resulted from the writings of Fr Émile Petitot. The lake is drained into the North Arm of Great Slave Lake by **Rivière la Martre**. *See also* Wha Ti.

Marwayne AB A village (1952) northwest of Lloydminster, it was named in 1906 after W. Creasey Marfleet's farm. He arrived here in 1901, and named the farm after his family name, and after his hometown, Wainfleet All Saints, Lincolnshire, England, near the North Sea coast.

Maryborough ON A township in Wellington County, it was named in 1840 after William Wellesley-Pole, 1st Baron Maryborough (1763–1845), a brother of the Duke of Wellington. His daughter, Mary Charlotte Anne, was the wife of Sir Charles Bagot, governor of the province of Canada, 1841–3. The title was from the Irish town of Maryborough (now Port Laoise) in County Queens (now County Laois), Ireland.

Maryfield SK Southeast of Moosomin and near the Manitoba border, this village (1907) was named after the school district, which had been created in 1883 and called after John Young McNaught's farm. When he had left Scotland, he had promised his sister Mary that he would name his farm after her. He had moved to Harrison Hot Springs, BC, in 1881, but provided $10 for the building of the school. Postal officials adopted the name in 1896 and the Canadian Northern Railway accepted it in 1907.

Mary Henry, Mount BC In the Muskwa Ranges of the Rocky Mountains, this mountain (2972 m) was named in 1931 by surveyor K.F. McCusker after botanist Mary Gibson Henry, who led an expedition to the area that year.

Maryhill ON In Waterloo Region, east of Waterloo, it was called *Freiburg* from 1851 to 1941, when it was renamed after the Virgin Mary, whose shrine was on the nearby hill.

Mary March's Brook NF Rising west of Badger, this brook flows south into the northeast end of Red Indian Lake. It was named after Demasduit, called Mary March, the wife of Beothuk chief Nonosbawsut. She had been captured, and died in captivity 8 January 1820.

Marystown NF A town (1951) on the east side of the Burin Peninsula, its present area was known as *Mortier Bay* until the first of the twentieth century. The name first occurred in the 1845 census, and was possibly given in honour of the Virgin Mary.

Marysvale NF On the southwest side of Conception Bay, this place was known as *Turks Gut* until 1919, when it was renamed Marysvale. The cove here retained the name Turks Gut, and a nearby pond is still called Turks Gut Long Pond. Presumably the gut was called after Barbary pirates who plundered Newfoundland's coasts in the seventeenth century.

Marysville NB In the city of Fredericton, northeast of the city centre, it was an incorporated town from 1886 to 1973. Its post office (1868–1964) was named by lumberman Alexander Gibson after his wife, Mary.

Marysville ON On Wolfe Island and in Frontenac County, this community was named in the 1820s for Mary Hitchcock, who was the postmistress of Wolfe Island from 1845 to 1877. Marysville was officially accepted as the community name in 1977, while maintaining Wolfe Island as the postal name. There is another **Marysville** in Hastings County, 55 km to the west, which was named in 1851 after the local Catholic parish of St. Mary's.

Mary Vaux, Mount AB This mountain (3027 m) near Maligne Lake was named in 1908 by climber Mary Schaffer after Mary Vaux Walcott (1870–1940) of Philadelphia, who performed extensive glaciological studies in the mountains of Canada.

Mascouche QC A town (1971) on the **Rivière Mascouche**, north of Laval, it had been a parish municipality from the mid-1800s. The river's name, which occurred in records as early as 1772, means 'little bear' in Algonquin.

Maskinongé QC A village (1931) in the regional county of **Maskinongé**, midway between Joliette and Trois-Rivières, it took its name from the parish municipality of *Saint-Joseph-de-Maskinongé*, created in 1845. In the early 1700s, the religious parish of Saint-Joseph-de-Maskinongé was formed here. The regional county was established in 1982. The river's name, which occurred as early as 1672 as *Masquinongé* in reference to a seigneury, means 'big pike'.

Masset BC A village (1961) at the north end of Graham Island, the largest of the Queen Charlotte Islands, its post office was named in 1909 after a Haida village on the shore of **Masset Harbour**, 3 km to the northwest. Two years earlier it had been named *Graham City* after Benjamin Graham, the president of Graham Steamship, Coal and Lumber Company. After adopting the name Masset, the older place became *Old Masset*, and later was named Haida. The Haida believe the name Masset came from the burial of a white officer called Massetta on a small island, which they called *Mahsht* after him. Geologist George M. Dawson recorded the island's name in 1878 as *Maast*.

Massey ON A town (1904) in Sudbury District, west of Espanola, its post office was named *Massey Station* in 1889, after Hart Almerrin Massey (1823–96), founder of the Massey Manufacturing Company in Newcastle in 1870, which was relocated to Toronto nine years later. The postal name was shortened to Massey in 1939.

Massey Island NT One of the Parry Islands in the Queen Elizabeth Islands, this island (423 km^2) was named in 1961 after Charles Vincent Massey (1887–1967), Canada's first native-born governor general, 1952–9.

Masson-Angers QC A town (1992) in the Communauté urbaine de l'Outaouais, southwest of Buckingham, it was named after consulting the residents of the town of Masson, which included the area of Angers. From 1975 to 1980, the two had been part of the town of Buckingham. Masson had been incorporated as a village in 1897 and a town in 1966. The township of Masson was proclaimed in 1894 and named after Louis-François-Rodrigue (or Roderick) Masson (1833–1903), member of the House of Commons, 1876–82, and lieutenant-governor of Québec, 1884–7. The **Angers** post office was named in 1869 after the birthplace in western France of Fr Eugène Trinquier, first parish priest of L'Ange-Gardien. Angers became a village in 1915.

Massueville QC Located southeast of Sorel, this village (1903) was named after its principal benefactor, Gaspard-Aimé Massue (1812–75) who donated land for a church, a convent, and a college.

Matachewan ON A township in Timiskaming District, southwest of Kirkland Lake, it was the site of the Hudson's Bay Company's Fort Matachewan trading post in 1850. Matachewan post office was opened in 1935. In Ojibwa, the word *matadjiwan* means 'meeting of the currents'.

Matagami QC Situated just south of **Lac Matagami** and in northwestern Québec, this town (1963) was created a village in 1961 to exploit the nearby rich reserves of copper and zinc. The lake was named in the late 1800s after the Algonquin for 'junction of waters', in reference to the Bell, the Waswanipi and the Allard rivers all flowing into it. The present Rivière Harricanaw was identified on a 1703 map as *R. Matagami*, although the lake is drained north to James Bay by the Rivière Nottaway.

Matane QC Located on the south shore of the St. Lawrence, northeast of Rimouski, this town (1937) was created the village of *Saint-Jérôme-de-Matane* in 1893. In 1603, Samuel de Champlain identified the **Rivière Matane** as *Mantanne*, derived from the Mi'kmaq *mtctan*, 'beaver pond'. The regional county of **Matane** was established in 1982.

Matapédia, Rivière QC Rising in **Lac Matapédia**, south of Matane, it flows 70 m south to join the Restigouche, west of Campbellton, NB. The name in Mi'kmaq has the sense of 'river that forks'.

Matawinie QC A regional county northwest of Joliette, it was created in 1982. It is centred on the **Rivière Matawin**, which in Algonquin means 'meeting of the waters'.

Matheson ON A township in Cochrane District, it was named in 1903 for Arthur James Matheson (d. 1913), member for Lanark South in the Legislative Assembly of Ontario, 1895–1913, and provincial treasurer, 1905–13. It forms part of the municipal township of **Black River-Matheson**. The post office in the community of Matheson was first named *McDougall Chute* in 1907, renamed *Matheson Station* in 1911, and changed to Matheson in 1949.

Matilda ON A township in the united counties of Stormont, Dundas and Glengarry, it was named in 1787 for Princess Charlotte Augusta Matilda (1766–1828), the eldest daughter of George III.

Matsqui BC A community in the north side of the city of Abbotsford, its post office was named in 1903. Until 1995 it was part of the district municipality of Matsqui, which had been created in 1892. **Matsqui**, **Matsqui Prairie**, and **Matsqui Island** in the Fraser River take their names from a Halkomelem word for 'easy portage', possibly referring to the ease of travelling overland to Sumas Lake (now dry), or to the Nooksak River, in present Washington State.

Mattagami River ON Rising in **Mattagami Lake**, at Gogama, it flows north through Timmins and Smooth Rock Falls to join the Moose River, upriver from Moosonee. The lake's two long arms provide the meaning of the name in Ojibwa and Cree, 'forked lake'.

Mattawa ON This town (1892) is in Nipissing District, and at the point where the **Mattawa River** flows into the Ottawa. The river's name in Algonquin and Ojibwa means 'meeting of waters'. The North West Company founded Mattawa House here in 1784, and it continued as a Hudson's Bay Company post from 1821 to 1915.

Mattice ON On the Missinaibi River and in Cochrane District, east of Hearst, this place was first called *Missinaibi*. With the coming of the railway in 1915, the post office was called Mattice, possibly after a member of an Eastern Ontario family. Since 1983, it has been part of the municipal township of **Mattice–Val-Coté**.

Matty Island NT Between Boothia Peninsula and King William Island, this island (477 km^2) was named on 23 May 1830 by James Clark Ross after the family that donated the expedition's silk colours. The colours were flown that day in honour of the first anniversary of the expedition's departure from London.

Maugerville NB A community on the north side of the Saint John River and opposite Oromocto, its post office was open from 1800 to 1812, and 1854 to 1966. It was named after **Maugerville Parish**, created in 1786. It took its name from Maugerville Township, established in 1765, and named after Halifax merchant Joshua Mauger (1725–88). Appointed Nova Scotia's agent in England in 1762, his name was the first on the list of grantees of the township.

Maurelle Island BC Between Vancouver Island and the mainland, northeast of Campbell River, this island was named in 1903 by Capt. John Walbran after Francisco Antonio Maurelle, the second-in-command of the *Sonora* in 1775, when Juan Francisco de la Bodega y Quadra undertook an exploratory voyage along the northwest coast. Until 1903 the island was considered part of *Valdes Island*, now divided among Maurelle, Sonora, and Quadra islands.

Maxville ON A village (1891) in the united counties of Stormont, Dundas and Glengarry, north of Cornwall, its post office was called *Macs Corners* from 1847 to 1852, after the many people with surnames beginning with Mac or Mc. When the railway arrived in 1880, postmaster John McEwan reopened the office as Maxville.

Mayerthorpe AB A town (1961) northwest of Edmonton, its post office was named in 1910 after first postmaster R.I. Mayer and a schoolteacher called Thorpe.

Maymont SK Southeast of North Battleford, the Canadian Northern Railway station and the post office at this village (1907) were named in 1905, because the construction of the railway was done in May, and it was the most elevated place of the line on the north side of the North Saskatchewan River, between the crossing northwest of Saskatoon and North Battleford.

Mayne Island BC One of the Gulf Islands, east of Saltspring Island, it was named about 1859 by Capt. George Richards after Lt (later Rear-Adm.) Richard Charles Mayne (d. 1892), who served on Richards's survey vessels, 1857–61.

Maynooth ON North of Bancroft in Hastings County, this place was called *Doyle's Corners*, *Tara*, and *Oxenden*, before it was named Maynooth in 1863. In Ireland, Maynooth is in County Kildare, west of Dublin.

Mayo QC A municipality in the regional county of Papineau, northeast of Buckingham, it was created in 1954, having been called the parish municipality of *Saint-Malachy* 100 years before. Mayo post office was open from 1866 to 1926.

Mayo YT A village (1984) on the Yukon River, southeast of Dawson, it was founded in 1902–3, and called after nearby **Mayo Lake**. The lake was named in 1887 by prospector Alexander MacDonald, after trader and steamboat captain Alfred S. Mayo (d. 1924).

McAdam NB A village (1966) southwest of Fredericton and near the Maine border, it was established in 1869 as a station on the St. Andrews and Quebec Railway. When a railway was built in 1870 across Maine to Saint John, this place's post office opened as *McAdam Junction*, and it was shortened to McAdam in 1940. It was named after lumberman John McAdam (1807–93), a member of the provincial legislature, 1854–66 and 1882–6, and a member of the House of Commons, 1872–4.

McArthur, Mount BC In Yoho National Park, north of Field, this mountain (3015 m) was named in 1886 by astronomer Otto Klotz after surveyor James Joseph McArthur (1856–1925), who surveyed the forty-ninth parallel through the Rocky Mountains in 1887, and undertook several other surveys. **Lake McArthur**, southeast of Field, was discovered and named by him.

McArthur Peak YT This peak (4344 m) in the St. Elias Mountains, and north of Mount Logan, was named in 1918 after James Joseph McArthur (1856–1925). He was the leader of the Canadian surveyors of the Alaska-British Columbia-Yukon boundary survey, 1901–24 and commissioner of the Yukon, 1917–24.

McBride BC A village (1932) on the Fraser River, southeast of Prince George, it was named in 1913 by the Grand Trunk Pacific Railway after Sir Richard McBride (1870–1917), who became premier of the province in 1903, at the age of 33.

McCoubrey, Mount BC In the Purcell Mountains, west of Invermere, this mountain (3216 m) was named *McCoubrey Mountain* in 1918 by Alexander Addison McCoubrey (1885–1942) of Winnipeg. A location engineer for the Canadian Pacific Railway, he made several ascents in the Purcells. It was earlier called *Peacock Mountain* on an 1915 map because pools of water on a glacier reflected like the eyes of a spreading peacock's tail. McCoubrey had also proposed Silverhorn or Silverspire sfor it. It was renamed Mount McCoubrey in 1960.

McCoubrey, Mount YT In the St. Elias Mountains, this mountain (3124 m) was named in 1970 by the Alpine Club of Canada after Alexander Addison McCoubrey, a surveyor and president of the club, 1932–4.

McCreary MB Located east of Riding Mountain National Park, the post office at this village (1964) was first called *Chamberlain* in August 1897. It was renamed *Elliott Station* two months later after homesteader George Elliott, but, because of similar names in Canada, it was changed to McCreary in 1899. Its Canadian Northern Railway station had been named in 1897 after William Forsythe McCreary, member for Selkirk in the House of Commons. He promoted the settlement of Ruthenians and Doukhobors in the area.

McDonalds Corners ON In Lanark County, west of Perth, its post office was named in 1853, a year after Alexander McDonald had settled here.

McGillivray ON A township in Middlesex County, it was named in 1830 after Simon McGillivray (1783–1840), a merchant in London, England and a director of the Canada Company, a British company set up in 1826 to develop the Huron Tract, a large area of what is now Southwestern Ontario. He promoted the merger of the North West Company with the Hudson's Bay Company in 1821.

McGregor ON In Essex County, south of Windsor, this community was named in 1881, probably for William McGregor (1836–1903), Liberal member of Parliament, 1874–8, and 1891–1900.

McKellar ON A township in Parry Sound District, it was named in 1869 after Archibald McKellar (1816–94), member of the Legislative Assembly of the province of Canada, 1857–67, and of the Ontario Legislature, 1867–75. **McKellar** post office was opened in the township in 1870.

McKerrow ON In Sudbury District, north of Espanola, this place was named *Espanola Station* in 1908. The post office was renamed in 1932 after Abitibi Pulp and Paper traffic manager, Jack McKerrow.

McKillop ON A township in Huron County, it was named in 1830 after James Mackillop, a director of the Canada Company, a British company established in 1826 to develop the Huron Tract, a large area of what is now Southwestern Ontario.

McLean SK East of Regina, the Canadian Pacific Railway station at this village (1966) was named about 1883. Its post office was opened on 1 April 1884. A claim that it was named for a Hudson's Bay Company post manager at Fort Pitt, north of Lloydminster, whose family had been taken prisoner in 1885 by the Cree, cannot be verified.

McLennan AB Northwest of Lesser Slave Lake, this town (1948) was named in 1915 after Dr J.K. McLennan, then secretary of the Edmonton, Dunvegan and British Columbia Railway, and later its vice-president. Its post office was also opened in 1915.

McLeod Lake BC Located north of Prince George, this lake was named about 1806 by Simon Fraser after chief trader Archibald Norman McLeod (d. 1839) of the North West Company. Previously the lake was called *Trout Lake*, where trader James McDougall had built Trout Lake House the previous year. The post office was first called *Fort McLeod* here in June 1952 and became **McLeod Lake** two months later to avoid confusing the place with Fort Macleod in Alberta.

M'Clintock Channel NT Between Victoria and Prince of Wales islands, this 400-km channel was named *Lady Franklin Channel* in 1859 by Capt. Allen Young. Lady Jane Franklin, the widow of Sir John Franklin, requested the naming of the channel after Francis (Sir Francis in 1860) Leopold McClintock. In 1859 McClintock had discovered the fate of Franklin's expedition in the area of King William Island. Long spelled 'M'Clintock' on maps, its name was confirmed by the Canadian Board on Geographical Names in 1950.

McLure BC On the east bank of the North Thompson River, north of Kamloops, this place's post office was named in 1920 after John McLure (d. 1933), who arrived here in 1906.

M'Clure Strait NT Separating Banks Island on the southwest, and Melville, Eglinton, and Prince Patrick islands on the northeast, this 300-km strait was named after Robert John Le Mesurier McClure (1807–73). McClure explored the islands and waters of the western Arctic extensively in 1850–1, but his vessel *Investigator* became frozen in ice on the Banks Island side of the strait, from 1851 to 1853. Spelled 'M'Clure' on many navigation charts and maps, the name was confirmed by the Canadian Board on Geographical Names in 1950.

McMasterville QC Located in the regional county of La Vallée-du-Richelieu and on the west bank of the Richelieu, east of Montréal, this village was created in 1917. It was named after William McMaster, the first president of Canadian Explosives, which became Canadian Industries Limited (C.I.L.) in 1927.

McNab ON A township in Renfrew County, it was named in 1825 after Archibald McNab (1775–1860), who was granted the township by the government of Upper Canada. Bitter strife with the tenants he had brought from Scotland in 1823 resulted in the government buying him out in 1842. After living for 10 years in Hamilton, he returned to Scotland to manage an estate in the Orkneys.

McNabs Island NS Situated in the entrance to Halifax Harbour, this island was called *Île de Chibouqueto* during the French regime. In 1749 it was named in honour of the Revd Frederick Cornwallis (1713–83), who was appointed the archbishop of Canterbury in 1768. The island was purchased by John McNab in 1783, and thereafter it was known by his name.

McNair, Mount NB South of Mount Carleton, this mountain was named in 1969 by director of surveys Willis F. Roberts after John Babbit McNair (1889–1968), premier, 1940–52, chief justice, 1952–65, and lieutenant-governor of New Brunswick, 1965–8.

McNaughton Lake NT Northwest of Baker Lake, this lake was named in 1969 by the Army Survey Establishment after Gen. Andrew George Latta McNaughton (1887–1966), commander of the Canadian forces in Britain, 1939–43, and chairman of the Canadian section of the International Joint Commission, 1950–62.

McWatters QC A municipality east of Rouyn-Noranda, it was established in 1981, having been called *Kinojévis* in 1979. McWatters Gold Mine was opened in 1932, named after prospector David McWatters, who had made its discovery two years earlier. The post office was called *McWaters* from 1936 to 1986, when it was corrected to McWatters.

Meadowbank PE A municipal community (1974) southwest of Charlottetown and on the north bank of the West River, it was named after the school district of *Meadow Bank*, established about 1877.

Meadow Lake SK Located on the west shore of **Meadow Lake**, this town (1936) was first settled in 1910, and its post office was opened 1 January 1911. Its development had been slow until the arrival of the Canadian Pacific Railway in 1936. The lake had been known to French traders as *Lac des Prairies*.

Meadows NF On the north side of Humber Arm, northwest of Corner Brook, this municipal community (1970) was established at **Meadows Point** in the late 1800s. The point may have been named because of its grassy cover, when it was encountered by the first explorers or hydrographic surveyors.

Meaford ON A town (1874) in Grey County, it was named in 1858 after Meaford Hall, the birthplace in Staffordshire, England, of John Jervis, Earl of St. Vincent (1735–1823). He defeated superior Spanish forces in 1797 off Cabo de São Vincente, Portugal. Earlier the place had been called *Peggy's Landing*, *Stephenson's Landing*, and *Purdytown*.

Meaghers Grant NS In the Musquodoboit River valley, northeast of Dartmouth, this place was named after Martin Meagher, a Loyalist from North Carolina, who received a 2024 ha (5000 a) grant on 7 June 1783. Its post office was opened in 1856.

Meath Park SK Northeast of Prince Albert, this village (1938) was developed 5 km to the south at Janow Corners, before 1932, when the Canadian Pacific Railway arrived at the present site. Its post office was opened in 1913, and was possibly named after County Meath in Ireland.

Medenagan Mountain BC In the Purcell Mountains, west of Invermere, this mountain (3295 m) was named *Meden-Agan Mountain* in 1971 by climber Curt Wagner after the message recorded at the oracle of Delphi and meaning 'nothing too much'. He named it after climbing Mount Delphine, 6 km to the southeast. The spelling was altered to Medenagan in 1973.

Medicine Hat AB This city (1906) was named in 1882 by Cpl Walter Johnson of the North-West Mounted Police. Johnson may have simply translated the Blackfoot *Saamis*, 'head-dress of a medicine man', in reference to the shape of a small hill. However, at least 13 legends relating to the naming of the place have evolved over the years. One story frequently told claims a Blackfoot medicine man was killed during a battle with the Cree, with his war bonnet floating down the South Saskatchewan, after which the Blackfoot fled. Other tales claim it was a Cree medicine man's hat, with the Cree fleeing. When some suggestions were made in 1910 to give the city a more prosaic name, such as Gasburg, or Smithville, the calamity was stifled after famous British writer Rudyard Kipling sharply rebuked the thought of such insolence.

Medonte ON Now part of the municipal township of **Oro-Medonte**, Simcoe County, it was created as a township in 1822. It took its name from Ojibwa *nin mâdondan*, 'I carry it away on my back on a portage-strap', referring to a portage route from Orillia to Georgian Bay.

Medstead SK North of North Battleford, a post office was opened near this village (1931) in 1911. Sylvester Perry, the first postmaster wanted to call it Medford, after Medford, WI, northwest of Wausau. Believing it could be confused with Melfort, postal authorities modified it to Medstead, and Perry accepted it. The Canadian Northern Railway built its line through Medstead in 1927.

Meductic NB A village (1966) on the south side of the Saint John River, and southeast of Woodstock, its post office was named in 1896 after the former Maliseet village of *Medoctec*. The name means 'the end', referring to the end of a portage from the Eel River to the Saint John River. Its former post office was called *Canterbury* from 1860 to 1897. The straight stretch of Mactaquac Lake here is called **Meductic Reach**. Before the lake was created in 1970, the rapids at the village was called *Meductic Falls*. *See also* Canterbury.

Meduxnekeag River NB Rising in the state of Maine, this river flows into the Saint John River at Woodstock. The name was derived from Maliseet *medukseneekik*, 'rough at the mouth'. After 1689 several different spellings were used until the Geographic Board confirmed the present spelling in 1901.

Meech, Lac QC Arguably the best known lake in the province, because of the ill-fated Meech Lake Accord (1987), it is located northwest of Hull. Congregational minister Asa Meech (1775–1849), a native of Brockton, MA, settled here in 1822. Following an 1870 survey, the lake's name was misspelled *Meach*, and this was confirmed in 1931, based on long-established use on maps. The spelling was corrected in 1982 by the Commission de toponymie du Québec.

Meelpaeg Lake NF In the south central part of the island of Newfoundland, this lake's name was derived from Mi'kmaq *Makpaq*, 'lake of many bays'.

Meighen Island NT In the Sverdrup Islands of the Queen Elizabeth Islands, this island (955 km^2) was named in 1916 by Vilhjalmur Stefansson after Arthur Meighen (1874–1960), who was then solicitor general in Sir Robert Borden's cabinet. He followed Sir Robert as the prime minister in 1920. Defeated the next year, he returned briefly as prime minister in 1926.

Mékinac QC A regional county, north of Shawinigan, it was named in 1982 after 15 places associated with the name, including **Lac Mékinac** and **Saint-Roch-de-Mékinac**. It means 'turtle' in Algonquin.

Melancthon ON A township in Dufferin County, it was named in 1821 after the Protestant reformer Philip Melanchthon (1497–1560), Martin Luther's co-worker. His German surname had been Schwarzerd, which an uncle translated into Greek *melan chthon*, 'black earth'.

Melbourne ON A community in Ekfrid Township, Middlesex County, southwest of London, it was called *Ekfrid* from 1832 to 1857, when it was changed to *Longwood*. That name was changed to *Wendigo* in 1882, and changed again five years later to Melbourne. It was likely named after William Lamb, 2nd Viscount Melbourne (1779–1848), British prime minister, 1834–41. Coincidentally, miller Walter Melburne lived in nearby Appin in the mid-1800s.

Melbourne QC A village midway between Sherbrooke and Drummondville, it was separated in 1860 from the parish municipality of Melbourne, erected five years earlier. Its post office was opened about 1834. The name of the township of Melbourne, given in 1795, and proclaimed in 1802, may reflect one of two English places, one in Hampshire, the other in Derbyshire.

Melfort SK The post office in this city (1980) was named in 1892 by Mrs Reginald Beatty, after the Campbell family home of Melfort House, at the head of Melfort Loch, 15 km south of Oban in Argyllshire, Scotland. The Canadian Northern Railway was built through it in 1909.

Melita MB A town (1906) in the Souris River valley, southwest of Brandon, Manchester was first proposed in 1884 for its post office, but that was refused because of duplication. The community was then inspired by the text of Acts 28:1, 'When we were safe on land, we found that the island was called Melita', in reference to Malta in the Mediterranean. The hardships of Paul's travels likely resonated in the difficulties faced by the settlers in carving out a living in the wilderness.

Melocheville QC A village in the regional county of Beauharnois-Salaberry, southwest of Montréal, it was called *Lac-Saint-Louis* from 1919 to 1953, when it was renamed after its post office, given in 1863, after pioneer merchant Joseph Meloche.

Melrose NF South of Catalina, this municipal community (1971) had been incorporated as a village in 1968. It was known as *Ragged Harbour* until 1904 when it was renamed, likely after Melrose, Roxburghshire, Scotland, near the birthplace of William MacGregor (1846–1919), governor of Newfoundland (1904–9).

Melville SK Founded in 1907–8, this city (1960) was established as a divisional point on the Grand Trunk Pacific Railway. It was named after Charles Melville Hays (1856–1912), the railway's general manager, 1896–1909, and its president, from 1909 until 16 May 1912, when he went down on the *Titanic*. Its post office was opened in 1908.

Melville Island NT The largest of the Parry Islands of the Queen Elizabeth Islands, this island (42,149 km^2) was named in 1819 by William (later Sir William) Parry after Robert Dundas, 2nd Viscount Melville (1771–1851), then first lord of the Admiralty. Parry named **Melville Peninsula**, which separates Foxe Basin from the Gulf of Boothia, in 1822. **Melville Sound**, south of Victoria Island and between the mainland and Kent Peninsula, was named in 1821 by John (later Sir John) Franklin.

Melville Lake NF The upper part of Hamilton Inlet, and east of Happy Valley-Goose Bay, this body of water was named after Henry Dundas, 1st Viscount Melville (1742–1811), first lord of the Admiralty 1804–5.

Memphrémagog QC A regional county centred on the town of Magog, it was named in 1982 after **Lac Memphrémagog**. The name of the lake, which extends for 44 km from Newport, VT, to Magog, is derived from Abenaki, and means 'at the great stretch of water'.

Memramcook River NB Rising east of Moncton, this river flows into Shepody Bay, west of Sackville. The name is derived from Mi'kmaq *Amlamkook*, 'variegated', or 'all spotted yellow', in reference to the rock with variegated colours at Cape Maringouin. The community of **Memramcook**, midway between Moncton and Sackville, was identified in 1755 as *Mameramecou* on a French map. Its post office was opened in 1845.

Meota SK Northwest of North Battleford, this village (1911) is located on the west shore of Jackfish Lake. In 1894 Arthur Mannix proposed Meotate from the Cree *mewasinota*, meaning 'good place to camp', or 'it is good here', for a post office near the present village. Postal officials shortened it to Meota. It was closed in 1897, and reopened in 1904 at the present-day Prince, 6 km to the southeast. After the Canadian Northern Railway was built through the area in 1910, Meota was restored the next year near its former location.

Mercier QC Located in the regional county of Roussillon, south of Châteauguay, this town was named in 1968, having been erected as the town of *Sainte-Philomène* four years earlier. *Sainte-Philomène* was the name of the parish municipality and the post office from the mid-1800s, but it was dropped because of its lack of commercial appeal, and because church authorities had declared that there was no certainty the saint ever existed. The new name reflected the nearby bridge crossing the St. Lawrence, which had been named in honour of Honoré Mercier (1840–94), premier of Québec, 1887–91.

Merigomish NS On **Merigomish Harbour** and east of New Glasgow, this place's name was derived from Mi'kmaq *malegomich*, possibly meaning 'merrymaking place'. Postal service was established here in 1838.

Merlin ON In Kent County, south of Chatham, this community was named *Smith's Corners* in 1855, after a local family. It was changed in 1868 to Merlin, after a place near Edinburgh, Scotland.

Mermaid PE East of Charlottetown and on the east bank of the Hillsborough River, this place's post office was named *Mermaid Farm* in 1870 by first postmaster John Farquharson. He named his farm after HMS *Mermaid*, which was commanded by Capt. William Johnston at the mouth of the river in 1764–5.

Merrickville ON A Loyalist from Massachusetts, William Merrick (1760–1844), settled on the Rideau River in 1793, and built mills at what became *Merrick's Mills*. Merrickville post office was opened in 1829 and the village was incorporated in 1860.

Merritt BC Located on the Nicola River, southwest of Kamloops, this city (1911) was laid out in 1906 by the Nicola, Kamloops and Similkameen Railway, and named after William Hamilton Merritt Jr, one of its promoters. Earlier names for the place were *Forksdale* and *Diamond Vale*. Its post office was opened in 1907.

Merritton ON Within St. Catharines since 1961, this urban centre was named after William Hamilton Merritt (1793–1862), an owner of the Welland Canal Loan Company, who had been a major promoter of a canal joining Lake Ontario to Lake Erie. It was called *Welland City* in 1829, but it was exchanged with *Merrittsville*, now the city of Welland, in 1858.

Mersea ON A township in Essex County, it was named in 1792 after Mersea Island, on the east coast of Essex, England.

Mersey River NS Draining several lakes in south central Nova Scotia, this river flows into the Atlantic at Liverpool. It was called *Liverpool River* until 1937, when it was renamed after the River Mersey in England, which empties into the Irish Sea at Liverpool.

Merville BC On Vancouver Island, between Courtenay and Campbell River, this place's post office was named in 1919 by returned soldiers from the First World War. They gave it after Merville, France, near the Belgian border, the site of Canada's first field headquarters during the war.

Mervin SK Northwest of North Battleford, the post office in this village (1920) was named in 1908 by postmaster Archie Gemmell, after his son Mervin. The Canadian Northern Railway arrived here in 1912.

Mesilinka River BC Rising west of Williston Lake, this river flows east into the Omineca Arm of the lake. In Sekani the word means 'stranger'.

Messines QC A municipality in the regional county of La Vallée-de-la-Gatineau, south of Maniwaki, it was named *Messine* in 1921 in honour of distinguished service by Canadian soldiers in capturing Messines in western Belgium during the First World War. The spelling was corrected to Messines in 1986, in conformity with the post office name, established in 1920. It had replaced the postal name *Burbidge*, given in 1907 after the federal deputy minister of justice, George Wheelock Burbidge.

Métabetchouan QC A town on the southeast shore of Lac Saint-Jean, it was created in 1975 through the merger of the village of *Saint-Jérôme* and the parish municipality of *Saint-Jérôme*. Its name was derived from the post office, opened in 1860, and the township, proclaimed in 1857. **Rivière Métabetchouan**, which flows north for 128 km, was first noted in the late 1600s. Its name in Montagnais may mean 'flowing down rapids to a lake'.

Meta Incognita Peninsula NT The most southeasterly peninsula of Baffin Island, it was named in 1957 by the Canadian Board on Geographical Names. The board was inspired by remarks made by Queen Elizabeth I about 1578, when she described the new-found lands of Martin (later Sir Martin) Frobisher as *meta incognita*, 'limits unknown'.

Metcalfe ON In Osgoode Township, Ottawa-Carleton Region, southeast of Ottawa, this community's post office was called *Osgoode* from 1838 to 1877, when it was changed in honour of Charles Theophilus Metcalfe, 1st Baron Metcalfe (1785–1846), governor of the province of Canada, 1843–6. **Metcalfe Township** in Middlesex County was named in 1847 after him.

Metchosin BC A district municipality (1984) southwest of Victoria, its name was mentioned as early as 1842 by Hudson's Bay Company chief trader James Douglas, when he chose the site for Fort Victoria to the east. Derived from Straits Salish, the name may relate to 'the oil of a beached dead whale'. Its post office was opened in 1881.

Meteghan NS Located on the French Shore, north of Yarmouth, this place derived its name from Mi'kmaq *muntoogun*, 'blue stone', referring to the blue-tinged stones along the shore of St. Marys Bay. Its post office was opened in 1851, and other offices opened later include **Meteghan River** (1871), **Meteghan Station** (1883), and **Meteghan Centre** (1911).

Methuen ON A township in Peterborough County, it was named in 1823 after a Scottish title of Charles Lennox, 4th Duke of Richmond and Baron Methuen (1764–1819), governor-in-chief of British North America, 1818–9. His title is from the village of Methven, west of Perth, Scotland. **Belmont and Methuen** comprise a single township municipality.

Métis-sur-Mer QC A village in the regional county of La Mitis, it was named in 1921, succeeding *Petit-Métis*, created as a village in 1850. Its name, and that of the regional county and Rivière Mitis, are derived from Mi'kmaq *miti sipo*, 'poplar river'. Well known in English as *Metis Beach*, its site has long been renowned as a place to enjoy the sun and the sand away from the searing summer heat of interior cities.

Metropolitan Toronto ON Created in 1953, this municipality then comprised the city of Toronto, and the boroughs of East York, Etobicoke, North York, Scarborough, and York. Later, all the boroughs, except East York, became cities.

Miami MB This place northwest of Morden was named by the post office department in 1878, likely after the Aboriginal tribe that was once based in the present states of Ohio and Indiana. The tribe's name in Delaware may mean 'all friends', or in Ojibwa, 'pigeon'. Miami County in southeastern Indiana had been named in 1834. Manitoba's Miami predates Miami in Florida by 18 years.

Michener, Mount AB In the upper reaches of the North Saskatchewan River, this mountain (2545 m) was named in 1979 after Daniel Roland Michener (1900–91), governor general of Canada, 1967–74. Noted for his fitness and vigour, he climbed to its summit in his eighty-third year, accompanied by two mountain guides.

Michipicoten ON A township in Algoma District, it was created in 1952 and named after a nearby river, bay, and island. In Ojibwa it means 'great bluff'. The township's main urban centre is Wawa.

Midale SK Midway between Estevan and Weyburn, this name of the Canadian Pacific Railway station at this town (1962) was created in 1893 from the surnames of Dr R.M. Mitchell and Ole Dale, who were pioneers here. Its post office was opened in 1903.

Middle Arm NF Located on the **Middle Arm** of Green Bay, a westerly extension of Notre Dame Bay, this municipal community (1966) was established in the late 1800s as a fishing and logging centre.

Middlechurch MB Situated on the west bank of the Red River, just north of Winnipeg, this place's post office was named *Middle Church* in 1871, because St Paul's Parish had the middle church between St Andrew's and St John's parishes. The post office was closed in 1963.

Middle Lake SK Southeast of **Middle Lake**, which is centrally located between the larger Basin Lake and Lake Lenore, the post office at this village (1963) was named in 1909. Its station was founded in 1929, when the Canadian Pacific Railway built a line from Humboldt to Prince Albert.

Middlesex ON Created from *Suffolk County* in 1796, it was named after Middlesex, England, now organized as Greater London.

Middleton NS This town (1909) in the Annapolis Valley is located midway between Kentville and Annapolis Royal. It was also in the centre of the former *Wilmot Township*, and was previously called *Wilmot Corner* and *Fowlers Corner*. Its post office was opened in 1855.

Middlewood NS Located south of Bridgewater, this place was first called *Bull Run*, after the American Civil War battle. It was renamed *Middleton* about 1890, because it was situated about halfway between Bridgewater and Liverpool. To distinguish it from Middleton, in the Annapolis Valley, its post office was called Middlewood in 1917.

Midhurst ON In Simcoe County, north of Barrie, this place was named in 1863 after Midhurst in Sussex, north of Chichester, England. It was earlier known as *Oliver's Mills*, after miller George Oliver.

Midland ON A town (1890) in Simcoe County, it was developed in 1872–3 by the Midland Railway of Canada. The first train arrived in 1879, and it became the main shipping centre at the south end of Georgian Bay. It had been called *Mundy's Bay* and *Aberdare*, but company president Baron Adolf von Hugel insisted that it be called Midland.

Midnight Dome YT Overlooking Dawson, this elevation was early called *Moosehide Hill* and *Mooseskin Mountain*, because of its shape, and the colour of a rockslide on its west side. On the evening of the longest day of the year, 21 June, people have been gathering on the dome since the early 1900s to watch the sun almost set, before it rises again.

Midway BC A village (1967) on the Washington State border, west of Grand Forks, it was first named *Boundary City* in 1893. It was renamed in 1894 by Capt. R.C. Adams after the Midway Pleasance at the 1893 Chicago World's Fair, and after its location halfway between the Rocky Mountains and the Pacific Ocean.

Milden SK Southwest of Saskatoon and midway between Rosetown and Outlook, the post office at this village (1911) was named by Charles Mills in 1907, deriving it from his surname and the surname of neighbour Robert Bryden. They were the first settlers here. The Canadian Pacific Railway arrived the following year.

Mildmay ON A village (1918) in Bruce County, south of Walkerton, it was named in 1867, possibly after Mildmay Park, near Islington in Greater London, England. For many years it was informally called *Mernersville*, after Samuel Merner, who laid out a street plan in 1867.

Milestone SK South of Regina, this town (1906) was established in 1893 by the Canadian Pacific Railway. It was named after C.W. Milestone, CPR superintendent of the extension of the Soo Line from North Portal to Moose Jaw. Its post office was opened in 1900.

Milford Station NS Between Elmsdale and Shubenacadie, this place's post office was named *Wickwire Station* in 1866, six years after the Intercolonial Railway was built to Halifax. It was renamed Milford Station in 1875, possibly after a mill at a crossing of the Shubenacadie River.

Milk River AB The post office in this town (1956), southeast of Lethbridge, was named in 1908, after the river, a tributary of the Missouri. The river was named in 1805 by American explorers Meriwether Lewis and William Clark, when they noted it looked like 'a cup of tea with the admixture of a tablespoon of milk'.

Millbank ON A community in Perth County, northeast of Stratford, it was named in 1851, after saw and grist mills were built on both banks of the Nith River by William Rutherford and John Freeborn.

Millbrook ON A village (1879) in Peterborough County, southwest of the city of Peterborough, it was named in 1846 after a mill built by John Deyell on Baxter Creek, a tributary of the Otonabee River.

Mille-Îles, Rivière des QC Uniting Lac des Deux Montagnes with the St. Lawrence and passing north of Île Jésus, this river was first noted in the Jesuit *Relation* of 1674 as *rivière Jésus*, with the additional observation that 'one only finds submerged shoals sprinkled with a thousand islands'.

Mille Lacs, Lac des ON A large lake northwest of Thunder Bay, it was named during the French regime, when its complex network of islands, peninsulas, bays, and coves gave it the appearance of a large number of small lakes.

Millertown NF A municipal community (1959) on the east side of Red Indian Lake, east of Buchans, it was named after Lewis Miller (1848–1909), who established a lumbering operation here in 1900. *See also* Lewisporte.

Millet AB Between Leduc and Wetaskiwin, the station in this town (1983) was named about 1892 by the Canadian Pacific Railway. Two stories have evolved to explain the name. One states it was given by Sir William Van Horne, who was an admirer of the paintings of Jean-François Millet (1815–75). A second suggests Van Horne asked Fr Albert Lacombe for a suggestion, and he proposed naming it after fur buyer August Millet, his frequent canoeing companion, who had been drowned in the Red Deer River.

Millhaven ON In Lennox and Addington County, southeast of Napanee, it had a government mill in 1784, and other mills were on both sides of *Mill Creek* by 1802. The post office was called *Mill Haven* in 1852. Subsequently it became Millhaven, and in 1951, the creek became **Millhaven Creek**.

Millidgeville NB In the north side of the city of Saint John, this place's post office was named in 1864 after merchant and legislator Thomas Millidge (1776–1838) and his son, shipbuilder Thomas Edward Millidge (1814–94). The name of the post office was spelled *Milledgeville* from 1864 to 1963.

Milltown NB Part of the town of St. Stephen since 1973, Milltown had been incorporated as a town in 1878. Known in the late 1700s as *Christie Town* and *Stillwater*, its post office was named Milltown in 1839.

Milltown-Head of Bay d'Espoir NF Near the head of Bay d'Espoir, this town (1952) comprises the separate population centres of **Head of Bay d'Espoir**, a fishing community established in about 1850, and **Milltown**, a sawmilling community created in the late 1800s. *See also* Espoir, Bay d'.

Mill Village NS This place, near the mouth of the Medway River, and northeast of Liverpool, was named after saw and grist mills built here in the late 1700s. Postal service was established in 1847. There is a second **Mill Village** west of Shubenacadie, where there was a mill located on Ryans Creek in the 1800s.

Millville NB East of Woodstock, this village (1966) was established about 1860 as a New Brunswick and Nova Scotia Land Company settlement. A Mr Hayes built a mill here, and Millville post office was opened in 1866.

Milton NS Located at falls on the Mersey River, upriver from Liverpool, this municipal community (1996) was the site of three sawmills in 1762. When postal service was provided in 1855, it was given its present name.

Milton ON Jasper and Sarah Martin built a grist mill in 1822 on Sixteen Mile Creek, northwest of Oakville, and the site was called *Martin's Mills*. At a meeting in 1837 it was renamed Milton, after English poet John Milton (1608–74), and after its many mills. The post office was *Milton West* from 1836 to 1962. It became a town in 1857, and annexed Nassagaweya Township and parts of Esquesing Township and the town of Oakville in 1974.

Milton, Mount BC This mountain (3185 m) in the Cariboo Mountains, south of Valemount, was named in 1863 by Dr Walter Cheadle after his travelling companion, William Wentworth Fitzwilliam, Viscount Milton (1839–77).

Miltonvale Park PE A municipal community (1974) northwest of Charlottetown, its name is a composite of **Milton** and **Springvale**. Milton's post office was called *Milton Station* in 1878. Noted on an 1830s plan, Milton may have been named after a place on South Uist, Scotland, where Scottish nationalist Flora Macdonald (1722–90) was born, or it may have been proposed after the many mills here. Springvale had been established as a school district about 1862.

Milverton ON In Perth County, north of Stratford, this village (1881) was called *West's Corners* from 1854 to 1871 after shoemaker Andrew J. West. In 1871 Presbyterian minister the Revd Peter Musgrave proposed renaming the post office after his birthplace of Milverton in Somerset, England, southwest of Bristol.

Mimico ON On the shore of Lake Ontario and at the mouth of **Mimico Creek**, this urban centre was named in 1857 after the creek, whose name in Mississauga is *Omimeca*, 'place of wild pigeons'. Incorporated as a village in 1911 and a town in 1917, it was annexed by the borough (now city) of Etobicoke in 1967.

Miminegash PE A municipal community (1968) on the northwestern coast of the province, its name may be derived from Mi'kmaq *m'negash*, 'portage place'. Its post office was opened in 1861, closed in 1921, and reopened in 1967. *Ebbsfleet* post office was opened here in 1894 and continued until 1967. The school district was called *West Port* in 1882, and changed to *Ebbsleet* in 1904. *Ebbsfleet* may have been named after the ebb tides and the local fishing fleet, replicating the same name in Kent, England.

Minaki ON In Kenora District, northwest of the town of Kenora, it was known as *Winnipeg River Crossing* in 1870. In 1912 Minaki post office was opened, and the National Transcontinental Railway built a luxury summer lodge here in 1914. In Ojibwa *minaki* means 'blueberry ground'.

Minas Basin NS With dimensions of 25 km by 38 km, this basin was called *Bassin des Mines* by Samuel de Champlain in 1604 after the discovery of copper at Cape d'Or, on the north side of **Minas Channel**. The latter, extending for 45 km from Cape Chignecto to Cape Sharp, is the southern arm of the Bay of Fundy. *See also* New Minas.

Mindemoya ON On Manitoulin Island, southwest of Little Current, this place was named in 1880 after **Mindemoya Lake**. The lake's name means 'old woman' in the Ojibwa, Algonquin, and Ottawa languages. There is a legend that the great hero Nanaboozhoo cut off an old woman's head, and tossed her into the lake, leaving the outline of a kneeling old woman.

Minden ON Both the township and the community in Haliburton County were named in 1858 after a town in Prussia, southwest of Hamburg, where the British and Prince Ferdinand of Brunswick's forces defeated the French in 1759. Both are in the municipal township of **Anson, Hindon and Minden**.

Minesing ON In Simcoe County, northwest of Barrie, this community was named in 1867 after the Ojibwa word for 'island'.

Mingan, Archipel de QC Located on the north shore of the St. Lawrence, east of Sept-Îles, this group of 47 islands, islets, rocks, and reefs became the **Mingan Archipelago National Park** (94 km^2) in 1984. The name was long thought to be Montagnais for 'wolf', but the current preference is that it is a Breton word for 'white stone'. The regional county of **Minganie**, which embraces the archipelago and the north shore of the Gulf of St. Lawrence all the way to Newfoundland's Labrador border, was created in 1982.

Miniota MB Located on the east bank of the Assinboine River, northwest of Brandon, this community was named after the rural municipality (1883) of **Miniota**. The municipality's name was adapted by W.A. Doyle, postmaster of nearby Beulah, from the Assiniboine *mini-ota*, 'plenty of water', after hearing a surveyor announce that some water he had drunk from a local spring creek was the best he had ever had. The post office had been called *Parkismo* in 1885, and was changed to Miniota five years later.

Ministers Island NB In Passamaquoddy Bay, northeast of St. Andrews, this island was named after Samuel Andrews (*c.* 1736–1818), a St. Andrews minister, who lived on the island after 1791. Sir William Van Horne purchased the island in 1890 and named it *Covenhoven* in honour of his father. Although it has also been called *Van Hornes Island*, the earlier name has prevailed. In 1977 the island was established as a provincial historic site.

Minitonas MB East of Swan River, the post office at this village (1901) was named in 1900, after a Cree word meaning 'home of the little god'.

Minnedosa MB A town (1883) on the Little Saskatchewan River, west of Neepawa, it was first called *Tanner's Crossing* after ferry operator John Tanner. Its post office was called Hallsford in 1876. In 1879 first postmaster J.S. Armitage, who built the first saw and grist mills there, suggested Minnedosa, derived from Sioux *minne duza*, 'fast water'. His wife's name was Minnie, and they named a daughter Minnedosa in 1880, the same year it was adopted for the post office. The river was identified as both *Rapid River* and Little Saskatchewan River in the early 1800s, but *Minnedosa River* was adopted for it in 1911. Local appeals persuaded the Manitoba names authority to restore Little Saskatchewan River in 1978.

Minto NB A village (1905) east of Fredericton, its post office was named in 1905 after Gilbert John Elliot, 4th Earl of Minto (1845–1914), governor general of Canada, 1898–1904.

Minto ON A township in Wellington County, it was named in 1840 after Gilbert Elliot Murray Kynynmound, 1st Earl of Minto and Viscount Melgund (1751–1814), governor general of India, 1806–13. His great-grandson, the 4th Earl of Minto, was governor general of Canada, 1898–1904 and viceroy of India, 1905–10. The name was taken from a village in Roxburghshire, Scotland.

Minto, Lac QC Situated at the head of Rivière aux Feuilles in northwestern Québec, and covering an area of nearly 600 km^2, it was named in 1898 by geologist Albert Peter Low after Gilbert John Elliot, Earl of Minto, who was named Canada's governor general that year.

Mira Bay NS At the mouth of the **Mira River**, north of Louisbourg, this bay's name may have a Portuguese origin, perhaps given for a place south of Oporto. Pierre Jumeau identified it as *Baye de Miray* on his 1685 map, and Mgr Joseph-Octave Plessis described the settlement here as *Miré* in his 1815 journal. The Mira River flows northeast into the bay.

Mirabel QC This town was named in 1973, succeeding the town of *Sainte-Scholastique*, created two years earlier through the amalgamation of eight municipalities, together covering almost 500 km^2. The name was chosen in 1972 through public consultation to describe the new international airport, opened three years later. A post office called Mirabel had served part of its area from 1880 to 1914. It is said to have been devised from the names Miriam and Isabelle, daughters of a Scottish settler. It could also have been from a small plum called a *mirabelle*.

Miramichi NB On 1 January 1995 the city of Miramichi was established by amalgamating the towns of Chatham and Newcastle, the villages of Douglastown, Loggieville, and Nelson-Miramichi, and the surrounding area. Long proposed, it took its name from the region commonly called The Miramichi. The region is centred on the **Miramichi River**, which rises as three main branches called the **Southwest**, the **Northwest**, and the **Little Southwest**

Miramichi rivers. **Miramichi Bay** may have been named by Jacques Cartier in 1535 after the Montagnais, living on the north shore of the Gulf of St. Lawrence, observed that he had been in *Maissimeu Assi*, 'Mi'kmaq land'. In 1632 Samuel de Champlain called it *Baie de Petit Misamichy*. The same year map-maker Jean Guérard identified it as *Petit Miramichi*, and this spelling was subsequently adopted.

Mirror AB East of Lacombe, this village (1912) was named in 1911 by the Grand Trunk Pacific Railway Railway after *The Daily Mirror* of London, England. In its 10 August 1911 edition, the newspaper ran a headline reporting 'Mirror — Alberta / Rising Township Named After London Journal'.

Miscou Island NB On the northeast coast of the province, its name was derived from Mi'kmaq *susqu*, 'wet bog'. It was identified by Samuel de Champlain in 1632 as *I de Miscou*.

Miscouche PE A municipal community (1983) northwest of Summerside, it had been incorporated as a village in 1957. It was named after nearby **Miscouche Cove**, whose name in Mi'kmaq, *Menisgotjg*, means 'little marshy place'. Called *Belle Alliance* about 1824, it was noted as *Mascouche* in 1835. Miscouche post office was opened in 1866.

Missaguash River NB Part of the boundary between New Brunwick and Nova Scotia, this river flows into Cumberland Basin. Its name was likely derived from Mi'kmaq *muskwash*, 'muskrat'. After 1731 there were many variant forms of the name, with the Geographic Board confirming the present spelling in 1901.

Mission BC A district municipality (1892) on the north side of the Fraser River, and east of Vancouver, it was named after the St. Mary's Indian mission established in 1861 by Oblate Fr Léon Fouquet. The urban centre on the Fraser River was called *Mission Junction* after the arrival of railways between 1885 and 1891, and was incorporated as the village of Mission in 1922. When it became the first incorporated town in the province in 1950, it adopted the name *Mission City*, the name it retained until 1973. *Mission City* was the name of the post office from 1891 to 1973.

Mississagi River ON Rising in Algoma District, this river flows south into Lake Huron's North Channel, just west of Blind River. The name was derived from Ojibwa *missi-sauk*, 'large outlet'.

Mississauga ON A city in Peel Region, it is on land surrendered in 1805 by the Mississauga to William Claus, deputy superintendent of Indian affairs. Much of the area of the present city was organized that year as *Toronto Township*, which became the town of Mississauga in 1967 and a city in 1974, when it annexed the towns of Port Credit and Streetsville. An Ojibwa band, the Mississauga were first encountered in 1634 by French explorer Jean Nicolet at the mouth of Mississagi River, in Algoma District. In the 1700s they occupied a large part of present Southern Ontario. They were located at the mouth of the Credit River from around 1720 to about 1850, when they moved to the New Credit Reserve on the Grand River.

Mississippi River ON Rising in Lennox and Addington County, this river flows east to the Ottawa River, close to the mouth of the Madawaska River, at Arnprior. The first reference to it was in 1816 by government surveyor William Graves, while the nearby Madawaska and Bonnechere rivers were recorded over a century before. The name means 'big river' in Cree and other Algonquian languages, but it is not appropriate, since it is smaller than the nearby Madawaska. Graves may have misinterpreted what an Algonquin said, such as *Mazinaw-sippi,* after the pictographs on the east side of Upper Mazinaw Lake, near the head of the river.

Mistassini QC Located on the east bank of the **Rivière Mistassini**, north of Lac Saint-Jean, this town (1947) was founded in 1930 as the village of Mistassini. It annexed the parish municipality of *Saint-Michel-de-Mistassini* (1897) in 1976. The river rises east of **Lac Mistassini**, which is drained west into James Bay by the Rivière Rupert. *Mistassini* means 'big rock' in Montagnais. The Cree village of **Mistissini**, at the extreme southeastern end of Lac

Mistassini, has had several names since the establishment of the Hudson's Bay Company post of Canadian House in 1775, including *Mistassini Post* and *Baie-du-Poste*.

Mitchell ON A town (1874) in Perth County, northwest of Stratford, it may have been named by the Canada Company about 1836–7 for Judge James Mitchell, who served in London from 1814 to 1844. Two of his daughters married company surveyors. Originally called Big Thames, Mitchell's post office was opened in 1847. The tale that Mitchell was a black innkeeper was probably wrong.

Mitchell, Mount NB South of the Nepisiguit River, this mountain was named in 1964 by Arthur F. Wightman, then the provincial names authority, after Peter Mitchell (1824–99), premier of New Brunswick, 1866–7, and a Father of Confederation.

Moira River ON Flowing into the Bay of Quinte, at Belleville, this river was known as *Sagonaska, Singleton's Creek*, and *Meyers' Creek*, before it was named Moira River in 1807 after Francis Rawdon-Hastings, who succeeded his father as the Earl of Moira in 1793. The title was taken from a village in County Down, Ireland, southwest of Belfast.

Moisie QC Situated at the mouth of the **Rivière Moisie**, east of Sept-Îles, this town was created in 1984 through the merger of the municipalities of Moisie, *De Grasse*, and *Rivière Pigou*. Under the form of *Moisy*, the river was identified on a 1685 map by Louis Jolliet. The river's name may be from the old French word *moise* or *moyse*, 'wet river bank', which describes the area of its marshy sand and clay plain, where there are several meanders and serpentine streams at its mouth.

Monarch Mountain BC In the Coast Mountains, and at the southern edge of Tweedsmuir Provincial Park, this mountain (3459 m) was named in the early 1930s by W.A.D. (Don) Munday in honour of Canada's head of state, George V.

Monashee Mountains BC Bounded by the Columbia River on the east, and by the Kettle, Shuswap, and the upper North Thompson rivers on the west, these mountains extend 440 km from the Washington State border to Valemount. The name is from Scots Gaelic *monadh sith*, 'mountain of peace'. It was given about 1880 by prospector Donald McIntyre to his mine in the Gold Range, south of Revelstoke. He coined the name after a particularly peaceful sunset over a nearby mountain.

Monastery NS East of Antigonish, this place was named in 1925 after the Trappist priory, established here in 1815, but abandoned in 1919. It was taken over by the Augustinian brothers in 1938 and turned into a monastery.

Moncton NB This city (1890) was named in 1854 after **Moncton Parish**, which had been established in 1786. The parish had been named after *Monckton Township*, created in 1765, and called after Gen. Robert Monckton (1726–82), who had captured Fort Beauséjour on the Chignecto Isthmus in 1755, and renamed it Fort Cumberland. Its post office was called *Bend of Petitcodiac* from 1833 to 1865, when it was named Moncton. In 1930 the city council voted to restore the spelling *Monckton*, but universal public opposition resulted in a reversal of the vote.

Monkland ON In the united counties of Stormont, Dundas and Glengarry, northwest of Cornwall, this community may have been named after Old Monkland, east of Glasgow, Scotland, or after Monkland, near Leominster, Herefordshire, England. The post office was called *Monckland* in 1862, possibly influenced by the name of Lord Monck, appointed the governor general a year earlier. It was called *Monckland Station* from 1888 to 1966, when it was changed to Monkland, the name of the police village.

Monkman Pass BC This pass (1082 m) through the Rocky Mountains, northeast of Prince George, was discovered in 1921–2 by hunter and trapper Alexander Monkman.

Monkton ON In Perth County, north of Stratford, this place's post office was named in

1858 by Edward Winstanley, possibly after his birthplace in England.

Monmouth Mountain BC In the Coast Mountains, northwest of Whistler, this mountain (3194 m) was named in 1924 after HMS *Monmouth*, sunk in action 1 November 1914 off Coronel, Chile.

Mono ON A township in Dufferin County, it was named in 1821, possibly by Lt-Gov. Sir Peregrine Maitland, and possibly after the Spanish for 'monkey'. It may, however, have been derived from another source: a daughter of Shawnee chief Tecumseh; an Aboriginal word for 'let it be so'; or 'little ironwood tree'; the Roman name for the Welsh Isle of Anglesey; the Gaelic word for 'hilly'; or from the Monach Islands or the Monadhliath Mountains in Scotland. **Mono Mills**, partly in the township, but mainly in the town of Caledon, Peel Region, was named in 1839 after mills built on the Humber River about 20 years earlier by Michael, Francis, and Daniel McLaughlin.

Montague ON A township in Lanark County, it was named in 1798 for Adm. Sir George Montague (1750–1829), who was a British naval commander during the American Revolution. Numogate, a community in the township, was named about 1870 by township clerk John Ferguson by rearranging the letters of Montague.

Montague PE A town (1917) on the **Montague River**, its post office was opened about 1854, and named *Montague Bridge*. Samuel Holland named the river in 1765, possibly after Montague Wilmot (d. 1766), governor of Nova Scotia, 1763–6. It may also have been named for George Brudenell, 4th Earl of Cardigan (1712–90), who was created the Duke of Montagu in 1766. In 1730 he had married the daughter of John, the Duke of Montagu.

Montcalm QC This regional county, situated between the cities of Saint-Jérôme and Joliette, was created in 1982, taking its name from the former county of Montcalm. It had been named after Gen. Louis-Joseph, Marquis de Montcalm (1712–59), commander of the French forces in New France, 1756–9, who lost his life at the Battle of the Plains of Abraham.

Monteagle ON A township in Hastings County, it was named in 1857 after Thomas Spring-Rice, Lord Monteagle (1790–1866), who had served as British secretary of the treasury, secretary for the colonies, and chancellor of the exchequer.

Montebello QC Located in the regional county of Papineau, east of Buckingham, this village was named in 1878 after the residence of Louis-Joseph Papineau. He had named it *Monte-Bello* in 1854 after Napoléon-Auguste Lannes, Duc de Montebello (1801–74), French minister of foreign affairs in 1839. Papineau met him in France during his exile from 1839 to 1845.

Mont-Joli QC The largest town in the regional county of La Mitis, northeast of Rimouski, it was created in 1945, having been a village since 1880. Its post office was called *Sainte-Flavie-Station* from 1877 to 1914. The name is descriptive of its pretty (*joli*) view from its location on top of a hill (*mont*).

Mont-Laurier QC At the end of the nineteenth century this town north of Buckingham was known as *Rapide-de-l'Orignal*, named, according to legend, after a prodigious leap by a moose (*orignal*) over a falls of the Rivière du Lièvre. Mont-Laurier was named in 1909 after Sir Wilfrid Laurier (1841–1919), prime minister of Canada, 1896–1911.

Montmagny QC This town on the south shore of the St. Lawrence, downriver from the city of Québec, was incorporated as a village in 1845 and as a town in 1883. The name honours Charles Huault de Montmagny (*c.* 1583–*c.* 1653), the first governor of New France, 1636–48, who was granted the seigneury of Rivière-du-Sud in the area of the present town in 1646. The regional county of **Montmagny** was established in 1982.

Montmartre SK Southeast of Regina, the post office at this village (1908) was named in 1894, taking its name from the famed 'mount of martyrs' in Paris, France. It was proposed by a number of French settlers from Paris who had arrived here in 1893, sponsored by a French colonization society called La Société foncière du Canada.

Montmorency, Chute QC At a height of 85 m, this pretty waterfall 13 km northeast of the city of Québec was named as early as 1608 by Samuel de Champlain after French Adm. Charles de Montmorency, to whom Champlain dedicated his 1603 book, *Des Sauvages*.

Montréal QC The largest city in Québec, it traces its roots to 1535 when Jacques Cartier named the 233-m elevation Mont Royal, either in honour of François I, or as a feature worthy of the King. The evolution of Mont Royal to Montréal may be explained by sixteenth century evidence that *royal* and *real* had the same meaning, and that map-maker Belleforest used an Italian translation of Cartier's writings to identify the mountain in 1575 as *Monte Real*, and the place as *Montreal*. In 1612 Champlain referred to the mountain as *montreal*, and 20 years later wrote about the *Isle de Mont-real*. In 1642 Paul de Chomedey de Maisonneuve founded the colony of *Ville marie en l'isle de Montréal*. By the early 1700s *Ville-Marie* was superseded by Montréal. The city of **Montréal-Nord**, on the north side of the island, was named in 1915. The post office of the town (1897) of **Montréal-Ouest**, southwest of Montréal, was called *Montreal West* from 1894 to 1913. The industrial town of **Montréal-Est** was founded in 1910 on the southeastern shore of the island. The **Communauté urbaine de Montréal**, composed of 29 municipalities, was established in 1971.

Montreal River ON Rising west of Kirkland Lake, this river flows into the Ottawa River, at the south end of Lake Timiskaming. It was named in 1686 by French military commander Chevalier de Troyes, on his way to Hudson Bay to seize British forts. A second **Montreal River** rises southwest of Chapleau, and flows southwest into Lake Superior. It was probably named in the early 1700s by French explorers and voyageurs in honour of the historic island in the St. Lawrence, where they planned their trips to the Northwest.

Mont-Rolland QC Located in the regional county of Les Pays-d'en-Haut, north of Saint-Jérôme, this village was founded in 1881 by papermaker Jean-Baptiste Rolland (1815–88). Created in 1918 as the parish municipality of *Saint-Joseph-de-Mont-Rolland*, the parish municipality's name was abbreviated to Mont-Rolland in 1967, following the postal name established in 1905. In 1981 it became a village after annexing the town of Mont-Gabriel.

Montrose BC A village (1956) east of Trail, its post office was named *Beaver Falls* in 1949. It was renamed in 1953 by Trail lawyer A.G. Cameron, after his hometown of Montrose, Angusshire, Scotland.

Mont-Saint-Hilaire QC Located on the east side of the Richelieu, and in the shadow of the Monteregian hill called **Mont Saint-Hilaire**, the town's roots may be traced to the mission of Saint-Hilaire, established in 1799, and the religious parish of Saint-Hilaire, created in 1827. The mountain was known as both *Belœil* and *Rouville* until 1916, when it was called *Montagne Saint-Hilaire*, with 'mont' being substituted five years later. Mont-Saint-Hilaire became a town in 1950, and annexed *Saint-Hilaire-sur-Richelieu* in 1966.

Mont-Tremblant QC One of the province's most popular all-season tourist destinations northwest of Montréal, the post office of this municipality (1940) was opened in 1902. It took its name from the 968-m mountain known in Algonquin as *Manitou Ewitchi Saga*, 'the mountain of the formidable spirit', or *Manitonga Soutana*, 'mountain of the spirits or of the devil', with the *manitou*, the god of nature, having a reputation of making mountains tremble. That was rendered first in French as *Montagne Tremblante*, but became Mont Tremblant in 1936.

Monument Peak BC In the Purcell Mountains, west of Invermere, this mountain (3094 m) was named in 1910 when geologists with the Geological Survey of Canada observed that its summit was shaped like a gravestone.

Moonbeam ON A community in the municipal township of **Moonbeam**, Cochrane

District, southeast of Kapuskasing, its railway station was named in 1913 after **Moonbeam Creek**, given earlier when a splash of moonlight was observed in it. Its post office was named in 1915.

Moore ON A township in Lambton County, it was named in 1829 by Lt-Gov. Sir John Colborne after Sir John Moore (1761–1809), one of his commanders in the Peninsular War, who was killed at La Coruña, in Spain. It had been named *St. Clair Township* about 1820. The post office at **Mooretown**, in the township, was called *Moore* from 1837 to 1906.

Moorefield ON In Wellington County, east of Listowel, this place's station was named in 1871 after Baptist minister George Moore (d. 1883), through whose property the Wellington, Grey and Bruce Railway passed.

Moore, Mount BC In the Coast Mountains, east of Mount Waddington, this mountain (3041 m) was named in 1960 after pioneer rancher Ken Moore, who had settled in the Tatlayoko Lake area about 1915.

Moose Creek ON A community in the united counties of Stormont, Dundas and Glengarry, northwest of Cornwall, and on a tributary of the South Nation River, it was named in 1864 after the creek, which had been observed to rarely freeze at a small waterfall, providing a place for moose to gather in the winter.

Moose Factory ON A Hudson's Bay Company trading post founded in 1671 on an island in **Moose River**, this place was captured by the French in 1686. It was recovered by the British in 1730, and, after it was rebuilt, Moose Factory became the company's main post in the James Bay area.

Moose Jaw SK A city (1903) on a bend of the **Moose Jaw River**, its station was established by the Canadian Pacific Railway in 1883 and the post office was named the same year. Although some fanciful stories have been told about the origin of the name, it is generally accepted that the configuration of the Moose Jaw River, as it flows northwest, then, within the city, flows east, and then north to the Qu'Appelle River, provides the origin of the name. First recorded by John Palliser as *Moose Jaw Creek* in 1857, its name may have been a translation of its Cree name. The Geographic Board called the place *Moosejaw* and the river *Moosejaw creek* in 1901, but accepted the city's name in 1931, and the river's name was authorized in 1967.

Moose River ON A large river with several major tributaries, it drains northeast into the south end of James Bay. Identified on a map in 1679 as *R. des Monsonis*, it was named after the Monsoni tribe, whose totem was the moose. The French also called it *Rivière des Orignaux*, 'moose river'. After the Treaty of Utrecht in 1713, the English called it Moose River.

Moosomin SK Established by the Canadian Pacific Railway in 1882, this town (1889), near the Manitoba border, was named after Moosomin, a Cree chief, who lived at Jackfish Lake, north of North Battleford. He remained loyal to the government of Canada during the 1885 North-West Rebellion.

Moosonee ON On the west bank of the Moose River, it was the site in 1903 of a Revillon Frères trading post. In 1932 it became the end of the line of the Temiskaming and Northern Ontario Railway. The name was devised from Moose River and *Rivière des Monsonis*, named after the Monsoni, the first Algonquian tribe met by French explorers and traders in the area of James Bay.

Moraviantown ON In Kent County, and on **Moravian Indian Reserve 47**, northeast of Chatham, this place was first established at *Fairfield* on the north side of the Thames River in 1792 by the Delaware. They were survivors of a formerly active Moravian church in the United States, who sought asylum in Canada. Two years after their settlement was destroyed in 1813 by American forces, *New Fairfield* mission was established on the south side of the Thames, and renamed Moraviantown.

Morden MB The name of this town (1903) originated in 1882, when the Canadian Pacific

Railway was built west, 20 km north of the North Dakota border. It was named after Alvey Morden (d. 1891), who had moved west from Bruce County, ON, in 1874. His name was adopted for the post office in 1884.

Morell PE A municipal community (1983) northeast of Charlottetown and at the mouth of the **Morell River**, it had been incorporated as a village in 1953. Its post office was called *Morell Station* in 1884, and was changed to its present form in 1916. The river was named after Jean-François Morel, who was living here in 1752.

Moresby Island BC The second largest of the Queen Charlotte Islands, it was named in 1853 by Cdr James Charles Prevost after his father-in-law, Rear-Adm. (later Adm. Sir) Fairfax Moresby (1786–1877), commander-in-chief of the Pacific station, 1850–3.

Morewood ON In the united counties of Stormont, Dundas and Glengarry, north of Chesterville, it was named in 1862 by first postmaster Alex McKay, because it was surrounded by dense woods.

Morice River BC Rising in **Morice Lake**, east of Kitimat, this river flows east and north to join the Bulkley River at Houston. It was named after Fr Adrien Gabriel Morice (1859–1938), an Oblate missionary among the Aboriginals of the north central part of the province. Between 1885 and 1905 Fr Morice explored, mapped, and named the geography centred on the basins of the Skeena and Nechako rivers, but was much distressed when the province did not consult him. He viewed the Morice River as the main source of the Bulkley River, rather than the smaller river now considered the head of the Bulkley, rising east of Houston, but authorities in Victoria recognized the latter as the natural valley of the Bulkley.

Morigeau, Mount BC This mountain (3155 m) in the Purcell Mountains, east of Kootenay Lake, was named in 1884 by geologist George M. Dawson after François Baptiste Morigeau, a native of Saint-Martin, QC, who settled in the East Kootenay about 1819.

Morin-Heights QC Located in the regional county of Les Pays-d'en-Haut, west of Sainte-Adèle, this municipality owes its name to Augustin-Norbert Morin (1803–65), member for Bellechasse in the Legislative Assembly of the province of Canada, 1830–8 and 1844–51. He developed a farm, a sawmill, and a flour mill here during the mid-1800s. In 1852 the township of Morin was proclaimed, with the township municipality of *Morin-Partie-Sud* being erected three years later. It became the municipality of Morin-Heights in 1950, adopting the name given several years before to its railway station.

Morinville AB A town (1911) north of Edmonton, its post office was named in 1893 after Fr Jean-Baptiste Morin (1852–1911), who founded the francophone community here two years earlier.

Mornington ON A township in Perth County, it was named in 1845 for Richard Wellesley, Marquess of Wellesley, 2nd Earl of Mornington (1760–1842). His younger brother, William, Baron Maryborough, was made 3rd Earl of Mornington in 1845. Both were brothers of the Duke of Wellington. Mornington is a village in Ireland's County Meath, near Drogheda.

Morpeth ON In Howard Township, Kent County, east of Chatham, this community was named *Howard* in 1832. When a new name was considered at a meeting in 1851, *Jamesville* was proposed, after pioneer James Coll. However, others preferred Morpeth, after the Earl of Morpeth, who was travelling in the area. Morpeth is in Northumberland, England, north of Newcastle upon Tyne.

Morrin AB A village (1920) north of Drumheller, it was named in 1911 by the Canadian Northern Railway. Locally it is believed it was named after the locomotive engineer who brought the first train through the community. Records of the Canadian National Railways suggest it may have been named after Dr Joseph Morrin (1794–1861), a native of Dumfriesshire, Scotland, who practised medicine in the city of Québec after

1815, and served as the city's mayor in the 1850s. The place was earlier called *Blooming Prairie*.

Morris MB A town (1883) on the Red River, south of Winnipeg, its post office was named in 1881 after Alexander Morris (1826–89), lieutenant-governor of Manitoba, 1872–6. The place had been called *Scratching River* in 1874. *Scratching River* had been shown on the 1865 Palliser expedition map, but it was officially called **Morris River** in 1901. It had been earlier called by the French *Rivière aux Gratias*, referring to thorns that tore clothes and slashed flesh.

Morris ON A township in Huron County, it was named in 1850 after William Morris (1786–1858), receiver general of the province of Canada, 1844–7, and president of the Executive Council, 1846–8.

Morrisburg ON Called *Williamsburg West* from 1830 to 1851, after the township of Williamsburgh, in the united counties of Stormont, Dundas and Glengarry, the post office in this village (1860) was renamed in 1851 after James Morris (1798–1865), Canada's first postmaster general, 1851–3.

Morriston ON This place in Wellington County, southeast of Guelph, was named in 1854 after merchant R.B. Morison, who arrived here in 1847. He was still the postmaster in 1905, when he provided the details about the naming of the place, but he did not comment on the misspelling.

Morrisville NF Near the head of Bay d'Espoir and east of St. Alban's, this municipal community (1971) was first known as *Lynch Cove*, presumably after a fisherman who drowned here. It was renamed in 1910 after Edward Patrick Morris (1859–1935), prime minister of Newfoundland, 1909–17.

Morro, Mount BC In the Rocky Mountains, northeast of Cranbrook, this mountain (2912 m) was named in 1964 after FSgt Frank P. Morro of Cranbrook, who was killed in action in Europe, 4 December 1943.

Morse SK East of Swift Current, the station at this town (1912) was established by the Canadian Pacific Railway in 1883. It was not developed as an urban centre until 1908, two years after its post office was opened. It may have been named after Samuel Morse (1791–1872), the American inventor of the telegraph and of the Morse code in the 1830s.

Mortlach SK West of Moose Jaw, the station at this village (1949) was named in 1883 by the Canadian Pacific Railway, likely after the parish of Mortlach in the area of Dufftown, Banffshire, Scotland. The post office was named in 1905, the same year the first house was built here.

Mosa ON A township in Middlesex County, it was named in 1821 by Lt-Gov. Sir Peregrine Maitland after the Latin and Spanish form of the River Meuse in Belgium. Prussian Gen. Gebhard von Blücher's forces were drawn up there before the historic Battle of Waterloo in 1815.

Moscow ON In Lennox and Addington County, northeast of Napanee, this community was named *Springfield* in 1854, but because of duplication, it was renamed that year to commemorate Napoléon Bonaparte's retreat from Moscow in 1812.

Moser River NS On the Eastern Shore, between Sheet Harbour and Sherbrooke, this place was founded in 1809 by Henry Moser Sr. He had migrated to the province from Germany in 1751. Mosers River post office was opened in 1873.

Mossbank SK Southwest of Moose Jaw, the post office of this town (1959) was named in 1909 by Robert Jolly, modifying Mossgeil, the name of a farm north of Mauchline, Ayrshire, Scotland. The Canadian Pacific Railway was built through it in 1914.

Mountain ON A township in Dundas County, it was named in 1798 after the Rt Revd Jacob Mountain (1750–1825), who became the first Protestant bishop of Quebec in 1793. The post office in the community of **Mountain** was named in 1888. *See also* South Mountain.

Mount Albert ON In the town of East Gwillimbury, York Region, northeast of Newmarket, this community was first called *Birchardtown* and *Newland*. After Prince Edward Albert, the Prince of Wales visited Newmarket in 1860, the residents of *Newland* decided to rename the place after him, and it was changed to Mount Albert in 1864.

Mount Brydges ON In Middlesex County, southest of London, this place's station was named in 1854 after Charles John Brydges (1827–89), managing director of the Great Western Railway, 1852–62, general manager of the Grand Trunk Railway, 1862–74, and general superintendent of government railways, 1874–9.

Mount Carmel NF On the north side of the mouth of Salmonier River, where it flows southwest into St. Mary's Bay, this urban centre, in the municipal community of **Mount Carmel-Mitchell's Brook-St. Catherine's**, was called *Salmonier North* until 1930, when it was renamed after the local Our Lady of Mount Carmel Catholic Church.

Mount Carmel PE Located west of Summerside and on the shore of Northumberland Strait, this place's first post office was called *Fifteen Point* in 1858, after a point in Lot 15, one of the province's 70 divisions. It was renamed Mount Carmel in 1899 after the local parish church, Notre-Dame-du-Mont-Carmel.

Mount Currie BC In the Lillooet River valley, northeast of Whistler, this Lillooet community was named after John Currie (d. 1910), who became a rancher at nearby Pemberton about 1885.

Mount Elgin ON A community in Oxford County, southeast of Ingersoll, its post office was named in 1851 after James Bruce, 8th Earl of Elgin (1811–63), governor of the province of Canada, 1847–54.

Mount Forest ON On the South Saugeen River and in Grey County, this town (1879) was named *Maitland River* in 1849, in the mistaken belief that it was at the head of that river. Four years later it was renamed Mount Forest, after the townland of Mountforest, near Gorey, in County Wexford, Ireland.

Mount Hope ON A community in Hamilton-Wentworth Region, south of Hamilton, it was named *Glandford* in 1847, and corrected to *Glanford* three years later, after the township. In the 1850s pioneers suggested renaming it Mount Hope, because it was at the high point between Lake Ontario and Lake Erie, and because of hopes for prosperity and temperance, but the name was not changed until 1913.

Mount Moriah NF On the west side of Corner Brook, this town (1971) was named after the adjacent rounded hill, called earlier from the reference in the Bible to *Moriah*, 'land of hills'. The place was formerly called *Giles Point*, likely called after Joseph Giles, who was living here in 1871.

Mount Pearl NF A city (1988) on the west side of St. John's, it was developed after 1829 by retired British naval officer James (later Sir James) Pearl (1790–1840), and called *Mount Cochrane* after Sir Thomas John Cochrane (1789–1872), governor of Newfoundland (1825–34). In 1837 he renamed it Mount Pearl. It was subdivided into building lots after Pearl's widow's death in 1860. The town of *Mount Pearl-Glendale* was incorporated in 1955, and Glendale was dropped in 1958.

Mount Pleasant ON In Brant County, south of Brantford, this place was named about 1800 by Loyalist Henry Ellis after his birthplace in Flintshire, Wales. The post office was named Mount Pleasant in 1823, but it was called *Mohawk* from 1836 to 1922, when it reverted to Mount Pleasant.

Mount Royal QC This city on the northwest side of Mont Royal was incorporated in 1912 as both Mont-Royal and Mount Royal, and both names continue as official. It is widely known in English as Town of Mount Royal, and by the initials TMR.

Mount Sir Richard BC In the Coast Mountains, northeast of Squamish, this mountain (2728 m) was named in 1930 after Sir Richard McBride (1870–1917), premier of British Columbia, 1912–15. It had been previously called *Pyramid Mountain.*

Mount Stewart PE A municipal community (1983) northeast of Charlottetown, it had been incorporated as a village in 1953. Its post office was named about 1867 after John Stewart (1758–1834), founder of the place, and author of *Account of Prince Edward Island* (1806).

Mount Uniacke NS Located midway between Halifax and Windsor, this place was named after the country residence of Richard John Uniacke (1753–1830), solicitor general of Nova Scotia, 1781–97, and attorney general, 1797–1830. The residence was built in 1813–15, taken over by the province in 1949, and opened to the public in 1952.

Mousseau, Lac QC Located above Lac Meech, northwest of Hull, this lake is well known as the site of the Harrington Lake residence of the prime minister of Canada. Among early settlers on the lake were John Hetherington (a son-in-law of Asa Meech), and Louis Mousseau. Although a survey plan of 1850 called it *Harrington's Lake*, and official maps had *Harrington Lake* until 1962, it was locally called Lac Mousseau and *Mousseau Lake.* It was officially called Lac Mousseau (Lac Harrington) in 1962, with *Lac Harrington* being dropped in 1987.

Muenster SK Located east of Humboldt, this village (1908) became the site of the first Benedictine monastery in Canada. It was established in 1903, and its post office was first called *St. Peter's Colony* on 1 January 1904. It was renamed 1 August 1904 after Münster, in what is now **Nordrhein-Westfalen**, Germany, whose name had been derived from the Latin *monasterium,* 'monastery'.

Mulgrave NS On the mainland side of the Canso Causeway, this town (1923) was first called *McNair's Cove* and *Wylde's Cove* before being named *Port Mulgrave* in 1859 after the 2nd Earl of Mulgrave (1797–1863), the father of George Augustus Augustine Phipps, 3rd Earl of Mulgrave, lieutenant-governor of Nova Scotia, 1858–63. *Port Mulgrave* post office, opened in 1860, and *Port Mulgrave Station* post office, opened in 1893, were united as Mulgrave post office in 1897.

Mulmur ON A township in Dufferin County, it was named in 1821. Suggestions that it was named after a son of Shawnee chief Tecumseh, or after the ninth-century Irish poet Maolmura are questionable. If it has a Gaelic origin, it may be derived from Maol Mhor, 'big hill', in Scotland, where there is an elevation called that on the island of Mull.

Mummery, Mount BC Southwest of Howse Pass, this mountain (3328 m), part of the **Mummery Group**, 1813–15, was named in 1898 by British climber Norman Collie after alpinist Albert Frederick Mummery (1855–96), who died climbing Nanga Parbat in the Himalayas.

Mundare AB A town (1951) east of Edmonton, its station was named in 1906 by the Canadian Northern Railway, after station agent William Mundare. Its post office was opened in the same year.

Munday, Mount BC Southeast of Mount Waddington, this mountain (3367 m) was named in 1928 after climber W.A.D. (Don) Munday, who led a survey party to the area of Mount Waddington in 1926–7.

Munson AB Between Drumheller and Morrin, this village (1911) was founded in 1911 by the Canadian Northern Railway, and was likely named after J.A. Munson of the Winnipeg law firm Munson Allan Laird and Davis.

Munster ON In Ottawa-Carleton Region, southwest of Ottawa, this community was named in 1864 after the province of Munster, in southwestern Ireland. Following residential development after 1970, it has been popularly called *Munster Hamlet.*

Murchison, Mount AB Near the head of the North Saskatchewan River, this mountain

(3333 m) was named in 1858 by Dr James (later Sir James) Hector after Sir Robert Impey Murchison (1792–1871). He was a noted English geologist, who had recommended Hector for the position of geologist with the Palliser expedition, 1857–60.

Murdochville QC This town, almost 100 km west of Gaspé, was incorporated in 1953. It was named after James Y. Murdoch, the first president of Noranda Mines, which started mining the copper ore here in 1950.

Murillo ON This place in Thunder Bay District, west of the city of Thunder Bay, was called *Murillo Station* in 1880, and shortened to Murillo 10 years later. It was given for the celebrated Spanish painter Bartolomeo Estaban Murillo (1617–82).

Murray ON A township in Northumberland County, it was named in 1792 after James Murray (1721/22–94), a senior officer at the Battle of the Plains of Abraham, 1759, military governor of the district of Quebec, 1760–3, and governor of Quebec, 1763–8.

Murray Harbour PE A municipal community (1983) south of Montague, and on the south side of **Murray Harbour**, it had been incorporated as a village in 1953. The municipal community (1983) of **Murray River**, west of Murray Harbour, had been incorporated as a village in 1955. It was first known as *Cambridge Mills* and then *Murray Mills*. Both the harbour and the river were named in 1765 by Samuel Holland after James Murray (1721/22–94), governor of Quebec, 1763–8.

Musgrave Harbour NF On the north coast of the island of Newfoundland, this town (1954) was first called *Muddy Hole*, until it was renamed in the 1860s after Sir Anthony Musgrave (1828–88), governor of Newfoundland, 1864–9.

Musgravetown NF A town (1974) on the west side of Goose Bay, an arm of Chandler Reach, which is a southwestern extension of Bonavista Bay, it was once known as *Goose Bay*. It was renamed before 1869 after Sir Anthony Musgrave, governor of Newfoundland, 1864–9.

Muskoka ON The district was created in 1851 and separated from Simcoe County in 1888. In 1970 it became the province's only district municipality, divided among three towns and three municipal townships. The municipal township of **Muskoka Lakes**, west of Gravenhurst and Bracebridge, was created in 1971. **Lake Muskoka** and the **Muskoka Lakes** recall Chief Yellowhead and his son William Yellowhead (1769–1864), also known as Mesqua Ukie and Musquakie, who were chiefs of the Ojibwa around Lake Simcoe and Lake Huron, and had their favourite hunting grounds between Lake Muskoka and Lake of Bays. The name Yellowhead was derived from Ojibwa *mesqua*, 'red', and *ahkees*, 'ground'. The **Muskoka River** is comprised of two main branches, the **North Branch** and the **South Branch**, which join at Bracebridge.

Muskwa River BC Rising in the Muskwa Ranges of the Rocky Mountains, this river flows east and then north into the Fort Nelson River, east of the town of Fort Nelson. In Cree *muskwa* means 'black bear'.

Musquodoboit River NS Rising southeast of Truro, this river flows into the Atlantic at **Musquodoboit Harbour**, east of Dartmouth. Derived from Mi'kmaq *mooskudoboogwek*, the name means 'rolling out in foam at its mouth'. **Middle Musquodoboit**, southeast of Stewiacke, had postal service in 1821, **Upper Musquodoboit**'s post office opened 10 years later, and **Musquodoboit Harbour**'s post office was established in 1855.

Myrnam AB A village (1930) northeast of Vegreville, its post office was named in 1908 after the Ukrainian words *myr nam*, 'peace to us'. Many of the early settlers came from Ukraine.

N

Nackawic NB A town (1976) on the Saint John River and at the mouth of **Nackawic Stream**, it was established in 1970 as a residential community for the families of employees of the new St Annes Pulp and Paper Company. The stream's name in Maliseet was *Nelgwaweegek*, possibly meaning 'straight stream', referring to the fact that its lower part, before flooding by the waters of Mactaquac Lake, was in a direct line with the Coac Reach of the Saint John River. There was a *Nackawick* post office at Upper Queensbury, 3 km to the east, from 1862 to 1914.

Nahanni Butte NT At the mouth of the South Nahanni River, where it flows into the Liard River, this settlement was named in 1958 after the nearby imposing hill. The butte, which is not in fact a flat-topped butte, rises to 1372 m. It was named before 1910 after the Nahanni, an Athapascan tribe in the northern Rocky Mountains and the southern Mackenzie Mountains. **Nahanni National Park Reserve** (4766 km^2) was established on both sides of the South Nahanni River in 1972.

Naicam SK Northwest of Humboldt, this town (1954) was established in 1920, when the Canadian Pacific Railway line was surveyed from Watson to Melfort. Its name was derived from the names of two railway contractors who built the line, Naismith and Campbell. Its post office was also opened that year.

Nail Pond PE On the northwest coast of the province, this place's post office was named in 1878. It was a municipal community from 1975 to 1996. The adjacent seaside pond was possibly named after an Acadian family, whose ancestors may have originated as Noil in the Netherlands. An 1833 petition referred to *Neal's pond.*

Nain NF On the Labrador coast, this town (1970) was named in 1771 by Moravian missionaries, who chose the site as their first mission among the Inuit. It was the first settlement in Labrador, and is now the most northerly one.

Nairn ON In Nairn Township, Sudbury District, southwest of the city of Sudbury, this place's post office was named *Nairn Centre* in 1897, after a railway engineer had named a station Nairn in 1886, for his birthplace in Nairnshire, Scotland, northeast of Inverness. The post office at **Nairn** in Middlesex County, northwest of London, ON, was named in 1857 after the same county.

Nakina ON A community in Thunder Bay District, north of Geraldton, it was founded in 1913 as a railway construction base. Of uncertain origin, its name in Cree may mean 'land covered with moss'.

Nakusp BC A village (1964) on the eastern shore of Upper Arrow Lake, south of Revelstoke, its post office was named in 1892. In the Okanagan language the name may mean 'lake narrowing', a description of the lake 10 km to the south.

Nalaisk Mountain NB Southeast of Mount Carleton, this mountain (785 m) was named in 1901 by naturalist William F. Ganong after the Maliseet *Nalaisk*, 'snake', their name for the Serpentine River.

Nampa AB A village (1958) southeast of the town of Peace River, it was first called *Tank* in 1916 by the Edmonton, Dunvegan and British Columbia Railway. Confusion with Frank, AB, prompted local merchant Robert 'Pa' Christian to suggest naming it after his native Nampa, ID, west of Boise.

Namur QC A municipality (1964) in the regional county of Papineau, it was settled in 1865 by Belgian Protestants from the province of Namur. Its post office was named in 1874.

Nanaimo BC The second largest city (1874) on Vancouver Island, it was founded in 1852 at

Winthuysen Bay to exploit the coal resources of the area. Hudson's Bay Company chief factor James Douglas noted that year that the bay was commonly called *Nanymo Bay*. The growing settlement was called *Colviletown* in 1854, after HBC governor Andrew Colvile, but its use was discontinued by 1860. The bay was named after a confederacy of Coast Salish called *Sneněymexw*, whose pronunciation approximated 'na-NY-mo'. In Halkomelem the name means 'the great and mighty people'.

Nanga Parbat, Mount AB, BC On the continental divide and west of Howse Pass, this mountain (3240 m) was named by British climber Norman Collie after Nanga Parbat in present-day Pakistan, the world's ninth highest peak. The latter was associated with climber Albert Mummery, who was killed while ascending it in 1896. The peaks surrounding Mount Nanga Parbat are part of the Mummery Group.

Nanisivik NT The site of a lead-zinc mine on northern Baffin Island, it was founded in 1976 as Canada's first commercial mine in the Arctic. In Inuktitut the name means 'place where people find things'.

Nantes QC Located in the regional county of Le Granit, northwest of Lac-Mégantic, this municipality was called *Whitton* in 1874, after the township of Whitton, proclaimed in 1863. The township was probably named after a suburb of London, England. It was renamed Nantes in 1957 in honour of settlers from Nantes, France, who arrived in this part of the province in the late 1800s.

Nanticoke ON Although mostly rural in character, Nanticoke was created a city in 1974 in Haldimand-Norfolk Region through the amalgamation of several municipalities, including the towns of Port Dover and Waterford and the village of Jarvis. Believed to mean 'crooked creek' in an Aboriginal language, it was taken from the small community of Nanticoke, on **Nanticoke Creek**, near the Lake Erie shore. In Maryland, the name of the Nanticoke nation is interpreted as 'tidewater people'.

Nanton AB Midway between Calgary and Fort Macleod, this town (1907) was first called *Mosquito Creek*. It was renamed in 1893 by the Canadian Pacific Railway after Sir Augustus Meredith Nanton (1860–1925), a partner of the Winnipeg financial office of Osler, Hammond and Nanton, and a director of the CPR. His firm had a contract with the CPR to lay out townsites along the line south of Calgary.

Napanee ON A town (1864) in Lennox and Addington County, it took its name in the 1820s from **Napanee River**, originally called *Appanea*, whose meaning is unknown, but it is not 'flour', as often claimed. In the 1780s the mill site on the south side of the river was developed by millwright Robert Clark and government agent James Clark, and became known as *Clarkville*. In 1799 Richard Cartwright of Kingston built a mill on the north side, and that area became known as *Cartwrightville*.

Napierville QC A village (1873) in the regional county of Les Jardins-de-Napierville, southwest of Saint-Jean-sur-Richelieu, its post office was named in 1832. It was named in honour of Lt-Gen. Napier Christie Burton (d. 1835). In 1766 his father, Gabriel Christie, had purchased the seigneury of Léry, which included the area of Napierville.

Napinka MB In the Souris River valley, southwest of Brandon, this community's post office was named in 1884. It was a village from 1908 to 1986. Its name may be from a Sioux word meaning 'double' or 'equal to two', because it was hoped that its location on one Canadian Pacific Railway line, and the terminus of another CPR line, would promote its growth. It may also be an Aboriginal word signifying 'mitts'.

Nappan NS At the mouth of the **Nappan River**, south of Amherst, this place's post office was named in 1842. The river's name was derived from Mi'kmaq *Nepan*, 'good place to gather wigwam poles'.

Naramata BC On the eastern shore of Okanagan Lake, north of Penticton, this place was established in 1907 by John Moore Robinson. He reported that a spiritualist derived the name from the wife of an Aboriginal chief called Narramattah, from which he trimmed the superfluous letters. He had earlier planned to call the place Brighton Beach.

Nashwaak River NB This river flows into the Saint John River, opposite the city centre of Fredericton. The name is derived from Maliseet *Nahwijewauk*, possibly meaning 'winding among others', referring to how it interlaces with the head of the Southwest Miramichi. Noted in a 1676 French grant as *Nachouac*, the name had several spellings until 1901, when Surveyor General Thomas Loggie used the present form on a map.

Nashwaaksis NB A neighbourhood in the city of Fredericton and on the north side of the Saint John River, it is located at the mouth of **Naskwaaksis Stream**, whose name means 'little Nashwaak'. Its post office was *Naswaaksis* from 1854 to 1857, when it was changed to the present form. It was an incorporated village from 1966 to 1973, when it was annexed by Fredericton.

Naskaupi River NF Rising adjacent to Smallwood Reservoir, this river flows southeast into Lake Melville, part of Hamilton Inlet. The river took its name from the Naskapi, an Algonquian tribe related to the Montagnais. Their name was taken from a derisive Montagnais word, while calling themselves Nanénot, 'true, real men'. Both the Naskapi and the Montagnais in Labrador are now called Innu, 'people'. The river is also known as *North West River*.

Nass River BC Rising in the Skeena Mountains, northeast of Stewart, this river flows southwest into Portland Inlet. The name was given by the Tlingit, and literally means 'stomach' in their language. They came from southern Alaska to obtain their supplies of salmon and eulachon (a fish valued for its rich oily flesh) at the mouth of the river. Capt. George Vancouver established the spelling in 1793, after consulting the Tlingit. The Nisga'a, who occupy its basin, call it *K'alii Aksim Lisims*.

Natashquan QC Situated on the north shore of the Gulf of St. Lawrence, and at the mouth of **Petite Rivière Natashquan**, just west of the mouth of **Rivière Natashquan**, this township municipality was erected in 1907, taking its name from the township, proclaimed in 1869. Explorer Louis Jolliet identified the larger river as *R. noutasquan* in 1684. In Montagnais *Nutahquaniu Hipu* means 'river where they caught the black bear'.

Nation River BC Flowing east into the Parsnip Reach of Williston Lake, this river was mentioned by Simon Fraser in 1806 as *River au Nation*. He said it was named after the nation of Big Men (the Sekani), who were different from the nation (the Carrier) at *Trout Lake* (renamed McLeod Lake about 1806).

Naughton ON Andrew McNaughton, the first judge in Sudbury, became lost in the woods in 1883, but found some railway engineers camped at this site, now in the town of Walden, southwest of Sudbury. First called *McNaughtonville*, it was shortened to Naughton, when the post office was opened in 1902.

Nauwigewauk NB A community on the Hammond River, southwest of Hampton, it was named in 1852 by the European and North American Railway after the Maliseet name for Hammond River. Its post office was open from 1852 to 1969.

Navan ON In the municipal township of Cumberland, Ottawa-Carleton Region, east of Ottawa, its post office was named in 1861, after the town of Navan (An Uaimh), in County Meath, Ireland, northwest of Dublin.

Navy Peak BC In the Coast Mountains, east of Mount Waddington, this mountain (2941 m) was named in 1985 by Lt-Cdr Dave Redmond, who led a party into the **Naval Anniversary Range** that year. These names were given in honour of the seventy-fifth anniversary of the founding of the Royal Canadian Navy.

Nechako River BC Rising in Tweedsmuir Provincial Park, north of Bella Coola, it flows northeast, and then east into the Fraser River at Prince George. In the Carrier language, *Nechako* means 'big river'. When the Kenney Dam was built in 1956 to provide power for aluminum production at Kitimat, it inundated several lakes (Natalkuz, Eutsuk, Ootsa, Whitesail, Tetachuck) to form **Nechako Reservoir**.

Necum Teuch NS On the Eastern Shore and west of Ecum Secum, this place's post office was established in 1855. The name of the adjacent **Necum Teuch Bay** was derived from Mi'kmaq *Noogoomkeak*, 'soft sand place'.

Neebing ON A municipal township in Thunder Bay District, on the west side of the city of Thunder Bay, it was named about 1860 after the Ojibwa for 'summer'.

Neepawa MB This town (1883), northeast of Brandon, was named in 1873, after the Ojibwa *nibiwa*, 'plenty' or 'abundance', and its post office was opened in 1882. The name was chosen by its promoters on account of the rich and productive agricultural land surrounding the town. It adopted the cornucopia (the horn of plenty) as its emblem.

Neguac NB Southwest of Tracadie, this village (1967) was first noted in various sources in the 1780s. Its post office was called *Niguac* in 1857 and was changed to its present form in 1873. The name may have been derived from Mi'kmaq *Annikeooek*, 'improperly situated', referring to Hay Island in Neguac Bay

Neilburg SK West of Battleford, the post office at this village (1947) was established in 1908 and named after first postmaster Clifford O'Neill. The Canadian Pacific Railway arrived here in 1923.

Neils Harbour NS Located on the northeastern coast of Cape Breton Island, this place was named in 1867 after pioneer settler Neil McLennan. In 1760 French writer Thomas Pichon referred to the site as *Quarachoque*.

Nelles Corners ON In the town of Haldimand, Haldimand-Norfolk Region, this place was named in 1876 after merchant John Hamilton Nelles, murdered in 1854 by a notorious gang. Two men were hanged, and a third imprisoned, but a man who may have been gang leader William Townsend was acquitted three years later.

Nelson BC This city (1897), on the West Arm of Kootenay Lake, was named in 1888 after Hugh Nelson (1830–93), lieutenant-governor of British Columbia, 1887–92. Previously it was called *Salisbury*, after the 3rd Marquess of Salisbury (1830–1903), prime minister of Great Britain for three terms between 1885 and 1902, and *Stanley*, after Frederick Arthur Stanley, Baron Stanley of Preston (1841–1908), governor general of Canada, 1888–93.

Nelson-Miramichi NB In the city of Miramichi and opposite Newcastle, it had been incorporated as a village in 1967. Its post office was called *Nelson* in 1842 after **Nelson Parish**, established in 1814, and named after Adm. Viscount Horatio Nelson (1758–1805), the victor in 1805 of the naval battle off Cabo Trafalgar. *South Nelson* post office was opened in 1853 and changed to Nelson-Miramichi in 1968.

Nelson, Mount BC This mountain (3294 m) in the Purcell Mountains, west of Invermere, was named in 1807 by geographer David Thompson after Adm. Lord Nelson.

Nelson River MB This 540-km river was named in 1612–3 by Thomas Button after Robert Nelson, the sailing master of the *Resolution*, who died near the river's mouth, where Button had spent the winter.

Nemiscau QC A Cree village on the north shore of **Lac Nemiscau**, east of Fort Rupert, it was an important fur trade site at various times between 1685 and the present. The name means 'where there is plenty of fish', in reference to the fact that five rivers flow into the lake, which is drained into James Bay through the Rivière Rupert.

Nepean ON A city in Ottawa-Carleton Region, it was created in 1978, when the region was established. Laid out as a township in 1792, it was named two years later, after Sir Evan Nepean (1751–1822), British under-secretary of state for the home office. The area of the city of Ottawa west of the Rideau River was originally part of Nepean Township.

Nepisiguit River NB Rising in the **Nepisiguit Lakes**, on the eastern flank of Mount Carleton, this river flows east, then north into Bathurst Harbour, and **Nepisiguit Bay**. It was named after the Mi'kmaq designation *Winpegijooik*, 'river that dashes roughly along'. Noted by Acadian Gov. Nicolas Denys in 1672 as both *Nepegiguit* and *Nepiziguit*, its present spelling was not established officially until 1935. In various spellings, Nepisiguit Bay applied to Bathurst Harbour until 1777, when mapmaker J.F.W. DesBarres applied it beyond the sand bars, which almost enclose the harbour.

Neptune Peak BC Located in the Selkirk Mountains, north of Revelstoke, this mountain (3185 m) was named in 1939, likely by N.E. McConnell after the Roman god of water, the equivalent of the Greek god Poseidon.

Nettilling Lake NT The largest lake (5063 km^2) on Baffin Island, its name was recorded by anthropologist Franz Boas in 1884. The name may be derived from Inuktitut *netsek*, in reference to the ringed seal, which has its only permanent freshwater habitat in this lake. In 1897 geologist Robert Bell was informed by his Inuk guide that the name meant 'flat floor', referring to the flat beds of rocks extending to Amadjuak Lake and Foxe Basin.

Neudorf SK Southwest of Melville, this village (1905) was founded by Austrian settlers. A Mr Rathgeber, asked to name the post office in 1895, called it after his home town north of Vienna and near the Czech border. The Canadian Pacific Railway was built through it in 1903.

Neustadt ON A village (1907) in Grey County, south of Hanover, it was founded by several German immigrants in 1855. Its post office was called *Newstead* in 1856, but it was amended the following year to Neustadt, German for 'new town'.

Neuville QC Located in the regional county of Portneuf, southwest of the city of Québec, this village was named in 1919 after the seigneury of Neuville, given its name in 1680 by Nicolas Dupont de Neuville (1632–1716), member of the King's council and keeper of the seals. The name may have come from a village in France, north of Paris, whose name means 'new estate'.

New Annan PE East of Summerside, this place was named before 1836 by William 'Squire' Jamieson after Annan, Dumfriesshire, Scotland. Its post office was opened in 1878.

New Bandon NB On the shore of Chaleur Bay, northeast of Bathurst, this community's post office was named in 1850 after **New Bandon Parish**. The parish was named in 1831 after Bandon, Cork, Ireland, from where many families in the parish emigrated in 1819. There is a second community called **New Bandon** on the Southwest Miramichi, midway between Boiestown and Doaktown.

Newboro' ON In the united counties of Leeds and Grenville, and on the Rideau Canal, it was first called *New Borough*, which was shortened to Newboro' for the post office in 1836, and for the incorporated village in 1876. Although the apostrophe is legally part of the name, it is rarely, if ever, used.

New Brunswick The province was separated from Nova Scotia in 1784, and named after the House of Brunswick of the British royal family. Sir William Alexander wanted to call the area New Alexandria in 1624. In 1784 British under-secretary of state William Knox proposed New Ireland, to complement New England and Nova Scotia, but Ireland was then out of favour with George III. Pittsylvania was also suggested as a compliment to William Pitt, the prime minister.

New Brunswick, Mount YT A mountain (3388 m) in the Centennial Range of the St.

Elias Mountains, it was named in 1967 after the province.

Newburgh ON A village (1859) in Lennox and Addington County, northeast of Napanee, it was named in 1839 by Isaac Brock Aylesworth after Newburgh on the Hudson River, in New York State. The postal name was called *Newburg* in 1846 and altered to Newburgh in 1862.

Newbury ON A village (1872) in Middlesex County, midway between London and Chatham, it was named in 1854 after a place in England, possibly Newbury in Berkshire, west of Reading.

New Carlisle QC Located on the north shore of Chaleur Bay, southwest of Gaspé, this municipality (1877) was founded by Loyalists in the late 1700s. The name, which appeared on maps as early as 1815, may have been given after Carlisle in Cumberland County, PA, which had been named after Carlisle in Cumberland, England.

Newcastle NB In the city of Miramichi since 1 January 1995, this urban centre had been incorporated as a town in 1899. The post office, opened in 1823, was named after **Newcastle Parish**, created in 1786, and possibly given by Benjamin Marston after Thomas Pelham Holles, Duke of Newcastle (1693–1768), prime minister of Great Britain, 1754–62.

Newcastle Village ON In the municipality of Clarington, Durham Region, its post office was named *Newcastle* in 1845, which became the name of the village in 1857. When regional government was introduced in 1974, the village, the town of Bowmanville, and the townships of Clarke and Darlington became the town of *Newcastle*, taking its name from Newcastle District, which existed from 1798 to 1850. The former village became Newcastle Village in 1985 to distinguish it from the town. *See also* Clarington.

Newdale MB Located on the Canadian Pacific Railway, northwest of Minnedosa, this place's post office, originally located 5 km to the south, was named by Edward Cook in 1881. The post office moved 5 km north, when the CPR line was built.

New Denmark NB Southeast of Grand Falls, this community was named after the homeland of Danish immigrants, who settled here in 1872. Its post office was opened in 1887.

New Denver BC A village (1929) on the eastern shore of Slocan Lake, north of Nelson, its name was proposed in 1892 by Thomas Latheen after Denver, CO. The year before it had been called *El Dorado* (after the mythical city of gold) by a surveyor called Perry, but when it turned out that silver and lead, not gold, comprised the mineral wealth of the area, a meeting was called to choose an alternate name.

New Dundee ON A community in Waterloo Region, southwest of Kitchener, its post office was named in 1852 after Dundee, Scotland, the birthplace of brothers John and Frederick Millar.

Newellton NS Located on Cape Sable Island, north of the town of Clark's Harbour, its post office was opened in 1889. There were people with the name Newell living here as early as 1812.

New Ferrolle Peninsula NF On the northwestern coast of the island of Newfoundland and north of Port au Choix, it was named by the Basques in the late 1500s *Ferrolgo Amuixco Punta*, probably after El Ferrol in Spain's Galicia, where the Basque whalers spent the winter. It was called 'new' to distinguish it from the original harbour of Ferrolle, called *Ferrol Çaharra*, at the present-day Plum Point to the north.

Newfoundland The Portuguese identified the island in the 1500s as *Tierra de Bacalaos*, 'land of the codfish'. In 1497 John Cabot called it the *new founde isle*, and by 1502 *New found launde* was being used in British documents. Giovanni da Verrazzano referred to it as *Terra Nova* in 1529. The coast of Labrador became part of Newfoundland in 1809, and its boundary with Québec was defined by the Imperial Privy Council in 1927. In 1964 the province designated its administration as the government of Newfoundland and Labrador.

Newfoundland, Mount YT In the Centennial Range of the St. Elias Mountains, this mountain (3670 m) was named in 1967 after the province.

New Germany NS Northwest of Bridgewater, this place on the LaHave River was settled in 1785 by immigrants from Germany. Its first post office in 1855 was *Chesleys Corner* and it was given its present name in 1901. In 1854 the post office 5 km to the east had been called *New Germany*, but it was changed to Barss Corner in 1882.

New Glasgow NS This town (1875) in Pictou County was founded in the 1780s and named by surveyor William Fraser. He perceived the East River of Pictou as the River Clyde and the new settlement as Glasgow. Postal service was established here in 1834.

New Glasgow PE Between Hunter River and North Rustico, this place was named in 1819 after Glasgow, Scotland. Its postal service began about 1834. The post office at nearby **New Glasgow Mills** was established in 1908.

New Hamburg ON In Waterloo Region, west of Kitchener, this urban centre's post office was named in 1851 after the German city. Created a village in 1858 and a town in 1966, it was annexed in 1973 by the municipal township of Wilmot.

New Harbour NF Northeast of Dildo, this place was named after its adjoining harbour, which may have been named because it was established about 1750 as the first harbour on the east side of Trinity Bay, developed from the base at Trinity on the west side.

New Haven-Riverdale PE West of Charlottetown, this municipal community (1974) is comprised of the localities of **New Haven**, **Riverdale**, and Churchill. New Haven's post office was named in 1872, possibly after Newhaven, a suburb of Edinburgh, Scotland. It may have been named because it was a shipping port for some years. Riverdale is located in the deep valleys of West River and Howells Brook. Its post office was opened in 1885.

Newington ON Northwest of Cornwall and in the united counties of Stormont, Dundas and Glengarry, it was named in 1862 by postmaster Jacob Baker after his birthplace of Stoke Newington, in north central London, England.

New Liskeard ON A town (1903) in Timiskaming District, it was called *Liskeard* in 1896 after a town in Cornwall, England, but it was renamed *Thornloe* that year after newly ordained Anglican bishop George Thorneloe, because of confusion with Leskard, in the present Durham Region. In 1898 George Paget, a native of Cornwall, persuaded the province to restore *Liskeard*, and 'New' was added three years later for the post office and the town.

New London PE A district northeast of Summerside, its post office was opened at Clifton in 1827. The name had been chosen in 1773 for a port 5 km to the north in present French River, which remains part of the New London district. The nearby **New London Bay** was named after the port in the 1830s. In 1765 Samuel Holland had called it *Grenville Bay*, after George Grenville (1712–70), prime minister of Great Britain, 1763–5.

New Lowell ON In Simcoe County, west of Barrie, its post office was named New Lowell in 1858 after Lowell, MA.

Newmarket ON A town (1880) in York Region, it was named about 1816, when William Roe established a 'new market' for Aboriginals to sell furs, and trade for other goods. The post office was opened in 1823.

New Maryland NB South of Fredericton, this community was named by a Mr Arnold, who had come from the American state of Maryland about 1817. It was first called *Maryland*, but New Maryland was in use by 1825. **New Maryland Parish** was established in 1850.

New Minas NS Between Kentville and Wolfville, the post office in this village (1968) was named in 1858 after the nearby Minas Basin.

New Norway AB Southwest of Camrose, the post office at this village (1910) was named in 1903 after the homeland of some of the early settlers.

Newport NS In 1761 Newport Township was established across the St. Croix River from *Pisiquid* (Windsor) at present-day Avondale. It was named by Lt-Gov. Jonathan Belcher after a friend, Lord Newport. Newport, the post office at Brooklyn, was established in 1831. Post offices were also opened at **Newport Corner** (1850), **Newport Landing** (1851), and **Newport Station** (1863).

Newport QC On the north shore of Chaleur Bay, south of Gaspé, this municipality was created in 1845, abolished two years later, and re-established in 1855, taking its name from the township of Newport, proclaimed in 1840. The name, also spelled *New Port* in the early days, had been given by the Loyalists, who settled here at the beginning of the nineteenth century, perhaps recalling a place in the American colonies.

New Richmond QC The most populous town on the north shore of Chaleur Bay, it was first called *Richmond* in the early years of the 1800s. The township of New Richmond was identified on a map in 1803, but was not proclaimed until 1842. It became a township municipality in 1845, was abolished two years later, and was re-established in 1855. The name was likely given by Loyalists after a place they left in the American colonies, such as Richmond, the borough embracing Staten Island, NY, or Richmond, VA.

New Ross NS In Lunenburg County, northwest of Chester, this place was first called *Sherbrooke* in 1855 after Sir John Coape Sherbrooke, lieutenant-governor of Nova Scotia, 1811–16. It was renamed in 1863 after the second title of Lt-Gov. Lord Mulgrave, who chose it after New Ross, in County Wexford, Ireland. It also honoured Capt. William Ross, who had settled there in 1816.

New Sarepta AB A village (1960) southeast of Edmonton, its post office was named in 1905. It is believed Moravian bishop Clement Hoyler named it after the settlement of Sarepta in Vohlynia, Russia. That name may have been given for the ancient Sarepta (later Sidon, now Sayda in Lebanon). The Alberta village was formerly called *Little Hay Lakes*.

New Tecumseth ON Created as a town in Simcoe County in 1992, it comprises the former town of Alliston, the villages of Tottenham and Beeton, and the municipal township of Tecumseth. 'New' was added to distinguish it from the town of Tecumseh in Essex County. *See also* Tecumseth.

Newton, Mount YT In the St. Elias Mountains, this mountain (4210 m) was named in 1890 by explorer and geologist Israel C. Russell after Henry Newton, a professor at the Columbia School of Mines, Columbia, MO, and a member of the United States Geological Survey.

Newton Robinson ON In Simcoe County, northwest of Newmarket, it was first called *Latimer's Corners* and *Springville*. It was renamed in 1852, after one of the Newtowns in Ireland, and after William Benjamin Robinson (1797–1873), Conservative member for Simcoe in the House of Assembly of Upper Canada, 1830–41, and in the Legislative Assembly of the province of Canada, 1844–57.

Newtonville ON In Durham Region, west of Port Hope, this community's post office was called *Newton* in 1835. When a railway was built from Montreal to Toronto in 1853, a station 4 km to the southeast was called Newtonville, which was adopted by the community by the 1870s, although the post office was only changed to Newtonville in 1946.

New Toronto ON On the shore of Lake Ontario and in the city of Etobicoke, this urban centre was developed as a new industrial town in 1890. Incorporated as a village in

1913, and a town in 1920, it became part of the borough (now city) of Etobicoke in 1967.

Newtown NF A neighbourhood (1996) in the town of New-Wes-Valley, it was first known as *Inner Pinchard's Islands*. It was named in 1874, and became a municipal community in 1954.

New Waterford NS An urban community in the regional municipality of Cape Breton since 1 August 1985, it had been incorporated as a town in 1913. First known as *Barrachois*, its post office was opened as *Dominion No. 12* in May 1909 and renamed New Waterford two months later. It was named by J.J. Hinchey, a native of the Irish city of Waterford.

New Westminster BC A city (1860) on the north bank of the Fraser River, southeast of Vancouver, its site was selected in 1858 by Col Richard C. Moody as an appropriate location for the capital of British Columbia. He proposed calling it Queenborough. In 1859 Gov. James Douglas countered with Queensborough, but the colonial secretary, W.A.G. Young, stated Victoria was the 'Queen's borough'. To ensure that a suitable name was chosen, Douglas consulted Sir Edward Bulwer-Lytton, who responded 'that Her Majesty has been graciously pleased to decide that the capital of British Columbia shall be called "New Westminster".' But in 1867, when the colonies of Vancouver Island and British Columbia were united, Victoria became the capital.

New-Wes-Valley NF A town (1996) on the northwest side of Bonavista Bay, it comprises three separate populated areas. Four years earlier it had been incorporated as the town of Badger's Quay-Valleyfield-Pool's Island-Wesleyville-Newtown. The new name is a composite of the names Newtown, Wesleyville, and Valleyfield.

New World Island NF Between Twillingate Island and the island of Newfoundland, this island may have an indirect connection with *Novus Mundus*, noted on the Girolamo da Verrazzano map of 1529. It was not occupied by permanent settlements until after 1783.

New Zealand PE Located northwest of Souris, this place was named in 1858 when settlers left Charlottetown in two ships, one travelling to New Zealand in the South Seas and the other to the eastern part of the province.

Niagara ON The regional municipality of Niagara was created in 1970 through the amalgamation of Lincoln and Welland counties, and the restructuring of the municipalities within the region. **Niagara Peninsula** is a misnomer, as it is really an isthmus, extending from Hamilton to Niagara-on-the-Lake. **Niagara Escarpment**, although really geological in structure, has acquired geographical significance between Queenston Heights and Tobermory.

Niagara Falls ON Arguably the most famous set of waterfalls in the world, they and the **Niagara River** are said to be named after the Neutral word *Onghiara*, 'neck', referring to the **Niagara Peninsula** being cut aross by a river. The city of **Niagara Falls** was created in 1903 through the amalgamation of the town of *Clifton* and the village of *Niagara Falls Village*. In 1963 the city annexed the municipal township of *Stamford*, and in 1970 absorbed parts of the municipal townships of *Crowland*, *Willoughby*, and *Humberstone*.

Niagara-on-the-Lake ON In 1792 Lt-Gov. John Graves Simcoe made *West Niagara* the capital of Upper Canada, and renamed it *Newark*, as it provided an ark of safety for the Loyalists, as well as recalling Newark upon Trent in Nottinghamshire, southwest of the city of Lincoln, England. Also in 1792 Niagara Township was renamed Newark Township by Simcoe, but the former name was reinstated in 1798. In 1970 the township was united with the town (1850) of *Niagara* to form the town of Niagara-on-the-Lake, which had been made the postal name in 1902.

Nichol ON A township in Wellington County, it was named in 1822 after Robert Nichol (*c.* 1780–1824), who was killed accidentally at Queenston on returning from his duties as a judge at *Niagara* (Niagara-on-the-Lake).

Nicholas Denys NB A community northwest of Bathurst, its post office was named in 1940, misspelling it after Acadian Gov. Nicolas Denys (1598–1688), who had lived in the area of present-day Bathurst after 1669.

Nickel Centre ON A town in the Sudbury Region, on the northeast side of Sudbury, it was created in 1973, when the town of Coniston, the townships of Falconbridge and Maclennan, and most of the township of Garson, were amalgamated.

Nicola River BC Rising west of Kelowna, this river flows west through **Nicola Lake** to enter the Thompson River, midway between Lytton and Ashcroft. It was named by fur traders after a strong Thompson chief called Nukuala (*c.* 1785–1865), interpreted as Nicola, but whose Aboriginal name was *Hwistemexe'quen,* 'walking grizzly bear'. An 1849 map identified *R. Nicholas* and *Lac de Nicholas.*

Nicolet QC A town at the mouth of **Rivière Nicolet**, where it flows north into the east end of Lac Saint-Pierre, it was erected in 1872. The river, identified on a 1641 map, was named after Jean Nicolet (or Nicollet) de Belleborne (*c.* 1598–1642), who had come to New France with Samuel de Champlain, and explored the area of the Great Lakes. The regional county of **Nicolet-Yamaska** was created in 1982.

Nictau Lake NB North of Mount Carleton, this lake is situated near the heads of several river systems. Its name in Maliseet means 'forks', referring to the confluence of the Little Tobique, Mamozekel, and Right Hand Branch Tobique rivers, where they unite to form the Tobique River.

Nictaux NS On the **Nictaux River**, south of Middleton, this place was named for the river, called after the Mi'kmaq word *niktawk*, 'river forks', referring to the confluence of the Nictaux with the Annapolis River. Among places with post offices opened in the area are **Nictaux Falls** (1848), **Nictaux West** (1889), and **Nictaux South** (1896).

Nigadoo NB A village (1967) northwest of Bathurst, its post office was named in 1889 after the **Nigadoo River**. An 1807 land petition referred to Nigadoo. The name of the river is derived from Mi'kmaq *anigadoo*, whose meaning is unknown.

Nigel, Mount AB A mountain (3211 m) in the area of Maligne Lake, it was named in 1898 by British climbers Hugh Stutfield and Norman Collie after their guide Nigel Vavasour.

Nine Mile River NS The community of Nine Mile River is located nine miles (15 km) north of the **Nine Mile River**'s confluence with the Shubenacadie River. There was postal service here from 1852 to 1959.

Ninette MB Southeast of Brandon, this community's post office was named in 1884. A contraction of Antoinette, it was chosen from the name of a heroine in a novel, for a French actress, or suggested by French-speaking Canadians, who had met the delegates on their way to Winnipeg to obtain a post office for their community.

Nipawin SK On the Saskatchewan River, east of Prince Albert, the post office in this town (1937) was named in 1910 after a Cree word meaning 'place where one stands', in reference to the site's broad outlook over the river and surrounding country.

Nipigon ON Now a municipal township in Thunder Bay District, and at the mouth of **Nipigon River**, it was known as *Red Rocks* in 1872, but it was renamed *Nepigon* in 1889, and changed to Nipigon in 1912. The river drains **Lake Nipigon**, derived from Ojibwa *anemebegong*, 'continous water', referring either to being unable to see the far side of the lake, or to the straight route from Lake Superior up the river to the lake.

Nipissing ON The district was named in 1858 after **Lake Nipissing**, and the township on the south shore of the lake was named in 1879 after the lake. The lake's name was derived from an Algonquin word meaning 'little body of water', referring to its size compared to the

Great Lakes, or in contrast to the vast area of Georgian Bay.

Nippers Harbour NF Located on the western side of Notre Dame Bay and southeast of Baie Verte, this municipal community (1964) was first settled in the late 1700s by English immigrants. Although its name may have been derived from a Nippard family on Fogo Island, as early as 1854 Bishop Edward Feild wrote that the 'Nipper is the largest and most formidable of the mosquitoes' here.

Niut Mountain BC In the Coast Mountains, northeast of Mount Waddington, this mountain (2911 m) was named in 1925 by surveyor V. Dolmage after a Chilcotin legend of a woman who had been banished with her husband Tzelos for cannibalism during a time of famine. Subsequently they acquired potatoes from Ashcroft and planted them along the tribe's migration route. They were readmitted to the tribe, and when they died they were turned into the mountains west of Tatlayoko Lake. Potato Range east of the lake recalls the same legend. **Niut Range**, southwest of the mountain, was named in 1910.

Nivelle, Mount BC Near the continental divide and at the head of the Elk River, it was named in 1918 after Gen. Robert Georges Nivelle (1856–1924), First World War commander of the French armies until 1917.

Niverville MB A town (1993) on the east side of the Red River, south of Winnipeg, its name was given to a station by the Canadian Pacific Railway in 1877 after Joseph Boucher de Niverville (1715–1804). He had set out from Montréal in 1750 to find the 'Western Sea', returning home four years later without achieving his goal. Postal service was provided in 1879.

Nobel ON A community in Parry Sound District, northwest of the town of Parry Sound, its post office was named in 1913 by Canadian Industries Limited after Alfred Bernhard Nobel (1833–96), who invented dynamite in 1866, and estabished the Nobel prizes for peace, physics, chemistry, medicine, and literature. The post office here was called *Ambo* from 1911 to 1913.

Nobleford AB A village (1918) northwest of Lethbridge, it was first called *Noble*, after Charles Sherwood Noble (1873–1957), a native of Iowa, who homesteaded in 1902 near Claresholm. He accumulated 14,164 ha (35,000 a) of land for the Noble Foundation, which became the largest and highest-yielding farmland in the British Empire. The post office was named in 1918.

Nobleton ON In King Township, York Region, this urban centre was named after merchant Joseph Noble, whose brother, Thomas, was appointed the first postmaster in 1851.

Noel NS On **Noel Bay** and on the south side of Cobequid Bay, this place's post office was named in 1832. There were also post offices at **Noel Shore** (1855), **Noel Road** (1883), and **North Noel Road** (1922). The bay was likely named after Acadian settler Noel Pinet.

Noëlville ON In Sudbury District, southeast of the city of Sudbury, this community was first called *Cosby*, after the township, in 1904. After being confused with Crosby in Southern Ontario, it was named in 1911 after lumberman and merchant Noël Desmarais.

Nokomis SK Located between Regina and Humboldt, this town (1908) was named in 1906 by postmistress Mrs Thomas Halstead, who admired the name in Longfellow's *Hiawatha*, and related it to the romantic landscape long occupied by the Aboriginals. At a major crossing of the Canadian Pacific Railway's line from Virden, MB to Humboldt, with the Grand Trunk Pacific Railway's line from Winnipeg to Saskatoon, the CPR wanted to call it Blaikie in 1907 and the GTP proposed Blakemore in 1908, but Nokomis was accepted by both.

Nonacho Lake NT South of Great Slave Lake, the name of this lake (698 km^2) was approved in 1926. Meaning 'big point lake' in Chipewyan, it was submitted as *Nonachoh Lake* by surveyor G.H. Blanchet. On an earlier map it had been identified as *Lady Grey Lake* and on an 1926 map it was called *The Big Lake*.

Nootka Sound BC A large bay on the west coast of Vancouver Island, it was named *King George's Sound* in 1778 by Capt. James Cook, but he changed it to Nootka Sound in the belief that it was the Aboriginal name for it. Cook had asked the Aboriginals what they called it, but they interpreted the sweep of his arms to refer to his trip around the bay, and they said something like *nootk sitl*, 'to go around'. Claims that it was named *Boca San Lorenzo* four years earlier by Spanish explorer Juan Perez appear to be unfounded. It is likely he was referring to another bay near Estevan Point, to the southwest, and later explorers equated it with Nootka Sound. **Nootka Island**, to the northwest of Nootka Sound, is the largest island on the west coast of Vancouver Island.

Nordegg AB West of Rocky Mountain House, this place's post office was named in 1914 after Martin Nordegg (1868–1948), who arrived in Canada in 1906 from Germany. He adopted the surname Nordegg in place of his family name, Cohn or Cohen, and became general manager of the Brazeau Collieries Ltd before the First World War.

Norfolk ON Lt-Gov. John Graves Simcoe named the county in 1792 after the English county. It was united in 1974 with Haldimand County to form the regional municipality of **Haldimand-Norfolk**. The municipal township of **Norfolk** was also created the same year by amalgamating the village of Port Rowan, the municipal townships of Houghton and North Walsingham, and parts of the municipal townships of Middleton and South Walsingham.

Norland ON In Victoria County, north of Lindsay, this place's name was proposed in 1862 by the Revd Bayard Taylor, after Nordland, a village in Africa, where he had been working. The 'd' was intentionally omitted, leading to the impression it was at the end of civilization in the county.

Normanby ON A township in Grey County, it was named in 1840 after Constantine Henry Phipps, Marquis of Normanby (1797–1863), secretary of state for the colonies in 1839.

Normandin QC This town northwest of Lac Saint-Jean had its beginnings in 1878, when a colony was planted here by businessmen living in the city of Québec. The township of Normandin was proclaimed nine years later and named after Joseph-Laurent Normandin, who explored the region in 1832 to determine the limits of the Domaine du Roy, 'the territory reserved exclusively for the King'. The town was formed in 1979 though the merger of the township municipality of Normandin (1899) and the village of Normandin (1926).

Norman Wells NT On the east bank of the Mackenzie River and west of Great Bear Lake, this town (1992) was developed in 1944 after the discovery of oil here. It took its name from *Fort Norman*, now known as Tulita, 75 km upriver at the mouth of Great Bear River.

Normétal QC Located in the regional county of Abitibi-Ouest, near the Ontario border, this municipality was incorporated in 1945. It had been named in 1931 by blending the words *nord* and *métallurgique*, an appropriate description of the copper and zinc mines developed here in 1925.

Norquay SK Northeast of Canora, the post office in this town (1963) was named in 1907 by Hugh Wylie, who greatly admired John Norquay (1841–89), premier of Manitoba, 1878–87.

Norquay, Mount AB This mountain (2522 m) on the north side of Banff, was named in 1904 after John Norquay, premier of Manitoba, 1878–87, who had climbed it in 1888.

Norris Arm NF Located on Norris Arm and at the southern end of the Bay of Exploits, this town (1971) was permanently settled after 1871. The arm, noted on an 1811 map by Capt. David Buchan, was named after pioneer settler James Norris, whom Lt John Cartwright mentioned in his book, *A Sketch of the River Exploits*, 1768.

Norris Point NF On the northwestern coast of the island of Newfoundland and in Bonne Bay, this municipal community (1960) was named after the point. The point and the adjacent Neddy Harbour were called after trapper Neddy Norris, who was located here about 1790.

North Saskatchewan River AB, SK Rising in the Columbia Icefield and adjacent to **Mount Saskatchewan** (3342 m), this river (1287 km) flows northeast to Edmonton, where it takes a generally eastern course to join the South Saskatchewan River, 50 km east of Prince Albert, SK. *See also* Saskatchewan River *and* South Saskatchewan River.

North Augusta ON A community in Augusta Township, united counties of Leeds and Grenville, north of Brockville, it was first known as *Bellamy's Mills* and *Bellamyville*, after millers Samuel and Hiram Bellamy. Its post office was named North Augusta in 1840.

North Battleford SK When the Canadian Northern Railway reached this site on the north bank of the North Saskatchewan River in 1905, opposite the historic town of Battleford, it wanted to call its station Riverview. However, much to the distress of residents of Battleford, the station and the post office were called North Battleford. Created as a village and then a town in 1906, it became a city in 1913.

North Bay ON A city (1925) in Nipissing District, it became a divisional point of the Canadian Pacific Railway in 1881. Although located over 35 km east of the Great North Bay, it was identified that year as the 'north bay' of Lake Nipissing, and the post office was opened in 1884.

North Bend BC In the Fraser Canyon, almost opposite Boston Bar, it was first called *Yankee Flats* and *Yankee Town*, but was renamed by the Canadian Pacific Railway. When the CPR built its transcontinental line from west to east through the province, it had to make a 150-km 'north bend' at Hope, with its easterly direction only restored at Ashcroft. The site of *Boston Bar* post office in 1884, it was renamed North Bend three years later.

Northbrook ON In Kaladar Township, Lennox and Addington County, north of Napanee, it was first known as *Flint's Mills and Flint*, after lumberman and politician Billa Flint. Its post office was called *Kaladar* in 1857. It was changed to Northbrook in 1890, possibly after Thomas George, 1st Earl of Northbrook (1826–1904), governor general of India, 1872–6. *See also* Flinton *and* Kaladar.

North Burgess ON Burgess Township was named in 1798, possibly for James B. Burgess, a British member of Parliament, who voted in 1791 for the Canada Bill. In 1841, the part of Burgess Township north of Big Rideau Lake was joined to Lanark County. *See also* South Burgess.

North Buxton ON In Kent County, south of Chatham, this place was founded in 1849 as a settlement of American blacks and named *Elgin Settlement*, after Lord Elgin, the governor of the province of Canada. It was renamed in 1874 after Sir Thomas Fowell Buxton, a British member of Parliament, who promoted the passing of the 1833 Emancipation Act. *See also* South Buxton.

North, Cape NS The most northerly point of Cape Breton Island, it was identified as *Le Cap de Nort* by Nicolas Denys in 1672 and *C. de Nort* by Vincenzo Coronelli on his map of 1698. *North Cape* occurred on English maps in the mid-1700s, and the present form was established by the chart-maker J.F.W. DesBarres in 1775.

North Cape PE The most northwesterly point of the province, it was named by Samuel Holland in 1765. Jacques Cartier referred to it as *Cap dez Sauvaiges* in 1534, as he observed an Aboriginal fleeing when his vessel approached the shore. Jacques Nicolas Bellin called it *Pointe du Nord* in 1744 and subsequent English references called it *North Point*, but North Cape is traditional usage.

North Cayuga ON Part of the town of Haldimand in Haldimand-Norfolk Region since 1974, the township was named in 1835 after the Cayuga tribe of the Six Nations confederacy. The name has many interpretations, including 'place where the canoes are drawn out', 'mountain rising from the water', and 'advising nation'. *See also* South Cayuga.

North Cowichan BC A district municipality (1873) on Vancouver Island, extending north from Duncan to Chemainus, it was named after the Cowichan River and Cowichan Bay. The Cowichan, a Salish tribe, which once occupied a large area on both sides of the Strait of Georgia, owe their name to an Island Halkomelem word meaning 'land warmed by the sun', referring to a rock formation on the side of Mount Tzuhalem, resembling a frog basking in the sun.

North Crosby ON Crosby Township was created in 1798, and probably named after Brass Crosby, Lord Mayor of London, a member of Parliament, and a friend of Lt-Gov. John Graves Simcoe. North Crosby was separated from South Crosby in 1806.

North Dorchester ON Dorchester Township, created in 1798 in Middlesex County, was named after Sir Guy Carleton, Baron Dorchester (1724–1808), governor of Quebec, 1768–78, commander-in-chief of British forces in North America, 1782–6, and governor-in-chief of British North America, 1786–96. It was divided in 1851, with North Dorchester remaining with Middlesex County, and South Dorchester transferring to Elgin County.

North Dumfries ON A township in Waterloo Region, it is the north half of Dumfries Township, named in 1816 by Niagara merchant William Dickson, after his birthplace of Dumfries, in Dumfriesshire, Scotland. *See also* South Dumfries.

North Easthope ON Easthope Township in Perth County was named in 1830 for Sir John Easthope (1784–1865), a director of the Canada Company, a British company established in 1826 to develop the Huron Tract, a large area of what is now Southwestern Ontario. It was divided into North Easthope and South Easthope townships in 1843.

North Elmsley ON Elmsley Township was named in 1798 after Upper Canada's chief justice John Elmsley (1762–1805). In 1841 the part of the township north of the Rideau River was attached to Lanark County, with South Elmsley Township remaining in the united counties of Leeds and Grenville.

Northern Peninsula NF Extending for 225 km from the head of White Bay to L'Anse aux Meadows, this name describes the 35-km-wide peninsula of the island of Newfoundland. Since the 1960s, it has become more commonly called the Great Northern Peninsula.

North Fredericksburgh ON Fredericksburgh Township was named in 1784, likely after Prince Augustus Frederick (1773–1843), sixth son of George III, with his next older brother honoured in Ernestown Township, to the east, and the next younger brother honoured in Adolphustown Township, to the west. As Prince Frederick, the second son, was commemorated in Osnabruck Township, and the townships named after the royal children were given in chronological order, it is unlikely this township was named after him. It was divided in 1858 into North Fredericksburgh and South Fredericksburgh townships in Lennox and Addington County.

North Gower ON Gower Township, created in 1798 in Grenville County, was divided into two townships in 1850. North Gower Township became part of Rideau Township in the regional municipality of Ottawa-Carleton in 1974. The original township was named after John Levenson Gower (1740–92), lord of the Admiralty, 1783–9. The post office in the community of **North Gower** was named in 1847.

North Hatley QC This village at the north end of Lac Massawippi, southwest of Sherbrooke, was named in 1897 after the municipal township of Hatley, erected in 1845. The township of Hatley had been proclaimed in 1803 and named after a village in Cambridgeshire, England.

North Head NB At the north end of Grand Manan Island, it was an incorporated village from 1966 to 1995, when it became an urban community in the village of Grand Manan. Its post office was opened in 1859. Its port connects the island with the mainland at Blacks Harbour.

North Kent Island NT Between Ellesmere and Devon islands, this island (590 km^2) was named *Kent Island* in 1853 by Sir Edward Belcher, after Prince Edward Augustus, Duke of Kent (1767–1820), the fourth son of George III, and father of Queen Victoria. It was renamed North Kent Island in 1939 based on a recommendation made over thirty years before by geologist Albert P. Low.

North Marysburgh ON Marysburgh Township in Prince Edward County was named in 1786 for Princess Mary (1776–1857), fourth daughter of George III and the wife of the Duke of Gloucester, her first cousin. North Marysburgh was separated from South Marysburgh in 1871.

North Monaghan ON A township in Peterborough County, it was until 1850 part of Monaghan Township, named in 1820 after County Monaghan, in north central Ireland. *See also* South Monaghan.

North Nahanni River NT Rising in the Mackenzie Mountains, this river flows into the Mackenzie River, 110 km downriver from Fort Simpson. It was named before 1910 after the Nahanni, an Athapascan tribe in the northern Rocky Mountains and the southern Mackenzie Mountains. *See also* Nahanni Butte *and* South Nahanni River.

North Pender Island BC One of the Gulf Islands, east of Saltspring Island, it and adjacent **South Pender Island** were jointly named *Pender Island* in 1859–60 by Capt. George Richards after Daniel Pender, who was then master of the survey vessel *Plumper*.

North Plantagenet ON The original Plantagenet Township was named in 1798 after the royal family surname from 1154 to 1485. North Plantagenet was separated from South Plantagenet in 1850, and both remained in the united counties of Prescott and Russell.

North Pole Mountain NB In north central New Brunswick, and adjacent to **North Pole Stream**, this mountain was named in 1964 by Arthur F. Wightman, then the provincial names authority. Inspired by the Christmas spirit inherent in the stream's name, Wightman also named mountains after St Nicholas, and after the eight reindeer (Blitzen, Comet, Cupid, Dancer, Dasher, Donder, Prancer, Vixen) in Clement Moore's 1823 poem *A Visit from Saint Nicholas*. The stream was named by lumbermen in the 1840s, possibly because it was the farthest north they had cut trees, or because the weather was especially cold.

North Portal SK Southeast of Estevan, this village (1903) was created in 1893, when the Soo Line connected with the Canadian Pacific Railway at the forty-ninth parallel. It took its name from the idea of being the entrance to north of the border. Its post office was opened in 1894.

North River NF On the west side of Conception Bay and south of Bay Roberts, this municipal community (1964) is located on the northwest bank of the river, and extends southwest as far as The Pond That Feeds the Brook. Both the place and the river were earlier known as *Northern Gut*.

North River PE A locality in the town of Cornwall since 1995, it had been created as a municipal community in 1983. It was named after a school district, established about 1864, and its post office, opened in 1872. They were named after the adjacent river, which was called *Rivière du Nord* by the French. Samuel Holland named the river in 1765 after John Yorke (1728–69), then lord commissioner of plantations. *Yorke River* remained official until 1966, when it was changed to *North (Yorke) River*, and changed again in 1971 to North River. During the 1800s North River was used interchangeably with *York River.* York Point, at the river's mouth, was called after the river's variant name in the mid-1800s.

North Rustico PE Northwest of Charlottetown, this municipal community (1983) had been incorporated as a village in 1954. Part of the Rustico district, its post office had been named in 1855. The post office at **South Rustico** was called *Rustico* from 1852 to 1967, when it was changed to South Rustico, but closed the following year. **Rusticoville**, between the two, had postal service from 1886 and 1947. *See also* Rustico Bay.

North Saanich BC A district municipality (1965) at the north end of the Saanich Peninsula, north of Victoria, it takes the second part of its name from the Saanich tribe. Its name in Straits Salish means 'elevated', in reference to Mount Newton (307 m), whose profile looks like a raised rump. *See also* Saanich *and* Central Saanich.

North Shore PE A municipal community (1974) north of Charlottetown, and extending for 6 km along the north shore of the province, it embraces the localities of Covehead Road, West Covehead, Stanhope, and Stanhope by the Sea.

North Sydney NS An urban community in the regional municipality of Cape Breton since 1 August 1985, it had been incorporated as a town in 1885. It was named after its location on the north side of Sydney Harbour. Postal service was established here in 1837.

Northumberland ON The county was named in 1792 by Lt-Gov. John Graves Simcoe after the county of Northumberland in the north of England. In 1850 it was united with Durham County for municipal purposes, but it became an independent county in 1974.

Northumberland Strait NB, NS, PE In the southern part of the Gulf of St. Lawrence, this strait was named about 1777 by map-maker J.F.W. DesBarres, after HMS *Northumberland*, the flagship of Adm. Lord Colville (1717–70). In 1765 Samuel Holland observed that 'the sea betwixt the island and the continent is frequently of a red hue...by many people called the Red Sea'. Lord Colville undertook a survey of Nova Scotia's North Shore in 1764. **Northumberland County**, NB, was named in 1785, likely after the strait.

North Vancouver BC A district municipality (1891) and a city (1907) on the north shore of Burrard Inlet, they were named for their location relative to the city of Vancouver. The site of the city was called *Moodyville* in 1872 after Sewell Prescott Moody (*c.* 1835–75), who bought a sawmill here in 1864. It was renamed North Vancouver in 1902.

Northwest Miramichi River NB A principal tributary of the Miramichi River, it rises in north central New Brunswick. Its name in Mi'kmaq, *Elmunokun*, means 'a beaver hole', in reference to Big Hole, at the mouth of the Big Sevogle River. Variant forms of the Aboriginal name appeared on maps between 1686 and 1755, but the lumbermen's descriptive name ultimately prevailed.

Northwest Territories When Canada acquired the Hudson's Bay Company's Rupert's Land and The North-Western Territory from Great Britain in 1870, the combined lands were called North-West Territories. Ten years later Great Britain transferred its rights to the Arctic Archipelago. Meanwhile parts of the present province of Manitoba (1870) and the district of Keewatin (1876) were separated from the territories, and in 1897 the territory of Yukon was taken from it. In 1905 the provinces of Alberta and Saskatchewan were severed from the newly named territory of the Northwest Territories, with Keewatin being restored to it as one of its four districts (the others were Ungava, Mackenzie, Franklin). In 1912 Ontario and Quebec took away parts of its territory by extending to the shores of Hudson and James bays. Nunavut will be separated from it in 1999. Meanwhile the residents of the remaining territory, primarily based in the Mackenzie River valley, are debating what to call it, with the preference apparently to retain the misnomer, Northwest Territories.

Northwest, Mount YT A mountain (3216 m) in the Centennial Range of the St. Elias Mountains, it was named in 1967 after the Northwest Territories.

North West River NF On the east side of the mouth of the Naskaupi River, where it flows into Lake Melville, northeast of Happy Valley-Goose Bay, this town (1958) was named after an earlier name of the river. A French trading post was established here in 1742, and a Hudson's Bay Company post was opened in 1836.

North Wiltshire PE South of Hunter River and in the Wiltshire district, its school district was called *New Wiltshire* from about 1846 to about 1875, when it became North Wiltshire, the name of the railway station. The post office opened as *New Wiltshire* in 1863, and was only changed to North Wiltshire in 1967, with this becoming the name of the municipal community in 1974. The school district was likely called after the English county.

North York ON Separated from York Township in 1922, North York Township became a borough in Metropolitan Toronto in 1967, and then a city in 1979. With a population of 558,000 (1996), it is second in size only to Toronto in the province.

Norton NB A village (1966) southwest of Sussex, its post office was called *Fingerboard* in 1840. It was changed to *Norton Station* in 1869 and became Norton in 1900. It was called after **Norton Parish**, created in 1795, and named after Norton, MA, near Taunton.

Norval ON A community in the town of Halton Hills, Halton Region, it was named in 1836 by Alexander McNab, possibly from a reference to Norval in John Hume's Scottish play, *Douglas*. Or possibly the McNab family came from Scotland in 1820 on the *Young Norval*.

Norway House MB Located on the eastern shore of Little Playgreen Lake, an enlargement of the main channel of the Nelson River, this Cree reserve has its roots in the building of a winter road from York Factory to the Red River Settlement in 1814 by Norwegian axemen employed by the Hudson's Bay Company. First located at the Lake Winnipeg outlet, 25 km to the south, it was moved to its present site in 1826 to avoid the possibility of its original location being destroyed by a flood. Its post office was opened in 1904.

Norwegian Bay NT Surrounded by Ellesmere, Devon, Cornwall, and Axel Heiberg islands, this water body was named *Norske Bugten* in 1900 by explorer Otto Sverdrup, after his fellow countrymen.

Norwich ON The township in Oxford County was named in 1795, when it was part of Norfolk County, and took its name from the largest city in Norfolk, England. When Oxford County was restructured in 1975, the municipal townships of *North Norwich* and *South Norwich*, created in 1856, the community of **Norwich**, incorporated in 1876, and the municipal township of *East Oxford* were united as the municipal township of Norwich.

Norwood ON A village (1877) in Peterborough County, east of the city of Peterborough, it was named in 1841 by Harriet Maria (Keeler) Groves, who was reading a book about a suburb called Norwood on the south side of London, England. She also noted that its site was in the woods, north of the family's former home of Colborne.

Notre Dame Bay NF On the north coast of the island of Newfoundland, this large bay, extending from Cape St. John, east of Baie Verte, to Joe Batt's Point on Fogo Island, was likely named by French explorers in the 1500s after the Virgin Mary. The bay was generally known to the Portuguese as *baia verde*, which has survived in the name Green Bay, a small bay on its southwestern side.

Notre-Dame-de-Bon-Secours QC Located east of Chambly, this municipality was created in 1989, having been made a parish municipality in 1869. The name celebrates the feast day on 24 May, the day of Our Lady of Perpetual Help. Its post office has been called *Village-Richelieu* since 1864.

Notre Dame de Lourdes MB A post office at this village (1963), south of Portage la Prairie, was opened in 1892. In the 1880s a seminary was founded here for the training of priests of the new order of Canons Régulaires, which was later broken up after a decree from Rome changed its rules. Lourdes in southwestern France is where the Virgin Mary revealed herself as the Immaculate Conception to Marie Bernarde Soubirous (St Bernadette) in 1858.

Notre-Dame-de-Lourdes QC North of Joliette, this parish municipality was named in 1925. It was given in honour of the sanctuary of Lourdes, in France's Hautes-Pyrénées, where the Blessed Virgin appeared to Marie Bernarde Soubirous (St Bernadette) in 1858. Its post office has been called Lourdes-de-Joliette since 1930.

Notre-Dame-du-Bon-Conseil QC This village, northeast of Drummondville, was named in 1957. The religious parish of Notre-Dame-du-Bon-Conseil was created in 1895, and the municipal parish of the same name followed three years later. The original parish may have been named after the Sisters of Our Lady of Good Counsel, which provided Christian education and instruction here.

Notre-Dame-du-Lac QC Located on the western shore of Lac Témiscouata, southeast of Rivière-du-Loup, it was created a parish municipality in 1871, became a village in 1949, and was elevated to town status in 1968. The religious parish had been named in 1869.

Notre-Dame-du-Laus QC Situated north of Buckingham, this municipality was created in 1946, taking its name from its post office, opened in 1878. The religious parish, established in 1873, was named after Laus in France's Hautes-Alpes, where there were some 600 apparitions of the Virgin Mary between 1664 and 1718.

Notre-Dame-du-Mont-Carmel QC A parish municipality between Trois-Rivières and Shawinigan, it was erected in 1859, a year after the religious parish was established. The name may be traced to an order of monks established in the twelfth century at the foot of Mount Carmel in the Holy Land, where an order of Carmelite nuns was permitted to affiliate in the fifteenth century.

Notre-Dame-du-Nord QC Located at the head of Lake Timiskaming, this municipality was established in 1928, taking its name from the religious parish, created in 1895.

Notre-Dame-du-Portage QC Situated on the south shore of the St. Lawrence, southwest of Rivière-du-Loup, this municipality was created in 1856 after the religious parish, named by Mgr Charles-François Baillargeon, who had a considerable devotion to the Immaculate Conception of the Virgin Mary. The portage is in reference to the point where the land route from Halifax reached the St. Lawrence in the 1780s.

Notre-Dame, Monts QC These mountains, part of the continental Appalachian chain, extend from Sherbrooke to Gaspé. Jacques Cartier may have named them on 15 August 1535, the feast day of the Assumption of the Blessed Virgin Mary. In 1544 pilot Jean Alfonse *dit* Fonteneau mentioned the *montz Nostre Dame* four times in his *Cosmographie*.

Nottawa ON In Simcoe County, south of Collingwood, its post office was named *Nottawa Mills* in 1854 after Nottawasaga Township and mills on the Pretty River. 'Mills' was soon dropped from the name.

Nottawasaga ON The township in Simcoe County was named in 1832 after the **Nottawasaga River**, which rises in Dufferin County, north of Orangeville. The river's name was taken from Ojibwa *Nahdowasaga*, 'outlet of the river of the Iroquois'. In 1994 the municipal township, the town of Stayner, the village of Creemore, and the municipal township of Sunnidale were merged to form the municipal township of Clearview. *See also* Wasaga Beach.

Nouvelle QC A municipality north of the mouth of the Restigouche River, it was named in 1953, recalling the names of the mission

(1834), the religious parish (1868), and the post office (1881). It was given in memory of Jesuit Fr Henri Nouvel (*c.* 1621–*c.* 1702), who was a missionary among the Aboriginals of the north shore of the St. Lawrence and the Gaspé from 1663 to 1669, after which he went to the Great Lakes area to minister in several missions for the following 32 years.

Novar ON In Parry Sound District, north of Huntsville, its post office was named in 1887 by miller Alexander McGillivray after his birthplace in Ross and Cromarty, Scotland, northeast of Dingwall.

Nova Scotia The province was named on 29 September 1621 in the grant to Sir William Alexander of the 'lands lying between New England and Newfoundland...to be known as Nova Scotia, or New Scotland'. At that time it included the areas of present-day Prince Edward Island (separated in 1769) and New Brunswick (1784).

Nova Scotia, Mount YT In the Centennial Range of the St. Elias Mountains, this mountain (3292 m) was named in 1967 after the province.

Nueltin Lake MB, NT On the sixtieth parallel, this lake (2033 km^2) was identified as *Nu-thel-tin-tu-eh* by geologist Joseph Burr Tyrrell in 1894. Tyrrell determined that in Chipewyan the name meant 'sleeping island lake'.

Nunavut This new territory was created in 1993, to come into being on 1 April 1999. It consists of the administrative regions of Keewatin, Baffin, and Kitikmeot, which together comprise all the former district of Keewatin, the northeastern part of the district of Mackenzie, and most of the Arctic Archipelago in the district of Franklin, except Banks and Prince Patrick islands, and parts of Victoria and Melville islands, and some smaller islands. In Inuktitut Nunavut means 'our land'.

Nyland, Mount BC In the Coast Mountains, southeast of Bella Coola, this mountain (2789 m) was named in 1962 after Pvt. Arthur W. Nyland of Bella Coola, who was killed in action on 19 February 1945.

Oak Bay BC A district municipality (1906) on the east side of Victoria, it was named after the adjacent bay, identified on Capt. Henry Kellett's 1847 chart.

Oak Bay NB East of St. Stephen, this bay's name was derived from Passamaquoddy *Wahquaeek*, 'head of the bay'. The name went from *Waughweig* in the 1780s through *Aouk Bay* to *Oak Point Bay* before 1800. **Waweig River** flows into the east side of the bay. There were post offices in the communities of **Oak Bay** (1847) and **Waweig** (1859), and both were closed in 1969.

Oak Island NS On the west side of Mahone Bay, this 1.3 km^2 island has been the legendary site of hidden treasure since 1795, when three men uncovered a shaft. Subsequent digging to 29.5 m and lowering a camera to a depth of 60 m have failed to find any buried treasure.

Oak Lake MB Located 10 km northeast of Oak Lake and west of Brandon, the Canadian Pacific Railway established a station here in 1882, at what is now the smallest (population 350) town (1907) in the province. It was earlier called *Flat Creek*. The lake had been known as *Lac des Chênes*, where there were real oak trees on the east side of the lake, not the usual scrub oak.

Oakland NS On the east side of the village of Mahone Bay, this place's post office was named in 1893 after the groves of oak trees along the shore of Mahone Harbour.

Oakland ON A township in Brant County, it was named in 1821 after the ridge of oak trees extending through the area. The community of **Oakland** was first called *Perth* and *Malcolm's Mills*, and its post office was named Oakland in 1840.

Oak Point MB A popular summer resort on the eastern shore of Lake Manitoba, its post office was opened in 1872. It was changed to *Radway* in 1896, but Oak Point was restored in 1918.

Oak Ridges ON In the town of Richmond Hill, York Region, its post office was named in 1851 after the large oak trees on a prominent morainic ridge extending from the Niagara Escarpment in the west to Rice Lake in the east.

Oakville MB Located between Winnipeg and Portage la Prairie, this community was settled in 1872–4, and became a Canadian Northern Railway station in 1888–9. Its post office was named in 1891, but confusion with Oakville, ON, led to changing the name in 1900 to *Kawende*, which means 'no name' in Ojibwa. It was changed back to Oakville in 1939.

Oakville ON On the Lake Ontario shore, this town (1857) in Halton Region was founded in 1827 by William Chisholm, and named in 1835 by his friend Robert Baldwin Sullivan, after the great white oaks that supported the local oak-stave industry. Sullivan was then the mayor of Toronto, and was later appointed to the Executive Council.

Oakwood ON In Victoria County, west of Lindsay, this community's post office was named in 1843 by Joseph Pearson after the extensive oak woods in the area.

Observatory Inlet BC Extending 70 km from the head of Portland Inlet, northeast of Prince Rupert, this inlet was named in 1793 by Capt. George Vancouver, who set up his chronometers at its mouth, and amended his latitude and longitude positions.

Ochre River MB A community on **Ochre River**, southeast of Dauphin, its post office was named in 1893. The Canadian Northern Railway was built through the settlement in 1896–7. The river, noted on the Palliser expedition map of 1865, has a bottom of yellow marl.

Odessa ON In Lennox and Addington County, west of Kingston, its post office was called *Mill Creek* in 1838, after mills built on what is now Millhaven Creek. It was renamed in 1855 after the city in present-day Ukraine, then under siege by the British during the Crimean War.

Odessa SK Southeast of Regina, this village (1911) was founded in 1904 by settlers from the the area of Odessa in Ukraine. The Canadian Northern Railway station was established here in 1907. Its post office was called *Magna* in 1908, *Odessa Station* in 1910, and simply Odessa in 1938.

Ogema SK West of Weyburn, this town (1913) was named in 1910, when the post office was opened, and the Canadian Pacific Railway's survey for a line had been laid out. Settlers here wanted to call it Omega, which is Greek for 'end', referring to the fact that it was at the end of the line. They were told Omega was already in use, so the letters were jumbled to form Ogema, which coincidentally is Cree for 'chief'.

Ogidaki Mountain ON An elevation rising to 665 m (2183 ft) in Algoma District, northeast of Sault Ste. Marie, it was named in 1966 after the Ojibwa for 'high ground'. It was then believed to be the highest point in Ontario, but in 1972 it was discovered that Ishpatina Ridge, north of Sudbury, at 685 m (2250 ft), was the highest point in the province.

Ogilvie Mountains YT In the west central part of the territory and northeast of the Yukon River, this chain of mountains was named *Ogilvie Range* in 1898 after William Ogilvie (1846–1912), and was changed to its present name in 1948. Ogilvie made the first survey of the Yukon-Alaska boundary in 1887–8 and laid out the townsite of Dawson in 1896. He was the commissioner of the Yukon from September 1898 to April 1901.

Ogre Mountain BC In the Coast Mountains, southeast of Bella Coola, this mountain (2880 m) was named in 1960 because it had the profile of Mont Ogre in Switzerland.

Oil City ON In Lambton County, southeast of Sarnia, this community's post office was named in 1874 after the oil boom, 15 years earlier. Never incorporated, its population is about 265 (1996).

Oil Springs ON A village (1864) in Lambton County, southeast of Sarnia, its post office was named in 1859, when North America's first commercial oil field was developed here.

Oka QC On the north shore of Lac des Deux Montagnes, west of Montréal, Oka describes both a municipality and a parish municipality, each with nearly identical populations of some 1660. The municipality was erected as the parish municipality of *L'Annonciation* in 1875, and the current name was adopted in 1953 to conform with the postal name, given in 1867. The parish municipality was called the municipality of *L'Annonciation-Partie-Nord* in 1918, and was changed to its present name in 1977. It annexed the town of *Oka-sur-le-Lac* (1942) in 1982. Oka is said to be derived from the name of an Algonquin chief, whose name was given after a variety of pike.

Okanagan Lake BC This lake (361 km^2), in the southern interior of the province, was noted in 1811 as *Ookanawgan* by geographer David Thompson. The origin of the name is uncertain. It may refer to the 'farthest point' on the **Okanagan River**, near **Okanagan Falls**, and south of Penticton, which was as far upstream that salmon ascended. The river (314 km) drains south into Washington State, where the Okanogan River joins the Columbia, 83 km south of the border. The Okanogan, an important division of the Interior Salish, occupied a large territory in the Okanagan and Columbia River valleys. In the late 1860s, Gov. Frederick Seymour of British Columbia decided on the spelling Okanagan to differentiate it from Okanogan south of the border.

Okotoks AB A town (1904) south of Calgary, it had been named *Dewdney* in 1894 by the Canadian Pacific Railway after Sir Edgar Dewdney, lieutenant-governor of the North-West Territories, 1882–8, minister of the interior, 1888–92, and lieutenant-governor of British Columbia, 1892–7. It was renamed in 1897 after the Blackfoot designation of a glacial erratic called the Big Rock, 8 km west of the town.

Olalla BC A community on Keremeos Creek, southwest of Penticton, its post office was named in 1900. It took its name from the Chinook jargon *olallie*, 'berries'. Saskatoon berries grew abundantly here.

Old Barns NS Situated on the south shore of Cobequid Bay, west of Truro, this name recalls two Acadian barns found here when the first English-speaking settlers arrived after 1760. The barns were torn down in the 1820s. Old Barns post office served the area from 1851 to 1968.

Old Crow YT A community at the junction of the **Old Crow** and Porcupine rivers, it was founded in 1911 by the Gwitch'in. They had moved in 1869 from Fort Yukon, on American territory in Alaska, to Old Rampart House, which they assumed was in British territory. When it was found in 1889 to be in Alaska, they moved again to Rampart House, just east of the border, and, after a smallpox epidemic, they moved yet again to the present site. The place was named after a respected chief called Tshim-Gevtik, which meant 'Walking Crow'. The bird after which it was named is really the northern raven.

Olden ON A township in Frontenac County, it was named about 1823 after John Olden, its first surveyor. He carved his name and the date 'June 12, 1831' in a tree.

Oldman River AB Rising in the Rocky Mountains west of Claresholm, this river flows south, and then east through Fort Macleod and Lethbridge. Joining the Bow River west of Medicine Hat, the two continue east as the South Saskatchewan River. At the point where Livingstone River joins the Oldman, northwest of Pincher Creek, was 'Old man's playing ground', marked by a cairn of rocks, each one placed there as a sign of good luck before its bearer entered the mountains. The mythical 'Old man' was called *Apistoki* by the Blackfoot, and *Wisukishak* by the Cree. *See also* Belly River.

Old Perlican NF On the southeast shore of Trinity Bay, northeast of Carbonear, this place's name was recorded as early as 1597 as *Parlican*, with *Le Vieux Perlican* occurring 100 years later. For a small island offshore, Jacques Nicolas Bellin had *I du Vieux Perlican* in 1754. **New Perlican** is 35 km southwest on the same shore. Recorded as early as 1669 as *Little Pernocan*, it was referred to as New Perlican eight years later. The meaning of 'perlican' as a designation for geographical features is possibly related to the 'pelican', perhaps an erroneous description of another bird encountered here.

Olds AB South of Red Deer, this town (1905) was named in 1892 by the Canadian Pacific Railway after George Olds, a CPR traffic manager. It was previously called *Lone Pine* and *6th Siding*.

Old Wives Lake SK Located 37 km southwest of Moose Jaw, this lake recalls a story of Cree hunters and their families being pursued in the early 1800s by the Blackfoot, after the Cree had had a successful hunt for buffalo meat. To allow the men, the younger women, and children to escape, the older women hatched a plan to give the appearance that the entire party remained camped at the lake overnight. However, the next morning the Blackfoot found that only the older women had tended the fires, and in their fury they killed them all. About 1900 the lake was named after Sir Frederick John W. Johnstone, who had hunted in the area. Nevertheless Old Wives Lake persisted in local use, and was officially restored in 1953.

O'Leary PE A municipal community (1983) in the western part of the province, it had been incorporated as a village in 1951. It was named after the road, which had been called after Michael O'Leary, who lived at the West Cape end of O'Leary Road in the first half of the 1800s. Its post office was called *O'Leary Station* in 1877, and received its present name in 1967.

Oliver BC A town (1991) on the Okanagan River, south of Penticton, its post office was named in 1921 after John Oliver (1856–1927), premier of the province, 1918–27. It was during his administration that an extensive irrigation project was undertaken in the area to provide land for returning First World War veterans. *See also* John Oliver, Mount.

Oliver ON A township in Thunder Bay District, and west of the city of Thunder Bay, it was named in 1874 after Adam Oliver (1823–82), a member of the Legislative Assembly of Ontario, 1867–75, and an owner of the Fort William (Thunder Bay) lumbering firm Oliver, Davidson and Company.

Omemee ON A village (1873) in Emily Township, Victoria County, southeast of Lindsay, it was first called *Williamstown*, *Emily*, and *Metcalfe*, before the post office was renamed Omemee in 1857, after the Omemee (Pigeon) subtribe of the Mississauga, who once hunted here.

Omerville QC A village in the regional county of Memphrémagog, on the north side of the town of Magog, it was erected in 1945. Its post office was open from 1950 to 1974. It was named after Magog butcher Omer Gaudreau, who acquired a farm here in 1932, divided it into village lots in 1945, and became the village's first mayor.

Omineca Mountains BC Located west of Williston Lake, this group of mountains was named before 1910 after the **Omineca River**, which rises in the Hogem Ranges, and flows east into the lake, opposite its Peace Arm. In Sekani its name may mean 'sluggish river', or 'lake-like'.

Ompah ON In Frontenac County, northwest of Perth, its post office was named in 1865 after the Algonquin for 'long step' or 'long portage', referring to the 5-km portage from the Mississippi River to Palmerston Lake.

Onanole MB Located south of Riding Mountain National Park and north of Minnedosa, this place's post office was named in 1928 on the suggestion of postmaster N.W. Tracy, whose house was located 'on a knoll'.

Onaping Falls ON A town in Sudbury Region, northwest of the city of Sudbury, it was created in 1973 by amalgamating the townships of Levack and Dowling, as well as the improvement district of Onaping, which had been laid out as a township in 1911, and named after **Onaping Lake**. The name may have been derived from the Ojibwa word *onapina*, 'I harness it', possibly referring to the Onaping Dam.

Onderdonk, Mount BC Located in the Selkirk Mountains, north of Revelstoke, this mountain (2694 m) was named in 1971 by mountaineers William L. Putnam and Roger W. Laurilla after Andrew Onderdonk (1848–1905), a railway engineer who supervised the construction of the Canadian Pacific Railway through the Fraser Canyon in the early 1880s.

100 Mile House BC This district municipality (1991) in the Cariboo was established in 1862 on the Cariboo Road, 100 miles (161 km) northeast of Lillooet. Its post office was opened in 1916, and it was created a village in 1965.

Oneida ON A township in the regional municipality of Haldimand-Norfolk, it was named in 1835 after the Oneida tribe of the Six Nations confederacy. Oneida may be an ancient name meaning 'upright stone', after a syenite boulder near their early home in central New York State.

Onondaga ON A township in Brant County, it was named in 1840 after the Onondaga nation, a member of the Six Nations confederacy, originally from the Lake Oneida and Syracuse area of New York State. The name means 'on the mountain'.

Onoway AB A village (1923) west of Edmonton, its post office was named in 1904. It may have been named after the singer Onaway in Longfellow's *Hiawatha*. Or it may be from the Ojibwa for 'good meadow', chosen by first postmaster W.P. Beaupre, because he could not call the post office Beaupre (meaning 'good meadow'), since Beaupré was already in use in Québec.

Onslow NS Located on the north side of Truro, this place was named after Onslow Township, established in 1759. The township was named after Arthur Onslow (1691–1768), speaker of the British House of Commons, 1727–54. Onslow post office was opened in 1855, with subsequent offices in the area called **Onslow Upper** (1854), **Central Onslow** (1866), **Onslow Mountain** (1884), **Onslow Station** (1884), and **Lower Onslow** (1884).

Ontario Ontario was first used for the lake in the Jesuit *Relation* of 1641. In 1656 map-maker Nicolas Sanson described it as *Ontario, ou Lac de St Louys*. René de Bréhant de Galinée in 1670 referred to it simply as *Lac Ontario*. Three years later explorer Louis Jolliet called it *Lac Frontenac ou Ontario*, and explorer René-Robert Cavelier de La Salle in 1684 described it only as *Lac Frontenac*, given in honour of the colonial governor. By the mid-1700s only *Lac Ontario* was shown on maps, with **Lake Ontario** being adopted by the British. In 1683 Fr Louis Hennepin said it meant 'beautiful lake'. The Huron word for 'lake' is *ontare*, and similiar words are used in the other Iroquoian languages. The main islands at the eastern end of Lake Ontario formed the first Ontario County in 1792. The second was created in 1849 from the east side of York County, and separated from it in 1853. In 1974 all of Ontario County, except Mara and Rama townships (now Ramara Township in Simcoe County), was transferred to the regional municipality of Durham. The province was officially named in 1867, and extended to its present limits in the north in 1912.

Ontario, Mount YT In the Centennial Range of the St. Elias Mountains, this mountain (3719 m) was named in 1967 after the province.

Ootsa Lake BC Part of the Nechako Reservoir, on the north side of Tweedsmuir Provincial Park, it derived its name from Carrier *yootsoo*, 'very low down', or 'way down towards the water'.

Ops ON A township in Victoria County, it was named in 1821, likely by Lt-Gov. Sir Peregrine Maitland after the Roman goddess of plenty and fertility. She was the wife of Saturn and the mother of Jupiter.

Orangeville ON A town (1873) in Dufferin County, it was named about 1845 for Orange Lawrence (1796–1861), a Connecticut native, who developed mills at the site and was the first postmaster in 1851. It was earlier called *The Mill*.

Or, Cape d' NS Extending into Minas Channel, this cape was named *Cap d'Or* by Samuel de Champlain in 1604, when he discovered outcroppings of copper here. He called the southern arm of the Bay of Fundy *Bassin des Mines*, which resulted in the names Minas Basin and Minas Channel.

Orford ON A township in Kent County, it was named in 1794 after the village of Orford, in Suffolk, England, and near the North Sea.

Orford, Mont QC Located northwest of the town of Magog, this 850-m mountain dominates the landscape of the township municipality of Orford, which was created in 1855. It took its name from the township, proclaimed in 1801, and named after a village in Suffolk, England. The area of the mountain was set aside as a 60 km^2 recreational park in 1933.

Orillia ON The township in Simcoe County was named by Lt-Gov. Sir Peregrine Maitland in 1822 after the Spanish word *orilla*, meaning 'bank' or 'shore'. In 1994 it was united with the village of Coldwater and the municipal township of Matchedash to form the municipal township of Severn. The city (1969) of **Orillia** was first known as *Invermara*, *The Narrows*, and *Newtown*, before its post office was called Orillia in 1836.

Orléans ON Divided between the city of Gloucester and the municipal township of Cumberland, Ottawa-Carleton Region, east of Ottawa, its post office was named in 1860, probably by first postmaster Théodore Besserer, after his birthplace on Île d'Orléans, downriver from the city of Québec.

Orléans, Île d' QC Comprising one of the continent's prettiest pastoral landscapes, this island was named by Jacques Cartier in 1536, likely after Henri II, Duc d'Orléans, the second son of François I. A year earlier he had called it *Isle de Bascuz* (recalling Bacchus, the Roman god of wine), because he had found strong vines here.

Ormstown QC Located southwest of Montréal, this village was incorporated in 1889, taking its name from its post office, opened in 1836. The origin of the name is uncertain, but the most widely accepted is that it was after the name of a son of merchant Alexander Ellice (1743–1805), who bought the seigneury of Beauharnois in 1795. However, Ellice did not have a son called Orm. In 1905 the village clerk said that it was chosen 20 years before for a station name from a road that was named in the 1820s after a noted person called Ormiston in Britain. It may have had a connection with Ormiston Corners in New York State.

Oro ON A township in Simcoe County, it was named in 1820 by Lt-Gov. Sir Peregrine Maitland, after the Spanish word for 'gold'. In January 1994 Oro was joined with Medonte to form the municipal township of **Oro-Medonte**.

Oromocto NB A town (1956) southeast of Fredericton, it was laid out in the 1950s to accommodate the headquarters and residential areas of Canadian Forces Base Gagetown, although Oromocto post office had been opened over a century earlier in 1843. **Oromocto River**, whose North and South branches rise in **Oromocto** and **South Oromocto** lakes respectively, has a name derived from Maliseet *Welamooktook*, 'fine river', in reference to easy navigation by canoe. In the late 1600s the river's name was usually spelled with variations of *Ramouctou*, with the present form being introduced by surveyors general Charles Morris and Thomas Sproule in the late 1700s.

Orono ON In the municipality of Clarington, Durham Region, this community was named in 1852, when a visitor from Maine proposed Orono during a public meeting. The town in Maine was named after Penobscot chief Orono (1688–1801), who sided with the colonists opposed to British rule during the American Revolution.

Orwell ON In Elgin County, west of Aylmer, this place's post office was named in 1856 after Orwell, VT, northwest of Rutland. Earlier, it was called *Catfish Corners* and *Temperanceville*.

Osgoode ON A township in Ottawa-Carleton Region, it was named in 1798 after William Osgoode (1754–1824), chief justice of Upper Canada, 1792–4, and of Lower Canada, 1794–1801. The post office in the community of Osgoode was called *Osgoode Station* in 1880, and was shortened to Osgoode in 1962. Osgoode Hall, the Toronto headquarters of the Law Society of Upper Canada, was constructed in 1829–32.

Oshawa ON A city (1924) in Durham Region, it was first known as *Skae's Corners*, after merchant Edward Skae. At a meeting in 1840, *Sydenham* was chosen in honour of Charles Edward Poulett Thomson, Baron Sydenham (1799–1841), governor-in-chief of British North America, 1839–41. Local trader Moody Farewell invited two Mississauga friends from Rice Lake to propose a more original name. They suggested *ajawi*, signifying 'crossing to the other side' or 'shore of a river or lake', and the name Oshawa evolved from it.

Osler SK North of Saskatoon, this town (1985) was named in 1890 after Sir Edmund Boyd Osler (1845–1924), member for West Toronto in the House of Commons, 1896–1917, president of the Dominion Bank from 1901, and a founder of the financial firm of Osler, Hammond and Nanton. The firm contracted to build the railway to Prince Albert.

Osnabruck ON A township in the united counties of Stormont, Dundas and Glengarry, it was named in 1787 after a title of the second son of George III, Frederick, the Duke of York and Albany (1763–1827). A year after his birth, he was appointed to the hereditary office of bishop of Osnabrück, which he abandoned in

1803. Osnabrück is a city in Germany's Niedersachsen.

Oso ON A township in Frontenac County, it was named in 1823 by Lt-Gov. Sir Peregrine Maitland after the Spanish word for 'bear'.

Osoyoos BC On the eastern shore of **Osoyoos Lake** and near the Washington State border, this town (1983) was established in 1865 as a customs post. It took its name from the Okanagan *sooyoos*, 'sand bar across', referring to a bar that almost cuts the lake in half. Peter O'Reilly of Hope may have suggested adding 'O' to the name to give it dignity.

Osprey ON A township in Grey County, it was named in 1822, likely by Lt-Gov. Sir Peregrine Maitland, after the osprey, a large bird of prey.

Otonabee ON A township in Peterborough County, it was named in 1820 after the **Otonabee River**, which flows southwest from Katchewanooka Lake through the city of Peterborough to Rice Lake. The name is from an Ojibwa word, such as *odonimaseebi*, meaning 'mouth water', in reference to the place where the river empties into Rice Lake through a delta.

Ottawa ON Queen Victoria chose Ottawa as the capital of Canada in 1857, after being advised by Gov. Gen. Sir Edmund Head and Henry Labouchere, secretary of state for the colonies. It had become a city two years earlier, having been incorporated as the town of Bytown in 1850. Ottawa was chosen in 1854 in honour of the two hundredth anniversary of the first flotilla of furs brought from the west side of Lake Michigan by the Ottawa and other Aboriginals to trade with the French. The townsite was developed in 1827 by Lt-Col John By on the instructions of Lord Dalhousie, governor-in-chief of British North America.

Ottawa-Carleton ON Created a regional municipality in 1969, it united the cities of Ottawa, Kanata, and Vanier, the municipal townships (later cities) of Nepean and Gloucester, the municipal townships of Goulbourn, Osgoode, Rideau, and West Carleton, the village of Rockcliffe Park, and the municipal township of Cumberland Township, previously in the united counties of Prescott and Russell.

Ottawa River ON; **Rivière des Outaouais** QC Samuel de Champlain referred to the river in 1613 as *Rivière du Nord* and *Grande rivière des Algommequins*. By the late 1600s it was called *Rivière des Outaouacs*, which was subsequently rendered as Rivière de l'Outaouais by the French, and Ottawa River by the British. The Ottawa, an Algonquian nation, were known to the Huron as *Andatahouats*, noted as early as 1623–4 by Récollet brother Gabriel Sagard, and as *Ondataouaouat*, *Outaouacs*, and *Outaouak*, in the Jesuit *Relations* of the mid-1800s. Ottawa has been interpreted as being from *adawe,* used in Algonquin, Cree, Ojibwa, and other Algonquian languages to mean 'trade' and 'buy and sell'.

Otter Lake QC A community in the regional county of Pontiac, north of Shawville, its post office was named in 1866 after a nearby lake. There are numerous lakes in Québec named after the otter, *Lutra canadensis*, prized for its glossy fur.

Otterburn Park QC Located on the east bank of the Richelieu, south of Mont-Saint-Hilaire, this place became a municipality in 1953 and a town in 1969. It was named after a Grand Trunk Railway station, which was given in 1871 for Otterburn, in Northumberland, England, the birthplace of Sir Joseph Hickson (1830–97), general manager of the GTR, 1874–97.

Otterburne MB A place on the Canadian Pacific Railway, south of Winnipeg, it was named about 1877 after Otterburn, in Northumberland, England, where James Douglas, 2nd Earl Douglas was killed in 1388, although his forces defeated Sir Henry Percy (Hotspur). Its post office was opened in 1879.

Otterville ON On **Big Otter Creek** and in Oxford County, northeast of Tillsonburg, it was first known as *Otter Creek Mills*. Its post office was named Otterville in 1837.

Ouimet Canyon ON Located northeast of Thunder Bay, it was named after the nearby Canadian Pacific Railway station of Ouimet. The station was given for Joseph-Aldéric Ouimet, who served in a military expedition sent to the West in 1885 to subdue the North-West Rebellion. He was the minister of public works in Ottawa, 1892–6.

Outaouais, Communauté urbaine de l' QC Embracing the cities of Gatineau, Hull, and Aylmer, and the towns of Buckingham and Masson-Angers, it was established in 1991. In 1971 the *Communauté regionale de l'Outaouais* had been established, which then also included the present regional county of Les Collines-de-l'Outaouais, the territory immediately north and west of Hull and Gatineau. In French, the river is known as Rivière de l'Outaouais. *See also* Ottawa River.

Outardes, Rivière aux QC A major river on the north shore of the St. Lawrence, it flows south parallel to Rivière Manicouagan for almost 400 km. Identified on the mid-1700 maps of Jacques Nicolas Bellin, its name was derived from the wild goose. It had been described as an *outarde* in 1535 by Jacques Cartier, presuming it was the equivalent of the European bustard, which is not native to North America.

Outhouse Point NB Extending into the Petitcodiac River, opposite Moncton, this point was likely named after Edwin Outhouse. An early settler in Coverdale Parish, his father was a Loyalist who had received a land grant in Sackville Parish.

Outlook SK A town (1909) south of Saskatoon and on the east bank of the South Saskatchewan River, it was founded by the Canadian Pacific Railway in 1908. Presumably one of the CPR men was impressed with the views up and down the river, and off toward the horizon, and proposed the name. Its post office was established that year as well.

Outram, Mount AB This mountain (3240 m), near the head of the North Saskatchewan River, and north of Howse Pass, was named in 1921 by the Alberta-British Columbia Boundary Commission after climber Sir James Outram (1864–1925), and grandson of Gen. Sir James Outram, who served in India. Settling in Victoria in 1900, he made many first ascents in the Rocky Mountains, including Mount Assiniboine.

Outram, Mount BC In the Cascade Mountains, southeast of Hope, this mountain (2454 m) was named about 1858 by Hudson's Bay Company factor Alexander C. Anderson after his uncle, Gen. Sir James Outram (1803– 63), one of the relievers of Lucknow, India, in 1857.

Outremont QC Located on the north side of Mont Royal, this city's name means literally 'beyond the mountain'. The Le Bouthillier family had called their residence Outremont before the place was incorporated as a village in 1875, and a town 20 years later.

Overton NS This place was named because it is 'over' on the west side of Yarmouth Harbour. It had its own post office from 1881 to 1917.

Owen Sound ON A city (1920) in Grey-County, it was called *Sydenham* in 1840 after the adjoining bay, which had been named after Charles Edward Poulett Thomson, Baron Sydenham (1799–1841), governor-in-chief of British North America, 1839–41. However, two decades before, Admiralty surveyor Henry W. Bayfield had named the harbour after Adm. Sir Edward Campbell Rich Owen (1771–1849). Navigators called the harbour *Owen's Sound*, and this name was chosen as the post office in 1846. *Sydenham* was named the county seat in 1854, but three years later it became the town of Owen Sound, with the post office changing to this form.

Oxbow SK Named in 1884 by Canadian Pacific Railway official Samuel Sproul, this town (1904), east of Estevan, is located at a pronounced 'ox bow' of the Souris River, giving the appearance of part of a yoke used to harness draft oxen.

Oxford NS This town (1904) on River Philip, east of Amherst, was first known as *Head of Tide*. It was renamed in 1871 after the practice of fording oxen across the river to pasture, replicating the name of the English university city. **Oxford Junction**, where the Intercolonial Railway routes to Halifax and Sydney separated, 5 km to the south, was named in 1893.

Oxford ON Created in 1798, the county was named after the English city and county west of London, following Lt-Gov. John Graves Simcoe's proposal in 1793 that Oxford be established up the Thames River from London. Oxford Township was also created 1798. In 1845 it was divided into the municipal township of *East Oxford* (part of Norwich Township since 1975), and the municipal township of *West Oxford* (part of South-West Oxford Township, also since 1975). On the north of the river, the municipal township of *North Oxford* was organized in 1850, but was merged with the municipal township of Zorra in 1975.

Oxford House MB Located on the north shore of **Oxford Lake**, at the point where Hayes River flows toward the north, it was the site of a Hudson's Bay Company post established by William Sinclair in 1798. Its post office was opened in 1950. The lake was called by the Cree *Pethapaw Nipi*, 'lake with bottomless hole', which had been misleadingly rendered as *Holy Lake*.

Oxford on Rideau ON Oxford Township in the united counties of Leeds and Grenville was named in 1798 after the English city and county west of London. To differentiate it from Oxford County, and East, North, and West Oxford townships, it became the municipal township of Oxford-on-Rideau in 1850. **Oxford Mills**, southwest of Kemptville, was named in 1852.

Oyama BC Midway between Kelowna and Vernon, this community's post office was named in 1906 after Prince Iwao Oyama (1842–1916), a Japanese field marshal. In 1894 he captured Port Arthur (now Lüshun in northeastern China) during the Sino-Japanese War.

Oyen AB A town (1965) east of Hanna, and near the Saskatchewan border, it was named in 1912 by the Canadian Northern Railway after settler Andrew Oyen (1870–1937). He arrived here in 1909 from the United States, having left Norway in 1887.

Oyster Bed Bridge PE On the Wheatley River, northwest of Charlottetown, it was named after 1840, when local residents petitioned for a bridge. In 1765 Samuel Holland wrote 'Oyster Banks' on his map at the mouth of the river. The place's post office was established in 1875.

P

Pabok QC This regional county, on the north shore of Chaleur Bay, and south of Gaspé, was named in 1981. The name, an earlier form of Pabos, occurred in a 1753 register of births, marriages, and burials. The municipality of **Pabos** was erected in 1876. Several meanings have been advanced for both Pabok and Pabos, with the most convincing being the Mi'kmaq word *papôg*, 'calm waters'.

Pacific Rim National Park BC Located on the west coast of Vancouver Island, this national park (389 km^2), created in 1970, comprises three distinct parts between Tofino, on the northwest, and Port Renfrew, on the southeast, a distance of 105 km. It also extends offshore to the 10-fathom line.

Pacquet NF On the north coast of the island of Newfoundland and midway between Baie Verte and La Scie, this municipal community (1962), on the northern arm of **Pacquet Harbour**, became a base for French migratory fishing after 1713. Derived from the French word *paquet*, the name may mean 'big wave', perhaps a common feature of the harbour.

Paddockwood SK Northeast of Prince Albert, this village (1949) was named in 1912 by a Mrs Pitts, after her birthplace of Paddock Wood in Kent, England, southwest of Maidstone. Its post office was opened the following year.

Pagoda Peak BC In the Coast Mountains, northeast of Mount Waddington, this mountain (3215 m) was named in 1947 by mountaineer Fred W. Beckey because of its resemblance to a Chinese pagoda.

Pain Court ON In Dover Township, Kent County, west of Chatham, its post office was named *Dover South* in 1860. At that time the place was also known as Pain Court, either because they had often become short of bread (*pain deviendrait court*), or the loaves were smaller than baked elsewhere. Pain Court became the postal name in 1911, but from 1916 to 1991 the words were fused as *Paincourt*.

Painswick ON In the town of Innisfil, Simcoe County, southeast of Barrie, this place's post office was named in 1871 after a place in Gloucestershire, England, south of the city of Gloucester.

Paipoonge ON A township in Thunder Bay District and on the southwest side of the city of Thunder Bay, it was named in 1889 after the Ojibwa word for 'winter'.

Paisley ON A village (1873) in Bruce County, northwest of Walkerton, it was founded in 1851. Three names were considered for the place: Orchardville, Forfar, and Roweville. A surveyor observed the close connections in Scotland of several local names in the county, and suggested Paisley, which was adopted in 1856.

Pakenham ON A township in Lanark County, it was named in 1823 after Gen. Sir Edward Michael Pakenham (1778–1815), who was killed at the Battle of New Orleans. The community of Pakenham was first known as *Harvey's Mills* and *Pakenham Mills*, before Pakenham post office was opened in 1832.

Palgrave ON A community in the town of Caledon, Peel Region, it was first known as *Buck's Town*. Postal authorities named it Palgrave in 1869 after the noted English literary figure Francis Turner Palgrave (1824–97).

Palliser Range AB North of Banff, this range commemorates Capt. John Palliser (1817–87), who led an expedition in the late 1850s to investigate the geography of Canada between the Red River and the Rocky Mountains, and between the forty-ninth parallel and the North Saskatchewan River. His resulting reports on the people and land capability set the direction for much of the agriculture, communication, and settlement policies and practices that followed. A vast area of semi-arid land in southwestern Saskatchewan and southeastern Alberta is unofficially known as the Palliser Triangle. **Palliser River**, a tributary of the Kootenay in British Columbia, rises northeast of Windermere.

Palmerston ON A town (1874) in Wellington County, northeast of Listowel, it was named in 1872 after Henry John Temple, 3rd Viscount Palmerston (1784–1865), prime minister of Great Britain 1855–65.

Pambrun, Mount BC In the Purcell Mountains, east of Kootenay Lake, this mountain (3063 m) was named in 1937 after Pierre Chrysologue Pambrun (1792–1840), a Hudson's Bay Company chief trader.

Pamdenec NB A neighbourhood in the village of Grand Bay, west of Saint John, its name was chosen for a Canadian Pacific Railway station in 1902 by historian W.O. Raymond and Maliseet-speaker Sabatis Paul, from the Maliseet for 'small mountain', to replace *Hillside*. Pamdenec was incorporated as a village in 1966, but was annexed by Grand Bay in 1972.

Panache, Lake ON Located southwest of Sudbury, it was commonly called *Lake Penage*, before the present name was authorized in 1927. Panache is the French word for 'feather' or 'plume'. The lake's outline resembles a feather with a main stem and many little branches.

Pangnirtung NT On the south side of Cumberland Peninsula, Baffin Island and at the mouth of **Pangnirtung Fiord**, this hamlet (1972) was established in 1921 as a Hudson's Bay Company post. The name in Inuktitut means 'place of many buck deer', although the occurrence here of deer and caribou since the beginning of the twentieth century has been rare.

Papineau QC The regional county was erected in 1983, succeeding the former county of Papineau. The latter had been named after the celebrated politician Louis-Joseph Papineau (1786–1871), who lived in exile in Paris from 1839 to 1845 for his role in the aborted Rebellion of 1837–8. The village of **Papineauville** was erected in 1896, taking the name from its post office, opened in 1855. There is a claim that it was named after Louis-Joseph's father, Joseph Papineau (1752–1841), who was granted the seigneury of Petite-Nation in 1801. There is another claim that Louis-Joseph's brother, Denis-Benjamin (1789–1854), also a prominent politician, founded the place about 1850, and named it after himself.

Paquetville NB A village (1966) southwest of Caraquet, its post office was named in 1878 after Fr Joseph-Marie Paquet (1804–69), the parish priest at Caraquet, 1848–69. It was established as a free-grant settlement in 1866.

Paradise NS In the Annapolis Valley, east of Bridgetown, this place's post office was called *Paradise Lane* in 1856 and was renamed Paradise in 1901. It was derived from the Acadian designation *Paradis Terrestre*, 'earthly paradise', used in the area as early as 1684.

Paradise NF The name of this town (1971), west of St. John's, may have been chosen in the 1890s to suggest it was an inviting place to live and raise a family. It may have been given by a Revd Mr Colley of Topsail or by a politician. In 1992 it annexed the area of St. Thomas, which was first called *Horse Cove*. It was renamed in 1922 in honour of Fr Thomas O'Connor. St. Thomas had been created a municipal community in 1977.

Parent QC A village (1947) on a line of the Canadian National Railways, northwest of La Tuque, its post office was named in 1915, three years after a division point had been established by the Transcontinental Railway here. It was named after Simon-Napoléon Parent (1855–1920), mayor of the city of Québec, 1894–1906, premier of Québec, 1900–5, and president of the Transcontinental Railway Company, 1905–11.

Parham ON In Frontenac County, north of Kingston, this community's post office was named in 1862, possibly after either Parham, in Suffolk, England, or Parham, in West Sussex.

Paris ON A town (1856) in Brant County, it was founded in 1829 by Hiram 'King' Capron, who named it three years later after the extensive plaster of Paris gypsum beds. Plaster of Paris was first recognized near Paris, France, as a valuable ingredient for luxuriant crop growth.

Parkdale PE A neighbourhood in the east side of Charlottetown, it had been incorporated as a village in 1950, and a town in 1973. In the 1880s its area was called *St. Avards*, after Joseph Avard (1761–1816), who had come from England to manage lands at Murray Harbour.

Parkers Cove NF On the western shore of Placentia Bay and northeast of Marystown, this municipal community (1966) may have been named after a migratory fisherman, or after surveyor William Parker (1743–1802), who prepared maps and the ship's log and journal in the 1760s for Capt. James Cook.

Parkers Cove NS North of Annapolis Royal and on the Bay of Fundy shore, this place was named after an early Parker family that had settled here. Its post office operated from 1871 to 1961.

Parkhill ON A town (1886) in Middlesex County, northwest of London, it was founded in 1860 and first named *Westwood* and *Swainby*. Its post office was named in 1864 after Parkhill in the parish of Logie, Ross and Cromarty, Scotland, the birthplace of resident Simon McLeod.

Parksville BC This city (1986) on the east coast of Vancouver Island, northwest of Nanaimo, was first called *Englishman's River* after the local river named for English settlers who had arrived in the late 1880s. Its post office was named in 1887, after first postmaster Nelson Parks, the earliest settler in the area.

Parrsboro NS On the north shore of Minas Basin, this town (1884) was named after John Parr (1725–91), governor of Nova Scotia, 1782–91. Its post office had been named *Parrsborough* in 1812, with the present form being adopted about 1895.

Parry Channel NT This channel, from the entrance of Lancaster Sound in the east, to the outlet of M'Clure Strait in the west, was named in 1956 by geologist Y.O. Fortier after Sir William Edward Parry (1790–1855). He commanded four Arctic expeditions between 1818 and 1825. By penetrating Lancaster Sound, Barrow Strait, and Viscount Melville Sound as far as 110°W on 6 September 1819, he was awarded £5000 by a King's order in council.

Parry Islands NT A group of islands in the Queen Elizabeth Islands, and embracing Melville, Byam Martin, Bathurst, Cornwallis, Prince Patrick, Eglinton, Mackenzie King, Brock, Borden, and Lougheed islands, it was called the *Parry Group* before 1910. It was named after Sir William Parry, who discovered and named some of the individual islands. In 1939, the group was given its present designation.

Parry Sound ON The harbour was named by Admiralty surveyor Henry W. Bayfield during his 1822–5 survey. The name recalled northern Canada's *Parry Sound* (now Viscount Melville Sound), named in 1819 after Sir William Parry, who took the first ship west to the one hundred and tenth meridian. The post office in the town (1887) of **Parry Sound** was named in 1865, and **Parry Sound District** was created five years later.

Parsnip River BC Rising in the Hart Ranges of the Rocky Mountains, this river flows northwest into Williston Lake. Before the creation of the lake in 1968, it formed one of the two main sources of the Peace River (the other was the Finlay). It took its name from giant cow parsnips growing along its banks.

Parson's Pond NF On the northwestern coast of the Northern Peninsula, this municipal community (1966) is located at the outlet of **Parsons Pond**. Parsons is a common surname on the coast, a Parsons family having settled in 1878 on Bonne Bay, 50 km to the south. It may have been called after the Revd Isaac Parsons, who visited the settlement. Capt. James Cook called the pond *Sandy Bay* in 1768.

Partridge Island NB In Saint John Harbour, this island was noted in a French grant of 1672 as *Isle au Perdrix*, 'partridge island'. By 1711 it was described by Nathaniel Blackmore as *Partridge I.*

Pasadena NF On the southeast side of Deer Lake, and northeast of Corner Brook, this town (1955) took its name in the late 1930s from a farm established in 1933 by Lawrence Earle. The farm was called after the city in southern California, where Earle's wife once had lived. The city's name had been devised in 1874 from Chippewa (Ojibwa) words meaning 'crown of the valley'.

Paspébiac QC Located on the north shore of Chaleur Bay, east of New Carlisle, this municipality's post office was opened in 1832, and the municipality was established in 1877. In 1672 Acadian governor and author Nicolas Denys mentioned a cape called *Paspec-biac* in his *Description géographique et historique*. In 1707 the seigneury of Paspébiac was granted to Pierre Haimard (1674–1724). The municipality of **Paspébiac-Ouest** was created in 1922.

Passamaquoddy Bay NB On the southwestern side of the province, this bay's name was derived from Passamaquoddy *Peskutumaquadik*, 'place where pollock are', which is also the source of the Aboriginal tribe's name. Occurring as early as 1675–7 in the Jesuit *Relations* as *Pessemouquote*, its present spelling was established by chart-maker J.F.W. DesBarres in 1776.

Patapedia River NB, QC Rising in Québec, this river flows southeast into the Restigouche River, west of Campbellton. It forms part of the boundary between New Brunswick and Québec, where it is officially **Rivière Patapédia**. Its name was derived from Mi'kmaq *Patawegeok*, which possibly means 'running through burnt land', or 'place to peel birch bark, and build a canoe'.

Patricia Portion ON In 1912 that portion of Canada north of the Albany River, west of James Bay, east of Manitoba, and south of Hudson Bay, was attached to Ontario, and named after Princess Patricia, the daughter of the Duke of Connaught, governor general of Canada, 1911–16. For judicial purposes, it was joined to the district of Kenora in 1927.

Patterson, Mount YT In the St. Elias Mountains, this mountain (3444 m) was named in 1970 after John Duncan Patterson (1864–1940), a former president of the Alpine Club of Canada.

Pattullo, Mount BC In the Coast Mountains and northeast of Stewart, this mountain (2729 m) was named in 1927 after Thomas Dufferin Pattullo (1873–1956), premier of British Columbia, 1933–41. Previously he had been an alderman in Dawson, YT, a mayor of Prince Rupert, and a minister of lands in British Columbia.

Paul Lake BC Located east of Kamloops, this lake's name was proposed in 1925 by surveyor C.H. Taggart, after Jean Baptiste Lolo, who was commonly called St Paul.

Paynton SK Northwest of Battleford, this village (1905) was first settled about 1885 by three North-West Mounted Police constables, including Peter Paynter. Its post office was opened in 1904, and the postmaster reported the next year that Paynter was farming 2000 a (809 ha), and was 'known for his kindly assistance to one & all'.

Peace River AB, BC One of Canada's major rivers, it rises in Williston Lake in British Columbia, and flows east, north, and then east again to enter the Slave River, north of Fort Chipewyan. The name was derived from a point near its mouth, where the Cree and Beaver reached an accord on their respective territories, which explorer Alexander Mackenzie described in his *Voyages from Montreal* (1801). The town (1919) of **Peace River** is located at the junction of the Peace and Smoky rivers, northeast of Grande Prairie. Known as *Peace River Crossing* until 1916, its development was assured when the Edmonton, Dunvegan and British Columbia Railway crossed the river here the previous year.

Peachland BC On the western shore of Okanagan Lake, southwest of Kelowna, this district municipality (1955) was named in 1898 by John Moore Robinson, probably after the first peach orchards planted here by the Lambley family. Earlier it was called *Camp Hewitt*, after Gus Hewitt, who prospected here about 1890.

Peary Channel NT The entrance between Ellef Ringnes and Meighen islands was named *Peary Strait* by Vilhjalmur Stefansson in 1917 after Adm. Robert Peary. The Geographic Board did not accept the proposal in 1921, but authorized Peary Channel in 1939. Peary led the United States Polar Expedition, 1898–1902, which explored the lands and waters on the east side of Ellesmere Island. Peary set out to reach the North Pole, getting as far as 84°17′27″N in 1902, reaching 87°06′N in 1906, and attaining the vicinity of the pole in 1909.

Peawanuck ON After the community of Winisk, near the mouth of Winisk River, was destroyed by ice and water in 1986, its residents were relocated 32 km upriver. The name of the new community was possibly derived from Cree *payukwunok*, 'in one place'.

Peck, Mount BC In the Rocky Mountains, north of Sparwood, this mountain (2921 m) was named in 1919 after H.M. Peck, a Geological Survey assistant, who was killed in action in Europe.

Peck, Mount BC In the Muskwa Ranges of the Rocky Mountains, this mountain (2750 m) was named in 1987 after Don Peck (1920–80), a long-time guide, outfitter, and rancher in Fort St. John, and earlier at Trutch. It had been named *Mount Stalin* in 1944 after Marshal Joseph Stalin (1879–1953), leader of the Union of Socialist Soviet Republics after 1925. It had been given in this area where other Allied leaders of the Second World War were honoured, and was changed as a result of several petitions to expunge the name of a man deemed to be a notorious criminal.

Peel ON Peel County was named in 1849 after Sir Robert Peel (1788–1850), who held many senior posts in the British government, including prime minister, 1834–5 and 1841–6. It was reorganized as a regional municipality in 1974, with three municipalities, the cities of Brampton and Mississauga, and the town of Caledon. **Peel Township** in Wellington County was named in 1835 after Sir Robert Peel, during his first term as prime minister.

Peel River NT, YT Rising at the confluence of the Ogilvie and Blackstone rivers, in north central Yukon, this river (684 km, to the head of the Ogilvie) flows northeast, then north, to flow into the Mackenzie River, north of Fort MacPherson, NT. It was named in 1826 by John (later Sir John) Franklin after Sir Robert Peel. Peel was then the secretary of the home department and later served as the British prime minister, 1834–5 and 1841–6.

Peel Sound NT Between Prince of Wales and Somerset islands, this channel was likely named in the 1830s after Sir Robert Peel, prime minister of Great Britain, 1834–5 and 1841–6.

Pefferlaw ON In the town of Georgina, York Region and east of Sutton, its post office was named in 1851 by Capt. William Johnson after the Gaelic for 'beautiful hills'.

Peggys Cove NS At the eastern entrance to St. Margarets Bay, southwest of Halifax, this fishing community may have been named after Margaret (Peggy) Rodgers (Rogers), who with her husband, William, migrated from Ireland about 1770, lived here for six years, and then moved to Caribou Island, north of Pictou.

Pelee ON A township, embracing **Pelee Island** in Essex County, it was named in 1868. The island was called after **Point Pelee**, named *Pointe Pelée* during the French regime, after the bare appearance on its east side. **Point Pelee National Park** (16 km^2) was established in 1918.

Pelham ON A town in the regional municipality of Niagara since 1970, it had been organized as a township in Lincoln County in 1790. It was named after Henry Fiennes Pelham Clinton, 9th Earl of Lincoln and 2nd Duke of Newcastle under Lyme (1720–94). His mother was the sister of two prime ministers, Sir Henry Pelham and Thomas Pelham-Clinton, the 3rd Duke of Newcastle, and his wife was Catherine Pelham, whose surname he adopted.

Pelly SK Located northeast of Canora, this village (1911) was named by the Canadian Northern Railway in 1910 after nearby Fort Pelly. The fort was established in 1824 by the Hudson's Bay Company and named after Sir John Henry Pelly (1777–1852), a governor of the company. Its post office was opened in 1910.

Pelly River YT Rising in the Mackenzie Mountains, south of Macmillan Pass, this river (608 km) flows northwest to join the Yukon at the site of Fort Selkirk. It was named in 1840 by Hudson's Bay Company trader Robert Campbell after Sir John Henry Pelly, a governor of the company. **Pelly Crossing** is a small community at the point where the Klondike Highway crosses the river. The ferry crossing was replaced by a bridge in 1958. The **Pelly Mountains**, which extend for 375 km along the northwest side of the river, were named in 1887 by geologist George M. Dawson.

Pemberton BC A village (1956) in the Lillooet River valley, northeast of Whistler, it was founded about 1911. Its post office was called *Agerton* in 1912, but was changed to Pemberton in 1931. It was named after Joseph Despard Pemberton (1821–93), who was engaged by the Hudson's Bay Company as the surveyor of Vancouver Island, 1851–59. After leaving the HBC, he was the surveyor general of Vancouver Island until 1864, and also conducted surveys on the mainland, including one in the Pemberton area.

Pembina River MB Rising in the area of Turtle Mountain Provincial Park, south of Boissevain, this river flows east, crossing into North Dakota, south of Morden. The name was derived from Ojibwa *nepemenah*, 'summerberry', an ingredient, along with lard and buffalo meat, in the preparation of pemmican.

Pembroke ON The province's smallest township municipality, it was named about 1835 after George Augustus Herbert, 11th Earl of Pembroke (1759–1827). The site of the present city (1971) of Pembroke was developed in the 1820s, with parts of it being called *Miramichi*, *Campbelltown*, *Lowertown*, *Moffat*, and *Sydenham*, before it became the village of Pembroke in 1857.

Pend-d'Oreille River BC This river rises in Pend Oreille Lake in Idaho, crosses into Pend Oreille County, WA, and flows north into British Columbia to enter the Columbia River, south of Trail. It was named after the Pend d'Oreille, an Aboriginal group now known as the Kalispell. They had been named Pend d'Oreille by French traders, who called the ornaments in their ears *pendants d'oreilles*.

Pendleton ON In the united counties of Prescott and Russell, north of Casselman, this place was named in 1860 after James Pendleton Wells, who was later appointed the sheriff of the united counties.

Penetanguishene NF On the north side of the city of St. John's, and part of the city since 1982, this place was established in 1942, and took its name from the town in Ontario. It is reported that Stan Condon had then found a board attached to a farmer's fence with the name inscribed on it, and thereafter it was called Penetanguishene.

Penetanguishene ON A town (1882) in Simcoe County, it was named in 1793 as the Georgian Bay terminus of a road from Lake Simcoe. In Ojibwa it means 'place of the white rolling sands'. Its post office was called *Penetanguichine* in 1829, and was changed to Penetanguishene in 1833. The name is often shortened to Penetang.

Penhold AB A town (1980) south of Red Deer, it was originally planned by the Canadian Pacific Railway in 1892 to call this station Essexville. However, when one of the railway officials accidentally dropped his pen, and it became stuck in their route map, one of them coined the name Penhold.

Pennant SK Northwest of Swift Current, this village (1912) was given its name in 1911, when the Canadian Pacific Railway built a line south of the South Saskatchewan River. It was named when a surveyor found a pennant (a triangular flag) nearby. The post office was called *Pennant Station* in 1912.

Pennfield NB East of St. George, this place's post office was opened in 1857. It was called after **Pennfield Parish**, established in 1786, and named by Quakers from Pennsylvania after William Penn (1644–1718), who founded the state in 1681.

Penobsquis NB East of Sussex, this place's name was coined in 1857 by Moses H. Perley for a European and North American Railway station. It had been called *Stones Brook*, after a local stream. Perley blended the name from Maliseet *penobsq*, 'stone', and *sips*, 'brook'. The post office, called *Upper Sussex* in 1852, was renamed Penobsquis in 1865.

Pense SK Between Regina and Moose Jaw, this village (1904) was named by the Canadian Pacific Railway in 1883 after Edward John Barker Pense (1848–1910). While he was president of the Canadian Press Association, he followed the laying of the CPR's track, and was later publisher of the Kingston *Whig*. Pense post office was also opened in 1883.

Penticton BC A city (1948) at the south end of Okanagan Lake, its post office was named in 1889 after Tom Ellis's ranch. He choose the name from the Okanagan word *pentaktin*, 'always place', likely suggesting the site was permanently occupied. It is also said its name came from *snpnpiniyatn*, 'place where the deer net was used', from the extinct language of the Nicola-Similkameen. The townsite was laid out in 1892.

Percé QC Located at one of the continent's prettiest land and sea settings, this town overlooks the hauntingly beautiful **Rocher Percé,** widely known in the English-speaking world as Percé Rock. It was noted by Samuel de Champlain in 1603 as *Isle Percée*, but one of its two arches collapsed in 1845. Percé was created a township municipality in 1845 and became a town in 1971 when it annexed the municipalities of Barachois (1953), Bridgeville (1933), Cap-d'Espoir (1935), and *Saint-Pierre-de-la-Malbaie-N°2* (1876).

Percy ON A township in Northumberland County, it was named in 1798 after Lady Elizabeth Seymour, who became Baroness Percy after the death in 1786 of her husband, the Duke of Northumberland.

Perdue SK West of Saskatoon, this village (1909) was established in 1908 by the Canadian Pacific Railway, and named after Judge W.E. Perdue of Winnipeg. Postal service was also established in 1908.

Péribonka QC Situated at the mouth of **Rivière Péribonka**, on the north side of Lac Saint-Jean, this municipality was erected in 1926, and its post office was opened in 1898. The river, which rises east of Lac Mistassini, was noted as early as 1679. In Montagnais the name expresses the idea of a 'river cutting through sand'.

Perkins QC A community in the regional county of Les Collines-de-l'Outaouais, north of Gatineau, it was established as a village in 1866 and its post office was opened the same year. Lumberman John Adams Perkins had arrived here in 1845. In 1975 the village was annexed by the new municipality of Val-des-Monts.

Perth ON Created as a district in 1847 and a county in 1850, it was named in honour of Perthshire settlers from Scotland, the first to occupy its eastern townships of North Easthope and South Easthope.

Perth ON A town (1853) in Lanark County, it was founded in 1816 by Scottish settlers, mainly families of disbanded officers and soldiers, and took its name from the Scottish city of Perth on the River Tay. It was called *Perth-on-Tay* in 1816, and shortened to Perth four years later.

Perth-Andover NB In 1966 the communities of *Perth*, on the east side of the Saint John

River, and *Andover*, on the west, were united as the village of Perth-Andover. Perth post office had been opened in 1857, south of the village, and called after **Perth Parish**, created in 1833, and named by Lt-Gov. Sir Archibald Campbell after his birthplace of Perth, Scotland. *Perth Centre* post office had been opened in the village in 1879, and became Perth in 1902, after the older office's name was changed to Coronation. Andover post office had been named in 1846, after Andover Parish, established in 1833, and possibly given for Andover, Hampshire, England.

Pétain, Mount AB, BC On the continental divide and at the head of the Kananaskis River, this mountain was named in 1922 after Marshal Henri Philippe Pétain (1856–1951), a military hero in France during the First World War and commander-in-chief of the French armies, 1917–18.

Petawawa ON A township in Renfrew County, it was named in 1857 after the **Petawawa River**, a major tributary of the Ottawa River. The river's name is said to mean in Algonquin 'a noise is heard far away'. **Canadian Forces Base Petawawa** was established as a military base in 1904. The village of **Petawawa**, near the mouth of the river, was incorporated in 1961.

Peterborough ON The county was created in 1849 and named after the town of **Peterborough** (a city since 1905). The post office was named in 1825 after Peter Robinson (1785–1838), who organized the settlement here of 2024 Irish immigrants that year. Robinson was an Upper Canadian politician, commissioner of Crown lands and surveyor general of woods.

Peter, Mount BC In the Purcell Mountains, southwest of Invermere, this mountain (3338 m) was named in 1924 after alpinist Peter Kerr, who climbed in the area in 1914, and was killed in France during the First World War.

Petersburg ON In Waterloo Region, west of Kitchener, this place's post office was named in 1842 after farmer and blacksmith Peter Wilker (d. 1889).

Peterview NF On the south side of **Peters Arm** of the Bay of Exploits and at the mouth of **Peters River**, northeast of Grand Falls-Windsor, this town (1962) may have been named after a Peters family, established here in the early 1800s. It was called *Peters Arm* until about 1949. Capt. David Buchan referred to Peters Arm in 1811, when he dropped anchor here.

Petitcodiac NB A village (1966) southwest of Moncton and near the head of the **Petitcodiac River**, its first post office in 1848 was called *Head of Petitcodiac,* and was renamed Petitcodiac in 1865. The river's name was derived from Mi'kmaq *Epetkutogoyek*, 'river that bends around back'. Identified as *Petcoucoyec* by map-maker Jean-Baptiste-Louis Franquelin in 1686, its present form occurred on a map of 1799.

Petit-de-Grat NS On **Petit-de-Grat Island**, east of Isle Madame, this place's name was derived from the Basque word for 'fishing station', with the French *petit* denoting it as small. Its post office, opened in 1866, was called *Petit-de-Grat Bridge* between 1894 and 1953.

Petite Rivière NS At the mouth of **Petite Rivière**, south of Bridgewater, this place's post office was called *Petite Rivière Bridge* in 1852. The river's name, meaning 'little river' in English, may have been given by the French during the early years of the seventeenth century.

Petitot River AB, BC, NT Rising in northwestern Alberta, this river flows northwest through northeastern British Columbia to empty into the Liard River at Fort Liard, NT. It was named after Fr Émile Petitot (1838–1917), a missionary in the Mackenzie River valley, 1862–86, who produced a dictionary and a grammar of Dene languages, and other important studies of the Dene.

Petit-Rocher NB A village (1966) northwest of Bathurst, its post office was called *Madisco* in

1850. It was given its present name, which means 'little rock', in 1877.

Petrolia ON A town (1874) in Lambton County, southeast of Sarnia, its post office was named *Petrolea* in 1859, after the discovery of oil and the building of a refinery. Through a clerical mistake, the name was spelled Petrolia when the place was incorporated as a village in 1866. Postal authorities relented 40 years later, and respelled the office's name Petrolia.

Petty Harbour-Maddox Cove NF Located 10 km south of St. John's Harbour, this town (1969) is at the head of **Petty Harbour**. The harbour may have been named *Petit Havre* in French, in contrast to the larger harbour at St. John's. It may have had a Basque name before that. As early as 1675 there were people with variant spellings of the surname Maddox on the east coast of the Avalon Peninsula.

Philipsburg QC A village in the regional county of Brome-Missisquoi, southwest of Cowansville, it was incorporated in 1846, having been noted in 1815 by surveyor Joseph Bouchette as *Phillipsbourg*. Its post office from 1812 to 1922 was called *Philipsburg East*. The village was named after the son of John Ruiter (Johannes Ruyter), who had purchased land here in 1809.

Phillips, Mount AB, BC On the continental divide, northwest of Mount Robson, this mountain (3249 m) was named in 1910 after climber and guide Donald 'Curley' Phillips (1884–1938), who was the best authority on the mountains in the area of Jasper.

Piccadilly NF On the north side of Port au Port Penunsula, this place's name is an English interpretation of a local description of a hill, *Pic à Denis*, after an Acadian who had settled here. The road leading to Abrahams Cove, on the south side of the peninsula, is called **Piccadilly Slant**, with 'slant' being an old English word for a slope.

Pickering ON The township, originally called *Edinburgh*, was renamed in 1792 after the town of Pickering in North Yorkshire, England, northeast of the city of York. It was part of York County until 1849, when it was transferred to Ontario County. On the creation of the regional municipality of Durham in 1974, the township was divided between the towns of Ajax and **Pickering**. The community of *Pickering*, where a post office had been opened in 1829, became part of the town of Ajax in 1974, and was renamed **Pickering Village** in 1987 to differentiate it from the town of Pickering.

Pickle Lake ON A municipal township in Kenora District, northeast of Sioux Lookout, it was developed as a mining site in the 1930s. Its post office was first called Pickle Crow in 1935, but was known as Central Patricia from 1937 to 1977, when it was changed to Pickle Lake.

Picton ON A town (1850) in Prince Edward County, it was named in 1837 after Lt-Gen. Sir Thomas Picton (1758–1815), who served in the Peninsular War in Spain, 1810–14, and was killed at the Battle of Waterloo. In 1820, its post office was called *Hallowell*, after the township, but was changed to Picton in 1840.

Pictou NS This town (1873) is located on the north side of **Pictou Harbour**, and in **Pictou County**. It was settled in 1767 by Pennsylvanians and in 1773 by Scots from ports on the River Clyde. It was named after the harbour, identified in 1662 by Nicolas Denys as *Rivière de Pictou*. The name is derived from Mi'kmaq *piktook*, 'explosion of gas'. The three rivers entering the harbour are called **West River of Pictou**, **Middle River of Pictou**, and **East River of Pictou**. Pictou County was named in 1835. The community of **Pictou Landing**, across the harbour from Pictou, received postal service in 1867. **Pictou Island** in Northumberland Strait was named in the 1760s by chart-maker J.F.W. DesBarres.

Picture Butte AB A town (1960) north of Lethbridge, its post office was named in 1912 after petroglyphs on a butte. The butte was subsequently excavated for its gravel, and became the site of a subdivision.

Piedmont QC Established in 1923, this municipality adjacent to Saint-Sauveur-des-

Monts and north of Saint-Jérôme was named after its post office, which had been named in 1875 for its location at the foot (*pied*) of Mont Olympia.

Pierceland SK Near the Alberta border and west of Meadow Lake, the post office at this village (1973) was named in 1932 after a pioneer settler. To the north is an enlargement of Waterhen River called **Pierce Lake**.

Pierrefonds QC In the Communauté urbaine de Montréal and fronting on the Rivière des Prairies, west of the city of Montréal, Pierrefonds was incorporated as a village in 1904. It was called after the residence of Joseph-Adolphe Chauret (1854–1918), member of the Legislative Assembly, 1897–1908. Built in 1902, the residence's design was inspired by a castle in Pierrefonds, near Compiègne, northeast of Paris. The parish municipality of *Sainte-Geneviève*, created in the mid-1800s (Chauret had been its secretary-treasurer from 1880 to 1904), became the city of Pierrefonds in 1958, with the village of *Pierrefonds* becoming the town of Sainte-Geneviève the following year.

Pierreville QC A village in the regional county of Nicolet-Yamaska, near where the Rivière Yamaska flows into Lac Saint-Pierre, it was created in 1887, taking its name from the post office, opened in 1853. It had been severed from the parish municipality of *Saint-Thomas-de-Pierreville*, created in 1855. The name has its roots in the seigneury of *Pierreville*, created in 1723, whose name may have been derived from the lake's name.

Pierson MB Located near the southwest corner of the province, this place's Canadian Pacific Railway station was named in 1886 after Jan Lodewijk Pierson (1854–1945) who was then a partner in the Amsterdam firm of Adolphe Boissevain and Company, which sold shares of the CPR in the European financial markets. Its post office was opened in 1892. In 1925 Pierson wrote a book about Sir William Van Horne and the CPR.

Pigeon River ON Part of the boundary separating Ontario from Minnesota, this river was known during the French regime as *Rivière aux Groseilles*, 'gooseberry' or 'currant river', and *Rivière aux Tourtes*, 'wild pigeon river'. Pigeon River was adopted about 1826.

Pikangikum ON A community on the east shore of **Pikangikum Lake**, and in Kenora District, north of Red Lake, it was named in 1953. In Ojibwa the name means 'dirty water narrows'.

Pilkington ON A township in Wellington County, it was named in 1851 after Robert Pilkington (1765–1834), an officer with the Royal Engineers, 1790–1802. In 1799 he bought the area of the township, and subsequently arranged for some emigrants from Northamptonshire and Warwickshire to settle here.

Pilkington, Mount AB On the continental divide, west of Howse Pass, this mountain (3285 m) was named in 1898 after climber Charles Pilkington (1850–1919).

Pilley's Island NF Located on **Pilley's Island** and in Notre Dame Bay, this municipal community (1975) embraces the entire island, which is connected by a bridge to the island of Newfoundland. Several Pelly, Pilly, and Pilley families were living along the north coast in the 1840s, but the island was not settled until after 1887.

Pilot Butte SK This town (1980) 15 km east of Regina was established in 1882 by the Canadian Pacific Railway, and named after one of three nearly flat-topped hills, which provided extended views of the area's flat land. The post office was opened in 1903.

Pilot Mound MB A village (1904) west of Morden, its post office had been named in 1880 after an Aboriginal burial mound, which acted as a guide for travellers. When the Canadian Pacific Railway was built through the area in 1884, the post office was moved 3 km southeast to the line.

Pinawa MB A local government district (1962) on the Winnipeg River, northeast of

Winnipeg, its post office was named in 1910. In Ojibwa, its name means 'quiet waters'. Atomic Energy of Canada constructed the Whiteshell Research Establishment here in 1960.

Pincher Creek AB A town (1906) southwest of Lethbridge, it was named about 1878 after a tributary of the Oldman River. There are four different stories on the name's origin, all involving the loss of a pair of pinchers an older form of pincers). One states some American Fur Company prospectors mislaid a pair of pinchers here in 1867, and Joseph Kipp was sent back to recover them. A second story relates the loss of a pair of pinchers by a group of prospectors, and the North-West Mounted Police found them rusty and old in 1875. A third tale suggests Leonard Harnois lost them about 1870. A fourth story has prospectors being killed by Aboriginals, and the pinchers were found later at the grisly site.

Pincourt QC In the regional county of Vaudreuil-Soulanges and on the west side of Île Perrot, this town's post office was named in 1932, taking its name from the short pines (*pins courts*) in nearby woods. Becoming a village in 1950, it was incorporated as a town in 1959.

Pine, Cape NF Extending into the Atlantic Ocean, southwest of Trepassey, this cape was called *c de pena*, 'cape of sorrow', by the Portuguese in the early 1500s. By the 1670s, maps portrayed it as *C de Pine* and *C Pine*.

Pine Falls MB Situated 11 km above the mouth of the Winnipeg River, this place was founded by the Manitoba Pulp and Paper Company in 1925. Its post office was located at the site of Fort Alexander from 1879 to 1881, when it was moved upriver to its present location, on the south side of the falls.

Pinehouse SK On the western shore of **Pinehouse Lake**, and east of Île-à-la-Crosse, the post office of this northern village (1983) was known as *Snake Lake* until 1955. The previous year the residents requested the renaming of the lake and the post office.

Pine River BC Rising in the Rocky Mountains, this river flows northeast into the Peace River, opposite the district municipality of Taylor. Several species of pine trees, including the white, the lodgepole, the limber, and the whitebark support a thriving timber industry on the eastern flank of the mountains. **Pine Pass** (874 m), at the head of the river, is the lowest of the passes through the Rockies and was discovered in 1808 by one of Simon Fraser's men.

Pine River MB Located on the **North Pine River**, northwest of Dauphin, this place was founded by the Canadian Northern Railway about 1904. Its post office was called *Pine River Station* from 1904 to 1915, when 'Station' was dropped from the name.

Pinette PE In the municipal community of Belfast and southeast of Charlottetown, this place's post office was named in 1877 after the **Pinette River**. The river was called after Noel Pinet, who had arrived here in 1738 from Minas Basin in Nova Scotia.

Pintendre QC A municipality in the regional county of Desjardins, south of Lévis, its post office was named in 1900. The municipality was established in 1986, having been earlier called *Saint-Louis-de-Pintendre*. The name probably has its roots in the area's soft white pines (*pins tendres*).

Pinware NF On the north side of the Strait of Belle Isle, and at the mouth of the **Pinware River**, this municipal community (1971) and the river were named after a rock in **Pinware Bay**, called by French fishermen *Pied Noir*, 'black foot'.

Pipestone MB South of Virden, this place's post office was named in 1884. It is located south of **Pipestone Creek**, which rises south of Grenfell, SK. The name of the creek, shown on the 1865 Palliser expedition map, is said to have been transferred by Aboriginals from Pipestone, in the southwestern part of Minnesota.

Pitt Island BC South of the mouth of the Skeena River, this island (1375 km^2) took its name from *Pitt Archipelago*, given in 1792 by

Capt. George Vancouver after William Pitt the younger (1759–1806), prime minister of Great Britain, 1783–1801 and 1804–6. The present name was established in the 1920s.

Pitt Meadows BC A district municipality (1914) at the confluence of the Pitt and Fraser rivers, and between Port Coquitlam and Maple Ridge, it was named after the river. Noted in 1827 as *Pitt's River* by James McMillan, the founder of Fort Langley, the river was likely named after William Pitt the younger (1759–1806), prime minister of Great Britain for 20 years between 1783 and 1806.

Pittsburgh ON A township in Frontenac County, it was named in 1787 after William Pitt (1759–1806), prime minister of Great Britain, 1783–1801 and 1804–6.

Placentia NF A town (1945) on the east side of **Placentia Bay**, it was the site of an important land base for the Basque fishery in the late 1500s. It was likely named by the Basques after their seaport of Plentzia, in the Basque country of Spain. It was renamed *Plaisance* by the French, and served as their capital on the island of Newfoundland after 1624. The present spelling was adopted after 1713, when Newfoundland became British territory.

Plaisance QC A municipality in the regional county of Papineau, midway between Thurso and Papineauville, it was erected in 1931, having been called *Coeur-Très-Pur-de-la-Bienheureuse-Vierge-Marie-de-Plaisance* in 1900. The post office was named in 1882, possibly for a small house built about 1803 at the mouth of the Rivière de la Petite-Nation by Joseph Papineau, the local seigneur. The name of the house may have been derived from its attractive setting, or after a similar place in France.

Plamondon AB A village (1965) northwest of the town of Lac La Biche, its post office was named in 1909 after postmaster Joseph Plamondon (1861–1923), who homesteaded here the year before.

Plantagenet ON A village (1963) in the united counties of Prescott and Russell, its post office was named in 1838 after North Plantagenet Township. *See also* South Plantagenet.

Plaster Rock NB Southeast of Grand Falls and on the Tobique River, this village (1966) was named in 1900 by Henry Day after the red gypsum hills and river banks. In 1841 Edmund Ward described the area as an 'immense body of Plaister of Paris'.

Plattsville ON In Oxford County, northeast of Woodstock, this place was named in 1855 after miller and first postmaster Edward Platt (d. 1881).

Pleasant Grove PE A municipal community (1980) northeast of Charlottetown, its first post office was called *Suffolk Road* in 1870. Postmaster W.W. Duck renamed it in 1885 after someone observed a 'pleasant grove' of beech trees here.

Pleasantville NS On the LaHave River, southeast of Bridgewater, this place's post office was opened in 1881. It was named after its attractive setting.

Plenty SK Northwest of Rosetown, this village (1911) was founded by the Canadian Pacific Railway in 1910. Presumably it was named by a railway engineer, because the only crops that he encountered on his survey between Rosetown and Kerrobert were at Plenty. The post office had been called *Boraston* in 1907 and was renamed Plenty in 1911.

Plessisville QC Created as a village in 1855 and as a town 100 years later, it is located northeast of Victoriaville. The name was proposed by Fr Charles Trudelle after Mgr Joseph-Octave Plessis (1763–1825), eleventh bishop of the city of Québec, who became its first archbishop in 1818. The parish municipality of **Plessisville** was established in 1946 through the merger of the township municipalities of *Somerset-Nord* and *Somerset-Sud*.

Plevna ON In Frontenac County, northwest of Sharbot Lake, this community's post office was named *Buckshot* in 1872, after Buckshot Creek and nearby Buckshot Lake. Local residents did not like the name, but quarrelled over a suitable substitute. When postal authorities threatened to impose a name in 1877, Sam Barton Sr proposed Plevna, after a place (now Pleven) in Bulgaria, where many battles and quarrels had taken place.

Plumas MB This place northeast of Neepawa was first called *Richmond* in 1879. Because of confusion with other places called Richmond in Canada, postmaster James Anderson, who had arrived here in 1878 from Plumas County, CA, proposed renaming it Plumas in 1888. In Latin the name means 'feather'. Plumas County has many birds with beautiful plumage.

Plum Coulee MB A village (1901) east of Winkler, it was named by Mennonites in the 1880s after the normally dry bed (coulee) of **Plum Creek**. In the 1800s its banks had wild plum trees. Its post office was opened in 1890.

Plymouth NS On the south side of Stellarton, this community's post office was called *Plymouth Road* in 1884. First called *Irishtown* and *Churchtown*, it may have been named after Plymouth, England. There is another **Plymouth** between Yarmouth and Wedgeport.

Plympton NS On the east side of St. Marys Bay, and southwest of Digby, this community was first called *Everette* after pioneer Jeremiah S. Everette. It was named Plympton in 1885 by Judge Alfred Savary after his family's ancestral home in Massachusetts.

Plympton ON In Lambton County, this township was named in 1834 by Lt-Gov. Sir John Colborne, after Plympton, near Plymouth, Devon, England, not far from the childhood home of his wife, Elizabeth Yonge, Lady Colborne.

Poboktan Mountain AB In Jasper National Park, southeast of Maligne Lake, this mountain (3323 m) was named by geologist Arthur P. Coleman, after the Stoney word for 'owl'.

Pocologan NB East of St. George, this place's post office was opened in 1883, closed in 1898, reopened in 1929 and closed again in 1948. It took its name from **Pocologan Harbour** and **Pocologan River**. In Passamaquoddy *pekelugan* means 'enclosed harbour', or 'it opens out across'.

Pohénégamook QC A town surrounding **Lac Pohénégamook**, adjacent to the most northerly point of Maine, it was created in 1973 through the merger of three villages, Saint-Éleuthère (1903), Saint-Pierre-d'Estcourt (1922), and Sully (1916). The name was taken from the township, proclaimed in 1872, which had been given after the lake, whose name was derived from an Abenaki word meaning 'winter camp'.

Poilus, Mont des BC In Yoho National Park, this mountain (3161 m) was named *Mount Habel* in 1900 by British climber Norman Collie after German geographer Jean Habel. It was renamed by order in council in 1916 to remove an obvious German name from the Canadian Rockies. The new name was given in honour of the heroic foot soldiers of the French Army during the First World War.

Point de Bute NB A community east of Sackville, its post office was opened as *Pointe de Bute* in 1886. Settled by New Englanders in 1761, they adapted the name from *Pont à Buot*, a bridge over the Missaguash River. Pierre Buhot was a French-speaking settler here about 1700.

Pointe-du-Lac QC A municipality on the north side of Lac Saint-Pierre, where the St. Lawrence continues its journey to the Atlantic, it was created in 1978 through the merger of the municipal parishes of *La Visitation-de-la-Pointe-du-Lac* (1855), and *Pointe-du-Lac* (1928). The prominent sandy point was mentioned as early as 1656.

Pointe-à-la-Croix QC A municipality opposite the city of Campbellton, it was created in 1970, having been called the township municipality of *Mann* in 1845. Its post office was called *Cross Point* from 1846 to 1952. The name recalls a cross, probably erected here by the Mi'kmaq.

Pointe au Baril ON In Parry Sound District, northwest of the town of Parry Sound, this place's post office was named in 1892 after a whiskey barrel that had been mounted on a pole as a beacon. The barrel of whiskey had been lost many years before, and was found by French traders, who consumed its contents, before putting the barrel on the pole. The postmaster in 1905 recalled having seen the barrel in 1873. **Pointe au Baril Station**, on the Canadian Pacific Railway line, 10 km to the east, was named in 1923.

Pointe-au-Père QC Located just east of Rimouski, this town was created in 1988. Noted as early as 1696 as *Sainte-Anne-de-la-Pointe-au-Père*, it took its name from the celebration of the first mass on 8 December 1663 by Fr Henri Nouvel on the south side of the St. Lawrence. Well known as *Father Point*, its post office was called this from 1863 to the early 1970s, when it became Pointe-au-Père.

Pointe-au-Pic QC In the regional county of Charlevoix-Est, just west of La Malbaie, this village was incorporated in 1876. Its post office was named six years earlier after a peak (*pic*) on a point here, noted as *Point au Pic* in 1776 by chart-maker J.F.W. DesBarres. Widely known as a restful place to get away from the continent's summer heat, it has been frequently called Murray Bay by English-speaking visitors.

Pointe-aux-Trembles QC Located in the regional county of Portneuf, between Neuville and Donnacona, this parish municipality was called *Saint-François-de-Sales* in 1855, although the popular name was Pointe-aux-Trembles. In the early 1700s the religious parish was identified as *Saint-François-de-Sales*, with Pointe aux Trembles describing the nearby point covered with aspen (*trembles*) and birch. In 1912 the town of **Pointe-aux-Trembles** was incorporated at the northeast end of Île de Montréal, but was annexed by the city of Montréal in 1982. The trees have long disappeared from the point.

Pointe-Calumet QC A village near the east end of Lac des Deux Montagnes, it was erected in 1953. The word *calumet* means 'peace pipe', with a local tradition that the Abenaki and the Iroquois settled their differences by exchanging peace pipes on the point about 1720.

Pointe-Claire QC A city west of Montréal, it was created as a municipality in 1845, abolished two years later, and re-established in 1855. It became the village of *Saint-Joachim-de-la-Pointe-Claire* in 1854, and the town of Pointe-Claire in 1911. Its post office was named in 1835, taking its name from the point, identified on maps as early as 1686.

Pointe-des-Cascades QC Located at the east end of the regional county of Vaudreuil-Soulanges, where the St. Lawrence and Ottawa rivers unite, this village was created in 1961. The name reflects the fast water at that point, noted on maps of the late 1600s as *Cascades*. The post office was called *Cascades Point* from 1893 to 1951, when it was changed to the present name.

Point Edward ON A village (1878) in Lambton County, where Lake Huron empties into the St. Clair River (and where it is surrounded on the south and east by the city of Sarnia), it was called *Huron* in the early days, but it was renamed in 1865, after a visit in 1860 by Albert Edward, the Prince of Wales.

Pointe-Fortune QC A village on the Ontario border, northwest of Rigaud, it was incorporated in 1881. Its post office was called *Point Fortune* from 1851 to 1954, when the present form was adopted. The point was named after Col William Fortune, a native of County Wexford, Ireland, who had lived in South Carolina before coming to Canada. He was granted large areas of land on both sides of the Ottawa River, and was living here as early as 1796. He surveyed many of the townships of Eastern Ontario.

Pointe-Lebel QC Located at the mouth of Rivière Manicouagan, southwest of Baie-Comeau, this village was incorporated in 1964. The point was named after Norbert Lebel, the first settler here in 1902.

Pointe-Sapin NB On Northumberland Strait and north of Kouchibouguac National Park, this place's post office was called *Point Sapin* in 1873 after the local fir trees. It was renamed Pointe-Sapin in 1962.

Pointe-Verte NB Northwest of Bathurst, this village (1966) was named after a small point extending into Chaleur Bay. The post office was called *Green Point* in 1883 and was renamed Pointe-Verte in 1938.

Point La Nim NB West of Dalhousie, this place was called after a point extending into Restigouche River. Its post office, called *Point La Nin* in 1855, was renamed Point La Nim in 1862. The name was taken from Mi'kmaq *ananimkik*, possibly meaning 'lookout place'.

Point Tupper NS Adjacent to Port Hawkesbury, this place's post office was named in 1891. Ferdinand B. Tupper received a grant at the point in 1823.

Pokemouche NB South of Caraquet, this place was named after **Pokemouche River**, whose name in Mi'kmaq may mean 'salt water extending inward'. Its post office was opened as Pokemouche in 1845, but was renamed *Pockmouch* in 1854. The original spelling was restored in 1940.

Pokesudie NB On **Pokesudie Island** and east of Caraquet, this place's post office was called *Pokesudi* in 1912. It was changed to its current spelling in 1955. The island's name was derived from Mi'kmaq *booksadadek*, possibly meaning 'narrow passage between rocks'.

Pokiok NB On the Saint John River and at the mouth of **Pokiok Stream**, this place's post office was called *Poquiock* from 1854 to 1917. The stream's name, taken from Maliseet *Pokweok*, 'runs out through narrows', was recorded in 1762 by map-maker Joseph Peach as *R. Bogwiack*.

Pollard's Point NF Situated on the southwest side of White Bay and northeast of Deer Lake, this place was settled by Pollard families in the mid-1800s. It was named in the 1950s after the point. Droverville had been rejected that year for a postal name, because Sam Drover, after whom it was to be named, was still living. Drover then represented the White Bay district in the House of Assembly.

Pomquet NS In 1761 five Acadian families returned to Acadia from St-Malo, France, and settled at Pomquet. The name was derived from Mi'kmaq *pogumkek*, 'sandy beach'. Its post office, called *Pomquet Chapel* in 1864, was renamed Pomquet in 1900. There were other post offices at **Pomquet Forks** (1856), **Pomquet Station** (1897), and **Pomquet River** (1911).

Pond Inlet NT A hamlet (1975) on the north side of Baffin Island, it was established in 1922 as a Royal Canadian Mounted Police post. It took its name from the inlet that separates Baffin and Bylot islands. The inlet was named in 1818 by John (later Sir John) Ross, after John Pond (1767–1836), who was then the British astronomer royal.

Ponoka AB Midway between Red Deer and Wetaskiwin, this town (1904) was first called *Siding 14* in 1891 by the Canadian Pacific Railway. It was named Ponoka in 1897 after the Blackfoot word *ponokaii*, 'elk'.

Pont-Rouge QC A village in the regional county of Portneuf, west of the city of Québec, it was incorporated as the village of *Sainte-Jeanne-de-Pont-Rouge* in 1911 and adopted the present name seven years later. The bridge (*pont*), after which the village was named, had been built over the Rivière Jacques-Cartier in 1838, when it was entirely painted red (*rouge*).

Ponteix SK Southeast of Swift Current, this town (1957) was named in 1913 by the Canadian Pacific Railway after the home town in France of Fr Albert Royer, who had founded the mission of Notre Dame d'Auvergne just to the north. The post office had been called *Notre Dame d'Auvergne* in 1908, and was renamed Ponteix in 1914.

Pontiac QC A regional county extending along the Ottawa River from west of Quyon to

Rapides-des-Joachims, and covering an area of nearly 13,900 km^2 in western Québec, it was named in 1983 after the former county of Pontiac (1853). The county was named after noted Ottawa chief Pontiac (*c.* 1712–69), who supported the French during the Seven Years' War (1756–63). The municipality of **Pontiac**, embracing the original townships of Onslow and Eardley, was created in 1975. It has been part of the regional county of Les Collines-de-l'Outaouais since 1991.

Pontypool ON In Victoria County, south of Lindsay, this community's post office was named in 1881 after Pontypool, in Wales, northeast of Cardiff.

Pool's Island NF A neighbourhood (1996) in the town of New-Wes-Valley, it is located offshore from Badger's Quay on **Main Pool's Island**. The island may have been named after a fisherman.

Poplar Point MB Located northeast of Portage la Prairie, this place's post office was opened in 1871. It took its name from a bluff (a Western Canadian word for 'grove') of poplar trees growing further out from the belt of trees along the Assiniboine River.

Porcher Island BC This island, south of Prince Rupert, was named about 1867 by Capt. Daniel Pender after Cdr Edwin Augustus Porcher (d. 1878), who served on the west coast, 1865-8.

Porcupine ON On the north side of **Porcupine Lake** and in the city of Timmins, 12 km east of the city centre, it was named in 1910 after an island that looked like a porcupine in nearby **Porcupine River**. Sometimes called the 'Golden City', gold mining began here in 1907.

Porcupine Plain SK This town (1968), southeast of Tisdale, was founded by the Canadian National Railways in 1929. Richard Cooper proposed naming its post office Porcupine, but because that name was already in use in the Timmins area of Ontario, it was amended to Porcupine Plain that year.

Porcupine River YT Rising in the Ogilvie Mountains, north of Dawson, this major river (721 km) crosses the Yukon-Alaska border, west of Old Crow and flows into the Yukon River at Fort Yukon. It was possibly named in 1844 by John Bell, a chief trader of the Hudson's Bay Company.

Porquis Junction ON In the town of Iroquois Falls, southwest of the town centre, it was named in 1910, when a branch line of the Temiskaming and Northern Ontario Railway was built east to the new Abitibi Pulp and Paper company town of Iroquois Falls, and another west to Porcupine. 'Porquis' was created by blending the names Porcupine and Iroquois Falls.

Portage-du-Fort QC A small village in the regional county of Pontiac, west of Shawville, it was incorporated in 1863, 16 years after its post office was opened. Several possible origins have been suggested, the most likely being a place where supplies were kept for shantymen portaging around a series of rapids.

Portage la Prairie MB This city (1907) on the Assiniboine River, 25 km south of Lake Manitoba, acquired its name from the practice of fur traders, travellers, and Aboriginals crossing the flat land from the river to the lake. Pierre Gaultier de La Vérendrye noted in his 1739 journal that the Assiniboine carried their canoes over the portage on their way to the English posts on Hudson Bay. Its post office was opened in 1871, nine years before the Canadian Pacific Railway arrived.

Port Alberni BC At the head of **Alberni Inlet**, this city (1912) was amalgamated with the city (1945) of **Alberni** in 1967. The first post office was *Sayward-Alberni* in 1886, which was changed to Alberni in 1891, and closed in 1962. *New Alberni* post office opened on its south side in 1900, and was changed to Port Alberni in 1910. The inlet was named *Canal de Alberni* in 1791 by Lt Francisco Eliza, after Don Pedro Alberni, captain of a company of soldiers accompanying Eliza. *Alberni Canal* was adopted in 1861 by Capt. George Richards. Its narrow 45-km-long channel was reminiscent of other 'canals' (e.g., Portland, Gardner) on the west coast, but the Geographic Board changed it to Alberni Inlet in 1945.

Port Alice BC On the eastern shore of Neroustsos Inlet, an extension of Quatsino Sound, on the northwest coast of Vancouver Island, this village (1965) was founded in 1917 by the Whalen Pulp and Paper Company. The Whalen brothers, who established the company, named the place after their mother, Alice.

Port Anson NF On Sunday Cove Island, in Notre Dame Bay, this municipal community (1961) was earlier called *Sunday Cove Tickle* and *Newtown* before the present name was adopted in 1903. It is said that local resident Josiah Ryan proposed Port Handsome to the Nomenclature Board.

Port Arthur ON A main urban sector of the city of Thunder Bay, it began in the late 1850s as a transfer point between Lake Superior and the West. Col Garnet Wolseley, who was sent to put down the North-West Rebellion, named the place *Prince Arthur's Landing* in 1870 after Queen Victoria's third son, Prince Arthur, Duke of Connaught (1850–1942), who became governor general of Canada, 1911–6. Its post office was called Thunder Bay in 1869, but changed to Port Arthur in 1883, 11 years after it was incorporated as a village. It became a town in 1884 and a city in 1907. Port Arthur, Fort William, and parts of two neighbouring townships were amalgamated in 1970 to become the city of Thunder Bay.

Port au Bras NF Located on the western shore of Placentia Bay and northeast of Burin, the narrow coves at this municipal community (1971) were likely named *Port au Bras* ('port of arms') by fishermen from Saint-Malo, France. The name was adopted by English fishermen in the late 1700s.

Port au Choix NF On the northwestern coast of the island of Newfoundland, this town (1966) is located on a narrow isthmus joining Point Riche Peninsula to the mainland. Its name was derived from the Basque *Portuchua*, a description of a little harbour on the outside of the peninsula. The present Back Arm, on the inside of the peninsula, was called *Portuchua Çaharra*, 'old little port'. Capt. James Cook adopted the French form of the name in 1767.

Port au Port Bay NF Protected by the **Port au Port Peninsula**, this bay was named by the Basques in the late 1500s *Ophorportu*, 'port of relaxation', suggesting a place of calm waters. Its name, rendered as Port au Port by the French, was transferred to the 500-km^2 triangular peninsula, and adopted by Capt. James Cook in 1767.

Port au Port West-Aguathuna-Felix Cove NF This town (1975) was created by joining three separate communities. **Aguathuna** was first known as *Jack of Clubs Cove*. After a petition in 1911, it was renamed by the Most Revd Michael Howley after a Beothuk word he believed meant 'white rock', but it more likely meant 'grindstone'. **Felix Cove** was called after the cove, named about 1850 after French settler Harry Felix.

Port Blandford NF At the head of Clode Sound, a southwesterly extension of Bonavista Bay, this town (1983) was named in the 1890s after Capt. Samuel Blandford (1840–1909), who was a member of the House of Assembly, 1889–93, and a legislative councillor, 1893–1909. Another seaman of note was Capt. Darius Blandford, who had lived at Greenspond.

Port Burwell ON A village (1949) in Elgin County, it was surveyed in 1830 by Col Mahlon Burwell (1783–1846). A native of New Jersey, Burwell settled in the area in 1810, and served three terms in the House of Assembly of Upper Canada between 1812 and 1840.

Port Carling ON In the district municipality of Muskoka, northwest of Bracebridge, this resort centre was named in 1868 after John (later Sir John) Carling (1828–1911), then Ontario's minister of public works and agriculture, who was then on a fishing trip in the area.

Port-Cartier QC Situated on the north shore of the St. Lawence, southwest of Sept-Îles, this town was established in 1959 by Quebec Cartier Mining as its trans-shipment point for its iron ore from Lac Jeannine near Gagnon, and Mont Wright near Fermont. Previously it was the site of *Shelter Bay*, where the Ontario Paper Company produced paper for the *Chicago Tribune*, but the mill was closed in 1955.

Port Clements BC On the eastern shore of Masset Inlet and on Graham Island, one of the Queen Charlotte Islands, this village (1975) was first called *Queenstown*. When postal authorities declined to accept that name, it was named in 1914 after Herbert Sylvester Clements (1865–1939), member for Kent West, ON, 1904–8, Comox-Atlin (1911–7), and Comox-Alberni (1917–21) in the House of Commons.

Port Colborne ON In Niagara Region and at the Lake Erie entrance to the Welland Canal, this city (1966) was named in 1832 by William Hamilton Merritt after Sir John Colborne (1778–1863), lieutenant-governor of Upper Canada, 1828–36.

Port Coquitlam BC A city (1913) at the confluence of the Pitt and Fraser rivers, it took its name from the adjacent district municipality of **Coquitlam**. The latter was named after the Coquitlam, a Salish tribe. Their name means either 'small red salmon', or 'stinking with fish slime', in reference to their butchering salmon for their masters, the Kwantlen, a Cowichan tribe.

Port Credit ON At the mouth of the **Credit River**, its post office was named Port Credit in 1842. Incorporated as a village in 1914 and a town in 1961, it was annexed by the city of Mississauga in 1974.

Port-Daniel QC Located on the north shore of Chaleur Bay, southwest of Gaspé, this municipality was created in 1990 through the merger of the township municipalities of *Port-Daniel-Partie-Est* and *Port-Daniel-Partie-Ouest*, both established in 1882. The township of Port-Daniel, proclaimed in 1839, was called after the harbour known as Port Daniel, which had been named after Dieppe sea captain Charles Daniel (d. 1661), a contemporary of Samuel de Champlain. He had made several ocean crossings to Acadia and New France, spending some 10 years in the New World.

Port Dalhousie ON Annexed by the city of St. Catharines in 1961, this urban centre had been named in 1829, when it became the Lake Ontario outlet of the Welland Canal. It was given in honour of Gen. George Ramsay, 9th Earl Dalhousie (1770–1838), who promoted the construction of the canal, while he was governor-in-chief of British North America, 1820–8.

Port de Grave NF Located on the west side of Conception Bay and on **Bay de Grave**, this place's name was taken from the French *baie de grève*, 'bay with shingle beach', where they dried their cod. The shingle beach is not at the present site of Port de Grave, but at the town of Clarkes Beach, at the head of Bay de Grave.

Port Dover ON An urban centre in the city of Nanticoke, Haldimand-Norfolk Region, it was first called *Dover* and *Dover Mills*, after the famous port in Kent, England. Destroyed by American marauders during the War of 1812, it was called Port Dover after it was rebuilt.

Port Edward BC Near the mouth of the Skeena River and south of Prince Rupert, this district municipality (1991) was established in 1908 as a possible rail terminus on the west coast, and named after Edward VII. It became an important fish canning centre, but the last cannery closed in 1983.

Port Elgin NB A village (1922) northeast of Sackville, its post office was named in 1854 after James Bruce, 8th Earl of Elgin (1811–63), governor of the province of Canada, 1847–54. It was previously called *Gaspereau*, after the Gaspereau River, and the French Fort Gaspereau.

Port Elgin ON A town (1949) in Bruce County, it was named *Normanton* in 1854, possibly after James Ellis, 3rd Earl Normanton (1818–96). The earl's wife was related to Viscount Bury, who had named several of Bruce County's townships. The residents did not like the name, and had it changed in 1873, when it was incorporated as a village. It was named after James Bruce, 8th Earl of Elgin, governor of the province of Canada, 1847–54.

Porters Lake NS East of Dartmouth, this place's post office was named in 1855 after the adjacent lake. William Porter, one of the grantees of land at the lake in 1784, built a sawmill at the head of the lake in 1790. He had been deputy commissary general of musters to the foreign troops serving with the British armies during the American Revolution.

Port Franks ON In Lambton County, northeast of Sarnia, it was named in 1853 after a Capt. Franks, who had delivered some railway ties here, when the the Grand Trunk Railway was being built from Stratford to Sarnia.

Port Hammond BC In the district municipality of Maple Ridge, west of Haney, this place was named in 1885 after farmer John Hammond and civil engineer William Hammond. The townsite was developed when the Canadian Pacific Railway was built on the north shore of the Fraser River.

Port Hardy BC On the northeastern coast of Vancouver Island, this district municipality (1966) was founded in 1904. It took its name from **Hardy Bay**, named in 1860 by Capt. George Richards after Vice-Adm. Sir Thomas Masterman Hardy (1769–1839), the captain of Lord Nelson's *Victory* at the Battle of Trafalgar, 1805.

Port Hastings NS At the Cape Breton side of the Canso Causeway, this place's post office was named *Plaister Cove* in 1832. It was renamed Port Hastings in 1871, likely after Sir Charles Hastings Doyle (1804–83), lieutenant-governor of Nova Scotia, 1867–73.

Port Hawkesbury NS A town (1889) on the Cape Breton side of the Strait of Canso, its first post office was named *Ship Harbour* in 1823. It was renamed in 1868, likely after Robert Banks Jenkinson, Baron Hawkesbury and Lord Liverpool (1770–1828), prime minister of Great Britain, 1812–27. He succeeded his father as Baron Hawkesbury in 1796.

Port Hood NS A community on the west coast of Cape Breton Island, it was known as *Juste-au-Corps* during the French regime. Samuel Holland called its harbour *Port Barrington* in 1767, but it was renamed Port Hood by chart-maker J.F.W. DesBarres about 10 years later. He called it after Adm. Samuel Hood (1724–1816), who served in the Caribbean Sea during the American Revolution. Port Hood received postal service in 1829.

Port Hope ON A town (1850) in Northumberland County, it was known as both *Smith's Creek* (after Elias Smith) and *Toronto* in 1817. Because of confusion with Toronto post office (after Toronto Township) at present-day Erindale in the city of Mississauga, a special meeting, called in 1819, resulted in renaming it after Hope Township, named in 1792 after Col Henry Hope (*c.* 1746–89), lieutenant-governor of Quebec, 1785–9.

Port Hope Simpson NF Located on the south bank of the Alexis River and in southeastern Labrador, this municipal community (1973) was established as a lumbering centre in 1934 by J.O. Williams, of Wales, and named after Sir John Hope Simpson (1868–1961), commissioner of natural resources, 1934–6.

Portland ON A community on Big Rideau Lake and in the united counties of Leeds and Grenville, its post office was named in 1833 after William Henry Cavendish Bentinck, 3rd Duke of Portland (1738–1809), British secretary of the home department and for the colonies in the 1790s. The title was taken from the Isle of Portland in Dorset, offshore from Weymouth. **Portland Township** in Frontenac County was named in 1798 after the same duke.

Portland Canal BC This narrow 145-km-long channel, extending from the north side of Wales Island to Stewart, became part of the boundary between British Columbia and Alaska in 1903. It was named in 1793 by Capt. George Vancouver after William Henry Cavendish Bentinck, 3rd Duke of Portland (1738–1808), British home secretary, 1794–1801. Capt. Vancouver gave the name

Brown Inlet to the present **Portland Inlet**, extending for 50 km from the south side of Wales Island to the mouth of the Nass River. The present name was adopted in 1920, after it was found to be in local use.

Port Lambton ON In Lambton County, northwest of Wallaceburg, this place was first called *Lambton Village*. It was renamed Port Lambton in 1871, when the post office was opened.

Port La Tour NS This harbour, southwest of Shelburne, was named after Charles de Saint-Étienne de la Tour (1593–1666), governor of Acadia, 1651–66. A post office was opened in 1855 in the community of **Port La Tour**.

Port Maitland NS North of Yarmouth, its post office was first called *Maitland* in 1851 after Sir Peregrine Maitland, lieutenant-governor of Nova Scotia, 1828–34. It was given its present name in 1884.

Port Maitland ON At the mouth of the Grand River and in the town of Dunnville, Haldimand-Norfolk Region, it was named in 1846 after Lt-Gov. Sir Peregrine Maitland, who had once visited the site in the 1820s.

Port McNicholl ON In Tay Township, Simcoe County, this community was named in 1911 for David McNicholl, then vice-president and general manager of the Canadian Pacific Railway. It had become the eastern terminus of the Great Lakes Steamship service two years earlier. Incorporated as a village in 1917, it was annexed by Tay Township in 1994.

Port McNeill BC A town (1984) on the northeastern coast of Vancouver Island, it was named after **McNeill Bay**. The bay was named by the Hudson's Bay Company in 1837 after Capt. William Henry McNeill (1803–75), who served with the HBC from 1832 to 1863. He was chief factor at the company's Fort Rupert, now in the district municipality of Port Hardy.

Port Moody BC At the head of Burrard Inlet, east of Vancouver, this city (1913) was named in 1879 after the adjacent bay, when it was projected it would become the western terminus of the Canadian Pacific Railway. The bay was named **Port Moody** in 1860 by Capt. George Richards after Lt-Col Richard Clement Moody (1813–87), who served as commissioner of lands and works in the new colony of British Columbia, 1858–63.

Port Morien NS On **Morien Bay**, southwest of Glace Bay, this place's post office was called *Cow Bay* in 1855. It was given its present name in 1894. It may have been derived from the Portuguese designation for the nearby cape, now called Northern Head. The French called it *Cap de Mordienne* and the bay was described as *Baye de Mordienne*.

Port Mouton NS Southwest of Liverpool, the place's post office was opened in 1855. It was named after the bay, described as **Port Mouton** by Pierre du Gua de Monts in 1604, when a sheep (*mouton*) jumped overboard and was drowned, but was recovered for its meat. Sir William Alexander called it *St. Luke's Bay* in 1623, and chart maker J.F.W. DesBarres assigned the name *Gambier Harbour* to it in 1776, but the original designation prevailed.

Portneuf QC In the regional county of **Portneuf**, it was created as a municipal parish in 1863, which became a village in 1914 and a town in 1961. Its post office was opened in 1817. The name has its roots in a seigneury granted in 1647 to Jacques Leneuf (later Portneuf) de La Poterie (1606–*c.* 1685), which was located at the mouth of a river called about 1641 *port neuf*, meaning 'new port'. The regional county was created in 1982.

Port Perry ON In the township of Scugog, Durham Region, this urban centre was named in 1852 after Peter Perry (1792–1851), a Whitby merchant, who had laid out lots here in 1848. He had been the member for Lennox and Addington in the House of Assembly of Upper Canada, 1825–36.

Port Renfrew BC On the southwestern coast of Vancouver Island, this place was first called *Port San Juan*, after *Puerto San Juan*, given in 1790 by Spanish Sub-Lt Manuel Quimper.

Confusion with the San Juan Islands, in Washington State, led to renaming the post office in 1895 after Baron Renfrew of Selkirk, who had proposed settling some Scottish crofters in the San Juan River valley.

Port Rexton NF Situated on the northwest side of Trinity Bay and southwest of Bonavista, this municipal community (1969) was formerly called *Ship Cove* and *Robin Hood*. The name was chosen in 1910, because the first male child born here was a son of John Rex, a native of Longham, Dorset, England, who had settled here in the 1790s.

Port Robinson ON In the city of Thorold, Niagara Region, this place was called *Port Beverley* in 1829, when the Welland Canal had been built south to the Welland River. It was named after Sir John Beverley Robinson (1791–1863), chief justice of Upper Canada, 1829–62, and a director of the canal company. In 1834 it was renamed Port Robinson.

Port Rowan ON In Haldimand-Norfolk Region, southwest of Simcoe, it was named in 1845 after William (later Sir William) Rowan (1789–1879), secretary to Lt-Gov. Sir John Colborne, 1832–8, and commander-in-chief of the British forces in Canada, 1848–55.

Port Ryerse ON A community in the city of Nanticoke, Haldimand-Norfolk Region, south of Simcoe, it was named in 1854 after Col Samuel Ryerse (1752–1812), who received a large grant of land in 1795, and subsequently held a number of public offices. Ryerse's name was really Ryerson, but it was misspelled when he received his militia commission, and he decided to keep the new spelling. It also distinguished him from a rebel branch of the family.

Port Saunders NF Situated on Ingornachoix Bay, southeast of Port au Choix, this town (1956) was called after the port, named in 1767 by Capt. James Cook after Adm. Sir Charles Saunders (*c.* 1715–75). From July to October 1752 Saunders served as the commodore for the protection of Newfoundland's fishery. He commanded the British naval forces in the capture of the city of Québec in 1759.

Port Stanley ON A village (1874) in Elgin County, south of London, it was named about 1824, when Edward George Geoffrey Smith Stanley, the 14th Earl of Derby (1799–1869), visited Col Thomas Talbot at his home at Port Talbot, and the post office was opened five years later. Lord Stanley was later prime minister of Great Britain for three terms, and was the father of Frederick Arthur Stanley, the 16th Earl of Derby (1841–1908), Canada's governor general from 1888 to 1893.

Port Sydney ON On the south end of Mary Lake and in the town of Huntsville, district municipality of Muskoka, this place's post office was named *Mary Lake* in 1871. Three years later it was changed to honour Albert Sydney-Smith (1846–1925), who laid it out in village lots in 1872.

Portugal Cove South NF East of Trepassey, this municipal community (1963) was officially called this in 1913 to distinguish it from Portugal Cove in Conception Bay. It was identified in 1610 as *B of Portingal*, and in 1625 as *B Portugall*. It may have been named as early as 1500 by Gaspar Corte-Real. Both names reveal the extensive Portuguese connection with the fishery around Newfoundland.

Portugal Cove-St. Philip's NF On the east side of Conception Bay and northwest of St. John's, this town was established in 1992 by amalgamating the town (1977) of Portugal Cove and the town (1977) of *St. Phillips*. Portugal Cove was named after the adjacent cove, noted in 1675 by map-maker Henry Southwood. *St. Phillips* was first called *Broad Cove*. It was renamed in 1905 by Canon Smith, rector of St Philip's Catholic Church in Portugal Cove, 1886–1921.

Port Union NF South of Catalina and on the northwest side of Trinity Bay, this town (1961) was founded in 1914 by Sir William Ford Coaker. He established the headquarters of the

Fishermen's Protective Union and the Fishermen's Union Trading Company here in 1916, and the place was named that year.

Port Weller ON In the city of St. Catharines since 1961, this urban centre was named in 1913 after engineeer John Laing Weller, who worked on the construction of the fourth Welland Canal, 1907–17.

Port Williams NS Located west of Wolfville, this village (1951) was named in 1856 after Sir William Fenwick Williams (1800–83), a native of the province, who endured the siege of Kars, in present-day Turkey, in 1855, and was the lieutenant-governor of Nova Scotia, 1865–7. Its post office was opened in 1858.

Postville NF On the Labrador coast and on the northwest side of Kaipokok Bay, this municipal community (1975) was founded by the Revd William Gillett, a Pentecostal pastor, in 1941. It was named after a post established here in 1830 by D.R. Stewart, and taken over by the Hudson's Bay Company in 1837. Although the HBC moved its post nearer to the mouth of the bay in 1879, the original site continued to be called 'the Post'.

Potton QC A township municipality in the regional county of Memphrémagog, west of Lac Memphrémagog, it was established in 1845, taking its name from the township, proclaimed in 1797. Shown on a map two years earlier, it was named after a village in Bedfordshire, north of London, England.

Pouce Coupe BC A village (1932) on the Pouce Coupé River, southeast of Dawson Creek, its post office was named in 1912, after the river. The river is reported to be named after a Sekani trapper who was nicknamed Pouce Coupé ('cut thumb') by French-speaking voyageurs, because his thumb had been lost through an accident with a gun. Explorer Simon Fraser made a reference to him in 1806. Locally it is believed the real origin was a Beaver chief whose name was Pooscapee in his language, with voyageurs rendering it with the homonym Pouce Coupé. Although the village's name is pronounced 'poos-KOO-pee', it does not have an 'é' accent.

Pouch Cove NF On the southeast side of Cape St. Francis, and north of St. John's, this town (1970) was first noted in 1805, and was recorded by Joseph B. Jukes in 1840 as *Pouche Cove*. Pronounced like 'pooch', the name may be derived from a French family name, although locally there is no accepted explanation for it.

Powassan ON A town (1904) in Parry Sound District, south of North Bay, it was named in 1891 after the Ojibwa for 'bend in the river', because there are some sharp bends in the South River, just west of the town.

Powell River BC A district municipality (1955) on the mainland coast and at the west end of Malaspina Strait, it was named after the river, which drains **Powell Lake**. The river and lake were named in 1881 by Lt-Cdr V.B. Orlebar after Dr Israel Wood Powell (1836–1915), when they made a tour of the coast. Powell was a superintendent of education and a superintendent of Indian affairs in the province.

Powerview MB This village (1951) on the Winnipeg River, northeast of Winnipeg, was created as a residential community outside the Manitoba Pulp and Paper Company's planned community of Pine Falls, created in 1925. Its post office was opened in 1963.

Pownal PE On **Pownal Bay**, southeast of Charlottetown, this place's post office was established as *Lot 49* about 1854, renamed *Pownall* in 1871, and given its present form in 1940. The bay was named *Pownall Bay* in 1765 by Samuel Holland after John Pownall, secretary of the board of trade and plantations, 1758–61. In 1779 chart-maker J.F.W. DesBarres spelled it **Pownal Bay**, and this ultimately became the preferred form.

Prairies, Rivière des QC Separating the Île de Montréal and the Île Jésus, it was known to Samuel de Champlain in 1610. In both his writings, and in the Jesuit *Relation* of 1637, there is a reference to a courageous young man from Saint-Malo called des Prairies. According to the Jesuit *Relation*, it was named after him.

Preeceville SK North of Canora, this town (1946) was named in 1912 by the Canadian Northern Railway after Fred Preece, who homesteaded its site in 1904. His mother fed the men who laid the track. The post office was also opened in 1912.

Prelate SK Just east of the town of Leader and northwest of Swift Current, the site of this village (1913) was established by the Canadian Pacific Railway in 1913. Its post office had been opened in 1912, 5 km south of the village centre. A 'prelate' is a bishop or abbot in the Anglican and Catholic churches.

Premier Range BC In the Cariboo Mountains, west of Valemount, this range's peaks exceeding 3048 m (10,000 ft) were reserved in 1927 to honour Canadian prime ministers not previously commemorated. Peaks were then named after prime ministers Sir John Thompson, Sir John Abbott, and Sir Mackenzie Bowell. The names of Richard Bennett, Arthur Meighen, and Mackenzie King followed in 1962, and Louis St-Laurent was added in 1964, and Lester Pearson in 1972. Also in the group are mountains named for British prime minister Stanley Baldwin (1927), British Columbia premier John Oliver (1927), and pre-Confederation co-premier of Canada Sir Allan MacNab (1974). Eight peaks over 3048 m remain unnamed officially.

Prescott ON The county was named in 1798 after Maj.-Gen. Robert Prescott (1725–1815), governor-in-chief of Canada, Nova Scotia, and New Brunswick, 1797–9, although retaining the official title until 1807. It became part of the united counties of **Prescott and Russell** in 1822.

Prescott ON A separated town (1868) in the united counties of Leeds and Grenville since 1849, it was laid out in 1810 by Maj. Edward Jessop. It was named in 1817 for Maj.-Gen. Robert Prescott, governor-in-chief of Canada, Nova Scotia, and New Brunswick, 1797–9.

President Range BC In Yoho National Park, north of Field, it was named in 1907 by the Alpine Club of Canada after the senior officials of the Canadian Pacific Railway. As their surnames had already been used elsewhere, the highest peaks were called The President (3124 m) and The Vice President (3063 m).

Preston NS Located east of Dartmouth, this place was named in 1886, after Preston Township, established in 1784. The latter may have been named after Preston, Lancashire, England, or after Maj. Charles Preston, who served with Sir Guy Carleton in the 1770s in the defence of British North America. Descendents of slaves expelled from Jamaica settled in Preston in 1796, but migrated four years later to Sierre Leone in Africa. In 1816 the British brought 924 African-American slaves to Halifax, and they were located in Preston and **North Preston**.

Preston ON Part of the city of Cambridge since 1973, it was first called *Erb's Mills* and *Cambridge Mills*, before it was named in 1833 by William Scollick after his home town in Lancashire, England.

Prévost QC A municipality in the regional county of La Rivière-du-Nord, north of Saint-Jérôme, it was created in 1977 through the amalgamation of the villages of Shawbridge (1909) and Prévost (1927). Shawbridge post office, opened in 1860, became Prévost in 1981, after having been a summer office since 1823. It may have derived its name from Wilfrid Prévost (1832–98), member for Deux-Montagnes in the House of Commons, or for his son, Jean-Benoît-Berchmans Prévost, or his nephew, Jules-Édouard Prévost, who were both elected officials.

Price QC A village in the regional county of La Mitis, east of Mont-Joli, its post office was opened in 1891. The village was incorporated as *Priceville* in 1926, but because of confusion with Princeville, near Victoriaville, it was renamed Price in 1945. It was named after prominent lumberman William Evan Price (1827–80), who served in both the House of Commons, 1872–4, and the Legislative Assembly of Québec, 1875–80.

Primrose Lake AB, SK On the Alberta-Saskatchewan boundary and north of Cold Lake, this lake was named by boundary surveyors on 19 April 1909 on the anniversary of Primrose Day, and in commemoration of Benjamin Disraeli, Earl of Beaconsfield, who died on that day in 1881.

Prince PE This county was named in 1765 by Samuel Holland after George Augustus Frederick, Prince of Wales (1762–1830), who became George IV in 1820.

Prince Albert SK In 1866 the Revd James Nisbet established a Presbyterian mission on the North Saskatchewan River, and named it after the consort of Queen Victoria. Prince Albert had died five years earlier. A post office was established in the city (1904) in 1879. **Prince Albert National Park** (3875 km^2), northwest of the city, was created in 1927.

Prince Albert, Mount BC In the Royal Group, east of Invermere, this mountain (3209 m) was named in 1913 after Prince Albert (1895–1952), who was crowned George VI in 1936.

Prince Albert Peninsula NT In northwestern Victoria Island, this peninsula was first called *Prince Albert Land*, possibly by explorer Richard Collinson in 1852 after the consort of Queen Victoria. It was redesignated in 1905. **Prince Albert Sound**, which extends into Victoria Island between Wollaston and Diamond Jenness peninsulas, was named by Collinson in 1852.

Prince Charles Island NT In northern Foxe Basin, this island (9521 km^2) was discovered in 1948 by a Royal Canadian Air Force aerial survey. It was officially named the following year, after Prince Charles, who had been born 14 November 1948. It was the second last major Canadian island to be identified, the last being Stefansson Island, in the western Arctic, in 1952.

Prince Edward ON The county was named in 1792 after Prince Edward Augustus, Duke of Kent (1767–1820), fourth son of George III, and father of Queen Victoria. When he was commander-in-chief of the British forces in North America, he took a tour from Quebec to Niagara Falls, and visited the county named after him on his return.

Prince Edward Island The province was named in 1798 by the Legislative Assembly after Prince Edward, Duke of Kent (1767–1820), who then commanded the troops in Halifax. It was confirmed 5 February 1799 by George III. After 1759 it had been called *St. John's Island*, a translation of *Isle de Saint Jean*, given by French explorers, and recorded by Samuel de Champlain in 1604. **Prince Edward Island National Park** (18 km^2), on the north shore of the province, was established in 1937.

Prince Edward Island, Mount YT In the Centennial Range of the St. Elias Mountains, this mountain (3737 m) was named in 1967 after the province.

Prince Edward, Mount BC In the Royal Group, east of Invermere, this mountain (3200 m) was named in 1913 after Prince Edward (1894–1972), who was crowned Edward VIII in 1936. On abdicating the throne that year, he was appointed the Duke of Windsor.

Prince George BC This city (1915), at the confluence of the Nechako and Fraser rivers, was established in 1807 as Fort George by Simon Fraser of the North West Company and named after George III. The fort was closed by the Hudson's Bay Company in 1915. With the Grand Trunk Pacific Railway's planned arrival from Edmonton after 1910, it chose a site north of Fort George, and called it Prince George, after the Duke of Kent (1902–42), a younger brother of George VI (whose forename was Albert). During the first civic elections in 1915, the ratepayers indicated a preference for Prince George over Fort George by a vote of 153 to

13. **Mount Prince George** (2880 m) in the Royal Group, east of Invermere, was named in 1913 after the same prince.

Prince Gustav Adolf Sea NT Surrounded by Ellef Ringnes, King Christian, Lougheed, Mackenzie King, and Borden islands, this body of water was named by Otto Sverdrup during his 1898–1902 expedition after the crown prince of Sweden.

Prince Henry, Mount BC In the Royal Group, east of Invermere, this mountain (3227 m) was named in 1913 after Prince Henry, the Duke of Gloucester (1900–74), a brother of Edward VIII and George VI.

Prince John, Mount BC In the Royal Group, east of Invermere, this mountain (3236 m) was named in 1913 after Prince John (1905–19), a brother of Edward VIII and George VI.

Prince of Wales Island NT Between Victoria and Somerset islands, this island (33,339 km^2) was named in 1851, possibly by explorers Erasmus Ommanney, Sherard Osborn, and William Browne, after Albert Edward, Prince of Wales (1841–1910). As Edward VII he was the king from 1901 to 1910. **Prince of Wales Strait**, between Victoria and Banks islands, was named by Robert McClure in 1852 after the same prince.

Prince Patrick Island NT One of the Parry Islands of the Queen Elizabeth Islands, this island (15,848 km^2) was named about 1852 after Arthur William Patrick Albert, Duke of Connaught and Strathearn (1850–1942), third son of Queen Victoria. He served as governor general of Canada, 1911–16.

Prince Regent Inlet NT Between Somerset Island and Boothia Peninsula on the west, and Brodeur Peninsula on the east, this 300-km inlet was named in 1819 by William (later Sir William) Parry, on the anniversary of the birth of George Augustus Frederick (1762–1830), eldest son of George III. He was appointed Prince Regent in 1811 and became George IV in 1820.

Prince Rupert BC A seaport just north of the mouth of the Skeena River, this city (1910) was founded by the Grand Trunk Pacific Railway as the western terminus for its intercontinental line. In 1906 the GTP decided to offer a prize of $250 for the best name with no more than 10 letters and three syllables. From 5000 entries the suggestion from Eleanor MacDonald of Winnipeg to name it after the first governor of the Hudson's Bay Company was accepted. Since her proposal had 12 letters, the GTP gave two additional first prizes for the name Port Rupert.

Princess Mary Lake NT Located west of Baker Lake, the name of this lake (471 km^2) was proposed in 1958 by the Royal Canadian Corps of Signals after Princess Mary, the Princess Royal (1897–1965), the eldest daughter of Queen Victoria.

Princess Mary, Mount BC In the Royal Group, east of Invermere, this mountain (3084 m) was named in 1913 after Princess Mary, the Princess Royal, the sister of Edward VIII and George VI.

Princess Mountain BC In the Coast Mountains, southeast of Bella Coola, this mountain (2941 m) was named in 1955 by a member of the Alpine Club of Canada, who observed that the glaciers surrounding it reminded him of a princess's white gown radiating from its summit.

Princess Royal Island BC The largest of the islands along the province's Inside Passage, this island (2251 km^2) was named in 1788 by Capt. Charles Duncan, after his trading ship *Princess Royal*.

Princeton BC A town (1978) at the junction of the Tulameen and Similkameen rivers, midway between Hope and Penticton, it was named in 1860 by Gov. James Douglas after Prince Albert Edward, the Prince of Wales (1841–1910), who visited eastern North America that year. It was earlier called *Vermilion Forks* and *Red Earth Forks* (after a local red ochre source valued by Aboriginals), *Similkameen* (after the river), and *Allison's* (after pioneer rancher and miner John Fall Allison).

Princeton ON In Oxford County, east of Woodstock, this community was named by Thomas Watson, who came from Princeton, NJ in 1793, but returned there about 1802, after quarrelling with Upper Canada administrator Peter Russell over a land patent issued to his father by Lt-Gov. John Graves Simcoe. The post office was opened in 1836.

Princeville QC A town (1964) located northeast of Victoriaville, it had been created a village in 1857. The adjacent parish municipality of **Princeville** was named in 1969, having been called *Saint-Eusèbe-de-Stanfold* in 1855. *Stanfold* post office was opened here in 1849 and was changed to Princeville in 1914. The name was given in honour of an early settler, Pierre Prince.

Prince William NB On the Saint John River, west of Fredericton, this community's post office was named in 1845 after **Prince William Parish**. The parish, established in 1786, had been created as a township in 1783, and named by Edward Winslow and Daniel Murray after Prince William, third son of George III, and the future William IV. He was the patron of the King's American Dragoons, which received land grants in the parish.

Prior Peak AB, BC On the continental divide, southwest of Howse Pass, this mountain (3270 m) was named in 1924 after Edward Gawler Prior (1853–1920), lieutenant-governor of British Columbia, 1919-20.

Pritchard BC On the south bank of the South Thompson River, east of Kamloops, this place's post office was named in 1911 after farmer Walter Pritchard, who had settled here in 1904. Earlier it was called *Pemberton Spur* after sawmill owner Arthur G. Pemberton.

Procter BC On the south side of the West Arm of Kootenay Lake, east of Nelson, this place was named in 1906 after Nelson real estate speculator and first postmaster Thomas Gregg Procter, who was the manager of the Kootenay Valley Company.

Prophet River BC Rising in the Muskwa Ranges of the Rocky Mountains, this river flows east and then north into the Muskwa River, south of the town of Fort Nelson. It was likely named for a prophet of the Beaver nation, such as a recent one called Notseta or an earlier one known as Decutla.

Prospect NS On the Atlantic coast, southwest of Halifax, this place's name may have originated in Mi'kmaq *nospadakun*, 'herb mixed with tobacco'. Between 1672 and 1744 it appeared in various forms, including *Passepec*, *Paspek*, *Paspee*, and *Prospec*, before occurring as Prospect in 1754. Its post office was opened in 1852.

Proton ON A township in Grey County, it was named in 1827, likely by Lt-Gov. Sir Peregrine Maitland after the Greek word for 'the first'.

Provost AB A town (1952) near the Saskatchewan border and southeast of Wainwright, it was named in 1907 by the Canadian Pacific Railway, after the equivalent of a chief magistrate of a Scottish burgh. Its post office was opened the following year.

Prud'homme SK A village (1922) northeast of Saskatoon, its post office was named in 1922 after Mgr Joseph H. Prud'homme, bishop of Prince Albert and Saskatoon. It had been called *Lily* when the Canadian Northern Railway was built through it in 1904, and from 1906 to 1923 its post office was known as *Howell*.

Ptolemy, Mount AB, BC On the continental divide, south of Crowsnest Pass, this mountain (2813 m) was named in 1915 by surveyor Arthur O. Wheeler, because it resembled the head and chest of a man lying prone and gazing upwards, reminding him of Ptolemy, an astronomer and geographer of the second century AD.

Pubnico NS At the head of **Pubnico Harbour**, southeast of Yarmouth, this place's name was derived from Mi'kmaq *pogomkook*, 'dry sandy place'. Its post office was called *Pubnico Harbour* in 1850, *Pubnico Head* in 1900, and its present form in 1918. On each side of the 15-km harbour are several communities, with each incorporating the name Pubnico. With postal opening dates they are **East Pubnico** (1860), **Middle East Pubnico** (1899), **Centre East Pubnico** (1925), **Lower East Pubnico** (1873), **West Pubnico** (1874), **Upper West Pubnico** (1923), **Middle West Pubnico** (about 1899), and **Lower West Pubnico** (1873).

Pugwash NS At the mouth of **Pugwash River** and on the Northumberland Strait, east of Amherst, the first post office in this village (1948) was called *Waterford* in 1825. It was renamed after the river in 1828. The name of the river was derived from Mi'kmaq *pagweak*, 'shallow water' or 'shoal'. In 1957, Pugwash native and American industrialist Cyrus Eaton convened the first of his Pugwash Conferences here. In 1995, the promotion of peace and understanding by the Pugwash Movement among the world's nations was rewarded with a Nobel Peace Prize.

Pukaskwa River ON Rising west of Wawa, this river flows west into Lake Superior. Adopted in 1922, the name follows the spelling on a 1918 department of interior map, although its pronunciation, 'PUK-a-saw', suggests a name more like *Pukoso River,* shown on an 1878 Canadian chart of the lake. The name may have been taken from Ojibwa *Pagisowinakak*, 'bathing place'. **Pukaskwa National Park**, established in 1971, has an area of 1878 km^2.

Punkeydoodles Corners ON In Waterloo Region, midway between Kitchener and Stratford, this locality has one of the province's most amusing names. It is said that busy, prosperous farmers of German descent resented a lazy farmer among them, who grew only pumpkins, earning the nickname 'Punkeydoodle'. The name is also said to be a garbled rendition of the song 'Yankee Doodle'. For the record, Punkeydoodles Corners had a post office for six hours on 26 June 1982.

Punnichy SK Northeast of Regina, this village (1909) was named in 1908 by the Grand Trunk Pacific Railway as part of its alphabetical order of names, with Quinton and Raymore following to the west. Its post office was opened in 1909. It is said that its name was a Siouan nickname given to pioneer settler Mr Heuback, because his bald head fringed by a ring of hair reminded them of a chicken without feathers.

Purcell Mountains BC These mountains, part of the Columbia Mountains, extend north from the Idaho-Montana border to a point north of Golden, and from the upper valleys of the Columbia and the Kootenay rivers on the east to Kootenay Lake and Duncan and Beaver rivers on the west. They were named in 1859 by Dr James (later Sir James) Hector after Dr Goodwin Purcell (1817–76), who had taught him therapeutics and medical jurisdiction in Edinburgh.

Puslinch ON A township in Wellington County, it was named in 1835 after the Devon, England birthplace of Elizabeth Yonge, the wife of Sir John Colborne, lieutenant-governor of Upper Canada, 1828–36.

Puvirnituq QC A northern village on the east shore of Hudson Bay, and at the mouth of the **Rivière de Povungnituk**, the present spelling was adopted in 1995, having been erected as the northern village of *Povungnituk* in 1989. It took its name from the Inuit community, first identified on maps about 1945. In Inuktitut the name means 'it smells of rotting meat'.

Quaco Head NB Extending into the Bay of Fundy at St. Martins, this point was named after the Mi'kmaq *goolwahgahweek*, 'place of the hooded seal'. Map-maker Jean-Baptiste-Louis Franquelin noted it as *Ariquaki* in 1686. **Quaco Bay**, behind the head, was identified on a 1762 map.

Quadeville ON In Renfrew County, southeast of Barry's Bay, this place's post office was first called *Strathtay* in 1893. In 1907 it was renamed for pioneer miller August Quade, but the cancellation hammer had *Quadville*. It was corrected in 1921 to Quadeville.

Quadra Island BC At the north end of the Strait of Georgia and just north of Campbell River, this island was named in 1903 by the Geographic Board of Canada after Nootka governor Juan Francisco de la Bodega y Quadra (1744–94). He negotiated with Capt. George Vancouver a friendly transfer of the area of Vancouver Island in 1792 to the British Crown. Capt. Vancouver called the latter island *Quadra and Vancouver's Island*, but the addition of Quadra's name fell into disuse. The present Quadra Island plus Maurelle and Sonora islands was formerly believed to be part of a single island called *Valdes Island*.

Qualicum Beach BC This town (1983) is located on the east coast of Vancouver Island, northwest of Nanaimo. Its post office was named in 1913 after the **Qualicum River**, whose name in the Nanaimo language means 'place of the dog (or chum) salmon'.

Qu'Appelle SK A town (1904) east of Regina and 25 km south of Fort Qu'Appelle, its post office was founded as *Troy* in 1882 on the Canadian Pacific Railway. It was renamed *Qu'Appelle Station* two years later, and changed to *South Qu'Appelle* in 1902, and finally became Qu'Appelle in 1911. It took its name from the **Qu'Appelle River**, which rises northwest of Moose Jaw, and flows east to join the Assiniboine at St. Lazare, MB. The name is said among whites to be from a legend of a Cree brave travelling far from home, who, hearing his lover's voice, asked 'who calls?' On returning home he found she had died, with her last words calling for him. This legend may have come from a poem by Métis poet Pauline Johnson. Among the Cree there is a legend of an old woman travelling upriver with four dogs, but when one would not go on, she beat it to death. Every summer after that the dog could be heard calling. The lakes at Fort Qu'Appelle are called Calling Lakes, with one known as Katepwa Lake, *katepwa* meaning 'calling' in Cree. *See also* Fort Qu'Appelle.

Quaqtaq QC A northern village on the northwest shore of Ungava Bay, it had been called *Koartak* in 1950, *Notre-Dame-de-Koartac* in 1961, and *Koartac* in 1965. In Inuktitut the name means 'intestinal worms', making the name *Notre-Dame-de-Koartac* a rather weird designation during the francization of northern Québec place names in the early 1960s.

Quathiaski Cove BC This community on Quadra Island, opposite Campbell River, was named in 1899 after the Comox word meaning 'island in the mouth', in reference to a small island in the cove.

Quatsino Sound BC On the northwest coast of Vancouver Island, this inlet was identified on an Admiralty chart of 1849 as *Quatsinough Sound*, and the present form was adopted in 1920. It was called after the Koskimo, a Kwakiutl tribe, whose name may mean 'people of the north country' in the Kwakwala language. The post office in **Quatsino**, a community at the head of the sound, was named in 1896.

Québec After the 1763 Treaty of Paris, the name Quebec was adapted to describe the former territory of New France, roughly centred on the St. Lawrence River, extending north to Lake Nipissing and south to the forty-fifth parallel. In 1774 it was extended north to the limits of the watershed of the St. Lawrence, west beyond the Great Lakes to the lands between the Ohio and Mississippi rivers, and east to include the north shore of the Gulf of St. Lawrence and the present area of Labrador. In 1791, the territorial name of Quebec was replaced by Lower Canada and Upper Canada. The two were united as the province of Canada in 1841, with each division being called Canada East and Canada West, although Upper and Lower Canada continued in both popular and official use. With the passing of the British North America Act in 1867, the name Quebec was restored as the name of the province centred on the lower St. Lawrence, with its northern limits following the heads of rivers flowing into the river and the gulf. These limits were extended north to the central part of the present province in 1898, and all of the land area of the territory of Ungava was added in 1912. Its area was reduced by the Imperial Privy Council in 1927 by awarding the area of present Labrador to Newfoundland.

Québec QC Arguably North America's prettiest city, Québec occupies a significant place in the hearts not only of its residents, but among Québécois wherever they live. In 1985 it was added to the list of world heritage sites by UNESCO. When Jacques Cartier arrived at the site in 1535 its Iroquioan-speaking people called it *Stadacone*. In 1601 its site was identified as *Quebecq,* the description in Algonquin and Abenaki of the narrow, relatively blocked, channel of the St. Lawrence between the city and the south shore. Samuel de Champlain founded the city in 1608, and built his 'Abitation' here, making it the focus of the vast French political and commercial empire based on the St. Lawrence and the Great Lakes. The Communauté urbaine de Québec, comprising 13 municipalities, was created in 1970.

Quebec, Mount YT This mountain (3749 m), in the Centennial Range of the St. Elias Mountains, was named in 1967 after the province.

Queen Bess, Mount BC In the Coast Mountains, southeast of Mount Waddington, this mountain (3290 m) was named in 1935 after Queen Elizabeth I, in commemoration of the signing of the international law of the freedom of the seas, initiated by the Queen on the return to England of Sir Francis Drake in 1580.

Queen Charlotte Islands BC These islands were named in 1787 by Capt. George Dixon after one of his two ships, the *Queen Charlotte.* The ship had been named after the consort of George III. The community of **Queen Charlotte**, at the south end of Graham Island, was called **Queen Charlotte City** in the 1800s, and this remains a second official name. **Queen Charlotte Sound**, now designating that part of the Pacific Ocean between Vancouver Island and the Queen Charlotte Islands, was given to the wide entrance between Vancouver Island and the mainland in 1786 by S. Wedgborough after the consort of George III. This entrance is now called **Queen Charlotte Strait**, which originally designated the channel uniting Johnstone Strait to Queen Charlotte Sound.

Queen Elizabeth Foreland NT On the north side of the entrance to Frobisher Bay, this point was named in 1576 by Martin (later Sir Martin) Frobisher after Queen Elizabeth I.

Queen Elizabeth Islands NT All the islands in the Arctic Archipelago, north of Parry Channel, were named Queen Elizabeth Islands in 1954 after Queen Elizabeth II.

Queen Elizabeth Ranges AB Surrounding Maligne Lake, southeast of Jasper, these mountain ranges were named in 1953 by the Geographic Board of Alberta in honour of the coronation of Elizabeth II.

Queen Mary, Mount BC In the Royal Group, east of Invermere, this mountain (3245 m) was named in 1913 after Queen Mary (1867–1953), the consort of George V, and the mother of Edward VIII and George VI.

Queen Mary, Mount YT In the Icefield Ranges of the St. Elias Mountains, this mountain (3886 m) was named in 1935 by American climber Bradford Washburn in honour of the silver jubilee of George V and Queen Mary.

Queen Maud Gulf NT Southeast of Victoria Island and southwest of King William Island, this gulf was named in 1904 by Roald Amundsen after Queen Maud of Norway, the third daughter of Edward VII.

Queens NB To the west of Kings County, this county was named in 1785, likely as an expression of loyalty to George III and Queen Charlotte.

Queens NS Created as a county in 1762, and named after Queen Charlotte, the consort of George III, it became the region municipality of Queens on 1 April 1996.

Queens PE This county was named in 1765 by Samuel Holland after Queen Charlotte (1744–1818), the consort of George III.

Queensborough ON In Hastings County, northeast of Madoc, this place was named in 1854 by Daniel Thompson after Queensborough, east of Drogheda, Ireland. A native of Drogheda, Thompson came to Canada in 1841.

Queensland NS On St. Margarets Bay, west of Halifax, this place's post office was first called *North Shore St. Margarets Bay* in 1885. It was renamed Queensland in 1914, possibly in memory of Queen Victoria.

Queenston ON On the west bank of the Niagara River and in the town of Niagara-on-the-Lake, it was founded in 1789 by Robert Hamilton and named about 1810. Possibly it was named after either Queen Charlotte, the wife of George III, or John Graves Simcoe's regiment, the Queen's Rangers.

Queensville ON A community in the town of East Gwillimbury, York Region, north of Newmarket, it was named in 1843 after its location on Queen Street, itself named after Queen Victoria. The post office was opened in 1851.

Quesnel BC This city (1928) south of Prince George and at the junction of the Quesnel and Fraser rivers, was named in the early 1860s after the river. **Quesnel River** (203 km) was named in 1808 by explorer Simon Fraser after his North West Company clerk Jules-Maurice Quesnel (1786–1842), who remained in New Caledonia for three more years. The river drains the 90-km-long **Quesnel Lake**.

Quetico Provincial Park ON Set aside as a forest reserve in 1909, it became a provincial park in 1913. The name of **Quetico Lake** may have been derived from Ojibwa *Gwe ta ming*, 'bad, dangerous', with the concept of clinging to the shore for safety rather than going straight across. Or it may also have come from an earlier Cree word for a good spirit.

Quidi Vidi Harbour NF On the northeast side of the city of St. John's, this harbour was identified in 1669 as *Kitty-vitty* by Devon surgeon James Yonge. Between then and the late 1800s another two dozen spellings were used for the name, but none reveals its meaning. In 1971 writer Edgar R. Seary suggested it might have been derived from a French or Channel Islands family name, such as Quédville, Quidville, Quiédville, or Quetteville, or from Quetteville, near Honfleur, France.

Quill Lake SK Located east of Humboldt and north of Big Quill and Little Quill lakes, the post office of this village (1906) was named by first postmaster R.A. Gordon in 1904. When the Canadian Northern Railway arrived in the area the next year, it named its station *Lally* after a company official, but agreed to use Quill Lake when the post office was moved 3 km to its line. The Aboriginals collected pelican feathers at the lakes for their arrows.

Quincy Adams, Mount BC In the St. Elias Mountains, 4 km east of Fairweather Mountain, this mountain (4133 m) was named in 1923 after John Quincy Adams (1767–1848), president of the United States, 1825–9. As secretary of state in 1825, he negotiated the first treaty with Russia, which subsequently led to the establishment of the northwestern boundary between American and Canadian territory.

Quinte, Bay of ON The long channel separating Prince Edward County from the mainland, the bay's name likely has Mohawk roots, possibly after *Yo-ya-da-do-konthe,* the sanctuary home of the Thunderbird. It is also possible it is from *kenta* or *kenhenta*, 'meadow', 'prairie', or 'stopping place'.

Quinton SK North of Regina, this village (1910) was named in 1908 by the Grand Trunk Pacific Railway after one of its engineers. It was assigned between Punnichy and Raymore as part of its alphabetical list of names. Its post office was opened in 1909.

Quirpon NF At the northeast end of the Northern Peninsula, this place (pronounced 'kar-poon') was named after **Quirpon Harbour** and **Quirpon Island**, offshore in the entrance to the Strait of Belle Isle. Jacques Cartier was forced to spend two weeks in 1534 in the harbour of *Le Karpont*. It may have been named after Havre Le Kerpont, between St. Malo and the Île de Bréhat in Bretagne, France.

Quispamsis NB A town (1982), northeast of Saint John, it was named in 1857 by the European and North American Railway after the Maliseet *Quispem Sis*, 'little lake', in reference to Ritchie Lake. Its post office served the area from 1883 to 1931.

Quyon QC In the municipality of Pontiac since 1975, and at the point where **Rivière Quyon** flows into the Ottawa River, this community had been incorporated as a municipality in 1875. Before then the name was spelled both *Quio* and *Couillon*. The current spelling had been recommended by etymologist Edward Van Cortlandt of Ottawa, as best preserving its sound. The name likely means 'river with a sandy bottom' in Algonquin. The suggestion that it was derived from a lacrosse-type game called *des couillons* has the marks of a folk tale evolving from the similar pronunciations of two words.

Raanes Peninsula NT On the west side of Ellesmere Island, this 90-km peninsula was named before 1910 after Oluf Raanes, a member of the Sverdrup expedition, 1898–1902. He and Ivar Fosheim explored the west side of the island in 1901.

Rabbit Lake SK Northeast of North Battleford, the post office of this village (1928) was first named *Round Stone* in 1910, but was renamed after a small lake in 1926, two years before the Canadian Northern Railway arrived.

Race, Cape NF The most southeasterly point of Newfoundland, it was identified on early maps of the 1500s as *capo raso* ('flat cape'), a Portuguese description of the flat cliffs here. It may also have been named after Cabo Raso at the mouth of the River Tagus, the last point seen after sailing from Lisbon.

Radcliffe ON A township in Renfrew County, it was named in 1859 after Thomas Radcliff (1794–1841), who fought with the Duke of Wellington during the Peninsular War in Spain. He settled in Middlesex County in 1832, and was appointed to the Legislative Council of Upper Canada in 1839.

Radisson SK Southeast of North Battleford, the first post office of this town (1913) was called in 1903 *Great Bend*, after the sharp left-angled turn of the North Saskatchewan River, while the growing community was known as *Goodrich*, after homesteader J.S. Goodrich. In 1905 the Canadian Northern Railway named its station after French explorer Pierre-Esprit Radisson (*c.* 1640–1710), and its post office name was also changed that year.

Radisson QC Located adjacent to the Barrage LG-Deux, a dam at the point where La Grande Rivière continues its course to James Bay, it was named in 1973 by the Société de développement de la Baie James. The name was chosen to honour Pierre-Esprit Radisson, the noted explorer and fur trader, who was one of the founders of the Hudson's Bay Company in 1670.

Radium Hot Springs BC In the upper Columbia River valley, north of Windermere, this village (1990) was first called *Sinclair Hot Springs*. It took its name from adjacent Sinclair Canyon, given after trader James Sinclair (1806–56), who brought an Oregon-bound party of emigrants through the canyon in 1841. The place was renamed Radium Hot Springs in 1915 after the high radioactivity in the springs.

Radville SK Southwest of Weyburn, this town (1913) was founded by the Canadian Northern Railway in 1910, at a point where Conrad Paquin had settled. After reviewing various suggestions for the station's name, the last syllable of Paquin's first name was fused with 'ville' to form Radville. *Wallindale* post office had been opened nearby in 1909, but two years later it was moved into Radville, and renamed.

Radway AB A community northeast of Edmonton, its post office was named in 1910 after merchant and postmaster Orland S. Radway, who had settled here the year before, and first proposed calling it *Radway Centre*. His father, Frankford O. Radway, homesteaded in the area in 1897. It was an incorporated village from 1943 to 1996.

Rae-Edzo NT At the head of the North Arm of Great Slave Lake and northwest of Yellowknife, this hamlet (1971) is divided into two distinct communities, 24 km apart. **Rae** is the site of Fort Rae, established here by the Hudson's Bay Company in 1904, having been located 8 km to the southeast in 1852 by Dr John Rae. **Edzo** was created in 1965 as a site with better drainage, but most of the Dogrib continued to live at the older location. The name Edzo recalls a Dogrib chief who worked out a peace settlement with the Yellowknife, a

Chipewyan tribe, in the early nineteenth century.

Rae Isthmus NT A 75-km isthmus at the southwest end of Melville Peninsula, it was named after Dr John Rae (1813–93), who explored the area of the isthmus in 1846–7.

Rae Lakes NT On an isthmus between Faber Lake and **Rae Lake**, northwest of Yellowknife, this settlement was named in 1978. Rae Lake was named after Dr John Rae, who established the Hudson's Bay Company's post of Fort Rae, northwest of Yellowknife, in 1852.

Raglan ON A township in Renfrew County, it was named in 1857 after Fitzroy James Henry Somerset, 1st Baron Raglan (1788–1855), who commanded the British forces during the Crimean War, and died there of cholera.

Ragueneau QC Located in the regional county of Manicouagan, southwest of Baie-Comeau, this parish municipality was created in 1951, although some sources indicate it had been incorporated in 1933, 13 years after the township of Ragueneau had been proclaimed. The township was named after Jesuit Fr Paul Ragueneau (1608–80), who served as a missionary in New France from 1636 to 1662.

Rainbow Lake AB A town (1966) in the northwestern part of the province, it was named after an enlargement of Hay River, 23 km to the southeast. The name is a translation of the Dene description of its crescent shape. The post office was opened in 1967.

Rainville QC A municipality in the regional county of Brome-Missisquoi, northwest of Cowansville, it was known as the township municipality of *Farnham-Partie-Ouest* from 1855 to 1962. It was renamed after Théophile Rainville, the first francophone mayor, who served for several terms between 1887 and 1905.

Rainy Lake; Rainy River ON The boundary between Ontario and Minnesota goes through Rainy Lake and Rainy River, and the latter flows west into the Lake of the Woods. The French forms Lac à la Pluie and Rivière à la Pluie appear in records only after 1759. By 1800, they were being rendered as Rainy Lake and Rainy River. The names may be rooted in the Ojibwa expression *tekamammaouen,* 'it rains all the time', referring to the spray of the falls at Fort Frances. Explorer Pierre Gaultier de la Vérendrye wrote *Lac Tekemamihouenne* in 1738. Rainy River post office was opened in 1886. However, the site of the town (1904) was called *Beaver Mills* in 1898, and was only changed to **Rainy River** in 1903. **Rainy River District** was separated from Thunder Bay District in 1914.

Raleigh NF Near the northern end of the Northern Peninsula and north of St. Anthony, this municipal community (1973) was first known as *Ha Ha* and *Ha Ha Bay*, after the adjacent bay whose name refers to a narrow isthmus that separates it from Pistolet Bay. It was renamed in 1914 in honour of Sir Walter Raleigh (*c.* 1552–1618), who supported Humphrey Gilbert, his half-brother, in his colonizing and fishing plans in Newfoundland. *See also* Ha! Ha!, Baie des, *and* Saint-Louis-du-Ha! Ha!, QC.

Raleigh ON A township in Kent County, it was named in 1794 after the town of Rayleigh, in Essex, England, east of London.

Raleigh, Mount BC In the Coast Mountains, east of Bute Inlet, this mountain (3131 m) was named in 1933 by George G. Aitken, provincial member of the Geographic Board of Canada, after Sir Walter Raleigh. He had supported an expedition of his half-brother, Sir Humphrey Gilbert, who planned a voyage in the late 1500s to trade on the west coast of North America, but who died at sea near Newfoundland. *See also* Gilbert, Mount.

Ralston AB Northwest of Medicine Hat and on the Suffield Experimental Station, it was named in 1949 after Col James Layton Ralston (1881–1948), minister of national defence, 1940–4.

Ramara ON In 1994 Rama and Mara townships were united to become the municipal township of Ramara in the reorganized Simcoe County. Rama Township was named in 1820, possibly after the Spanish word for 'branch', or after a town in the Holy Land, north of Jerusalem. Mara Township was also named in 1820, possibly after the Spanish word for 'tide', or the biblical Marah, where Moses sweetened the bitter water.

Ramea NF This town (1951) is on Northwest Island, one of the **Ramea Islands**, located off the southwestern coast of the island of Newfoundland. The islands may have been named after the French *rameau*, 'branch', in reference to the narrow channels among the islands; after *ramée*, 'leafy branches', perhaps alluding to the fact that they had vegetation, when most of the islands along the coast were devoid of plants; or after Le Ramée, on Guernsey, the largest of the Channel Islands.

Ramore ON Located in Cochrane District, 40 km northwest of Kirkland Lake, its post office was first called *Claybelt* in 1909. It was renamed in 1915 after a prospector, who had been the first white man to die in the area.

Ramsay ON As township in Lanark County, it was named in 1823 after Gen. George Ramsay, Earl of Dalhousie (1770–1838), governor-in-chief of British North America, 1820–8.

Rankin Inlet NT At the head of **Rankin Inlet**, on the west coast of Hudson Bay, this hamlet (1975) was established as a nickel-mining settlement in 1955. The inlet was named after Lt John Rankin, who served in 1741–2 on HMS *Furnace*, which, under the command of Christopher Middleton, explored the west coast of Hudson Bay.

Rapid City MB Situated on the Little Saskatchewan River, north of Brandon, this town (1883) was almost called Saskatchewan City, but the name was deemed too lengthy. As the river was also called *Rapid River* (an English equivalent of Saskatchewan) in the 1800s, the name Rapid City was approved unanimously at a public meeting in 1877, and the post office was opened two years later. Earlier the place was called *Ralstons Colony*, after John Ralston, who had come from Burritts Rapids, ON, south of Ottawa, and brought many settlers with him.

Rapides-des-Joachims QC A municipality on the north shore of the Ottawa River, northwest of Pembroke, it was erected in 1955, taking its name from its post office, opened in 1853. The office was named after rapids in the Ottawa River, which were submerged by a hydroelectric dam in 1951. The first reference to the rapids was as *Rapides des Joachims de l'Estang* in 1686. Subsequent references referred to *Portages des Deux Joachims*, but it is uncertain who the two Joachims might have been. There may have been a connection with Chevalier de Troyes, who undertook a difficult trip in 1686 up the Ottawa and overland to Hudson Bay. He had married Marie Petit de L'Estang five years earlier. In English, the place's name is pronounced 'duh-swish-uh'.

Ratcliff, Mount BC In the Coast Mountains, southeast of Bella Coola, this mountain (3033 m) was named in 1948 by surveyor Maj. F.V. Longstaff after Walter E. Ratcliff, a pioneer settler in the area.

Rathwell MB Southwest of Portage la Prairie, this place's Canadian Pacific Railway station was named in 1887 after early settler John Rathwell. He had arrived there five years before from Innisville, ON, near Perth. The post office was called *Brunton* in 1888, after a place in Yorkshire, England, and was renamed Rathwell in 1890.

Rawdon ON A township in Hastings County, it was named in 1798 after Francis Rawdon-Hastings, 1st Marquess of Hastings and 2nd Earl of Moira (1754–1826), governor general of Bengal, 1812–23.

Rawdon QC This village (1919) in the regional county of Matawinie, west of Joliette, is surrounded by the township municipality of **Rawdon**, which was created in 1845, taking its name from the township, identified on a 1795 map, and proclaimed in 1799. It may have been named after Francis Rawdon-Hastings, 1st Marquess of Hastings and 2nd Earl of Moira (1754–1826), a noted military leader and a capable administrator. He became Baron Rawdon in 1783. There is a Rawdon in West Yorkshire, northwest of Leeds.

Ray, Cape NF The most southwesterly point of the island of Newfoundland, it was named by the Portuguese *Cabo do Rei*, 'king's point'. The Basques rendered this in the late 1500s as *Cap d'Array* and *Cadarrai*. In 1767 Capt. James Cook identified the cape's name as Cape Ray, and the nearby river as Grand Codroy River.

Raymond AB Southeast of Lethbridge, this town (1903) was named in 1902 by Jesse W. Knight after his eldest son Raymond. Knight planned the town as the centre of a 10,522-ha (26,000-a) farm devoted to sugar beet production for the North Western Coal and Navigation Company Limited.

Raymond, Mount NB Southeast of Mount Carleton, this mountain was named in 1903 by naturalist William F. Ganong after historian William Odber Raymond (1853–1923).

Raymore SK Located north of Regina, this town (1963) was named by the Grand Trunk Pacific Railway in 1908, with its name following the alphabetical arrangement favoured by the company, with Quinton to the east and Semans to the west. Its post office was also opened that year.

Rayside-Balfour ON A town in the regional municipality of Sudbury, northwest of the city of Sudbury, it was created in 1973 by amalgamating Rayside Township with part of Balfour Township. James Rayside and William Douglas Balfour were Liberal members in the Legislative Assembly of Ontario, 1882–94.

Razorback Mountain BC In the Coast Mountains, northeast of Mount Waddington, this mountain (3180 m) was named in 1924–5 by geologist V. Dolmage after its long serrated crest.

R.C.A.F. Peak BC In the Coast Mountains, northwest of Pemberton, this mountain (2880 m) was named in the 1950s after the Royal Canadian Air Force.

Rear of Leeds and Lansdowne ON This municipal township was created in 1850 from the interior portions of the townships of Leeds and Lansdowne. They were named in 1788, the one for Francis Osborne, 5th Duke of Leeds (1751–99), secretary of state for foreign affairs, 1783–91, and the other for William Petty Fitzmaurice, 2nd Earl of Shelburne and Marquess of Lansdowne (1737–1805), prime minister of Great Britain, 1782–3. *See also* Leeds County *and* Lansdowne.

Rear of Yonge and Escott ON A township in the united counties of Leeds and Grenville, it was created in 1854 from the interior portions of the townships of Yonge and Escott. The former was named after George Yonge (1732–1812), British secretary of war, 1782–94, and the latter after the Yonge family estate near Honiton in Devon, England.

Red Bay NF On the north side of the Strait of Belle Isle, this bay was the site of Basque whaling stations in the last half of the 1500s. Its name is derived from the red cliffs bordering the harbour. It was called *Havre des Buttes* by the French. The municipal community (1973) of **Red Bay** is on the east side of the bay.

Redcliff AB On the west side of Medicine Hat, the post office in this town (1912) was named in 1910 after the red shale cliffs along the banks of the South Saskatchewan River, south and east of the townsite.

Red Deer AB A city (1913) on the **Red Deer River**, midway between Calgary and Edmonton, it was first known as *Red Deer Crossing* on the Calgary-Edmonton Trail, with Red Deer post office opening in 1884. When the Canadian Pacific's Calgary and Edmonton Railway crossed the river in 1891, 7 km downstream, the settlement was relocated. The river, which joins the South Saskatchewan just east of the Alberta-Saskatchewan border, was known in Cree as *Waskasioo*, 'elk', which Scottish fur traders confused with the red deer of their homeland.

Red Indian Lake NF In the central part of the island of Newfoundland, this lake was the area where the last of the Beothuk were found. By 1829 disease and slaughter had annihilated them. Because they extensively used powdered hematite to paint their bodies, canoes, and other items, they were given the nickname 'Red Indians'. As they were the first North American Aboriginals encountered by Europeans, the nickname was ultimately assigned to all North American Aboriginals.

Red Lake ON In northwestern Ontario, north of the town of Kenora, the lake was known to the Ojibwa as *Miskwa Sagaigon*, 'blood-red lake'. Legendary hunters are said to have killed a great animal suspected to be *Matchee Manitou*, 'evil spirit', leaving the lake dyed blood-red. Gold was first discovered in 1897, and after rich deposits were found in 1925, a gold rush ensued. Red Lake post office was opened the next year, and the place became a municipal township in 1960.

Red River MB Known as the Red River of the North south of the forty-ninth parallel, where it forms the boundary between the states of Minnesota and North Dakota, it flows north through Winnipeg to Lake Winnipeg. In Ojibwa, it was called *Miskwagama Sipi*, 'red water river', likely after the red-brown silt carried by it. In 1738–9 explorer Pierre Gaultier de La Vérendrye called it Rivière Rouge, which remains one of its official names.

Red Rock ON A municipal township at the mouth of Nipigon River in Thunder Bay District, south of Nipigon, it was founded in 1915, and named after the dark red outcroppings in the area. There was a Hudson's Bay Company fort called Red Rock House at Nipigon in 1821.

Redvers SK Northeast of Estevan, this town (1960) was founded in 1900 by the Canadian Pacific Railway and named after Gen. Sir Redvers Buller (1839–1908). He had taken part in an expedition to the Red River in 1870, and had commanded the British forces during the South African War, 1899–1900. The post office was opened in 1902.

Redwater AB Northeast of Edmonton and on the south side of the **Redwater River**, the post office in this town (1950) was opened in 1909. The river is identified on geographer David Thompson's map of 1814 as *Vermilion River*, but was renamed because of the frequent occurrence of that name in Western Canada.

Regina SK As the Canadian Pacific Railway surveyed its route in 1881 across the southern prairies, it reached Pile of Bones Creek, south of the Qu'Appelle River. As the proposed site of the capital of the North-West Territories, the Marquess of Lorne had earlier proposed calling it *Victoria* after the Queen, the mother of his wife, Princess Louise. To avoid duplication, however, he chose the Latin equivalent of queen, used in her title. The post office opened on the first day of 1882. In 1883 the capital was moved from North Battleford. The town (1980) of **Regina Beach** is located to the northwest, on the southwest shore of Last Mountain Lake.

Reidville NF Located northeast of the town of Deer Lake, this municipal community (1975) was founded in the early 1930s by Thomas Reid, a native of Norris Point, on Bonne Bay, adjacent to the northwestern coast of the island of Newfoundland.

Reliance NT At the east end of Great Slave Lake, this settlement is at the site of the Hudson's Bay Company's Fort Reliance, built in 1833 as winter quarters for Lt George (later Sir George) Back. He named it as an expression of trust in the Creator.

Reliance Mountain BC In the Coast Mountains, east of Mount Waddington, this mountain (3140 m) was named in 1863 by Alfred Waddington. He projected the route of a railway through the valley of the Homathko River, on the west side of the mountain, and possibly relied on its summit as a reference point.

Remillard Peak BC Located in the Selkirk Mountains, north of Revelstoke, this mountain (3051 m) was named in 1939 after Louis Norbert Remillard, who had prospected for gold from 1910 to 1930 near the head of Goldstream River.

Remote Mountain BC In the Coast Mountains, northwest of Mount Waddington, this mountain (3033 m) was named in 1933 by W.A.D. (Don) Munday after the apparent difficulty of reaching it.

Rencontre East NF Located on the north side of Fortune Bay and northeast of Belleoram, this municipal community (1972) was named after **Rencontre Island**, the largest of the islands in front of its harbour. The island's name means 'meeting place', because French migratory fishermen met here in the eighteenth century. There is another place called **Rencontre West**, which is located on the southwestern coast, southwest of the town of St. Alban's.

Renews-Cappahayden NF This municipal community (1967) is located on the east coast of the Avalon Peninsula. The name **Renews** may be traced to the early sixteenth century description of **Renews Harbour** by the Portuguese as *ronhoso*, 'scabby', in reference to rocks being covered with shells and seaweed. **Cappahayden**, 8 km south of Renews Harbour, was first named *Broad Cove*. It was renamed in 1913, possibly by the Most Revd Michael Howley, after the townland of Cappahayden, County Kilkenny, Ireland.

Renforth NB A village (1966) northeast of Saint John, it was named in 1903 by the Intercolonial Railway after James Renforth, the stroke of the Tyne crew of Newcastle upon Tyne, England, who died 23 August 1871, after a race in Kennebecasis Bay against the Paris crew of Saint John. Its post office, named *Chalet* in 1911, after a Young Men's Christian Association building, was renamed Renforth in 1922.

Renfrew ON The county was named about 1825 after Renfrewshire, Scotland, southwest of Glasgow, and, as in Scotland, beside Lanark County. The town (1895) of **Renfrew**, first called *Second Chute*, after rapids on the Bonnechere River, was named in 1848, when the post office was opened here. It had been briefly known as *Renfrewville* in the 1840s.

Renous NB On the west bank of the Southwest Miramichi River and at the mouth of the Renous River, its post office was named *Renous Bridge* in 1856, and renamed Renous in 1911. The river was called after Mi'kmaq chief Sock Renou, whose family name may have been Renaud.

Repentigny QC This city (1956) is located in the regional county of L'Assomption, at the point where the rivières de l'Assomption and Mille-Îles join the St. Lawrence. The name may be traced back to the parish municipality of *Notre-Dame-de-l'Assomption-de-Repentigny*, created in 1855, which was the name of the religious parish, founded in 1699. The name was given for Pierre Legardeur de Repentigny (*c.* 1608–48), who was granted a seigneury here in 1647.

Repulse Bay NT Located at the south end of Melville Peninsula and at the northwest side of Hudson Bay, this hamlet (1978) was established in the early 1960s. It was named after the bay, which was given its name in 1741 by explorer Christopher Middleton, who was disappointed in not finding the Northwest Passage.

Reserve Mines NS Between Sydney and Glace Bay, this community was named in 1873. It reflected the decision of the General Mining Association to hold the coal seams here in reserve, while mining continued elsewhere in the area.

Resolute NT On the south coast of Cornwallis Island, this settlement was established in 1947 as the site of a joint American/Canadian High Arctic weather station. The bay was named after HMS *Resolute*, which may have wintered there in 1850–1. The settlement is commonly called Resolute Bay.

Resolution Island NT Southeast of Baffin Island and at the entrance to Hudson Strait, this island (1015 km^2) was named in 1612 by explorer Thomas Button after his ship *Resolution*.

Resplendent, Mount BC South of Mount Robson, this mountain (3426 m) was named in 1910 by British climber Norman Collie after observing its brilliant colours in the first rays of the sun.

Restigouche River NB, QC Formed at the confluence of the Kedgwick and Little Main Restigouche rivers, this river flows northeast into Chaleur Bay, and forms the boundary with Québec, below the mouth of the Patapedia River. Its name was derived from Mi'kmaq *lustagooch*, probably 'good river for canoeing', a similar meaning to that given for Aroostook, and for *Woolastook*, the Maliseet name for the Saint John River above tide. Map-maker Jean-Baptiste-Louis Franquelin recorded it in 1686 as **Ristigouche**, which continues to be the official form in Québec.

Reston MB Southwest of Virden, this community's first school was named in 1886 after a place in Berwickshire, Scotland, northwest of Berwick upon Tweed. Subsequently the post office (1890) and the Canadian Pacific Railway station (1900) were named Reston, with the school being renamed Lanark.

Revelstoke BC A city (1899) on the Columbia River, its site was first called *The Eddy*, after a swirling current that had eroded the west bank of the river, and *Second Crossing* by Canadian Pacific Railway surveyors. The site on the east bank of the river was laid out by surveyor Arthur Stanley Farwell (1841–1908) in 1880, and named *Farwell*. When the CPR arrived, it avoided Farwell's exorbitant price for his land by locating its station on higher land to the west, and naming it after Edward Charles Baring, 1st Lord Revelstoke (1828–97). His Baring Brothers banking house in London bought a $15 million CPR bond to prevent a financial disaster in the final months of building the CPR's line through the province. *Farwell* post office, opened in February 1886, was renamed Revelstoke in June 1886. There is access by road to the top of **Mount Revelstoke** (2027 m), which rises above the city in **Mount Revelstoke National Park** (263 km^2), established in 1914.

Reversing Falls NB At the mouth of the Saint John River and within the city of Saint John, this rough water in the river has its flow reversed every 13 hours, when the tide rises in the Bay of Fundy. Its name was suggested in 1882 by poet Sir Charles G.D. Roberts, when he wrote about its reversible character.

Rexton NB On the south side of the Richibucto River, opposite the town of Richibucto, this village (1966) was named *Kingston* in 1850 by shipbuilders Holderness and Chelton, after Kingston upon Hull, England. To reduce confusion with other places called Kingston, Dr J.W. Doherty proposed the name Rexton in 1901, using the Latin for 'king'.

Rhein SK Northeast of Yorkton, the post office at this village (1913) was named in 1911 by hardware merchant Emil Meugersing after the German river on which his home town of Köln (Cologne) was located. The area was settled by Germans, who had immigrated from the area of the Volga River in Russia.

Rhodes, Mount BC In the Clemenceau Icefield of the Rocky Mountains, southeast of Athabasca Pass, this mountain (3063 m) was named in 1927 by alpinist James Monroe Thorington after Sir Cecil Rhodes (1853–1902), who had made a fortune developing the diamond mines at Kimberley, South Africa and gave most of it to fund scholarships.

Rice Lake ON A long lake separating Northumberland and Peterborough counties, it was known to the Mississauga as *Pamadusgodayong*, 'lake of plains', because the Mohawk had earlier cleared the land for corn. In the 1830s settler John Langton wrote about the wild rice beds giving the lake the appearance of being a large grass plot.

Richard Bennett, Mount BC In the Premier Range, west of Valemount, this mountain (3190 m) was named in 1962 after Viscount Richard Bedford Bennett (1870–1947), prime minister of Canada, 1930–5.

Richards Island NT In the Mackenzie Delta, this island (2165 km^2) was named in 1826 by John (later Sir John) Richardson after a governor of the Bank of England.

Richardson Mountains NT, YT Located between Fort MacPherson, NT and Old Crow, YT, these mountains were named in 1825 by John (later Sir John) Franklin after John (later Sir John) Richardson (1787–1865), a surgeon and natural historian on Franklin's expeditions in 1819–20 and 1825–7. In 1848 he commanded an expedition in search of Franklin, who had disappeared three years earlier.

Richelieu QC On the east side of the Rivière Richelieu, opposite Chambly, this town (1968) was part of the parish municipality of *Notre-Dame-de-Bon-Secours* until 1869, when it became the village of Richelieu. The river was named after Fort Richelieu, built at its mouth in 1642. The fort's name was given in honour of Armand-Jean du Plessis, Cardinal de Richelieu (1585–1642), chief minister of Louis XIII. He created the Compagnie des Cent-Associés in 1627 to manage the development of New France during its early years.

Richer MB Northeast of Steinbach, this place was named in 1900 after H.I. Richer, who was the reeve of the rural municipality of Ste. Anne for many years.

Richibucto NB A town (1985) on the north bank of the **Richibucto River**, it was named *Liverpool* in 1826, when it was chosen as the shire town of Kent County. It was renamed Richibucto in 1832, and this was chosen as its postal designation the following year. The source of the name was Mi'kmaq *elsedabooktook*, which may mean 'runs back bay place', in reference to **Richibucto Harbour**. Ten km east of the town is the community of **Richibucto-Village**, whose post office was opened in 1872.

Richmond BC A city (1990) embracing Lulu and Sea islands, south of Vancouver, it had been incorporated as a district municipality in 1979, and, before that, a township in 1879. Australian Hugh McRoberts established Richmond View Farm here in 1862, with a daughter having suggested its name after a favourite place in Australia, possibly the town of Richmond, near Sydney.

Richmond NS This county, on southwestern Cape Breton Island, and including Isle Madame and adjacent islands, was named in 1835 after *Richmond Island*, the name given by Samuel Holland in 1767 to Isle Madame. He had named it after Charles Lennox, 3rd Duke of Richmond (1734–1806), who held many official positions, including British ambassador to France.

Richmond ON A township in Lennox and Addington County, it was named in 1786 after Charles Lennox, 3rd Duke of Richmond (1734–1806), who held many official positions, including ambassador to France. His nephew, Charles Lennox, 4th Duke of Richmond, was governor-in-chief of British North America, 1818–9.

Richmond ON Located in Ottawa-Carleton Region, 28 km southwest of Ottawa, this community was named in 1818 for Charles Gordon Lennox, 4th Duke of Richmond and Lennox, who had been appointed governor-in-chief of British North America in that year. While on a tour of settlements in Lower and Upper Canada the following summer, he was bitten by a pet fox in *William Henry* (now Sorel, QC), and after visiting Richmond, he became afflicted with hydrophobia, and died on 28 August 1819 in a barn 5 km to the north.

Richmond ON In Bayham Township, Elgin County, east of Aylmer, this community was named after Charles Lennox, 4th Duke of Richmond, governor-in-chief of British North America, 1818–9. However, its post office from 1829 to 1971 was *Bayham*.

Richmond PE A municipal community (1979) northwest of Summerside, it was named after **Richmond Bay**, the name given in 1765 by Samuel Holland for the whole of Malpeque Bay, but in later years applied to the west side of it. He named the bay after Charles Lennox, 3rd Duke of Richmond.

Richmond QC A town (1882), situated midway between Sherbrooke and Drummondville, its post office was opened in 1820, and named after Charles Lennox, 4th Duke of Richmond, governor-in-chief of British North America, 1818–9. It had been erected as a village in 1863.

Richmond Hill ON A town (1957) in York Region, it was first called *Miles Hill* and *Mount Pleasant*. Although legend has the Duke of Richmond stopping here in 1819 for a rest, it is more likely it was named in 1836 by schoolteacher Benjamin Barnard after Richmond Hill, Surrey, England.

Richmound SK West of Swift Current and near the Alberta border, this village (1947) was established by the Canadian Pacific Railway in 1924–5. Its post office had been opened in 1912, 5 km west of the village's site. It was named by Mrs Charlie Wilde after the area's rich earth and the rolling terrain.

Rideau ON A municipal township in Ottawa-Carleton Region, it was formed in 1974 by amalgamating the municipal townships of North Gower and Marlborough, plus Long Island, formerly split between the municipal townships of Gloucester and Osgoode. The Rideau River forms its southern and eastern boundary.

Rideau River; **Rideau Canal** ON In 1613 Samuel de Champlain observed the beauty of the 13-m **Rideau Falls**, noting that they looked like a curtain (*rideau*) behind which Aboriginals passed without getting wet. *Rivière du Rideau* only appeared on maps after 1694, and the English form appeared a century later. Among the well-known **Rideau Lakes**, southwest of Perth, are **Big Rideau**, **Lower Rideau**, and **Upper Rideau** lakes. The **Rideau Canal**, with 45 sets of locks, was built between 1827 and 1832, under the direction of Lt-Col John By.

Ridgetown ON Situated on a ridge separating the waters draining west into the Thames River and east into Lake Erie, this town (1881) in Kent County, east of Chatham, was first settled in the 1820s, and its post office was named in 1853.

Ridgeville ON In the town of Pelham, Niagara Region, this community was named in 1865 by Jonas Steele, the first postmaster, and John B. Crow, after its location on a southern ridge of the Short Hills.

Ridgeway ON In the town of Fort Erie, west of the town centre, its post office was named in 1873 from its location on Ridge Road, which divides the drainage flowing northeast to the Niagara River and south to Lake Erie.

Riding Mountain National Park MB Occupying rugged terrain between Neepawa and Dauphin, this 2973-km^2 park was created in 1929. The name relates to the trails followed by Aboriginals on horseback, when crossing the densely treed landscape. The 750-m mountain west of McCreary was first noted as *Riding or Dauphin Mountain* on an 1858 Crown lands map. **Riding Mountain** post office was opened south of McCreary in 1892.

Rigaud QC In the regional county of Vaudreuil-Soulanges, this town (1911) had been incorporated as a village in 1880, and its post office was opened in 1835. It owes its name to the grant of a seigneury in 1732 to Pierre de Rigaud de Vaudreuil de Cavagnial (1698–1778), governor of New France, 1755–60, and to François-Pierre de Rigaud de Vaudreuil (1703–79), governor of Trois-Rivières, 1749. It was in honour of the latter that the place was named.

Rigolet NF On the north side of Hamilton Inlet and northeast of Happy Valley-Goose Bay, this municipal community (1977) took its name from the French word *rigolet*, 'small stream', a description of The Narrows of the inlet there. A trading post was established here in 1788 by Pierre Marcoux, a native of the city of Québec.

Rimbey AB A town (1948) northwest of Red Deer, its post office was named in 1903 after postmaster James Rimbey, and his two brothers, Sam and Ben. In 1901 Sam Rimbey (1868–1952) had led a group of 200 pioneers from Kansas to the Blindman River valley.

Rimouski QC The largest city in the province east of the city of Québec, it is located on the south shore of the St. Lawrence, at the mouth of the **Rivière Rimouski**. Incorporated in 1869 as the town of *Saint-Germain-de-Rimouski*, it adopted its current name in 1920, taking the postal name assigned in 1832. The river was noted as early as 1688, when the seigneury of Rimouski was granted to Augustin Rouer de Villeray et de La Cardonnière. It was derived from Mi'kmaq *animouski*, 'place of the dog'. The village of **Rimouski-Est** was erected in 1939. The regional county of **Rimouski-Neigette** was created in 1982, succeeding the former county of Rimouski.

Riondel BC On the eastern shore of Kootenay Lake, northeast of Nelson, this place was named in 1907 after Comte Édouard Riondel, president of the French-owned Canadian Metal Company, which bought the Bluebell Mine here in 1905.

Ripley ON A community in Bruce County, west of Walkerton, its post office was named in 1857, after Ripley, Derbyshire, England, north of the city of Derby. Postmaster Paul McInnes persuaded the postal authorities in 1873 to call it *Dingwall*, after a place in Scotland, northwest of Inverness. However, its railway station was called Ripley that year, and six years later Ripley was restored as the postal name. It was annexed in 1994 by the municipal township of Huron.

Ripon QC A village in the regional county of Papineau, northeast of Buckingham, it was separated in 1923 from the township municipality of Ripon, which had been erected in 1861. The township was identified on a map in 1795, but was not proclaimed until 1855. The name recalls a town in West Yorkshire, England, north of Leeds.

Ripples NB East of Fredericton, this place's post office was called *Little River* from 1856 to 1903, when it was renamed for the rapids in Little River.

Riske Creek BC In the Chilcotin, southwest of Williams Lake, this place's post office was opened in 1912. It took its name from the creek, where some time before 1872 a Polish gentleman called L.W. Riskie had established a farm.

River Bourgeois NS On Cape Breton Island, southwest of St. Peters, this community's name was derived from the time of French occupation before 1758. Its post office was opened in 1838.

River Denys NS Northeast of Port Hawkesbury, this place's post office was named *River Dennis Station* in 1892, corrected to *River Denys Station* in 1926, and 'Station' was dropped from the name in 1967. The river flows into **Denys Basin**, part of Bras d'Or Lake. It may have been named after an Aboriginal chief who lived at its mouth, having adopted the surname of Nicolas Denys (1598–1688), governor of Acadia, 1653–87, and author of two books on the history and geography of the North American coasts. Nicolas Denys resided at *Saint-Pierre* (St. Peters) until 1669, when he moved to *Nipisiguit* (Bathurst, NB). On his 1672 map he had named Sydney Harbour *R. de Denys*.

Riverhead NF At the head of St. Mary's Harbour, and northwest of Trepassey, this municipal community (1966) was first noted in 1836. There is another **Riverhead**, at the head of Harbour Grace, first recorded in the census of 1901.

River Hebert NS Southwest of Amherst and on the **River Hébert**, the post office in this village (1950) was named in 1851. The river may have been named after Louis Hébert (*c.* 1575–1627), an apothecary, who accompanied Pierre du Gua de Monts to Port Royal in 1604.

Riverhurst SK Located near the southeastern bank of Lake Diefenbaker, northwest of Moose Jaw, the post office at this village (1916) was named in 1914, a year before the Grand Trunk Pacific Railway arrived. Its name was formed from the names of two nearby closed post offices, *Riverside* (1910–11) and *Boldenhurst* (1910–16).

River John NS East of Tatamagouche and at the mouth of the **River John**, where it flows into Amet Sound, it received postal service in

1838. The river's name was derived from the French *Rivière Jaune*, 'yellow river'. Chartmaker J.F.W. DesBarres introduced the name River John in the 1770s. **Cape John** is on the east side of the entrance to Amet Sound.

River of Ponds NF On the northwestern coast of the Northern Peninsula and south of Port au Choix, this municipal community (1970) was settled about 1870 and named after the river, which drains several ponds. The river may have named by Capt. James Cook in 1770. Earlier during that century French migratory fishermen knew it as the *Rivière des Roches* ('rocky river').

Rivers MB On the Little Saskatchewan River, northwest of Brandon, this town (1910) was established in 1907–8 by the Grand Trunk Pacific Railway, and named after the chairman of its board of directors, Sir Charles Rivers Wilson.

Riverside-Albert NB In Albert County, south of Moncton, this village united the separate communities of *Albert* and *Riverside* in 1966. *Hopewell Corner* post office, after the parish, was opened on the north side of Shepody River in 1854, and was called *Albert* in 1882, after the county. On the south side of the river, *River Side* post office was opened in 1867, and was renamed Riverside in 1932.

Rivers Inlet BC On the mainland coast, north of Vancouver Island, this 50-km fiord was named in 1792 by Capt. George Vancouver after George Pitt, 1st Baron Rivers (1721–1803), a British writer and politician.

Riverton MB North of Gimli, this village (1951) was first known as *Icelandic River*, after the river, which rises west of Arborg. It was changed to Riverton in 1914, when the Canadian Pacific Railway made it the northern terminus of a line on the western shore of Lake Winnipeg.

Riverton NS Located on the East River of Pictou and south of Stellarton, this place was first called *Fish Pools*. The residents had it changed to Riverton in 1877.

Riverview NB A town on the Petitcodiac River, opposite Moncton, it took its name from Riverview Heights, a postal designation in the central part of the town from 1949 to 1957, and a village from 1966 to 1973. In 1973 the villages of Riverview Heights, Gunningsville, and Bridgedale were united with adjacent parts of Coverdale Parish to form the town of *Coverdale*, which was changed to Riverview the following year. The parish, established in 1828, was named after *Coverdale River*, given before 1788 to present Little River. The original name of the river may have been called after Bible translator Miles Coverdale.

Rivière-Beaudette QC This municipality in the regional county of Vaudreuil-Soulanges is at the mouth of **Rivière Beaudette**, which rises in Ontario north of Cornwall. The name has its roots in the discovery during the 1600s of a wire-frame bed (*baudet*) in a burned shack on nearby **Pointe Beaudette**. The present municipality was created in 1990 through the amalgamation of the parish municipality of Rivière-Beaudette (1887) and the village of Rivière-Beaudette (1978).

Rivière-Bleue QC Situated in the regional county of Témiscouata, southeast of Rivière-du-Loup, and on the border with Maine, this municipality was created in 1975 through the merger of the village (1914) and the parish municipality (1920) of *Saint-Joseph-de-la-Rivière-Bleue*. Its post office was opened in 1910 and named after a river rising west of Cabano, on the western shore of Lac Témiscouata, and flowing southwest into the Rivière Saint-François.

Rivière-du-Gouffre QC In the regional municipality of Charlevoix, near Baie-Saint-Paul, this municipality was erected in 1921. The river was noted by Samuel de Champlain in 1608. Among the explanations of the name, the one favoured relates to landslides, with the river becoming turbulent (*engouffrant*) when a clay bank collapses into it.

Rivière-du-Loup QC The name of this town on the south shore of the St. Lawrence was noted as early as 1673 in reference to the grant of a seigneury to Charles Aubert de La Chesnaye. The *Rivière-du-Loup-en-Bas* post office was opened in 1828, but when the place was incorporated as a village in 1850 it was called *Fraserville*, and it continued until 1874 as the name of the town. It had been named after Alexander Fraser (1763–1837), who bought the seigneury in 1802. The name was changed to Rivière-du-Loup in 1919. There are three explanations for the name: harbour seals, called 'loups marins' by Jacques Cartier; an Aboriginal nation called the Loups or Mahicans; and a vessel called *Le Loup* (meaning 'the wolf'), which may have wintered here about 1660. The regional county of **Rivière-du-Loup** was erected in 1982, succeeding the former county of Rivière-du-Loup.

Rivière-Malbaie QC A municipality north of La Malbaie and in the regional county of Charlevoix-Est, it was named after the **Rivière Malbaie**, given in 1608 by Samuel de Champlain, because the bay at its mouth was too difficult (*mal*) for safe anchorage.

Rivière-Ouelle QC Situated in the regional county of Kamouraska, east of La Pocatière, this place was known as the parish municipality of *Notre-Dame-de-Liesse-de-la-Rivière-Ouelle* in 1855, and the name was shortened to its present form in 1983. Noted on a 1641 map as *R. Hoel*, and mentioned in the 1672 grant of a seigneury as *rivière Houelle*, the river may have been named after Louis Houel (or Ouel, sometimes written Houël), who performed several duties in New France, and was a friend of Samuel de Champlain.

Rivière-Verte NB A village (1966) southeast of Edmundston, its post office was called *Bellefleur* from 1884 to 1906, and *Green River Station* from 1906 to 1935, when it became Rivière-Verte. *Green River* post office, opened in 1863, became *Lynch* in 1906, and *Prime* in 1910. The river is officially called both Green River and **Rivière-Verte**.

Robert-Cliche QC A regional county centred on Beauceville, south of the city of Québec, it was named in 1982 after Robert Cliche (1921–78), a native of Saint-Joseph-de-Beauce, and a distinguished writer, teacher, judge, and politician.

Robert's Arm NF A town (1954) on the south side of Notre Dame Bay, it was first known as *Rabbits Arm*. By 1911 it was called Robert's Arm, possibly after local resident John Roberts.

Roberts Creek BC On the mainland shore of the Strait of Georgia, northwest of Gibsons, this community was named in 1904 after Thomas Roberts, who had built a cabin at the mouth of the creek in 1889. The first postmaster was J.F. Roberts, followed in 1908 by F.J. Roberts, and in 1910 by L.H. Roberts, who continued until 1930.

Robert Service, Mount YT Located at the head of the North Klondike River, east of Dawson, this mountain was named in 1968 by geologist Dirk Tempelman-Kluit after Robert William Service (1874–1958), the 'poet of the Yukon', and author of several collections of poems, including *Songs of a Sourdough*, *Ballads of a Cheechako*, and *Rhymes of a Rolling Stone*.

Robertsonville QC A village in the regional county of L'Amiante, just east of Thetford Mines, it was erected in 1909, taking its name from the railway station, opened in 1883. It was named after Joseph Gibb Robertson (1820–99), first president of the Quebec Central Railway, who also served as treasurer of the province, and postmaster of Sherbrooke.

Robertville NB Northwest of Bathurst, this place was established as a free-grant settlement in 1879. Its post office was opened as *Dumfries* in 1884, and was renamed in 1890, after Fr François-Antoine Robert (1820–88), parish priest at Petit-Rocher, 1866–88.

Roberval QC A town (1903) on the western shore of Lac Saint-Jean, it had been established as a village in 1884. The post office was opened in 1862 and a year later the township of Roberval was proclaimed. The name was given after Jean-François de La Rocque de Roberval (*c.* 1500–60), lieutenant-general of New France, 1541–3.

Robichaud NB A community east of Shediac, its post office was opened in 1878. It was named after Hippolite Robichaud, its postmaster until 1899.

Roblin MB West of Dauphin and near the Saskatchewan border, this town (1962) was first called *Goose Lake* after a small lake south of the town, but it was renamed by the Canadian Northern Railway in 1904 after Sir Rodmond Palen Roblin (1853–1937), premier of Manitoba, 1900–15.

Robson BC On the north side of the Columbia River, opposite Castlegar, this place was first called *Sproat's Landing* after the local gold commissioner, Gilbert Sproat. It became *East Robson* in 1890, with Robson post office opening two years later. **Robson West** is on the south bank and in the west end of present-day Castlegar. Both were named after John Robson (1824–92), premier of British Columbia, 1889–92.

Robson, Mount BC North of the Yellowhead Pass, this highest mountain (3954 m) in the Canadian Rockies was likely named after Hudson's Bay Company officer Colin Robertson (1783–1842), who led an HBC expedition of 10 canoes, 10 officers and 100 voyageurs to the Peace River country, 1819–20. It may have been named about 1820 by HBC party leader Ignace Giasson, after Robertson had sent him and a party of fur traders to explore the west side of the Yellowhead Pass.

Rocanville SK A town north of Moosomin, the post office in this town (1967) was named in 1884 after Auguste Henry Rocan Bastien, who had been born in 1842 in Saint-Vincent-de-Paul, QC (in the present city of Laval), and baptized Honoré Rocan. The office had been located 5 km to the northeast, and was moved to the line of the Canadian Pacific Railway in 1903.

Roche Percee SK Located between Estevan and the North Dakota border, this village (1909) received its name from a long sandstone rock with a large hole through it. Its post office was opened here in 1905, having been the name of the post office at *Coalfields*, 5 km to the northeast, 1896–7.

Roches Point ON Situated on the east shore of Cook's Bay, Lake Simcoe, and in the town of Georgina, York Region, north of Newmarket, this place was the first location of Keswick post office in 1836. It was renamed *Roach's Point*, after settler James Roche in 1870, and changed to *Roche's Point* in 1906. The present form was adopted in 1950.

Rochester ON A township in Essex County, it was named in 1792 after the town of Rochester, in Kent, England, near the town of Chatham, east of London.

Rockcliffe Park ON Adjacent to the city of Ottawa, this village (1925) was laid out by engineer Thomas Coltrin Keefer in 1864. He named it after Rockcliff House, built by Duncan MacNab in 1838, and acquired by Keefer in 1864 for his mother-in-law, Anne McKay. Her late husband Thomas McKay had built Rideau Hall, the unofficial title of the residence of governors general since 1868.

Rock Forest QC This town (1983) southwest of Sherbrooke had been established as a municipality in 1921. Its post office was named in 1872, two years after Parker Nagle called the place Rock Forest, after a residence he had owned in his native Ireland.

Rockglen SK South of Moose Jaw and near the Montana border, this town (1957) was named after a valley surrounded by rocky hills. Its post office was named in 1926, when the Canadian Pacific Railway was built south from Assiniboia.

Rockingham NS In Halifax and on the west side of Bedford Basin, this place was named *Rockingham Station* in 1886 after the Rockingham Club, established by Sir John Wentworth, when he was lieutenant-governor of Nova Scotia, 1792–1808. He named the club after his patron, Charles Watson-Wentworth, Marquess of Rockingham (1730–82), British statesman and prime minister, 1765–6, and 1782.

Rock Island QC Part of the town of Stanstead on the Vermont border, south of Magog, it had been an incorporated town from 1957 to 1995. At the beginning of the nineteenth century it was known as *Kilborn's Mills*. Charles Kilborn had constructed a canal across one of the bends of the Tomifobia River and created a small rocky island. Rock Island post office was opened in 1863.

Rockland ON A town (1908) on the Ottawa River and in the united counties of Prescott and Russell, east of Ottawa, it was named in 1869 by postmaster William Cameron Edwards. He had built a sawmill the previous year, after acquiring timber limits in the Gatineau Valley.

Rockwood ON A community in Wellington County, northeast of Guelph, it was called *Strange's Mills* and *Brotherstown*, before Rockwood Academy founder William Wetherald changed it to Rockwood in 1853, after its rocky and well-treed area.

Rockyford AB Northeast of Calgary, this village (1919) was named in 1914 by the Canadian Northern Railway's chief surveyor Beaumont after a narrow crossing of Serviceberry Creek, just south of the village site. The 'rocky ford' was well known to Aboriginals, ranchers, and surveyors.

Rocky Harbour NF Located on the western coast of Newfoundland and at the mouth of Bonne Bay, the shoreline of this municipal community (1966) is comprised of rocky ledges. The harbour was occupied by fishermen in the early 1800s.

Rocky Mountain House AB A town (1939) west of Red Deer, its post office was named in 1912. The North West Company had established Rocky Mountain House here in 1799, and it was continued intermittently by the Hudson's Bay Company from 1821 to 1875.

Rocky Mountains AB, BC Within Canada, this division of the Cordilleran Region's eastern system extends 1375 km northwest from the United States border to the Liard River, west of Fort Nelson. The system is approximately 100 km wide, with the **Rocky Mountain Trench** to the west and the **Rocky Mountain Foothills** to the east. Called *Shining Mountains* in 1716 by Hudson's Bay Company governor James Knight, they were described in 1752 by explorer Jacques Legardeur de Saint-Pierre as *Montagnes de Roches*, although he had not seen them. Two years later Hudson's Bay Company fur trader Anthony Henday was the first European to see them, although he did not mention them in his journal.

Roddickton NF At the head of Canada Bay and on the east coast of the Northern Peninsula, this town (1953) was named in 1906 in honour of surgeon Sir Thomas George Roddick (1846–1923), a native of Harbour Grace. He was the dean of medicine at McGill University, 1901–8, and a member of Parliament, 1896–1904. The site of the town was earlier known as *Easter Brook*.

Rodney ON In Elgin County, southwest of St. Thomas, this urban centre was first known as *Stewart's Mills* and *Centreville*, before it was renamed in 1865 after Sir George Brydges Rodney (1718–92). Local resident Lachlan MacDougall had proposed the name several years before to honour the English admiral's distinguished naval career. Created a village in 1907, Rodney was annexed by the municipal township of Aldborough in 1994.

Roes Welcome Sound NT This 300-km channel, between Southampton Island and the mainland, was named *Roe's Welcome* in 1631 by Luke Foxe after Sir Thomas Roe (1580–1644), a diplomat who supported Foxe's voyage of discovery. Several eighteenth and nineteenth century maps identified it as *Sir Thomas Roe's Welcome*, but the Geographic Board in 1902 named it Roes Welcome Sound.

Rogers Pass BC This route through the Selkirk Mountains was discovered in 1882 by Maj. Albert Bowman Rogers (1829–89) for the transcontinental route of the Canadian Pacific Railway, and was named after a promise made by CPR director James J. Hill to name a useful pass after him. He framed the $5000 cheque given to him by the CPR's general manager William Van Horne. **Mount Rogers**, a three-peaked mountain north of the pass, was named in 1890 by climber Carl Sulzer.

Rogersville NB A village (1966) southeast of Newcastle (Miramichi), its post office was called *Forest Station* in 1876, and was renamed *Carleton Station* the following year, after Carleton Parish. It was renamed in 1884 after Mgr James Rogers (1826–1903), Catholic bishop of Chatham, 1860–1902.

Roland MB Founded in 1889 on the Canadian Northern Railway line from Morris to Somerset, this place was named by the company after lumber dealer Roland MacDonald, who provided room and board for its surveyors. Its post office had been called *Lowestoft* in 1884 and was changed to Roland in 1890.

Rollo Bay PE West of Souris, this place's post office was named about 1854 after the nearby bay. The bay was named in 1765 by Samuel Holland, after Andrew Rollo, 5th Baron Rollo (1703–65). He accepted the surrender of the island (then called *Isle de Saint Jean*) by the French in 1759.

Romney ON A township in Kent County, it was named in 1794 after an ancient port in Kent, England, and on the Strait of Dover.

Roosevelt, Mount BC In the Muskwa Ranges of the Rocky Mountains, this mountain (2972 m) was named in 1944 by provincial minister of lands E.T. Kenny after United States President Franklin Delano Roosevelt (1882–1945). It was one of the names given in the area to honour Allied leaders of the Second World War.

Root, Mount BC In the St. Elias Mountains, 9 km northeast of Fairweather Mountain, this mountain (3901 m) was named in 1908 after Elihu Root (1845–1937), who was then the secretary of state in the administration of United States President Theodore Roosevelt.

Rosalind AB A village (1966) southeast of Camrose, its post office was named in 1905, drawing upon the two local school district names, Montrose and East Lynne. Montrose may have been derived from Montrose, Angusshire, Scotland, northeast of Dundee. East Lynne came from the title of a book an early settler had read.

Rose Blanche-Harbour Le Cou NF This town (1983) is situated on the southwestern coast of the island of Newfoundland and east of Channel-Port aux Basques. **Rose Blanche**'s name was derived from the French description of the rosy white granite here. **Harbour Le Cou** has a harbour shaped like a figure '8', with the narrow part having been called by the French *le cou*, 'the neck'.

Rosemary AB Northeast of Brooks, this village (1951) was named in 1914 by the Canadian Pacific Railway after Rosemary Millicent (1893–1930), the wife of the 3rd Earl of Dudley and daughter of the 4th Duke of Sutherland. The duke had acquired extensive lands in the Brooks area.

Rosemère QC Situated on the north shore of Rivère des Mille-Îles, north of Montréal, this town (1958) was created a parish municipality in 1947. Named Rosemere as a Canadian Pacific Railway station in 1880 after the abundance of wild roses and the old English word for a lake, its post office was open from 1901 to 1960. The parallel French form, Rosemère was adopted in 1940.

Rosemont ON On the boundary of Dufferin and Simcoe counties, its post office was named in 1861, possibly after wild roses blooming on the nearby hills. In providing its spelling, the post office department may have been unfluenced by Rosemont, a present-day neighbourhood in east central Montréal.

Roseneath ON In Northumberland County, northeast of Cobourg, its post office was named in 1858, after Rosneath, Dunbartonshire, Scotland, northwest of Glasgow, the birthplace of Mr and Mrs James Campbell, who had settled on the east side of Rice Lake 20 years earlier.

Rosenfeld MB Located southwest of Morris, this community was founded by Mennonites and named in 1882 when the Canadian Pacific Railway arrived. In German, the name means 'rose field'.

Rosetown SK Southeast of Saskatoon, this town (1911) was settled in 1904-5, and its post office was opened in 1907. It was named after pioneer settler James Rose. The lines of the Canadian Pacific and Canadian Northern railways crossed here in 1908.

Rose Valley SK Between Tisdale and Wadena, this town (1962) was named about 1915, when Martin Nelson proposed it at a meeting of grain growers. The post office opened 1 January 1916 and the Canadian Pacific Railway station was established here in 1924.

Ross ON A township in Renfrew County, it was established in 1830, and was possibly named by its surveyor and settlement agent John McNaughton, after Ross-shire, part of the Scottish Highlands county of Ross and Cromarty. It may have been given in honour of Adm. Sir James Clark Ross (1800–62), who joined his uncle Sir John Ross on six Arctic voyages.

Rossburn MB Southwest of Riding Mountain National Park, this village (1913) was named in 1880 after R.R. Ross (d. 1904), who had arrived here in 1879 from Molesworth, ON, northwest of Listowel.

Rosseau, Lake ON Jean Baptiste Rousseau (1758–1812), a fur trader who lived near the mouth of the Humber River, acted as an interpreter during discussions, agreements, and treaties with the Ojibwa in the areas of Lake Simcoe and the Muskoka Lakes. In 1835 Lt John Carthew of the Royal Navy, who took part in an exploration from the Severn River to Lake Nipissing, reported *Rousseau's Lake* was a large lake studded with beautiful islands. Geologist Alexander Murray called it *Lake Rousseau* in 1853, and surveyor Vernon Wadsworth referred to it as Lake Rosseau in 1860. The community of **Rosseau**, on the north shore of the lake, was named in 1866. It was an incorporated village from 1926 to 1971, when it was merged with the municipal township of Muskoka Lakes.

Rossignol, Lake NS The largest lake in the southwestern part of the province, this lake was named after Jean Rossignol, whose ship and furs were confiscated in 1604 in present-day Liverpool Harbour by Pierre du Gua de Monts. Four years later Rossignol proved he was legally trading on the coast and recovered his ship and compensation for the seized furs.

Rossland BC A city (1897) west of Trail, it was founded in 1893 by Ross Thompson, who called it *Thompson*. Likely because of possible confusion with Thompson's Landing on Lower Arrow Lake, the post office department established its post office as Rossland in 1894.

Ross River YT Located at the junction of the Ross and Pelly rivers, this place was named about 1903 by storekeeper Tom Smith. The river was named in 1843 by Hudson's Bay Company trader Robert Campbell after HBC chief factor Donald Ross.

Rosthern SK Midway between Saskatoon and Prince Albert, this town (1903) had its beginnings about 1890 when the line of the Qu'Appelle, Long Lake and Saskatchewan Railway and Steamboat Company was built from Regina to Prince Albert. The line became part of the Canadian Northern Railway in 1906. The place may have been named after a man called Ross, who drowned in the creek flowing through the townsite. The post office was opened in 1893.

Rothesay NB A town (1956) northeast of Saint John, it was named in 1860 by Robert Thompson after the Prince of Wales, Duke of Saxony, Cornwall, and Rothesay, the future Edward VII. He visited New Brunswick that year and sailed to Fredericton from a wharf nearby. Rothesay was first used as a European and North American Railway station. The post office had been named *Kennebecassis Bay* in 1853, but became Rothesay in 1866.

Rougemont QC In the regional county of Rouville, between Chambly and Granby, this village was separated from the parish municipality of *Saint-Michel-de-Rougemont* in 1914. Its post office was opened in 1852, taking its name from the adjacent 390-m **Mont Rougemont**. It possibly owes its name to Étienne Rougemont, a captain with the Carignan-Salières Regiment, who arrived in New France in 1665.

Rouleau SK Southwest of Regina, this town (1907) was established by the Canadian Pacific Railway in 1893 and named after Charles-Borromée Rouleau (1840–1901). He was appointed a stipendiary judge of the North-West Territories in 1883, serving until 1887, when he became a judge in the district of Alberta.

Rouville QC A regional county between Chambly and Granby, it was named in 1982 after the former county of Rouville, which had been named after Jean-Baptiste Hertel de Rouville (1668–1722), a military officer who was granted the seigneury of Rouville in 1694.

Rouyn-Noranda QC Both a regional county (1981) and a city (1986), they owe their origins to the discovery of gold and copper on the shore of Lac Osisko in 1911, and a gold rush a decade later. **Noranda**, a contraction of 'north' and 'Canada', became a company mining village, and was incorporated as a town in 1926. **Rouyn**, named after the township (1916) developed in the early 1920s on the south side of Noranda, was erected as a town in 1927. The township had been named after Jean-Baptiste Rouyn, a captain who served with the Royal-Roussillon Regiment in 1759–60.

Rowley Island NT At the north end of Foxe Basin, this island (1090 km^2) was named in 1941 after Graham M. Rowley, the first non-Aboriginal to explore the area. Rowley was subsequently the secretary coordinator of the Advisory Committee on Northern Development in the department of northern development and natural resources.

Roxboro QC In the Communauté urbaine de Montréal and on the south shore of the Rivière des Prairies, this town was erected in 1914. The reason for its choice of name is uncertain. It may have been given after a place in England, after an old farm of a military officer, or after a limestone quarry, suggesting a 'rocks borough'.

Roxborough ON A township in the united counties of Stormont, Dundas and Glengarry, it was named in 1798 after John Ker, the 3rd Duke of Roxburghe (1740–1804), a close friend of George III. The title of Lord Roxburghe had been granted in 1600 by James VI of Scotland to Robert Ker, and was given after the Scottish border county of Roxburghshire.

Roxton Falls QC Located in the municipal county of Acton, midway between Drummondville and Granby, this village was created in 1863, when it was separated from the township municipality of **Roxton**. The latter was established in 1855, taking its name from the township, noted on a 1795 map, and proclaimed in 1803. The township was named after a village in Bedfordshire, England, west of the city of Cambridge. **Roxton Pond**, a postal name since 1864, became a village in 1886, and also became the name of a parish municipality in 1985, succeeding *Sainte-Pudentienne*.

Royal Geographical Society Islands NT Between Victoria and King William islands, this group of islands was named in 1905 by Roald Amundsen after the Royal Geographical Society of London, England.

Royal Roads BC This body of water, between Victoria on the east and Albert Head on the west, was named *Royal Bay* in 1846 by Capt. Henry Kellett, possibly because of the royal names on each side honouring the Queen and her consort. As a good location to anchor ships, it became known as Royal Roads, the name in general use on marine charts by the first of the twentieth century.

Royston BC On the east coast of Vancouver Island, southeast of Courtenay, this place was named in 1910 by William Roy, who had settled here in 1890, and by land developer Frederick B. Warren, who had come from Royston, Cambridgeshire, England.

Rundle, Mount AB A landmark in Banff, this mountain (2949 m) was named in 1885 by Dr James (later Sir James) Hector after Methodist missionary Robert Terrill Rundle (1811–96). Rundle joined the Hudson's Bay Company in 1839 and was sent to its Saskatchewan district. He travelled through the West extensively, and, in either 1844 or 1847, would have seen the mountain named after him. He returned to England in 1847.

Rupert, Rivière QC Draining Lac Mistassini in northwestern Québec, this river flows west for 483 km, emptying into the southeast corner of James Bay. It was named in 1668 after Prince Rupert (1619–82), a son-in-law of Charles I and a first cousin of Charles II. He was appointed the first governor of the Hudson's Bay Company. The trading post at its mouth was first called Fort Charles, after Charles II, and then called Rupert House in 1670. The Cree village was known as *Fort Rupert* until 1978, when it was changed to Waskaganish.

Rusagonis NB On **Rusagonis Stream**, south of Fredericton, this place's post office was called *Rusagornis* (following local pronunciation) from 1853 to 1930, when it became Rusagonis until 1936. A nearby community's post office was known as *Rusagornis Station* from 1871 to 1930, when it became **Rusagonis Station** until 1945. The stream's name was derived from Maliseet *tesegwanik*, possibly 'meeting with the main stream' or 'flat branch'.

Rushoon NF Located on the western shore of Placentia Bay, and northeast of Marystown, this municipal community (1966) was noted in French records during the 1700s as *Rashoon*. Its name may have been derived from *roches jaunes*, 'yellow rocks', a description of the rocky shore.

Russell MB The post office at this town (1913), west of Riding Mountain National Park, was called *Shell River* in 1880. It was renamed in 1889 by Col C.A. Boulton, after Gen. Lord Alexander George Russell, commander of the Canadian forces, 1883–8.

Russell ON The county and the township in the united counties of **Prescott and Russell** were named in 1798 after Peter Russell (1733–1808), administrator of Upper Canada from the departure of Lt-Gov. John Graves Simcoe in 1796 until the appointment of Peter Hunter as lieutenant-governor in 1799. **Russell** post office was named in 1848, with developer William Duncan as the postmaster. Four years later, his son, John, the next postmaster, renamed the community *Duncanville*, but the postal name was not changed. When it became a police village in 1898, Russell was restored.

Russell Island NT North of Prince of Wales Island, this island (940 km^2) was named in 1819 by William (later Sir William) Parry after John Russell, 1st Earl Russell (1792–1878).

Rustico Bay PE René Rassicot, a native of Normandie, settled on the north side of the island in 1724. By 1752 Joseph de la Roque referred to the bay as *Grand Racico*. In 1765 Samuel Holland called it *Harris Bay*, with *Grand Rastico* as a secondary name, but the name Rustico Bay soon took precedence. The bay is enclosed by **Rustico Island**. It had been previously known as *Gotteville Island*, *Peters Island*, *McAuslin Island*, and *Robinsons Island*. *See also* North Rustico.

Ruthven ON In Essex County, west of Leamngton, this place was first called *Inkermanville* after the 1854 battle in the Crimea. Its post office was opened in 1860, with Hugh Ruthven as the first postmaster.

Rycroft AB A village (1944) north of Grande Prairie, it was near the first site of Spirit River. In 1920 the Spirit River post office was moved 8 km to the west, with the original site being named *Roycroft* after Robert Henry Rycroft, an early settler and justice of the peace. It was corrected to Rycroft in 1931. Three other residents, W.S.O. English, 'Doc' Calkin, and George Garnett, wanted their names chosen, but Rycroft's was picked from a hat.

Ryley AB A village (1910) southwest of Vegreville, its post office was named in 1908, after George Urquhart Ryley, land commissioner of the Grand Trunk Pacific Railway. A station was established here in 1909.

S

Saanich BC Immediately north of Victoria and at the base of the **Saanich Peninsula**, this district municipality (1906) derived its name from the Saanich tribe. The tribe's name in Straits Salish means 'elevated', in reference to Mount Newton (307 m), whose profile looks like a raised rump. The deep 23-km bay on the west side of the peninsula is called **Saanich Inlet**. *See also* Central Saanich *and* North Saanich.

Sabine Peninsula NT A northern extension of Melville Island, this peninsula was named in 1820 by William (later Sir William) Parry after Edward (later Sir Edward) Sabine (1788–1883), the astronomer on Parry's expedition in search of the Northwest Passage.

Sable Island NS Located in the Atlantic Ocean, 300 km southeast of Halifax, this island was named on a mid-1500s map of eastern North America as *Isola dell rena*. A map by Guillaume Levasseur in 1601 has *Y sable*, and Samuel de Champlain's 1613 map has *isle de sable*.

Sachs Harbour NT On the west coast of Banks Island, this hamlet (1986) was established in 1953 and named two years later when a meteorological station and post office were opened. The harbour was named in 1947 after the schooner *Mary Sachs*, which was beached here in 1914, while carrying provisions for an expedition.

Sackville NB A town (1903) on the Tantramar River, its post office had been named in 1837. It took its name from **Sackville Parish**, created in 1786, and established earlier in 1772 as Sackville Township. The township was called after George Sackville-Germain (1716–85), who became 1st Viscount Sackville in 1782.

Sackville NS North of Halifax, this suburban area unites a number of residential communities. It takes its name from Fort Sackville, constructed at the head of Bedford Basin in 1749. It was named after George Sackville-Germain (1716–85), who became 1st Viscount Sackville in 1782. In 1745 he had distinguished himself at the Battle of Fontenoy in France. Postal service was established in 1884 at **Lower Sackville**, **Middle Sackville**, and **Upper Sackville**.

Sacré-Cœur QC A municipality (1973) northwest of Tadoussac, it became the parish municipality of *Sacré-Cœur-de-Jésus* in 1915. A village of the same name was incorporated in 1937, and the two of them were merged as Sacré-Cœur in 1973. Its post office was called *Dolbeau* in 1884, and was renamed Sacré-Cœur-Saguenay in 1927.

Saguenay, Rivière QC Draining a basin of 88,000 km^2, this river (160 km long, but 698 km to the head of Rivière Péribonka) enters the St. Lawrence at Tadoussac. Jacques Cartier drew attention in his *Relation* of 1535–6 to the river, the royal road, and the land of the Saguenay. The name, referring to the territory, and not to the river, means 'place where the water flows out', likely from Montagnais or Algonquin.

St. Adolphe MB On the east bank of the Red River, south of Winnipeg, this place's post office was named in 1893 after Adolphe Turner, who generously subscribed to the erection of the Catholic church. Two years before it had been called *Dubuc*.

Saint-Adolphe-d'Howard QC A municipality, northwest of Saint-Jérôme, it was created in 1939, having been called the township municipality of *Howard* in 1883. The township was named in 1873 after Sir Frederick Howard, colonial commissioner at the end of the eighteenth century. The name of the municipality honours Fr Adolphe Jodoin (1836–91), who served a mission here, 1878–82.

Saint-Agapit QC A municipality in the regional county of Lotbinière, southwest of the city of Québec, it was formed in 1979 through the merger of the parish municipality of *Saint-Agapit-de-Beaurivage* (1867) and the village of *Saint-Agapitville* (1911). The name had been given after St Agapetus (d. 536), the fifty-seventh Pope, 535-6.

St. Agatha ON In Waterloo Region, west of the city of Waterloo, this community's post office was named in 1852 after St Agatha parish and church, dedicated in the 1830s. St Agatha was a third-century Sicilian martyr.

St. Alban's NF On the west side of Bay d'Espoir and near the southwestern coast of the island of Newfoundland, this town (1953) was named in 1915 by Fr Stanislaus St Croix, after the third century English martyr St Alban. St. Albans, Hertfordshire, England, northwest of London, is also named after him. It was formerly called *Ship Cove*.

St. Albert AB A city (1976) on the northwest side of Edmonton, it was named in 1861 by Mgr Alexandre Taché after Fr Albert Lacombe (1827–1916). He was a missionary among the Aboriginal and Métis communities in the present area of Alberta from 1852.

St-Albert ON In the united counties of Prescott and Russell, southwest of Casselman, its post office was named in 1880 after a church built here four years before. The church had been put under the protection of St Albert, while taking the name of the first parish priest, Fr Albert Philion.

Saint-Albert-de-Warwick QC A parish municipality in the regional county of Arthabaska, southwest of Victoriaville, it was erected in 1864. Its mission, established three years earlier, was named in honour of St Albertus Magnus (*c.* 1200–80), a scholar and theologian.

Saint-Alexandre QC Located southeast of Saint-Jean-sur-Richelieu, it was established in 1988 through the merger of the village and the parish municipality of Saint-Alexandre. The parish was set up in 1850, and named for a fourth century bishop of Alexandria, Egypt.

Saint-Alexandre QC In the regional county of Kamouraska, southwest of Rivière-du-Loup, this parish municipality was founded in 1851, taking its name from Mgr Alexandre-Antonin Taché, who was born nearby, and became archbishop of Saint-Boniface in Manitoba in 1853, and after local settler Alexandre Thériault. Its post office, **Saint-Alexandre-de-Kamouraska**, was opened in 1854.

Saint-Alexis-des-Monts QC In the regional municipality of Maskinongé, northwest of Louiseville, this parish municipality was created in 1984 through the amalgamation of the municipality of *Belleau* (1973) and the parish municipality of *Saint-Alexis* (1877). The Saint-Alexis-des-Monts post office was opened in 1876 and named after pioneer settler Alexis Bélanger.

Saint-Alphonse QC A parish municipality southwest of Granby, it was named in 1875 after the parish, erected six years earlier. The parish honoured the founder of the Redemptorists, St Alphonso Maria de' Liguori (1696–1787), patron saint of Fr Joseph-Alphonse Gravel (1843–1901), secretary of the diocese of Saint-Hyacinthe, 1875–86.

Saint-Alphonse-Rodriguez QC A municipality northwest of Joliette, it was erected in 1991, having been called *Bienheureux-Alphonse-de-Rodriguez* since 1855. Alfonso de Rodríguez (1531–1617), a Spanish brother of the Company of Jesus, was canonized in 1888.

Saint-Amable QC Located northeast of Longueuil, this municipality (1984) had been created a parish municipality in 1921. The religious parish was erected in 1913 and named after Fr Amable Prévost (1757–1820), who served in the area of Chambly and Belœil, 1807–20.

Saint-Ambroise QC A municipality northwest of Jonquière, it received its current status in 1982. It had been created a village in 1971 by merging the village (1917) and the municipality (1902) of the same name. The religious parish, founded in 1883, was named after a parish in Jonquière, which had been called after

Fr Ambroise-Martial Fafard (1840–99), pastor of the cathedral in Chicoutimi, 1880–9.

Saint-Ambroise-de-Kildare QC Located northwest of Jonquière, this parish municipality was created in 1855, taking its name from the religious parish, established in 1839. It was named after St Ambrosius (*c.* 340–97), bishop of Milan, and a noted doctor of the church. The township of Kildare was erected in 1803, taking its name from the town and county southwest of Dublin, Ireland.

Saint-André NB A village (1967) north of Grand Falls, its first post office was called *Chambord* in 1884 after Henry V, Comte de Chambord (1820–83). It was renamed **St-André-de-Madawaska** in 1912.

Saint-André-Avellin QC A village in the regional county of Papineau, northeast of Buckingham, it was named in 1911 when it was severed from the parish municipality of **Saint-André-Avellin**. The parish municipality and the post office were both established in 1855. The religious parish was created in 1851, taking its name from André Trudeau, who surveyed lots in the area in 1832, and from St Andrea Avellino (1521–1608), who was canonized in 1712.

Saint-André-d'Acton QC This parish municipality, southwest of Acton Vale, was erected in 1864. It was named after St Andrew, the younger brother of Simon Peter. The township of Acton was named after a suburb of London, England, west of the city centre.

Saint-André-Est QC Located south of Lachute, it was created as the village of *St. Andrews* in 1958, and the present name was adopted 20 years later. The Anglican parish of St Andrews was established here in 1822, three years after the *St. Andrews East* post office was opened. The parish municipality of **Saint-André-d'Argenteuil** was erected in 1855.

St. Andrews NB In Charlotte County and at the mouth of the St. Croix River, this town (1903) was named after **St. Andrews Parish**, created in 1786. Its post office was opened in 1817. Capt. William Owen identified *St. Andrews Point* on his 1770 chart. The point may have been named after a priest called St-André, or a priest may have celebrated mass here on the day of St Andrew. The town is commonly known as St. Andrews by the Sea.

St. Andrews ON In the united counties of Stormont, Dundas and Glengarry, north of Cornwall, it was first settled by Highland Scots in 1786. The original St. Andrew's Catholic Church, now the parish hall, was built between 1789 and 1801, and is Ontario's oldest stone structure. The post office was called St. Andrews in 1830, with 'West' being added to distinguish it from *St. Andrews East*, QC, renamed Saint-André-Est in 1978.

Saint-Anicet QC On the south shore of Lake St. Francis, southwest of Salaberry-de-Valleyfield, this parish municipality was established in 1885. The post office was opened in 1851, taking its name from the mission, created in 1810. St Anicetus was the Pope from *c.* 155 to *c.* 166.

St. Anns NS On Cape Breton Island and at the head of **St. Anns Harbour**, its post office was named in 1842. The harbour and **St. Anns Bay** were named after *Havre de Sainte-Anne*, which recalled *Fort Sainte-Anne*, established here in 1629 by Capt. Charles Daniel.

Saint-Anselme QC A village in the regional county of Bellechasse, southeast of Lévis, it was erected in 1920, when it was severed from the parish municipality of **Saint-Anselme**. It had been named after the parish, established in 1827. St Anselm (1033–1109), a scholar and theologian, was an archbishop of Canterbury.

St. Anthony NF A town (1945) on the northeastern coast of the Northern Peninsula, it was the site of a French fishing harbour in the 1700s, and became an English-speaking settlement in the mid-1800s. The harbour may have been named by Jacques Cartier in the spring of 1534. On 13 June of that year he had named *H. de St Antoine*, on the north shore of the estuary of the St. Lawrence River, between Havre Saint-Pierre and Sept-Îles.

Saint-Antoine NB A village north of Moncton, it was incorporated in 1966 as *St. Anthony*, and changed to its present name in 1969. Its post office was called *St. Anthony* from 1873 to 1928, when it was changed to **St-Antoine-de-Kent**. It may have been named after a French mission located on the Richibucto River about 1700, which was possibly given after the French seminary of St-Antoine-de-Padoue, where Fr Chrestien LeClercq, a Recollet missionary to the Mi'kmaq, was ordained about 1668.

Saint-Antoine QC Located just south of Saint-Jérôme, it was incorporated as a town in 1967, having previously been the village (1956) and the parish of *Saint-Antoine-des-Laurentides* (1940). It was named after Fr Antoine Labelle (1833–91), parish priest of Saint-Jérôme, 1868–91, who actively promoted the settlement of the lands to the north.

Saint-Antoine-de-Lavaltrie QC A parish municipality south of Joliette, it was created in 1855. The name was given in honour of St Anthony of Padua (1195–1231), a native of Portugal, who is claimed as a patron saint by Padua, Italy.

Saint-Antoine-de-Richelieu QC Situated on the west bank of the Richelieu, downriver from Belœil, the parish municipality was created in 1890, and named after the parish, founded in 1741 in honour of St Anthony of Padua (1195–1231), a doctor of the church. The current municipality was erected in 1982.

Saint-Antonin QC Located southeast of Rivière-du-Loup, this parish municipality was created in 1856, and named after Louis-Antonin (or Antoine) Proulx (1810–96), the first parish priest, 1840–54.

Saint-Apollinaire QC In the regional county of Lotbinière, southwest of the city of Québec, this municipality was created in 1974 through the merger of the parish municipality of Saint-Apollinaire (1855) and the village of *Francœur* (1919). Its post office was named in 1858. St Apollonaris (d. *c.* 200) was a bishop and martyr who lived at Ravenna, Italy.

Saint-Athanase QC Situated south of Iberville and across the Richelieu from Saint-Jean-sur-Richelieu, this parish municipality was named in 1845. The religious parish was named in 1828, likely after St Athanasius (*c.* 295–373), bishop of Alexandria, Egypt, 328–73.

Saint-Augustin-de-Desmaures QC A parish municipality in the Communauté urbaine de Québec, west of Sainte-Foy, it was erected in 1845, taking its name from the parish, founded in 1679. The name recalls Augustin de Saffray de Mézy, governor of New France, 1663–5, and Jean Juchereau de Maure, a grantee of the seigneury of de Maure (or Saint-Augustin) about 1681.

Saint-Basile NB On the Saint John River, southeast of Edmundston, this village was incorporated in 1966. Its post office was called *Upper St. Bazil* from 1852 to 1810, when it was renamed *Upper St. Basil*. It was *St. Bazil* from 1911 to 1938, when it was changed to the present name. **Saint-Basile Parish** was created in 1850, taking its name from St Basil the Great (*c.* 330–79), noted for organizing Greek monasticism.

Saint-Basile-le-Grand QC A town on the west bank of the Richelieu, east of Montréal, it was erected as a town in 1969. The religious parish was created in 1870, followed by the parish municipality the next year. It was named after local farmer Basile Daigneault, whose own name recalled the celebrated St Basil the Great (*c.* 330–79), who organized monastic life in Greece.

Saint-Bernard QC A municipality on the west side of the Rivière Chaudière, north of Sainte-Marie, it was created in 1987 through the amalgamation of the village (1959) and the parish municipality (1845) of Saint-Bernard. The religious parish was named in 1821 after St Bernard (1090–1153), the abbot of Clairvaux, France, and a doctor of the church.

St. Bernard's NF On the eastern shore of Fortune Bay and northeast of Marystown, this municipal community (1967) was named in 1915 after the local Catholic church, built here about 1890. It was formerly called *Fox Cove*.

St. Boniface MB Part of the neighbourhood of St. Boniface-St.Vital in the city of Winnipeg since 1972, this well-known French-speaking enclave was named in 1818 by Mgr Joseph-Norbert Provencher. He chose the name in honour of the English missionary to the Germans in the eighth century. It had been incorporated as a town in 1883, and a city in 1908.

Saint-Boniface-de-Shawinigan QC A village southwest of Shawinigan, it was erected in 1918, and the parish municipality of *Saint-Boniface-de-Shawinigan* (1956) was annexed to it in 1961. St Boniface (*c.* 673–754), a native of Devon, England, evangelized Germany.

St. Brendan's NF Situated on Cottel Island in Bonavista Bay, this municipal community (1953) was first called *Cottel Island*. It was renamed in the early 1900s after the Irish saint, who some 500 years before Christ undertook a legendary voyage in a leather boat across the Atlantic in search of the 'promised land of the saints'.

St. Bride, Mount AB This mountain (3315 m) northeast of Lake Louise, and directly south of Mount Douglas, was named in 1898 after the patron saint of the Douglas family.

St. Bride's NF Situated on the southeastern shore of Placentia Bay, this municipal community (1972) was called *Distress* as early as the 1830s, likely after the difficulty of finding a safe haven on the Cape Shore during rough seas. It was renamed about 1870 after St Brigid (453–523), a patron saint of Ireland.

St. Brieux SK Southwest of Melfort, this village (1913) was established in 1904 by Fr Paul Le Flock, who brought Breton settlers from Saint-Brieuc, Bretagne, France, west of Saint-Malo. Its post office was opened in 1905.

Saint-Bruno QC A municipality in the regional county of Lac-Saint-Jean-Est, south of Alma, it was created in 1886 and annexed the village of Saint-Bruno (1910) in 1975. The religious parish of Saint-Bruno was erected in 1884, and named after Fr Bruno-Élisée Leclerc (1838–1907), who was then pastor of Notre-Dame–d'Hébertville, recalling St Bruno (*c.* 1030–1101), the founder of the Carthusian order.

Saint-Bruno-de-Montarville QC Located on the south side of **Mont Bruno**, east of Montréal, this city was erected in 1958, 110 years after its post office had been opened. The religious parish of Saint-Bruno was founded in 1842, adapted from the name of the seigneur of Montarville, François-Pierre Bruneau. The seigneury was granted to Pierre Boucher de Boucherville, the younger (1653–1740), with its name having been constructed from the words *montagne* and *Boucherville*. First called *Colline de Montarville*, Mont Bruno came into general use for the mountain after 1842.

Saint-Calixte QC In the regional county of Montcalm, northeast of Saint-Jérôme, this municipality was named in 1954, having been the township municipality of Kilkenny since 1855. Its post office was called **Saint-Calixte-de-Kilkenny** in 1877. St Calixtus was the sixteenth Pope, 217–22. The township of Kilkenny was proclaimed in 1832 after the county and town of Kilkenny, southwest of Dublin, Ireland.

St. Catharines ON A city (1876) in Niagara Region, it was named about 1796 by Queenston merchant Robert Hamilton for his wife Catharine Askin Robertson, who died that year. It is often misspelled, because of the Catholic diocesan cathedral of St. Catherine's on Catherine Street, and because the post office was *St. Catherines* from 1817 to 1849, when it was respelled after the name of the town (1845). In the early years it was called *The Twelve*, after Twelve Mile Creek, and *Shipman's Corners*, after tavern keeper Paul Shipman.

Saint-Célestin QC A village in the regional county of Nicolet-Yamaska, southeast of Trois-Rivières, it replaced *Annaville* in 1991. *Annaville* had been named in 1896 after a relic of St Ann, brought from Rome by the founder of the parish of Saint-Célestin, Mgr Calixte Marquis. The parish was named after St Celestine V (d. 1296), the Pope in 1294.

Saint-Césaire QC In the regional county of Rouville, midway between Chambly and Granby, this town was erected in 1962, having been a village since 1859. Its post office was opened in 1832. The municipal parish was established in 1857, and the religious parish was founded in 1822. St Caesarius (*c.* 470–542) was a bishop of Arles, and primate of Gaul and Spain.

St. Charles ON In Sudbury District, southeast of the city of Sudbury, this community's post office was named in 1902 after the parish of Saint-Charles-Borromée, established in 1893. St Charles Borromeo (1538–84) was a sixteenth century archbishop of Milan.

Saint-Charles QC A village in the regional county of Bellechasse, southeast of the city of Québec, it was created in 1915 from the parish municipality of *Saint-Charles-Borromée*, established in 1855. St Charles Borromeo (1538–84) was a sixteenth century archbishop of Milan and the principal author of the 1566 Roman catechism.

Saint-Charles-Borromée QC In the regional county of Joliette and north of the city of Joliette, this municipality was created in 1986, having been a parish municipality from 1855. St Charles Borromeo was the founder in 1578 of the Oblates of the Blessed Virgin and St Ambrose, which preceded the familiar Oblates of Mary Immaculate, established in 1816.

Saint-Charles-de-Drummond QC Located on the east side of Drummondville, this municipality was established in 1988, and its post office had been opened in 1950. The parish was erected in 1868, and named after St Charles Borromeo, who was an energetic reformer of the church during the time of the Council of Trent, 1545–63.

Saint-Charles-de-Mandeville QC Located north of Joliette, this municipality was established in 1905, two years after the religious parish was created. The parish was named after either Fr Charles-Olivier Gingras or Fr Charles Turgeon. In 1905 **Mandeville** post office was opened here, taking its name from Maximillien Mandeville, who settled on the north shore of Lac Maskinongé about 1824.

Saint-Christophe-d'Arthabaska QC This parish municipality, southeast of Victoriaville, was named in 1845 after the religious parish, created in 1846. The parish was named after Jean-Chrysostome (also known as Christophe and Christo) Marcoux, who hunted in the area before 1830. It also recalls the name of St Christopher, the patron saint of travellers, especially drivers.

Saint-Chrysostome QC A village encircled by the parish municipality of **Saint-Jean-Chrysostome**, south of Montréal, it was established in 1902, 20 years after its post office was opened. St John Chrysostom (*c.* 347–407) was a patiarch of Constantinople and a scripture scholar.

St. Chrysostome PE Located northwest of Summerside and on the shore of Egmont Bay, this place's school district was established about 1854 and the post office was opened in 1893. It was formerly known as *Joe League Village*, after Joseph Arsenault, known for his precision in measuring distances.

St. Clair, Lake ON Between Lake Huron and Lake Erie, this body of water was named by René-Robert Cavelier de La Salle and Fr Louis Hennepin on 12 August 1679, the feast day of St Clair of Assisi. The **St. Clair River** comprised the first part of the *Rivière du Détroit* during the French regime, and was later called *Huron River*, before St. Clair River was given to it in the nineteenth century. The village of **St. Clair Beach**, east of Windsor, was incorporated in 1914.

St. Claude MB Southwest of Winnipeg, this village (1963) was named in 1891 by the Order of Canons Régulaires, based at nearby Notre Dame des Lourdes. They named it after a town in France, near the Swiss border, northeast of Lyon. The first settlers were French and Swiss.

St. Clements ON A community in Waterloo Region, northwest of Waterloo, its post office was named in 1853 after the first church built in the area. The church was named after Clement I, the pontiff of the Catholic Church for about 10 years, around 100 AD.

Saint-Colomban QC A parish municipality west of Saint-Jérôme, it was created in 1845, taking its name from the religious parish, established in 1835–6. The parish was named after St Colomban (543–615), an Irish missionary, who revived the monastic life in France, Switzerland, and Italy.

Saint-Constant QC Located south of Montréal, adjacent to the Kahnawake Reserve, this town (1973) had been estabished as a parish municipality in 1855, and its post office had been opened a year earlier. There is some uncertainty about the origin of the name, with the possibility of a connection with Constant Le Marchand de Lignery (or Ligneris) (*c.* 1663–1731), whose son, Jacques Le Marchand de Lignery, pastor at Laprairie, 1731–75, served the mission of Saint-Constant for a few years.

St. Croix NS East of Windsor and on the **St. Croix River**, this place's post office was named *Ste Croix* in 1855. The river forms the shape of a cross where it enters the Avon River, opposite Windsor.

St. Croix River NB Forming part of the boundary between Maine and New Brunswick, this river was named *Rivière saincte Croix* by Pierre du Gua de Monts in 1604. The St. Croix, Oak Bay, and Waweig River portray the shape of a cross. By the late 1700s, the estuary retained the name St. Croix, but the river above tide, and rising in Maine, was called the *Scoodic*, with the river on the boundary above Grand Falls Flowage being called the *Chiputneticook*. By 1885, the St. Croix was recognized as the boundary river rising in the Chiputneticook Lakes and flowing into Passamaquoddy Bay.

Saint-Cyrille-de-Wendover QC A municipality northeast of Drummondville, it was erected in 1982, having been the village of *Saint-Cyrille* from 1905. The Saint-Cyrille-de-Wendover post office was opened in 1875. It was named after first settler Cyrille Brassard, whose first name recalls St Cyril (*c.* 313–86), a bishop of Jerusalem and a doctor of the church. The township of Wendover was proclaimed in 1805, taking its name from a village in Buckinghamshire, England, northwest of London.

Saint-Damase QC A village south of the city of Saint-Hyacinthe, it was separated in 1952 from the parish municipality of **Saint-Damase**. The latter was erected in 1845, taking its name from the religious parish, organized in 1823. The post office was opened in 1837. The name was chosen by Mgr Joseph-Octave Plessis after St Damasus (*c.* 305–84), the Pope from 366 to 384.

Saint-Damien-de-Buckland QC A parish municipality in the regional county of Bellechasse, southeast of the city of Québec, it was erected in 1891. The parish of Saint-Damien was created in 1848, taking its name from St Damian, a celebrated physician in Asia Minor, who was beheaded about 303. The township of Buckland was proclaimed in 1806, and was named after a place in England, possibly Buckland in Buckinghamshire.

Saint-David-de-Falardeau QC Located north of Chicoutimi, this municipality was established in 1948, taking its name from the religious parish, created in 1937. It was named after Fr David Roussel (1835–98), pastor of Sainte-Anne-de-Chicoutimi after 1870. The township of Falardeau was proclaimed in 1920 and named after noted Québec-born painter Antoine-Sébastien Falardeau (1822–89), who spent most of his adult years in Italy.

Saint-Denis-de-Brompton QC Created in 1935, this parish municipality northwest of Sherbrooke was named after its founder, Fr Joseph-Denis Bellemare, a local parish priest. The township of Brompton was named in 1795 and proclaimed in 1801, possibly after a town in North Yorkshire, England, north of Northallerton.

Saint-Dominique QC Located east of Saint-Hyacinthe, this municipality was created in 1969 through the amalgamation of the village (1914) and the parish municipality of Saint-Dominique (1845). The parish was established in 1833. The post office, called Saint-Dominique in 1853, was renamed **Saint-Dominique-de-Bagot** in 1876. The name may have been chosen by the third seigneur of Saint-Hyacinthe, Hyacinthe-Marie Delorme in memory of his brother-in-law, Pierre-Dominique Debartzch, or after the latter's son, also Pierre-Dominique Debartzch, who acquired part of the seigneury in 1811.

Saint-Donat QC Located east of Mont-Tremblant, this municipality was established in 1953. It had been created in 1874 as the parish municipality of **Saint-Donat-de-Montcalm**, the name of the post office opened five years later. It may have been named after Fr Donat Coutu, the first pastor here in 1874, although recent evidence indicates Fr Alexis-Henri Coutu was the first pastor, and there was never a Fr Donat Coutu who had served in the parish.

Sainte-Adèle QC Created a town in 1973, this popular all-season resort north of Saint-Jérôme was previously a parish municipality (1855), which annexed the village of *Sainte-Adèle* (1922) in 1955. The religious parish was created in 1852, taking its name from Adèle Raymond, the wife of Augustin-Norbert Morin, who donated land for the church.

Ste. Agathe MB On the west bank of the Red River, south of Winnipeg, this community was named in 1872 by Mgr Noël Richot after his home town of Sainte-Agathe-des-Monts, northwest of Montréal. Its post office was named in 1878. The site was originally called *Pointe à Grouette*, possibly referring to fallow land, which is *guéret* in French.

Sainte-Agathe QC A village in the regional county of Lotbinière, southwest of the city of Québec, it was named in 1914 after the parish municipality of **Sainte-Agathe** (1857). The Sainte-Agathe post office was opened in 1858. St Agatha was martyred in either Palermo or Catania, Italy, in the year 251.

Sainte-Agathe-des-Monts QC A popular tourist destination in the mountains north of Saint-Jérôme, this town was incorporated in 1915, having been erected as a village in 1896. The adjacent village of **Sainte-Agathe-Sud** was named in 1964. The municipality of **Sainte-Agathe-Nord** was created in 1991, succeeding the parish municipality of *Sainte-Agathe* (1863). They recall a third-century Sicilian martyr.

Ste. Anne MB A village (1963) north of Steinbach, its name was given in 1864 by Fr Jean-Marie Le Floc'h, a native of France's Bretagne, where there was a special devotion to the mother of the Virgin Mary. It was earlier known as *Pointe des Chênes* and *Grande Pointe des Chênes*, and the village's post office was called *Ste. Anne des Chênes*. The oaks (*chênes*) have long disappeared from the banks of the Seine River, which flows into the Red River in Winnipeg.

Sainte-Anne-de-Beaupré QC A town northeast of the city of Québec, it was formed in 1973 through the merger of the village (1906) and the parish municipality of *Sainte-Anne-de-Beaupré* (1920). The name, which honours the mother of the Virgin Mary, has its roots in the early seventeenth-century phrase *le beau pré*, 'the pretty meadow', occurring in such names as the town of Beaupré and the regional county of La Côte-de-Beaupré.

Sainte-Anne-de-Bellevue QC Located at the extreme west end of the Île de Montréal, this town was established in 1895, having been incorporated as a village in 1878, a year before the *Sainte-Anne-de-Bout-de-l'Île* post office (1835) was changed to agree. The beautiful views over both Lac Saint-Louis and Lac des Deux Montagnes account for the ultimate part of the name.

Sainte-Anne-de-la-Pérade QC On the north shore of the St. Lawrence, northeast of Trois-Rivières, this municipality was created in 1989 through the fusion of the parish municipality of *Sainte-Anne-de-la-Pérade* (1845) and the village of *La Pérade* (1912).

Sainte-Anne-de-Madawaska NB Southeast of Edmundston, this village (1966) was first known as *André Settlement*. Its first post office was *Quisibus* in 1879, named after the Rivière Quisibis. It was renamed Ste-Anne-de-Madawaska in 1897. The church dedicated to St Ann was built in 1872, and **Sainte-Anne Parish** was created five years later.

Sainte-Anne-des-Monts QC On the south shore of the St. Lawrence and in the shadow of the Monts Chic-Chocs, this town (1968) is at the mouth of the **Rivière Sainte-Anne**, noted as early as a 1709 map. The religious parish originated in 1815, with the post office following in 1853, and the municipality two years later. Fr Jean-Baptiste Sasseville arrived here at the beginning of the nineteenth century, and may have dedicated the parish after his native Sainte-Anne-de-la-Pocatière.

Sainte-Anne-de-Sorel QC Located on the east side of the city of Sorel, the parish municipality was established in 1877, a year after the religious parish was set up. The post office was opened in 1879.

Sainte-Anne-des-Plaines QC Situated in the regional county of Thérèse-De Blainville, northeast of Blainville, this town had been created as a parish municipality in 1845. The latter had been named after the religious parish of Sainte-Anne-des-Plaines (1816), which had been created as Sainte-Anne-de-Mascouche in 1787. The word 'Plaines' in the name has its roots in the seigneury of Belle Plaine, granted to Louis Lepage de Sainte-Claire in 1731, and in the meaning of Algonquin *mascouche*, 'level plain'.

Sainte-Anne, Mont QC A well-known all-season leisure destination, just north of Sainte-Anne-de-Beaupré, this 820-m mountain acquired its name in the early 1800s from the parish of Sainte-Anne, which had been named in the mid-1600s after the mother of the Virgin Mary.

Sainte-Brigitte-de-Laval QC Located north of the city of Québec, this municipality has its roots in the religious parish, established in 1863. It was part of the seigneury granted to Mgr François de Laval (1623–1708), the first bishop of Québec, 1674–88. Settled after 1830 by Irish immigrants, they identified their community with St Brigid (*c.* 455–*c.* 528), a patron saint of Ireland.

Sainte-Catherine QC A town (1973) on the south side of the St. Lawrence, directly south of Montréal, it was created the parish municipality of *Sainte-Catherine-d'Alexandrie-de-Laprairie* in 1937, a year after the religious parish was erected. St Catherine of Alexandria was once considered a fourth-century martyr, but her name has been dropped from the official calendar of saints.

Sainte-Catherine-de-la-Jacques-Cartier QC Located on the Rivière Jacques-Cartier, and in the regional county of La Jacques-Cartier, west of the city of Québec, this municipality received this name in 1984. It was erected as the parish municipality of *Sainte-Catherine* in 1855, taking its name from the religious parish of Sainte-Catherine-de-Fossambault, erected in 1824.

Sainte-Claire QC A municipality in the regional county of Bellechasse, southeast of the city of Québec, it was erected in 1855, having been established 10 years earlier as *Sainte-Claire-de-Joliette*, but abolished in 1847. In 1977 it annexed the municipality of *Louis-Joliette* (1926). The land embraced the seigneury of Jolliet, granted to explorer Louis Jolliet in 1697. His wife was Claire-Françoise Byssot, perhaps the source of the name Sainte-Claire. St Clare of Assisi (1194–1253) was the founder of the order of the Poor Nuns.

Sainte-Croix QC A village in the regional county of Lotbinière, southeast of the city of Québec, it was severed from the parish municipality of **Sainte-Croix** in 1921. The latter, established in 1845, has its roots in the seigneury of Sainte-Croix, reserved in 1637 for the education of girls by the religious orders in the city of Québec, and taken over by the Ursulines in 1647. A mission and a parish were established in the early 1700s, and Samuel de Champlain referred to the place in 1613, noting it was originally further upriver at the present Pointe Platon.

Sainte-Félicité QC A village east of Matane, it was separated from the parish municipality of **Sainte-Félicité**, created in 1868, four years after the post office was opened. The site had been called Pointe-au-Massacre, likely because many shipwrecks happened here, but was renamed in 1860 by Mgr Baillargeon. St Felicity was a second-century Roman martyr.

Sainte-Foy QC A city in the Communauté urbaine de Québec, it was created a parish municipality in 1845, became a town in 1949, and annexed the parish municipality of *L'Ancienne-Lorette* in 1971. The name has its roots in the seigneury of Sainte-Foy, granted in 1637 to Pierre Puiseaux, which possibly took its name from a French village.

Sainte-Geneviève QC A town in the Communauté-Urbaine-de-Montréal, on the south shore of the Rivière des Prairies, it was erected in 1959, having been created in 1935 as the village of *Sainte-Geneviève-de-Pierrefonds*, and before that, the village of Sainte-Geneviève in 1860. *See also* Pierrefonds.

Sainte-Hélène-de-Breakeyville QC Situated south of the city of Québec, this parish municipality was created in 1909. The name recalls lumberman John Breakey (1846–1911), and his wife, Helen Henderson. The post office is called **Breakeyville**.

Sainte-Hélène, Île QC The principal site of Man and His World (Expo 67), this island was called *Saincte Elaine* in 1611 by Samuel de Champlain. Although he did not say so, the island was likely named after Hélène Bouillé (1598–1654), whom he married the previous year.

Sainte-Jeanne-d'Arc QC Located north of Lac Saint-Jean and east of Mistassini, this village was erected in 1922, two years after a mission was established here. St Joan of Arc (*c.* 1412–31) was canonized on 16 May 1920.

Sainte-Julie QC Located midway between the St. Lawrence and the Richelieu, east of Montréal, this city was erected in 1971, having been created a parish municipality in 1855. The religious parish was established in 1850 in honour of St Julia, a fifth-century martyr, and for Julie Gauthier *dit* Saint-Germain, whose husband donated land in 1849 for the chapel.

Sainte-Justine QC A municipality in the regional county of Les Etchemins, it was erected in 1993, having been a parish municipality from 1891, and the township municipality of *Langevin* from 1869. The name was given after Marie-Justine Têtu (1833–82), wife of Sir Hector-Louis Langevin (1826–1906), a Father of Confederation and minister of public works in the government of Canada, 1879–91. The township of Langevin had been proclaimed in 1862.

St. Eleanors PE Part of the city of Summerside since 1 April 1995, it was previously incorporated as a village in 1956 and as a municipal community in 1983. It may have been named after Eleanor Sanksey, the housekeeper of Maj. Harry Compton. Maj. Compton had settled nearby in 1804, but 10 years later moved to France. Hubert Compton, Harry's nephew, claimed it was after Harry's daughter, but it would appear that her name was Charlotte.

St. Elias, Mount YT This mountain (5489 m) is on the Yukon-Alaska boundary, and at the point where the Alaska panhandle begins its long, sinuous boundary with Canada south to the 54°40′ parallel. Cape St. Elias, on the tip of Kayak Island in Alaska, was named on 16 July 1741 by Vitus Bering, in honour of the saint's day. Bering, a Dane, was sailing on behalf of the Russian government. It would appear that Bering himself did not name the mountain, but that it was named by eighteenth century mapmakers. The **St. Elias Mountains** extend for 500 km from Alaska through the southwestern corner of the Yukon and across the most northwesterly part of British Columbia. They were called the *St. Elias Range* and the *St. Elias Alps* in the late 1800s, with the present name being confirmed in 1918.

Saint-Élie-d'Orford QC A municipality in the regional county of Sherbrooke, west of the city of Sherbrooke, it was established in 1992, having been created a parish municipality in

1899. The parish was created in 1889, six years after Fr Alfred-Élie Dufresne (1826–91) opened a mission here. The post office was called *Glen Iver* in 1885 and was changed to the present name in 1931.

Sainte-Madeleine QC A village midway between Mont-Saint-Hilaire and Saint-Hyacinthe, it was established in 1919, and its post office had been opened in 1876. The parish of Sainte-Marie-Madeleine was created in 1870, taking its name from St Mary Magdalen, the repentant sinner and follower of Jesus Christ.

Sainte-Madeleine-de-Rigaud QC Located on the Québec-Ontario border, and surrounding the town of Rigaud, this parish municipality was erected in 1855, and its mission had been set up about 1802. As well as honouring St Mary Magdalen, it was also named after Louise-Madeleine Chaussegros de Léry, the wife of Michel Chartier de Lotbinière, who had acquired the seigneury of Rigaud in 1763.

Sainte-Marie QC A town on both banks of the Rivière Chaudière, midway between the cities of Québec and Saint-Georges, its present limits were established in 1978 through the merging of the town (1958) of Sainte-Marie (erected a village in 1913) and the parish municipality (1845) of *Sainte-Marie*. Its name was given for Marie-Claire de Fleury de La Gorgendière (1708–97), whose mother in 1728 became the wife of Thomas-Jacques Taschereau (1680–1749), who had acquired the seigneury of Sainte-Marie in 1736.

Sainte-Marie-de-Monnoir QC A parish municipality between Chambly and Granby, it was established in 1845, taking its name from the religious parish, created in 1801. Claude de Ramezay was granted the seigneury of Monnoir in 1708, taking the name from a fief he owned in France. *See also* Marieville.

Sainte-Marie-Saint-Raphaël NB On Île Lamèque, northeast of Shippagan, this village (1986) was established by amalgamating the communities of **Sainte-Marie-sur-Mer** and **Saint-Raphaël-sur-Mer**. The former was a postal name from 1947 to 1970. The latter's post office was first called *Hachi* in 1915. It was renamed Saint-Raphaël-sur-Mer in 1946. Raphael Chiasson had a land grant here.

Sainte-Marthe-du-Cap-de-la-Madeleine QC This municipality was created in 1916, when it was separated from the parish municipality of *Sainte-Marie-Madeleine-du-Cap-de-la-Madeleine*. It was named after St Martha of the Gospels.

Sainte-Marthe-sur-le-Lac QC Located on the north shore of the Lac des Deux Montagnes, this town was erected in 1973, having become a municipality in 1960, when the parish (1956) was separated from the town of Saint-Eustache. The parish was named after St Martha of the Gospels.

Sainte-Martine QC A municipality in the regional county of Beauharnois-Salaberry, it was created in 1991, having been a parish municipality from 1855. The religious parish was established in 1829 and named after St Martina, a third-century Roman martyr.

Sainte-Mélanie QC Located in the regional county of Joliette, north of the city of Joliette, this municipality was established in 1845, with *Daillebout* being its post office from 1836 to 1881, when Sainte-Mélanie was adopted. It was named after Charlotte-Mélanie Panet (1794–1872), the wife of Marc-Antoine-Louis Lévesque, who donated land to erect a chapel in 1830. The name also recalls St Melanie the younger (383–439) of Rome, who became a nun in Jerusalem.

Saint-Émile QC Located northwest of Charlesbourg, in the Communauté urbaine de Québec, this town (1993) had been incorporated as a village in 1956, and as a municipality in 1929. The parish had been erected in 1925, but the post office had been called **Saint-Émile-de-Québec** in 1904. The parish was named after Fr Joseph-Nazaire-Émile Bédard, the pastor here from 1925 to 1940.

Sainte-Pétronille QC Located at the southwestern end of Île d'Orleans, this village was erected in 1980, having been called *Beaulieu* in 1874, the name of its post office, opened three years earlier, and continued until 1991. The religious parish was established in 1879. St Petronilla was a first-century martyr. *Beaulieu* was given for Jacques Gourdeau de Beaulieu, who in 1652 married the widow of the first seigneur.

Sainte-Rosalie QC On the the east side of the city of Saint-Hyacinthe, this village was incorporated in 1949, having been detached from the parish municipality of **Sainte-Rosalie**, erected in 1845, abolished two years later, and re-established in 1855. The original parish was created in 1832, and was named in honour of the patron saint of Palermo, Sicily, St Rosalia (d. 1160). Noted Québec women with this name include the mother of Louis-Joseph Papineau's, Rosalie Cherrier, and his sister, Marie-Rosalie, the wife of Jean Dessaulles, who was the proprietor of the seigneuries of Dessaulles and Saint-Hyacinthe when the parish was established.

Ste. Rose du Lac MB A village (1920) east of Dauphin, and south of Dauphin Lake, its post office was named in 1894 by Fr Eugène Le Coq.

Sainte-Thècle QC A municipality in the regional county of Mékinac, northeast of Shawinigan, it was created in 1989 through the amalgamation of the village (1909) and the parish municipality (1874) of the same name. The religious parish was erected in 1873 and the post office was opened six years later. The name recalls the first-century martyr St Thecla, who studied with St Paul.

Sainte-Thérèse QC A city in the regional county of Thérèse-De Blainville, north of Montréal, it achieved its present status in 1916, having been incorporated as the village of *Sainte-Thérèse-de-Blainville* in 1849. The religious parish was created in 1825 and was named in honour of Marie-Thérèse Dugué de Boisbriand (1671–1744), the daughter of Michel-Sidrac Dugué de Boisbriand, who was granted the seigneuries of Mille-Îles and Blainville in 1683.

Saint-Étienne-de-Lauzon QC A municipality southwest of the city of Québec, this municipality was erected in 1982, having been called simply *Saint-Étienne* in 1957. The religious parish of Saint-Étienne-de-Lauzon, in the original seigneury of Lauzon, was established in 1860, and named after Fr Étienne Baillargeon (1807–70), pastor of Saint-Nicolas, 1838–70. Its post office was called *Baillargeon* from 1862 to 1937, when the current name was adopted.

St. Eugène ON In the united counties of Prescott and Russell, southeast of Hawkesbury, this place's post office was named in 1861. A chapel was built here in 1854 and named after the first Catholic bishop of Ottawa, Joseph-Eugène-Bruno Guigues (1805–74).

Saint-Eustache QC A city in the regional county of Deux-Montagnes, northwest of Montréal, it was established with its present area in 1972 by annexing the parish municipality of *Saint-Eustache-sur-le-Lac* (1912). The religious parish was founded in 1768, and named after Eustache-Lambert Dumont, who was granted an extension of the seigneury of Mille-Îles in 1749. He built a flour mill here in 1762. His son Eustache-Louis, who developed the seigneury, is considered the real founder of the place.

Sainte-Véronique QC A village east of Mont-Laurier, it was named in 1984, having been called the township municipality of *Turgeon* in 1904. The religious parish of Sainte-Véronique was established in 1894, and named after St Veronica, who had a special devotion to the Holy Face of Christ, because she had wiped his brow when he was being taken to the cross on Golgotha. The township was proclaimed in 1900 and named after Fr Adrien Turgeon (1846–1917), rector of Collège Sainte-Marie in Montréal.

Sainte-Victoire-de-Sorel QC Situated south of Sorel, this parish municipality was erected in 1845, abolished two years later, and re-established in 1855, two years after its post office was opened. The religious parish was created in 1842 and was named in honour of

Queen Victoria, who had ascended the throne in 1837.

Saint-Fabien QC Located southwest of Rimouski, this parish municipality was erected in 1855, with the religious parish having been established in 1828. The post office was opened in 1856. The name recalls St Fabian, who was the Pope from 236 to 250.

Saint-Félicien QC A town northwest of Lac Saint-Jean, it took on its present form in 1976 with the merger of the town (1946) and the municipality (1882) of *Saint-Félicien*. The post office was named in 1875 after St Felicianus, a martyr who was beheaded at the beginning of the fourth century by Roman Emperor Maximian.

St. Felix PE On the south side of Tignish, this municipal community (1977) was established as a school district in 1846. The origin of the name is not recorded in the official records.

Saint-Félix-de-Valois QC A village north of Joliette, it was erected in 1926 from the township municipality of the same name, established in 1845, abolished two years later, and re-established in 1855. The religious parish was organized in 1840, taking its name from St Felix of Valois (*c*. 1127–1212), a founder of the Order of Trinitarians in 1197.

Saint-Flavien QC Located southwest of the city of Québec, adjacent to Laurier-Station, this village was established in 1912 when it was detached from the parish muncipality of **Saint-Flavien-de-Sainte-Croix**, created in 1855. The religious parish was set up in 1834 and named after Mgr Pierre-Flavien Turgeon (1787–1867), fourteenth bishop of Québec, 1850–67.

St. Francis, Cape NF North of St. John's, this peninsula was called by the Portuguese in the early 1500s *C de S francisco*. Explorer Gaspar Corte-Real may have named it after St Francis of Assisi.

St. Francis, Lake ON; **Lac Saint-François** QC An enlargement of the St. Lawrence River, downriver from Cornwall, it was named in 1656 in honour of St Francis Xavier (1506–52), who was canonized in 1622. It may have been given more directly after François de Lauson, a seigneur near Montréal, or after Fr François Le Mercier, the superior of the New France missions, who took some Jesuit missionaries to Huronia in 1656.

St. Francis River; Rivière Saint-François NB, QC Rising in **Lac Saint-François**, 20 km southeast of the city of Rivière-du-Loup, this river flows south through Lac Pohénégamook. From the lake's outlet it forms the international boundary, with Maine on the west, and Québec and New Brunswick on the east. About 1687 Mgr Jean-Baptiste de Saint-Vallier called it *Petite rivière St-François*.

Saint-François-de-Madawaska NB Southwest of Edmundston, this village (1966) is located on the Saint John River. Its first post office was *Webster's Creek*, opened in 1847, with Augustin Webster as the first postmaster. In 1874 it was called *Winding Ledges*, after rapids in the Saint John River It was given its present name in 1901 after **Saint-François Parish**, created in 1877, and named after the nearby river.

Saint-François-du-Lac QC A village in the regional county of Nicolet-Yamaska, near the mouth of the Rivière Saint-François, it was separated in 1917 from the parish municipality of **Saint-François-du-Lac**. The latter was created in 1855, having been called *Saint-François-du-Lac-Saint-Pierre* ten years earlier. The religious parish dates from 1714, a few years after the Jesuits, who had a mission called *Saint-François-de-Sales* at the mouth of the Chaudière, near the city of Québec, established a mission among the Abenaki on Lac Saint-Pierre.

Saint-François, Rivière QC A major river in the southeastern part of the province, it flows through Sherbrooke and Drummondville before emptying into Lac Saint-Pierre. Often called the St. Francis River in English, it was named in the mid-1600s after François de Lauson, who mentioned *Rivière Saint-François des*

Prés in granting the seigneury of Saint-François to Pierre Boucher de Grosbois in 1662.

Saint-Fulgence QC A municipality on the north side of the Saguenay, east of Chicoutimi, it was created in 1973 through the merger of the village (1947) and the parish municipality (1873) of *Saint-Fulgence*. A mission was established here in 1843 and named after St Fulgentius (*c.* 467–*c.* 533), appointed bishop of Ruspe in North Africa, 507 AD.

Saint-Gabriel QC Located north of Joliette, this town was erected in 1967, having been the village of *Saint-Gabriel-de-Brandon* since 1892. The village had been separated from the parish municipality of *Saint-Gabriel-de-Brandon*, which was erected in 1855. The religious parish was called Saint-Gabriel-du-Lac-Maskinongé in 1837, but was changed three years later to Saint-Gabriel-de-Brandon. The Angel Gabriel announced the births of John the Baptist and Jesus. The township of Brandon, noted on a map in 1795, and proclaimed in 1827, was named after the town of Brandon, in Suffolk, England, northeast of Cambridge.

Saint-Gabriel-de-Valcartier QC Adjacent to Canadian Forces Base Valcartier, northwest of the city of Québec, this municipality was created in 1862, having been called a parish municipality in 1855. Ten years before it had been called the municipality of *Valcartier*, contracted from 'vallée de la rivière Jacques-Cartier'. The seigneury of Saint-Gabriel was granted to Robert Giffard in 1647.

Saint-Gédéon QC A village midway between Lac-Mégantic and Saint-Georges, it was erected in 1950 when it was separated from the parish municipality of **Saint-Gédéon** (1912). The religious parish of Saint-Gédéon-de-Beauce was established in 1910, having been a mission since 1889. St Gideon was a judge who delivered Israel from the desert hordes in the eleventh and twelfth centuries BC.

St. George NB Near the mouth of the Magaguadavic River, this town (1904) was surveyed in 1786 and called *Magaguadavic*. Its post office was called St. George in 1829, taking its name from **Saint George Parish**, established in 1786.

St. George ON A community in Brant County, north of Brantford, it was named in 1835 after first postmaster George Stanton. The post office was called *St. George Brant* until it was closed in 1970.

St.-Georges MB A community on the Winnipeg River, northeast of Winnipeg, it was named St. George in 1902, after the patron saint of England, and was subsequently given a perceived, but improperly spelled, name in French. It is a residential community upriver from the Manitoba Pulp and Paper Company's planned community of Pine Falls, created in 1925.

St. George's NF Located on the southeast side of **St. George's Bay**, this town (1965) was first called *South Side* and *Little Bay*. It was renamed St. George's in 1904, when Mgr Neil McNeil moved the seat of the area's diocese to *South Side*. The bay have been named as early as 1534, when Jacques Cartier sailed across it. **Cape St. George**, the southwesterly point of the Port au Port Peninsula, may have been named at the same time.

Saint-Georges QC In 1990 the towns of Saint-Georges (1948) and *Saint-Georges-Ouest* (1943) were united to form the city of Saint-Georges, the most populous municipality in the Beauce valley. The religious parish, created in 1835, was named in honour of merchant Johann Georg Pfozer (1752–1848), who acquired the seigneury of Aubert-Gayon in 1808, and recruited 189 German settlers to immigrate to Saint-Georges.

Saint-Georges QC A village in the Saint-Maurice valley, opposite Grand-Mère, its name was adopted in 1919, two years after the religious parish was founded. From 1927 to 1951 its post office was *Saint-Georges-de-Champlain*. St George, the patron saint of England, was a fourth-century warrior and martyr, reputed to have rescued a maiden from a dragon in Palestine.

St. Georges Bay NS Northeast of Antigonish, this 35 km^2 bay was named in the 1770s by chart-maker J.F.W. DesBarres after the patron saint of England. It was renamed *George Bay* in 1902 by the Geographic Board, but the original name was restored in 1967.

Saint-Gérard QC A village midway between Sherbrooke and Thetford Mines, it was named in 1924 after the religious parish of Saint-Gérard-Majella, established in 1905 by Bishop Larocque of Sherbrooke, who assisted at the canonization of St Gerard Majella (1726–55) in Rome the previous year. From 1886 to 1924 the village was called *Lac-Weedon*, while the post office was known as *Lake Weedon* from 1869 to 1911, when it was changed to Saint-Gérard. The township of Weedon, shown on a 1795 map and proclaimed in 1822, was named after Weedon, Buckinghamshire, England.

Saint-Gérard-Majella QC A parish municipality, south of Joliette, it was adopted in 1982, having been created as *Saint-Gérard-Magella* in 1906. St Gerard Majella (1726–55) was canonized in 1904 in Rome.

Saint-Germain-de-Grantham QC A village west of Drummondville, it was separated from the parish municipality of **Saint-Germain-de-Grantham** in 1946. The religious parish of Saint-Germain was created in 1858, with the Saint-Germain-de-Grantham post office opening in 1867, followed by the parish municipality in 1871. The parish was named after early settler Germain Sylvestre, whose name likely recalled the sixth-century Paris bishop. The township of Grantham, shown on a 1795 map, was proclaimed in 1800, and named after a town in Lincolnshire, England.

Saint-Grégoire-de-Greenlay QC North of Sherbrooke and across the river from Windsor, this village was erected in 1947. *Greenlay*, the post office from 1903 to 1973, was named after Louisa Stevens Greenlay (d. 1895), the widow of Lazenby Greenlay, a native of East Yorkshire, England, and the first European to settle in the township of Brompton. The mission of Saint-Grégoire was established in 1931 and named after Pope Gregory VII (1015–85).

Saint-Grégoire-le-Grand QC A parish municipality east of Saint-Jean-sur-Richelieu, it was erected in 1855. It was named after Grégoire Bourque, who donated land in 1805 for a church site, as well as in honour of Pope Gregory I (*c.* 540–604), known as Gregory the Great, who was the pontiff from 590 to 604.

Saint-Guillaume QC Located directly west of Drummondville, this village was named in 1902, when it was separated from the parish municipality of **Saint-Guillaume**. The religious parish of Saint-Guillaume was erected in 1833, and named in honour of Charles William (the English for Guillaume) Grant, Baron de Longueuil (1782–1848), proprietor of the township of Upton, 1806–48. The post office was named **Saint-Guillaume-d'Upton** in 1862. The township of Upton was proclaimed in 1800, and possibly named after Upton, near Grantham (the name of an adjoining township), in Lincolnshire, England.

Saint-Guillaume-de-Granada QC A municipality, just south of the city of Rouyn-Noranda, it was created in 1978, having been established as a religious parish in 1935. It was named after St William the Great (*c.* 755–812), a prominent political and military figure during the time of Charlemagne. **Granada**, its post office since 1936, recalls the Granada Gold Mines, which started developing the site in 1927. The company's name also recalls the rich lead mines of the Spanish province of Granada.

Saint-Henri QC Situated south of the city of Lévis, the present municipality was established in 1976 through the annexation of the municipality of *Rivière-Boyer*. The municipality of Saint-Henri had been erected the previous year by merging the village (1913) and the parish municipality of **Saint-Henri** (1855). The name recalls the sixth bishop of Québec, Mgr Henri-Marie Dubreuil de Pontbriand (1708–60).

Saint-Hilaire NB Southwest of Edmundston, this village (1967) is located on the Saint John River. Its post office was called *Baker's Creek* in 1848, renamed *St. Hilaire* in 1873, closed in 1947, and reopened as Saint-Hilaire in 1968 for two years. **Saint-Hilaire Parish** was named in 1877 after the church, consecrated in 1868 by Mgr James Rogers, who named it after the former bishop of Tours in France. Albertine was the railway station and a second post office (1893–1968) in Saint-Hilaire.

Saint-Hippolyte QC A parish municipality north of Saint-Jérôme, it was created in 1951, having been called the township municipality of *Abercrombie-Est* in 1855. The post office was opened as **Saint-Hippolyte-de-Kilkenny** in 1870. St Hippolytus (*c.* 165–*c.* 235), was a Roman theologian and martyr. The township of Kilkenny was proclaimed in 1832, and named after the Irish city and county. The township of Abercrombie, established in 1842, recalls Gen. James Abercromby (1706–81), whose corps was defeated in 1758 by Montcalm's French forces at Fort Carillon, near present-day Ticonderoga, NY.

Saint-Honoré QC A municipality north of Chicoutimi, it was created in 1973 through the amalgamation of the village (1953) and the parish municipality (1914) of *Saint-Honoré*. The religious parish was organized in 1911 and named in honour of Honoré Petit (1847–1922), member for Chicoutimi in the House of Commons, 1892–7.

Saint-Hubert QC A city east of Montréal, it was incorporated in 1958 and the following year annexed the city of *Laflèche*, which had been called the town of *Mackayville* in 1947. **Saint-Hubert** was erected as a parish municipality in 1860, taking its name from St Hubert (d. 727), a Belgian bishop and patron saint of hunters.

Saint-Hyacinthe QC Situated on the Rivière Yamaska, east of Montréal, this city attained its present area in 1976 through the amalgamation of the towns of Saint-Hyacinthe (1850) and *La Providence* (1965), and the villages of *Saint-Joseph* (1898) and *Douville* (1947). It owes its name to Jacques-Hyacinthe-Simon Delorme *dit* Lapointe (*c.* 1718–78), who acquired the seigneury of Maska in 1753. The religious parish of Saint-Hyacinthe-le-Confesseur was created in 1757, recalling the founder of the Dominican Order in Poland in the thirteenth century.

Saint-Isidore NB A village (1991) northwest of Tracadie, it was established about 1867 as a free-grant settlement, and its post office was opened in 1878. The name may been given after the patron saint of farmers.

St. Isidore ON A village in the united counties of Prescott and Russell, southwest of Hawkesbury, its post office was first called *Kerry* in 1863. It was renamed St. Isidore in 1882, three years after a church dedicated to St Isidore, a noted theologian of the early Catholic church, was erected. The post office name was altered to *St. Isidore de Prescott* in 1930. When the village was incorporated in 1989, the modifier was dropped.

Saint-Isidore QC Located south of Lévis, this municipality was established in 1962, having been the parish muncipality of *Saint-Isidore-de-Lauzon* from 1855. The religious parish was erected in 1829, and named after St Isidore (*c.* 560–636), a distinguished theologian and primate of Spain, 610–36 AD.

Saint-Isidore QC Located south of Montréal and south of the Kahnawake Indian Reserve, this parish municipality was erected in 1842, taking its name from the religious parish, set up in 1833, and named for St Isidore (*c.* 560–636), the patron saint of farmers. The post office, called *Saint-Isidore-de-Laprairie* in 1853, is now **Saint-Isidore-de-La Prairie**.

Saint-Jacques NB A village (1966) northwest of Edmundston, its post office was first called *Silverstream* in 1870. It was renamed St-Jacques in 1902, after **Saint-Jacques Parish**, created in 1877. The church was called after the French translation of the forename of Mgr James Rogers, bishop of Chatham.

Saint-Jacques QC A village southeast of Joliette, it was detached from the parish municipality of *Saint-Jacques-de-l'Achigan* in 1912, with the latter becoming **Saint-Jacques** in 1920. The religious parish of Saint-Jacques-de-l'Achigan was created in 1831 and Saint-

Jacques post office was opened four years later. The parish was named after Fr Jacques Degeay (1717–74), long-time pastor of L'Assomption, who took a special interest in the plight of the Acadians. The Algonquin word *achigan*, as reflected in the nearby Rivière de l'Achigan, means 'bass'.

St. Jacobs ON In Waterloo Region, north of the city of Waterloo, this community's post office was named in 1852 after miller Jacob Snider, his son Jacob, and Jacob Eby.

St. James MB A neighbourhood in the urban community of **St. James-Assiniboia**, it has been part of the city of Winnipeg since 1972. Incorporated as a city in 1956, it took its name from the Church of England parish established here in 1853.

St. Jean Baptiste MB On the west bank of the Red River, south of Morris, this community was named in 1876. Also that year the parish church was built and named after the patron saint of French-speaking Canadians.

Saint-Jean-Baptiste QC A parish municipality southeast of Mont-Saint-Hilaire, it was named in 1855 after the religious parish of Saint-Jean-Baptiste-de-Rouville, founded in 1797. The seigneury of Rouville was granted in 1694 to Jean-Baptiste Hertel de Rouville (1668–1722).

Saint-Jean-Baptiste-de-Nicolet QC Located in the regional county of Nicolet-Yamaska, east of the town of Nicolet, this parish municipality was erected in 1845, abolished two years later, and re-established in 1855. The religious parish of Saint-Jean-Baptiste-de-Nicolet was created in 1831 and named after Jean-Baptiste Poulin de Courval (1657–1727), an important Trois-Rivières merchant, who facilitated the setting up of the parish.

Saint-Jean-Chrysostome QC A town south of the city of Québec and near the mouth of the Rivière Etchemin, it was established in 1965, having been created a parish municipality in 1845, abolished two years later, and re-established in 1855. The religious parish was created in 1828. Although identified with St John Chrysostom, the patriarch of Constantinople, 398–407 AD, it also honoured Sir John Caldwell (1775–1842), a prominent politician and industrialist, who managed the seigneury of Lauzon.

Saint-Jean-de-Dieu QC Located south of Trois-Pistoles and northeast of Rivière-de-Loup, this municipality was erected in 1947 and its post office was opened in 1881, taking its name from the religious parish, established in 1873. A patron of hospitals and the sick, St John of God (1495–1550) founded his first hospital in 1537 in Granada, Spain.

Saint-Jean-de-Matha QC Situated northwest of Joliette, this municipality was established in 1993, having been a parish municipality since 1855. The religious parish had been set up three years earlier, and named after St John of Matha (*c.* 1160–1213), a co-founder with St Felix of Valois of the Order of Trinitarians in 1197.

Saint-Jean, Lac QC With an area of over 1000 km^2, this well-known lake at the head of the Saguenay was named in 1652 by Fr Jean de Quen in honour of his patron saint. In his *Relation* of 1647 he had called it *Lac Piouagamik*, meaning 'shallow lake' in Montagnais. In 1603 Samuel de Champlain had described it as *Mer du Nord*, and in 1544 navigator Jean Fonteneau *dit* Alfonse referred to it as *Mer du Saguenay*.

Saint-Jean-Port-Joli QC Situated on the south shore of the St. Lawrence, midway between Montmagny and La Pocatière, this municipality was established in 1857, having been created a parish municipality two years earlier. The religious parish was created in 1721 in the seigneury of Port-Joly, and dedicated to St John the Baptist. The seigneury at the mouth of **Rivière Port Joli** was granted to Noël Langlois-Traversy in 1677. The river's name describes the fine harbour at its mouth.

Saint-Jean-sur-Richelieu QC Located on the west bank of the Richelieu, the present name of this city was adopted in 1978, having been incorporated as the town of *Saint-Jean* in 1856. In 1665 Jean-Frédéric Phélypeaux erected a fort here, and called it Fort Saint-Jean. The religious parish established here in 1667 was called Saint-Jean-l'Évangeliste, which became a parish municipality in 1855, and was united with the city in 1970. It was also officially known as *St. Johns*, but this was dropped in 1962.

Saint-Jérôme QC A city northwest of Montréal, it was erected in 1881, having been incorporated as a village in 1856. The religious parish of Saint-Jérôme-de-la-Rivière-du-Nord was founded in 1832, and was named after St Jerome (*c.* 347–420), noted for his translations of scripture and commentaries.

Saint John NB This city was named and incorporated in 1785, making it the oldest incorporated city in the country. It was named after the river, identified as *Rivière saincte-Jean* by Pierre du Gua de Monts on 24 June 1604, the day dedicated to St John the Baptist. In 1783 the central part of the city had been called *Parr Town* after John Parr (1725–91), governor of Nova Scotia, 1782–91. The west part was called *Carleton*, after Sir Guy Carleton, Lord Dorchester, governor-in-chief of British North America, 1786–96. In 1889 the city annexed the city of *Portland* (1883), and in 1966, it annexed the city and the parish of *Lancaster*, and part of the parish of Simonds. Although the 'saint' in the city's name is officially spelled in full, most people living in the province outside the city prefer to spell the river's name as St. John River. But the names authorities, having decreed that names of common origin should be spelled the same, made **Saint John River** official.

St. John's NF **St. John's Harbour** is supposed to have been entered on 24 June 1497, perhaps by explorer John Cabot, who sailed on behalf of the Bristol merchants that year. A Portuguese map of 1519–20 described it as *R de Sam joham*, and a letter of 1527, written by British seaman John Rut, called it *Haven of St. John*. In the mid-1500s the French called the harbour *baie de Saincte Jehan*. By 1689 the *English Pilot* used the present form of the harbour's name. St. John's became the seat of the government of Newfoundland in 1832 and its first incorporated city in 1888.

Saint-Joseph NB A village (1966) on the Memramcook River, northwest of Sackville, its post office was named in 1869. A common name in New Brunswick, another **Saint-Joseph** and a **Saint-Joseph-de-Kent** are both in Kent County, and **Saint-Joseph-de-Madawaska** is north of Edmundston.

Saint-Joseph-de-Beauce QC Situated on the east bank of the Chaudière, northwest of Saint-Georges, this town was erected in 1965, having been incorporated as a village in 1889. The religious parish of Saint-Joseph-de-la-Nouvelle-Beauce was set up in 1835, became the parish municipality of *Saint-Joseph* in 1855, and was amended to **Saint-Joseph-de-Beauce** in 1957. In 1747 Joseph de Fleury de La Gorgendière (1676–1755) acquired a seigneury in the Beauce valley, with his name being adopted for it. He had recruited the settlers in 1738 and built the first chapel the following year.

Saint-Joseph-de-la-Rive QC In the regional county of Charlevoix, east of Baie-Saint-Paul, this village was erected in 1931, and its name replaced *Les Éboulements-en-Bas* as its postal designation the next year. There are two explanations for calling the religious parish Saint-Joseph in 1931: one, that it is after church architect Joseph Archer, and two, that prayers by sailors to Christ's father helped bring them safely ashore. *Rive* means 'shore'.

Saint-Joseph-de-la-Sorel QC A town on the west bank of the Richelieu, opposite the city of Sorel, it was erected in 1942, 35 years after the village was detached from the parish municipality of **Saint-Joseph-de-la-Sorel**. The religious parish was established in 1875, taking its name from Mgr Joseph La Rocque (Laroque) (1808–87), then bishop of Saint-Hyacinthe. Pierre de Saurel (1628–82) was granted the seigneury of Saurel (Sorel) in 1672.

Saint-Joseph-du-Lac QC A parish municipality in the regional county of Deux-Montagnes, it is located west of Saint-Eustache and inland from Pointe-Calumet. The religious parish of Saint-Joseph was created in 1853 and became the parish municipality of Saint-Joseph-du-Lac three years later, when the post office was opened. In the late 1700s the name Saint-Joseph described a hillside (*côte*) in the area of the present municipality.

Saint-Jovite QC A popular tourist destination south of Mont-Tremblant, this town was incorporated in 1986, having been erected as a village in 1917. The parish municipality of **Saint-Jovite**, established in 1960, had been set up as the united townships of *De Salaberry-et-Grandison* in 1881. Charles-Michel d'Irumberry de Salaberry (1778–1829) successfully blocked an American invasion in 1813. Sir Charles Grandison was a hero in a 1753 novel by British writer Samuel Richardson.

Saint-Lambert QC A city in the regional county of Champlain, on the east side of the St. Lawrence opposite Montréal, it was established in its present area in 1969 when it annexed the town of *Préville*, erected in 1948. Established as a Protestant parish in 1855 and a municipality in 1857, it became a village in 1892, and a town six years later, with its post office opening in 1904. In the seventeenth century the shoreline was known as Côte-Saint-Lambert, likely named after Raphaël-Lambert Closse (*c.* 1618–62), commander of Ville-Marie (Montréal), 1655–7. The Protestant parish may have been named in honour of reformer Francis Lambert (1486–1530).

Saint-Lambert-de-Lauzon QC A parish municipality on the east bank of the Rivière Chaudière, south of the city of Québec, it was established in 1855, taking its name from the religious parish, erected in 1851. The name recalls surveyor Pierre Lambert, who laid out plans for the seigneury of Lauzon (1828) and the city centre of Lévis (1849). Jean de Lauson (1584–1666) was granted the seigneury of Lauzon (Lauson, Loson) in 1636.

St. Laurent MB On the southeastern shore of Lake Manitoba, this place's mission was named in 1862 by Fr Joseph Charles Camper, after the third century martyr, St Lawrence, who was put to death by Emperor Valerian. Its post office was named in 1883.

Saint-Laurent QC A city in the Communauté urbaine de Montréal, northwest of Montréal, it was erected in 1893, and annexed the parish municipality of *Notre-Dame-de-Liesse* in 1964. It has its roots in the name Côte-Saint-Laurent, created by official decree in 1722. It took its name from St Lawrence, a martyr, who may have been roasted on a gridiron by Emperor Valerian in 258.

St. Lawrence NF On the southeastern end of the Burin Peninsula, this town (1949) was named after **Great St. Lawrence Harbour**. The harbour may have been named by Richard Clarke, during Sir Humphrey Gilbert's 1583 voyage, because he found salmon as plentiful there as in the St. Lawrence River. Or it may have been called after a parish in Jersey in the late 1500s by fishermen from the Channel Islands.

St. Lawrence River ON; **Fleuve Saint-Laurent** QC On 10 August 1535 Jacques Cartier named a small bay, on the north shore of the **Gulf of St. Lawrence** after St Lawrence, whose feast day is on that date. Both Cartier and Samuel de Champlain referred to the river and its mighty estuary as *Rivière du Canada*, but ultimately the bay's name was extended to both the gulf and the river. **St. Lawrence Islands National Park** (4 km^2), comprising 18 islands, several islets, and a small part of the mainland in the area of the Thousand Islands, east of Kingston, ON, was created in 1904.

St. Lazare MB On the east bank of the Assiniboine River, opposite the mouth of the Qu'Appelle River, this village (1949) had been established as an Oblate mission in 1880. When the Oblates decided to abandon the mission in 1894, Archbishop Alexandre Taché ordered them to keep it open. Its resurrection prompted its naming after Lazarus, who had been raised from the dead. Its post office was opened in 1895.

Saint-Lazare QC A parish municipality in the regional county of Vaudreuil-Soulanges, just west of the towns ofVaudreuil and Dorion, it was established in 1875, and its post office was called **Saint-Lazare-Vaudreuil** three years later. Located beside the parish of Sainte-Marthe, its religious parish was named after Lazarus, the brother of Martha and Mary, who was miraculously raised from the dead.

Saint-Léolin NB A village (1978) west of Caraquet, its first post office was called *Saint-Joseph* in 1888. It was renamed *Pépère* in 1903, and became Saint-Léolin the following year.

Saint-Léonard NB A town (1920) on the Saint John River, northwest of Grand Falls, its first post office was called *Grand River*, after a nearby river, in 1847. It was called *St. Leonard Station* in 1883 and was changed to **St-Léonard** in 1925. The place was named after Leonard Reed Coombs, who settled here about 1840.

Saint-Léonard QC A city in the Communauté urbaine de Montréal, northeast of Montréal, it was created in 1962, having been established as the town (1915) and the parish municipality of *Saint-Léonard-de-Port-Maurice* (1886). The religious parish was erected in 1885, taking its name from Côte-Saint-Léonard, which described a hillside in 1706. That name may have been given after St Leonard of Port Maurice (1676–1751), who was canonized in 1867, or after Sulpician missionary Léonard Chaigneau (1663–1711), who persuaded several pioneers to settle at Côte-Saint-Léonard.

Saint-Léonard-d'Aston QC A village in the regional county of Nicolet-Yamaska, north of Drummondville, it was separated in 1912 from the municipality of **Saint-Léonard-d'Aston**. The religious parish was established in 1857, and named after the sixth-century founder of the Saint-Léonard-de-Noblat monastery near Limoges, France.

Saint-Liboire QC Situated east of Saint-Hyacinthe, this village was created in 1919 when it was detached from the parish municipality of **Saint-Liboire**, established in 1860, the same year the religious parish was set up and the post office was opened. The name was given for Fr Liboire-Henri Girouard (1798–1876), who participated actively in the founding of the parish.

Saint-Lin QC A municipality in the regional county of Montcalm. northeast of Saint-Jérôme, it was established in 1855, having been erected as a parish municipality in 1845, but abolished two years later. In 1847 its post office was opened. The name recalls St Linus, successor to St Peter as the pontiff of the church, from 67 to 76 AD.

Saint-Louis-de-France QC Located north of Trois-Rivières, this town was erected in 1993, taking its name from the religious parish, established in 1903. The name may have been given in honour of Mgr Louis-François Laflèche (1818–98), bishop of Trois-Rivières from 1870. It may also have been adapted from the name of the national French church in Rome, Saint-Louis-des-Français.

Saint-Louis-du-Ha! Ha! QC A parish municipality southeast of Rivière-du-Loup, it was named in 1874, with its mission having been established in 1860. The first part of its name may have been given after pioneer Louis Marquis, or after either Fr Louis-Antoine Proulx (1810-96) of Rivière-du-Loup, or Fr Louis-Nicolas Bernier (1833–1914), pastor at Notre-Dame-du-Lac when the parish was named. The last part of the name has nothing to do with laughter, but is an old French word for an unexpected barrier, or dead end. The dead end is 8 km east in Lac Témiscouata, where canoeists were forced to portage by land to Notre-Dame-de-Portage, 80 km to the west. *See also* Ha! Ha!, Baie des.

St. Louis PE Northwest of Alberton, this municipal community (1883) had been incorporated as a village in 1964. The school district was called St. Louis about 1890, having been earlier called *Smith Road* and *Union Road*. The post office, opened as *Kildare Station* in 1882, was renamed St. Louis in 1896.

St. Louis SK On the south bank of the South Saskatchewan, south of Prince Albert, this village (1959) was founded as *Boucher Settlement* in 1882. Its post office was called *Boucher* in 1888, after Jean Baptiste Boucher, a native of St. Boniface, MB. It was changed to St. Louis in 1897, in honour of Albert and Romuald St-Louis, who had settled here in 1884. Albert returned to Québec in 1895 and Romuald went to Winnipeg, where he became a director of the Federal Life Insurance Company.

Saint-Louis-de-Kent NB A village (1966) northwest of Richibucto, its first post office was called *Palmerston* in 1859, taking its name from the parish. The parish was renamed Saint-Louis in 1866, and the postal designation was changed to St-Louis-de-Kent in 1879.

Saint-Louis, Lac QC Situated west of Montréal and at the point where the Ottawa River joins the St. Lawrence, its name first appeared in the Jesuit *Relations* of 1656. Earlier Samuel de Champlain had named the rapids at present Lachine *Le Grand Sault St Louis,* after a young man in the service of Pierre du Gua de Monts called Louis, who had been drowned in the rapids.

Saint-Luc QC Located north of Saint-Jean-sur-Richelieu, this town was erected in 1963, having been incorporated as the municipality of Saint-Luc in 1855. The name was chosen in 1799 by Mgr Pierre Denaut, who wanted to assure that each of the authors of the four Gospels had a parish south of Montréal.

St. Lunaire-Griquet NF On the northeastern coast of the Northern Peninsula, this municipal community (1958) is comprised of two separate places. Both were developed fishing stations when Jacques Cartier visited them in 1534. **St. Lunaire** is known locally as St. Leonard's. **Griquet** may have been named after the French word *grecqué* ('sawtooth'), because the surrounding hills look like the teeth of a saw.

Saint-Malachie QC Located southeast of Lévis, this parish municipality was called this in 1948. It had been established in 1874 as *Saint-Malachie-de-Frampton*, taking its name from a mission opened here in 1841. It was settled by Irish, who likely suggested the name after St Malachy (1094–1148), archbishop of Armagh, the primatial seat of Ireland. The township of Frampton, shown on a 1795 map, and proclaimed in 1806, may have been named after a place near Boston, in Lincolnshire, England.

Saint-Malachie-d'Ormstown QC Located adjacent to the town of Ormstown, south of Salaberry-de-Valleyfield, this parish municipality was named in 1855. It was named after St Malachy (1094–1148), archbishop of Armagh, who reformed the church in Ireland.

St. Malo MB Located midway between Emerson and Steinbach, this place's post office was named in 1890, likely after the French port of Saint-Malo, from where Jacques Cartier sailed. It may have been inspired by an early settler there called Malot, or after settler Louis Malo in nearby St. Pierre, who had been teased about living in St. Malo, with this name being chosen for the new post office.

Saint-Marc-des-Carrières QC A village in the regional county of Portneuf, southwest of the city of Québec, it was incorporated in 1918. Its post office had been named seven years before, after having been called *Poiré* (1863) and *Châteauvert* (1901). The religious parish was set up in 1901, taking its name from St Mark's Basilica in Rome and the local limestone quarries.

St. Margarets Bay NS West of Halifax, this bay was identified on a 1615 map as *B de s margarita* and on a 1640 map as *B S Marguerite*. It was anglicized in the mid-1700s. J.F.W. DesBarres portrayed it on a chart of the 1770s as *Charlotte Bay alias Margarets*, but his effort to name the bay after Queen Charlotte was ignored. **Head of St. Margarets Bay** was given postal service in 1860.

Saint-Martin QC Located south of Saint-Georges and near the head of the Beauce valley, this parish municipality was founded in 1912, taking its name from the religious parish of Saint-Martin-de-Tours, also known as Saint-

Martin-de-la-Beauce. St Martin of Tours (*c.* 316–97) is venerated by the valley's francophone residents. Its post office has been called **Bolduc** since 1882. It was named after Joseph Bolduc (1847–1924), member for Beauce in the House of Commions, 1876–84, and then a senator until 1922.

St. Martins NB On the Bay of Fundy, east of Saint John, this village (1966) was named in 1844 after **Saint Martins Parish**. The parish had been named in 1786, possibly after St. Martins in Maryland, or after Martin Head, 33 km to the northeast. In the 1800s the place was commonly called *Quaco*, after the adjacent bay and nearby head.

St. Mary's NF A municipal community (1966) on **St. Mary's Harbour**, and on the east side of **St. Mary's Bay**, it was named before 1836. In the mid-1550s the bay was identified on maps as *Sa Maria* and *B St marie*. It was likely named after **Cape St. Mary's**, at the southern side of Placentia Bay, shown on a map of 1536 as *Cabo de Sancta Maria*.

St. Marys ON A separated town (1864) in Perth County, it was named in 1844 for Elizabeth Mary Jones, wife ofThomas Mercer Jones, commissioner of the Canada Company, and daughter of Bishop John Strachan. She subscribed £10 for the new stone school then being constructed.

St. Marys Bay NS Separating the Nova Scotia mainland from Digby Neck, and Long and Brier islands, it was called *Baye Ste Marie* by Pierre du Gua de Monts in 1604.

St. Marys River ON Joining Lake Superior to Lake Huron at Sault Ste. Marie, this river was named soon after 1641 by Jesuit missionaries, who dedicated their mission to the Virgin Mary, and named the place and the falls on the St. Marys River, *Sainte Marie du Saut*.

Saint-Mathieu-de-Belœil QC Located between Belœil and Sainte-Julie, this municipality was established in 1992, having been a parish municipality from 1855. The religious parish was founded in 1772, and named after Mathieu Camin LaTaille (1725–82), its first parish priest.

Saint-Mathias-de-Richelieu QC A parish municipality on the east side of the Richelieu, opposite Chambly, it was given this name in 1988, having been the municipality of *Saint-Mathias* since 1855. The religious parish, called Saint-Olivier in 1772, after Mgr Jean-Olivier Briand, then bishop of Québec, was renamed in 1828 by Mgr Joseph-Octave Plessis after the apostle who replaced Judas after the Ascension of Christ.

Saint-Maurice, Rivière QC Draining almost 43,000 km^2, it rises in Réservoir Gouin, and flows some 300 km south to the St. Lawrence at Trois-Rivières. It was identified on Guillaume Levasseur's 1601 map as *3 Rivieres*, a description of the river's mouth. During the early 1700s, it was renamed after Maurice Poulin de La Fontaine (*c.* 1620–*c.* 1676), the seigneur and king's attorney of Trois-Rivières in the late 1600s. The parish municipality of **Saint-Maurice**, northeast of Trois-Rivières, was founded in 1855, three years after its post office opened.

St. Maur, Mount BC In the Purcell Mountains, east of Kootenay Lake, this mountain (2863 m) was named in 1953 after Lady Susan St. Maur, Duchess of Somerset.

Saint-Michel QC A parish municipality in the regional county of Les Jardins-de-Napierville, south of Montréal, it was erected in 1855, two years after the religious parish was organized, taking its name from the Archangel Michael. Since 1880 its post office has been called **Saint-Michel-de-Napierville**.

Saint-Michel-des-Saints QC Located north of Joliette, and northeast of Mont-Tremblant, this municipality was incorporated in 1979, when the parish municipality of *Saint-Michel-des-Saints* was amalgamated with the united townships of *Masson-et-Laviolette*. The mission was established in 1863, taking its name from the Spanish Trinitarian St Michael of the Saints (1591–1625).

Saint-Narcisse QC A parish municipality east of Shawinigan, it was established in 1859, with the religious parish having been organized in 1851. It took its name from St Narcissus (*c.* 106–212), the third bishop of Jerusalem.

Saint-Nazaire QC Located in the regional county of Lac-Saint-Jean-Est, northeast of Alma, this municipality was named in 1988, having been the township municipality of *Taché* since 1906. The religious parish was organized in 1890, named in honour of Mgr Louis-Nazaire Bégin (1840–1925), second bishop of Chicoutimi, 1888–91, archbishop of Québec 1898–1925, and a cardinal from 1914. The post office was called *Saint-Louis-Nazaire* from 1908 to 1927, when it was renamed Saint-Nazaire-de-Chicoutimi. The township of Taché was proclaimed in 1883, taking its name from Pascal (baptised Paschal-Jacques) Taché (1757–1830), proprietor of part of the seigneury of Kamouraska after 1790.

Saint-Nicéphore QC A municipality southeast of Drummondville, it was created in 1944. The religious parish was set up in 1916, and named after Fr Nicéphore Lessard (1879–1951), who was then the vicar of Saint-Frédéric-de-Drummondville. St Nicephorus (*c.* 758–*c.* 829) was a patriarch of Constantinople.

St. Nicholas PE A municipal community (1991) west of Summerside, its name was first noted in 1845 in the *Journal of the House of Assembly*. Postal service was provided by Muddy Creek from 1870 to 1917.

Saint-Nicolas QC Situated on the south shore of the St. Lawrence, southwest of the city of Québec, this town was incorporated in 1962, having been a parish municipality since 1845. The creation of a religious parish dates from 1660, with the name Saint-Nicolas being assigned to it in 1694. The name was given in honour of the parish of Saint-Nicolas-de-la-Ferté in the diocese of Chartres, Normandie, France. St Nicholas was a celebrated fourth-century bishop, possibly in Asia Minor, who became the patron saint of schoolchildren and sailors and the progenitor of Santa Claus.

Saint-Noël QC A village in the regional county of La Matapédia, southwest of Matane, it was named in 1945, having been the village of *Saint-Moïse* since 1906. The religious parish of Saint-Noël was organized in 1951, taking its name from Fr Noël Chabanel (1613–49), one of the Canadian martyrs, who was canonized in 1930.

St. Norbert MB Founded in 1857, this neighbourhood on the south side of Winnipeg, west of the Red River, was named after Mgr Joseph-Norbert Provencher (1787–1853), bishop of the North-West, 1847–53.

Saint-Ours QC Located on the east bank of the Richelieu, south of Sorel, it became a town in 1866, having been created a village in 1847, and a municipality two years earlier. The name was derived from the seigneury of Saint-Ours, which Pierre de Saint-Ours (1640–1724) had acquired through marriage in 1668. He was a captain in the Carignan-Salières Regiment, who had arrived in Canada in 1665, and encouraged the settlement of the seigneury.

Saint-Pacôme QC A municipality in the regional county of Kamouraska, east of La Pocatière, it was established in its present territory in 1980 through the merger of the village (1926) and the parish municipality (1855) of *Saint-Pacôme*. It recalls the monastic life of an Egyptian abbot of the fourth century.

Saint-Pamphile QC Located on the Maine border, due east of Montmagny, this place became an incorporated town in 1963, having been made a village in 1889. The religious parish was founded about 1870 and was named after Pamphile-Gaspard Verrault (1832–1906), member for L'Islet in the Legislative Assembly, 1867–78. He had been named after St Pamphilus (*c.* 240–309), a doctor of the church, teacher, and translator.

Saint-Pascal QC A town in the regional county of Kamouraska, northeast of La Pocatière, it was erected in 1966, having been made a village in 1939. It had been established as the parish municipality of *Saint-Paschal-de-Kamouraska* in 1855, with its post office being

Saint-Paschal from 1851 to 1916, when it was changed to Saint-Pascal. The religious parish was named in 1827 in honour of Paschal-Jacques Taché (1786–1833), who became the seigneur of Kamouraska in 1830 on the death of his father.

Saint-Patrice-de-la-Rivière-du-Loup QC This parish municipality, incorporated in 1855, was founded as a mission at Rivière-du-Loup in 1683, and its first chapel was built in 1792. The name honours the patron saint of Ireland, St Patrick (*c.* 385–*c.* 461).

Saint-Patrice-de-Sherrington QC A parish municipality in the regional county of Les Jardins-de-Napierville, it was erected in 1855. The religious parish was set up in 1848, and named after St Patrick, who converted Ireland to Christianity in the fifth century. Sherrington was proclaimed as a township in 1809, and named after the village of Sherington, in Buckinghamshire, England, northwest of London.

St. Paul AB North of the North Saskatchewan River and northeast of Edmonton, this town (1936) was the site of the mission of *St. Paul des Cris* in 1866, but it was abandoned eight years later. In 1896 Fr Albert Lacombe and Fr Adéodat Therien had the mission reopened. *St. Paul des Cris* post office was established in 1899 and its name was shortened in 1929.

Saint-Paul QC Located south of the city of Joliette, this municipality was given its name in 1954, having been called *La Conversion-de-Saint-Paul* in 1855. That was the name of the religious parish set up in 1716, which was later called *Saint-Paul-de-Lavaltrie*. In 1851 its post office was called *Saint-Paul-d'Industrie*, with *L'Industrie* having been an early name for Joliette. Four accounts of the conversion of Paul appear in the New Testament.

Saint-Paul-d'Abbotsford QC Located in the regional county of Rouville, northwest of Granby, this parish municipality was established in 1855. The St Paul Episcopal Church was constructed here in 1822, and the Catholic parish of Saint-Paul-d'Abbotsford was set up in 1855. The post office was called *Yamaska Mountain* in 1825, and renamed *Abbotsford* four years later after Joseph Abbott (1790–1862) and his wife, Harriet Bradford, the parents of Canada's fourth prime minister, Sir Joseph J.C. Abbott.

St. Paul Island NS North of Cape Breton Island and in Cabot Strait, this island was identified on maps from the mid-1500s to the late 1600s as *I S Pol* and *I S Paul*.

St. Pauls NF At the south side of a bar separating **St. Pauls Inlet** from **St. Pauls Bay**, and surrounded by Gros Morne National Park, this municipal community (1968) was settled after 1870. The origin of its name is not recorded in official records.

St. Peters NS Located on a narrow isthmus of Cape Breton Island, separating the Atlantic Ocean from Bras d'Or Lake, this village (1940) was founded as a Portuguese fishing station in 1521 and called *San Pedro*. From 1650 to 1713 it was known as *Saint-Pierre*, where Gov. Nicolas Denys had his headquarters until 1669. The French renamed it *Port Toulouze* in 1713 after Comte de Toulouse, a son of Louis XIV. Postal service was established at St. Peters in 1844. St. Peters Canal has connected the ocean to the lake since 1869.

St. Peters Bay PE Northwest of Souris and at the head of **St. Peters Bay**, this municipal community (1883) had been incorporated as the village of *St. Peters* in 1953. The bay was named *havre St Pierre* in 1721 by Louis Denys de la Ronde after Comte de Saint-Pierre, equerry to the Duchesse d'Orléans at the French court.

Saint-Philippe QC Located east of Candiac and southeast of Montréal, this parish municipality was erected in 1855. Set up as a mission in 1750, it became a religious parish in 1841. Its post office, called Saint-Philippe in 1852, was renamed *Saint-Philippe-de-La Prairie* in 1876. The parish was named after the Apostle Philip.

Saint-Pie QC A village southeast of Saint-Hyacinthe, it was incorporated in 1904, having been detached from the parish municipality of **Saint-Pie**, which was created in 1855. The religious parish was set up in 1828 and was likely named after St Pius V (1504–72), who introduced several reforms in church practices. When the parish was named, Pius VIII (1761–1830) was the Pope. The place had been named *Village-Bistodeau* about 1817, after merchant Joseph Bistodeau (1768–1856). In 1837 its post office was called **Saint-Pie-de-Bagot**. The former county of *Bagot* had been named after Sir Charles Bagot (1781–1843), governor of the province of Canada, 1841–3.

Saint-Pierre QC A town in the Communauté urbaine de Montréal, adjacent to Montréal-Ouest and Lachine, it was incorporated in 1908, having been called the village of *Saint-Pierre-aux-Liens* in 1894, during the time when the religious parish was organized. The name recalls an event in the life of St Peter, when he was miraculously released from the chains of King Herod.

Saint-Pierre QC A village on the west side of the regional county of Joliette, it was erected in 1922 as *Saint-Pierre-de-Joliette* and was modified to its present form seven years later. It is usually referred to as Village Saint-Pierre, to distinguish it from other places called Saint-Pierre.

St. Pierre-Jolys MB Located west of Steinbach, this village (1947) was named after the parish, established in 1872. Fr J.M. Jolys was the first parish priest. The post office was first called *Joly* in 1883, then *Laurier* in July 1897, and followed by *St. Pierre* in October, 1897, which was changed in 1922 to St. Pierre Jolys.

Saint-Pierre, Lac QC An enlargement of the St. Lawrence, extending for 35 km downriver from Sorel, it was probably named by Samuel de Champlain on 29 June 1603, the feast day of St Peter, although he made no reference to it by name until six years later. It may have been named *Lac d'Angoulême* by Jacques Cartier in the period 1535–40, as the name appeared on several maps during the rest of that century. Angoulême, northeast of Bordeaux, is the chief town of the French department of Charente. It had been the ancient capital of the county of Angoulême.

Saint-Polycarpe QC A municipality in the regional county of Vaudreuil-Soulanges, northwest of Salaberry-de-Valleyfield, it was created in 1988 through the merger of the village (1887) and the parish municipality of Saint-Polycarpe. The religious parish was founded in 1818, and named after St Polycarp (*c.* 69–*c.* 167), bishop of Smyrna in Asia Minor, and a martyr.

Saint-Prime QC A municipality west of Lac Saint-Jean and between Roberval and Saint-Félicien, it was established in its present area in 1968 through the amalgamation of the village (1923) and the municipality (1873) of Saint-Prime. The religious parish was founded in 1863 and the post office was opened in 1872, taking the name from Fr Prime Girard (1829–76), its first pastor and after St Primus, a third-century martyr.

Saint-Prosper QC A municipality in the regional county of Les Etchemins and east of Saint-Georges, it was established in 1929, having been called the township municipality of *Watford-Ouest* in 1886. Its post office was called *Saint-Prosper-de-Dorchester* in 1882, as it was then in the county of *Dorchester*. It was named after Fr Prosper-Marcel Meunier, who built a chapel here in 1882. The township of Watford was shown on a map of 1795, but not proclaimed until 1864, taking its name from a town in Hertfordshire, England, northwest of London.

Saint-Quentin NB Midway between Grand Falls and Campbellton, the first post office in this village (1966) was called *Anderson Siding* in 1912. It was renamed in 1919 after an Allied victory in early October 1918 at St-Quentin, France, near the Belgian border.

Saint-Raphaël QC A village in the regional county of Bellechasse, east of the city of Québec, it was severed from the parish municipality (1855) of Saint-Raphaël in 1921. The religious parish was organized in 1851 and named after Fr François-Raphaël Paquet (1762–1836), first pastor of the mission.

Saint-Raphaël-de-l'Île-Bizard QC Embracing the whole of Île Bizard in the Communauté urbaine de Montréal, this parish municipality was established in 1855, taking its name from the religious parish set up in 1839. Saint-Raphaël was proposed by Denis-Benjamin Viger after a college in the old manor house of Vaudreuil, where he studied. The island was named after Jacques Bizard (1642–92), its first seigneur in 1678.

Saint-Raymond QC In the regional county of Portneuf, west of the city of Québec, this town was created in 1957, having been detached as a village from the parish municipality of Saint-Raymond in 1898. The latter was created in 1845, abolished two years later, and re-established in 1855. Its religious parish was organized in 1842 as Saint-Raymond-Nonnat, in honour of St Raymond (1203–40), a Catalan cardinal noted for freeing Christian slaves.

Saint-Rédempteur QC A town (1981) near the mouth of the Rivière Chaudière, south of the city of Québec, it had been established as a village in 1919. It took its name from the religious parish, created then as Très-Saint-Rédempteur, 'The Most Holy Redeemer', one of the loftiest titles given to Jesus Christ.

Saint-Rémi QC Located in the regional county of Les Jardins-de-Napierville, south of Montréal, this town was established in 1949 and absorbed the village (1859) of Saint-Rémi in 1975. The municipality of Saint-Rémi had been created in 1855, and the post office was opened in 1831, taking its name from the religious parish of Saint-Rémi-de-Lasalle (La Salle) in 1828. The name was given in honour of Daniel de Rémy de Courcelle (Corcelles) (1626–98), governor of New France, 1665–72. He had a special devotion for St Remi (*c.* 437–*c.* 533), who was the bishop of Reims, France, from 459.

Saint-Roch-de-l'Achigan QC Located midway between Saint-Jérôme and Joliette and on the **Rivière de l'Achigan**, this parish municipality acquired its current name in 1957, having been erected as the municipality of *Saint-Roch* in 1855. Ten years earlier it had been organized as the parish municipality of Saint-Roch-de-l'Achigan (abolished in 1847), taking its name from the religious parish, founded in 1787. The name was given after Paul-Roch de Saint-Ours de l'Eschaillon (1747–1814), seigneur of L'Assomption from 1781, who donated land for the church. In Algonquin, *achigan* means 'bass'.

Saint-Romuald QC A town (1982) on the south shore of the St. Lawrence, opposite the city of Québec, it had been incorporated as the town of *Saint-Romuald-d'Etchemin* in 1963, and attained city status in 1965 when it annexed the municipality of Saint-Télesphore. The religious parish was organized in 1853 and named after St Romuald (952–1027), the founder of the monastic order of Camaldolese Benedictines in 1012.

Saint-Sauveur-des-Monts QC A popular resort destination in the mountains north of Saint-Jérôme, this village was severed from the parish municipality of **Saint-Sauveur** in 1926. The religious parish of Saint-Sauveur was officially erected in 1853 and two years later became the municipality of **Saint-Sauveur**, meaning 'Holy Saviour', a title of Jesus Christ.

St. Sepulchre, Mount BC In the Tower of London Range of the Muskwa Ranges of the Rocky Mountains, this mountain (2819 m) was named in 1960 by Capt. M.F.R. Jones after the regimental chapel of the Royal Fusiliers in London. A party of British and Canadian fusiliers climbed it.

St. Shott's NF The most southerly place in Newfoundland, it took its name from the old French word *chinchette*, 'small harbour with shoals'. On a 1544 map the adjacent cape was called *cap de chincete*. In 1775 Capt. James Cook and Michael Lane referred to the place as *St. Shot or Chink Hole*. In 1849 Joseph B. Jukes used its present form in his journal.

Saint-Siméon QC A village on the north shore of the St. Lawrence, opposite Rivière-du-

Loup, it was established in 1911 when it was detached from the parish municipality of Saint-Siméon. The latter and the religious parish were created in 1869, taking their names from St Simeon, bishop of Jerusalem in the first century and a close relative of Jesus Christ.

St. Stephen NB This town (1871), on the St. Croix River, was first called *Morris Town* in 1784 by Charles Morris Jr and Moses Gerrish, after Charles Morris Sr (1711-81), surveyor general of Nova Scotia, 1759-81. It was later called *Dover Hill*, resulting in Calais being built across the river in Maine. It was also known as *Scoodic*, after the name given to that part of the St. Croix, when the post office was named in 1825 after **Saint Stephen Parish**. The parish was created in 1786, and called after surveyor Stephen Pendleton, with its name complementing the other local 'saint' names for parishes, including Saint Andrews and Saint George. For a few months in 1973 the town was known as *St. Stephen-Milltown*, when the two towns were amalgamated, but Milltown was soon dropped as part of the official name.

St. Thomas ON A city (1881) in Elgin County, it was named in 1825 after Col Thomas Talbot (1771–1853). 'Saint' was added to Thomas out of respect for Talbot, who organized the settlement of thousands of acres south of London, during the first half of the nineteenth century.

Saint-Thomas QC A municipality on the east side of the city of Joliette, it attained its current status in 1993, having been established as a parish municipality in 1855. The religious parish was founded about 1840, and named in honour of Fr Thomas-Léandre Brassard (1805–91), the pastor of an adjoining parish. In 1845 it had been incorporated as the municipality of *Saint-Thomas-de-North-Jersey*, which described a concession road there. The place was also known as *Saint-Thomas-de-Voligny*, after an officer who had donated land for the church, presbytery, and cemetery.

Saint-Thomas-d'Aquin QC A parish municipality northwest of Saint-Hyacinthe, it was created in 1893, two years after its post office was opened. Served by a Dominican mission between 1889 and 1891, it was named after St Thomas Aquinas (1225–74), a celebrated theologian of the thirteenth century.

Saint-Timothée QC Located east of Salaberry-de-Valleyfield, this municipality was created in 1990 through the amalgamation of the village (1919) and the parish municipality (1855) of Saint-Timothée (originally erected in 1845, and abolished two years later). The religious parish of Saint-Timothée-de-Beauharnois was organized in 1820 and named after St Timothy, a disciple of St Paul during the first century.

Saint-Tite QC A town northeast of Shawinigan, it was severed from the parish municipality of **Saint-Tite** in 1910. The latter was erected in 1863, taking its name from the religious parish, established four years earlier, the same year the post office was opened. St Titus was a companion of St Paul during the first century.

Saint-Tite-des-Caps QC Located in the regional county of La Côte-de-Beaupré, northeast of Sainte-Anne-de-Beaupré, this municipality was established in 1872. Serving as a mission since 1853, the religious parish was organized in 1876, ten years after the post office was opened. At the time that it was named, Pope Pius IX had entered the name of St Titus, St Paul's disciple, into the Roman calendar.

Saint-Ubalde QC In the regional county of Portneuf, northeast of Trois-Rivières, this municipality was established in 1973 through the union of the village of Saint-Ubalde (1920) and the parish municipality of *Saint-Ubald*. The form, without the final 'e', was adopted for the post office (1874) in 1930. The religious parish of Saint-Ubald was organized in 1866, and named after Ubald (Ubalde) Gingras (1824–74), brother-in-law of Mgr Charles-François Baillargeon, archbishop of Québec, 1867–70, who was the first sacristan of Saint-Ubald.

Saint-Vallier QC In the regional county of Bellechasse, between Lévis and Montmagny, this municipality was established in 1993, having been incorporated as a village in 1958. The religious parish of Saint-Vallier was organized in 1714 and named after Jean-Baptiste de la Croix de Chevrières de Saint-Vallier (1653–1727), second bishop of Québec.

Saint-Victor QC Located west of Saint-Georges, this village was created in 1955, when it was detached from the municipality of **Saint-Victor-de-Tring**. The municipality was established in 1870, taking its name from the religious parish, organized in 1848. Prior to that it was part of the township municipality of Tring. The post office was called Tring from 1852 to 1865, then *Saint-Victor-de-Tring* until 1937, when it became Saint-Victor-de-Beauce. The parish was named after St Victor, the fourteenth Pope, 186–197 AD. The township of Tring, named after a town in Hertfordshire, England, northwest of London, was identified on a 1795 map and proclaimed in 1804. *See also* Tring-Jonction.

St. Vincent ON A township in Grey County, it was named in 1840 after John Jervis, Earl of St. Vincent (1735–1823), who achieved a great naval victory in 1797 over a larger Spanish fleet off Portugal's Cabo de São Vincente.

St. Vincent's-St. Stephens-Peter's River NF On the east side of St. Mary's Bay, this improvement district was created in 1971. **St. Vincent's** was known as *Holyrood* as early as the 1830s and later as *Holyrood South*, but it was renamed in 1910 to avoid confusion with Holyrood on Conception Bay. **St. Stephens** was called *Middle Gut* until 1945, when it was renamed after Fr Stephen O'Driscoll of St. Mary's. The origin of the name **Peters River** is not recorded in official the records.

St. Vital MB A neighbourhood on the south side of Winnipeg and on the east side of the Red River, it is part of the urban community of **St. Boniface-St. Vital**. It was named about 1860 by Mgr Alexandre Taché, after Mgr Vital-Justin Grandin (1829–1902), coadjutor bishop of St. Boniface, 1857–71 (based in Île-à-la-Crosse, in present-day Saskatchewan), and bishop of St. Albert, 1871–1902 (northwest of Edmonton).

St. Walburg SK A town (1953) northeast of Lloydminster, its post office was named in 1910 after Walburga Musch, a woman widely respected for her charitable work in the community. Her name was given in honour of St Walburga, an eighth-century English missionary who served in Germany at the time of St Boniface. The post office was moved 4 km north in 1920 to the site of *Clansman*, with the Canadian Northern Railway arriving here the following year.

Saint-Wenceslas QC A village in the regional county of Nicolet-Yamaska, it was separated from the municipality of **Saint-Wenceslas** in 1922. The municipality was established in 1864, and named after the religious parish, erected in 1857, and named after St Wenceslas (*c.* 907–35), a duke of Bohemia and a pious man, who was murdered by his brother.

St. Williams ON A community in Delhi Township, Haldimand-Norfolk Region, it was first known as *Walsingham* after the original township, *Cope's Landing*, after settler William Cope, and *Neal's Corners*, after the Revd George Neal, who owned a farm here. It was renamed in 1869 after William Gillaspey, an early settler much admired for his upright Christian principles.

Saint-Zacharie QC Near the Maine border and in the regional county of Les Etchemins, east of Saint-Georges, this municipality was created in 1990 by amalgamating the village (1961) and the parish municipality (1932) of Saint-Zacharie. The latter was formerly the township municipality of *Metgermette-Nord*, organized in 1886. The religious parish was set up in 1873, and named after Fr Zacharie Lacasse (1845–1921), who had promoted the settlement of the area. The township, proclaimed in 1885, was named after a river and a lake whose names in Abenaki express the ideas of 'disaster' or 'misfortune'.

Saint-Zotique QC A village in the regional county of Vaudreuil-Soulanges and on the north shore of Lake St. Francis, it was created in 1913 and annexed the parish municipality (1855) of Saint-Zotique in 1967. Mgr Ignace Bourget obtained a relic of the martyr St Zoticus during a European journey in 1849.

Salaberry-de-Valleyfield QC A city at the east end of Lake St. Francis, it was created in 1874. The first part was given in honour of Lt-Col Charles-Michel d'Irumberry de Salaberry (1778–1829), who defeated the superior American forces at Châteauguay in 1813. The second recalls The Valleyfield Paper Mills, established here by Thomas F. Miller in 1854, the same year the **Valleyfield** post office was opened.

Salisbury NB West of Moncton, this village (1966) was named in 1846 after **Salisbury Parish**, created in 1787. The parish may have been called after *Salisbury Cove*, a name given about 1777 by chart-maker J.F.W. DesBarres to present Rocher Bay, in the Bay of Fundy. The cove may have been named in honour of John Salisbury, who accompanied Gov. Edward Cornwallis in Nova Scotia in the mid-1700s.

Salluit QC Located near the southern coast of Hudson Strait, this northern village was given its present spelling in 1979, having been earlier spelled *Saglouc* (1962), *Notre-Dame-de-Sugluc* (1961), and *Sugluk* (1947). The meaning of the name is uncertain, although it has been suggested it means 'sparse' in Inuktitut.

Salmo BC A village (1946) south of Nelson, it was named in 1897 after the **Salmo River**, which rises east of Nelson and flows south to join the Pend d'Oreille River, southeast of Trail. The river was called the *Salmon River* as early as 1859. *Salmo* is Latin for 'salmon'.

Salmon Arm BC A district municipality (1970) east of Kamloops and on the **Salmon Arm** of Shuswap Lake, it had been incorporated as a city in 1912. Its post office was named in 1890. It had its beginnings when a projected route for an intercontinental railway was surveyed in 1871 by Walter Moberly.

Salmon Cove NF Northeast of Carbonear, this town (1974) was referred to by this name in the 1680s. *The English Pilot* of 1689 stated that 'a River in the said Cove runs up, in which are a store of salmon'.

Salmon River NS This place is located east of Truro on the **Salmon River**, which flows west through Truro into Cobequid Bay. Although the province once had four post offices called Salmon River, this place, with some 1900 people, was not one of them.

Saltair BC On the east coast of Vancouver Island, between Ladysmith and Chemainus, this community was named in 1912 after the salty air of the Strait of Georgia.

Saltcoats SK Southeast of Yorktown, this town (1910) was first called *Stirling* in 1886 by the Manitoba and North Western Railway after the home town of the Allan brothers of the Allan Steamship Lines. When that name was found in use elsewhere, it was renamed by Sir Hugh Allan after a town in Ayrshire, Scotland, southwest of Glasgow. Its post office was opened in 1888.

Saltspring Island BC The largest (180 km^2) of the Gulf Islands, north of Victoria, it was named *Salt Spring Island* by the Hudson's Bay Company in the early 1800s. In 1859 Capt. George Richards substituted *Admiral Island*, after Rear-Adm. Robert L. Baynes, commander of the Pacific station, 1857-60, but it did not supplant the earlier name. In 1910 the Geographic Board of Canada adopted the current name, fusing the words 'salt' and 'spring', a practice the board preferred for multiple-word specifics of geographical names. Preferred local usage remains Salt Spring Island.

Salvage NF A fishing community on the outer end of Eastport Peninsula, its name may refer to a marine incident or tragedy. It was visited by migratory fishermen from England as early as 1676.

Sambro NS On the Atlantic Ocean, south of Halifax, this place was identified by Samuel de

Champlain on his 1632 map as *Sesambre*. It may have been named after Cesambre, an island adjacent to St-Malo, France. Postal service was provided at Sambro in 1860.

Sampson, Mount YT This mountain (3353 m) in the St. Elias Mountains, was named in 1970 by the Alpine Club of Canada after Herbert E. Sampson (1871–1962), president of the club, 1930–2.

Sandringham NF On the Eastport Peninsula and east of Glovertown, this municipal community (1968) was founded in 1939 for the resettlement of many nearby communities. It was likely named after the summer residence of the Royal family in northwestern Norfolk, England.

Sandwich South ON Sandwich Township in Essex County was named in 1792 for the English borough of Sandwich in Kent. The town of *Sandwich* was formed in 1858, but Windsor annexed it in 1935. The municipal townships of *Sandwich East*, Sandwich South, and *Sandwich West* were created in 1854, but only Sandwich South remains a separate municipality. *See also* LaSalle.

Sanford MB Situated on the La Salle River, southwest of Winnipeg, this community's post office was called *Mandan* in 1900, but was renamed in 1906 after Senator W.E. Sanford (1838–99).

Sango Bay NF On the inner side of Davis Inlet, its area had been chosen in 1996 as a relocation site for the Innu community of Davis Inlet, on an adjoining island. In the language of the Innu the name means 'sandy plateau'. It is pronounced 'shango'.

Sangudo AB A village (1937) northwest of Edmonton, its post office was named in 1912, created as an acronym from the names of settlers and other local names: Sutton or Sides, Albers, Nanton (where Mrs Albers formerly lived), Gaskell, United, Deep Creek (the first proposal), and Orangeville (the school district name).

Sanikiluaq NT At the north end of Flaherty Island, one of the Belcher Islands, this hamlet (1976) was established in 1970 and named after the first Inuk to hunt seals there. Before Sanikiluaq was founded, the Inuit of the Belcher Islands had previously lived at two separate places called North Camp and South Camp.

Sarawak ON A township in Grey County, it was named in 1855 after the state of Sarawak (now a part of Malaysia) by Viscount Bury, superintendent of Indian affairs in Canada. In 1841 the rajah of Sarawak ceded the state to English soldier and explorer Sir James Brooke, a relative of the viscount.

Sardis BC In the district municipality of Chilliwack, south of its main urban centre, this community was named in 1888 by Mrs A.S. Vedder, who opened her Bible at *Revelations* 3:4, and was inspired by the following verse: 'Thou hast a few names even in Sardis which have not defiled their garments; and they shall walk with me in white: for they are worthy.'

Sarnia ON The township was named in 1829 by Sir John Colborne, after the Roman name for Guernsey in the Channel Islands, where Colborne had been lieutenant-governor, 1821–8. In 1836 a growing community on the St. Clair River was named *Port Sarnia*. In 1860 'Port' was dropped from the postal name to conform with the name of the town, incorporated three years earlier. The township became the town of *Clearwater* in 1988. It was annexed by the city (1914) of Sarnia on 1 January 1991, which existed as the city of *Sarnia-Clearwater* for one year, before reverting to Sarnia.

Saskatchewan The district of Saskatchewan was established in 1882 in the North-West Territories, and embraced a central section (between Saskatoon and Prince Albert), a section of present-day Manitoba, northwest of Lake Winnipeg, and a narrow strip of present-day Alberta. The limits of the present province were decided in 1905, by adding most of the district of Assiniboia and the east half of the district of Athabaska, which extended north to the sixtieth parallel.

Saskatchewan, Mount YT In the Centennial Range of the St. Elias Mountains, this mountain (3472 m) was named in 1967 after the province.

Saskatchewan River MB, SK Rising as the North Saskatchewan River in the Rocky Mountains, and as the South Saskatchewan River east of Lethbridge, AB, it officially becomes Saskatchewan River 45 km east of Prince Albert, SK. It was derived from the Cree *Kisiskatchewan*, 'swift current'. The present spelling was established in 1813 by geographer David Thompson, who had then extended it to the mouth of the Nelson River.

Saskatoon SK John Neilson Lake founded a temperance colony on the South Saskatchewan River in 1882, and planned to call it Minnetonka (Siouan for 'big water'), possibly after a western suburb of Minneapolis, MN. When a branch of reddish purple berries was brought to him, he was told the Cree called the berries *misaskwatomina*, 'fruit of the much wood'. It sounded like Saskatoon to him and he pronounced it the name of the new colony. It became a town in 1903 and a city in 1906.

Saturna Island BC The most southeasterly of the Gulf Islands, it was named in 1791 by José Maria Narvaez after his Spanish ship *Saturnina*. As early as 1862 the name was being rendered as Saturna.

Saugeen ON A township in Bruce County, it was named in 1850 after the **Saugeen River**. *Saugeen* is an Ojibwa word meaning 'river mouth'. Admiralty surveyor Henry W. Bayfield spelled the river's name *Saugink* in 1822, but by the late 1840s it was spelled Saugeen. The river rises north of Dundalk in Grey County, and has numerous tributaries, including the **South Saugeen**, the **North Saugeen**, the **Rocky Saugeen**, the **Beatty Saugeen**, and the Teeswater rivers. Based on the knowledge of Chief Newash at Owen Sound, surveyor Charles Rankin encountered the *Saugin* in 1837 at the site of Mount Forest. John McDonald resurveyed the Rankin route to Owen Sound in 1841, and changed the name of the Saugeen at Mount Forest to the Maitland, deeming it to be that river's headwaters. In 1844 a survey party led by Casimir (later Sir Casimir) Gzowski explored downriver from Mount Forest to Hanover, finding that it was clearly the Saugeen.

Saugstad, Mount BC In the Coast Mountains, southeast of Bella Coola, this mountain (2905 m) was named in 1948 by surveyor Maj. F.V. Longstaff after the Revd Mr Saugstad. He and Thorwald Jacobsen brought Norwegian settlers to the area about 1892.

Saulnierville NS On the French Shore, southwest of Digby, this place's post office was called *Souvier Wharf* in 1860. Four years later it was renamed Saulnierville, after several Saulnier families, whose ancestors had settled here after the 1755 expulsion of the Acadians.

Sault-au-Mouton QC A village (1947) located on the north shore of the St. Lawrence, across from Rimouski, its post office had been opened in 1906. The name, which occurred as *Saut a mouton* on Jean Guérard's 1631 map, describes the foaming froth, which looks like the white wool of sheep (*mouton*), at the foot of the falls near the mouth of the **Rivière du Sault au Mouton**.

Sault Ste. Marie ON The site of the city (1912) was called by the Ojibwa *Pawating*, meaning 'turbulent waters', and Étienne Brûlé named it *Saut de Gaston* in the early 1600s after Louis XIII's brother. In 1641 Jesuit missionaries dedicated their mission to the Virgin Mary, and called the place *Sainte Marie du Sault*. The North West Company opened a trading post here in 1783, and continued the French name, slightly modified, for the post. When the post office was established in 1846 the historic trading-post name was maintained. In the seventeenth and eighteenth centuries the French word *sault* was used as an equivalent for 'falls' and 'rapids', but was gradually replaced in everyday language and place names by *chute*.

Savona BC On the south shore of Kamloops Lake, west of Kamloops, this place was originally founded on the north shore by François Saveneux, a native of Corsica, who became known as Francis Savona. He established a ferry in 1859 across the Thompson River, where it flows out of the lake. When the Canadian Pacific Railway passed along the south shore in 1882, its station here was called *Port Van Horne*, after William (later Sir William) Van Horne, but it was soon superseded by Savona. In 1881 the post office was called *Savona's Ferry*, and was changed to Savona in 1896.

Sawyerville QC Located east of Sherbrooke, this village was established in 1892 and its post office had been opened in 1853. Josias Sawyer (d. 1837) crossed the border from Vermont about 1792 to inspect the site, and subsequently built sawmills on the Rivière Eaton. He donated land for a school in 1822.

Sayabec QC Situated west of Lac Matapédia and southwest of Matane, this municipality was erected in 1982 by merging the parish municipality of *Sainte-Marie-de-Sayabec* (1894) and the village of Sayabec (1951), incorporated as *Saindon* in 1917. The latter was the railway station name, given after Fr Joseph-Cléophas Saindon (1866–1941), who had been appointed the pastor here in 1896. The post office was named Sayabec in 1900, derived from the **Rivière Sayabec**, which means 'obstructed river' in Mi'kmaq.

Sayward BC On the northeastern coast of Vancouver Island, northwest of Campbell River, this village (1968) was named in 1911 after lumberman William Parsons Sayward, who moved to Victoria from California in 1858. Its post office had been called *Port Kusam* in 1899.

Scarborough ON A city in Metropolitan Toronto, its area was first called *Glasgow Township* in 1791. Two years later, after Elizabeth Simcoe noted how the impressive bluffs reminded her of the magnificent cliffs of Scarborough in North Yorkshire in England, her husband renamed it. The township became a borough in 1967 and a city in 1983.

Scatarie Island NS Off the east coast of Cape Breton Island, this island's name is derived from Basque *Escatari*, 'steep coast', or 'line of cliffs'. The island had postal service from 1884 to 1957.

Sceptre SK Northwest of Swift Current and south of the South Saskatchewan River, this village (1913) was named in 1913 by the Canadian Pacific Railway, possibly after the staff carried by a prelate, with the village to the west being called Prelate. Its post office was also opened in 1913.

Schefferville QC A town (1955) north of Labrador City, it was a centre of iron ore mining from 1953 to 1982, when its population tumbled from 5000 to a little over 100. Although the town was officially dissolved in 1986, its status was restored four years later as market centre for nearby Montagnais and Naskapi communities. It was named after Mgr Lionel Scheffer (1903–66), first apostolic vicar of Labrador, 1946–66.

Schomberg ON In York Region, west of Newmarket, this community was first called *Brownsville*, after Thomas Brown had built a mill here in 1836. Because of possible confusion with simliar names, the post office department declined to accept Brownsville in 1861. Thomas R. Ferguson, a member of the Legislative Assembly and a staunch Orangeman, proposed Schomberg in 1862, after the Duke of Schomberg (1615–90), commander-in-chief of the British forces at the Battle of the Boyne in Ireland, where he was killed.

Schreiber ON When the Canadian Pacific Railway arrived at this site on the north shore of Lake Superior in 1885, its station was named after Collingwood Schreiber (1831–1918), who became the CPR's chief engineer in 1880. He was appointed the deputy minister of the federal department of railways and canals in 1892 and was knighted in 1916.

Schumacher ON In Timmins, east of the city centre, it was named in 1912 after Fred Schumacher, who was a wealthy manufacturer of patent medicines in Ohio. He had come to the Porcupine area that year, invested wisely in property, and made a fortune when his lots were sold for gold exploration. Subsequently he retired to a life of seclusion in Columbus, OH.

Scotch Village NS Located near the mouth of the Kennetcook River and east of Windsor, this place was named after pioneer settler John Smith, who came from Scotland. Its post office was opened in 1858.

Scotland ON A community in Brant County, southwest of Brantford, it was first known as *Malcolm's Mills*, after Loyalist Finlay Malcolm, whose ancestors were Scottish. It was renamed in 1852 by Janet Graham, when the post office was opened.

Scots Bay NS North of Wolfville and east of Cape Split, this place is located on the shore of **Scots Bay**. The bay was first called *Scotch Bay*, after some Scottish immigrants were driven ashore, where they stayed for the winter, before proceeding to their destination. When its post office was opened in 1855, it was called *Scotts Bay*, and *Scotts Bay Road* post office was opened to the south in 1877.

Scotsburn NS Southwest of Pictou, this place's post office was called *Rogers Hill* in 1845 after John Roger, who had settled here in the late 1700s. It was renamed in 1874 after the birthplace of Hugh Ross in Ross and Cromarty, north of Inverness, Scotland.

Scotstown QC A town (1892) located between Sherbrooke and Lac-Mégantic, its post office was opened in 1873. It was named after John Scott, director of the Glasgow Canadian Land and Trust Company, established in Scotland in 1873. He acquired land in the area of Scotstown and built a sawmill to serve the settlers, but he was dismissed in 1875, and the company ceased operations 21 years later.

Scott ON Part of the municipal township of Uxbridge, Durham Region, since 1974, Scott Township had been named in 1820 after Thomas Scott (1746–1824), chief justice of Upper Canada, 1806–16.

Scott QC A village (1978) in the regional county of La Nouvelle-Beauce, north of Sainte-Marie, it replaced the parish municipality of *Saint-Maxime* (1895). Its station was called *Scott-Jonction*, established when the Lévis and Kennebec Railway had been built in the Beauce valley between 1868 and 1875. It was named after Charles Armstrong Scott (d. 1893), a contractor who had built the first section of tracks.

Scott SK A town (1910) southwest of Battleford, it was named in 1908 after Frank Scott, the treasurer of the Grand Trunk Pacific Railway. Postal service also arrived here that year. With a population of 118 in 1991, it tied Fleming as the province's smallest town.

Scott, Cape BC The most northwesterly point of Vancouver Island, it may have been named in 1786 by trader James Strange after David Scott, a member of a Bombay (India) syndicate, who financially supported Strange's expedition to trade on the northwestern coast of North America.

Scott, Mount AB, BC On the continental divide and northeast of Athabasca Pass, this mountain (3300 m) was named in 1913 by climbers Arnold L. Mumm and Edward Howard after British naval officer and explorer Robert Falcon Scott (1868–1912). He reached the South Pole in 1912, but died returning to his base camp.

Scoudouc NB Northeast of Moncton, this place's post office was called *Scadouc* in 1890. It was renamed Scoudouc in 1932, after the **Scoudouc River**, which flows north into Shediac Bay. The river's name was derived from Mi'kmaq *omskoodook*, whose meaning is unknown.

Scugog ON Named as a municipal township in 1851, it was united with the village of Port Perry and the municipal townships of Cartwright and Reach in 1974 to form an enlarged municipal township of Scugog in Durham Region. Its name was probably derived from Mississauga *sigaog*, 'waves leap over a canoe', referring to the flooding of the

Scugog River valley in 1834, when William Purdy built a dam at Lindsay, which created both **Lake Scugog** and **Scugog Island**.

Seabright NS On the east side of St. Margarets Bay and west of Halifax, this place's post office was called *Hubley Settlement* in 1896. It was renamed in 1902 after the Seabright Hotel.

Seaforth ON A town (1874) in Huron County, southeast of Goderich, it was named in 1855 by Barrie lawyer and developer James Patton, possibly after the Seaforth Highlanders, a Scottish regiment raised by Francis H. Mackenzie, Baron Seaforth and Mackenzie (1754–1815). The title may have been taken from Loch Seaforth, on the east side of Lewis, in the Outer Hebrides.

Seal Cove NB At the south end of Grand Manan Island, this urban community was an incorporated village from 1966 to 1995, when it became part of the village of Grand Manan. Its post office was named in 1875. Both the hooded seal and the harp seal have played significant roles as sources of food and clothing in the economic development of the Atlantic coast.

Seal Cove NF A municipal community (1972) on the northwest side of Fortune Bay, and west of Harbour Breton, it was settled about 1837. The town (1958) of **Seal Cove**, on the eastern shore of White Bay, and west of Baie Verte, was occupied about 1840. A third place called **Seal Cove**, in the town of Conception Bay South and on the southeastern shore of Conception Bay, was settled in the early 1800s.

Searletown PE North of Borden-Carleton, this place was settled by James Searle in 1767. It was called *Searle Town* as early as 1843. His grandson, James Searle Mann, a native of Philadelphia, moved here in 1854 and arranged to have the Searletown post office opened the next year.

Seattle, Mount YT In the St. Elias Mountains, this mountain (3069 m) was named in 1890 by explorer and geologist Israel C. Russell after the city of Seattle, WA.

Sebringville ON In Perth County, northwest of Stratford, it was named in 1852 after miller and first postmaster P.A. Sebring. John Sebring, a native of Vermont, had built a sawmill here in 1834 and a store a few years later.

Sechelt BC A district municipality (1986) on the **Sechelt Peninsula**, northwest of Vancouver, it was named after the community of **Sechelt**, whose post office was opened in 1896. The community, on a narrow isthmus separating **Sechelt Inlet** from the Strait of Georgia, was known as *Se-shalt* by the Seechelt tribe of the Coast Salish. The tribe's name may have been derived from *chatelech*, 'climb over', in reference to a tree that had to be climbed over to get to the beach. The Indian government district of **Sechelt** was created in 1988.

Sedgewick AB A town (1966) southeast of Camrose, its post office was named in 1907 after Robert Sedgewick (1848–1906), deputy minister of justice, 1888–93 and justice of the Supreme Court of Canada from 1893. A native of Scotland, he practised law in Halifax from 1873 to 1888.

Sedley SK Southeast of Regina, the post office at this village (1907) was named in 1904 by Dr R.J. Blanchard after his brother Sedley Blanchard, a Winnipeg lawyer, who had died of typhoid in 1886.

Seeleys Bay ON In the united counties of Leeds and Grenville, northeast of Kingston, it was settled in 1825 by trader Justus Seeley (d. 1828), and later by his son John. It was first called *Coleman's Corners*, and William Coleman became the first postmaster of *Seely's Bay* in 1851. The name was changed to Seeleys Bay in 1950.

Seldom-Little Seldom NF On the south side of Fogo Island, the two names of this town (1972) may relate to few fishing schooners seldom willing to negotiate the treacherous Stag

Harbour Run to the west, resulting in the only access to its harbour being from the east. Seldom was formerly called *Seldom Come By*.

Selkirk MB On the west bank of the Red River, north of Winnipeg, the post office in this town (1882) was named in 1876 after Thomas Douglas, 5th Earl of Selkirk (1771–1820), who established the Red River Colony in 1812.

Selkirk ON In the southeastern part of the city of Nanticoke, it is in the area where the Earl of Selkirk received the deed for a large grant of land in 1808, although he may never have paid the Six Nations confederacy for it, before establishing the Red River Colony in Manitoba three years later. In 1836 its post office was called *Walpole*, after the township, but was renamed Selkirk in 1855.

Selkirk, Mount BC In the Rocky Mountains, northeast of Invermere, this mountain (2938 m) was named about 1886 by geologist George M. Dawson after Thomas Douglas, 5th Earl of Selkirk (1771–1820), who founded the Red River Colony in present-day Manitoba in 1811.

Selkirk Mountains BC Extending from the Washington State border on the south to the Big Bend of the Columbia River on the north, and between the Columbia River on the west, and Kootenay Lake and Duncan and Beaver rivers on the east, these mountains were first called *Nelson's Mountains* by geographer David Thompson about 1811, after Adm. Viscount Horatio Nelson. In 1821 the Hudson's Bay Company named them after Thomas Douglas, 5th Earl of Selkirk (1771–1820).

Selwyn Lake NT, SK On the sixtieth parallel, this lake (593 km^2) was named before 1910 after Alfred Richard Cecil Selwyn (1824–1902), director of the Geological Survey of Canada, 1869–95.

Selwyn, Mount BC In the Selkirk Mountains, south of Rogers Pass, this mountain (3360 m) was named in 1901 after Alfred Richard Cecil Selwyn (1824–1902), director of the Geological Survey of Canada, 1869–95.

Selwyn Mountains YT These mountains, a division of the Mackenzie Mountains, extend for 650 km along the west side of the territorial boundary with the Northwest Territories. They were named in 1901 by geologist Joseph Keele after Alfred Richard Cecil Selwyn, director of the Geological Survey of Canada, 1869–95.

Semans SK North of Regina, this village (1908) was named in 1908 by the Grand Trunk Pacific Railway after the birth name of the wife of one of its officials. The post office was also opened the same year.

Senneterre QC Located northeast of Val-d'Or, this town (1956) had been incorporated as a village in 1948, the same year the post office was opened. The township of Senneterre was surveyed between 1911 and 1913 and proclaimed in 1916. It took its name from an officer of the Languedoc Regiment in Montcalm's army, who fought at the Battle of Sainte-Foy in 1760.

Senneville QC Created as a village in 1895, this place at the west end of the Communauté urbaine de Montréal, north of Sainte-Anne-de-Bellevue, was named after Jacques Le Ber de Saint-Paul de Senneville (1633–1706), who shared the rear fief of Boisbriand after 1679, and built a fort, store, and mill. He was born at Senneville-sur-Fécamp, Seine-Maritime, France. *See also* Val-Senneville.

Sept-Îles QC The metropolis of the north shore of the St. Lawrence, this city is situated on a natural harbour, with seven islands sited in front of its entrance. It was incorporated in 1951, and since 1981, has been the principal urban centre of the regional county of Sept-Rivières. The post office was called *Seven Islands* in 1886, and was renamed Sept-Îles in 1931. In 1535 Jacques Cartier referred to the bay by the name of *Sept Ysles*.

Sept-Rivières QC A regional county on the north shore of the Gulf of St. Lawrence, it was named in 1981 after its seven rivers: Trinité, Penticôte, aux Rochers, Sainte-Marguerite, Moisie, Pigou, and Manitou.

Serra Peaks BC In the Coast Mountains, northeast of Mount Waddington, these peaks (3642 m) were named in 1933 by climber W.A.D. (Don) Munday, because the peaks had a sharp serrated edge along their crest.

Seul, Lac ON A large lake between Lake Nipigon and Lake of the Woods, it was called in Ojibwa *Opasiccocan*, 'white pine at the narrows', but which French traders interpreted as *Lac Seulment*, possibly because of disappointment in not finding that it led to the Western Sea. It was then shortened to Lac Seul, the form adopted by English fur trader Edward Umfreville in his 1784 diary.

70 Mile House BC This community in the Cariboo was named in the 1860s, because it was judged to be 70 miles (113 km) north of Lillooet on the Cariboo Road. Its post office was opened in 1908.

Severn ON A municipal township in Simcoe County, it was created in 1994 by merging the village of Coldwater, the municipal townships of Orillia and Matchedash, and parts of the municipal townships of Medonte and Tay. It was named after the **Severn River**, which drains Lake Couchiching into Georgian Bay and forms the western part of the Trent-Severn Waterway. The river was possibly named by Admiralty surveyor Henry W. Bayfield in the 1820s.

Severn River ON A main river of northwestern Ontario, it flows northeast for 575 km to Hudson Bay. In 1631 it was named the *New Severn River* by explorer Thomas James, after England's River Severn.

Sexsmith AB A town (1979) 20 km north of Grande Prairie, its original post office was located 5 km farther south, and named after David Sexsmith, who had arrived here first in 1898, moved to Edmonton in 1901, and returned in 1912. The townsite, first known as *Benville*, after John Bernard 'Benny' Foster, was surveyed in 1915. It was named Sexsmith the next year by the Edmonton, Dunvegan and British Columbia Railway, and the post office was transferred here.

Seymour ON A township in Northumberland County, it was named in 1798 after Lady Elizabeth Seymour, the widow of the Duke of Northumberland. After the duke's death in 1786, she became Baroness Percy.

Seymour, Mount BC In the district municipality of North Vancouver, this mountain (1455 m) was named in 1951 after Frederick Seymour (1820–69), governor of the colony of British Columbia, 1864–9. **Seymour River** flows south, on the west side of the mountain, into Burrard Inlet. **Seymour Inlet**, on the mainland coast, opposite the north end of Vancouver Island, was also named after him. **Seymour Narrows** in Discovery Passage, northwest of Campbell River, was named in 1846 by Cdr George T. Gordon after Rear-Adm. Sir George Francis Seymour (1787–1870), commander of the Pacific station, 1844–8.

Shackleton, Mount BC In the Clemenceau Icefield and west of Howse Pass, this mountain (3330 m) was named in 1923 by surveyor Arthur O. Wheeler after Sir Ernest Shackleton (1874–1922), who had led British Antarctic expeditions in 1907–9 and 1914–16.

Shakespeare ON In Perth County, east of Stratford, this community was settled by David Bell in 1832 and its post office was named *Bell's Corners* in 1848. Four years later, it was named after William Shakespeare, by either first postmaster Alexander Mitchell or Stratford notary John James Edmonstoune Linton.

Shallow Lake ON A village (1911) in Grey County, northwest of Owen Sound, it was first known as *Stoney Creek*. In 1891 Shallow Pond was proposed after a nearby shallow water body, where marl was being excavated for the production of cement. Because Shallow Pond sounded too dull, it was named Shallow Lake.

Shanadithit Brook NF This brook, which flows into the northwest side of Red Indian Lake, was named in 1956 after Shawnadithit, a Beothuk woman who was captured in 1823, but died nine years later in St. John's. She may have been the last survivor of her people.

Shand, Mount BC In the Coast Mountains, northeast of Mount Waddington, this mountain (3095 m) was named in 1947 by mountaineer Fred W. Beckey after climber William Shand Jr (1818–46), who had been killed in a car accident. Shand had climbed in the St. Elias Mountains in the early 1940s.

Shannon QC A municipality (1947) northwest of the city of Québec, its post office served the area from 1905 to 1968. It was settled early by Irish immigrants, suggesting it might have been named after Ireland's River Shannon, but it would appear the name has its roots in a family name, because local church registers of the 1830s noted the deaths of Richard and Simon Shannon.

Shannonville ON A community in Hastings County, east of Belleville, it was named in 1833 after the *Shannon River* (now the Salmon River). It had been named after Ireland's River Shannon, from where several Tyendinaga Township families originated.

Shanty Bay ON In Simcoe County, northeast of Barrie, this place was named in 1832 after several shanties erected to house a settlement of blacks, sponsored by the government of Upper Canada.

Sharbot Lake ON In Frontenac County, north of Kingston, this community was settled in 1827 by Francis Sharbot, an Aboriginal, and its post office was named in 1877. The lake, divided into two equal sections, was first called *Crooked Lake* and was identified as *Chabot Lake* in 1856.

Sharon ON In York Region, north of Newmarket, it was named by Children of Peace community founder David Willson (1778–1866) in 1841, inspired by the reference to the rose of Sharon in the Bible's Song of Solomon. It was earlier called *Hope* and *Davidtown*.

Shaughnessy, Mount BC In Glacier National Park, north of Rogers Pass, this mountain (2807 m) was named in 1900 after Thomas George Shaughnessy, 1st Baron Shaughnessy (1853–1923), president of the Canadian Pacific Railway, 1899–1918.

Shaunavon SK A town (1914) southwest of Swift Current, it was named by the Canadian Pacific Railway in 1913 on the suggestion of Thomas George Shaughnessy, 1st Baron Shaughnessy, president of the CPR, 1899–1918. He proposed it after 'an area in the old country', although he was a native of Milwaukee, WI. Its post office was opened the same year.

Shawinigan QC A city (1958) in the Saint-Maurice valley, north of Trois-Rivières, it had been incorporated as the village of *Shawinigan Falls* in 1901, and became a town the following year. The name had many spellings until the formation of the Shawinigan Water and Power Company in 1898. Its origin points to the idea of 'portage', perhaps 'portage over a pointed crest' in Abenaki. The town of **Shawinigan-Sud** (1961), on the east side of the river, had been created a village in 1953. Before that it was called *Almaville*, derived from *Alma Mater*, in reference to the Virgin Mary, and meaning 'generous mother'.

Shawnigan Lake BC At the north end of **Shawnigan Lake**, southeast of Duncan, this community's post office was named in 1894. The lake may have been named in the 1850s by blending the surnames 'Shaw' and 'Finnegan'.

Shawville QC A village in the regional county of Pontiac, it was named in 1874 after James Shaw (1818–77), who had settled in the township of Clarendon in 1843. In 1856 he was appointed the postmaster of *Clarendon Centre*, which had been established in 1852. The first name proposed for the village was Daggville, after the pioneer Dagg family, but Shaw's gift of almost a hectare of land to the village ensured the perpetuation of his name.

Shedden ON A community in Elgin County, southwest of St. Thomas, its Canada Southern Railway station was named in 1872 after railway contractor and Toronto and Nipissing Railway president John Shedden (1825–73). He was accidentally killed the following year at Cannington in Victoria County. Its post office was named *Corseley* in 1875, but was renamed Shedden in 1883. There is also a municipal township in Algoma District, west of Espanola, named after John Shedden.

Shediac NB A town (1903) northeast of Moncton, its post office was opened about 1829. It took its name from the **Shediac River**, which flows east into **Shediac Bay**, north of the town. The river's name was derived from Mi'kmaq *Esedeiik*, 'running far back', possibly in reference to the indentation formed by the bay as part of Northumberland Strait, or to the portage route to the Petitcodiac River. Map-maker Pierre Jumeau called it *R. Chedaik* in 1685. In 1832 Surveyor General Thomas Baillie wrote 'Shediac river by the Acadians Gidaic'.

Sheet Harbour NS This place is on the Eastern Shore and at the head of **Sheet Harbour**. The harbour was first known as *Port North*, likely given in the 1770s by chart-maker J.F.W. DesBarres after Lord North, prime minister of Great Britain, 1770–82. The present name came into use about 1818, taking its name from a prominent white cliff at the harbour's entrance, which appeared to look like a white sheet spread out to dry. The place was first called *Campbelltown*, likely after Lord William Campbell (*c.* 1730–78), governor of Nova Scotia, 1766–73. It was also known as *Manchester*, before Sheet Harbour post office was opened in 1852.

Sheffield ON A township in Lennox and Addington County, it was named in 1798 after John Baker Holroyd, Lord Sheffield (1734–1821), who served in many British government ministries.

Sheffield ON In the town of Flamborough, Hamilton-Wentworth Region, this place's post office was named in 1837 by the Revd John A. Cornell after Sheffield, England.

Sheho SK Northwest of Yorkton, this village (1905) was named in 1892, after **Sheho Lake**, 3 km to the north, itself named by pioneer settler George Hill after the Cree word for 'prairie chicken'.

Shelburne NS On the South Shore and at the head of **Shelburne Harbour**, this town (1907) was called *Port Razoir* during the French regime, because the V-shaped harbour had the appearance of a partly opened razor. It was called *Port Roseway* by New Englanders (with Roseway River being the present name of the river, rising 45 km to the north in Roseway Lake). In 1783 Gov. John Parr named it after William Petty Fitzmaurice, 1st Marquess of Lansdowne and Earl of Shelburne, 1737–1805, prime minister of Great Britain, 1782–3.

Shelburne ON A town (1976) in Dufferin County, it was first called *Jelly's Corners*. William Jelly wanted to call it Tandragee, after a town in County Armagh, Ireland, where his ancestors had lived. However, at a session of the Legislative Assembly in the city of Québec in 1865, it was named after Henry Petty-Fitzmaurice, 3rd Marquess of Lansdowne and Earl of Shelburne (1780–1863), who had served in several government ministries.

Shellbrook SK A town (1948) just south of a stream called **Shell Brook**, west of Prince Albert, its post office was named *Shell Brook* in 1894 and was changed to Shellbrook in 1948.

Shell Lake SK West of Prince Albert, the post office at the village (1940) and the rural municipality were both named in 1913 after a lake to the southwest. The Canadian Northern Railway arrived in 1927.

Sherbrooke NS Located on St. Marys River, this place was the site of Fort Sainte-Marie, built in the 1650s by the French (and thus providing the name for the river). Postal service was established in 1829, when the office was named after Sir John Coape Sherbrooke (1764–1830), lieutenant-governor of Nova Scotia, 1811–16, and governor-in-chief of British North America, 1816–18.

Sherbrooke PE A municipal community (1983) northeast of Summerside, its school district was named in 1905, possibly after Sherbrooke, NS. Its post office was called *Grangemount* from 1909 to 1912.

Sherbrooke QC This city acquired its present status in the mid-1800s, having been incorporated as a village in 1823. Its post office was opened in 1819, taking its name from Sir John Coape Sherbrooke (1764–1830), governor-in-chief of British North America, 1816–18. His father, William Coape, had added the name Sherbrooke on his marriage to Sarah Sherbrooke. The Abenaki knew the place as *Shacewanteku*, 'place where vapour rises'. The French first called it *Grandes-Fourches*, in reference to the forks of the Saint-François and the Magog.

Sherridon MB Located northeast of Flin Flon and on the east side of Kississing Lake, this place was first called *Kississing*, a Cree word meaning 'cold'. In 1929 the Sherritt Gordon Mining Company opened the zinc and copper mines here, and the place was named after Carl Sherritt and Peter Gordon, who had discovered the property. Its population rose to 2000, but declined after 1951, when the operations were moved to Lynn Lake, and fewer than 200 are now living here.

Sherwood PE Part of the city of Charlottetown since 1 April 1995, it had been created a village in 1960 and a municipal community in 1983. Its post office from 1910 to 1912 was called *Sherwood Station*, after the nearby Sherwood Cemetery, named in 1891, but established as Charlottetown Cemetery before 1880.

Sherwood Park AB A residential area in Strathcona County, east of Edmonton, it was named in 1957 after the Sherwood Forest in Nottinghamshire, England. The name was likely chosen to elicit a sense of woodsy surroundings in a rural landscape.

Sheshatsheit NF On the west side of the mouth of the Naskaupi River and northeast of Happy Valley-Goose Bay, this place was founded by the Innu (Naskapi and Montagnais) in the 1950s, and separated from North West River in 1979. In their language, the name means 'narrow place', a description of the narrows between Grand Lake, a widening of the river, and Lake Melville.

Shilo MB Now the site of a Canadian Forces base east of Brandon, the Canadian Northern Railway station established here in 1905 was named after a Jewish pedlar.

Shippagan NB A town (1958) on **Baie de Shippegan**, east of Caraquet, its post office was called *Shippegan* 1850–2, 1862–1900, and 1955–83, and *Shippigan*, 1852–62, and 1900–55. The preference by Acadians for the spelling Shippagan for the town's name was acknowledged in 1980, but most of the associated names, such as Baie de Shippegan, **Shippegan Gully**, and **Shippegan Channel**, retain the older spelling. The bay's name was derived from Mi'kmaq *Sepaguncheech*, 'duck passage' and was called *Cibaguen* on a 1656 map. Shippegan Gully connects the bay to the Gulf of St. Lawrence.

Shipshaw QC A municipality (1930) north of Jonquière, its present area was established in 1977, when it absorbed the village of *Saint-Jean-Vianney* (1952), which had been wiped out by a landslide six years earlier. Shipshaw post office had been opened in 1904. The name in Montagnais means 'blocked river', probably in reference to an earlier landslide into the **Rivière Shipshaw**. The religious parish of Saint-Jean-Vianney was named in 1928 after St Jean Baptiste Marie Vianney (1786–1859), the famed Curé d'Ars, who was said to have many curative powers, and was canonized in 1925.

Shipton QC A township municipality in the regional county of Asbestos and surrounding the towns of Asbestos and Danville, it was erected in 1845. It took its name from the township, shown on a 1795 map, and proclaimed in 1801. The name possibly came from a place in Buckinghamshire, England, northwest of London.

Shoal Lake MB Located northeast of Minnedosa, this village (1909) was named after the rural municipality of Shoal Lake (1881), which was named after a shallow lake to the south. There was a North-West Mounted Police station at the south end of the lake as early as 1876. When the Manitoba and North

Western Railway was built in 1909, the place was moved to the north end of the lake.

Shubenacadie NS Situated on the **Shubenacadie River** and between Halifax and Truro, this place's post office was named in 1836. Derived from Mi'kmaq *Segubunakade*, the river's name means 'place where ground nuts grow'.

Shulaps Peak BC In the **Shulaps Range** of the Coast Mountains, northwest of Lillooet, this mountain (2772 m) was named in 1913 after the Lillooet word for 'ram' of the bighorn sheep.

Shuniah ON In 1873 the Ontario government united several townships to create the municipal township of Shuniah at the head of Lake Superior. Although it only comprises the townships of McGregor and McTavish now, it extends for 50 km northeast of the city of Thunder Bay. The name may have been derived from Ojibwa *joniiag*, meaning 'silver', possibly in reference to Silver Islet, at the south end of the Sibley Peninsula.

Shuswap Lake BC This lake with its four long arms, north of the district municipality of Salmon Arm, was named after the Shuswap, an Interior Salish tribe. The meaning of the tribal name is obscure. The **Shuswap River** (185 km) rises just south of Revelstoke, but takes a convoluted route south, then north, before entering the lake at Sicamous.

Sibbald, Mount YT In the St. Elias Mountains, this mountain (3063 m) was named in 1970 by the Alpine Club of Canada after Andrew S. Sibbald (1888–1945), president of the club, 1934–8.

Sibley Peninsula ON This prominent peninsula shelters Thunder Bay from the open waters of Lake Superior. It comprises the township of **Sibley**, named in 1873 after Maj. Alexander Sibley, of Detroit, who bought the silver mine at Silver Islet in 1869. *Sibley Provincial Park* became Sleeping Giant Provincial Park in 1992. As seen from the city of Thunder Bay, the peninsula portrays the profile of The Sleeping Giant, with The Head, The Adam's Apple, and The Chest as distinctive features.

Sicamous BC A district municipality (1989) northeast of Salmon Arm, its post office was named in 1887 In Shuswap the name means 'narrow' or 'squeezed in the middle', in reference to the delta of Eagle River having created Mara Lake, rather than leaving it as an arm of Shuswap Lake.

Sidney BC A town (1967) near the north end of the Saanich Peninsula, north of Victoria, it was named in 1892 after **Sidney Island**, offshore in Haro Strait. The island, called *Sallas Island* about 1850 by the Hudson's Bay Company, was renamed in 1859 by Capt. George Richards, likely after Frederick W. Sidney, who was also a surveyor with the British Navy. The origin of the name *Sallas Island* is not known.

Sidney MB Midway between Portage la Prairie and Brandon, this place was named in 1881 by the Marquess of Lorne after Sidney Austin, a correspondent for the London *Graphic*, who followed the progress of the Canadian Pacific Railway to the West.

Sidney ON A township in Hastings County, it was named in 1787 after Thomas Townshend, 1st Viscount Sydney (1732–1800), British home secretary, 1782–9, lord of the Admiralty, 1790–3, and lord of the treasury, 1793–1800.

Sifton MB North of Dauphin, this place was founded in 1898 by the Canadian Northern Railway and named after William M. Sifton, a railway contractor and a resident of Minitonas. He was an uncle of Sir Clifford Sifton (1861–1929), who aggressively promoted the settlement of Western Canada. Sir Clifford objected to the name, so the townsite was registered as *Lemberg*, the capital of Galicia (now L'viv in Ukraine). However, Sifton was not supplanted.

Signal Hill NF This distinctive landmark (153 m) at the entrance of St. John's Harbour had a signal cannon placed on it in the late 1500s. Subsequently flags were flown to warn ships about hazards and bad weather. In 1901

Guglielmo Marconi received the first transatlantic radio signals on this hill from Cornwall, England.

Sikanni Chief River BC Rising in the Muskwa Ranges of the Rocky Mountains, this river flows northeast into the Fort Nelson River, east of the town of Fort Nelson. It was named after Makenunatane, a Beaver prophet who was also known as the Sikanni Chief.

Sillery QC A town in the Communauté urbaine de Québec, between the cities of Québec and Sainte-Foy, it recalls the name of Noël Brulart de Sillery (1577–1640), a French nobleman whose name was given to the first mission here in 1637, because he financially supported its establishment on the St. Lawrence River shore.

Silverton BC A village (1930) on the eastern shore of Slocan Lake, it was laid out in 1893, and named *Four Mile City*, after a nearby creek. Its post office was named in 1894 after Silverton, CO. The latter is in San Juan County, in the southwestern part of the state.

Silverwood NB In the west end of the city of Fredericton and on the south bank of the Saint John River, it was established as a residential area in 1961. Incorporated as the village of Springhill in 1968, it was changed to Silverwood in 1969, and annexed by Fredericton in 1973. Springhill post office was named in 1859 after a residence established by Chief Justice George D. Ludlow in 1786, which he named after the residence of New York Gov. Cadwallader Colden.

Simcoe ON The county was named in 1798 for John Graves Simcoe (1752–1806), Upper Canada's first lieutenant-governor. He was appointed in September 1791 and arrived at the capital, Newark (Niagara-on-the-Lake), the following spring. He left the province in 1796 and resigned two years later. **Lake Simcoe** was named in 1793 after the lieutenant-governor's father, Capt. John Simcoe, who died in 1759, on his way up the St. Lawrence River to take part in the capture of the city of Québec.

Simcoe ON A town (1878) in Haldimand-Norfolk Region, its site was visited in 1795 by Lt-Gov. John Graves Simcoe. Some merchants at the north end of the place wanted to call its post office Wellington in 1829, after the victor at the Battle of Waterloo, but Aaron Culver's proposal to honour Simcoe was adopted. Culver had been granted land here by Simcoe.

Similkameen River BC Rising in Manning Provincial Park, near the Washington State border, this river (84 km in British Columbia) flows north through Princeton, and then passes southeast through Keremeos. It crosses the border west of Osoyoos and joins the Okanogan River in Washington State, just south of the border. It was named after a designation of several Okanogan bands, which came from the extinct language of the Nicola-Similkameen.

Simpson SK South of Watrous, this village (1911) was named in 1910 by the Canadian Pacific Railway after Sir George Simpson (1792–1860), governor of the Hudson's Bay Company, 1821–60. The post office was opened in 1911.

Simpson Pass AB This pass (2107 m) west of Banff, was named in 1859 by Dr James (later Sir James) Hector for Sir George Simpson (1792–1860), the long-serving governor of the Hudson's Bay Company. **Simpson River** drains the area of the pass into the Vermilion River, in British Columbia, and is adjacent to **Simpson Ridge**.

Sintaluta SK On the main line of the Canadian Pacific Railway, east of Regina, this town (1907) was first called *Carson* in 1885. Its post office was renamed Sintaluta three years later, after the Assiniboine word for 'at the end of the fox's tail'. The area was then known as *Red Fox Valley*.

Sioux Lookout ON A town (1912) in Kenora District, northeast of Dryden, it was first called *Knowlton*, when the Grand Trunk Pacific Railway arrived here in 1909. It was renamed *Graham* in 1911 after George P. Graham, minister of railways and canals. Two years later it was changed to Sioux Lookout, in memory of a defeat of the Sioux by the Ojibwa in the late 1700s or early 1800s.

Sioux Narrows ON A municipal township in Kenora District, southeast of Kenora, it is centred on a narrow passage separating Regina Bay from Whitefish Bay. Named in 1938, it commemorates a victory by the Ojibwa over the Sioux.

Sir Alexander, Mount BC In the Rocky Mountains, east of Prince George, this mountain (3270 m) was named in 1914 by alpinist Samuel Prescott Fay after Sir Alexander Mackenzie (1764–1820), who had completed the first crossing of North America, north of Mexico, in 1793.

Sir Donald, Mount BC In the **Sir Donald Range** of the Selkirk Mountains, south of Rogers Pass, this peak (3297 m) was originally called *Syndicate Mountain*, after the financiers who had ensured that the Canadian Pacific Railway's bonds were covered during the building of the CPR through the mountains. By order in council in 1887, it was renamed after Sir Donald Alexander Smith (1820–1914), who was elevated to the peerage as 1st Baron Strathcona and Mount Royal in 1897.

Sir Douglas, Mount AB, BC At the southeastern end of Banff National Park, this mountain (3406 m) was named in 1916 by the Alberta-British Columbia Boundary Commission after Gen. Sir Douglas Haig (1861–1928), commander of the British Expeditionary force in France, 1915–19.

Sir Francis Drake, Mount BC On the east side of Bute Inlet, northeast of Campbell River, this mountain (2695 m) was named in 1939 by surveyor R.P. Bishop after Sir Francis Drake, who had navigated up the west coast of North America, possibly as far as Vancouver Island, in 1579.

Sir John Abbott, Mount BC In the Premier Range, west of Valemount, this mountain (3411 m) was named in 1927 after Sir John Joseph Caldwell Abbott (1821–93), prime minister of Canada, 1891–2.

Sir John Thompson, Mount BC This mountain (3250 m) in the Premier Range, west of Valemount, was named in 1927 after Sir John Sparrow Thompson (1844–94), prime minister of Canada, 1892–4.

Sir Mackenzie Bowell, Mount BC West of Valemount in the Premier Range, this mountain (3280 m) was named in 1927 after Sir Mackenzie Bowell (1823–1917), prime minister of Canada, 1894–6.

Sir Robert, Mount BC In the Coast Mountains, northeast of Terrace, this mountain (2407 m) was named in 1916 after Sir Robert Laird Borden (1854–1937), prime minister of Canada, 1911–20.

Sir Sandford, Mount BC In the **Sir Sandford Range** of the Selkirk Mountains and south of Rogers Pass, this mountain (3530 m) was named in 1902 by surveyor Arthur O. Wheeler after Sir Sandford Fleming (1827–1915), engineer-in-chief of the Canadian Pacific Railway. He supervised the surveys for an intercontinental rail line through both the Kicking Horse and Yellowhead passes.

Sir Wilfrid Laurier, Mount BC The highest peak in the Premier Range, west of Valemount, this mountain (3520 m) was named in 1927 after Sir Wilfrid Laurier (1841–1919), prime minister of Canada, 1896–1911.

Siva Mountain BC In the Pantheon Range of the Coast Mountains, northeast of Mount Waddington, this mountain (2981 m) was named in 1964 after the third deity of the Hindu triad. Powers of reproduction and dissolution are attributed to him.

Skeena River BC Rising in the Skeena Mountains, north of Hazelton, this river (579 km) was identified as *Ayton's River* in 1788 by Capt. Charles Duncan. It was also known as *Simpson's River*, after Capt. Æmilius Simpson, and *Babine River*, after a tributary of the Skeena.

The river's name is derived from the Tsimshian for 'water out of the clouds'.

Skidegate BC On **Skidegate Indian Reserve** and on the southeastern coast of Graham Island, one of the Queen Charlotte Islands, this place was noted as early as 1805 by Samuel Patterson as *Skittagates*. In the 1790s a principal chief of the Haida was called Skiteiget, whose name meant 'red paint stone'. The bay to the southwest is called **Skidegate Inlet** and **Skidegate Channel** separates Graham Island from Moresby Island.

Skihist Mountain BC In the Coast Mountains, south of Lillooet, this mountain (2972 m) was named about 1895 after the Thompson Salish word *sk-haest*, 'split rock'.

Skinners Pond PE Northwest of Alberton, this place's post office was opened in 1856. It took its name from the sea-level pond, which may have been named after a Capt. Skinner, who was shipwrecked here, or derived from *Étang des Peaux*, 'skin pond'.

Skookumchuck BC Located in the Kootenay River valley, north of Cranbrook, this place's post office was opened in 1915. In Chinook jargon the name means 'swift water'.

Skookum Jim, Mount YT This mountain, in the Ogilvie Mountains, north of Dawson, was named in 1973 on the seventy-fifth anniversary of the discovery of gold on Bonanza Creek by Skookum Jim Mason (d. 1916), George Carmack, and Tagish (later Dawson) Charlie.

Slade, Mount BC In the Purcell Mountains, west of Invermere, this mountain (3218 m) was named *Slade Mountain* in 1914 after a prospector associated with the Parradice mine. Three years earlier climber E.W. Harnden had called it *Boulder Peak*. It was renamed Mount Slade in 1960.

Slaggard, Mount YT In the St. Elias Mountains, this mountain (4663 m) was named in 1958 after Joseph R. Slaggard, a prospector, who discovered copper on the nearby White River.

Slave Lake AB At the point where the Lesser Slave River drains from the eastern end of Lesser Slave Lake, the post office at this town (1965) was called *Sawridge* until 1922, when it was changed to Slave Lake.

Slave River AB, NT This river (415 km), part of the Mackenzie River system, drains Lake Athabasca north to Great Slave Lake. It was named after the Slavey, a Dene tribe, which has several communities around the lake and in the valleys of the Slave, Mackenzie, Hay, and Liard rivers.

Slocan BC At the south end of **Slocan Lake**, this village (1958) was an incorporated city from 1901 to 1958. Its post office was opened as *Slocan City* in 1896, but it was changed that year to Slocan. The Canadian Pacific Railway continued to use Slocan City after 1958. **South Slocan** is at the mouth of the **Slocan River**, where it flows into the Kootenay River, northeast of Castlegar. The community of **Slocan Park** is 10 km upriver from the mouth. In Okanagan *slocan* means 'pierce in the head', in reference to the practice of spearing salmon.

Smallwood Reservoir NF Described as the third largest man-made lake (6527 km^2) in the world, it was created in 1965 after the Churchill River was dammed at Churchill Falls, and several other rivers were diked. It was named after Joseph Roberts Smallwood (1900–91), the first premier of the province of Newfoundland, 1949–72, whose vision led to the construction of the huge Churchill Falls hydroelectric power project in Labrador.

Smeaton SK Northwest of Nipawin, this village (1944) was named in 1930 by the Canadian Pacific Railway after Senator R. Smeaton White. Its post office was opened in 1931.

Smith ON A township in Peterborough County, it was named in 1821, likely for Samuel Smith (1756–1826), a member of the Executive

Council from 1813, and administrator of Upper Canada during brief periods in 1817 and 1820. It is less likely that it was named for Sir David William Smith (Smyth) (1764–1837), surveyor general of Upper Canada until 1802, who returned to England in 1805.

Smithers BC A town (1967) on the Bulkley River, northeast of Prince Rupert, it was founded in 1913 by the Grand Trunk Pacific Railway and named after Sir Alfred Waldron Smithers (1850–1924), chairman of the railway's board of directors.

Smithfield ON In Northumberland County, west of Trenton, this place was named in 1853 after first postmaster Abijah Smith (1789–1870).

Smiths Falls ON Loyalist Thomas Smyth (1768–1832) received large grants in the upper Rideau Valley after the American Revolution. He built a sawmill at *Smyth's Falls* in 1823, but neither he nor any member of his family ever lived here. He lost his land in 1825 through a foreclosure. The growing community was briefly called *Wardsville*, after miller Abel Russell Ward (1796–1882). Rideau Canal plans in 1827 had Smiths Falls, and this spelling, without an apostrophe, was subsequently adopted locally. It became a separated town in 1902.

Smiths Cove NS East of Digby, this place was first called *Lower Clements*, after Clements Township. It was named Smiths Cove in 1856 after Loyalist Jeremiah Smith.

Smith Sound NT At the north end of Baffin Bay and between Ellesmere Island and Kalaallit Nunaat (Greenland), this water body was named in 1616 by William Baffin and Robert Bylot after Sir Thomas Smith (1558–1625). A London merchant, Sir Thomas was the first governor of the Company of Adventurers for the Discovery of the North-West Passage.

Smithville ON A community in Niagara Region, west of St. Catharines, it was first called *Griffintown* by Richard Griffin, who settled here in 1786. His son, Smith, named the place in 1831, in memory of his mother, whose maiden name was Mary Smith.

Smokey, Cape NS A dominating elevation on the east side of Cape Breton Island and just south of Ingonish, it was identified as *Cap Enfumé* on maps of 1600 and 1602, and on Samuel de Champlain's 1632 map. The name was anglicized in the late 1700s.

Smoky Lake AB A town (1962) northeast of Edmonton, its post office was opened in 1909, taking its name from a nearby lake. Known in Cree as *Kaskapite Sakigan*, 'smoky lake', because vapours often obscure the opposite shore, the lake was identified in 1810 by Alexander Henry the younger as *Lac qui Fume*.

Smoky River AB Rising east of Mount Robson, this river (380 km) flows northeast to join the Peace at the town of Peace River. Known as *Kaskapite Sipi*, 'smoky river', it received its name from smouldering coal beds on its banks in the Grande Cache area.

Smooth Rock Falls ON A town (1929) in Cochrane District, southeast of Kapuskasing, it was founded in 1917 and named after a large rock in the Mattagami River, which was washed smooth after a dam had been constucted here in 1915.

Snafu Creek YT A tributary of Lubbock River and between Jakes Corner and Atlin, BC, this creek was named in 1949–50 by Canadian Army engineers after the Second World War phrase 'Situation Normal, All Fouled Up'. *See also* Tarfu Creek.

Snag YT This locality, southwest of Dawson and on **Snag Creek**, a tributary of the White River, was named in 1942 when a landing strip and a weather station were located here. The creek had been named in 1899 by geologist Alfred Brooks and topographer William Peters of the United States Geological Survey. Canada's coldest temperature, -81°F, or -62.8°C, was recorded here on 3 February 1947.

Snelgrove ON In the city of Brampton, Peel Region, north of the city centre, this community was first called *Edmonton* in 1851 after the district in north central London, England. To avoid confusion with the new settlement that would become the capital of Alberta, it was renamed in 1859 after J.C. Snell, who won a contest by getting the most mail during a certain period.

Snowbird Lake NT Near the sixtieth parallel and southeast of Great Slave Lake, this lake (490 km^2) was possibly named in 1894 by geologist Joseph Burr Tyrrell. He explored the rivers and lakes from the Saskatchewan River northeast to near present-day Arviat on the Hudson Bay coast.

Snow Dome AB, BC On the continental divide and at the centre of the Columbia Icefield, this summit (*c.* 3520 m) was first named *Dome* by British climber Norman Collie in 1899. In 1819 the Alberta-British Columbia Boundary Commission determined it was the hydrographic apex of North America, with waters draining from it to the Pacific, the Arctic, and the Atlantic.

Snow Lake MB Located east of Flin Flon, this town (1976) was founded in 1945 by Howe Sound Exploration Company Ltd to develop the gold deposits, which were discovered by Lew Parres in 1927 at **Snow Lake**. He named the lake because he found its water as soft as melted snow.

Snow Road Station ON In Frontenac County, north of Sharbot Lake, this place's post office was named in 1890 after the road constructed through the area in 1859, which had been surveyed two years earlier by John Allen Snow (1824–88). *Snow Road* was a station on the Kingston and Pembroke Railway from 1883 to 1965, and the tracks were lifted five years later. However, over thirty years later the place is still known as Snow Road Station.

Sointula BC Located on Malcolm Island on the northeastern side of Vancouver Island, this place was founded by the Kalevan Kansa Colonization Company in 1901. It was designed as a Finnish co-operative community, with its name in Finnish meaning 'harmony'. Within four years the venture failed, because of poor business practices, but the place remained occupied.

Sombra ON A township in Lambton County, it was named in 1822 by Lt-Gov. Sir Peregrine Maitland after the Spanish word for 'shade', because he found the area heavily wooded. **Sombra** post office was opened in 1851.

Somerset MB A village (1962) northwest of Morden, it was named in 1881 by settlers from Somerset, England. There is a tall story that the name really came from an incident where a Mr Stevenson was somersaulted into a creek from his democrat, a double-seated carriage. The creek was then called *Somersault Creek*, which eventually became **Somerset Creek**.

Somerset Island NT North of Boothia Peninsula and between Prince of Wales Island and Brodeur Peninsula of Baffin Island, this island (24,786 km^2) was named *North Somerset* in 1820 by William (later Sir William) Parry, after his native county in England. The Geographic Board amended it to its present name in 1905.

Somervell, Mount BC In the Clemenceau Icefield of the Rocky Mountains, southeast of Athabasca Pass, this mountain (3120 m) was named in 1927 by alpinist Alfred J. Ostheimer III after Dr Theodore Howard Somervell (1890–1975), who was a British climber in the Himalayas as well as a physiologist.

Sonora Island BC Between Vancouver Island and the mainland, north of Campbell River, this island was named in 1903 by Capt. John Walbran after the Spanish schooner *Sonora*, which was under the command in 1775 of Juan Francisco de la Bodega y Quadra, who undertook an exploratory voyage along the northwestern coast. Until 1903 the island was considered part of *Valdes Island*, which was then divided among Sonora, Maurelle, and Quadra islands.

Sooke BC On **Sooke Inlet**, southwest of Victoria, this community's post office was named in 1864. It took its name from the Sooke tribe of the Coast Salish, which was almost annihilated in 1848 by neighbouring tribes. The meaning of their name is not available in published records.

Sophiasburgh ON A township in Prince Edward County, it was named in 1788 for Princess Sophia (1777–1848), fifth daughter and twelfth child of George III.

Sorcerer Mountain BC In the Selkirk Mountains, northwest of Rogers Pass, this mountain (3166 m) was named in 1907 by surveyor Percy Carson. In 1900 botanist Charles Shaw had called it *The Hub of the Big Bend*. Carson did not explain what had possessed him to give the name.

Sorel QC Situated at the junction of the Richelieu with the St Lawrence, this city's site was called Fort Richelieu in 1642. In 1792, Gov. Frederick Haldimand issued the municipal charter for *William Henry*, after buying the ancient seigneury of Saurel (Sorel). Five years earlier Prince William (who became William IV in 1830) had visited the province. In 1814 the post office was called *William Henry*, which became the town's name in 1848. It was renamed Sorel in 1860, and its post office adopted it two years later. The seigneury was granted in 1672 to Pierre de Saurel (1628–82), an officer with the Carignan-Salières Regiment. Within a few years the spelling Sorel superseded the original form.

Sorrento BC Located on Shuswap Lake, northwest of Salmon Arm, this place was established in the early 1880s by J.R. Kinghorn, an eastern businessman, in anticipation of the arrival of the Canadian Pacific Railway. The view over the lake reminded him of the view of the Isle of Capri from Sorrento, Italy, where Kinghorn had gone for his honeymoon. The railway bypassed it, but when the post office was opened in 1913, it was called Sorrento.

Souris MB A town (1903) at the junction of Plum Creek and **Souris River**, southwest of Brandon, its post office was called *Plum Creek* before 1882, when it was renamed after the river. The river rises northwest of Weyburn, SK, crosses into North Dakota south of Carnduff, SK, flows back into Canada south of Melita, MB, and empties into the Assiniboine, 35 km southeast of Brandon. Alexander Henry the younger recorded it as *Rivière à la Souris* in 1799, and geographer David Thompson noted it as *Mouse Rivulet* in 1814. In North Dakota it has been commonly called Mouse River. *Souris* means 'mouse' in French.

Souris PE A town (1910) near the east end of the province, its post office was opened about 1830. It was called *Souris East* from 1867 to 1967, when the original form was restored. Although it was claimed as early as 1816 that the adjacent **Souris Harbour** and **Souris River** were named after plagues of mice (*souris*), and French maps of the mid-1700s called the harbour *Havre à la Souris*, it might be more likely that the harbour may have been called *Havre de l'Échourie*, 'barred harbour', an apt description of the harbour. During the French regime, present South Lake, east of Souris, was identified as *Havre de l'Échourie*. The municipal community of **Souris West**, on the west side of the mouth of the Souris River, was established in 1972.

Southampton NS Between Amherst and Parrsboro, this place's first post office was called *Lawrence Factory* in 1873. It was renamed three years later, possibly after Southampton, England.

Southampton ON At the mouth of the Saugeen River, this town (1904) in Bruce County was first called *Saugeen* in 1851. Later that year, James Hervey Price, commissioner of Crown lands, named it after the seaport city on England's south coast.

Southampton Island NT At the north end of Hudson Bay, this island (41,214 km^2) was named in 1631 by explorer Luke Foxe after Henry Wriothesley, 3rd Earl of Southampton (1573-1624), a politician and soldier.

South Brook NF A town (1965) at the head of Halls Bay, a southwesterly extension of Notre Dame Bay, it is located at the outlet of

South Brook, which rises just north of Millertown Junction. Between 1945 and 1965, South Brook and Springdale had been incorporated as a rural district.

South Burgess ON Burgess Township was named in 1798, possibly for James B. Burgess, a British member of Parliament, who voted for the Canada Bill in 1791. In 1841, that part of the township south of Big Rideau Lake was united with Bastard Township in the united counties of Leeds and Grenville. *See also* North Burgess.

South Buxton ON In Kent County, south of Chatham, this place was founded in 1849 as a settlement of American blacks, and named *Elgin Settlement*, after Lord Elgin, the governor of the province of Canada. It was renamed in 1874 after Sir Thomas Fowell Buxton, a British member of Parliament, who promoted the passing of the 1833 Emancipation Act. *See also* North Buxton.

South Cayuga ON Part of the town of Haldimand in Haldimand-Norfolk Region since 1974, the township was named in 1835, after the Cayuga tribe of the Six Nations confederacy. The name has many interpretations, including 'place where the canoes are drawn out', 'mountain rising from the water', and 'advising nation'.

South Crosby ON Crosby Township was created in 1798, and probably named after Brass Crosby, Lord Mayor of London, member of Parliament, and a friend of Lt-Gov. John Graves Simcoe. South Crosby was separated from North Crosby in 1806, with both remaining in the united counties of Leeds and Grenville.

South Dorchester ON Dorchester Township, created in 1798 in Middlesex County, was named after Sir Guy Carleton, Baron Dorchester (1724–1808), governor of Quebec, 1768–78, commander-in-chief of British forces in North America, 1782–6, and governor-in-chief of British North America, 1786–96. It was divided in 1851, with North Dorchester remaining in Middlesex County and South Dorchester transferring to Elgin County.

South Dumfries ON A township in Brant County, it is the south half of Dumfries Township, named in 1816 by Niagara merchant William Dickson after his birthplace of Dumfries, in Dumfriesshire, Scotland.

South Easthope ON Easthope Township in Perth County was named in 1830 for Sir John Easthope (1784–1865), a director of the Canada Company, a British company established in 1826 to develop the Huron Tract, a large area of what is now Southwestern Ontario. It was divided into North Easthope and South Easthope townships in 1843.

South Elmsley ON Elmsley Township was named in 1798 after Upper Canada's chief justice John Elmsley (1762–1805). In 1841 that part of the township south of the Rideau River was attached to the united counties of Leeds and Grenville, and North Elmsley Township was transferred to Lanark County.

Southern Harbour NF On the northeastern shore of Placentia Bay and south of Arnold's Cove, the bay at this town (1968) was identified in 1770 by hydrographer Michael Lane as **Southern Harbour**.

Southesk, Mount AB This mountain (3120 m) in Jasper National Park, east of Maligne Lake, was named in 1859 after James Carnegie, 9th Earl of Southesk (1827–1905). He travelled through the Rocky Mountains at the time that the Palliser expedition was undertaking its extensive surveys in Western Canada.

Southey SK North of Regina, this town (1980) was named in 1905 by a Mr Chandler after English poet, author, and critic Robert Southey (1774–1843), whose two-volume *The Life of Nelson* remains a classic of historical writing. The town's main street is called Keats, with other streets honouring Burns, Byron, Cowper, and Browning. The Canadian Pacific Railway arrived in 1907.

South Fredericksburgh ON Fredericksburgh Township was named in 1784, likely after Prince Augustus Frederick (1773–1843), sixth son of George III. Ernestown Township, to the east, was called after his next older brother, and Adolphustown Township, to the west, was named in honour of his next younger brother. As Prince Frederick, the second son, had been commemorated in Osnabruck Township, and the townships named after the 12 royal children were given in chronological order, it is unlikely this township was named after him. It was divided in 1858 into North Fredericksburgh and South Fredericksburgh townships in Lennox and Addington County.

South Gloucester ON Near the south boundary of the city of Gloucester, Ottawa-Carleton Region, its post office was named in 1852.

South Gower ON Gower Township was created in 1798 in Grenville County and was divided in 1850, with South Gower Township becoming part of the united counties of Leeds of Grenville. The original township was named after John Levenson Gower (1740–1792), lord of the Admiralty from 1783 to 1789. *See also* North Gower.

South Henik Lake NT West of Arviat, this lake (512 km^2) was named on an eight-mile map sheet, produced in 1942. That year anthropologist Diamond Jenness proposed *Henik*, a contraction of the Inuktitut *Henningyouyouak*, 'the large lake that lies over to one side'. The Hydrographic and Map Service of the department of mines and resources perceived there were two lakes, calling them South and **North Henik** lakes. The Geographic Board approved the two names in 1945.

South Marysburgh ON Marysburgh Township in Prince Edward County was named in 1786 for Princess Mary (1776–1857), fourth daughter of George III, and the wife of the Duke of Gloucester, her first cousin. South Marysburgh was separated from North Marysburgh in 1871.

South Monaghan ON A township in Peterborough County, it was originally part of Monaghan Township, named in 1820 after County Monaghan, in north central Ireland. *See also* North Monaghan.

South Mountain ON A community south of Mountain and in Mountain Township, united counties of Stormont, Dundas and Glengarry, northwest of Morrisburg, its post office was named in 1851.

South Nahanni River NT Rising in the Mackenzie Mountains, this river (563 km) flows southeast into the Liard River, 150 km upriver from Fort Simpson. It was named after the Nahanni, an Athapascan tribe, whose name means 'people of the west'. **Nahanni National Park** (4,766 km^2) was established in 1972. It extends for 210 km on both sides of the river.

South Nation River ON Rising near Brockville, this river flows northwest to the Ottawa River. Known as the *Rivière de la Petite Nation* during the French regime, it was called *Petite Nation River* and *Nation River* during the 1800s. In 1908 it was called South Nation River to distinguish it from the Rivière de la Petite Nation, which flows south into the Ottawa, opposite the mouth of the South Nation. However, *Nation River* is preferred locally.

South Ohio NS North of Yarmouth, this place's post office was named in 1873. It and two other communities called **Ohio** in the province owe their names to the movement in the 1820s to the Ohio Territory, west of the Allegheny Mountains. As migration to the territory took place, there were simultaneous moves into the back country of the province, where their new lands were called Ohio.

South Plantagenet ON The original Plantagenet Township was named in 1798 after the royal family surname from 1154 to 1485. This municipal township was separated from the municipal township of North Plantagenet in 1850, and both remained in the united counties of Prescott and Russell.

South Porcupine ON At the south end of **Porcupine Lake** and in the city of Timmins, east of the city centre, this urban centre was founded in 1909. It was destroyed by fire in 1911 and subsequently rebuilt. The lake was named after an island shaped like a porcupine in the nearby **Porcupine River**.

Southport PE In the town of Stratford, across the Hillsborough River from Charlottetown, its post office served the area from 1860 to 1967. Created as a municipal community in 1983, it was united with Bunbury, Cross Roads, and Keppoch-Kinlock to form the town of Stratford in 1995.

South River NF A town (1966) at the head of Bay de Grave and on the west side of Conception Bay, it is located at the outlet of the **South River**. It was known as *Southern Gut* until the beginning of the twentieth century.

South River ON A village (1882) in Parry Sound District, its post office was named in 1882 after the **South River**, which flows north into the South Bay of Lake Nipissing.

South Saskatchewan River AB, SK The Bow and Oldman rivers unite 70 km west of Medicine Hat and become the South Saskatchewan. It flows generally east through Diefenbaker Lake, and north through Saskatoon to unite with the North Saskatchewan to form the **Saskatchewan River**. In 1814 geographer David Thompson labelled the entire South Saskatchewan as Bow River. *See also* Saskatchewan River.

South-West Oxford ON A township in Oxford County, it was created in 1975 by amalgamating the village of Beachville and the municipal townships of West Oxford and Dereham.

Spalding SK A village (1924) south of Melfort, its post office was named in 1906 by Mrs J.W. Hutchison after Spalding, Lincolnshire, England, north of Peterborough. When the Canadian Pacific Railway was extended north from Watson in 1921, the post office was moved 3 km southwest to the rail line.

Spallumcheen BC A district municipality (1892) north of Vernon, surrounding the city of Armstrong, it was named after the *Spallumcheen River*, a former name of the Shuswap River. The name was derived from Okanagan *spalmtsin*, 'flat area along the edge'. In the form of **Spillamacheen**, the name was transferred east to the area of the Purcell Mountains, where it identifies a river, a mountain, a range, and a glacier.

Spaniards Bay NF Located at the head of **Spaniard's Bay** and on the western side of Conception Bay, this town (1965) was named after the bay, noted by pioneer settler John Guy in 1610. The name may have been given by English fishermen, who sometimes identified Portuguese and Basque fishermen as Spaniards.

Spanish ON In Algoma District, west of Espanola, this place's post office was first called *Spanish River Station* in 1887. The river had been named in the 1820s by Admiralty surveyor Henry W. Bayfield. There are many legends about the origin of the name, most involving the Ojibwa capturing a Spaniard near the Gulf of Mexico, resulting in a family called Espaniel being given the river as its hunting grounds. *See also* Espanola *and* The Spanish River.

Spanish Bank BC On the northwestern side of Vancouver and at the outer limits of Burrard Inlet, it was here on 22 June 1792 that Capt. George Vancouver met the Spanish sea captains Dionisio Alcalá Galiano and Cayetano Valdés. *See also* English Bay.

Sparta ON In Elgin County, southeast of St. Thomas, this community was named in 1834 after Sparta in north central New Jersey, which recalls the Greek city state, noted for its austerity and strictness. Its post office was opened in 1841.

Sparwood BC In the valley of the Elk River and extending up Michel Creek toward the Alberta border, this district municipality was established in 1966 by amalgamating Sparwood, Natal, *Michel*, and *Middletown*. Sparwood's post office was opened in 1903, taking its name from the good quality of spars obtained from the trees here for the construction of the Crows Nest Pass Railway.

Spear, Cape NF The most easterly point of Canada, it extends into the Atlantic east of St. John's. It was named in Portuguese *Cauo de la spera*, 'place of waiting'. Identified on an 1505–8 map, it may have been the point where Portuguese vessels waited for the lost expeditions of Gaspar and Miguel Corte-Real. It became an important rendezvous point in the sixteenth century for explorers and fishermen on the Grand Banks.

Spencerville ON A community in the united counties of Leeds and Grenville, north of Prescott, it was first known as *Spencer's Mills* after miller Peleg Spencer. His son David surveyed village lots in 1841, when the post office was opened.

Spences Bridge BC On the Thompson River, northeast of Lytton, this place was first called *Cook's Ferry*, after Mortimer Cook, who started the ferry here in 1862. Three years later a toll bridge was built by Thomas Spence. The *Spence's Bridge* colonial post office opened in 1868, followed by a Canadian office in 1872. The apostrophe was dropped in the 1950s.

Spirit River AB A town (1951) north of Grande Prairie, its post office was named in 1920, near the present site of Rycroft on the Spirit River, 7 km to the east. It was moved to its current location in 1920 when the Edmonton, Dunvegan and British Columbia Railway was built that far. The river was known in Cree as *Chepe Sipi*, 'ghost river', and *Ghost River* appeared on some early maps of the Peace River country.

Spiritwood SK A town (1965) west of Prince Albert, its post office was named in 1923 by postmaster Rupert Dumond, after his home town east of Jamestown, ND.

Split, Cape NS Extending into the Minas Channel, this cape had been called *Cap Fendu*, 'split cape', during the French regime. Its name may have been the ultimate source of the name 'Fundy'.

Springdale NF A town (1965) on the western side of Halls Bay, a southwesterly extension of Notre Dame Bay, it was settled in the 1870s. It was first known as *Little Wolf Cove* and *Island Rock Cove*, and the two places became Springdale in 1898, when the post office was opened.

Springfield ON A village (1877) in Elgin County, east of St. Thomas, it was named in 1863 after the many springs in the fields near the head of Catfish Creek.

Springhill NS Southeast of Amherst, this town (1889) was first settled in the 1820s. Its post office was opened in 1852 and named after the springs on its hillside. Coal had been discovered there in 1834, but mining was not undertaken in a large scale until 1873.

Spring-Rice, Mount AB, BC On the continental divide and south of the Columbia Icefield, this mountain (3275 m) was named in 1918 by the Alberta-British Columbia Boundary Commission after Sir Cecil Arthur Spring-Rice (1859–1918), British ambassador to Washington, 1913–18.

Springside SK Located northwest of Yorkton, the post office in this town (1985) was named in 1904 after the good quality of spring water in the area.

Springwater ON A municipal township in Simcoe County, it was created in 1994 by merging the village of Elmvale and the municipal townships of Flos and Vespra. A contest had been held to choose a suitable name, with the bland Springwater being accepted, because no one objected to it.

Spruce Grove AB A city (1986) west of Edmonton, its post office was named in 1894 after a nearby grove of spruce trees. Its Grand Trunk Pacific Railway station was established in 1910, slightly to the west.

Spryfield NS In the south side of Halifax, this place's post office was named in 1882. It owes its name to Capt. William Spry, who acquired 607 ha (1500 a) of land in the area in the 1760s. On returning to England in 1783, he advertised the sale of 'Spryfield Farm'.

Spuzzum BC In the Fraser Canyon, between Yale and Boston Bar, this place's name was derived from the Thompson Salish *spozem*, 'little flat (land)'. Its post office was named in 1897.

Spy Hill SK A village (1910) east of Esterhazy, its post office was named in 1888 after a nearby hill used by the Cree to keep a watch on enemies approaching from various directions. A legend has a spying Sioux being discovered and killed near the hill, with his body being placed on the hill for the birds and animals to devour, resulting in the name *Kappa Gammaho*, 'spy hill'.

Squamish BC A district municipality (1972) at the mouth of **Squamish River**, where it flows south into Howe Sound, its post office was named in 1892. The office was renamed *Newport Beach* in 1912 to provide the place with a more suitable name. When the change was not accepted, the Pacific Great Eastern Railway held a contest in 1914 to select a better name. The winning name was Squamish. The river's name was derived from the Squohomish, a Coast Salish tribe, whose name possibly means 'strong wind'.

Stafford ON A township in Renfrew County, it was named in 1843 after Staffordshire in England, the home county of Sir Charles Bagot (1781–1843), governor of the province of Canada, 1841–3. *See also* Admaston *and* Bromley.

Standard AB A village (1922) east of Calgary, it was named in 1911 by the Canadian Pacific Railway, possibly after the Danish flag, the oldest European 'standard', dating from the thirteenth century. A Danish settlement had been established here two years earlier. Its post office was opened in 1912.

Stand Off AB Located on the Blood Indian Reserve, southwest of Lethbridge, it was named *Fort Stand Off* in 1871 by American whisky traders, who claimed they could not be pursued and charged by a United States marshal. The name was retained in 1882 by the North-West Mounted Police. Its post office was opened in 1896.

Stanhope PE North of Charlottetown, and adjacent to Prince Edward Island National Park, this place's post office was called **Stanhope by the Sea** in 1961. It took its name from *Stanhope Cove*, the name given in 1765 to present Covehead Bay by Samuel Holland. Holland had called it after William Stanhope, Viscount Petersham (1719–79).

Stanley NB A village (1966) northwest of Fredericton, it was named in 1834 by the New Brunswick and Nova Scotia Land Company after Edward George Geoffrey Smith Stanley, 14th Earl of Derby (1799–1869). One of the organizers of the company, he was prime minister of Great Britain three times. Its post office was opened in 1847.

Stanley ON A township in Huron County, it was named in 1835 after Edward Geoffrey Smith Stanley, 14th Earl of Derby (1799–1869), then a joint secretary of state for the colonies. Lord Derby had three brief terms as prime minister. His son, Lord Stanley of Preston, was governor general of Canada, 1888–93. *See also* Port Stanley.

Stanley Baldwin, Mount BC In the Premier Range, west of Valemount, this mountain (3249 m) was named in 1927 by Dr John D. MacLean, acting premier of the province, after Stanley Baldwin, 1st Earl Baldwin of Bewdley (1867–1947). Prime minister of Great Britain during three terms between 1924 and 1937, he toured British Columbia in 1927.

Stanley, Mount BC In the Clemenceau Icefield of the Rocky Mountains, southeast of Athabasca Pass, this mountain (3090 m) was named in 1927 by alpinist James Monroe Thorington after Henry Morton Stanley (1841–1904), an American newspaper reporter who found explorer David Livingstone in Africa in 1871.

Stanstead QC A town on the Vermont border, it was created in 1995 when the town (1957) of Rock Island, the village (1857) of Stanstead Plain and the village (1873) of Beebe Plain were amalgamated. The township municipality of **Stanstead** was incorporated in 1845.

The township, shown on a 1795 map, and proclaimed in 1800, was possibly called after Stanstead, Suffolk, England.

Starbuck MB On the La Salle River, west of Winnipeg, this community was named by a Canadian Northern Railway contractor in 1885 after Starbuck, MN, northwest of Minneapolis.

Star City SK A town (1921) midway between Melfort and Tisdale, its post office was named in 1902 after postmaster Walter Starkey, who homesteaded nearby in 1900.

Stavely AB Between Nanton and Claresholm, south of Calgary, this town was first called *Oxley*, after the Oxley Ranching Company, established here in 1882 by Alexander Staveley Hill, judge advocate of the British fleet. The company was named after Oxley Manor, Wolverhampton, West Midlands, England. When the town was incorporated in 1912, the second 'e' was dropped. Its post office had been opened in 1903.

Stayner ON An urban area in Simcoe County, southeast of Collingwood, its post office was called *Nottawasaga Station* in 1855, and it was incorporated as the village of *Dingwall* two years later. Both were renamed in 1864 after Thomas Stayner, deputy postmaster general, 1848–9, who donated land for a church. His son, Sutherland Stayner, purchased land in the area. Stayner was a town from 1888 to 1994, when it was united with the village of Creemore and the municipal townships of Nottawasaga and Sunnidale to form the municipal township of Clearview.

Steele, Mount YT In the St. Elias Mountains, this mountain (5072 m) was named in 1909 by surveyor James J. McArthur after Maj.-Gen Samuel (later Sir Samuel) Benfield Steele (1849–1919). Steele began his career with the North-West Mounted Police in 1874 and commanded the NWMP posts in the territory during the Klondike Gold Rush.

Steeves, Mount NB South of the Nepisiguit River, this mountain was named in 1964 by Arthur F. Wightman, then the provincial names authority, for William Henry Steeves (1814–73). He was a Father of Confederation, and one of the original 12 senators from New Brunswick.

Stefansson Island NT On the northeast side of Victoria Island, this island (4,463 km^2) was recognized in 1947 by a Royal Canadian Air Force aerial survey as being separated from Victoria Island by Goldsmith Channel. It was named in 1952 after explorer Vilhjalmur Stefansson (1879–1962), who had investigated Victoria Island in 1910–11 in search of Inuit settlements. In 1917 explorer Storker Storkerson named *Leffingwell Island* in this area, after Ernest de Koven Leffingwell, joint leader of the 1906–7 Leffingwell-Mikkelson Arctic expedition.

Steinbach MB A town (1946) southeast of Winnipeg, it was founded by Mennonite immigrants in 1874 and named after their former home in Ukraine. In German the name means 'stone brook'.

Stellarton NS A town (1889) south of New Glasgow, its postal way office was called *Albion Mines* in 1842, after coal mining had started here in 1827. It was renamed on 1 February 1870 after the top quality Stellar coal mined here.

Stephen ON A township in Huron County, it was named in 1835 after James Stephen Jr, who was then British under-secretary of state for the colonies.

Stephen, Mount BC In Yoho National Park, southeast of Field, this mountain (3199 m) was named in 1886 after George Stephen (1829–1921), president of the Canadian Pacific Railway, 1880–8. Created a baronet in 1886, he was promoted to the peerage in 1891, when he took the title of Baron Mount Stephen.

Stephenville NF Located on the north side of St. George's Bay, and on the west coast of the island of Newfoundland, this town (1952) was named in the early 1870s after St Stephen's Catholic Church, given by Catholic missionary

Fr Thomas Sears after Stephen Le Blanc (also called White), the first child baptized here. It was earlier called *Indian Head*. The town (1958) of **Stephenville Crossing** is located to the east, where the former railway crossed the mouth of the St. George's River.

Stettler AB A town (1906) east of Red Deer, its post office was named *Blumenau* in 1905, as it was the centre of a German-Swiss colony. Later that year, the Canadian Pacific Railway was built through the colony. The post office was moved 5 km east to the townsite, and renamed in 1906 after Carl Stettler (1861–1919), who had homesteaded here three years earlier.

Steveston BC On the southwestern point of Lulu Island and in the city of Richmond, this place was named in 1890 after William H. Steves, the first postmaster. He was the eldest son of Manoah Steves, who had come from the Moncton area of New Brunswick in 1877.

Stewart BC A district municipality (1976) at the head of Portland Canal and beside the state of Alaska, it was settled in 1902 by brothers John and Robert Stewart. Robert was appointed the first postmaster in 1905.

Stewart, Mount AB This mountain (3312 m) east of the Columbia Icefield, was named in 1902 after Louis B. Stewart (1861–1937), a professor of surveying and geodesy at the University of Toronto, who surveyed the boundaries of Banff National Park. His father, George A. Stewart (1830–1917), also a surveyor, was the first superintendent of the park.

Stewart River YT Rising in the Selwyn Mountains, this river (644 km) flows west to join the Yukon River, 80 km upstream from Dawson. It was named in 1849 by Hudson's Bay Company trader Robert Campbell after James G. Stewart, his assistant.

Stewarttown ON In the town of Halton Hills, Halton Region, west of Georgetown, its post office was called *Esquesing*, after the township, from 1832 to 1918. John and Duncan Stewart built mills here in the 1820s, and the place was called *Stewartstown*, *Stewarton*, and *Stewart Town*. The present form was confirmed in 1965.

Stewiacke NS The name of this town (1906) is derived from the **Stewiacke River**, which flows into the Shubenacadie River here. The river's name in Mi'kmaq is *Esiktaweak*, 'oozing slowly from still water'. Its post office was called *Lower Stewiacke* in 1836 and was only changed to Stewiacke in 1913.

Stikine River BC Rising in the Cassiar Mountains, southwest of Fort Nelson, this river (539 km) flows west, and then southwest, crossing the Alaska border northwest of Stewart. In Tlingit the name means 'the river', in the sense of being a big river.

Stirling AB A village (1901) southeast of Lethbridge, it was named in 1899 after John A. Stirling, managing director of the London-based Trusts, Executors and Securities Corporation, a large shareholder in the Alberta Railway and Coal Company.

Stirling ON A village (1858) in Hastings County, northwest of Belleville, its post office was called *Rawdon*, after the township, from 1832 to 1852. It was then changed to Stirling, because its site recalled the countryside of Stirlingshire in Scotland.

Stittsville ON In Goulbourn Township, Ottawa-Carleton Region, southwest of Ottawa, this community was named in 1854 after first postmaster Jackson Stitt, who had arrived here in 1818.

Stockholm SK West of Esterhazy, the Canadian Pacific Railway station was erected at this village (1905) in 1903 and its post office was named by first postmaster Alex Stenberg the following year. The area was settled by Swedes in the 1880s.

Stoneham-et-Tewkesbury QC Located north of the city of Québec, this united townships municipality was erected in 1845, abolished two years later, and re-established in

1855. The township of Stoneham was founded in the late 1700s by the Revd Philip Toosey, an Anglican priest from Stoneham, Suffolk, England, and was proclaimed in 1800. The township of Tewkesbury was also proclaimed in 1800, and named after the home town of its first settler, Kenelm Chandler, in Gloucestershire, England.

Stonewall MB Northwest of Winnipeg, the post office in this town (1908) was named in 1878. The first postmaster, O.P. Jackson, named it after his brother, the Hon. S.J. Jackson, who had donated the land for the post office. The latter was nicknamed 'Stonewall' because he walked with an erect military bearing during his political campaigns, recalling the famed Confederate general, Thomas 'Stonewall' Jackson (1824–63), whose troops stood like a stone wall at the first Battle of Manassas at Bull Run, VA, 1861.

Stoney Creek ON A city (1984) in Hamilton-Wentworth Region, it was likely named after the creek that flows north into Lake Ontario. There are claims it was named after trapper Jim Stoney, or after Anglican priest Edmund Stoney. Its post office was called *Stony Creek* in 1827, but it was changed to Stoney Creek about 1832, because this form of the name was closely identified with the battle that took place here during the War of 1812, which became well known as a turning point for the British forces.

Stoney Point ON In Essex County, east of Windsor, this community was named after the nearby point, called *Pointe aux Roches* during the French regime. Stoney Point post office was named in 1865, but it was changed to Pointe-aux-Roches in 1950. However, Stoney Point continued as the police village's name. In 1996 the municipal township of Tilbury requested equal recognition of both names for the post office and the community.

Stony Mountain MB Situated between Winnipeg and Stonewall, this place's post office was named in 1880, having been called *Rockwood* in 1873. The nearby mountain was identified on Henry Y. Hind's map of 1858.

Stony Plain AB A town (1908) west of Edmonton, its post office was named in 1893 after the surrounding land. The plain may have been originally called after the Stoney, a variant name of the Assiniboine, but Dr James (later Sir James) Hector observed in 1858 that it 'well deserves the name from being covered with boulders which are rather rare in general in this district or country'.

Stony Rapids SK On the Fond du Lac River, east of Lake Athabasca, this northern hamlet (1992) was named after the boulder-strewn fast water of the river. It was settled by trappers in the 1920s and its post office was opened in 1937.

Storkerson Peninsula NT The northeastern extension of Victoria Island, this 260-km peninsula was named in 1961 by geologist Raymond Thorsteinsson after Storker Storkerson, who investigated the north coast of Victoria Island in 1915 and 1917. A member of the Leffingwell-Mikkelson Arctic expedition of 1906–7, Storkerson was a Norwegian sailor, who had remained in Alaska to study the geography of the North.

Stormont ON The county was named in 1792 after David Murray, 7th Viscount Stormont and 2nd Earl of Mansfield (1727–96), who held many official posts, including justice general of Scotland. The title was taken in 1621 from Stormont Loch in Perthshire, Scotland. In 1852 it became part of the united counties of **Stormont, Dundas and Glengarry**.

Storrington ON A township in Frontenac County, it was named in 1845 after the English village where Sir Henry Smith (1812–68), speaker of the Legislative Assembly of the province of Canada, 1858–61, was born. Storrington is in West Sussex, northwest of Brighton.

Stouffville ON In York Region, east of Aurora, this community was named in 1832 after miller Abraham Stouffer (1776–1851), who settled here in 1804. In 1971 it was united with Whitchurch Township to form the town of **Whitchurch-Stouffville**.

Stoughton SK A town (1960) east of Weyburn, it was founded by the Canadian Pacific Railway in 1904 and named after John Stoughton Dennis (1820–85), the son of Joseph Dennis and Mary Stoughton. He was Canada's first surveyor general, 1871–8 and deputy minister of the interior, 1878–81. Its post office was also opened in 1904.

Straffordville ON A community in Elgin County, east of Aylmer, its post office was named in 1851 after Sir John Byng (1772-1860), who had become the Earl of Strafford in 1847. He served during the Peninsular War in Spain, 1811–14 and was the commander-in-chief in Ireland, 1828–31.

Strasbourg SK A town (1907) north of Regina, its post office was named *Strassburg* in 1886 after the chief city of Alsace, then part of Germany. It was called *Strassburg Station* in 1905. After Alsace was restored to France following the First World War, the Canadian town's post office was renamed *Strasbourg Station* in 1919, and was shortened to Strasbourg in 1956.

Stratford ON A city (1885) in Perth County, its post office was named in 1835. Possibly the London, England-based Canada Company instructed its Canadian co-commissioner Thomas Mercer Jones to call it that. Or it may have been inspired by a sign portraying William Shakespeare that Jones gave to William Sargint to hang outside his inn, which he had built in 1832.

Stratford PE A town (1 April 1995), south of Charlottetown, it was created through the amalgamation of the municipal communities of Southport, Bunbury, Cross Roads, and Keppoch-Kinlock. During the previous December the local residents voted to call it Waterview, but widespread dislike of that name led to a second ballot, which endorsed the second choice, Stratford. It was named after the main road in Southport and Cross Roads, which had been the name of a school district, founded in Southport in 1858.

Strathclair MB Located northeast of Minnedosa, this place was the location of a Hudson's Bay Company trading post, taking its name from the Gaelic word *strath*, 'valley' and the last syllable of surveyor Duncan Sinclair's name. Its post office was opened in 1880.

Strathmore AB A town (1911) east of Calgary, it was named by the Canadian Pacific Railway in 1884 after Claude Bowes-Lyon, 13th Earl of Strathmore (1824–1904).

Strathroy ON A town (1870) in Middlesex County, west of London, it was named in 1832, likely after Straughroy (also called Strathroy), County Tyrone, Ireland. Its post office was opened in 1851.

Streetsville ON In the city of Mississauga, Peel Region, this urban community was developed in 1818–19 by Timothy Street, and its post office was opened in 1828. It became a village in 1857 and a town in 1962, but was annexed by the city of Mississauga in 1974.

Strickland, Mount YT This mountain (4212 m) in the St. Elias Mountains was named about 1918 after North-West Mounted Police inspector D'Arcy Strickland (d. 1908), who commanded the White Pass detachment in 1898.

Strome AB A village (1910) southeast of Camrose, its first post office was called *Knollton* after postmaster Mac Knoll. It was renamed in 1906 after Stromeferry, on Loch Carron, east of the Isle of Skye, and in Ross and Cromarty, Scotland.

Stroud ON In the town of Innisfil, Simcoe County, southeast of Barrie, this place was named in 1873 by member of Parliament William C. Little after a town near his birthplace in Gloucestershire, England.

Stuart Lake BC Northwest of Prince George, this lake was known to Simon Fraser in 1806 as *Sturgeon Lake*. It was soon named after John Stuart (1780–1847), who spent the winter of 1806–7 at McLeod Lake, to the northeast, as a clerk of the North West Company. He became a partner of the NWC in 1813, and, after the merger with the Hudson's Bay

Company in 1821, he served as a chief factor in New Caledonia for three years.

Stukely-Sud QC A village midway between Granby and Magog, it was detached in 1935 from the municipality of **Stukely-Sud**, which had been created in 1855. Ten years earlier the latter had been erected as the township municipality of Stukely-Sud. Well known as Stukely South, the name was derived from the township of Stukely, proclaimed in 1800, and named after Stewkley, in Buckinghamshire, England, northwest of London.

Stupendous Mountain BC In the Coast Mountains, southeast of Bella Coola, this mountain (2728 m) was named in 1793 by Alexander (later Sir Alexander) Mackenzie. On descending from the Interior Plateau into the Bella Coola River valley he was much impressed with this mountain, whose summit disappeared into the clouds.

Sturgeon Falls ON A town (1895) in Nipissing District, west of North Bay, it is located on the **Sturgeon River**. After the transcontinental Canadian Pacific Railway arrived in 1882, the Sturgeon Falls post office was opened.

Sturgeon Point ON A village in Victoria County, north of Lindsay, it is on a peninsula where **Sturgeon Lake** divides into two large arms. Incorporated in 1899, it is Ontario's smallest village in terms of population, having only 103 permanent residents in 1996.

Sturgis SK North of Canora, the post office of this town (1951) was named in 1908 by postmaster F. Brooks after his home town northwest of Rapid City, SD.

Stutfield Peak AB On the northwest side of the Columbia Icefield, this mountain (3459 m) was named in 1899 after climber Hugh Edward Millington Stutfield (1858–1929), who made several ascents in the area of the icefield.

Sudbury ON The site of the city (1930) was named in 1883 by Canadian Pacific Railway superintendent of construction James Worthington. It was given in honour of his wife's birthplace in Suffolk, England, west of Ipswich. **Sudbury District** was created from Nipissing District in 1894. In 1973 the regional municipality of **Sudbury** was severed from the district, and divided into the city of Sudbury, and the towns of Capreol, Nickel Centre, Valley East, Rayside-Balfour, Onaping Falls, and Walden.

Suffield AB Northwest of Medicine Hat, this place was named in 1883 by the Canadian Pacific Railway after Charles Harbord, 5th Baron Suffield (1830–1914). He was a brother-in-law of Lord Revelstoke, a major financier of the CPR's construction in British Columbia. The nearby Suffield Experimental Station is a military training ground, where biological warfare studies are undertaken.

Sullivan ON A township in Grey County, it was named in 1840 after Robert Baldwin Sullivan (1802–53), then the commissioner of Crown lands in Upper Canada.

Sullivan QC Situated north of Val-d'Or, this municipality was named in 1972, after its post office, Sullivan Mines, had been opened in 1935. The place was called *Carrièreville* (meaning 'quarry town') from 1937 to 1967, when the mine closed. Jos Sullivan discovered gold here in 1911, and mining began in 1934.

Sulphur Mountain AB This mountain (2451 m), overlooking Banff, was named in 1916 after the distinctive odour of the hot springs at its base.

Summerford NF A municipal community (1971) on New World Island, it was first called *Farmer's Arm*. It was renamed before 1911 to avoid confusion with *Farmer's Arm*, near Twillingate, now called Gillesport.

Summerland BC A district municipality (1906) on the west side of Okanagan Lake and north of Penticton, it was founded in 1902 by developer John Moore Robinson. It was named after the pleasant sunny climate of the Okanagan Valley.

Summerside PE A city (1 April 1995), it had been incorporated as a town in 1877. In creating the city it annexed the neighbouring municipal communities of St. Eleanors and Wilmot, and part of Sherbrooke. It was named in 1852 after Summerside House, established by Joseph Green in 1840. The name of the inn was suggested by Maj. Harry Compton, because he described the site as the 'sunny side' of the island.

Summerville NS On the east bank of the Avon River, north of Windsor, this place was once called *Black Rock* after a rock in the river. Its post office was named in 1866.

Sunbury NB Southeast of Fredericton, this county was created in 1765 to cover most of the western part of present New Brunswick. Twenty years later it was confined to the present area, centred on Oromocto. It was named after George Montagu Dunk, 2nd Earl Halifax, Viscount Sunbury (1716–71). He held several positions in the British government, including president of the board of trade and lord privy seal.

Sunbury ON In Frontenac County, northeast of Kingston, its post office was named in 1864, probably by member of Parliament Sir Henry Smith after a district in Greater London, England, west of London's city centre.

Sunderland ON In Durham Region, west of Lindsay, this community's post office was called *Brock*, after the township, from 1836 to 1868, when it was renamed after Sunderland, on the North Sea coast of County Durham, southeast of Newcastle upon Tyne.

Sundre AB This town (1956), northwest of Calgary, was named in 1909 after the birthplace in Norway of first postmaster N.T. Hagen, who had arrived here three years earlier, and bought a store.

Sundridge ON A village (1889) in Parry Sound District, north of Huntsville, its post office was named in 1879. First postmaster John Paget had suggested *Sunridge*, a description of the place's setting at the north end of Lake Bernard. The postal authorities may have been inspired by Sundridge in Kent, England, southeast of London.

Sunnyside NF Located at the north end of the Avalon Isthmus, and at the head of Bull Arm, an extension of Trinity Bay, it was formerly called *Bay Bulls Arm*. It was renamed in 1930, presumably because it was on the northern (sunny) side of the arm.

Sunwapta Peak AB A mountain (3315 m) southeast of Jasper, it was named in 1892 by climber Arthur P. Coleman after the Stoney words for 'turbulent water'. The **Sunwapta River** rises on the north flank of Mount Athabasca and flows northwest into the Athabasca River.

Superior, Lake ON Called *Grand Lac* by Samuel de Champlain in 1632, this lake was identified on later maps as *Lac de Tracy ou Svperievr*, the former in honour of Alexandre de Prouville, Marquis de Tracy (*c.* 1602–70), commander-in-chief of French forces in America, 1665–7. The lake's name describes its position as the most elevated of the Great Lakes.

Surrey BC On the south side of the Fraser River, southeast of Vancouver, this city (1993) had been created a district municipality in 1879. It was named that year after the English county of Surrey. Its location south of New Westminster likely inspired the name, as Surrey in England is south of the city of Westminster in Greater London.

Sussex NB This town (1904) in the Kennebecasis River valley was founded in the early 1800s. It was named after **Sussex Parish**, established in 1786, and possibly called after Sussex, NJ. Its post office was named *Sussex Vale* about 1810, and was changed to Sussex in 1898. The post office of the adjacent village (1966) of **Sussex Corner** was named in 1860.

Sutton ON In the town of Georgina, York Region, northeast of Newmarket, this urban centre was first known as *Bourchier's Mills*, after miller James O'Brien Bourchier (1797–1872). Bourchier was appointed the first postmaster of

Georgina in 1831. The post office was renamed **Sutton West** in 1885, possibly after Sutton on Hull, a suburb of Kingston upon Hull in East Yorkshire, but the place is usually called just Sutton.

Sutton QC A town (1962) in the regional county of Brome-Missisquoi and on the west flank of the **Monts Sutton**, southeast of Cowansville, it had been incorporated as a village in 1896. The township municipality of **Sutton** was organized in 1845, abolished two years later, and re-established in 1855. The post office was opened in 1837 and named after the township, shown on a 1795 map and proclaimed in 1802. It may have been named after Sutton, Bedfordshire, England (adjacent to Potton, the name of the township east of Sutton in Québec), and reinforced by places called Sutton in four of the New England states.

Sverdrup Islands NT In the Queen Eizabeth Islands, the group comprising Axel Heiberg, Amund Ringnes, Ellef Ringnes, Cornwall, King Christian, and Meighen islands was named before 1910 after Norwegian Capt. Otto Neumann Sverdrup (1855–1930). From 1898 to 1902 Sverdrup led an expedition to southern Ellesmere Island and the islands west of it.

Swan Hills AB A town (1967) in the **Swan Hills**, north of Whitecourt, was named in 1960 after an oil field had been developed here. The hills are said to be have been named after huge swans, which make thunderous noises when they flap their wings.

Swan Lake MB Northwest of Morden, this place's post office was named in 1881 after a lake 7 km to the southwest. White swans had a habitat at the lake, which is an enlargement of the Pembina River.

Swannell Ranges BC Located west of Williston Lake, these ranges were named in 1950 after Frank Cyril Swannell (1879–1969), who performed several land surveys throughout the province. **Mount Swannell**, south of the Nechako Reservoir, was named after him in 1947.

Swan River MB Located on the **Swan River**, which rises west of the Porcupine Hills in Saskatchewan, this town (1908) took its name from *Swan River House*, a fur trading post founded in the late 1700s. During 1875 it was the legislative capital of the North-West Territories.

Swastika ON In the town of Kirkland Lake, Timiskaming District, this place was named in 1911, five years after brothers Bill and Jim Dusty found gold at a nearby lake, and named the mine after a visitor's good luck charm. Because of the symbol's later association with Adolph Hitler, the Ontario government tried, during the Second World War, to rename the place *Winston*, after Winston Churchill, but local residents protested the desecration of their name.

Swift Current NF At the mouth of **Swift Current**, which flows into the north part of Placentia Bay, this place, formerly called *Piper's Hole*, was renamed about 1920. There is a legend attached to Piper's Hole that ghosts could be seen in the area, accompanied by the mournful sound of bagpipes.

Swift Current SK A city (1914) on the main line of the Canadian Pacific Railway, its post office was named in 1883. Its name is a translation of the Cree name for **Swift Current Creek**, *Kisiskatchewan*, also the Cree name for the Saskatchewan River.

Sydenham ON A township in Grey County, it was named in 1842 after Charles Edward Poulett Thomson, Baron Sydenham (1799–1841), governor-in-chief of British North America, 1839–41. **Sydenham River** rises near Williamsford, south of Owen Sound, and flows north into Owen Sound Harbour.

Sydenham ON In Frontenac County, north of Kingston, this community was first named *Loughborough* in 1836, after the township, but was changed in 1839 to honour Lord Sydenham, governor-in-chief of British North America, 1839–41.

Sydenham River ON Rising north of London at Ilderton, this river flows southwest, parallel to the Thames, to enter the Snye River (Chenail Écarté), downriver from Wallaceburg, and ultimately into Lake St. Clair. It was named for Lord Sydenham, governor-in-chief of British North America, 1839–41.

Sydney NS A metropolitan area in the regional municipality of Cape Breton since 1 August 1995, it had been incorporated as a town in 1886 and as a city in 1904. It was named after **Sydney Harbour**, which had been named in 1784–5 by J.F.W. DesBarres, then governor of Cape Breton, after Thomas Townshend, 1st Viscount Sydney (1733–1800), British home secretary, 1782–9, lord of the Admiralty, 1790–3, and lord of the treasury, 1793–1800. The harbour had been called *Dartmouth Harbour* in 1767 by Samuel Holland, replacing *Spanish Bay*. The latter had been named *Riviere aux Espagnols* in the late 1600s, perhaps reflecting not only Spanish fisherman coming ashore here, but Portuguese and Basque as well. The community of **Sydney River**, southwest of Sydney, was named in 1892. **Sydney Forks**, also southwest of Sydney, and at the confluence of **Sydney River** and Meadows Brook, was named in 1874.

Sydney Mines NS On the north side of Sydney Harbour, this urban community has been in the regional municipality of Cape Breton since 1 August 1995. It had been incorporated as a town in 1889.

Sylvan Lake AB Located on the south shore of Sylvan Lake, west of Red Deer, the post office in this town (1946) was named in 1907. The lake was once called *Snake Lake*. Geographer David Thompson recorded it as *Methy Lake* in 1814. The present name is from the Latin *silvanus*, 'forested'.

T

Taber AB A town east of Lethbridge, its station was named in 1894 by the Canadian Paci-fic Railway. Its post office was called *Tabor* in 1904 after Mount Tabor in the Holy Land. The townsite was first registered as *Tabor*, but was changed to Taber when the town was incorporated in 1907. Although claims had been made that its origin came from the first part of 'tabernacle,' Judge J.A. Jackson of Lethbridge reported in 1928 that it was not true. Horace Tabor (1830–99), an appointed United States senator for Colorado in the 1880s, may have visited the place, leading to the erroneous conclusion that it was named after him.

Tabusintac NB At the mouth of the **Tabusintac River**, southwest of Tracadie, this place's post office was named *Tabisintac* in 1840, changed to *Tabucintac* in 1857, and given its present spelling in 1931. The river's name in Mi'kmaq may mean 'two entrances', referring to the main river and the entrance to French Cove. Map-maker Pierre Jumeau called it *R. tabochemkek* in 1685.

Tadoussac QC Situated on the north side of the mouth of the Saguenay, this village has one of the prettiest settings in Québec. Its name in Montagnais, *Totouskak*, means 'breasts', in reference to two rounded sandy hills to the west of the village. In 1535 Jacques Cartier visited the site, which became a major trading centre during the next two centuries. Its post office was opened in 1851. The village was erected in 1899 and the parish municipality was established in 1937. The latter was dissolved 12 years later because its population had declined below 500.

Taghum BC On the north side of the Kootenay River, this place's post office was named in 1924 after the Chinook jargon for 'six', as it is approximately six miles (10 km) west of Nelson. The post had been called *Williams Siding* in 1906 after the first postmaster, J. Williams.

Tagish Charlie, Mount YT Located in the Ogilvie Mountains, in northern Yukon, this mountain was named in 1973 on the seventy-fifth anniversary of the discovery of gold on Bonanza Creek, a tributary of the Yukon River, by Tagish (later Dawson) Charlie (d. 1908), George Carmack, and Skookum Jim Mason. A Tagish, he lived at Carcross and died after a fall from a railway bridge there.

Tagish Lake BC, YT This lake, southeast of Whitehorse, extends south from **Tagish**, YT, for 100 km and connects with Atlin Lake through Taku Arm. It was named in 1887 by geologist George M. Dawson after the Tagish tribe. Four years earlier American Army Lt Frederick Schwatka had called it *Lake Bove* after Lt Bove of the Italian Navy, but Dawson restored the Aboriginal designation. The Tagish, who originally spoke an Athapaskan dialect, now speak the language and follow the customs of the coastal Tlingit.

Tahsis BC A village (1979) at the head of **Tahsis Inlet**, an arm of Nootka Sound, and west of Campbell River, its post office was established in 1938. The name of the inlet is a Nootka word meaning 'trail at beach', referring to the practice of coming up the inlet by boat, and then proceeding by a trail to the head of the north-flowing Nimpkish River, leading to the east side of Vancouver Island.

Takakkaw Falls BC On the Yoho River and in Yoho National Park, this set of falls (380 m) is the second highest in Canada, after Della Falls on Vancouver Island. It was named in 1897 by Sir William Van Horne. In Cree *takakkaw* means 'it is magnificent'.

Takla Lake BC Located east of Hazelton and north of Babine Lake, this lake was named in 1914. It was previously called *Tacla Lake* after the Carrier word for 'lower end of the lake', in reference to the original site of Bulkley House at the southeastern end of the lake.

Talbotville Royal ON In Elgin County, northwest of St. Thomas, it was first called *Five Stakes* in 1811, after five survey stakes marking the junction of five roads in Southwold Township. First postmaster and innkeeper John Allworth named it in 1853 after Col Thomas Talbot (1771–1853). Talbot, who was born at Malahide Castle, north of Dublin, was the private secretary to Lt-Gov. John Graves Simcoe, 1792–4. In 1801 he was granted 2023 ha (5000 a) of land south of London, and eventually quadrupled his land holdings, while promoting British immigration to the area.

Talchako Mountain BC In the Coast Mountains, southeast of Bella Coola, this mountain (3063 m) was named about 1933 by climber W.A.D. (Don) Munday after the **Talchako River**, named in 1922 by the Geographic Board. The name may be a reference to the devil in either the Carrier or the Bella Coola language.

Taloyoak NT On the west side of the south end of Boothia Peninsula, this hamlet (1981) had been called *Spence Bay* in 1964. The Inuit name, in use for many generations, was adopted in 1992. In Inuktitut the name means 'caribou blind', in reference to a pair of stone piles along a path, where hunters speared caribou. The bay was named in 1831 by John (later Sir John) Ross, after a relative.

Tamworth ON In Lennox and Addington County, north of Napanee, its post office was named in 1848 by miller Calvin Wheeler after Tamworth in Staffordshire, England, north of Birmingham. Wheeler admired Sir Robert Peel, who represented Tamworth in Parliament.

Tancook Island NS On Big Tancook Island, in the entrance to Mahone Bay, this place's post office was opened in 1866. The island's name was derived from Mi'kmaq *Uktankook*, 'facing the open sea'. **Little Tancook Island** is located on its east side.

Tangier NS On the Eastern Shore, this place received postal service in 1864. The adjacent harbour was likely named after the shipwreck of the schooner *Tangier* in 1830.

Tanquary Fiord NT Extending north from Greely Fiord, in central Ellesmere Island, this fiord was named in 1915 by geologist W. Elmer Ekblaw after Maurice Cole Tanquary, zoologist with the United States 'Crocker Land' expedition, 1913–16.

Tantallon NS On the northeastern shore of St. Margarets Bay and west of Halifax, this place's post office was named in 1896, likely after Tantallon Castle, east of Edinburgh, Scotland, once the Douglas stronghold, which was destroyed in 1651. **Upper Tantallon** post office was open from 1914 to 1947.

Tantallon SK On the north bank of the Qu'Appelle River, southeast of Esterhazy, this village (1904) was founded in 1893 by the Revd James Moffat Douglas (d. 1920) and its post office was opened in 1897. Douglas named it after his ranch, which had been called after Tantallon Castle, east of Edinburgh. His ancestors built it as a stronghold, but it was destroyed in 1651.

Tantalus, Mount BC In the **Tantalus Range** of the Coast Mountains, northwest of Squamish, this mountain (2606 m) and the range were named about 1953 by climber Dr Neal M. Carter because the unnamed mountains rising to the east in the area of Mount Garibaldi tantalized early climbers and recalled to him the Greek legend of King Tantalus. The legend described food and drink that was just beyond the king's grasp.

Tantramar River NB Rising east of Moncton, this river flows south through the **Tantramar Marshes**, and empties into Cumberland Basin, south of the town of Sackville. The name was derived from the French *tintemarre*, 'thundering noise', in reference either to the flocks of geese or to the rushing tide. Map-maker Dugald Campbell referred to it in 1799 as *Tantaramar*, and it was not until 1901 that the current spelling was confirmed.

Tara ON A village (1880) in Bruce County, southwest of Owen Sound, it was called *Eblana* in 1862, after an ancient name of Dublin, Ireland, but was changed to Tara two years later, after the site of the ancient royal palaces, northwest of Dublin.

Tarfu Creek YT A tributary of Lubbock River and between Jakes Corner and Atlin, BC, this creek was named in 1949–50 by Canadian Army engineers after a variation of the Second World War phrase, which provided the name Snafu Creek. Tarfu means 'Things Are Really Fouled Up'.

Taschereau QC A village in the regional county of Abitibi-Ouest, midway between Amos and La Sarre, it was detached in 1980 from the municipality of **Taschereau**, which had been erected in 1926. Until then it had been called the township municipality of *Privat*, established in 1919, and named after the township, proclaimed three years earlier. The village was named after Louis-Alexandre Taschereau (1867–1952), premier of Québec, 1920–36. The township was named after Marc-Antoine de Privat, an officer in the Languedoc Regiment, who was wounded at the Battle of the Plains of Abraham, 1759.

Taseko Mountain BC In the Coast Mountains, northwest of Pemberton, this mountain (3062 m) was named about 1923 after the **Taseko River** and the **Taseko Lakes**, both named in 1911. In Chilcotin *taseko* means 'mosquito river'.

Tasiujaq QC A northern village near the mouth of the Rivière aux Feuilles, northwest of Kuujjuaq, it was incorporated in 1980. The name in Inuktitut means 'it looks like a lake'.

Tatamagouche NS A village (1950) situated on the south side of **Tatamagouche Bay** and west of Pictou, its name was derived from Mi'kmaq *Takamegoochk*, 'barred across the entrance by sand'. Postal service was provided in 1738. An effort by early Scottish settler Wellwood Waugh to have it renamed Southampton failed.

Tathlina Lake NT Southwest of Hay River, this lake (573 km^2) was officially named in 1923. In 1875 Fr Émile Petitot made a reference to the lake, stating its meaning (possibly in Slavey) was 'river flowing out of a corner of a lake', in reference to its location, where Kakisa River flows north from the lake.

Tatlow, Mount BC In the Coast Mountains, east of Mount Waddington, this mountain (3062 m) was named in 1911 after Robert Garnett Tatlow, provincial minister of finance, who was killed two years earlier when his horse bolted, throwing him from his carriage.

Tatshenshini River BC, YT A tributary of the Alsek River, this river rises near Chilkat Pass and flows north into Yukon, where it turns south to cut its way through the St. Elias Mountains. It was officially approved by the Geographic Board in 1898, but its meaning is not provided in the names records. In 1993 the British Columbia government reserved this northwestern part of the province as **Tatshenshini-Alsek Wilderness Park**.

Tavistock ON In Oxford County, southeast of Stratford, this urban centre was first called *Freiburg*, after a city in Baden, Germany, and *Inkerman*, after a Crimean War battle. In 1857 postal authorities renamed it after the town of Tavistock, Devon, England, and on the west edge of the Dartmoor. In 1975 it was joined with East Zorra Township to form the municipal township of **East Zorra-Tavistock**.

Tay ON A township in Simcoe County, it was named in 1822 after a pet dog of Lady Sarah Maitland (1792–1873), the wife of Lt-Gov. Sir Peregrine Maitland. In 1994, it annexed the villages of Victoria Harbour and Port McNicholl. *See also* Tiny, *and also* Flos *under* Springwater.

Taylor BC A district municipality (1989) on the Peace River, southeast of Fort St. John, it was named in 1906 after homesteader D.H. (Herbie) Taylor, a former Hudson's Bay Company trader.

Tecumseh ON A town (1921) in Essex County, east of Windsor, it was named *Ryegate* in 1870. Six years later it was renamed after Tecumseh Road, the old route from Windsor to Chatham, named earlier after the famous Shawnee chief, who fought on the British side during the War of 1812, dying near present-day Thamesville in 1813.

Tecumseth ON Established as a township in Simcoe County in 1822, it was named after the great Shawnee chief, who was killed in 1813 at the Battle of Moraviantown. The spelling followed that of a vessel called *Tecumseth*, sunk in 1819 at Penetanguishene. In 1991 it was united with the town of Alliston and the villages of Tottenham and Beeton to form the town of **New Tecumseth**.

Teepee Mountain BC In Rocky Mountains, northeast of Cranbrook, this mountain (3118 m) was named in 1933 by climber Katie Gardiner after the shape of its summit. She and Lillian Gest made its first ascent that year.

Teeswater ON A village (1874) on the **Teeswater River** and in Bruce County, southwest of Walkerton, it was named in 1856. The river had been named four years earlier after the River Tees in England, which forms the boundary between Yorkshire and County Durham. Named *Mud River* in the late 1840s, it had earlier been called *Yokasipi*, from Ojibwa *ahtayahkosibbi*, 'drowned lands river'.

Teeterville ON In Delhi Township, Haldimand-Norfolk Region, northwest of Simcoe, its post office was named in 1864 after miller George Teeter.

Telegraph Creek BC On the Stikine River, where **Telegraph Creek** flows into it from the north, this place's post office was named in 1899. The creek was named in 1866, when the projected Collins Overland Telegraph, to connect the United States with Russia, was surveyed along the creek. The project was abandoned when the transatlantic cable was successfully laid that year, but several names, including Telegraph Creek, remain as evidence of the project.

Telkwa BC In the Bulkley River valley, south of Smithers, this village (1952) is located at the mouth of the **Telkwa River**. Its post office was opened in 1910. In Carrier the name possibly means 'frog'.

Temagami ON A muncipal township in Nipissing District, north of North Bay, it is located at the east end of the Northeast Arm of **Lake Temagami**. The Hudson's Bay Company's Temagamang Post was located at the lake in 1834. The site was named *Timagami* in 1903, when the Temiskaming and Northern Ontario Railway was built through it. The lake's name was spelled *Timagami* until 1968, when the Ontario names authority was persuaded to accept the preferred spelling, but the postal name was not changed to Temagami until 1974. The name means 'deep lake' in Ojibwa.

Témiscaming QC Located on the east bank of the Ottawa River, northeast of North Bay, this town was created in 1920 as *Kipawa*, which had annexed *South Temiscaming* and *Lumsden's Mill*. The next year, however, it was renamed Témiscaming, after the municipality of **Témiscaming**, erected in 1888. Its name was derived from the nearby Lac Témiscamingue.

Témiscamingue QC A regional county in northwestern Québec, it was established in 1981, taking its name from the former county. It had been named after **Lac Témiscamingue**, taken from Algonquin *Timi-s-timing*, 'there is deep and shallow water'. In the lower 80 km of its 122-km length, it is narrow and quite deep, but the north part is much wider and relatively shallow. In Ontario the official spelling is Timiskaming.

Témiscouata QC A regional county southeast of Rivière-du-Loup, it was named in 1992 after the former county, which was called after the 40-km-long **Lac Témiscouata**. Noted in 1683 as *Cecemiscouata* and in 1746 as *Temisquata*, the name in Mi'kmaq means 'deep lake'.

Temple, Mount AB Northwest of Banff, this mountain (3543 m) was named in 1884 by geologist George M. Dawson after Sir Richard Temple (1826–1902), who led an expedition to the Rockies that year.

Terence Bay NS On the Atlantic Ocean, southwest of Halifax, this place was early known by a variety of names, including *Terrants Bay*, *Turns Bay*, *Turner Bay*, and *Twins Bay*. Postal service was provided at *Turns Bay* in 1856, but it was renamed Terence Bay that year.

Terrace BC A city (1987) on the Skeena River, east of Prince Rupert, it was established as a village in 1927 and a district municipality in 1960. It was settled in 1905 by George Little, who laid out a townsite in 1911 and called it *Littleton*. The post office department declined to accept the name because of similar names elsewhere in Canada. Little then proposed Terrace after a series of four benches on the north side of the Skeena, and the post office was opened on 1 January 1912.

Terrace Bay ON A municipal township on the north shore of Lake Superior and in Thunder Bay District, east of Schreiber, it was founded in 1947 and named after a series of flat terraces separated by escarpments and cliffs. They had developed after the last ice age, as the water level of Lake Superior receded.

Terra Cotta ON In the town of Caledon, Peel Region, northwest of Brampton, this community's post office was called *Salmonville* in 1866. It was renamed Terra Cotta in 1890 after the hard red clay used in the making of bricks.

Terra Nova River NF Rising in the east central part of the island of Newfoundland, it flows into Bonavista Bay, at Glovertown. It took its name from the Latin rendering of the island's name, used by Girolamo da Verrazzano in 1529. **Terra Nova National Park** (397 km^2) was established in 1957.

Terrebonne QC Situated on the north shore of the Rivière des Mille-Îles, north of Montréal, the town of Terrebonne was established in 1860, having being erected as a village seven years earlier and a parish municipality in 1855. The seigneury of Terrebonne (Terbonne) was granted to André Daulier Des Landes in 1673. It was named because the fertility of the land (*terre*) was equally good (*bonne*) in every concession.

Terrenceville NF Located at the northeastern head of Placentia Bay, this town (1972) was first known as *Head of Fortune Bay* and *Fortune Bay Bottom*. It was renamed in 1905, likely after Sir John Terence Nicholls O'Brien (1830–1903), governor of Newfoundland, 1889–95.

Terry Fox, Mount BC In **Mount Terry Fox Provincial Park**, south of Yellowhead Pass, this mountain (2651 m) was named in 1981, after Terrance Stanley (Terry) Fox (1958–81), who set out from St. John's, NF, in 1980 to run across the country to raise funds for cancer research. Running with one artifical leg, his Marathon of Hope came to an end at Thunder Bay, ON, when the renewed cancer forced him to curtail the heroic run. Since 1981, annual Terry Fox runs have been organized to promote cancer research.

Teslin YT This village, located on the eastern shore of Teslin Lake, became the site of Tom Smith's trading post in 1903, and its post office was opened 10 years later. The lake extends south for 125 km from Johnsons Crossing, YT, into British Columbia. The **Teslin River** (393 km) rises in British Columbia and drains the lake north to the Yukon River, downriver from Lake Laberge. The Tagish called it *Nasathane*, 'no salmon'. In 1881 Michael Byrne, of the Western Union Telegraph Company, was told it was called *Hootalinkwa*. American Army Lt Frederick Schwatka called it *Newberry River*, after an American professor, but it was ignored. In the Athapaskan language spoken here in the 1800s, possibly Tagish or Nahane, *Teslintoo* meant 'long, narrow water'.

Tête Jaune Cache BC At the west end of the Yellowhead Pass and on the Fraser River, northwest of Valemount, this name, meaning 'yellow head's hiding place', recalls a mixed-blood Iroquois called Pierre Bostonais (d. 1827). This yellow-headed trader, who worked with both the North West Company and the Hudson's Bay Company, also provided the name Yellowhead Pass.

Teulon MB North of Stonewall, this place was named in 1895 when the Canadian Northern Railway was built north to Arborg. It was named after the wife of pioneer settler C.C. Castle.

Texada Island BC Located in the Strait of Georgia, south of Powell River, this island (351 km^2) was named *Isla de Texada* in 1791 by Spanish Capt. José Maria Narvaez after Spanish Rear-Adm. Felix de Tejada (or Texada).

Thamesford ON On the Middle Thames River and in Oxford County, west of Woodstock, this place was named when the post office was opened in 1851.

Thames River ON Rising near Tavistock, Oxford County, this river flows southwest through Woodstock, London, and Chatham to Lake St. Clair. During the French regime it was called *Rivière à la Tranche*, 'cutting river', possibly because at its mouth it appeared to be cutting through tall trees. It was renamed in 1792 after England's River Thames by Lt-Gov. John Graves Simcoe. Its main branches, the **North Thames** and the **Middle Thames** rivers, rise near Stratford.

Thamesville ON A village (1873) in Kent County, northeast of Chatham, it was named in 1832 when the post office was opened, and village lots were laid out 20 years later on the southeast side of the Thames River. Lots were also laid out on the northwest side, and named *Tecumseh*, after the great Shawnee chief, who died nearby in 1813. When a railway was built through *Tecumseh* in 1854, its station was called Thamesville, after the post office was transferred here.

Thedford ON A village (1877) in Lambton County, northeast of Sarnia, its post office was named in 1867 by Nelson Southworth, a native of Thetford, VT. He had donated a site in 1859 for a railway station, which was then called *Widder*, after a nearby place. A misinterpretation of handwriting resulted in the spelling error. The village of Thedford in central Nebraska was named after the Ontario village.

Thelon River NT Rising in Whitefish Lake, east of Great Slave Lake, this river (904 km) flows northeast to Baker Lake and Chesterfield Inlet. In Chipewyan the name means 'whitefish'. In Inuktitut the river is called *Arkinlinik*, 'wooded river'.

Theodore SK A village (1907) northwest of Yorkton, it was named in 1893 after Theodora Seeman, the wife of a wealthy landowner from London, England. They had spent one summer with their family on their ranch, 5 km from the village site. Earlier the place had been called *New Denmark*.

The Pas MB Located at the point where the Pasquia River flows into the Saskatchewan River, the site of this town (1912) has a long history relating to exploration, fur trading, and resource development. Its name has its roots in the Cree *opasquaow*, 'narrows between wooded banks'. During the period of French exploration, the name was rendered as *Le Pas*, coincidentally part of the family name of the wife of explorer Pierre Gaultier de La Vérendrye.

Thérèse-De Blainville QC This regional county, north of Rivière des Mille-Îles, comprises part of the former county of Terrebonne. Its name, given in 1982, united the names of Marie-Thérèse Dugué de Boisbriand (1671–1744) and her son-in-law, Louis-Jean-Baptiste Céleron de Blainville (1696–1756), inheritors of parts of the seigneury of Mille-Îles.

The Spanish River ON A municipal township in Sudbury District, it extends 80 km from Espanola on the east to the mouth of the Spanish River on the west. It was created in 1975 by amalgamating the townships of Salter, May, Harrow, Victoria, and Hallam.

Thessalon ON A town (1892) at the mouth of the **Thessalon River**, Algoma District, southeast of Sault Ste. Marie, its post office was named *Thessalon River* in 1874 and shortened to Thessalon in 1881. The river was described by explorer René de Bréhant de Galinée in 1669 as *R de Tessalon*. The name was derived from Ojibwa *Neyashewan*, 'point of land', referring to the 2-km-long **Thessalon Point**.

Thetford Mines QC Known as the city of white gold and the world capital of asbestos, this city was established in 1905. Its name had been chosen for the post office in 1881 after the township of Thetford, proclaimed in 1802, and named after Thetford, Norfolk, England. In 1892 the area of the present city had been incorporated as the village of *Kingsville*, named after William King (d. 1892), a wealthy owner of the mines.

Thicket Portage MB On the Hudson Bay Railway, south of Thompson, this place is on a narrow isthmus separating Landing and Wintering lakes. In Cree it is called *Sagaskawskow Uniga*, 'heavily wooded crossing place'.

Thlewiaza River NT Draining Nueltin Lake east into Hudson Bay, this river's name in Chipewyan means 'small fish' or 'whitefish'.

Thompson MB This city (1970) was created in 1957 and named after International Nickel Company president John F. Thompson. In 1956 large deposits of nickel had been discovered in the area, and smelting, concentrating, and refining of nickel, copper, cobalt, and precious metals were undertaken five years later.

Thompson River BC This river (489 km to the head of the **North Thompson River**) was named on 20 June 1808 by Simon Fraser after geographer David Thompson (1770–1857), who explored the West extensively but is not known to have seen the river. Five years later, Thompson named the Fraser River after Simon Fraser.

Thorburn NS Located east of New Glasgow, this place's first post office was called *Vale Colliery* in 1873, after the mine had opened the previous year by the Vale Coal Iron and Manufacturing Company, based in Montréal. It was renamed in 1886, uniting the Scandinavian *Thor*, 'god of thunder' and the Scottish word *burn*, 'brook'.

Thorhild AB A village (1949) northeast of Edmonton, its post office was opened in 1914, 7 km east of its present location. When the Alberta and Great Waterways Railway was built in 1914 to Lac La Biche, the post office was moved to the line. It is said that first postmaster M.G. Jardy was inspired by lightning and thunder on a nearby hill, and created the name from the hill and the Norse god Thor.

Thornbury ON A town (1887) in Grey County, east of Owen Sound, its post office was named in 1853 by surveyor William Gifford after Thornbury, Gloucestershire, England, north of Bristol.

Thorndale ON A community in Middlesex County, northeast of London, its post office was named in 1859 after the Irish residence of settler James Shanly, a native of County Meath.

Thornhill ON Divided down the middle between the city of Vaughan and the town of Markham in York Region, this urban centre's post office was named in 1829 after miller Benjamin Thorne (1794–1848), who had settled here in 1820. Thorne, a prosperous importer and exporter, became one of two presidents of the Bank of Montreal in 1842. He was bankrupted after the repeal of Britain's Corn Laws in 1846.

Thornloe ON A village (1916) in Timiskaming District, northwest of New Liskeard, its post office was named in 1909 after the Rt Revd George Thorneloe, the Anglican bishop of Algoma.

Thornton ON In Simcoe County, southwest of Barrie, this place's post office was named in 1854 after Sir Edward Thornton (1766–1852), who had been the British ambassador in Washington.

Thorold ON A city (1975) in Niagara Region, it embraces part of the area of the township of Thorold, named in 1798 in Lincoln County, after Sir John Thorold (1734–

1815), who represented Lincolnshire in the British House of Commons. Thorold post office was named in 1826.

Thorsby AB A village (1949) southwest of Edmonton, its post office was named in 1908 after the Norse god Thor and the Norse word *by*, 'village'.

Thousand Islands ON In the St. Lawrence River, between Kingston and Brockville, they were described as *Mille Isles* in 1721 by French historian Pierre-François-Xavier Charlevoix, although he concluded that there were only about 500 islands. Various counts have ranged from 1149 to 1800, but only 367 have official names, with 241 of these on the Canadian side of the border.

Three Hills AB A town (1929) northeast of Calgary, its post office was established in 1904. There are three small hills running in a line from southeast to northwest.

Three Mile Plains NS This place is located three miles (5 km) southeast of Windsor and in a flat physiographic area. Its post office was opened in 1865, closed in 1873, and reopened in 1900. An effort to name the place Mapleton in 1901 failed.

Thrums BC This community northeast of Castlegar was named in 1906 after a 'Scottish village', described by Sir James Barrie in *A Window in Thrums* (1889). Perhaps he had Thrumster, 7 km southwest of Wick, Caithness, in mind, or another village to which he attached this fictitious name.

Thunder Bay ON A city on **Thunder Bay**, at the head of Lake Superior, it was created in 1970 through the amalgamation of Fort William, Port Arthur, and parts of the municipal townships of Neebing and Shuniah. Only three names were on the ballot, with Thunder Bay receiving 15,821 votes, Lakehead, 15,302, and The Lakehead, 8477. Although the majority of voters preferred a form of 'Lakehead', the ballot was skewed against it. However, Thunder Bay was soon fully accepted. **Thunder Bay District** was created in 1871.

Thurlow ON A township in Hastings County, it was named in 1787 after Edward Thurlow, 1st Baron Thurlow (1732–1806), who held many British ministerial posts, including solicitor general, attorney general, and lord chancellor.

Thurlow Islands BC Between Vancou-ver Island and the mainland, north of Campbell River, these islands were named in 1792 by Capt. George Vancouver after Edward Thurlow, 1st Baron Thurlow, who had been appointed British chancellor of the exchequer in 1778. The two islands are officially known as **East Thurlow** and **West Thurlow** islands.

Thurso QC A town in the regional county of Papineau, east of Buckingham, it was created in 1963, having been organized as a village in 1886. The post office was named in 1853 after the town of Thurso, in the north of Scotland, from where some pioneer settlers emigrated to western Québec in 1807.

Tide Head NB A village (1966) west of Campbellton, its post office was named *Head of Tide* in 1873 after the head of the tide of the Restigouche River. The post office was renamed Tide Head in 1920.

Tidnish NS On the shore of Northumberland Strait, and near the New Brunswick border, this place's post office was named in 1861 after the **Tidnish River**. The name of the river was derived from Mi'kmaq *tedeneche*, 'straight across', in reference to a straight entrance into the river. *See also* Tignish, PE.

Tiedemann, Mount BC In the Coast Mountains, directly east of Mount Waddington, this mountain (3848 m) was named in 1926 by climber W.A.D. (Don) Munday after surveyor Hermann Otto Tiedeman (1821–91). In 1862 he had surveyed a railway route for Alfred Waddington to connect Vancouver Island with the rest of Canada. In 1875 Canadian Pacific Railway surveyors had named **Tiedemann Glacier**, which extends southeast from the mountain.

Tignish PE Near the northwestern end of the province, this municipal community (1983) had been incorporated as a village in 1952. It was named after the **Tignish River**, whose name in Mi'kmaq may have been derived from *Tedeneche*, 'straight across', referring to the direct entrance to the harbour through **Tignish Run**. *See also* Tidnish, NS.

Tilbury ON **Tilbury West Township** in Essex County was named in 1792 and **Tilbury East Township** in Kent County was named in 1794, both after the town of Tilbury, in Essex, England. **Tilbury North Township** in Essex County was taken from Tilbury West in 1891. The town (1910) of **Tilbury** in Kent County had been incorporated as the village of *Tilbury Centre* in 1887, and was shortened to Tilbury in 1896.

Tilley AB A village (1940) southeast of Brooks, it was named in 1894 by the Canadian Pacific Railway, after Sir Samuel Leonard Tilley (1818–96), a Father of Confederation, a minister of customs and of finance in Ottawa, and lieutenant-governor of New Brunswick. His brother, Sir Malcolm Tilley, was a CPR director.

Tilley Ridge NB South of the Nepisiguit River, this ridge was named in 1964 by Arthur F. Wightman, then the provincial names authority, after Sir Samuel Leonard Tilley (1818–96), a Father of Confederation and lieutenant-governor of New Brunswick, 1873–8 and 1885–93.

Tillsonburg ON A town in Oxford County, its post office was named in 1836 after George Tillson, a native of Massachusetts, who settled here in 1825, and built a sawmill and a forge. It was incorporated as the town of *Tilsonburg* in 1872, with the spelling being corrected by an act of the Ontario Legislature in 1902.

Tilting NF Located on Fogo Island and southeast of Joe Batt's Arm, this municipal community (1975) was earlier known as *Tilton Harbour* after the tilts (shacks) built here for migratory fishermen. In 1906 the present name was adopted to distinguish the place from Tilton, on the west side of Conception Bay.

Tilton NF Located on the northwestern side of Spaniard's Bay and southwest of Harbour Grace, this local improvement district (1979) was formerly called *Tilts*, in reference to the temporary huts used by early fishermen. Soon after 1900, the Revd Dr William Pilot, the superintendent of Church of England and Salvation Army schools, proposed renaming it Tilton.

Timberlea NS West of Halifax, this place was first known as *Nine Mile River*. Its post office was opened as *Bowser Station* in 1905, named for hotel keeper Angus Bowser. It was renamed Timberlea in 1917 after the meadows and an active lumbering and sawmilling industry in the area from the mid-1800s.

Timiskaming ON Created as a district in 1902, it was named after Lake Timiskaming. The lake received its name from Algonquin *Timi-s-timing*, 'there is deep and shallow water'. In the lower 80 km of its 122-km length, it is narrow and quite deep, but the north part is much wider and relatively shallow. Locally, the spelling *Temiskaming* is commonly used. In Québec, the official spelling is Témiscamingue for the lake and the regional county, and Témiscaming for the town.

Timmins ON The largest incorporated city (1973) in area (3000 km^2) in Canada, it was founded in 1911 by Noah and Henry Timmins, former Mattawa storekeepers. They made a fortune in silver mining at Cobalt. Noah Timmins (1867–1936) became the president of Hollinger Consolidated Gold Mines Ltd.

Tincap ON In the united counties of Leeds and Grenville, northwest of Brockville, this community was named when the first schoolhouse had a roof-end cap covered with tin. *Spring Valley* post office was located here until 1912, when it was renamed Tincap. Spring Valley is the name of an adjacent community.

Tintagel BC East of Burns Lake, this place was named in 1936 after Tintagel, Cornwall, England, which is located on the Bristol Channel, northwest of Plymouth. A stone from Tintagel Castle is mounted on a monument here.

Tiny ON A township in Simcoe County, it was named in 1822 after a pet dog of Lady

Sarah Maitland (1792–1873), the wife of Lt-Gov. Sir Peregrine Maitland. *See also* Tay, *and also* Flos *under* Springwater.

Tisdale SK This town (1920) east of Melfort was named in 1904 by the Canadian Northern Railway after Frederick W. Tisdale, an agent of the Canadian Northern. Its post office opened the same year. It was earlier called *Doghide*, after the river flowing south through the town. The postmaster asked the chief geographer of Canada in 1905 whether Doghide River could be renamed, as 'you would confer a great favor on me and also all the people of Tisdale and community', but it was never changed.

Tiverton NS On the northeast end of Long Island, and southwest of Digby, the first post office at this village (1969) was called *Petite Passage* in 1862 after the narrow Petit Passage, separating the island from Digby Neck. It was renamed Tiverton in 1883 after the birthplace of teacher Thomas Mildon in Tiverton, Devon, England.

Tiverton ON A village (1878) in Bruce County, northeast of Kincardine, its post office was named in 1860 after a town in England's Devon, north of Exeter.

Tobermory ON In Bruce County and at the north end of the Bruce Peninsula, this community was first known as *St. Edmunds* after the township, and *Bury* after Viscount Bury, the superintendent of Indian affairs in the 1850s, who organized the purchase of the peninsula from the Ojibwa. It was renamed in 1882 after Tobermory, Island of Mull, Scotland.

Tobique River NB Rising in the west central part of the province, this river flows southwest to enter the Saint John River at Perth-Andover. It was named after a Maliseet chief, possibly Noel Toubic (*c.* 1706–67), who once lived at its mouth, but died at Kingsclear, west of Fredericton. In Maliseet it was called *Naygoot* or *Naygootcook*, probably meaning 'flat banks beside it'. Some early maps combined both names, such as *R. Tobed Nigaurlegoh* on Charles Morris's map of 1765. Dugald Campbell called it *Tobique's River* on his 1784 map.

Toby, Mount BC In the Purcell Mountains and on the east side of Kootenay Lake, this mountain (3222 m) was named in 1888 after Dr Levi Toby, a physician and prospector, who had settled nearby in 1864.

Tofield AB A town (1909) southeast of Edmonton, it was named in 1908 by the Grand Trank Pacific Railway after the local school district. It had been named after Dr J.H. Tofield, who settled briefly at the site of the town in 1896, before moving to near Fort Saskatchewan, and then returning to Tofield in 1903. The name fits the alphabetical arrangement used by the GTP, with Ryley and Shonts to the southeast.

Tofino BC This district municipality (1984), on the east side of Clayoquot Sound and on the west coast of Vancouver Island, was named after **Tofino Inlet**, an easterly arm of the sound. The inlet was named in 1792 by Dionisio Alcalá Galiano and Cayetano Valdés after Spanish hydrographer Vincente Tofiño de San Miguel (1732–95).

Togo SK On the Manitoba border, southeast of Canora, this village (1906) was founded in 1904 by the Canadian Northern Railway and named after Vice-Adm. Heihachiro Togo (1848–1934), whose Japanese Navy had defeated the Russian forces during the Russo-Japanese War, 1904–5. The post office was opened in 1905.

Toledo ON A community in Kitley Township, united counties of Leeds and Grenville, northeast of Brockville, it was first known as *Koyl's Bridge*, *Kitley Corners*, and *Chamberlain's Corners*. *Kitley* post office, opened in 1832, was renamed Toledo in 1856 after either the city in Spain or the city in Ohio.

Tombstone Mountain YT Northeast of Dawson, at the head of **Tombstone River**, this mountain (2499 m) was named *Mount Campbell* in 1896 by surveyor William Ogilvie after Hudson's Bay Company trader Robert

Campbell. However, miners, being unaware of that name, gave it the present name after a great shaft of black rock, which rises 1600 m above the surrounding terrain.

Tompkins SK On the main Canadian Pacific Railway line, southwest of Swift Current, this village (1910) was named in 1883 after railway contractor Thomas Tompkins. Its post office was named in 1904.

Toodoggone River BC Rising northwest of Williston Lake, this river is a tributary of the Finlay River. It was earlier called *Two Brothers Creek* and was known to the Sekani as *Thudegade*. In Sekani the word *toodoggone* may mean 'water's arms' or 'eagle nest'. **Toodoggone Peak** (2350 m) is situated on the north side of the river.

Topley BC In the Bulkley River valley, east of Houston, this place's post office was opened in 1921 and named after pioneer settler William J. Topley.

Torbay NF This town (1972), north of St. John's, is located on **Tor Bay**. In 1612 pioneer settler John Guy referred to it as *Torrebay*. It would appear that the bay may have been named by English fishermen, because of its resemblance to Tor Bay, adjacent to Devon, England.

Tornado Mountain AB, BC On the continental divide, north of the Crowsnest Pass, this mountain (3099 m) was named about 1917 by Richard W. Cautley of the Alberta-British Columbia Boundary Commission because it seemed to be the centre of thunderstorm activity. It was originally called *Gould Dome* in 1858 by surveyor Lt Thomas Blakiston. Gould Dome was then transferred 4 km to the southeast to a lower peak (2894 m).

Torngat Mountains NF; **Monts Torgat**, QC The Torngat Mountains extend north between Ungava Bay and Labrador Sea. In Inuktitut the name suggests a terrifying spirit which controls all the spirits of the land, sea, sky, wind, and clouds. Mont d'Iberville (QC) and Mount Caubvick (NF) comprise the adjoining summits, about 15 m apart, of the same mountain, the highest in the Torngats.

Toronto ON The capital of the province and Ontario's largest city, its name was derived from Mohawk *tkaronto,* 'trees standing in the water', referring to ancient fish weirs in The Narrows between Lakes Simcoe and Couchiching. The oft-quoted 'place of meeting' is erroneous. The name was transferred (1680 to 1759) through *Lac de Taronto* (Lake Simcoe), *Passage de Taronto* (Holland River to the mouth of the Humber), *Rivière Taronto* (Humber River), and Fort Rouillé, also called Fort Toronto. Its site was known as Toronto until 1793, when Lt-Gov. John Graves Simcoe renamed it York, after the Duke of York's victory in Flanders. It was renamed Toronto when it was incorporated as a city in 1834. Most of the present city of Mississauga was named Toronto Township in 1805. Toronto Gore Township, named in 1819, was merged with the city of Brampton in 1974.

Torquay SK A village (1923) west of Estevan, it was named in 1913, when the Canadian Pacific Railway built a line west to Minton. A railway official is reported to have said that its well water reminded him of Torquay, on the Devon coast, east of Plymouth, England.

Torrington AB Southeast of Red Deer, this village (1964) was named in 1930 by the Canadian Pacific Railway. It may have been named after Torrington, Devon, England.

Torryburn NB In the northeast side of the city of Saint John, this community's post office was named in 1910 after Torryburn House, noted on an 1851 plan. The house may have been named after Torryburn, on the north side of the Firth of Forth, in Scotland.

Tosorontio ON Part of the municipal township of **Adjala-Tosorontio** in Simcoe County since 1994, this township was named in 1822. It is believed to have been named after a famous Mohawk chief, possibly John Deserontyon, also known as Odeserundiye (*c.* 1740–1811), who led the Mohawk to the Bay of Quinte in 1784. *See also* Deseronto.

Tottenham ON In the town of New Tecumseth, Simcoe County, this urban centre was named in 1858 after first postmaster Alexander Totten. It also replicates the name of the well-known district of London, England.

Tracadie NS On the south side of St. Georges Bay and east of Antigonish, this place received postal service in 1834. It took its name from the **Tracadie River**, derived from Mi'kmaq *tulakadik*, 'camping ground'. *See also* Grand Tracadie, PE.

Tracadie-Sheila NB A town (1992) in the northeastern part of the province, it is comprised of the former town (1966) of Tracadie and the former village (1978) of Sheila. Its post office was called *Tracady* in 1845. It was renamed Tracadie six years later, after the **Big Tracadie River**. The river's name was derived from Mi'kmaq *Tulakadik*, 'camping ground'. Samuel de Champlain referred to it as *Tregate* in 1604, and map-maker Pierre Jumeau called it *R. tracadi* in 1685. In 1880 Surveyor General Thomas Loggie identified it with its present name, but in 1948 the Canadian Board on Geographical Names authorized *Tracadie River*. The locally used name was endorsed in 1968. The post office in Sheila was named in 1898 after Sheila Foster, whose husband established the Tracadie Mills here in the 1890s.

Tracy NB A village (1966) south of Fredericton, its post office was called *Tracey Station* in 1871 and was renamed Tracy in 1925. Jeremiah Tracy (1744–1812), a native of Maine, was the first settler here. His son Jeremiah Tracy Jr built a mill on Yoho Stream, and the latter's son Jeremiah ('Boss') Tracy became a well-known lumberman.

Tracy QC The town of Tracy is located on the west side of the mouth of the Richelieu, opposite Sorel. It was created in 1954 when it was separated from the parish municipality of *Saint-Joseph-de-Sorel*. It was named after Alexandre de Prouville de Tracy (*c.* 1596–1670), who built Fort Richelieu at the mouth of the river in 1665.

Trail BC A city (1901) on the Columbia River, its post office was named *Trail Creek* by Eugene S. Topping in 1891. Its name was shortened to Trail in 1897. The creek was the Columbia River end of the Dewdney Trail, which was cut through from Hope by Edgar Dewdney in 1860.

Tranquille BC On the north side of the Thompson River and at the west end of the city of Kamloops, this place's post office was named in 1916 after the **Tranquille River**. The river was called after a chief who had died in 1841. He was known to the French traders as being quiet (*tranquille*) and patient.

Transcona MB This urban community in the eastern part of the city of Winnipeg had been an incorporated city before 1972. In 1910 its site had been chosen by the Grand Trunk Pacific Railway for its repair shops, with the name, chosen through a contest, being a blend of Transcontinental Railway and Strathcona. Lord Strathcona drove the last spike of the first transcontinental line, the Canadian Pacific Railway, at Craigellachie, BC, in 1885.

Travellers Rest PE On the east side of Summerside, its post office was opened in 1827, and was named after a public house established here about 1810 by a Mr Baker. The first postmaster, John Townsend, may have called his inn Travellers Rest in 1826.

Traytown NF On the north side of Terra Nova National Park and east of Glovertown, this municipal community (1978) was originally called Troytown. This name may have been derived from Troytown in Dorset, England, where the name is often equated with disorder or maze. Perhaps the maze of passages and channels in Bonavista Bay suggested the name. *See also* Triton.

Treherne MB Located southwest of Portage la Prairie, the post office in this village (1948) was named in 1880 after pioneer settler George Treherne. Six years later the Canadian Pacific Railway built a line from Winnipeg to Souris, and the site of the present village was relocated to the line.

Trenton NS This town (1911) on the north side of New Glasgow, was named in 1882 by

Harvey Graham after Trenton, NJ. A noted manufacturing centre, it became the site of a steel plant that year, and glass, paint, railway car, and woodworking operations were built here later.

Trenton ON A city in Hastings County and at the mouth of the **Trent River**, it was first known as *River Trent, Port Trent, Trent Port,* and *Trentown*, before the village was incorporated as Trenton in 1853. The river, named after the River Trent in central England, rises in Rice Lake, although its watershed drains the Kawartha Lakes north of Peterborough. The **Trent Canal** permitted navigation from Lake Ontario to Lake Simcoe in 1918, and the **Trent-Severn Waterway** was completed to Georgian Bay two years later.

Trepassey NF Located at the head of **Trepassey Bay** and near the southeastern end of the Avalon Peninsula, this town (1967) was named after the bay. The bay was shown on a map as *trepasse* about 1555, with the present form coming into use by the early 1600s. Edgar R. Seary, author of *Place Names of the Avalon Peninsula* (1971), concluded that it was likely a transfer name from Baie des Trépassés, north of Pointe du Raz, on the southwestern side of Bretagne, France.

Tring-Jonction QC This village (1918) was established in 1895, when the Quebec Central Railway built a branch line from here to Lac-Mégantic. The post office was called *Tring Junction* from 1895 to 1936, when the present name was adopted. The township of Tring, shown on a 1795 map and proclaimed in 1805, was named after a town in Hertfordshire, England.

Trinity Bay NF Separating the north side of the Avalon Peninsula from the rest of the island of Newfoundland, this bay was called *Baia de s ciria* in the early 1500s after an obscure seventh-century saint of Portugal, and may have been given about 1500 by explorer Gaspar Corte-Real. It was only in the early 1600s that it acquired its present name. The municipal community (1969) of **Trinity**, on the west side of the bay, was called 'the best and largest Harbour in all the land' by the 1689 *English Pilot. See also* Centreville-Wareham-Trinity.

Triton NF A town (1980) on the east side of **Triton Island**, which is in the western side of Notre Dame Bay, was formerly the rural district (1961) of *Triton-Jim's Cove-Card's Harbour.* The island may have been named after the expression 'troytown', used in the West Country (Devon) of England to describe a maze, perhaps in reference to the narrow tickles and difficult harbour entrances around the island. The island was settled by 1840 and Triton occurred in the census of 1845. See also *Traytown.*

Trochu AB A town (1962) southeast of Red Deer, it was first known as *Trochu Valley* after Armand Trochu (1857–1930), who led a group of French Army officers to Alberta in 1905, and established the St Ann Ranch Trading Company. When the place became a village in 1911, its name was shortened to Trochu. Most of the officers returned to France when the First World War broke out, and Trochu himself returned in 1917.

Trois-Pistoles QC On the south shore of the St. Lawrence and southwest of Rimouski, this town was established in 1916. It took its name from the **Rivière des Trois Pistoles**, which appeared on the 1631 map by Jean Guérard. The real story about the naming of the river has been lost in the mists of time. Perhaps it might relate to a tale of a sailor losing a silver goblet, valued at three *pistoles* (gold coins), while dipping for fresh water, or it may refer to three firearms.

Trois-Rivières QC Located at the mouth of the Rivière Saint-Maurice, where the islands of La Poterie and Saint-Quentin separate the three outlets of the river, this city has had a long and vibrant history. The parish of Immaculée-Conception-des-Trois-Rivières was set up in 1615, and was formally erected in 1678. The post office was called *Three Rivers* from 1763 to 1906, when it was changed to its present form. The city was incorporated as both Trois-Rivières and *Three Rivers* until 1937, when the English form was abolished. *Three Rivers* remained in popular use among English-language speakers and the media until about 1980. Since then the English-language media (radio, television, newspapers, magazines) have used Trois-Rivières.

Trout Creek ON A town (1913) in Parry Sound District and on a tributary of the South River, south of North Bay, it was first known as *Little Bend of the South River*, *Melbourne*, and *Barkerton*, before it was named Trout Creek in 1890.

Trout River NF A municipal community (1966) on the west coast of the island of Newfoundland and southwest of Gros Morne National Park, it was settled about 1830. It was noted in the census of 1857. The river rises in the Long Range Mountains and flows through the Upper Trout River Pond to the Gulf of St. Lawrence.

Troy NS Located on the west side of Cape Breton Island, northwest of Port Hawkesbury, this place's post office was named in 1892. It may have been called after the ancient city in Turkey.

Truro NS This town (1875) was named in 1759, after Truro, Cornwall, England, and its post office was opened in 1812. During the French regime, the Acadian settlement was called *Cobequid*, after the nearby Cobequid Bay.

Trutch, Mount AB, BC This mountain (3210 m), southwest of Howse Pass, was named in 1920 after Sir Joseph Trutch (1826–1904), lieutenant-governor of British Columbia, 1871–6.

Tryon River PE Flowing into the Northumberland Strait, southeast of Borden-Carleton, this river was named in 1765 after William Tryon (1725–88), a former governor of New York, who later became lieutenant-governor of South Carolina. Surveyor Gen. Samuel Holland had served under him during the American Revolution. *Tryon River* post office was opened in 1827 and became **Tryon** about 1864. Another post office was opened at **North Tryon** in 1872.

Tsar Mountain BC In the Clemenceau Icefield and south of Athabasca Pass, this mountain (3424 m) was named in 1920 by surveyor Arthur O. Wheeler because he was struck by its majestic domination of its surroundings.

Tsiigehtchic NT A charter community (1993) at the confluence of the Arctic Red River with the Mackenzie River, its name replaced *Arctic Red River* in 1994. The name in Gwich'in means 'mouth of the iron river', in reference to the reddish colour of the Arctic Red River.

Tuckersmith ON A township in Huron County, it was named in 1830 after Martin Tucker Smith (1803–80), a director of the Canada Company, a British company established in 1826 to develop the Huron Tract, a large area of what is now Southwestern Ontario.

Tugaske SK A village (1909) northwest of Moose Jaw, it was founded by the Canadian Pacific Railway in 1907 and named after the Cree word for 'flat land'. Its post office was opened the following year.

Tuktoyaktuk NT On the shore of Beaufort Sea and northeast of Inuvik, this hamlet (1970) was known as *Tuktuyaktok* from 1928 to 1936, when the Hudson's Bay Company opened its Port Brabant post here and named it after HBC chief factor Angus Brabant. *Port Brabant* post office was opened in 1948, but was renamed Tuktoyaktuk in 1950. In Inuvialuktun the word *tuktuujaartuq* means 'rock caribou place'. There is a legend that a shaman went looking for food for the hungry Inuvialuit. He turned two caribou, escaping into the sea, into protruding rocks. The hamlet is commonly called both 'Tuk' and 'Tuktuk'.

Tuktut Nogait National Park NT Located on the south shore of Amundsen Gulf, east of Inuvik, this national park (28,000 km^2) was established in 1996. In Inuvialuktun the name means 'calving ground of the young caribou'.

Tulameen BC Located at the forks of **Tulameen River** and Otter Creek, northwest of Princeton, this place's post office was named in 1907. The river, which joins the Similkameen at Princeton, was named after the Thompson Salish word for 'red earth'. A steep bank near Princeton was an excellent source of red ochre.

Tulita NT At the confluence of the Great Bear and Mackenzie rivers, this hamlet (1984) was formerly known as *Fort Norman*. It was renamed Tulita on 1 January 1996 after the Slavey word for 'where the waters meet'. Named by the Hudson's Bay Company in 1810, after either Alexander Norman McLeod or Archibald Norman McLeod, *Fort Norman* was an important transportation centre.

Tumbler Ridge BC This district municipality was created in 1981 to provide for community facilities for coal miners and their families in the Rocky Mountain Foothills, 112 km southwest of Dawson Creek. Its town centre was developed in the following three years. The ridge, on the east side of the town, rises some 375 m above the site of the municipality's town centre.

Tungsten NT In the Mackenzie Mountains and near the Yukon border, this place was founded in 1961 by the Canadian Tungsten Mining Corporation. From 1968 to 1986 the mining community was served by Tungsten post office.

Tupper, Mount BC This mountain (2816 m) in the Selkirk Mountains, north of Rogers Pass, was first known as *Hermit Mountain*. It was renamed in 1895 after Sir Charles Tupper (1821–1915), prime minister of Canada from 1 May 1896 to 8 July 1896.

Turnberry ON A township in Huron County, it was named in 1850 after Turnberry Castle, southwest of Ayr, Scotland, and at the mouth of the Firth of Clyde. The castle had long associations with the ancestors of James Bruce, 8th Earl of Elgin, governor of the province of Canada, 1847–54.

Turner Valley AB A town (1930) south of Calgary, its post office was named in 1926 after Robert and James Turner, who had homesteaded in the Sheep River valley in 1886.

Turtle Mountain MB Situated south of Boissevain and on the North Dakota border, this range of hills exhibits the outline of a turtle. Alexander Henry the younger mentioned it in 1806, and the maps of David Thompson (1814) and Peter Fidler (*c.* 1830) identified it. Robert Douglas, in his *Place-Names of Manitoba* (1933), claimed the turtle was the dove or pigeon, not the tortoise, but North Dakota state historian Dana Wright, in Penny Ham's *Place Names of Manitoba* (1980), asserted that it was the tortoise.

Turtleford SK Northwest of North Battleford, the post office at this town (1983) was opened in 1913. It was named after a convenient crossing of the Turtlelake River, where the Canadian Northern Railway crossed it in 1914.

Tuscarora ON A township in Brant County, it was named in 1840 after the Tuscarora nation, which migrated in 1722 from North Carolina to New York to become the sixth of the Six Nations confederacy. The name means 'shirt wearer'.

Tusket NS East of Yarmouth and on the **Tusket River**, this place's postal service was established in 1837. The river's name was derived from Mi'kmaq *Neketaouksit*, 'great forked tidal river'.

Tusk Peak BC In the Clemenceau Icefield, northeast of Revelstoke, this peak (3360 m) was named in 1920 by the Alberta-British Columbia Boundary Commission after its 'sharp cone of rock'.

Tuxedo MB A neighbourhood in the city of Winnipeg, south of the Assiniboine River, it was established in 1908 by the Tuxedo Park Company Limited, which developed wealthy residential lots. Incorporated as a town in 1908, it was annexed by Winnipeg in 1972.

Tuzo, Mount AB, BC This mountain (3245 m) on the continental divide, south of Lake Louise, was named in 1906 after Henrietta Tuzo, later Wilson (1880–1955), who was the first to ascend it.

Tweed ON A village (1891) in Hastings County, north of Belleville, its post office was named in 1852 by miller James Jamison, after the River Tweed, which forms part of the boundary between England and Scotland.

Tweedsmuir Provincial Park BC In the Coast Mountains, northeast of Bella Coola, this park was named in 1936 after John Buchan, 1st Lord Tweedsmuir (1875–1940), governor general of Canada, 1935–40.

Twillingate NF Situated on the west side of **South Twillingate Island**, this town (1962) was named after the two Twillingate islands. They took their name from Pointe de Toulinguet at the entrance to the harbour of Brest, Bretagne, France. In 1992 the town annexed Durrell, which had been incorporated as a town in 1971, when it had joined together the settlements of Durrell's Arm, Hart's Cove, Jenkin's Cove, Gillesport, and Blow-Me-Down. Durrell's Arm was likely named after an early fishing family. Twillingate also annexed the municipal community (1986) of Bayview, which had been an amalgamation of Manuel's Cove and Gillard's Cove.

Two Hills AB A town (1955) northeast of Vegreville, its post office was named in 1913 after two distinctive hills southwest of the town. Alexander Henry the younger commented in 1808 on *les Deux Grosses Buttes*, where he rested for an hour.

Tyendinaga ON A township in Hastings County, it was named in 1800 after Joseph Brant (1742–1809), whose Mohawk name was Thayendanegea. After the American Revolution, he led many Mohawk to the area of the Bay of Quinte, and later to the Brantford area on the Grand River.

Tyndall MB Located southeast of Selkirk and adjacent to the village of Garson, this place was founded in 1877 by the Canadian Pacific Railway, and named after British physicist John Tyndall (1820–93). In 1900 William Garson developed the limestone quarries, with the phrase 'Tyndall stone' subsequently becoming well known as a handsome component of public buildings across the country.

Tyne Valley PE A municipal community (1983) northwest of Summerside, its post office was opened in 1872. The name had been proposed by James Rogers for a school district, likely after the River Tyne, in Northumberland, England.

Tyrrell, Mount YT Southwest of Dawson, this mountain (1447 m) was named in 1906 on a Geological Survey map after geologist Joseph Burr Tyrrell (1858–1957), who extensively explored the Yukon and the Northwest Territories.

U

Ucluelet BC On the northwest side of Barkley Sound and on the west coast of Vancouver Island, this village (1952) was named in 1894 after the Nootka word meaning 'people of the sheltered bay', in reference to a protected landing place for canoes.

Uigg PE Located west of Montague, this place was named about 1841 by schoolmaster Donald Macdonald after Uig, Isle of Skye, Scotland. Its area was settled by Scots in 1829.

Ulu Mountain YT In the St. Elias Mountains, southeast of Mount Kennedy, this mountain (3097 m) was named in 1970 after the Inuit woman's knife, which was used that year as the symbol of the Arctic Winter Games, held in Whitehorse.

Umingmaktok NT This settlement, on Baychimo Harbour, became the site in 1964 of the Hudson's Bay Company's Bathurst Inlet trading post. The post was called Baychimo Harbour, but confusion with Fort Chimo (now Kuujjuaq) in Québec persuaded the HBC to call it Bathurst Inlet. As the former site at the mouth of Burnside River, 95 km to the south, was still called Bathurst Inlet, a new name was needed. Umingmaktok, named in 1975, took its name from an Inuktitut word meaning 'musk-ox place'.

Umiujaq QC A northern village on the eastern shore of Hudson Bay, it was founded in the mid-1980s by Inuit from Kuujjuarapik, who chose to move away from the busier centre, 160 km to the south, in order to be nearer to natural surroundings and wildlife. In Inuktitut, the name means 'it has the form of a loaf of bread', 'looks like a turned-over ship', or 'it seems like a beard', all in reference to rounded grassy hills on the bar separating the bay from Lac Guillaume-Delisle.

Ungava Bay NT Administratively part of the Northwest Territories, this 250-km bay, surrounded on the east, south, and west by the province of Québec, has a width of 275 km at its mouth, from Cap Hopes Advance, QC on the west to Killiniq Island, NF, NT on the east. Of uncertain meaning in Inuktitut, Ungava may mean 'toward the open water', 'far away', 'unknown land', 'place to the south', or 'place visited by white whales'. The vast peninsula to the west of the bay, with a base of 520 km between Inukjuaq and the mouth of Rivière aux Feuilles, is called Péninsule d'Ungava. In 1879 the area of present-day Québec north of the St. Lawrence River watershed was conveyed by the British government to Canada. The government of Canada called the area the provisional district of **Ungava**. It transferred Ungava to Québec in 1912.

Union Bay BC Southeast of Courtenay and on the east coast of Vancouver Island, this place's post office was named in 1894. It was the coal shipping port for the Union Coal Company, based at Cumberland.

Union Road PE A municipal community (1977) north of Charlottetown, its post office was opened in 1875. The road may have been named because it united Charlottetown with the north shore.

Unionville ON In the town of Markham, York Region, west of the town centre, its post office was named in 1849 after Ira White's Union Mills, opened in 1841, a year following the Act of Union of 1840 had united Upper and Lower Canada into the single province of Canada.

Unity SK A town (1919) southwest of Battleford, its Grand Trunk Pacific Railway station was named in 1909 after Unity, WI, southeast of Wausau. Its post office was opened the following year.

Unwin, Mount AB This mountain (3268 m), on the west side of Maligne Lake, was named in 1908 by climber Mary Schaffer after guide Sidney Unwin, who was killed during the First World War.

Upper Arrow Lake BC The upper of the two Arrow lakes, this narrow 65-km-long widening of the Columbia River was named in the early 1800s after *Arrow Rock*. The rock was, before the flooding of the lake behind the Keenleyside Dam, a steep cliff on the east side of the Lower Arrow Lake where the arrows of Aboriginal braves were lodged. Arrows that remained lodged in the rock were believed to bring good luck to the brave.

Upper Island Cove NF A town (1965) on the north side of Spaniard's Bay and on the west side of Conception Bay, it had been referred to as *Haylinscove* in 1697, and *Island Cove* in 1773. The cove is on the north side of Big Island, which is almost joined to the shore.

Upper Mazinaw Lake ON Near the head of the Mississippi River system, north of Kaladar, this lake was considered part of *Mazinaw Lake* until 1981, when it was divided at The Narrows into Lower and Upper Mazinaw lakes. *Mazinaw* is an Algonkian word meaning 'picture', referring to pictographs on the face of **Mazinaw Rock**, on the east shore of the lake.

Upsalquitch River NB Rising in its Northwest and Southeast branches, north of Mount Carleton, this river flows north into the Restigouche River, southwest of Campbellton. Its name was derived from Mi'kmaq *Apsetkwechk*, 'little river', in reference to its small size compared to the Restigouche. Shown on a map in 1786 as *Upsatquitch*, a map of 1803 by William Vondenvelden and Louis Chartrand did not have the 't' crossed, resulting in the spelling that has come down to the present.

Uranium City SK Near the north shore of Lake Athabasca, this place was founded in 1952 as an uranium mining community and became both a town and a municipal district in 1954. Three years after the mining and milling operations were closed in 1982, and most residents had moved away, its status was changed to a northern settlement.

Urbainville PE Situated northwest of Summerside, this place was named after early settler Urbain Arsenault. The post office was called *Urbinville* from 1896 to 1917, but the school district, first called *Portage* and *Egmont Bay Road*, had been renamed Urbainville in 1901.

Usborne ON A township in Huron County, it was named in 1830 after Henry Usborne (*c.* 1780–1840), a director of the Canada Company, a British company established in 1826 to develop the Huron Tract, a large area of what is now Southwestern Ontario.

Utopia, Lake NB Northeast of St. George, this lake was named in 1784 by either Capt. Peter Clinch or Lt-Gov. Thomas Carleton, when it was discovered that a plan of land grants extended under the lake. Getting to them would have been as impossible as achieving the ideal of Sir Thomas More's *Utopia* (1515–16). A second plan in 1829 stated that it was a 'Reserve to make good the deficiency caused by the Lake Eutopia'.

Uxbridge ON A municipal township in Durham Region, it was created in 1974 through the merging of the town of Uxbridge and the municipal townships of *Scott* and Uxbridge. The original township was named in 1798 after Uxbridge, in Greater London, England, west of the city centre. The post office in the urban centre of **Uxbridge** was named in 1832.

V

Val-Bélair QC A town in the Communauté urbaine de Québec, northwest of the city of Québec, it was erected in 1973 through the merger of the town of *Bélair* (1965) and the municipality of *Val-Saint-Michel* (1933). Bélair was an informal name as early as 1733 of the seigneury of Guillaume-Bonhomme (1682), in the area of the present town, and the name of another seigneury (also called Pointe-aux-Écureuils) to the southwest near Portneuf.

Val Caron ON In the town of Valley East, Sudbury Region, north of the city of Sudbury, this community was named in 1944 after Fr Hormidas Caron, the first Jesuit missionary in the region, who founded the parish of Blezard Valley in 1901.

Val-Côté ON In Cochrane District, southeast of Hearst, this place was named *Côté Siding* in 1927 after first postmaster G. Côté. It was changed to Val-Côté about 1935. It became part of the municipal township of **Mattice-Val-Côté** in 1983.

Valcourt QC Situated northwest of Sherbrooke, this town was established in 1974, having been erected as a village in 1929. It is surrounded by the township municipality of **Valcourt**, organized in 1965, which had been created the municipal township of *Ély* in 1855. Valcourt post office was opened in 1864, taking its name from a perceived short valley (*val court*) here. The township of Ely was identified on a 1795 map, and proclaimed in 1802, probably acquiring its name from the village of Ely, northeast of Cambridge, England.

Val-David QC North of Saint-Jérôme and adjacent to Sainte-Anne-des-Monts, this village was established 1921. It was named after Louis-Athanase David (1882–1953), secretary of the province of Québec, 1919–36, and his father, Laurent-Olivier David (1840–1926), member for Berthier in the House of Commons, 1887–99, and a prolific writer on the careers of noted Québécois.

Valdes Island BC In the Gulf Islands, between Gabriola and Galiano islands, this island was named in 1859 by Capt. George Richards after Spanish Cdr Cayetano Valdés, who explored the shores of the Gulf of Georgia with Cdr Dionisio Alcalá Galiano in 1792. The present Quadra, Maurelle, and Sonora islands had been jointly known as *Valdes Island* until 1903.

Val-des-Monts QC Located in the regional county of Papineau, north of Gatineau, this municipality dates from 1975 when four municipalities, including Perkins (1960), were amalgamated. Embracing 455 km^2, its landscape is marked by numerous lakes, streams, valleys, and hills.

Val-d'Or QC Located in the regional county of **Vallée-de-l'Or** (1981), the area of this city was established in 1968 through the amalgamation of the towns of Val-d'Or (1937) and Bourlamaque (1934) with the municipality of *Lac-Lemoine* (1958). Val-d'Or had been incorporated as a village in 1935, following a gold rush comparable to the Yukon's Klondike. Bourlamaque took its name from the township, noted on a 1911 map, and proclaimed in 1920. It was called after a noted officer of the French regime, François-Charles de Bourlamaque (1716–64), who wrote *Mémoire sur le Canada* (1762). *Lac-Lemoine*, the first post office in 1935, was named after missionary Georges Lemoine (1860–1912), author of a French/Montagnais dictionary.

Valemount BC A village (1962) in the Rocky Mountain Trench and near the watershed of the drainage basins of the Fraser and Columbia rivers, its post office was named *Cranberry Lake* in 1913. It was renamed *Swift Creek* five years later, and then was called Valemount in 1927. Its location, at 900 m, is framed by mountains rising over 2600 m on the east and west.

Val Gagné ON A community in Cochrane District, south of Iroquois Falls, its post office was called *Nushka Station* in 1911, but the place was destroyed by fire in 1916. It was rebuilt, and named Val Gagné in 1920, after Fr Wilfrid Gagné, who died at Matheson during the 1916 fire, which swept a large area of Northern Ontario.

Valhalla Ranges BC Between Lower Arrow and Slocan lakes, these ranges in the Selkirk Mountains were named *Valhalla Mountains* in 1875 by geologist Reginald W. Brock after the hall of the immortal souls of Norse heroes. The Geographic Board substituted the word 'ranges'.

Vallée-Jonction QC Located in the Beauce valley at an important Canadian Pacific Railway junction, with lines leading west to Thetford Mines and south to Saint-Georges, this municipality was formed in 1989 through the fusion of the village of Vallée-Jonction (1949) and the parish municipality of *L'Enfant-Jésus* (1900). Its post office was first called *Beauce-Jonction* in 1883, and was changed to Vallée-Jonction in 1922.

Valley East ON In Sudbury Region, north of the city of Sudbury, this town (1973) was created a municipal township in 1969 by amalgamating the municipal township of Blezard with the municipal township of Capreol and Hanmer. On the creation of Sudbury Region four years later, parts of three adjoining townships were added to it. The name decribes the agricultural valley in the heart of the region.

Valleyfield NF A neighbourhood (1996) in the town of New-Wes-Valley, it was first called *North West Arm*. It became known as Valleyfield by 1891.

Valleyfield PE A municipal community (1974) west of Montague, it embraces the localities of Head of Montague, Kilmuir, Commercial Cross, and Brooklyn. Valleyfield was named about 1864 by the Revd Alexander Munro and its post office was opened in 1871.

Valleyview AB A town (1957) east of Grande Prairie, its post office was named in 1929 for the view of the Little Smoky River valley and the valleys of the river's tributaries, the Red Willow and Sturgeon creeks.

Val Marie SK South of Swift Current and near the Montana border, this village (1926) was founded in 1910 and named by Fr Passaplan after its location in the valley of the Frenchman River, and after the Virgin Mary. The post office was opened two years later and the Canadian Pacific Railway arrived in 1924.

Valparaiso SK Midway between the towns of Tisdale and Star City, this village (1924) was named in 1904 by postmaster George E. Green, possibly after the city of Valparaiso in northwestern Indiana, itself recalling the seaport city in Chile. Green had first proposed Beaver Forks, but that was rejected because it was too similiar to other names in Canada.

Val Rita ON In Cochrane District, northwest of Kapuskasing, it was named in 1925 after the parish church, dedicated to St Rita. Since 1983, it has been part of the municipal township of **Val Rita-Harty**.

Val-Senneville QC Northeast of Val-d'Or, this municipality was created in 1980, and its post office had been opened in 1945. The names were taken from a lake and a river named Senneville, with both of them recalling the township of Senneville, proclaimed in 1916, and named after a lieutenant in the Languedoc Regiment of Montcalm's army. From 1940 to 1973 the place had been named the mining village of *Pascalis* after the adjacent township, proclaimed in 1916, and called after a captain in the La Reine Regiment in 1760. *See also* Senneville.

Vananda BC On the east shore of Texada Island and southwest of Powell River, this place was named in 1897 by copper-mine developer Edward Blewett, after his son Van Anda Blewett. The son was named after New York journalist Carl Van Anda.

Vancouver BC The first settlement at the site of Vancouver was made in the 1860s, with the names *New Brighton*, *Hastings*, and *Gastown* being attached to developing urban centres. In 1870 *Granville* was selected by residents in honour of George Leveson-Gower, 2nd Earl of Granville (1815–91), British secretary for the colonies, 1868–70, and foreign secretary, 1880–5. In 1882 pioneers John Morton and Sam Brighouse, anticipating the arrival of the Canadian Pacific Railway, laid out a street plan in present downtown Vancouver and called the proposed development the city of *Liverpool*. After the CPR chose Coal Harbour in Burrard Inlet as its West Coast terminus in 1884, another townsite was surveyed by the CPR's L.A. Hamilton. He asked William (later Sir William) Van Horne what the new seaport should be called. Van Horne replied that 'this eventually is destined to be a great city in Canada. We must see that it has a name that will designate its place on the map of Canada. Vancouver it shall be, if I have the ultimate decision.' No doubt he was guided by the knowledge that both he and Capt. Vancouver had Dutch roots. In the spring of 1886 Vancouver was incorporated as a city, and *Granville* post office was renamed Vancouver.

Vancouver Island BC This island (31,285 km^2) was named the *island of Quadra and Vancouver* in 1792 by Capt. George Vancouver, after he and Spain's Nootka Gov. Juan Francisco de la Bodega y Quadra agreed at the head of Tahsis Inlet that an appropriate feature should be named jointly after them. Subsequent marine charts had *Quadra and Vancouver's Island*, but Hudson's Bay Company traders soon shortened the name to *Vancouver's Island*. By the mid-1800s it was called simply Vancouver Island.

Vancouver, Mount YT On the Yukon-Alaska boundary and in the St. Elias Mountains, this mountain (4785 m) was named in 1874 by William H. Dall, of the Western Union Telegraph Company, after Capt. George Vancouver.

Vanderhoof BC A district municipality (1982) west of Prince George, it was named in 1914 by the Grand Trunk Pacific Railway after Chicago publicity agent Herbert Vanderhoof. In 1908 the GTP had joined the Canadian government, the Canadian Pacific Railway, and the Canadian Northern Railway to encourage Americans to migrate north of the forty-ninth parallel.

Vanguard SK A village (1912) southeast of Swift Current, it was named in 1912 by the Canadian Pacific Railway, because it was then the foremost location of the line, with anticipation of advancing further. Its post office was also opened that year.

Van Horne Range BC In Yoho National Park, northwest of Kicking Horse Pass, this range of mountains was named in 1884 after William (later Sir William) Cornelius Van Horne (1814–1915), who managed the construction of the Canadian Pacific Railway in the West, 1882–5 and was the president of the CPR, 1888–99.

Vanier ON Canada's smallest city in area (2.79 km^2), it is totally surrounded by the city of Ottawa. In 1908 it became the village of *Eastview*, taking its name from its view of the Parliament Buildings from its eastern limits. It became a town in 1912 and a city in 1950. In 1969 it was renamed after Georges Philias Vanier (1888–1967), governor general of Canada, 1959–67.

Vanier QC A town in the Communauté urbaine de Québec and on the north side of the Rivière Saint-Charles, it is surrounded by the city of Québec. It was established in 1966 on the occasion of the fiftieth anniversary of the founding of the town of *Québec-Ouest*, and named in honour of Georges Vanier, a native of Montréal and governor general of Canada, 1959–67.

Vanier, Île NT West of Bathurst Island and between Massey and Cameron islands, this island (1126 km^2) was named in 1961 after Georges Vanier, governor general of Canada, 1959–67. In 1963, a small island between Massey and Vanier islands was named Île Pauline, after the governor general's wife. **Mount Vanier** in northern Ellesmere Island was named in 1969 by glaciologist Geoffrey Hattersley-Smith.

Vanier, Mount YT Located 50 km west of Whitehorse, this mountain (1845 m) was named in 1967 by the Whitehorse Lions Club after Georges Vanier, governor general of Canada, 1959–67.

Vankleek Hill ON A town (1897) in the united counties of Prescott and Russell, south of Hawkesbury, it was first settled about 1798 by Loyalist Simon Vankleek (1743–1827). Its post office was named in 1831 after him, his son Simon (1767–1865), and his grandsons Peter and Barnabus.

Vanscoy SK Southwest of Saskatoon, the post office in this village (1919) was named in 1908 by the Canadian Northern Railway after pioneer settler Vern Vanscoy. The post office opened the following year.

Varennes QC Located on the eastern shore of the St. Lawrence River, northeast of Montréal, this town was created in 1972 through the amalgamation of the village of Varennes (1848) and the parish municipality of *Sainte-Anne-de-Varennes* (1855). The parish of Sainte-Anne-de-Varennes had its first pastor in 1692, taking its name from the seigneury of Varennes, granted in 1672 to René Gaultier de Varennes (*c.* 1635–89), who was then governor of Trois-Rivières.

Vars ON In the municipal township of Cumberland, Ottawa-Carleton Region, this community was known from 1881 to 1886 as *Bearbrook Station*. It was renamed by Fr Casimir Guillaume, a native of Vars in France's Hautes-Alpes, who had come to Canada in 1859.

Vaudreuil QC Located in the regional county of **Vaudreuil-Soulanges** (1982), this town was created in 1963 through the merging of the village of Vaudreuil (1850) and the parish municipality of *Saint-Michel-de-Vaudreuil* (1855). The seigneury of Vaudreuil was established in 1702 and granted to Philippe de Rigaud de Vaudreuil, Marquis de Vaudreuil (*c.* 1643–1725), governor of New France, 1703–25.

Vaughan ON A city in York Region, it was created a township in 1792 and named after Benjamin Vaughan, a British negotiator in 1783 of the treaty of peace with the United States. It became a town in 1971 and a city in 1991.

Vauxhall AB A town (1961) north of Taber, it was founded in 1910 by the Canada Land and Irrigation Company. Its post office was named in 1920 after Vauxhall, a district between Lambeth and Battersea in London, England, on the south side of the River Thames. Presumably the name was chosen to inspire English investment in the company's lands.

Vaux, Mount BC In Yoho National Park, south of Field, this mountain (3319 m) was named in 1858 by Dr James (later Sir James) Hector after William Sandys Wright Vaux (1818–85), an antiquarian and numismatist at the British Museum.

Vavenby BC On the North Thompson River, northeast of Kamloops, this place's post office was opened in 1910. Its name was suggested by settler Daubney Pridgeon after his birthplace of Navenby, south of the city of Lincoln, in Lincolnshire, England. The post office department misread his handwriting.

Vegreville AB East of Edmonton, this town (1906) was named in 1895 by French-speaking settlers from Kansas after Fr Valentin Végréville (1829–1903), a missionary in Western Canada after 1852. He became an accomplished linguist in Cree, Chipewyan, and Assiniboine, and was a director of the Collège Saint-Boniface in present-day Winnipeg.

Ventnor ON Located in the united counties of Leeds and Grenville, 18 km southeast of Kemptville, this place was first known as *Adams*, after four Adams families living there. *Adams* and *Adamsville* were suggested as a post office name in 1865, but it was named after a place on England's Isle of Wight.

Verchères QC On the southeast shore of the St. Lawrence and northeast of Montréal, this municipality dates from 1971 when the village of Verchères (1913) and the parish municipality of *Saint-François-Xavier-de-Verchères* (1855) were united. Its post office had been opened in 1827.

The name has its roots in the seigneury of Verchères, granted in 1672 to François Jarret de Verchères (1641–1700), an officer with the Carignan-Salières Regiment.

Verdun QC In the Communauté urbaine de Montréal, southwest of Montréal, this city was established in 1907, having been incorporated in 1875 as the village of *Rivière-Saint-Pierre*, a name in use as early as 1830. Verdun is not named after the city in eastern France, made famous during the First World War, but after the small village of Saverdun, south of Toulouse, in southern France. Zacharie Dupuy (*c.* 1608–76), a native of Saverdun, was granted some land in 1662 along the shore of the St. Lawrence and called it Verdun.

Veregin SK A village (1912), between Canora and Kamsack, it was named by the Canadian Northern Railway in 1904 after Peter Veregin (1859–1924), who led about 8000 pacifist Doukhobors from Russia to Canada in 1899. Its post office was opened as Verigin in 1906.

Verendrye, Mount BC On the western border of Kootenay National Park, this mountain (3086 m) was named in 1884 by George M. Dawson after Pierre Gaultier de Varennes de La Vérendrye (1685–1749), who explored the West, but may have only got as far as the Black Hills of South Dakota.

Vermilion AB A town (1906) between Vegreville and Lloydminster, it was named by the Canadian Northern Railway in 1906 after the **Vermilion River**, which rises south of Vegreville and flows into the North Saskatchewan, northeast of the town. Its post office from 1905 to 1906 was called *Breage*, after a place in southwestern Cornwall, England.

Vermilion Pass AB, BC This mountain pass (1651 m), west of Banff, was named by Dr James (later Sir James) Hector in 1858, after the red, yellow, and orange ochre beds in the mineral springs adjacent to **Vermilion River**, 9 km south of the pass. The river is one of the headwaters of the Kootenay River.

Verner ON In Nipissing District, northwest of Sturgeon Falls, this place's post office was named about 1883 after the family name of the wife of Archer Baker, general superintendent of the Canadian Pacific Railway.

Vernon BC This city (1892), located near the north end of the Okanagan Valley, was first known as *Priest's Valley* after Fr Paul Durieu, who had built a cabin here. In 1887 it was renamed after Forbes George Vernon, who, with his brother Charles, bought the nearby Coldstream Ranch in 1869. He was elected to the provincial legislature and served as chief commissioner of lands and works.

Vernon ON A community in Osgoode Township, Ottawa-Carleton Region, southeast of Ottawa, it was named in 1862 at a public meeting. The reason for choosing the name is unknown locally. It may have been derived from the name Mount Vernon, the Virginia home of United States President George Washington.

Verona ON In Frontenac County, north of Kingston, this community was first known as *Buzztown*, because it was a boisterous place. Called Verona in 1858 after the city in Italy, the name may have been inspired by a local Italian hotel keeper.

Vernon River PE Located northwest of Montague, this place's post office was opened in 1827. The river was named in 1765 by Samuel Holland after Adm. Edward Vernon (1684–1757), who is remembered for diluting his sailors' rum, which was called grog after his grogram (grosgrain) coat.

Verret NB On the southwest side of Edmundston, the post office in this village (1978) was opened in 1908 and closed in 1948. The first postmaster was Joseph Verret.

Verulam ON A township in Victoria County, it was named in 1823 after James Walter Grimston, Earl of Verulam (1775–1845), a brother-in-law of the prime minister, Lord Liverpool.

Veteran AB A village (1914) east of Stettler, it was named in 1911 by the Canadian Pacific Railway. It was named during the year of the coronation of George V, reflecting the King's long service during the reign of his mother, Queen Victoria. Its post office was called *Wheat Belt* from 1910 to 1913.

Vibank SK Southeast of Regina, the post office at this village (1911) was named *Elsas* in 1908, but it lasted only a month, when it was renamed *Ubank*. After another two months, it was called Vibank. The meaning of Vibank is not available in the official names records.

Victoria BC This city (1862), at the southeast end of Vancouver Island, was founded in 1843 by James Douglas as the site of a Hudson's Bay Company depot. Chief trader Charles Ross built the fort, which he believed was to be named after Prince Albert. However, the HBC's northern department decided on 10 June of that year that it was to be called Fort Victoria and a terse message from London confirmed it. The townsite was laid out in 1851–2 and in 1868 it was chosen as the capital of the colony of British Columbia. Over 300 populated places and physical features have been named after her in Canada.

Victoria NB This county, in the area of Grand Falls, was named in 1844 after Queen Victoria. No individual has been more honoured in Canada's place names than the Queen.

Victoria NF A town (1971) north of Carbonear, it was founded in the early 1800s as a place to cut wood in the winter. It was first named *Victoria Village* about 1860 after Queen Victoria.

Victoria NS This county on Cape Breton Island was named in 1851 after Queen Victoria.

Victoria ON The county was named in 1851 after Queen Victoria (1819–1901), the sovereign of Great Britain, Ireland, Canada, and her other realms across the seas from 1837, and Empress of India from 1867.

Victoria PE On **Victoria Harbour**, southwest of Borden-Carleton, this municipal community (1983) had been incorporated as a village in 1952. Its post office was named in 1870 after Queen Victoria. **Victoria Cross**, west of Montague and **Victoria West** on Egmont Bay, west of Tyne Valley, were also named after her.

Victoria Beach NS On the east side of Digby Gut and north of the town of Digby, this place was named in 1875 after Queen Victoria.

Victoria, Grand lac QC Located near the head of the Ottawa River, southeast of Val-d'Or, it has an area of 108 km^2. It was named in the 1860s after Queen Victoria.

Victoria Harbour ON In Tay Township, Simcoe County, east of Midland, this community's post office was named in 1872 after Queen Victoria. It was first known as *Hogg's Bay*, after the bay, named for early settler John Hogg.

Victoria Island NT Canada's second largest island (217,291 km^2), it was named in 1838 by Hudson's Bay Company trader Thomas Simpson after the Queen, who was crowned the year before.

Victoria Lake NF Southwest of Red Indian Lake, this lake was likely named after Queen Victoria. The lake is drained northeast into Red Indian Lake by **Victoria River**.

Victoria, Mount AB, BC This mountain (3464 m) on the continental divide, and west of Lake Louise, was named in 1886 by surveyor James J. McArthur after Queen Victoria.

Victoria Peak BC Located on Vancouver Island, west of Campbell River, this mountain (2163 m) was originally called *Mount Victoria* after the queen, but was designated a peak in 1887 by geologist George M. Dawson.

Victoriaville QC In the regional county of Arthabaska, almost midway between Montréal and Québec, it was created a village in 1860 and named after Queen Victoria. In 1890 it was reorganized as a town. It became the city of *Victoriaville-Arthabaska* in the fall of 1993 by amalgamating the towns of Victoriaville and Arthabaska and the parish municipality of *Sainte-Victoire-d'Arthabaska*. The following summer the city's name was abbreviated to Victoriaville.

Vienna ON In Haldimand-Norfolk Region, south of Tillsonburg, its post office was named in 1836 by Capt. Samuel Edison after the birthplace of either one of his ancestors or of his wife. Samuel was the grandfather of the inventor Thomas Alva Edison.

View Royal BC This town (1988) west of Victoria was named in the 1950s after its fine view of Royal Roads, the wide bay to the south.

Viking AB A town (1952) southeast of Edmonton, its post office was named in 1904 by Norwegian settlers after their Norse ancestors. When the Grand Trunk Pacific Railway passed between Viking and *Harland* in 1909, it planned to call its station Meighen, as part of its alphabetical arrangement of names. The settlers disliked a name that sounded like 'mean,' and persuaded the GTP to accept Viking. At that time Arthur Meighen, who later became the prime minister of Canada, was solicitor general in Sir Robert Borden's cabinet.

Ville-Marie QC Located in the regional county of Témiscamingue, this town was erected in 1962, having been incorporated as a village in 1897. In 1863 its name was given to the mission of the Oblates of Mary Immaculate after their patron, the Virgin Mary. Its post office was called *Baie-des-Pères* in 1891, after the Oblate fathers, but was renamed Ville-Marie in 1898.

Vilna AB North of the North Saskatchewan River and midway between Edmonton and Cold Lake, this village (1923) was named by the Canadian Northern Railway in 1919 after Vilna, then in Poland, but now known as Vilnius, the capital of Lithuania. The first post office was 3 km to the east, and was called *Villett*. It was moved to Vilna in 1919 and renamed after the station. Polish homesteaders arrived here in 1918–19.

Vineland ON In the town of Lincoln, Niagara Region, west of St. Catharines, this place's post office was named in 1894 after its rich vineyards in the best part of the region's soft fruit section.

Virden MB Located west of Brandon, near the Saskatchewan border, this town (1904) was first known as *Gopher Creek*, but was named *Manchester* in 1882 by the Canadian Pacific Railway, after the 8th Duke of Manchester. Because there were other places called Manchester in Canada, it was renamed Virden the following year. It may have been named after the duke's country residence in England, which may have been called after a place in Germany, the home of the duke's wife.

Virgin Arm NF On New World Island and south of Twillingate, this place may have been named because the land around its narrow arm was not settled until the 1870s, while most of the island had been occupied much earlier.

Virginia Falls NT On the South Nahanni River, this waterfall (90 m) was named in 1928 by American explorer Fenley Hunter after his new-born daughter, Virginia.

Virginiatown ON In Timiskaming District, east of Kirkland Lake, its post office was named in 1938 after Virginia Webster, wife of George B. Webster, the president of Kerr-Addison Mines.

Viscount SK In anticipation of the arrival of the Canadian Pacific Railway's tracks in 1909, this village was named in January 1908 after William Conyngham Plunket, 1st Baron Plunket (1764–1854), a noted Irish lawyer and judge, and a spokesman for Irish rights. When he was appointed lord chancellor of Ireland in 1830, he became Baron Plunket of Newton, and was thereafter called Viscount Plunket. The next station to the east was called Plunkett.

Viscount Melville Sound NT Part of Parry Channel, and between Victoria and Melville islands, this water body was named in 1819 by William (later Sir William) Parry after Robert Dundas, 2nd Viscount Melville (1771–1851), then first lord of the Admiralty.

Vita MB Northeast of Emerson, near the Minnesota border, this community's post office was called *Szewczenko* in 1907 after Ukrainian poet and painter Taras Shevchenko. Because of its perceived awkward spelling by postal authorities, it was renamed the following year after the Latin word for 'life'. Efforts to restore *Szewczenko* in the 1960s failed, but the Szewczenko Collegiate serves the community.

Vittoria ON A community in Delhi Township, Haldimand-Norfolk Region, south of Simcoe, its post office was named in 1819 to honour the Duke of Wellington's 1813 victory at Vitoria, in northern Spain.

Voisey Bay NF The site of intensive mineral exploration in 1995–6, this bay on the Labrador coast, 35 km south of Nain, had been named after Amos Voisey (d. 1887). A native of Plymouth, England, he settled on the bay about 1850. The former settlement of *Voisey's Bay* was first recorded in the census in 1921, but by the 1950s, most residents had relocated to Nain and Postville.

Vonda SK This town (1907) northeast of Saskatoon was named in 1905 by the Canadian Northern Railway after Vonda Warman, the daughter of American journalist Cy Warman, who followed the progress of the railway from Winnipeg to Edmonton. The post office was opened as *Vauuder* the following year, but was quickly corrected. *See also* Warman.

Vulcan AB Midway between Calgary and Lethbridge, this town (1921) was named by the Canadian Pacific Railway in 1910 after the Roman god of fire and metalworking. Postal service was provided the same year.

Vuntut National Park YT Located in north central Yukon, north of Old Crow, this national park (4345 km^2) was created in 1994 to protect valuable wetlands, fossils, the habitat of the Porcupine caribou herd, and other wildlife. Its name in Gwitch'in means 'crow flats'.

Wabamun AB A village (1980) on the north shore of **Wabumun Lake**, west of Edmonton, its post office was named in 1903. Identified as *White Lake* on the Palliser expedition map of 1865, the lake's present name was derived from the Cree for 'mirror'.

Wabana NF This town (1950) on Bell Island and in Conception Bay, was named in 1895 by Thomas Cantley of the Nova Scotia Steel and Coal Company, devising it from the Mi'kmaq words *wâbun*, 'dawn', or 'east land', and *aki*, 'place'. He chose the name because the mine was the most easterly in North America.

Wabasca-Desmarais AB North of Lesser Slave Lake, this hamlet was named in 1882 by amalgamating the hamlets of **Wabasca** and **Desmarais**. The post office at Wabasca was named in 1908 after the **Wabasca River**, whose name in Cree means 'grassy narrows'. The post office at Desmarais was named in 1927 after Fr Alphonse Desmarais (1850–1940), the first Catholic missionary at Wabasca in 1891.

Wabigoon ON On the north shore of **Wabigoon Lake** and in Kenora District, southeast of Dryden, this community's post office was named in 1897 after the lake, whose name in Ojibwa means 'white feather', possibly referring to white lilies on the shore of the lake.

Wabowden MB A community on the Hudson Bay Railway, southwest of Thompson, its station was named *Bowden* in 1928, after W.A. Bowden, then chief engineer of the department of railways and canals in Ottawa. Because there was a station already called Bowden in Alberta, his initials were added to differentiate the two stations.

Wabush NF Southeast of Labrador City and on **Wabush Lake**, this town (1967) was named after the lake. It was established in the 1950s when the iron ore was developed by Wabush Mines Ltd, and the original townsite was informally called 'city of Wabush' in the mid-1960s. The lake's name was derived from Montagnais *waboz*, 'rabbit'.

W.A.C. Bennett Dam BC This 180-m-high dam on the Peace River, above Hudson's Hope, was completed in 1967 and officially opened by William Andrew Cecil Bennett (1900–79), premier of British Columbia, 1952–72.

Waddington, Mount BC The highest mountain (4016 m) in the Coast Mountains, northwest of Bute Inlet, it was named in 1928 after Alfred Penderell Waddington (1801–72). A Victoria businessman from 1858, he set out three years later to create a railway route to the Cariboo gold fields, which would cling to the steep shoreline of Bute Inlet, and then follow the Homathko and Chilco rivers to Alexandria on the Fraser. After 19 of his workers were murdered in 1864 by Chilcotin men, he abandoned the project. From 1867 until his death he promoted a transcontinental railway scheme, which would have followed the Bute Inlet route. **Waddington Peak** (2630 m), in the Rocky Mountains, south of the Yellowhead Pass, named *Mount Waddington* in 1918 by surveyor Arthur O. Wheeler, after Alfred Waddington, was given its present name in 1951.

Wadena SK This town (1912) southwest of Humboldt, was named in 1905 by the Canadian Northern Railway on the suggestion of the Tolen family, which had come from Wadena, MN, in the north central part of the state. The post office was also opened that year.

Wager Bay NT On the northwestern coast of Hudson Bay, this bay was named in 1742 by explorer Christopher Middleton after Sir Charles Wager (1666–1743). Middleton was seeking an outlet to the 'Western Sea'.

Wainwright AB A town (1910) southwest of Lloydminster, it was named in 1908 by the Grand Trunk Pacific Railway after William

Wainwright (1840–1914), who was then second vice-president of both the Grand Trunk and the Grand Trunk Pacific railways. The post office had been called *Denwood*, 1907–8. Denwood became the post office of the Wainwright Regional Training Area of the department of national defence in 1961.

Wakaw SK South of Prince Albert and at the west end of **Wakaw Lake**, the post office in this town (1953) was named in 1905 after the Cree name of the lake, *Wakawkomah*. Meaning 'crooked lake', it is an apt description of the crescent-shaped lake.

Wakefield QC A community on the west bank of the Gatineau and in the municipality of La Pêche, it was an incorporated village from 1917 to 1975. Also in 1975 the township municipality of **Wakefield** (1845) was merged with La Pêche. The township was identified on a 1795 map, but it was not proclaimed until 1843. It was named after Wakefield in West Yorkshire, England, south of Leeds.

Waldeck SK A village (1913) east of Swift Current, its post office was named in 1906 by the Revd Klaas Peters, who had joined some Mennonites four years earlier when they emigrated from Westphalia in Germany. It was named after Pierre-Marie-René Waldeck-Rousseau (1846–1904), president of France, 1899–1902, who promoted the right of religious freedom and association.

Walden ON A town in Sudbury Region, southwest of the city of Sudbury, it was created in 1973 when the town of Lively, the townships of Denison, Dieppe, Drury, Graham, Louise, Lorne, and Waters, and parts of the townships of Dowling and Balfour were amalgamated. The name was blended from letters in the names Waters, Lively, and Denison, replicating the name of the pond in Massachusetts, where writer Henry Thoreau lived in the mid-1800s.

Waldheim SK North of Saskatoon, this town (1967) was settled by German immigrants in the 1880s, and its post office was named in 1900. In German the name means 'forest home'. There is a town called Waldheim in Sachsen, Germany, between Leipzig and Dresden.

Wales Island NT In Committee Bay, between Melville and Simpson peninsulas, this island (1137 km^2) was first called *Prince of Wales Island*, possibly by Dr John Rae in 1854, after Albert Edward, Prince of Wales (1841–1910). It was renamed in 1926 to avoid confusion with the larger island to the northwest, named in 1851.

Walhachin BC On the south bank of the Thompson River, west of Kamloops, this place was named in 1909 by American real estate developer C.E. Barnes after the Thompson Salish for 'close to the edge'. Wealthy British were attracted by the Marquess of Anglesey to live here, but various misadventures, and the return of many men to fight for Britain during the First World War, doomed the project.

Walkerton ON A town (1871) in Bruce County, it was named *Brant*, after the township, in 1852. It was renamed five years later after miller Joseph Walker (1801–73) and his son William, who laid out the place in village lots. Joseph Walker died on Manitoulin Island, having moved there in 1871.

Walker, Mount AB This mountain (3303 m) at the head of the North Saskatchewan River, was named in 1897 by British climber Norman Collie after Horace Walker (1838–1908), president of the Alpine Club of England in 1890.

Walkerville ON A neighbourhood on the east side of the city of Windsor, its post office was opened in 1869, 10 years after Hiram Walker (1816–99) established a distillery here. Incorporated as a town in 1890, it was amalgamated with Windsor in 1935.

Wallace NS At the mouth of the Wallace River, east of Amherst, this place was first called *Ramsheg*, after a Mi'kmaq designation. It was renamed in 1810 after Michael Wallace (*c.* 1744–1831), then treasurer and deputy surveyor general of the king's woods in Nova Scotia, and its postal service was established in 1828. Other post offices opened in the area include **Wallace River** (1848), **Head of Wallace Bay** (1851), **Head of Wallace Bay North** (1856), **Wallace Bridge** (1864), **Wallace Ridge** (1864), **Wallace Grant** (1884), **Wallace Station** (1892), and **Wallace Highlands** (1903).

Wallace ON A township in Perth County, it was named in 1849 after Baron Thomas Wallace, a promoter of free trade, who served in the government of Lord Goderich, 1827–8.

Wallaceburg ON A town (1896) in Kent County, it was named in 1837 by first postmaster Hugh McCallum in honour of the great Scottish hero, Sir William Wallace (1270–1305), who was hanged by the English in London.

Wallacetown ON In Elgin County, southwest of St. Thomas, this community's post office was named in 1852 by an early Scottish settler after the Scottish national hero, Sir William Wallace.

Walsh, Mount YT On the Yukon-Alaska boundary and in the St. Elias Mountains, this mountain (4505 m) was named in 1900 by surveyor James J. McArthur after James Morrow Walsh (1840–1905), commissioner and superintendent of the North-West Mounted Police in the territory, from the fall of 1897 to the spring of 1898.

Waltham QC A community in the regional county of Pontiac, it takes its name from the township of Waltham, identified on a 1795 map, but not proclaimed until 1849. Its post office has been called **Waltham Station** since 1898, although the Canadian Pacific Railway ceased serving the place in 1959. The township was named after a village in England, possibly Waltham in Kent.

Walton NS On the south shore of Minas Basin and northeast of Windsor, this place was named in 1837 after Halifax lawyer and Baptist newspaper editor James Walton Nutting (*c.* 1787–1870). Born nearby at Cheverie and a major land owner in the area, he was the son of Mary Walton Nutting.

Wanham AB A village (1958) northeast of Grande Prairie, it was likely named in 1916 by engineer J.B. Prest of the Edmonton, Dunvegan and British Columbia Railway after Wonham Manor, in Surrey, England. The post office was opened in 1918. Its site may have had a name in Cree meaning 'warm winds'.

Wapella SK On the main line of the Canadian Pacific Railway, midway between Moosomin and Whitewood, the post office in this town (1903) was named in 1883 after the Cree word for 'water under ground', in reference to the ease of drilling for water.

Wapusk National Park MB Located on the shore of Hudson Bay, southeast of Churchill, this national park (11,475 km^2) was established in 1996 to protect the denning areas of polar bears and the habitat of other wildlife. In Cree the name means 'white bear'.

Warburg AB A village (1953) southeast of Edmonton, its post office was named in 1916 after an ancient castle in Varberg, Sweden, south of Göteborg. The first settlers were from Sweden.

Warden QC Located east of Granby and just north of Waterloo, it was organized in 1916, taking its name after the post office, which replaced *Knowlton* in 1860. It was named after settler Jack Warden. *Knowlton* had been given in 1851 after *Knowlton Falls*, and the flour mill operated here by Col Paul Knowlton at the beginning of the 1800s. *See also* Knowlton.

Ward Hunt Island NT On the northwestern side of Ellesmere Island, this island was named in 1875 by explorer George (later Sir George) Nares after George Ward Hunt (1825–77), then the first lord of the Admiralty.

Wardner BC Southeast of Cranbrook, in the Kootenay River valley, this place was founded in 1896 by James F. Wardner. He also established Wardner in Idaho, a suburb of Kellogg.

Wardsville ON A village (1867) in Middlesex County, southwest of London, its post office was named *Mosa,* after the township, in

1832. It was renamed Wardsville in 1856 after John Ward, who settled here in 1810.

Ware BC On the Finlay River, northeast of Fort St. John, this place was first called Whitewater Post by the Hudson's Bay Company in 1927. When its post office was opened in 1938, it was renamed *Fort Ware*, probably after William Ware, manager of HBC's British Columbia district, 1927–32.

Warfield BC A village (1952) on the west side of Trail, it was founded in 1939 and named after an associate of F.A. Heinze, who constructed the first smelter at Trail.

Warkworth ON A community in Northumberland County's Percy Township, southwest of Campbellford, it was first called *Upper Percy*. It was named Warkworth in 1857 by postmaster Israel Humphries after Warkworth in Northumberland, England, north of Newcastle upon Tyne.

Warman SK In 1905 the Winnipeg-Edmonton line of the Canadian Northern Railway was built north of Saskatoon, crossing a railway constructed from Regina to Prince Albert in 1890. First called *Diamond*, after the crossing, it was renamed by the Canadian Northern after American journalist Cy Warman, who was following the progress of the railway. *See also* Vonda.

Warner AB This village (1908) southeast of Lethbridge was named in 1906 after American land agent A.L. Warner, who, with his partner, O.W. Kerr, encouraged settlers to take up land from Lethbridge to Coutts, along the line of the Alberta Railway and Irrigation Company.

Warren MB Located northwest of Winnipeg, this place was founded in 1904 and named after Canadian Northern Railway chief clerk A.E. Warren. Its post office was opened in 1907 and called *Hanlan*. It was renamed **Warrenton** in 1912.

Warren Grove PE A municipal community (1983) northwest of Charlottetown, it was named after the former school district of *Warrens Grove*, established before 1880.

Warren, Mount AB A mountain (3300 m) southeast of Maligne Lake, it was named in 1911 by climber Mary Schaffer after William Warren (1885–1943), her head packer and guide and, after 1915, her second husband.

Warsaw ON In Dummer Township, Peterborough County, northeast of Peterborough, this community was first called *Dummer's Mills* and *Choate's Mills*, after Thomas George Choate. In 1839, Choate named the post office after Warsaw, NY, where he had once lived. Warsaw, NY, is near Buffalo and directly south of Warsaw, ON.

Warwick ON A township in Lambton County, it was named in 1834 by Lt-Gov. Sir John Colborne after the Earl of Warwick, who was instrumental in obtaining a military commission for Colborne.

Warwick QC Situated south of Victoriaville, this town was founded in 1956, having been created a village in 1866, when it had been separated from the township municipality of **Warwick**. The latter was established in 1864, taking its name from the township, proclaimed in 1804, and likely named after Warwick, the principal borough of Warwickshire, near Coventry, England.

Wasa BC North of Cranbrook, this place was named in 1902 by Nils Hanson after Vaasa, on the west coast of Finland. **Wasa Lake**, to the north, was formerly called *Hanson Lake*.

Wasaga Beach ON A town (1974) in Simcoe County and at the mouth of the Nottawasaga River, its post office was named in 1923. It is one of Ontario's most popular summer resorts, and its beach, part of **Wasaga Beach Provincial Park**, during some years is one of the longest and widest on the continent, depending on the water level of Georgian Bay.

Wascana Creek SK Rising 100 km southeast of Regina, this creek flows through the capital city and into the Qu'Appelle River at Lumsden. In Cree *wascana* means 'pile of bones', in reference to the bleached buffalo bones found at the site of Regina in the 1880s. *See also* Regina.

Waseca SK Southeast of Lloydminster, this village (1911) was named by the Canadian Northern Railway in 1905, taking its name from the Cree for 'hill of the swan', referring to a hill to the south of the village. Its post office was opened in 1907.

Washademoak Lake NB Extending for 37 km from the mouth of the Canaan River southwest to the Saint John River, southeast of Grand Lake, this lake was named after the Maliseet word *wasetemoik*, whose meaning is unknown. In 1685 map-maker Pierre Rameau identified it as *Quaigesmock*. The present form occurred in the *Land Journal* of 1785.

Washago ON In Ramara Township, Simcoe County, north of Orillia and at the outlet of Lake Couchiching, it was named in 1868 by Ojibwa chief Bigwin, at the request of miller Henri St George. He created the name from *washagomin*, 'green and sparkling waters', referring to the clear water of the Severn River, before the Black River joins it to the north.

Washburn, Mount BC In the Rocky Mountains, west of Sparwood, this mountain (3039 m) was named in 1964 after F/O Dean J. Washburn of Fernie, who was killed in action in Europe on 24 December 1944.

Waskada MB In the southwestern part of the province, this village (1948) was created by the Canadian Pacific Railway in 1883. The name may be from Assiniboine *wastadaow*, 'better further on', because the Assiniboine wanted homesteaders to move beyond Turtle Mountain and leave them alone. The postmaster in 1905 stated that the post office was named in 1884 by the post office department, and that he was told it meant 'the best of everything'.

Waskaganish QC Situated at the mouth of the Rivière Rupert, this Cree village was reconstructed in 1987. Nine years earlier, when it had been incorporated, it was officially called *Waskaghegaganish* in Cree, *Fort Rupert* in French, and *Rupert House* in English. The English and French forms were given in 1670 after Prince Rupert, the first governor of the Hudson's Bay Company. In Cree, Waskaganish means 'little house', referring to the fort.

Waskatenau AB A village (1932) northeast of Edmonton, its post office was named in 1919 after **Waskatenau Creek**. In Cree *waskatenau* means 'opening in the banks', referring to a cut in a ridge through which the creek flows to the North Saskatchewan River.

Waskesiu SK A popular tourist destination on the southeastern end of **Waskesiu Lake** in Prince Albert National Park, its post office was named *Waskesiu Lake* in 1929. In Cree *wawaskesiu* means 'red deer'.

Waswanipi QC Located at the outlet of **Lac Waswanipi**, west of Chibougamau, this Cree village was named in 1978, two years after it was constructed. The name literally translates as 'fishing by torch water'.

Waterdown ON A community in the town of Flamborough, Hamilton-Wentworth Region, north of Hamilton, its post office was named in 1840 after the plunging of Grindstone Creek over Great Falls, at the Niagara Escarpment.

Waterford ON In the city of Nanticoke, Haldimand-Norfolk Region, north of Simcoe, this community was first known as *Averill's* (or *Avery's*) *Mills*, *Sovereign's Mills*, and *Loderville*, before it was named in 1826 after Waterford on the Hudson River, north of Albany, NY.

Waterloo ON The township was named in 1817 after the British-led victory in 1815 at Waterloo, Belgium. The county was created in 1849 and was reorganized in 1972 as the regional municipality of **Waterloo**, with three cities and four townships. In 1973 the township was divided among the cities of Cambridge, Kitchener, and **Waterloo**, and the municipal township of Woolwich. The city of Waterloo was named in 1831 when its post office was opened, and it became a village in 1857, a town in 1876, and a city in 1948.

Waterloo QC Situated between Granby and Magog, this town was established in 1890, having been incorporated as a village in 1867. The place had been named in 1815 in honour of Wellington's victory in Belgium, and its post office was opened in 1836.

Waterton Lakes AB These lakes, which include **Upper**, **Middle**, and **Lower Waterton** lakes, were named in 1858 by surveyor Thomas Blakiston after Charles Waterton (1782–1865), a noted English naturalist, who had visited the northwest of the United States. **Waterton Lakes National Park** (526 km^2) was named in 1895. It became part of the first international peace park in 1932, when it was joined with Glacier National Park in Montana.

Waterville NS In the Annapolis Valley, west of Kentville, this place's first post office was called *Pineo Village* in 1858. The office was renamed in 1873, 18 months after the residents had voted to call it Waterville. It has been part of the village of Cornwallis Square since 1977.

Waterville QC Erected as a town in 1965, and having been created a village in 1876, this place south of Sherbrooke was named in 1852 when its post office was opened. Having an excellent mill site on the Rivière Coaticook, it was earlier called *Pennoyer's Mills*, *Hollister's Mills*, and *Ball's Mills*.

Watford ON A village (1873) in Lambton County, its post office was named in 1854 after the city of Watford, in Hertfordshire, England, northwest of London.

Watrous SK Southeast of Saskatoon, this town (1909) was named by the Grand Trunk Pacific Railway in 1908. It was given in honour of Frank Watrous Morse, vice-president and general manager of the company. Its choice followed the GTP practice of naming its stations in alphabetical order, with Venn to the east and Xena to the west.

Watson SK East of Humboldt, this town (1908) was founded by the Canadian Northern Railway in 1905 and named after Senator Robert Watson of Portage la Prairie, Manitoba, who owned the site of the station. *Vossen* post office was opened in the area in 1904, but was changed to Watson in 1906.

Watson Lake YT This town was established in 1942, on the building of the Alaska Highway through the southern part of the territory, and the construction of an airport with two 1524-m (5000-ft) runways. It took its name from the lake, 16 km to the west, where Frank Watson, a Yorkshireman, settled in the spring of 1898. The lake was previously called *Fish Lake*.

Waubaushene ON In Tay Township, Simcoe County, east of Midland, this community was variously named *Baushene*, *Wabashene*, and *Wabaushene*, before the present spelling was introduced in the late 1860s. It was derived from an Ojibwa word meaning 'place of the meeting of the rocks', after two rocks that once obstructed navigation in the entrance to the harbour.

Waverley NS A community in the regional municipality of Halifax and on Lake William, north of Dartmouth, its post office was named in 1863 after Waverley Cottage. The cottage was built after 1847 by Charles P. Allen, which he called after Sir Walter Scott's novel, *Waverley* (1814).

Wawa ON In Michipicoten Township, Algoma District, north of Sault Ste. Marie, this community's post office was first called *Wa-Wa* in 1899, but it was closed in 1908. It was opened again as Wawa in 1933 after the Algoma Central and Hudson Bay Railway arrived from Sault Ste. Marie. In Ojibwa, the name means 'wild goose', referring to **Wawa Lake**, where wild geese congregate in the spring and fall. It was renamed *Jamestown* in 1951 after Algoma Steel president Sir James Dunn, and Wawa was restored in 1960.

Wawanesa MB The post office in this village (1908), southeast of Brandon, was first proposed in 1890 as Sipweske, after a Cree word, but objections were raised by the temperance movement. It was then named after *Wawonaissa*, which had been a post office from 1882 to 1883, 3 km to the southwest. That name was taken from Wawonaissa, the whippoorwill in Henry Wadsworth Longfellow's *Hiawatha*.

Wawota SK A town (1975) southwest of Moosomin, its post office was named in 1884. In Assiniboine the name may mean 'lots of snow' or 'deep snow'.

Webbwood ON A town (1906) in Sudbury District, west of Espanola, its post office was named in 1889 after Andrew Webb, its first settler.

Webequie ON On Winisk Lake, Kenora District, northeast of Thunder Bay, Webequie post office operated here from 1941 to 1944, reopened in 1964 as *Webekwei*, and became Webequie again in 1966. In Cree, it means 'place of many ducks'.

Webster, Mount NB South of the Nepisiguit River, this mountain was named in 1969 by director of surveys Willis F. Roberts after John Clarence Webster (1863–1950), a historian and antiquarian.

Wedgeport NS Southeast of Yarmouth and on the west side of the Tusket River, this community was first called *The Wedge* and *Wedge Port*. Its post office was called *Tusket Wedge* in 1860 and was changed to its present name in 1910. It was an incorporated town from 1910 to 1947.

Weedon Centre QC A village northeast of Sherbrooke, it was established in 1887, when it was separated from the township municipality of **Weedon**, which had been erected in 1855. The township, identified on a 1795 map, but not proclaimed until 1822, took its name from Weedon, Buckinghamshire, England, northwest of London.

Weldon SK Northwest of Melfort, the post office at this village (1914) was named on 1 October 1895 after Weldon Ellis, the son of postmaster George Ellis, who had settled here in 1883. The post office actually opened in the winter of 1896, a few months before Weldon died.

Welland ON The county was created in 1851, taking its name from the **Welland River**. In 1970, it and Lincoln County were united to form the regional municipality of Niagara. The river had been named in 1792 by Lt-Gov. John Graves Simcoe after the Welland River in Lincolnshire, England. *Chippawa Creek*, its previous name, still has strong local usage more than 200 years later. The **Welland Canal**, first opened in 1829 and named after the river, has never drawn water from it to replenish the canal. The second canal was opened in 1845, the third in 1887, and the current route in 1932. In 1849, the site of the present city (1917) was named *Merrittsville*, after William Hamilton Merritt, and was changed to Welland in 1858. *See also* Merritton.

Wellesley ON A township in Waterloo Region, it was named in 1840 after Richard Wellesley, 2nd Earl of Mornington and Baron Wellesley (1760–1842), eldest brother of the Duke of Wellington and governor general of India, 1798–1805. The post office in the community of **Wellesley** was named in 1851.

Wellington NS On the south shore of Shubenacadie Grand Lake and north of Dartmouth, this place was first called *Fletcher's Bridge*. Its post office was called *Fletchers Station* in 1866 and was renamed *Wellington Station* in 1902. The latter name was chosen by Lt-Gen. John W. Laurie (1835–1912), after the Duke of Wellington. *See also* Fletchers Lake.

Wellington ON The county was named in 1851 after Arthur Wellesley, Duke of Wellington (1769–1852), who won major battles in Spain during the Peninsular War, and at Waterloo, Belgium. He was prime minister of Great Britain, 1828–30.

Wellington ON A village (1862) in Prince Edward County, it was first called *Smokeville* after Daniel Reynolds, nicknamed Old Smoke by Aboriginals. In 1830 it was renamed after the Duke of Wellington, then the British prime minister.

Wellington PE A municipal community (1983) northwest of Summerside, it had been incorporated as a village in 1959. Its post office was named *Wellington Station* about 1875 after the Duke of Wellington. About 1855 its school district was called *Quagmire* after the Long Swamp, which extends east for about 7 km toward Miscouche. **Wellington Centre**, 4 km to the north, was the area of *Wellington* school district about 1864. *Quagmire* post office opened here about 1853, but was changed to *Wellington* in 1871 and was closed in 1913.

Wells BC Just west of historic Barkerville and east of Quesnel, this place was founded in 1932 when Fred M. Wells discovered the Cariboo Gold Quartz Mine. By 1940, 3000 people were living here, but 50 years later its population stood at 240.

Wells Gray Park BC In the east central part of the province, this park was established in 1939 and named after Arthur Wellesley (Wells) Gray, provincial minister of lands, 1933–44. The lake embraces the headwaters of the Clearwater River, and Mahood, Murtle, Azure, Hobson, and Angus Horne lakes.

Welsford NB Northwest of Saint John, this place's post office was named by Robert Bayard in 1859 after Maj. Augustus Frederick Welsford (1811–55) a native of Windsor, NS, who was killed 8 September 1855 scaling the Great Redan fortification of the city of Sevastopol, during the Crimean War.

Welsford NS East of Pictou, this place's post office was named in 1888 after Maj. Augustus Frederick Welsford, a native of Windsor, NS, who was killed 8 September 1855 during the Crimean War. There is another **Welsford** in the Annapolis Valley, north of Berwick, where the *Welsford Road* post office was opened in 1892.

Welshpool NB Located on Campobello Island, this place was named in 1835 by David Owen after Welshpool, in northern Wales. Its first post office was called *Campo Bello* in 1837. It was renamed *Welchpool* in 1924, and was changed in 1940 to Welshpool.

Welwyn SK The post office at this village (1907) on the Manitoba border, northeast of Moosomin, was named in 1902 by J. Wake after Welwyn in Hertfordshire, England, north of London. Welwyn Garden City in Hertfordshire was designed as a model of ideal urban planning in 1920.

Wembley AB A town (1980) west of Grande Prairie, its station and post office were named in 1924 by the Lake Saskatoon Board of Trade after Wembley, England, the site of the British Empire Exhibition, held in 1924–5 in the northwestern part of Greater London.

Wemindji QC Located on the eastern shore of James Bay, this Cree village was erected in 1978, having earlier been called *Paint Hills* and *Nouveau-Comptoir*. In Cree, the name means 'ochre mountain'.

Wentworth ON This county was named in 1816 after Sir John Wentworth (1737–1820), lieutenant-governor of Nova Scotia, 1792–1808, and the brother of Annabella Wentworth Gore, the wife of Lt-Gov. Francis Gore. In 1974 the county was reorganized as the regional municipality of Hamilton-Wentworth.

Wentworth Valley NS Northwest of Truro, this valley provides the route for the Trans-Canada Highway through the Cobequid Mountains. It was named after Sir John Wentworth (1737–1820), lieutenant-governor of Nova Scotia, 1792–1808.

Wesleyville NF A neighbourhood (1996) in the town of New-Wes-Valley, it was first known as *Coal Harbour*. It was renamed in 1884 after John Wesley (1703–91), the founder of Methodism. Incorporated as a town in 1945, it had become part of the town of Badger's Quay-Valleyfield-Pool's Island-Wesleyville-Newtown in 1992.

Westbank BC On the west bank of Okanagan Lake, opposite Kelowna, this community's post office was named in 1902, on the suggestion of John Davidson, who had arrived here 10 years earlier.

Westbrook ON In Frontenac County, northwest of Kingston, this place's post office was named *West Brook* in 1860, after the west branch of Collins Creek, and was changed in 1950 to Westbrook.

West Carleton ON A municipal township in the regional municipality of Ottawa-Carleton, it was created in 1974, when the three municipal townships of Huntley, Fitzroy, and Torbolton, in the western part of Carleton County, were amalgamated.

Westchester Station NS Northwest of Truro, this place's post office was called *Greenville* in 1870. It was changed in 1895 after a name given to the area in the 1780s by Loyalists from Westchester, NY, northeast of New York City.

Westcock NB South of Sackville, this place was named in 1854 after the Micmac word *oakshaak*, whose meaning is unknown. It was spelled several ways in the 1700s, including *Worshcock*, *Westqua*, and *West Coup*, before the present form appeared on a plan in 1792.

Western Bay NF On the west side of Conception Bay, northeast of Carbonear, this place is located on a small inlet, with the point called **Western Bay Head** on its south side. The two names originated during a hydrographic survey by Michael Lane in 1774. Before then they were called *Green Bay* and *Point Prime*, the latter because it was the first point seen by mariners on rounding Cape St. Francis, on the east side of Conception Bay.

Western Shore NS This place was named for its location on Mahone Bay. Although settled as early as 1784, it did not receive its own post office until 1924.

Westfield NB A village (1966) northwest of Saint John, its post office was opened as *Mouth of Neripis* in 1846, renamed *Westfield Station* in 1876, changed back to *Mouth of Neripis* in 1879, and finally called Westfield in 1895. It took its name from **Westfield Parish**, created in 1786, which was possibly called after Westfield, MA, or Westfield, NJ, or for its location on the west side of the Saint John River.

West Garafraxa ON A township in Wellington County, it was separated from East Garafraxa Township in 1869, and the latter was added to Dufferin County in 1874. There are many guesses as to the meaning of Garafraxa, with none of them more convincing than any other.

West Hawkesbury ON A municipal township in the united counties of Prescott and Russell, it was separated in 1884 from Hawkesbury Township. The latter was named in 1798, either for Charles Jenkinson (1727–1808), Baron Hawkesbury and Lord Liverpool, or for his son, Robert Banks Jenkinson (1770–1828), created Baron Hawkesbury in 1796. The son, who became Lord Liverpool in 1808, was prime minister of Great Britain, 1812–27.

West Lincoln ON A municipal township in Niagara Region, it was created in 1970 by amalgamating the municipal townships of Caistor, Gainsborough, and South Grimsby, all formerly in Lincoln County.

Westlock AB A town (1947) northwest of Edmonton, its original location 7 km to the west was first named *Edison*, after early settler Jack Edgson, who had arrived there in 1902. When it was moved to its present site in 1912, it occupied the homesteads of two settlers, Westgate and Lockhart. The postal name was devised from their surnames.

West Lorne ON In 1872 the Canada Southern Railway named a station in Elgin County's Aldborough Township after the German chancellor, Prince Otto von Bismarck (1815–98). The next year, the post office was called West Lorne after Archibald McKillop's Lorne Mills, named after the Marquess of Lorne, who married Queen Victoria's daughter Princess Louise in 1871, and became the governor general in 1878. The station was renamed West Lorne in 1907 after one of McKillop's shipments was left at Bismarck in Lincoln County by mistake. That year the village was incorporated as well.

West Luther ON Luther Township, named in 1831 after Protestant reformer Martin Luther (1483–1546), was divided in 1879, with East Luther transferring to Dufferin County in 1883. West Luther remained in Wellington County.

Westmeath ON A township in Renfrew County, it was named in 1830 after the east central county in Ireland. *Bellowston* post office was opened on the Ottawa River in 1832 and named after first postmaster Caleb Strong Bellows. It was changed to Westmeath in 1837, when it was moved downriver to *Goddard's Corners*, which also adopted the name Westmeath.

West Montrose ON In Woolwich Township, Waterloo Region, north of Waterloo, this place was named in 1866 by A.L. Anderson, a native of Montrose, Angusshire, Scotland. 'West' was added to distinguish it from another place called Montrose in the Niagara Peninsula.

Westmorland NB In southeastern New Brunswick, this county was named in 1785. Located northwest of Cumberland County, NS, it may have been named after the county in England, which is also adjacent to Cumberland.

Westmount NS On the west side of Sydney River, opposite Sydney, this place received postal service as *Westmount North* in 1925, but it was closed when a second office opened as *Westmount South* in 1954.

Westmount QC A city in the Communauté urbaine de Montréal, west of Mont Royal, it was named in 1895, having been established as the village of *Côte-Saint-Antoine* in 1879, and, before that, the village of *Notre-Dame-de-Grâce* in 1874. With an area of only 3.96 km^2, it is one of the smallest municipalities in the province.

West Nissouri ON Nissouri Township was named in 1818 and attached to Oxford County in 1821. West Nissouri Township was separated from it in 1852 and transferred to Middlesex County. It has been speculated that the name Nissouri may be similar to Missouri, with the meaning of 'gurgling' or 'struggling waters'.

Weston ON An urban centre in the city of York, it was named in 1843 by miller James Farr after his home town in England. Incorporated as a village in 1881 and a town in 1915, it was merged in 1967 with the municipal township of York, which became the borough of York in 1970 and a city in 1983.

Westport NF On the east side of White Bay and southwest of Baie Verte, this municipal community (1967) is comprised of the former settlements of *Western Cove*, *Pound Cove*, and *Wiseman's Cove*. The name was adopted in the early 1900s.

Westport NS On Brier Island, southeast of Digby, postal service was established in this village (1946) in 1835. It was named for its location on the west side of Grand Passage, opposite Freeport, with which it is connected by ferry.

Westport ON A village (1904) in the united counties of Leeds and Grenville, southwest of Perth, it was named in 1841 after its location at the west end of the Upper Rideau Lake, and on the Rideau Canal system.

West Redonda Island BC Between Vancouver Island and the mainland, northeast of Campbell River, this island was considered by Spanish explorers Dionisio Alcalá Galiano and Cayetano Valdés y Bazan in 1792 to be part of *Isla Redonda*, which included present West Redonda Island. The two names were endorsed in 1950. In Spanish *redonda* means 'round'.

West River PE One of the three rivers flowing into Charlottetown Harbour, it was known during the French regime as *Rivière de l'Ouest*. In 1765 Samuel Holland called it *Elliot River*, after Edward Eliot, 1st Baron Eliot (1727–1804), lord commissioner of trade and plantations in 1760. Both names competed for acceptance, until West River was endorsed in 1970. *See also* Eliot River.

West Road (Blackwater) River BC Rising in the Chilcotin, this river (227 km) flows east to join the Fraser between Prince George and Quesnel. It was named *West Road River* in 1793 by Alexander (later Sir Alexander) Mackenzie, because he had followed its route to reach the Pacific Ocean, near Bella Coola. Known locally as the Blackwater River, its official designation is West Road (Blackwater) River.

West Royalty PE In the city of Charlottetown since 1 April 1995, it had been incorporated as a community in 1983. This locality is part of the original Charlottetown Royalty, one of the 70 territorial divisions laid out in 1765 by Samuel Holland.

West St. Modeste NF Located on the north side of the Strait of Belle Isle and on the west side of Pinware Bay, the name of this municipal community (1975) may have been given by French explorers, because the name of an eighth-century abbess sounded like *Semadet*, a Basque name given in the 1500s, and still used informally across Pinware Bay at the unoccupied site of *East St. Modeste*.

West Vancouver BC This district municipality was separated from the district municipality of North Vancouver in 1912, although the name had already come into general use. It embraces an area of 990 km^2.

Westville NS This town (1894) on the west side of Stellarton, was first called *Acadia Village*, after the Acadian Coal Company, which started mining operations here in 1866. It was renamed Westville on 25 February 1868, likely because it was west of the other Pictou County coal mining towns.

West Wawanosh ON Wawanosh Township was named in 1840 for Chief Joshua Wawanosh of the Chippewas on the Sarnia Reserve, who was among chiefs who transferred a large area of Southwestern Ontario to the British Crown in 1827. It was divided in 1867 into the municipal townships of East Wawanosh and West Wawanosh, with both remaining in Huron County.

West Williams ON Williams Township, created in Middlesex County in 1830, was named after William Williams, a deputy governor of the Canada Company, a British company established in 1826 to develop the Huron Tract, a large area of what is now Southwestern Ontario. It was divided into the municipal townships of East Williams and West Williams in 1860.

Westwold BC Southeast of Kamloops, this place was known as *Grande Prairie* as early as 1826. Its post office was called *Adelphi* in 1900, but it never really supplanted *Grande Prairie*. When *Adelphi* was dropped in 1926, *Grande Prairie* was no longer available for a postal name. Long time resident L.R. Pearse recommended Westwold, 'wold' being an English word for a high plain, a description of the land 'west' of the railway station.

West Zorra ON Zorra Township was divided into the municipal townships of East Zorra and West Zorra in 1850. On the restructuring of Oxford County in 1975, West Zorra became the municipal township of Zorra, and East Zorra became the municipal township of East Zorra-Tavistock.

Wetaskiwin AB A city (1906) south of Edmonton, it was named by the Calgary and Edmonton Railway about 1891 after the Cree *Witaskiwinik*, 'place of peace', referring to the nearby Peace Hills, where the Cree and Blackfoot made peace about 1867.

Weyburn SK This city (1913) was named in 1893 by the Canadian Pacific Railway, after either a construction contractor who laid the steel, or his brother-in-law. Its post office was opened in 1895.

Weymouth NS On the Sissiboo River, southwest of Digby, this village (1924) was settled in the 1760s by fishermen from Weymouth, MA. Its post office was called *Weymouth Bridge* in 1865 and became Weymouth in 1907. It was earlier called *Sissiboo*, but the name was discarded because it was considered uncouth. The first Weymouth post office had been opened in 1833 nearer the river's mouth, but it became **Weymouth North** in 1907.

Whalley BC The principal centre of the city of Surrey, it was founded in 1925 by Arthur Whalley. Whalley post office was opened in 1948, changed to *North Surrey* in 1966, and renamed Surrey three years later.

Whapmagoostui QC Located at the mouth of the Grande rivière de la Baleine, this Cree village shares its site with the northern (Inuit) village of Kuujjuarapik. In Cree, the name means 'river of the white whale'. *See also* Baleine, Grande rivière de la, *and* Kuujjuarapik.

Wha Ti NT Northwest of Yellowknife and on Lac la Martre, this community, previously known as *Lac la Martre*, was renamed 1 January 1996. Its name in Slavey and French means 'marten lake'. *See also* Martre, Lac la.

Wheatley ON A village (1914) in Kent County, northeast of Leamington, it was named in 1864 after Richard Wheatley, the father-in-law of first postmaster William Buchanan. Wheatley had settled here in 1832 and had recently died.

Wheeler, Mount BC In the Selkirk Mountains, south of Rogers Pass, this mountain (3386 m) was named in 1904 after Arthur Oliver Wheeler (1860–1945), who had surveyed in the area of Glacier National Park, 1900–2. He was the British Columbia representative on the Alberta-British Columbia Boundary Commission, 1913–25.

Whistler BC A resort municipality (1975) northeast of Squamish, it was called after **Whistler Mountain**. The mountain was named in 1965 after the whistling marmots on its slopes. It had been called *London Mountain* in 1932.

Whitbourne NF Southwest of St. John's, this town (1968) was the site of the Grange estate of Sir Richard Bond, prime minister of Newfoundland, 1900–9. He proposed the name in 1889 after Sir Richard Whitbourne (*c.* 1565–1628), who was involved in the fishery off Newfoundland as early as 1579, and governed a colony at Aquaforte for Sir William Vaughan, 1618–20. The place was earlier called *Harbour Grace Junction*, after a rail line constructed north to Harbour Grace in the 1880s, and *Davenport Junction*, after Daniel Davenport, who contracted to build the line.

Whitby ON Whitby Township was named in 1792 after the seaside town of Whitby, in England's North Yorkshire. The centre of the town was first known as *Hamer's Corners* and *Perry's Corners*, with the port on Lake Ontario beng called *Windsor*. The two areas were united as the town of Whitby in 1854. In 1974, on the creation of Durham Region, the municipal township of Whitby was merged with the town of Whitby.

Whitchurch-Stouffville ON Whitchurch Township was named in 1798, probably after one of the several places with that name in England and Wales. When York Region was created in 1971, most of the township, part of the municipal township of Markham, and the village of Stouffville became the town of Whitchurch-Stouffville.

White City SK East of Regina, this village (1967) was established in 1960 as a residential community. Although the name may have been given because it simply had a nice sound, it may been given in recollection of the modernistic White City at the 1893 Chicago World's Fair.

Whitecourt AB A town (1971) on the south bank of the Athabasca River, northwest of Edmonton, it was named in 1909 after the first postmaster, Walter White, who previously carried the mail from his former home in Greencourt, 33 km to the southeast.

White Fox SK A village (1941) on the **White Fox River**, north of Nipawin, its post office was named in 1920, 10 years before the Canadian Pacific Railway was extended west toward Prince Albert.

Whitehorn Mountain BC Northwest of Mount Robson, this mountain (3395 m) was named in 1911 by surveyor Arthur O. Wheeler after its sharp, snow-covered peak.

Whitehorse YT The capital and only city in the territory, it was established in 1900 at the head of navigation on the Yukon River. It was named after the **Whitehorse Rapids**, which likely received its name from the first miners in 1880–1, who perceived the manes of horses in its white waters. The first townsite was laid out in 1899 on the east side of the river, but it developed the following year on the west side, when the White Pass and Yukon Railway was completed from Skagway. The place was first called *Closeleigh* after the Close brothers, of London, England, who financed the construction of the railway. But by 21 April 1900 the railway officials renamed it Whitehorse, because it had become well known around the world. The post office and town were called *White Horse*, which was changed to the one-word form in 1957.

White Lake ON On the east shore of **White Lake** and in Renfrew County, southwest of Arnprior, this place was named in 1848. The lake was shown on some early maps as *Wabalac*, a blend of Algonquin for 'white' and French for 'lake'.

Whitemantle Mountain BC In the Coast Mountains, south of Mount Waddington, this mountain (2969 m) was named about 1927 by climber W.A.D. (Don) Munday because there was a glacier near its summit.

Whitemouth MB Located on the **Whitemouth River**, which flows into the Winnipeg River, 18 km downriver from the community on the main line of the Canadian Pacific Railway, this place's post office was named in 1877. The river was identified on the Palliser expedition map of 1865.

White Pass BC A pass (889 m) on the British Columbia-Alaska boundary, it is the summit of the famous route from Skagway to the Klondike. It was named in 1887 by surveyor William Ogilvie after Thomas White (1830–88), minister of the interior, 1885–8.

White River ON In Algoma District, midway between Sault Ste. Marie and Thunder Bay, this municipal township is where the Canadian Pacific Railway and the Trans-Canada Highway cross **White River**, which flows west into Lake Superior. Called *Snowbank* by the CPR in the mid-1880s, the station was renamed White River in 1887.

White River YT Rising west of the Yukon-Alaska boundary and in the St. Elias Mountains, this river (265 km) was named in 1850 by Hudson's Bay Company trader Robert Campbell, after its colour, caused by suspended volcanic ash. Volcanic eruptions had occurred about 1000 AD, and ash was deposited to the depth of 30 m in some places.

White Rock BC A city (1957) just north of the forty-ninth parallel and southeast of Vancouver, it was named after a large rock on the beach, painted white to guide mariners. There is an Aboriginal legend that a chief hurled the rock across the Strait of Georgia and where it landed he agreed to live with the girl he loved.

Whitesaddle Mountain BC In the Coast Mountains, northeast of Mount Waddington, this mountain (2972 m) was named in 1933 by mountaineer Henry S. Hall Jr of Cambridge, MA. A glacier flowing down its face reminded him of a saddle, which he deemed appropriate for this cattle-ranching country.

White Tower, The BC In the Tower of London Range of the Muskwa Ranges of the Rocky Mountains, this mountain (2819 m) was named in 1960 by Capt. M.F.R. Jones after one of the towers of the Tower of London. A party of British and Canadian fusiliers climbed it.

Whiteway NF On the southeast side of Trinity Bay, this municipal community (1975) was first known as *Witler's Bay* and *Witless Bay*. It was renamed in 1912 after Sir William Vallance Whiteway (1828–1908), who served three terms as prime minister of Newfoundland between 1878 and 1897.

Whitewood SK A town (1892) on the main line of the Canadian Pacific Railway, east of Regina, it was founded in 1882. Its post office was called *Whitewood Station* from 1883 to 1895, when it was shortened to Whitewood. A bluff (grove) of white poplars inspired the name.

Whitney ON In Nipissing District and at the eastern entrance of Algonquin Park, this community's post office was named in 1895 after Edward Canfield Whitney (1844–1924), the president of the St Anthony Lumber Company.

Wholdaia Lake NT Southeast of Great Slave Lake, this lake (608 km^2) was identified on Samuel Hearne's map of 1772. In Chipewyan the name means 'icefish'.

Whonnock BC On the north shore of the Fraser River and in the district municipality of Maple Ridge, the place's post office was called *Whonock* in 1885. It was renamed Whonnock in 1969 after a local petition. In Halkomelem the name means 'place where there are (always) humpback salmon'.

Whycocomagh NS Located at the head of St. Patricks Channel of Bras d'Or Lake, its name was derived from Mi'kmaq *Wakogumaak*, 'beside the sea', or 'end of the bay'. Postal service was established here in 1834.

Wiarton ON A town (1894) in Bruce County, northwest of Owen Sound, its post office was named in 1868 after Wiarton Place, the birthplace near Maidstone, Kent, England, of Sir Edmund Walker Head (1805–68), governor of the province of Canada, 1854–61.

Wickham QC South of Drummondville, this municipality was created in 1957, having been incorporated as the village of *Wickham-Ouest* in 1922. In 1867 it had been erected as the municipality *Wickham-Ouest*, although the post office from 1874 to 1954 was called *Wickham West*. The township of Wickham was shown on a map in 1795, and was proclaimed in 1802, taking its name from a place in England, possibly Wickham in Berkshire, west of London.

Wikwemikong ON On Manitoulin Island, southeast of Little Current, this community was founded about 1825 by the Ottawa, on their return from exile on the west side of Lake Michigan. The post office was opened in 1890. The name in Ottawa means 'beaver bay'.

Wilberforce ON A township in Renfrew County, it was named in 1851 after William Wilberforce (1759–1833), a distinguished British statesman, who strongly opposed slavery.

Wilberforce Falls NT On the Hood River and west of Bathurst Inlet, this waterfall (49 m) was named in 1821 by John (later Sir John) Franklin after William Wilberforce (1759–1833), the noted philanthropist.

Wilcox SK Directly south of Regina, this village (1907) was founded by the Canadian Pacific Railway in 1902. Prior to that year new settlers had to haul their possessions from Milestone, 14 km to the southeast. The settlers agreed to name a station after CPR official Albert Wilcox, if the CPR would provide a regular stop. It is the location of the Athol Murray College of Notre Dame, founded in 1927 by Fr Athol Murray.

Wilkie SK A town (1910) southeast of Battleford, it was first known in 1906 as *Glen Logan*. The Canadian Pacific Railway was built through in 1908, with plans to locate its Wilkie station 8 km east of Unity, and the present site was to be called Adanac (Canada spelled backward). When water was not found at Wilkie, the CPR switched the two names. Wilkie was named after D.R. Wilkie, president of the Imperial Bank of Canada and a promoter of the CPR.

Williamsburgh ON A township in the united counties of Stormont, Dundas and Glengarry, it was named in 1787 after Prince William Henry (1765–1837), the third son of George

III. In 1830 he succeeded his brother George IV and served as William IV until his death. The post office in the community of **Williamsburg** (without an 'h') in the township was named *Williamsburg North* in 1841, changed the following year to *North Williamsburg*, and to Williamsburg in 1907.

Williamsford ON A community in Sullivan Township, Grey County, south of Owen Sound, its post office was first called *Sullivan* in 1847, but it was renamed Williamsford in 1886 after blacksmith Alfred Williams.

Williams Lake BC This city (1981) in the Cariboo was named as a colonial post office in 1861 after the lake to its southeast. The lake was named after Chief William of the Shuswap's Sugarcane band. When the Chilcotin killed 19 whites in 1864 during Alfred Waddington's railway survey, Chief William dissuaded other Aboriginals from taking part in the massacre.

Williamstown ON In the united counties of Stormont, Dundas and Glengarry, northeast of Cornwall, this community was founded in 1790 by Sir John Johnson, who named it after his father, Sir William Johnson (*c.* 1715–74), superintendent of the northern Indians in New York.

Willingdon AB A village (1928) northeast of Edmonton, it was named in 1928 by the Canadian Pacific Railway after Freeman Freeman-Thomas, 1st Marquess of Willingdon (1866–1941), governor general of Canada, 1926–31. Its post office was also named in 1928. **Mount Willingdon** (3373 m), north of Lake Louise, was named in 1927.

Williston Lake BC At the head of the Peace River, this reservoir (1660 km^2) was created in 1968. It was named after Ray Williston, the provincial minister of lands and forests in W.A.C. Bennett's Social Credit government. The reservoir is impounded by the W.A.C. Bennett Dam.

Willow Bunch SK Almost due south of Moose Jaw, this town (1960) was first settled by Métis from Manitoba about 1865. They found a coulee filled with red willows, which they called *Talle de Harre Rouge* (bunch of red willow). When the North-West Mounted Police arrived in 1877, the name was translated as *Red Willow Bunch*, and was shortened to Willow Bunch in 1895, when the post office opened.

Willowdale ON In the city of North York, this urban centre's post office was named in 1855 by David Gibson, who had migrated from Scotland in 1829. The reason he chose the name is not explained in official names records.

Wilmer BC In the upper Columbia River valley, north of Invermere, this place was first known as *Peterborough*. It was renamed in 1906 after Wilmer Charles Wells, member of the provincial legislature, 1899–1906, and minister of public works.

Wilmot NS In the Annapolis Valley, between Middleton and Kingston, this place's postal way office was named in 1837 after Wilmot Township. The township was named in the 1760s after Montagu Wilmot (d. 1766), lieutenant-governor of Nova Scotia, 1763–4, and governor, 1764–6.

Wilmot ON A township in Waterloo Region, it was named in 1825, likely after Sir Robert John Wilmot Horton (1784–1841), under-secretary for war and the colonies in Lord Liverpool's administration, who was also involved in the Canada Company, which developed the Huron Tract. He assumed the additional name of Horton when he married Anna Beatrix Horton in 1823. The township may also have been named after surveyor Samuel Street Wilmot (1774–1856), who was a government inspector of Crown and clergy lands.

Wilmot PE Part of the city of Summerside, it had been incorporated as a village in 1965 and became a municipal community in 1983. Called *Wilmot District* before 1880, it took its name from the **Wilmot River**. The river had been named *Wilmot Cove* in 1765 by Samuel Holland after Montagu Wilmot (d. 1766), governor of Nova Scotia, 1764–6.

Wilno ON In Renfrew County, east of Barry's Bay, it was settled by Polish immigrants in 1864, and named in 1885 after the birthplace of Fr Ludwick Dembski, which is now Vilnius, the capital of Lithuania.

Wilson, Mount AB This mountain (3260 m), east of the Columbia Icefield, was named in 1898 by British climber Norman Collie after Thomas (Tom) Edmonds Wilson (1859–1933), a Banff packer and guide.

Wilsons Beach NB On Campobello Island, this place was named after Robert Wilson, who settled on the island in 1766, and successfully rejected William Owen's claim to the entire island. Wilsons Beach post office was opened in 1859.

Winchester ON A township in the united counties of Stormont, Dundas and Glengarry, it was named in 1798 after the English city of Winchester, Hampshire. The village (1887) of **Winchester** was first called *Bates Corners*. Its post office was opened as *West Winchester* in 1855, because Chesterville's post office was called Winchester from 1845 to 1876. In the mid-1880s the station at *West Winchester* was called Winchester, leading to its post office dropping 'West'.

Winchester Springs ON In the united counties of Stormont, Dundas and Glengarry, south of Winchester, its post office was named in 1864. A natural spa was operated here from 1870 to 1904.

Windermere BC On the eastern shore of **Windermere Lake**, and in the upper Columbia River valley, this place was founded in 1887. In the early 1800s the lake was called *Kootenae Lake* and *Lower Columbia Lake* before it was named in 1883 by G.M. Sproat, because it resembled Lake Windermere in England's Lake District.

Windsor NS A town (1878) on the Avon River, it was named in 1784 after Windsor Township, created 20 years before and named after Windsor in Berkshire, England. The site was earlier called *Pisiquid* (*Pesiquid*) by the Acadians from the Mi'kmaq for 'place where the tidal flow forks'.

Windsor ON A city (1892) in Essex County, it was named in 1835 after the town in Berkshire, England, the site of the famous Windsor Castle. One hundred years later it annexed the towns of Sandwich, East Windsor, Walkerville, and Ojibway, and in 1966, the separate town of Riverside.

Windsor QC Located north of Sherbrooke, this town (1914) had been incorporated as the town of *Windsor Mills* in 1899, which had been erected as village in 1876. The township of Windsor was shown on a 1795 map, and proclaimed in 1802, taking its name from the town of Windsor, in Berkshire, England, west of London.

Windthorst SK South of Grenfell, this village (1907) was founded in 1907 by the Canadian Pacific Railway, and named after German parliamentarian Ludwig Windthorst (1812–91), who vigorously debated many policies of Prince Otto von Bismarck. The post office was also opened in 1907.

Winfield BC On the north side of Kelowna, this place was first called *Alvaston*, likely after a suburb of Derby, England. It was renamed in 1948 after Winfield Lodge, the residence of rancher Thomas Wood.

Wingham ON A town (1879) in Huron County, northeast of Goderich, its post office was named in 1862 after Wingham in Kent, England, east of Canterbury.

Winisk ON A Hudson's Bay Company post called Fort Weenisk was opened in 1820 at the mouth of the **Winisk River**. The post office was called *Weenusk* in 1945 and renamed Winisk two years later. Destroyed by a flood in 1986, it was relocated upriver, and named Peawanuck. In Cree, *winisk* means 'woodchuck'.

Winkler MB In 1892 the Canadian Pacific Railway wanted to name its station at this town (1954) after landowner Isaac Wiens, but he declined because such practice went against his Mennonite faith. Morden lumber merchant and legislator Valentine Winkler exchanged a quarter section at the station site with Wiens, and the CPR named it after him. Winkler built an elevator, flour mill, and hotel here.

Winlaw BC Located in the Slocan River valley, this place was named in 1903 after first postmaster John B. Winlaw, who had built a sawmill here three years earlier.

Winnipeg MB This name for the city first appeared on the the front page of the *Nor'Wester* of 24 February 1866, with the previous issue of the newspaper being headed by 'Red River Settlement, Assiniboia'. References to the town of Winnipeg occurred in 1870. The first post office was opened that year as *Fort Garry*, but it was changed to Winnipeg in 1873, the year the city was incorporated. The *Red River Colony* was established here by Lord Selkirk in 1812, and the Hudson's Bay Company's Fort Garry was erected 1821–2. The **Winnipeg River** drains the Lake of the Woods and several other lakes in northwestern Ontario and flows across the Manitoba border to **Lake Winnipeg**. In Cree and Ojibwa, the name of the lake means 'murky water', in reference to the cloudiness of the water where the river flows into the lake. The Jesuit *Relation* of 1640 has a reference to *Ounipigon*, 'dirty people'.

Winnipeg Beach MB On the shore of Lake Winnipeg and south of Gimli, this town (1914) was incorporated as a village in 1909, seven years after the Canadian Pacific Railway was built north along the shore of Lake Winnipeg. Its post office was opened in 1904.

Winnipegosis MB A village (1915) on the southwestern shore of **Lake Winnipegosis** and north of Dauphin, it was first called *Mossy Point*. Winnipegosis post office was opened in 1898 and the Canadian Northern Railway arrived a year later. A map of 1720 identified the lake as *Ouenpigouchib*, meaning 'little Winnipeg', or 'little murky lake'.

Winona ON In the city of Stoney Creek, Hamilton-Wentworth Region, this community was called *Ontario* in 1851. No one objected to calling it 'Ontario, Ontario' after Confederation in 1867, but it was changed to Winona, after the legendary first-born daughter of the Sioux.

Winsloe PE Part of the city of Charlottetown, this place had been incorporated as a municipal community in 1983. It took its name from John Hodges Winsloe, the proprietor of its area in the mid-1800s. The municipal community of **Winsloe South**, north of Winsloe and on the southern part of Winsloe Road, was established in 1983.

Winston Churchill Range AB Embracing an area of 325 km^2 in the Rocky Mountains, north of the Columbia Icefield, this range of mountains was named in 1965 for Sir Winston Leonard Spencer Churchill (1874–1965). He is remembered, among many accomplishments, for his perseverence as prime minister of the United Kingdom in preventing the Nazis from overrunning Great Britain during the Second World War.

Winstone, Mount BC In the Coast Mountains, northwest of Pemberton, this mountain (3124 m) was named in 1962 by Alpine Club of Canada climbers after Harry P. Winstone. He had been killed three years earlier while climbing Mount Argus in Strathcona Provincial Park, BC.

Winterton NF A town (1964) on the eastern shore of Trinity Bay, northwest of Carbonear, it was called *Sille Cove* in 1675, and *Sillee Cove* and *Scilly Cove* occurred in later records. In 1912 it was renamed after Sir James Spearman Winter (1845–1911), prime minister of Newfoundland, 1897–1900.

Wishart SK Northeast of Yorkton, the post office at this village (1937) was named in 1884 after homesteader Robert Wishart, who failed at farming here, and went to Dauphin, MB, in 1891–2, where he prospered. The Canadian Pacific Railway built a line from Foam Lake to it in 1928.

Witless Bay NF South of St. John's, this municipal community (1986) is located at the head of **Witless Bay**, whose name has engendered much speculation over the years. The present spelling was in use in the mid-1600s, but spellings such as *Whittles*, *Whitley's*, and *Whitles* during the latter part of the 1600s suggest the possibility of it being a personal name. Because the bay is quite open to the sea, its name might suggest a seaman would have to be deranged to consider trying to enter it during a heavy storm.

Wolfe Island ON Named *Grande Isle* by the French, it was renamed in 1760 after Maj.-Gen. James Wolfe (1727–59), who died while his British forces won the Battle of the Plains of Abraham. It became a township in 1792, and **Wolfe Island** post office was opened in Marysville in 1845.

Wolford ON A township in the united counties of Stormont, Dundas and Glengarry, it was created in 1797 and named after Wolford Lodge, John Graves and Elizabeth Simcoe's family home, in Devon, England.

Wolfville NS In the Annapolis Valley and on the Cornwallis River, this town (1893) was named in 1829 after Elisha De Wolf. He had been the postmaster of *Horton*, opened in 1786, and called after Horton Township. The site was also known as *Upper Horton* and *Mud Creek*.

Wollaston Lake SK With an area of 2681 km^2, this lake in northeastern Saskatchewan was named in 1821 by explorer John (later Sir John) Franklin after William Hyde Wollaston (1766–1828), a prominent English scientist.

Wolseley SK A town (1898) on the main line of the Canadian Pacific Railway, east of Regina, it was named in 1882 after Col (later 1st Viscount) Garnet Wolseley (1833–1913), leader of the Red River Expedition in 1870 to put down the first North-West Rebellion, and commander-in-chief of the British Army, 1895–1900.

Wolstenholme, Cap QC The most northerly point of the province, it was named in 1610 by Henry Hudson after Sir John Wolstenholme (1562–1639), who supported Hudson's search for the Northwest Passage. In 1744 map-maker Jacques Nicolas Bellin identified it as *Cap Saint-Louis*, and this name was still being shown on maps as recently as 1980, although Cap Wolstenholme had been endorsed as the sole official name in 1968.

Wood Buffalo National Park AB Canada's largest national park, it embraces an area of 27,840 km^2 in northern Alberta and in the Northwest Territories, south of Great Slave Lake. It was established in 1922 as a preserve for the wood bison, with plains bison being moved here between 1925 and 1927. On 1 April 1995 the city (now an urban service centre) of Fort McMurray and the Improvement District No. 143 were amalgamated to form the regional municipality of **Wood Buffalo**.

Woodbridge ON In the city of Vaughan, York Region, this urban centre was first called *Burwick* about 1837 after miller Rowland Burr. When a post office was applied for that year, *Burwick* was not accepted, because of possible confusion with Berwick in Eastern Ontario. It was then called *Vaughan* until 1855, when it was renamed after Woodbridge in Suffolk, England, northeast of Ipswich.

Wood Islands PE At the southeastern side of the province, this place is the terminus of a ferry to Caribou, NS. Its post offices were called *Wood Islands North* in 1889 and *Wood Islands West* in 1905. The two original islands, called *Isles à bois* during the French regime, were greatly altered after 1868, when the harbour facilities were developed.

Wood, Mount YT In the St. Elias Mountains, this mountain (4842 m) was named in 1900 by surveyor James J. McArthur after Zachary Taylor Wood (d. 1915), a North-West Mounted Police inspector in Dawson during the Klondike Gold Rush. He was later the commissioner of the NWMP.

Woodridge MB Located southeast of Steinbach, this place was named in 1899 by the Canadian Pacific Railway, after it crossed a ridge and built its station at the edge of woods. Its post office was opened in 1902.

Woods Harbour NS On the South Shore and southwest of Shelburne, this narrow harbour was named after the Revd Samuel Wood, who settled here shortly after 1772. Post offices were located at **Upper Woods Harbour** (1872), **Lower Woods Harbour** (1872), and **Central Woods Harbour** (1897).

Woods, Lake of the MB, ON First called *Lac aux Iles, lac des Assiniboines*, and *lac des Gens de la Pierre-Noire,* it was called *Lac des Bois* in 1730 by explorer Pierre Gaultier de La Vérendrye. This name was subsequently translated as the Lake of the Woods.

Woodstock NB A town (1856) on the west bank of the Saint John River, its post office was opened in 1830. It was named after **Woodstock Parish**, established in 1786. The parish may have been named after William Henry Cavendish Bentinck, 3rd Duke of Portland and Viscount Woodstock (1738–1809), who was briefly the prime minister of Great Britain in 1783, or it may have been after Woodstock, in Oxfordshire, England.

Woodstock ON A city (1901) in Oxford County, it was founded in 1832 by Capt. Andrew Drew. He and Rear-Adm. Henry Vansittart, who arrived two years later, chose the name Woodstock in 1835 after the village of Woodstock in Oxfordshire, England, northwest of the city of Oxford.

Woodville ON In Victoria County, northwest of Lindsay, this village (1884) was named in 1853 by postmaster, and later village clerk, John Calder Gilchrist. As it could be confused with other names, a plebiscite in 1877 resulted in Otago, a region of New Zealand's South Island, outpolling Woodville by one vote (46 to 45), but the change was not made.

Wooler ON In Northumberland County, northwest of Trenton, this place's post office was named in 1857 after Wooler in Northumberland, England, south of Berwick upon Tweed.

Woolley, Mount AB This mountain (3405 m), in the Winston Churchill Range, north of the Columbia Icefield, was named in 1898 by British climber Norman Collie after British alpinist and publisher Herman Woolley (1846–1920).

Woolwich ON A township in Waterloo Region, it was named in 1822 after Woolwich, Kent, England, beside Greenwich. When the regional municipality of Waterloo was formed in 1974, the municipal township of Woolwich annexed the town of Elmira and part of the municipal township of Waterloo.

Wotton QC A municipality north of Sherbrooke, it was organized in 1993 when it annexed the village (1919) of *Wottonville*. It had been established as the township municipality of Wotton in 1855, taking its name from the township, shown on a 1795 map, but not proclaimed until 1849. The township was named after a village in Surrey, England, southwest of London.

Wrigley NT On the east bank of the Mackenzie River, 190 km northwest of Fort Simpson, this settlement was named in 1915. It took its name from Fort Wrigley, a Hudson's Bay Company post founded in 1880 on the west bank, and named after its chief commissioner, Joseph Wrigley. Efforts in 1974 to rename the place Fort Wrigley were rejected, as it is not at the site of the original fort.

Wroxeter ON In Huron County, east of Wingham, this community was surveyed in 1858 by James Patton and named after his birthplace in Shropshire, England, southeast of Shrewsbury.

Wunnummin Lake ON Located on **Wunnummin Lake** in Kenora District, north of Thunder Bay, this community's post office was named in 1964. The lake's name is from the Cree word for 'red paint'.

Wycliffe BC Midway between Cranbrook and Kimberley, this place was named in 1906 after John Wycliffe (*c.* 1330–84), a church reformer and translator of the Bible.

Wyebridge ON In Simcoe County, south of Midland, this place's post office was named in 1861 after a bridge over the **Wye River**. Its twisting course reminded its namer of the crooked channel of the River Wye in western England.

Wyevale ON A community in Simcoe County, south of Midland, its station, where the North Simcoe Railway crossed the Wye River, was named in 1871. Its post office was opened in 1880.

Wynndel BC North of Creston and in the Kootenay River valley, it was named in 1910 after an early fruit grower in the valley.

Wynyard SK On the Canadian Pacific Railway, and south of the Quill Lakes, this town (1911) was named in 1909 by the wife of railway agent W.H. McNalley, after Wynyard, County Durham, England.

Wyoming ON A village (1873) in Lambton County, east of Sarnia, its post office was named in 1858 by the Great Western Railway after Wyoming County, northwest of Scranton, PA. That name was from the Delaware *maughwauame*, 'large meadows'. The Wyoming Territory in the American West was named 10 years later.

Y

Yahk BC In the Moyie River valley, east of Creston, this place's post office was named in 1905 after **Yahk Mountain**, 35 km to the northeast, at the head of **Yahk River**. The river flows south, becoming Yaak River in Montana, which flows into the Kootenai River, northwest of Libby. In the Kootenay language, the name may mean 'female caribou', or 'arrow', because of the way the river appears to be placed against the 'bow' of the Kootenai River.

Yale BC In the Fraser Canyon, north of Hope, this place was known as *The Falls* in the early 1800s after a rough stretch of the Fraser River upstream. In 1847 Ovid Allard established a Hudson's Bay Company post here and named it Fort Yale after James Murray Yale (*c.* 1798–1871), chief trader at Fort Langley, 1833–59. Until the Canadian Pacific Railway arrived in 1885, Yale was a busy trans-shipment point.

Yamachiche QC A municipality southwest of Trois-Rivières, it was established in 1987 when the village of Yamachiche (1887) and the parish municipality of *Sainte-Anne-d'Yamachiche* (1855) were united. The **Rivière Yamachiche** flows south into Lac Saint-Pierre. Its name in Abenaki means 'little whitefish'.

Yamaska QC Located near the mouth of the **Rivière Yamaska**, east of Sorel, this village was incorporated in 1968, having been erected as the village of *Saint-Michel* in 1867. Yamaska post office was opened in 1826. The village of **Yamaska-Est** was established in 1955. In Abenaki *yamaska* means 'there are rushes all around' at the mouth of the river.

Yarker ON In Lennox and Addington County, northeast of Napanee, this community was first called *Vader's Mills*, after miller David Vader. Recalling that Lt-Gov. John Graves Simcoe, and his son Henry, owned the site until 1840, it was proposed in 1859 to call the place's post office Simcoe Falls. Because of possible confusion with town of Simcoe, it was rejected. It was then named after Kingston mill operator George W. Yarker, who had mills at nearby Sydenham.

Yarmouth NS This town (1890), in southwestern Nova Scotia, was named as early as 1759 by New England planters and fishermen from Yarmouth, MA. Samuel de Champlain had named the harbour *Port Fourchu*, 'forked harbour,' in 1604. That name is reflected in Cape Forchu, at the western entrance to the harbour.

Yarmouth ON A township in Elgin County, it was named in 1792 after the town of Great Yarmouth, on the North Sea coast of Norfolk, England.

Yathkyed Lake NT Southwest of Baker Lake, this lake (1334 km^2) was visited in 1894 by geologist Joseph Burr Tyrrell. In Chipewyan, the name means 'snow'.

Yellow Grass SK Northwest of Weyburn, this town (1906) was presumably named by Canadian Pacific Railway surveyors in 1885, when they observed high grass waving in the setting sun, giving it a yellowish glow. The CPR station was named in 1893, but the post office did not open until 1896.

Yellowhead Pass AB, BC West of Jasper and south of Mount Robson, this pass (1131 m) on the continental divide was named in the early 1800s after a mixed-blood yellow-headed Iroquois called Pierre Bostonais (d. 1827). He was a trader for both the North West and Hudson's Bay companies. *See also* Tête Jaune Cache.

Yellowknife NT On the east side of the North Arm of Great Slave Lake, and at the mouth of **Yellowknife River**, this city (1970) was established in the 1930s, when a gold rush resulted in a boom town by 1940. The river was named after a Chipewyan band, whose name implied they made knives from copper. The band remained a distinct group at the beginning of the twentieth century, but by the 1960s it had been blended with the neighbouring Dogrib and Chipewyan bands.

Ymir BC South of Nelson, this place was named in 1897 by Daniel C. Corbin, president of the Nelson and Fort Shepherd Railway, after **Ymir Mountain**, in the Nelson Range, southeast of Nelson. About 1884 geologist George M. Dawson had named a group of mountains *Ymir Range*, after the legendary Norse father of the giants, but the name survived in the single mountain only.

Yoho National Park BC This national park (1313 km^2) on both sides of the Kicking Horse River, and embracing some of the most rugged alpine landscape in the Rocky Mountains, was named in 1886 after the **Yoho River**. In Cree, *yoho* is a cry of astonishment.

York NB Established in 1785, this county was named after Prince Frederick, the Duke of York and Albany (1763–1827), second son of George III. He was made the Duke of York in 1784.

York ON The county was named in 1792 by Lt-Gov. John Graves Simcoe after Yorkshire, in east central England. The township, embracing the area of present Metropolitan Toronto, earlier called *Dublin Township*, was also named in 1792. The county, less Metropolitan Toronto, became the regional municipality of **York** in 1979. The modern township of York and the town of Weston became the borough of **York** in 1970 in Metropolitan Toronto and a city in 1983. *See also* East York *and* North York.

York PE This municipal community (1986), northeast of Charlottetown, was first known as Little York in the 1830s, and **Little York** has been its post office since 1872. Emigrants from Yorkshire, England settled here

York Factory MB Located at the mouth of the Hayes River, this place was established as the site of a Hudson's Bay Company fort, built in 1684 by Pierre Radisson and Médard Chouart Des Groseilliers. It was named after James, Duke of York (1633–1701), third governor of the company, who became James II in 1685.

Yorkton SK In 1882 *York City* was established some 4 km northeast of the present site of the city (1928) by the York Farmers Colonization Company. The company had been formed in York County, ON, to help prospective settlers move to the North-West Territories. Its post office was named Yorkton in 1884 and was moved to the Manitoba and North Western Railway line in 1889.

Youbou BC On the north shore of Cowichan Lake, northwest of Duncan, this place's post office was opened in 1926. The name was created in 1914 by the Empire Lumber Company by uniting the first three letters of Yount and Bouten, surnames of the general manager and the president of the company.

Young SK Southeast of Saskatoon, this village (1910) was founded by the Grand Trunk Pacific Railway in 1908 and named after real estate agent F.G. Young. The name maintained the alphabetical arrangement of the GTP stations, with Xena to the east and Zelma to the west.

Youngstown AB A village (1936) southeast of Hanna, its post office was named in 1912 after pioneer settler Joseph Victor Young.

Yukon Yukon was established as a district in the North-West Territories in 1894. Four years later it was upgraded as a separate territory. It took its name from the **Yukon River**, which the Gwitch'in call *Yukunah*, 'great river'. The fifth longest river on the North American continent, its length, from the head of the Nisutlin River to the mouth in Alaska, is 3185 km. **Mount Yukon** (3231 m) in the Centennial Range of the St. Elias Mountains, was named in 1967 after the territory.

Z

Zealand NB Northwest of Fredericton, this community was first named *New Zealand* by Philip Crouse (*c.* 1761–1857), a native of Zealand in the Netherlands, who settled here about 1789. Its post office was opened as *New Zealand Station* in 1885 and was changed to Zealand in 1961. At the time of Crouse's death, he left a family of one wife, 18 children, 196 grandchildren, and 118 great-grandchildren.

Zealandia SK Southwest of Saskatoon, this town (1911) was first called *Brock* by the Canadian Northern Railway in 1908, but the local settlers objected. (Brock was later given to another station between Rosetown and Kindersley.) When choosing a post office for the area two years earlier, the settlers put forward several suggestions, and Thomas Englebrecht's proposal of Zealandia was chosen. He had immigrated a few months before from New Zealand on the SS *Zealandia.*

Zeballos BC At the head of **Zeballos Inlet**, west of Campbell River, this village (1952) was named in 1937 after the inlet. The latter was named in 1791 by Capt. Alexandro Malaspina, after Lt Ciriaco Cevallos, one of his Spanish officers.

Zenon Park SK Northeast of Tisdale, the post office in this village (1941) was named in 1916 after first postmaster Zenon Chamberland. He had a small recreation ground on his property for local settlers to gather on Sundays and holidays.

Zephyr ON In Uxbridge Township, Durham Region, southeast of Sutton, this place's post office was named in 1865 after the classical Greek *zephyros,* 'west wind', possibly because the site was quite breezy.

Zone ON A township in Kent County, it was named in 1821 after the Moravian Indian Reserve on the Thames River, which was then known as the 'Indian Zone'.

Zorra ON The original Zorra Township was named in 1819 by Lt-Gov. Sir Peregrine Maitland after the Spanish word *zorro*, 'female fox', and divided into the municipal townships of East Zorra and West Zorra in 1845. A new municipal township of **Zorra** in Oxford County was created in 1975 by amalgamating the municipal townships of East Nissouri, West Zorra, and North Oxford. *See also* East Zorra *and* West Zorra.

Zurich ON A village (1960) in Huron County, northwest of Exeter, it was founded in 1854 by Frederick K. Knell, who named it after the city of Zurich in his native Switzerland. Two years later, Knell was appointed its first postmaster.

INDEX

What is the current name of Coppermine, NT? Port Simpson, BC? Father Point and Murray Bay, QC? These names in the index guide the reader to the current official names: Kugluktuk, Lax Kwa'laams, Pointe-au-Père, and Pointe-au-Pic. This index also provides many official place names described in other entries, such as Alouette Lake, BC, under Lillooet, and Rusticoville, PE, under North Rustico. Physical feature names are listed under their specifics, that is, Mount Vixen, NB, is found under Vixen, Mount, and Fleuve Saint-Laurent, QC, is under Saint-Laurent, Fleuve.

Abbotsford, QC, 362
Aberdare, ON, 248
Acadia Village, NS, 434
Adams, ON, 419
Adamstown, ON, 159
Adamsville, ON, 2, 419
Adanac, SK, 437
Addington, ON, 213
Adelphi, BC, 434
Admiral Island, BC, 367
Agerton, BC, 295
Agoforte, NF, 12
Aguathuna, NF, 306
Aigle, Cap à l', QC, 60
Ainleyville, ON, 51
Aispe, NS, 16
Ajax, ON, vii, 4
Aklavik East Three, NT, 177
Alameda, SK, vii, 4
Alberni, BC, 305
Alberni Canal, BC, 305
Alberni Inlet, BC, 305
Albert, NB, 331
Albion, ON, 41
Albion Mines, NS, 390
Alcan Highway, BC, YT, 4
Alderlea, BC, 107
Alderman Jones Sound, NT, 183
Aldred, SK, 68
Algommequins, Grande rivière des, ON, QC, 287
Algoumequins, Isle des, QC, 7
Allison's, BC, 314
Allright Island, QC, 162
Almaville, QC, 375
Alnwick, ON, 6
Alouette Lake, BC, 217
Alvaston, BC, 439
Ambo, ON, 272
Amherst, ON, 81
Anderson Siding, NB, 363
Andover, NB, 296-7
André Settlement, NB, 347
Ange-Gardien, QC, 205
Angers, QC, 239
Anglois, Havre à l', NS, 223
Angoulême, Lac d', QC, 363
Anguille, Rivière à l', NS, 214
Annaville, QC, 343
Appanea, ON, 263
Apple Tree Landing, NS, 59
Arabasca Lake, NT, 152
Arbre, Anse, NF, 159
Archibald, MB, 233
Arcs, Lac des, AB, 45
Arcs, Rivière des, AB, 45
Ardal, MB, 12
Argall's Bay, NB, NS, 138
Armstong's Mills, ON, 74
Arnoldville, AB, 38
Arrowhead, Mount, BC, 105
Arrow Rock, BC, 224, 414
Arctic Red River, NT, 411
Artemesia, ON, 128
Arthabaska, QC, 422
Articougnesche, NS, 11
Artiwinipeck, QC, 155
Ascot-Nord, QC, 128
Aspé, B. d', NS, 16
Aspy Bay, NS, 102
Assibiboiles, rivière des, MB, 16
Assiniboine, Mount, AB, BC, viii, 16
Assiniboines, lac des, MB, ON, 442
Assomption, Rivière de l', QC, 208
Assumption Bay, NF, 85
Assumption, isle de l', QC, 11
Astleyville, AB, 142
Aultsville, ON, 176
Aurora, AB, 233
Averill's Mills, ON, 428
Avery's Mills, ON, 428
Avesta, MB, 120
Avocat, Havre à l', NS, 3
Avondale North, NF, 85
Awmik, ON, 151
Ayton's River, BC, 380

Babine River, BC, 380
Baccaro, Pte de, NS, 21
Baffin Island National Park, NT, 18
Bagot, QC, 363
Baie-des-Pères, QC, 422
Baie-du-Poste, QC, 253
Baillargeon, QC, 350
Baker's Creek, NB, 353
Baleine, Grande rivière de la, QC, 23
Bald Mountain, NB, 69
Balfour, ON, 324
Ball's Mills, QC, 429
Baltimore, SK, 43
Baptiste, Cap, NS, 40
Baptiste, Lake, ON, 203
Barachois, NS, 270
Barber's Creek, ON, 45
Barkerton, ON, 411
Barnett, AB, 200
Barr'd Islands, NF, 182
Barretts Cross, PE, 189
Bassin, QC, 216
Bates Corners, ON, 439
Bates' Mills, ON, 17
Bathurst Harbour, NB, 266
Bathurst Inlet, NT, 414
Battenburg, AB, 142
Baushene, ON, 429
Bauval, SK, 30
Bay Bulls Arm, NF, 395
Baychimo Harbour, NT, 414
Bay de Grave, NF, 307
Bay d'Espoir, NF, viii, 121
Bayham, ON, 328
Bay of Robbers, NF, 28
Bearbrook Station, ON, 419
Bear Creek, ON, 116
Beatty Saugeen River, ON, 369
Beauce-Jonction, QC, 417

Index

Beauceville, QC, xii, 29
Beaudette, Pointe, QC, 331
Beaulieu, QC, 350
Beaumont, NF, 225
Beaumont North, NF, 225
Beautiful Plains, MB, 13
Beaver River Settlement, NS, 231
Beaverton, BC, 30
Becaguimic, NB, 161
Bedeque, NS, 21
Bedford Bay, PE, 150
Beghultesse, NT, 10
Behring River, YT, 8
Bélair, QC, 416
Bellamy's Mills, ON, 79, 274
Bellamyville, ON, 274
Belle Alliance, PE, 252
Belleau, QC, 340
Bellefleur, NB, 332
Bellin, QC, 186
Bellowston, ON, 433
Bell River, BC, NT, YT, 216
Bellsmount, MB, 34
Bells Point, PE, 61
Belœil, Mont, QC, 255
Belvidere, BC, 118
Bend of Petitcodiac, NB, 253
Ben Eby's, ON, 194
Bennett Dam, W.A.C., BC, 424
Bentinck, ON, 109
Benville, AB, 374
Berlin, ON, 194
Bersiamiste, QC, 36
Bersimis, QC, 36
Beverley, ON, 98, 169
Bic, QC, 211
Bicroft, ON, 62
Bideford, PE, 116
Bideford River, PE, 116
Big Chance Cove, NF, 69
Biggar's Town, ON, 46
Big Glace Bay, NS, 143
Big Lake, The, NT, 273
Big Otter Creek, ON, 287
Big Rideau Lake, ON, 329
Big River, QC, 149, 196
Big Thames, ON, 253
Big Tooth, BC, 13
Big Tracadie River, NB, 409
Binbrook, ON, 143-4
Birchardtown, ON, 259
Birchy Bay North, NF, 28
Birchy Cove, NF, 93
Birchy Head, NF, 144
Bird Island Cove, NF, 117
Birdtail Creek, MB, 37
Birmingham, ON, 183
Bismarck, ON, 432
Black Brook, NB, 220
Black Diamond, AB, 73
Black Rock, NS, 395
Blackwater River, BC, 434
Blaikie, SK, 272
Blakemore, SK, 272
Blayney, AB, 25
Blitzen, Mount, NB, 276
Bloody Bay, NF, 145
Blooming Prairie, AB, 258
Blumenau, AB, 391
Bog End, SK, 214
Bois, Lac des, MB, ON, 442
Boldenhurst, SK, 330
Bolton Forest, QC, 111
Bomore, ON, 108
Bonaventure, QC, 38
Bonaventure, Île, QC, 42
Bonnington, SK, 188
Boraston, SK, 301
Borgne, Isle du, QC, 7
Bosanquet, ON, 13
Boston Bar, BC, 274
Boucher, SK, 359
Boucher Settlement, SK, 359
Boucherville, Îles de, QC, 44
Boundary City, BC, 248
Bountiful, AB, 24
Bourbon, Lac, MB, 67
Bourchier's Mills, ON, 395
Bourlamaque, QC, 416
Bove, Lake, BC, YT. 398
Bowden, MB, 424
Bowser Station, NS, 406
Bradalbane, PE, 46
Bradyville, ON, 6
Brandenburgh, ON, 135
Brant, ON, 425
Brant's Block, ON, 53
Bread and Cheese Cove, NF, 38
Breakers Point, BC, 122
Breakeyville, QC, 348
Brega, NF, 48
Brennan's Corners, ON, 135
Brennanville, ON, 135
Bretons, Terre des, NS, 60
Bridgeburg, ON, 131
Bridgedale, NB, 331
Bridgeport Mines, 102
Bridgewater, ON, 2, 59
Brighton, NF, 111
Brighton Beach, NF, 264
Brigus South, NF, 167
Britannia Settlement, AB, SK, 219
Broad Cove, NF, 310
Broad Cove Marsh, NS, 108, 177
Brock, ON, 395
Brock, SK, 446
Brome, QC, 199
Brotherstown, ON, 334
Brown's Cove, NF, 71
Brownsville, ON, 370
Broyle, Cape, NF, 61
Bruno, Mont, QC, 343
Brunton, NB, 323
Buck's Crossing, ON, 159
Buckshot, ON, 301
Buck's Town, ON, 290
Buckville, ON, 81
Buctouche, NB, 44
Buctouche, Baie de, NB, 44
Buell's Bay, ON, 49
Bull Run, NS, 248
Bumfrau, NB, 31
Bumfrow Brook, NB, 31
Burbidge, QC, 246
Burlington Bay, ON, 158
Burns Settlement, PE, 136
Burnt Bay, NF, 216
Burwick, ON, 441
Bury, ON, 407
Buttermilk Creek, NB, 129
Butternut Ridge, NB, 162
Buttes, Havre des, NF, 324
Buzztown, ON, 420
Byng Inlet North, ON, 48
Bytown, ON, vii, 287

Caisiqupet, PE, 65
Calder's Mills, ON, 31
Calling Lakes, SK, 317
Cambridge Mills, ON, 312
Cambridge Mills, PE, 261
Cameron Bay, NT, 112
Campbell, Mount, YT, 407
Campbellton, ON, 295
Campbelltown, NS, 376
Camp Borden, ON, 43
Camp Hewitt, BC, 294
Campo Bello, NB, 431
Canada Sea, NT, 173
Canal de Alberni, BC, 305
Canning's Cove, NS, 192
Cannington Manor, SK, 233-4
Canrobert, QC, 205
Canterbury, BC, 177
Cantley, NS, 129

Cantons de l'Est, QC, 122
Cap a Lee, NB, 61
Cap-Caissie, NB, 56
Cap-des-Caissie, NB, 56
Cape d'Or, NS, 285
Cape Bald, NB, 61
Cape Negro River, NS, 80
Caplen Bay, NF, 57
Cappahayden, NF, 326
Card's Harbour, NF, 410
Careaux, B., NS, 21
Caribou Crossing, YT, 62
Carleton, NB, 356
Carleton Siding, PE, 43
Carleton Station, NB, 335
Carling's Corners, ON, 123
Carol Lake, NF, 198
Carriboo Harbour, NS, 63
Carriboo Island, NS, 63
Carrièreville, QC, 394
Carronbrook, ON, 106
Carrot River Settlement, SK, 193
Carson, SK, 379
Cartwright, ON, 39
Cartwrightville, ON, 263
Cascades, QC, 303
Cascades Point, QC, 303
Cascumpec, PE, 5
Catalogne, Barachois de, NS, 66
Catfish Corners, ON, 286
Cat Harbour, NF, 225
Cathartic, ON, 64
Catherine, MB, 215
Cats Cove, NF, 85
Caubvik, Mount, NF, viii, 67
Caughnawaga, QC, 185
Cavenick's Point, NB, 58
Cawdor, ON, 185
Cayoosh Flat, BC, 217
Cedars, QC, 214
Cèdres, QC, 214
Central Argyle, NS, 13
Central Onslow, NS, 285
Central Patricia, ON, 298
Central Woods Harbour, NS, 442
Centreville, NS, 25
Centreville, ON, 334
Centreville, PE, 31
Chabot Lake, ON, 375
Chalet, NB, 326
Chamberlain, MB, 241
Chamberlain's Corners, ON, 407
Chambord, NB, 341
Chapel Cove, NF, 160
Charleston, ON, 56
Charlotte Bay, NS, 359
Chasteaux, Baye des, NF, 33
Châteauvert, QC, 359
Chat, Lac du, ON, 120
Chebucto, NS, 157
Chemahawin, MB, 110
Chênes, Lac des, MB, 281
Chênes, Portage des, ON, QC, 99
Cherry Creek, MB, 41
Chesleys Corner, NS, 268
Chigonaise River, NS, 34
Chinese Wall, The, NT, 3
Chinguacousy, ON, 45
Chippawa Creek, ON, 430
Chiputneticook River, NB, 345
Choate's Mills, ON, 427
Christie Town, NB, 249
Christieville, QC, 175
Christopher Hall Island, NT, 157
Church Point, NB, 53
Churchtown, NS, 302
Churchville, QC, 90
Chute-aux-Iroquois, QC, 198
Clansman, SK, 366
Clare, NS, 84
Clarendon Centre, QC, 375
Clarksville, NS, 167
Clarksville, ON, 32
Clarkville, ON, 263
Claybelt, ON, 323
Clearwater, ON, 368
Cleverville, AB, 69
Clifton, ON, 79, 270
Closeleigh, YT, 436
Clothier's Mills, ON, 188
Clyde River, PE, 174
Coal Banks, The, AB, 81, 215
Coalfields, SK, 333
Coalhurst, AB, 215
Cobequid, NS, 411
Cockburn's Corners, ON, 36
Colborne, NB, 71
Colebridge, ON, 104
Colebrooke, NB, 149
Coleman's Corners, ON, 372
Colombiers, Les, NF, 179
Comeau Bay, NS, 21
Comet, Mount, NB, 276
Confederation Bridge, NB, PE, 1
Congdon Settlement, NS, 36
Consumption Bay, NF, 85
Cook's Ferry, BC, 388
Cookville, ON, 116
Cope's Landing, ON, 366
Copper City, BC, 177
Coppermine, NT, 196
Cordova, Puerto de, BC, 87, 121
Cornabuss, ON, 236
Cornfield, SK, 112
Coronation, NT, 196
Corte Real, Tiera del, NF, 198
Cortez Island, BC, 234
Cosby, ON, 272
Cossette, MB, 178
Côte-Saint-Luc, QC, xii, 89
Côte-Saint-Antoine, QC, 433
Côté Siding, ON, 416
Cottel Island, NF, 343
County Line, NS, 56
County Line, PE, 118
Covenhoven, NB, 251
Coverdale, NB, 331
Coverdale River, NB, 331
Cow Bay, NS, 309
Cowichan Lake, BC, 202
Cowichan River, BC, 202
Cranberry Lake, BC, 416
Credit, ON, 120
Crooked Lake, ON, 375
Cross Lake, ON, 91
Cross Point, QC, 302
Crowland, ON, 270
Cumberland, NS, 9
Cupers Cove, NF, 93
Cupid, Mount, NB, 276
Curling, NF, 88
Currie's Landing, MB, 104
Curzon, SK, 7
Cut-Off Valley, BC, 80
Cyprès, Montagnes des, AB, SK, 93

Daggville, QC, 375
Daillebout, QC, 349
Dalrymple, SK, 93
Dancer, Mount, NB, 276
Daneville, SK, 204
Danish River, MB, 76
Danville, ON, 2
Dark Tickle, NF, 47
Darlington, ON, 45
Darlington Mills, ON, 45
Darmet, Isle, NS, 9
Dartmouth Harbour, NS, 397
Dasher, Mount, NB, 276
Dauphin Mountain, MB, 329

Index

Davenport Junction, NF, 435
Davidsville, NF, 230-1
Davidtown, ON, 375
Dayton, NF, 219
Delnorte, AB, 177
Demoiselles, Cap des, NB, 170
Dennison's Bridge, ON, 84
Dentiform Mountain, BC, 13
Denwood, AB, 425
Denys Basin, NS, 330
Des Joachims, QC, 323
Desmarais, AB, 424
Deux Grosses Buttes, les, AB, 413
Deux Joachims, Portages des, QC, 323
Devil's Lake, SK, 224
Dewdney, AB, 282
Diamond, SK, 427
Diamond Vale, BC, 246
Dickinson Landing, ON, 176
Dildo Pond, NF, 40
Dingle, ON, 51
Dingwall, ON, 330, 390
Disappointment, River, NT, 228
Distress, NF, 343
Dixon's Corners, ON, 17
Dock River, PE, 80
Dog Bay, NF, 171
Doghide, SK, 407
Dog River, PE, 80
Dolbeau, QC, 339
Dominion No. 6, NS, 103
Dominion No. 12, NS, 270
Donder, Mount, NB, 276
Doré, Lac, ON, 146
Dorset, Cape, NT, 61
Dorval, QC, xii, 104
Douglas, ON, 34
Douville, QC, 354
Dover, ON, 307
Dover Hill, NB, 365
Dover Mills, ON, 307
Dover South, ON, 290
Doyle's Corners, ON, 240
Drew's Mills, QC, 102
Dublin, ON, 444
Dumfries, NB, 332
Dummer's Mills, ON, 427
Duncan, BC, 172
Duncanville, ON, 338
Dunk Cove, NS, 166
Durhamville, NB, 180
Durhamville, ON, 39
Durrell, NF, 413
Durrell's Arm, NF, 413
Dyson Bay, NS, 143

East Baccaro, NS, 21
East Chezzetcook, NS, 74
Easter Brook, NF, 334
Eastern Townships, QC, 122
East Hants, NS, 159
East Jeddore, NS, 181
Eastman's Corners, ON, 13
Eastman's Springs, ON, 64
East Margaree, NS, 235
East Oxford, ON, 278, 289
East Pubnico, NS, 316
East River, PE, 167
East River of Pictou, NS, 298
East Robson, BC, 333
East St. Modeste, NF, 434
East Thurlow Island, BC, 405
Eastview, ON, 418
Ebbsfleet, PE, 250
Éboulements, QC, 214
Ebytown, ON, 194
Échourie, Havre de l', PE, 384
Ederna, Baya, NF, 42
Edinborough, ON, 117
Edinburgh, ON, 298
Edison, AB, 432
Edmonton, ON, 383
Edward River, PE, 80
Edwardsburg, ON, 62, 103, 104
Edzo, NT, 321
Egmont Harbour, NS, 181
Eidswold, AB, 103
Eisenhower Peak, AB, 66
Eisenhower, Mount, AB, 66
Ekfrid, ON, 244
Ekfrid Centre, ON, 12
El Dorado, BC, 267
Eldorado Mountain, BC, 105
Elgin Settlement, ON, 274, 385
Elizabethtown, ON, 49
Elliot River, PE, 433
Elliott's Mills, ON, 159
Elliott Station, MB, 241
Elmsdale, ON, 175
Elsa, YT, xii, 117
Elsas, SK, 421
Ély, QC, 416
Emily, ON, 284
Emmett, ON, 73
Enfumé, Cap, NS, 382
Englishman's River, BC, 292
English River, MB, 76
Erb's Mills, ON, 312
Escott, ON, 137, 324
Eskimo Point, NT, 15
Espagnols, Rivière aux, NS, 397
Espanola Station, ON, 241
Espoir, Bay d', NF, viii, 121
Espoir, Cap d', QC, 60
Esquimaux, Baie des, NF, 158
Esquimaux Point, QC, 163
Etechimins, Rivière des, QC, 72
Everette, NS, 302

Fairfield, ON, 256
Fairfield, SK, 12
Fairleigh, NB, 124
Fairport, ON, 105
Fairy Lake, NS, 187
Famish Cove, NF, 124
Famish Gut, NF, 124
Farilham, NF, 126
Farmer's Arm, NF, 394
Farmersville, ON, 17
Farrans Point, ON, 176
Farwell, BC, 327
Father Point, QC, 303
Felix Cove, NF, 306
Fendu, Cap, NS, 138, 388
Fenwick, NB, 35
Ferguson Manor, NB, 17
Ferme-du-Milieu, QC, 205
Ferris, ON, 111
Fifteen Point, PE, 259
Finch, ON, 36
Fingerboard, NB, 278
Fish Lake, YT, 429
Fish Pools, NF, 331
Fitzhugh, AB, 181
Fitzback, ON, 170
Five Stakes, ON, 399
Flat Creek, MB, 281
Flesherton Station, ON, 68
Fletcher's Bridge, ON, 430
Fletcher's Station, ON, 430
Flint's Mills, ON, 274
Floridablanca, Rio, BC, 135
Flos, ON, 90, 388
Flounders Bight, NF, 231
Flour Cove, NF, 129
Fontaine, NF, 136
Forest Station, NB, 335
Forfar, ON, 290
Forgetville, QC, 50
Forillon, NF, 126
Forked Lake, ON, 203
Forksdale, BC, 246
Fort Amherstburg, ON, 9
Fort Babine, BC, 21
Fort Birdtail, MB, 37
Fort Brisebois, AB, 56

Fort-Chambly, QC, 69
Fort-Chimo, QC, 196
Fort Edmonton, AB, 113
Fort Franklin, NT, 97
Fort Garry, MB, 440
Fort George, BC, 313
Fort Hope, BC, 170
Fort McLeod, BC, 242
Fort Norman, NT, 273, 412
Fort of the Forks, NT, 133
Fort Reliance, NT, 325
Fort Rupert, QC, 338, 428
Fort Sackville, NS, 31
Fort Sainte-Anne, NS, 341
Fort Simpson, BC, 210
Fort Stand Off, AB, 389
Fortuna, I de la, NF, 33
Fortune Bay Bottom, NF, 402
Fortune's Landing, BC, 118
Fort Victoria, BC, 421
Fort Ware, BC, 427
Fort Wrigley, NT, 442
Fort Yale, BC, 444
47 Mile House, BC, 80
Fosbery, MB, 209
Foulon, Anse au, QC, 134
Four Mile City, BC, 379
Fowlers Corner, NS, 248
Fox Cove, NF, 343
Français, Lac des, BC, 135
Francistown, ON, 123
Francœur, QC, 342
François-Babel, QC, 186
Françoise, La Baye, NB, NS, 137
Fraser Lake, BC, 131
Fraserville, QC, 332
Frederick Bay, NS, 9
Fredericksburg, ON, 97
Freiburg, ON, 238, 400
French, SK, 79
French Bay, NB, 22
French Island Harbour, NF, 129
Freneuse, Lac, NB, 149
Friedsburg, ON, 94
Frobisher Bay, NT, 178
Frobyshire, SK, 137
Frogmore, ON, 103
Frohlich, SK, 168
Fronsac, Détroit de, NS, 59
Fronsac, Passage de, NS, 59
Frontenac, Lac, ON, 285
Fume, Lac qui, AB, 382
Furnace Falls, ON, 226

Gabari, Baye de, NS, 139
Gagnon, NB, 22
Galtois, Havre le, NF, 141
Gambier Harbour, NS, 309
Garafraxa, ON, 34, 111, 432
Gasburg, AB, 243
Gaspa, C., NS, 16
Gaspé, QC, xii, 141
Gaspereau, NB, 307
Gastown, BC, 418
Gaviola, Punta de, BC, 139
Gaviota, Punta de, BC, 139
Gayside, NF, 28
Gens de la Pierre-Noire, lac des, MB, ON, 442
George IV's Coronation Gulf, NT, 88
George Bay, NS, 353
Georgina, ON, 221, 396
Ghost River, AB, 388
Gibbon, NS, 96
Gilchrist's Mills, ON, 187
Gillesport, NF, 394
Gitwangak, BC, 195
Glace, Baie de, NS, 143
Glanford, ON, 259
Glasgow, ON, 370
Glass Bay, NS, 143
Glendale, NF, 259
Glenelg, ON, 104
Glen Iver, QC, 349
Glen Logan, SK, 437
Glensmith Station, MB, 188
Glover's, PE, 189
Gobetick, NS, 81
Goddard's Corners, ON, 433
Goffs, NS, 124
Golden, ON, 23
Golden Ears Provincial Park, BC, 40
Good Hope, NT, 132
Goodwood River, PE, 116
Goose Bay, NF, 159, 261
Goose Lake, MB, 333
Gopher Creek, MB, 422
Gotteville Island, PE, 338
Gould Dome, AB, BC, 408
Grâce, Île de, QC, 154
Graham, ON, 379
Graham City, BC, 238
Grand Belle doune, NB, 32
Grande-Baleine, QC, 196
Grande Baye, NF, 33
Grande Île de Miscou, NB, 204
Grande Isle, ON, 441
Grande Pointe des Chênes, MB, 346
Grande Prairie, BC, 149, 434
Grande rivière de la Baleine, QC, 23
Grandes-Fourches, QC, 377
Grand Falls, NF, 76
Grand Lac, ON, 395
Grand lac Victoria, QC, 421
Grand Manan, NB, 149
Grand Racico, PE, 338
Grand Rastico, PE, 338
Grand River, NB, 358
Grand-Sault, NB, 149
Grand Sault St Louis, Le, QC, 359
Grand'Terre, NF, 230
Grangemount, PE, 376
Grangousier Hill, ON, 140
Grant, ON, 164
Grants Point, ON, 222
Granville, BC, 418
Gratias, Rivière aux, MB, 258
Grave, Bay de, NF, 307
Gravel River, NT, 187
Great Araubaska, AB, SK, 17
Great Bend, SK, 321
Great Colinet Island, NF, 83
Great Fish River, NT, 21
Great Garnish Barasway, NF, 140
Great Marten Lake, NT, 237
Great Northern Peninsula, NF, 275
Great St. Lawrence Harbour, NF, 357
Great Whale River, QC, 23, 196
Green Bay, NF, 432
Greene bay, NF, 27
Greenlay, QC, 353
Green Point, NB, 304
Green River, NB, 332
Greenville, NS, 432
Grenville Bay, PE, 268
Gretna Green, NB, 104
Griffintown, NS, 382
Griquet, NF, 359
Groseilles, Rivière aux, ON, 299
Gruenfeld, MB, 195
Guillaume-Delisle, Lac, QC, 155
Gunningsville, NB, 331
Gwillimbury, ON, 169

Habel, Mount, BC, 302 OK
Habitant Corner, NS, 59
Haggerty's Corners, ON, 32
Ha Ha, NF, 322
Ha! Ha!, Baie des, QC, 156

Ha Ha Bay, NF, 322
Haldimand, ON, 147
Halfway River, NS, 159
Halifax Road, NS, 167
Halifax Bay, PE, 31
Hallewood, ON, 157
Hall Island, NT, 92
Hallowell, ON, 298
Hall's Bridge, ON, 51
Hamer's Corners, ON, 435
Hamilton, MB, 158
Hamilton, ON, 81
Hamilton River, NF, 77
Hamilton's Mills, ON, 163
Hanlan, MB, 427
Hanson Lake, BC, 427
Happyland, SK, 210
Haran, C. au, NB, 61
Harang, C., NB, 61
Harbour Grace Junction, NF, 435
Harbour Le Cou, NF, 335
Hardy Bay, BC, 308
Hareng, Ance du, NS, 166
Harker, AB, 103
Harkness, SK, 175
Harper's Camp, BC, 170
Harricanaw River, ON, 161
Harrington Lake, QC, 260
Harris Bay, PE, 338
Harrison River, YT, 8
Hartes content, NF, 164
Hartt's Mills, NB, 136
Hartwell, QC, 73
Harver Deep, NF, 152
Harvey's Mills, ON, 290
Hastings, BC, 418
Hastings, ON, 229
Haven of St. John, NF, 356
Havre-Aubert, QC, 216
Havre Bertrand, NF, 160
Havre-Content, NF, 164
Havre de Grâce, NF, 160
Havre le Galtois, NF, 141
havre St Pierre, NS, 362
Haylinscove, NF, 415
Hayward, SK, 218
Hazard, Golfe de, QC, 155
Hazen, Lake, NT, 164
Head of Bay d'Espoir, NF, 249
Head of Chezzetcook, NS, 74
Head of Fortune Bay, NF, 402
Head of Hillsborough, PE, 167
Head of Jeddore, NS, 181
Head of Jordan River, NS, 183
Head of Petitcodiac, NB, 297
Head of Tide, NB, 405
Head of Tide, NS, 289
Head of Wallace Bay, NS, 426
Head of Wallace Bay North, NS, 426
Heatherwood, AB, 114
Hébert, Rivière, NS, 29
Heiberg Land, NT, 19
Hermit Mountain, BC, 412
Heron, C., NB, 61
Herring, C., NB, 61
Herring Cove, NS, 50
Herriot's Falls, ON, 202
Heureka Sund, NT, 122
Hick's Ferry, NS, 47
Highland, AB, 97
High Woods River, AB, 167
Hillside, NB, 291
Hodgkinson's Corners, ON, 20
Hogg's Bay, ON, 421
Hohenlohe, SK, 205
Holland, ON, 35
Holland Bay, PE, 65
Hollister's Mills, QC, 429
Holmstown, AB, 223
Holt City, AB, 202
Holy Lake, MB, 289
Holyrood South, NF, 366
Hootalinkwa, YT, 402
Hope, ON, 375
Hopevale, NF, 216
Hopewell Corner, NB, 331
Horse Cove, NF, 291
Horton, NS, 441
Horton Corner, NS, 189
Horton Point, NS, 19
House Harbour, QC, 162
Howard, ON, 257, 339
Howard Settlement, NB, 59
Howe Bay, PE, 11
Howell, SK, 315
Howe Sound, BC, 142
Howick, ON, 130
Hub of the Big Bend, The, BC, 384
Hull, QC, xii, 173
Hull-Partie-Ouest, QC, 73
Humbermouth, NF, 88
Humberstone, ON, 270
Humqui, Rivière, QC, 10
Huron, ON, 303

Icelandic River, MB, 331
Île Jésus, QC, 182
Îles de la Madeleine, QC, viii, 214
Iles, Lac aux, MB, ON, 442
Imbert, R, NS, 29
Inconnue, NT, 10
Indian Arm, NF, 58
Indian Head, NF, 391
Indian Point, NS, 193
Inkerman, ON, 400
Inkermanville, PE, 338
Inner Pinchard's Islands, NF, 270
Innuksuac, Rivière, QC, 177
Inoucdjouac, QC, 177
Invermara, ON, 285
Inverness, SK, 177
Irishtown, NS, 302
Iroquois, Point, ON, 178
Island Cove, NF, 164, 415
Isla Redonda, BC, 111, 433
Isle Royale, NS, 60
Islington, NF, 164

Jackanet River, BC, 135
Jacques-Cartier, QC, 221
Jacques-Cartier, Rivière, QC, 202
James Bay, NT, 22
Jamestown, ON, 429
Jamesville, ON, 257
Jaune, Rivière, NS, 331
Jelly's Corners, ON, 376
Jésus, Île, QC, 182
Jésus, rivière, QC, 249
Jim's Cove, NF, 410
Joachims de l'Estang, Rapides des, QC, 323
Joe League Village, PE, 344
John, Cape, NS, 331
Johnstone Lake, SK, 283
Johnstons Crossing, NS, 167
Joice's Corners, ON, 150
Joly, MB, 363
Joly, QC, 198
Jones River, YT, 8
Juste au Corps, NS, 177
Juste-au-Corps, NS, 308

Kakibonga, QC, 55
Kaladar, ON, 274
Kane's Landing, BC, 186
Kanmore, ON, 3
Katepwa Lakes, SK, 317
Kavanagh's Point, NB, 58
Kawende, MB, 281
Kelheau, MB, 71
Kendal, SK, 144
Kennebecasis Bay, NB, 336

Kent Island, NT, 276
Kenyon, ON, 108
Kerry, ON, 354
Keystone, AB, 47
Kicking Horse Flats, BC, 146
Kilborn's Mills, QC, 334
Kildare Station, PE, 358
Kildonan, MB, 111
Kilmar's Fort, BC, 21
Kinburn, NS, 229
King George's Sound, BC, 273
Kingshurst, NB, 111
Kingston, NB, 327
Kingsville, QC, 404
Kinojévis, QC, 243
Kipawa, QC, 401
Kississing, MB, 377
Kitley, ON, 407
Kitley Corners, ON, 407
Kitty-vitty, NF, 319
Kleena Kleene, BC, 195
Kneehill, AB, 62
Knollton, AB, 393
Knowlton, ON, 379
Knowlton, QC, 426
Knowlton Falls, QC, 426
Koartac, QC, 317
Koartak, QC, 317
Kolin, SK, 121
Kootenae Lake, BC, 439
Koyl's Bridge, ON, 407
Kribbs Mills, ON, 113
Kuntze's Corners, ON, 87

Laberge, SK, 117
L'Acadie, Rivière, QC, 198
Lac de Bois, NT, 41
Lac de Gras, NT, 151
Lac la Martre, NT, 237, 435
Lac la Plonge, SK, 30
La Conversion-de-Saint-Paul, QC, 362
Lac qui Fume, AB, 382
Lac, R du, NB, 18
Lac-Saint-Louis, QC, 245
Lac-Weedon, QC, 353
Lading Tickles, NF, 211
Ladle Tickles, NF, 211
Lady Franklin Channel, NT, 242
Lady Grey Lake, NT, 273
Laflèche, QC, 354
Lafontaine, ON, 78
Laggan, AB, 202
La Haye Point, NF, 140
Lake Harbour, NT, 191
Lake St. Charles, QC, 201
Lakeview, NF, 160
Lake Weedon, QC, 353
La Magdelene, QC, 229
Lambly's Landing, BC, 118
Lambton Village, ON, 309
Lancaster, NB, 356
L'Ancienne-Lorette, QC, 348
Langara, Punta de, BC, 154
Langevin, QC, 348
L'Annonciation, QC, 282
Lansdowne, ON, 137, 324
L'Anse au Sable, BC, 188
L'Anse Sauvage, NF, 210
La Pérade, QC, 346
La Providence, QC, 354
Latimer's Corners, ON, 269
Laurier, MB, 363
Laurier, QC, 70
La Visitation-de-la-Pointe-du-Lac, QC, 302
Lawrence Factory, NS, 384
Lazo de la Vega, Punta de, BC, 210
Leavings, AB, 151
Leeds, ON, 137, 324
Lees Creek, AB, 62
Leeville, SK, 16
Leffingwell Island, NT, 390
Leger Corner, NB, 101
Legere Corner, NB, 101
Le Karpont, NS, 320
Lemberg, MB, 378
L'Enfant-Jésus, QC, 417
Lennoxville, QC, xii, 213
Lenora Lake, SK, 202
Le Pas, MB, 403
Le Petit Courant, ON, 219
Les Éboulements-en-Bas, QC, 356
Les Laurentides, QC, 209
Les Petites Oies, NF, 146
Levesqueville, ON, 16
Lewisville, ON, 46
Liard, NT, 132
Liberal, AB, 120
Lievres, B. aux, NF, 160
L'Île-Dorval, QC, xii
Lillooet, NB, 75
Lind Island, NT, 181
L'Industrie, QC, 183, 362
Lisadel, ON, 130
Liskeard, ON, 268
Little Bay, NF, 352
Little Bend of the South River, ON, 411
Little Britain, MB, 220
Little Cataraqui River, ON, 67
Little Chance Cove, NF, 69
Little Dover, NS, 105
Little Falls, NB, 113
Little Forks, QC, 213
Little Garnish, NF, 140
Little Glace Bay, NS, 143
Little Glocester, NF, 134
Little Hay Lakes, AB, 269
Little Opeongo Lake, ON, 19
Little Pernocan, NF, 283
Little Placentia, NF, 13
Little Pond, PE, 11
Little Prairie, BC, 74
Little River, NB, 330
Little Saskatchewan River, MB, 151
Little Seldom, NF, 372
Little Southwest Miramichi River, NB, 251-2
Little Tancook Island, 399
Little Thames River, ON, 19
Littleton, BC, 402
Littleton, MB, 93
Little Toronto, ON, 150
Liverpool, BC, 418
Liverpool River, NS, 246
Lobo Junction, ON, 196
Loderville, ON, 428
Logans Tannery, NS, 226
London Mountain, BC, 435
Lone Pine, AB, 283
Long Island, NS, 136
Longuil River, NB, 114
Longwood, ON, 244
Longworth, SK, 1
Loon's Cove, NF, 216
Lopez de Aro, Canal de, BC, 160
Loranger, QC, 200
L'Or-Blanc, QC, 15
Lorraine, cap de, NS, vii, 28
Lot 49, PE, 311
Loughborough, ON, 396
Louis-Joliette, QC, 347
Lourdes-de-Blanc-Sablon, QC, 40
Lowell, ON, 225
Lower Allumette Lake, ON, QC, 8
Lower Argyle, NS, 13
Lower Bay du Vin, NB, 22
Lower Burton, NB, 54
Lower Caraquet, NB, 26
Lower Clements, NS, 382

Index

Lower Cove, NS, 136
Lower Columbia Lake, BC, 439
Lower Coverdale, NB, 47
Lower East Chezzetcook, NS, 74
Lower East Pubnico, NS, 316
Lower Garry Lake, NT, 140
Lower Horton, NS, 150
Lower L'Ardoise, NS, 207
Lower Onslow, NS, 285
Lower Rideau Lake, ON, 329
Lower Sackville, NS, 339
Lower Stewiacke, NS, 391
Lower Woods Harbour, NS, 442
Lowertown, ON, 295
Lower Waterton Lakes, AB, 429
Lower West Pubnico, NS, 316
Lowestoft, MB, 335
Lumley Inlet, NT, 137
Lumsden's Mill, QC, 401
Lunes Cove, NF, 216
Luther, ON, 150
Lynch, NB, 332
Lynch Cove, NF, 258

Macaza, Lac, QC, 203
Mackayville, QC, 354
MacMillan's Mill, ON, 120
Macpherson, AB, 35
Madamkeswick River, NB, 190
Maddox Cove, NF, 298
Madeleine, Îles de la, QC, viii, 214, 229
Madisco, NB, 297
Magaguadavic, NB, 352
Magdalen Islands, QC, 229
Magna, SK, 282
Magpie Mine, ON, 106
Mahonne, Baie de la, NS, 230
Main Point, NF, 231
Maitland, ON, 309
Maitland River, ON, 259
Malcolm's Mills, ON, 281, 371
Manasseh, ON, 150
Manchester, MB, 422
Manchester, NS, 376
Manchester, ON, 18
Mandan, MB, 368
Mandeville, QC, 344
Mann, QC, 302
Mansfield Island, NT, 234
Mapleton, ON, 218
Maquako, NB, 234
Maricourt, QC, 186
Marienquactacook, NB, 22
Mariposa, ON, 233
Marksville, ON, 167
Marshallville, NF, 216
Marshville, ON, 78
Marten Lake, NT, 237
Martin Lake, NT, 237
Martin's Mills, ON, 249
Martin's Point, NB, 58
Martre, Lac la, NT, 237, 435
Martre, Rivière la, NT, 237
Mary Lake, ON, 310
Maryland, NB, 268
Macs Corners, ON, 240
Mascouche, PE, 252
Matapédia, Rivière, QC, 203
Matsqui, BC, 1
Maurepas, I. de, NS, 229
Mauricie National Park, La, QC, 203
Mazinaw Lake, ON, 224, 415
Mazinaw Rock, ON, 224, 415
McAuslin Island, PE, 338
McCurdy's Village, ON, 32
McDame Creek, BC, 66
McGillivray's River, BC, 196
McIntosh's Corners, ON, 107
McNab's Corners, ON, 108
McNair's Cove, NS, 260
McNaughton Lake, BC, 191
McNaughtonville, ON, 264
McNeill Bay, BC, 309
Mcoun, SK, 60
Mcowan, SK, 60
Mégantic, QC, 200
Meighen, AB, 422
Melbourne, ON, 411
Melick, ON, 180
Meline, Isles de la, NF, 203
Menane, NB, 150
Menesetung, ON, 231
Mer Douce, ON, 142
Merligueshe, NS, 225
Mernersville, ON, 248
Merrittsville, ON, 430
Metcalfe, ON, 284
Metgermette-Nord, QC, 366
Methy Lake, AB, 397
Methy Lake, SK, 203
Meules, Île aux, QC, 60
Meyers' Creek, ON, 33, 253
Michel, BC, 387
Middle East Pubnico, NS, 316
Middle Gut, NF, 366
Middle Musquodoboit, NS, 261
Middle River, NS, 8
Middle River of Pictou, NS, 298
Middle Sackville, NS, 339
Middle St. Francis, NB, 77
Middleton, ON, 97
Middletown, BC, 387
Middle Waterton Lakes, AB, 429
Middle West Pubnico, NS, 316
Miles, SK, 218
Miles Hill, ON, 329
Millburn, ON, 108
Mill Creek, ON, 249, 282
Mille Isles, ON, 405
Mille Roches, ON, 221
Mill Point, ON, 100
Mill River, PE, 172
Milltown, NB, 365
Mill Town, ON, 31
Millville, ON, 159
Milton, ON, 31
Minto, Lac, QC, 251
Miramichi, ON, 295
Missinaibi, ON, 240
Missinipi, MB, 76
Mitchell's Brook, NF, 259
Mitis, Rivière, QC, 204
Mizonette, NB, 231
Moffat, ON, 295
Mohawk, ON, 259
Molstad, AB, 27
Monsonis, Rivière des, ON, 256
Montmagny, Isle de, QC, 182
Montréal, QC, viii, xii, 255
Montréal-Sud, QC, 221
Montrose, PE, 152
Moodyville, BC, 277
Moosehide Hill, YT, 248
Moose River, NS, 79
Mooseskin Mountain, YT, 248
Mordienne, Baye de, NS, 309
Mordienne, Cap de, NS, 309
Morien Bay, NS, 309
Morphy's Falls, ON, 63
Morris Town, NB, 365
Mortier Bay, NF, 238
Mosa, ON, 426
Mosquito Creek, AB, 263
Mosquito Point, NB, 236
Mostyn, SK, 215
Moulin-Bersimis, QC, 36
Moulin-Desbiens, QC, 99
Mountain Indian River, YT, 127
Mount Assiniboine, AB, BC, viii, 16
Mount Good Hope, BC, 146
Mount Lorne, YT, 222
Mount Pleasant, ON, 329

Mouse River, MB, 384
Mousseau, Lac, QC, 260
Mouth of Keswick, NB, 190
Mouth of Millstream, NB, 12
Mouth of Neripis, NB, 432
Muchalat, BC, 146
Mud Creek, NS, 441
Muddy Hole, NF, 261
Mud Lake, YT, 237
Mud River, ON, 401
Mundy's Bay, ON, 248
Munquart, NB, 26
Munsie's Corners, ON, 56
Murray, ON, 65
Murray Bay, QC, 203, 303
Muskoka Station, ON, 229

Nachouac, NB, 263
Nahanni National Park, NT, 385
Napierville, QC, 215
Narrows, NB, 57
Natal, BC, 387
Natashquan, QC, 11
Nation River, ON, 386
Nation, River au, BC, 264
Neal's Corners, ON, 366
Neal's pond, PE, 262
Neddy Harbour, NF, 274
Nelson's Falls, ON, 202
Nelson's Mountains, BC, 373
Nelsonville, QC, 90
Nerichac, NS, 13
Nerichat, NS, 13
New Aklavik, NT, 177
New Alberni, BC, 305
Newberry River, YT, 402
New Bideford, PE, 116
New Brighton, BC, 418
New Caledonia River, BC, 135
New Carlisle, ON, 39
Newcastle, ON, 147
New Denmark, SK, 403
New England, ON, 77
New Fairfield, ON, 256
New Gary, NB, 141
New Germany, NS, 25
New Glasgow River, PE, 174
New Harbour, NF, 84
New Hazelton, BC, 164
New Italy, NS, 179
New Johnstown, ON, 88
New Lanark, ON, 204
New Lancaster, ON, 204
Newland, ON, 259
New London, ON, 221
New Niagara, NB, 141
New Perlican, NF, 283
Newport Beach, BC, 389
New Severn River, ON, 374
Newstead, NF, 84
Newstead, ON, 266
Newtown, ON, 285
New Warren, SK, 19
New Wiltshire, PE, 278
New Zealand, NB, 446
Nicholsville, NF, 97
Nieven, SK, 152
Niganiche, NS, 176
Niganis, NS, 176
Nine Mile River, NS, 406
Nipisiguit, NB, 26
Nominingue, QC, 200
Noranda, QC, 337
Nord-Est, Rivière du, PE, 167
Nord, Mer du, QC, 355
Nord, Pointe du, PE, 274
Nord, Rivière du, ON, QC, 287
Nord, Rivière du, QC, 207
Nord, Rivière du, PE, 276
Norman, ON, 189
Normanton, ON, 307
Nort, Cap de, NS, 274
North Branch Muskoka River, ON, 261
North Camp, NT, 368
North Cape, NS, 274
North Cayuga, ON, 67
North Delta, BC, 98
North Douro, ON, 202
North East Arm, NF, 108
North East Margaree, NS, 235
Northern Gut, NF, 276
Northern Yukon National Park, YT, 179
North Fort St. John, BC, 133
North Lochaber, NS, 219
North Macmillan River, YT, 228
North Norwich, ON, 278
North Oxford, ON, 289
North Pine River, MB, 300
North Point, PE, 274
North Preston, NS, 312
North Saugeen River, ON, 369
North Sentinel, BC, 228
North Shore St. Margaret's Bay, NS, 319
North Somerset Island, NT, 383
North Surrey, BC, 435
North Thames River, ON, 403
North Thompson River, BC, 404
North Tryon, PE, 411
North Walsingham, ON, 205
North West Arm, NF, 417
North West Arm, NF, 53
Northwest Miramichi River, NB, 251
North West River, NF, 264
Northwest Territories, viii, 277
North-West Territories, viii, 277
North Williamsburg, ON, 438
Norway Pine Falls, ON, 39
Norwood, ON, 153
Notre-Dame-de-Grâce, QC, 433
Notre-Dame-d'Auvergne, SK, 304
Notre-Dame-de-Beauport, QC, 30
Notre-Dame-de-Bon-Secours, QC, 125
Notre-Dame-de-Bon-Secours-de-L'Islet, QC, 218
Notre-Dame-de-Bon-Secours, QC, 328
Notre-Dame-de-Koartac, QC, 317
Notre-Dame-de-la-Doré, QC, 201
Notre-Dame-de-l'Assomption-de-Repentigny, QC, 326
Notre-Dame-de-Liesse, QC, 357
Notre-Dame-de-Liesse-de-la-Rivière-Ouelle, QC, 332
Notre-Dame-de-Lorette, QC, 204
Notre-Dame-de-Rosaire, QC, 30
Notre-Dame-de-Sugluc, QC, 367
Nottawasaga, ON, 108
Nottawasaga Station, ON, 390
Nouveau-Comptoir, QC, 431
Nouvelle-Lorette, QC, 204
Numogate, ON, 254

Oak Lake, MB, 191
Oak Point, NS, 193
Old Chelsea, QC, 73
Old Bridgeport, NS, 102
Old Bridgeport Mines, NS, 102
Old Cove, NF, 71
Old Killaloe, ON, 190

Index

Old Masset, BC, 238
O'Leary Road, PE, 196
Oliver's Mills, ON, 248
Omega, SK, 282
Ontario, ON, 440
Onwa, SK, 214
Ophorportu, NF, 306
Oquapo, NB, 234
Orange, Baie l', NF, 152
Orangeville, NF, 219
Orchardville, ON, 290
Orignal, Pointe à, ON, 222
Orignal, Rivière de l', NS, 79
Orléans, Île d', QC, viii, 286
Orléans, Port d', NS, 44
Osborne, ON, 92
Osborne's, ON, 130
Osnaburg, NB, 136
Ossekeag, NB, 159
Otago, ON, 442
Ottawa, ON, vii, 287
Otter Creek Mills, ON, 287
Ouest, Rivière de l', PE, 433
Oxenden, ON, 240
Oxford Centre, ON, 176
Oxford-upon-the-Thames, ON, 176
Oxley, AB, 390
Oyster Harbour, BC, 201

Paint Hills, QC, 431
Paisley, ON, 56
Palestine, MB, 143
Palmer Road, NS, 17
Palmerston, NB, 359
Palmerston Depot, ON, 117
Pantagruel Bay, ON, 140
Parkismo, MB, 251
Parlican, NF, 283
Parr Town, NB, 356
Parson's Point, NF, 175-6
Pascalis, QC, 417
Passage Harbour, NF, 84
Payne River, QC, 186
Payne, Rivière, QC, 14
Peacock Mountain, BC, 241
Pécans, Rivière aux, MB, 127
Pêche, Lac la, QC, 206
Pêche, Rivière la, QC, 206
Peggy's Landing, ON, 243
Pelee, Point, ON, viii, 294
Pelham, ON, 130
Peliodore, Riviere, NS, 181
Pemberton Spur, BC, 315
Pender Island, BC, 276
Peninsula, ON, 234
Pennoyer's Mills, QC, 429
Pentangore, ON, 191
Pépêre, NB, 358
Perkins, QC, 416
Perry Lake, NT, 227
Perry's Corners, ON, 435
Perth, ON, 281
Peterborough, BC, 438
Petersham Cove, PE, 45
Peters Island, PE, 338
Peter's River, NF, 366
Pete's Point, NS, 235
Petite Passage, NS, 407
Petite-Nation, QC, 125
Petite Nation, Rivière de la, ON, 386
Petite-Rochelle, NB, 58
Petites-Fourches, QC, 213
Petit-Lac-Magog, QC, 96
Petit-Métis, QC, 247
Petit Miramichi, NB, 252
Petit-Sault, NB, 113
Pic, QC, 211
Pile of Bones Creek, SK, 325
Pinchgut, NF, 124
Pine Falls, MB, 131
Pineo Village, NS, 429
Piouagamik, Lac, QC, 355
Piper's Hole, NF, 396
Pisarinco, NB, 222
Pisiquid, NS, 159, 439
Plaisance, NF, 301
Plaister Cove, NS, 308
Pluie, Lac à la, ON, 322
Pluie, Rivière à la, ON, 322
Plum Creek, MB, 384
Point Brule, NS, 51
Pointe-à-Cavagnol, QC, 172
Pointe-aux-Esquimaux, QC, 163
Pointe aux Napraux, NB, 213
Pointe-aux-Roches, ON, 392
Pointe-de-l'Église, NS, 77
Pointe des Chênes, MB, 346
Pointe-des-Sauvages, NB, 58
Pointe-du-Lac, QC
Pointe-Gatineau, QC, 141
Point La Haye, NF, 140
Point Pelee National Park, ON, 294
Pointe à Grouette, MB, 346
Poplar Grove, AB, 177
Porcupine, Cape, NS, 40
porképic, Isle, NS, 9
Port Acadie, NS, 77
Portage de l'Oreille, MB, 110
Port aux Basques, NF, 70
Port Borden, PE, 43
Port Brabant, NT, 411
Port Fourchu, NS, 444
Port Haney, BC, 159, 234
Port Harrison, QC, 177
Port Kusam, BC, 370
Port Joli, Rivière, QC, 355
Portland, NB, 356
Port Loring, ON, 222
Port Mulgrave, NS, 260
Port North, NS, 376
Port-Nouveau-Québec, QC, 186
Port Radium, NT, 112
Port Razoir, NS, 376
Port Roseway, NS, 376
Port Rossignol, NS, 219
Port San Juan, BC, 309
Port Sarnia, ON, 368
Port Shubenacadie, NS, 231
Port Simpson, BC, 210
Port Toulouze, NS, 362
Port Trent, ON, 410
Portuchua, NF, 306
Portuchua Çaharra, NF, 306
Port Van Horne, BC, 370
Port Williams Station, NS, 153
Pound Cove, NF, 433
Poutrincourt, Cap de, NS, 40
Povungnituk, QC, 316
Povungnituk, Rivière de, QC, 316
Powerhouse, AB, 105
Prairies, Lac des, SK, 243
Prairie River, AB, 167
Prancer, Mount, NB, 276
President, The, BC, 312
Préville, QC, 357
Priest's Mill, ON, 6
Priest's Valley, BC, 420
Prime, NB, 332
Prime, Point, NF, 432
Prince Arthur's Landing, ON, 306
Prince of Wales Island, NT, 425
Princetown, PE, 232
Privat, QC, 400
Prince Henries Foreland, QC, 170
Providence, NT, 132
Prussia, SK, 210
Puerto de Cordova, BC, 87, 121
Puerto San Juan, BC, 309
Pultney, MB, 6

Punta de Lazo de la Vega, BC, 210
Purdy's Mills, ON, 217
Purdy's Rapids, ON, 217
Purdytown, ON, 243
Pyramid Mountain, BC, 260

Quaco, NB, 360
Quacopeck, NB, 234
Quadra and Vancouver's Island, BC, 418
Quadville, ON, 317
Quagmire, PE, 431
Quaigesmock, NB, 428
Qu'Appelle, SK, 133
Quarachoque, NS, 265
Québec, viii, 317-8
Québec-Ouest, QC, 418
Queenborough, BC, 270
Queenstown, BC, 307
Queen Victoria Island, NT, 27
Quévillon, Lac, QC, 211
Quinton's Point, NB, 58
Quisibus, NB, 347

Rabbit Creek, YT, 42
Rabbits Arm, NF, 332
Race, Cape, NF, viii, 321
Radway, MB, 281
Ragged Harbour, NF, 245
Ralphtown, MB, 189
Ralstons Colony, MB, 323
Ramea, QC, 229
Ramée, QC, 229
Ramouctou, NB, 286
Ramsay, ON, 8
Ramsayville, ON, 8
Ramsheg, NS, 425
Rapide-de-l'Orignal, QC, 254
Rapid River, MB, 251
Rashoon, NF, 338
Rat Portage, ON, 189
Rawdon, ON, 391
Rayside, ON, 20
Reagh's Cove, NS, 235
Reburn's Corners, ON, 56
Red Earth Forks, BC, 314
Red Fox Valley, SK, 379
Redonda, Isla, BC, 111, 433
Red River Colony, MB, 440
Red River, ON, 231
Red Sea, NB, NS, PE, 277
Redwillow, AB, 30
Red Willow Bunch, SK, 438
Refugee Cove, NF, 160
Remedy, Isle, NS, 9
Rendell, BC, 30
Resolution, NT, 133
Rhinebeck, ON, 28
Richards Station, NB, 187
Richmond, MB, 302
Richmond, QC, 269
Richmond, Baie de, QC, 155
Richmond Bay, PE, 232
Richmond, Golfe de, QC, 155
Richmond Gulf, QC, 155
Richmond Hill, ON, 234
Richmond Island, NS, 229
Rio Floridablanca, BC, 135
Ristigouche, Rivière, QC, 327
Ristook, NB, 14
Riverdale, ON, 176
Riverdale, PE, 268
River Debert, NS, 96
River Desert, QC, 233
Riverhead, NF, 158
Riverside Beach, NB, 111
Riverside, SK, 330
River View, SK, 115
Rivière Bersimis, QC, 36
Rivière-Boyer, QC, 353
rivière creuse, ON, QC, 96
Rivière de Bourq, NS, 96
R. de la Grande Baleine, QC, 23
Rivière de l'Ouest, PE, 433
Rivière-Duchesne, QC, 99
Rivière-du-Loup-en-Haut, QC, 223
Rivière Kénogami, QC, 189
Rivière-Saint-Pierre, QC, 420
Rivière Taronto, ON, 173
Robertson-et-Pope, QC, 100
Robin Hood, NF, 310
Rocher Bay, NB, 367
Roches, Montagnes de, AB, BC, 334
Roches, Rivière des, NF, 331
Rochester, ON, 33
Rocky Bay, NF, 64
Rocky Mountains Park Reserve, AB, 23
Rocky Saugeen River, ON, 369
Rogers Hill, NS, 371
Rolling Portage, ON, 172
Ronge, Lac la, SK, 207
Roseau Crossing, MB, 103
Rosebank, ON, 39
Roseneath, MB, 37
Rosenoll, AB, 38
Rosewood, ON, 169
Rosiers, Cap des, QC, 60
Rouge, Cap, QC, 61
Round Stone, SK, 321
Rousseau's Lake, ON, 336
Rousseaux's Mills, ON, 10
Rouville, Mont, QC, 255
Roweville, ON, 290
Royale, Isle, QC, 60
Royal Honey Harbour, ON, 169
Roycroft, AB, 338
Rupert, ON, 234
Rupert House, QC, 428
Rusticoville, PE, 277
Ryegate, ON, 401

Saanichton, BC, 68
Sables, Rivière aux, ON, 18
Sackville, NS, 31
Saglouc, QC, 367
Sagonaska, ON, 253
Saguenay, Mer du, QC, 355
Sailors Hope, PE, 11
saincte Croix, Rivière, NB, 345
Saincte Elaine, QC, 348
saincte-Jean, Rivière, NB, 356
Sainct Esperit, hable du, NF, 121
Saindon, QC, 370
St. Albans, ON, 169
Saint-Ambroise-de-la-Jeune-Loretteville, QC, 222
St. Andrews, QC, 341
St. Andrews East, QC, 341
St. Andrews North, MB, 220
Saint-Ange-Gardien, QC, 205
St. Anne's Point, NB, 136
St. Anthony, NB, 342
St Antoine, H. de, QC, 341
Saint-Antoine-de-Longueuil, QC, 221
St. Avard's, PE, 292
St. Azilda, ON, 20
Saint-Benôit-Joseph-Labre, QC, 10
Saint-Bernard-de-Lacolle, QC, 200
St. Catherine's, NF, 259
Saint-Charles-de-Lachenaie, QC, 199
Saint-Charles, Rivière, QC, 200
St. Clair, ON, 256
St. Clair City, MB, 37
Saint-Dunstan-du-Lac-Beauport, QC, 199
Sainte-Anastasie, QC, 226

Index

Sainte-Anastasie-de-Nelson, QC, 226
Sainte-Anne-de-la-Pocatière, QC, 206-07
Ste. Anne des Chênes, MB, 346
Sainte-Anne-des-Plaines, QC, 206
Sainte-Anne-de-Varennes, QC, 419
Sainte-Anne-du-Sault, QC, 95
Sainte-Anne-d'Yamachiche, QC, 444
Sainte-Anne, Havre de, NS, 341
Sainte-Anne, NB, 135
Sainte-Cécile-du-Bic, QC, 211
St. Edmunds, ON, 407
Sainte-Flavie-Station, QC, 254
Sainte-Geneviève, QC, 299
Ste Jeanne d'Arc, NB, 231
Sainte-Jeanne-de-Pont-Rouge, QC, 304
Sainte-Marie-de-Sayabec, QC, 370
Sainte Marie du Sault, ON, 369
Ste-Marie, Isle, NS, 229
Sainte-Marie-Madeleine, QC, 60
Saint-Émilien, QC, 99
Sainte-Philomène, QC, 245
Sainte-Pudentienne, QC, 337
Sainte-Rose-du-Dégelé, QC, 97
Sainte-Rose-du-Dégilis, QC, 97
Sainte-Scholastique, QC, 251
Sainte-Thérèse-de-Blainville, QC, 39
Sainte-Trinité-de-Contrecœur, QC, 86
Saint-Eusèbe-de-Stanfold, QC, 315
Saint-Évariste-Station, QC, 201
Sainte-Victoire-d'Arthabaska, QC, 422
Saint-Faustin, QC, 201
Saint-Faustin-Station, QC, 199
Saint-Félicien-Partie-Nord-Ouest, QC, 201
Saint-Félix-du-Cap-Rouge, QC, 61
Saint-Fernand-d'Halifax, QC, 35
Saint-François-de-Sales, QC, 303
Saint-François-Xavier-de-Verchères, QC, 419
Saint-Georges, QC, 195
Saint-Germain-de-Rimouski, QC, 330
Saint-Ignace, Cap, QC, 61
Saint-Jacques-des-Piles, QC, 149
Saint-Jacques-le-Majeur-de-Causapscal, QC, 67
Saint-Jean, QC, 206
Saint-Jean-Baptiste-de-l'Isle-Verte, QC 218
Saint-Jean-Chrysostome, QC, 344
Saint-Jean-de-Boischatel, QC, 41
Saint-Jean-Vianney, QC, 377
Saint-Jérôme, QC, 246
Saint-Jérôme-de-Matane, QC, 239
Saint-Joachim-de-la-Pointe-Claire, QC, 303
St. Johns, QC, 356
St. John's Island, PE, 313
St. John's River, ON, 173
Saint-Joseph, QC, 354
Saint-Joseph-d'Alma, QC, 8
Saint-Joseph-de-Deschambault, QC, 99
Saint-Joseph-de-la-Rivière-Bleue, QC, 331
Saint-Joseph-de-Maskinongé, QC, 238
Saint-Joseph-de-Sorel
Saint-Joseph-de-Soulanges, QC, 214
Saint-Joseph-de-Chambly, QC, 63
St-Jullian, Baie, NF, 179
S Laurent, Ance, NS, 28
S Laurent, C, NS, 28
Saint-Laurent, Fleuve, QC, 357
St. Lawrence, Cape, NS, vii, 28
St. Louis, Barachois, NS, 223
Saint-Louis, Baie, NF, 158
Saint-Louis, Cap, QC, 441
Saint-Louis-de-Kamouraska, QC, 185
Saint-Louis-de-Lotbinière, QC, 223
St. Louis de Moose Lake, AB, 43
Saint-Louis-de-Pintendre, QC, 300
St. Luke's Bay, NS, 309
Saint-Malachy, QC, 240
Saint-Maxime, QC, 371
Saint-Michel, QC, 444
Saint-Michel-de-Lachine, QC, 208
Saint-Michel-de-Mistassini, QC, 252
Saint-Michel-de-Rougemont, QC, 337
Saint-Michel-de-Vaudreuil, QC, 419
Saint-Moïse, QC, 361
Saint-Nicolas-Sud, QC, 35
Saint-Paul-d'Abbotsford, QC, 1
Saint-Paul-l'Ermite, QC, 212
St. Peters, NB, 26
Saint-Philippe-de-Clermont, QC, 79
St. Phillip's, NF, 310
St-Pierre, NB, 26
Saint-Pierre, NS, 362
Saint-Pierre, Détroit, QC, 180
Saint-Raphaël, NB, 349
Saint-Roch-de-Mékinac, QC, 244
Saint-Samuel-de-Gayhurst, QC, 199
Saints-Anges-de-Lachine, QC, 208
St. Stephens, NF, 366
Saint-Thomas-de-Pierreville, QC, 299
St. Thomas, ON, viii, 365
Salaberry, QC, 29
Salé, Rivière, MB, 208
Salisbury, BC, 265
Sallas Island, BC, 378
Salmon Cove, NF, 19
Salmonier North, NF, 259
Salmon River, ON, 375
Salmon River, NB, 8
Salmonville, ON, 402
Salt Pond, NF, 117
Salvage Bay, NF, 111
Sanderville, AB, 86
Sandhills, ON, 194
Sandville, ON, 67
Sandwich West, ON, 208
Sandy Bay, NF, 292
Sandyville, NF, 166
San Juan, Puerto, BC, 309
San Pedro, NS, 362
Santé, Cap, QC, 61
Sartigan, QC, 29
Saugeen, ON, 384
Sault-au-Cochon, QC, 130
Sault Ste. Marie, ON, viii, 369
Saunders Island, QC, 162
Saut de Gaston, ON, 369

Sauvages, Baie des, NF, 158
Sauvaiges, Cap dez, PE, 274
Sawridge, AB, 381
Sayward-Alberni, BC, 305
Scadouc, NB, 371
Schade's Mill, ON, 139
Scilly Cove, NF, 440
Scoodic, NB, 365
Scoodic River, NB, 345
Scotch Corners, ON, 108
Scott, ON, 415
Scott, C., NB, 61
Scratching River, MB, 258
Scugog, Lake, ON, 372
Second Chute, ON, 326
Second Crossing, BC, 327
Selby, ON, 202
Semadet, NF, 434
Seneca, ON, 56
Senneville, QC, 417
Serenity Mountain, BC, 64
Sesambre, NS, 368
Seven Islands, QC, 373
Seymour West, ON, 58
Shappee, Mount, AB, BC, 7
Sharpe's Brook, NS, 57
Shasheki Pass, BC, 75
Sheba's Paps, BC, 218
Sheila, NB, 409
Shell River, MB, 338
Shelter Bay, QC, 307
Shepherd's Falls, ON, 8
Sherbrooke, NS, 269
Sherbrooke, QC, xii, 377
Shiktehawk, NB, 48
Shinglebridge, ON, 39
Shining Mountains, AB, BC, 334
Ship Cove, NF, 44, 310, 340
Ship Harbour, NS, 308
Shipman's Corners, ON, 343
Shipman's Mills, ON, 8
Shippegan, Baie de, NB, 377
Shippegan Gully, NB, 212
Shippegan Island, NB, 204
Shippigan Gully, NB, 212
Shives Athol, NB, 17
Shoal Bay, NF, 105, 182
Shoal Brook, NF, 144
Shoreham, NS, 74
Sicotte, QC, 150
Sille Cove, NF, 440
Silverstream, NB, 354
Silver Stream, SK, 65
Simcoe Falls, ON, 444
Similkameen, BC, 314
Simpson, NT, 133
Simpson's River, BC, 380
Sinclair Hot Springs, BC, 321
Singleton's Creek, ON, 33, 253
Sir James Lancaster Sound, NT, 204
Sir Thomas Roe's Welcome, NT, 334
Sissiboo, NS, 434
Sisters, The, BC, 218
Skae's Corners, ON, 286
Skenesville, ON, 34
Slabtown, NS, 167
Slahaltkan, BC, 124
Slave Lake, AB, 215
Sleeping Giant Provincial Park, ON, 378
Sleeping Giant, The, ON, 378
Smith Road, PE, 358
Smith's Corners, ON, 246
Smith's Creek, ON, 308
Smithville, AB, 243
Smokeville, ON, 430
Smythe, ON, 116
Snake Creek, MB, 11
Snake Lake, AB, 397
Snake Lake, SK, 300
Snedden's Mills, ON, 39
Snowbank, ON, 436
Snowdrift, NT, 225
Somerset, PE, 193
Somerset-Nord, QC, 301
Somerset-Sud, QC, 301
Sore-foot Lake, AB, 232
Soulanges, QC, 214
Southampton, NS, 400
South Bay Ingonish, NS, 176
South Branch Muskoka River, ON, 261
South Camp, NT, 368
South Casselman, ON, 66
South Cayuga, ON, 67
South Durham, QC, 109
Southeast Arm, NF, 216
South Elmsley, ON, 220
Southern Gut, NF, 387
South Finch, ON, 127
South Hazelton, BC, 164
South Indian, ON, 217
South Lancaster, ON, 204
South Lochaber, NS, 219
South Macmillan River, YT, 228
South Moresby National Park Reserve, BC, 155
South Nelson, NB, 265
South Norwich, ON, 278
South Oromocto Lake, NB, 286
South Pender Island, BC, 276
South Qu'Appelle, SK, 317
South River, QC, 196
South Rustico, PE, 277
South Saugeen River, ON, 369
South Side, NF, 352
South Slocan, BC, 381
South Temiscaming, QC, 401
South Thames River, ON, 403
South Twillingate Island, NF, 413
South West Margaree, NS, 235
Southwest Miramichi River, NB, 251
South Westmeath, ON, 28
South-West Oxford, ON, 289
South Wiltshire, PE, 193
Souvier Wharf, NS, 369
Sovereen's Corners, ON, 97
Sovereign's Mills, ON, 428
Spalding-et-Ditchfield, QC, 137
Spallumcheen, BC, 118
Spanish Bay, NS, 397
Spanish River, The, ON, 403
Spark Point, QC, 208
Spillamacheen River, BC, 386
Sprague's Point, NB, 162
Springfield, ON, 258
Springfield-on-the-Credit, ON, 120
Spring Hill, AB, 167
Springhill, NB, 379
Springvale, PE, 250
Spring Valley, ON, 406
Springville, ON, 269
Sproat's Landing, BC, 333
Stadacone, QC, 318
Stalin, Mount, BC, 294
Stamford, ON, 270
Stanfold, QC, 315
Stanhope Cove, PE, 90
Stanley, BC, 265
Steamboat Landing, BC, 118
Stephenson's Landing, ON, 243
Stevens, NB, 2
Stevenstown, ON, 98
Stewarton, ON, 391
Stewart's Mills, ON, 334
Stewartstown, ON, 391
Stewart Town, ON, 391
Stewartville, AB, 117
Stillwater, NB, 249
Stirling, SK, 367
Stobart, SK, 106
Stone Mills, ON, 98

Index

Stones Book, NB, 296
Stoney Creek, ON, 374
Storm, C., NB, 61
Stormy Point, NB, 61
Storrington, ON, 177
Strange's Mills, ON, 334
Strathtay, ON, 317
Stringer, SK, 236
Strong's Corners, PE, 68
Stuart Lake, BC, 133
Stubbert, NS, 129
Studholm, NB, 12
Sturgeon Lake, BC, 393
Suffolk, ON, 248
Suffolk Road, PE, 301
Sugluk, QC, 367
Summerside, NF, 178
Summerside, NS, 96
Sunday Cove Tickle, NF, 306
Sunnyside, MB, 107
Swainby, ON, 292
Swan River, AB, 194
Swift Creek, BC, 416
Swift Harbour, NF, 70
Swints Harbour, NF, 70
Sydenham, ON, 109, 286, 288, 295
Sydney County, NS, 11
Symmes Landing, QC, 20
Szewczenko, MB, 423

Tandragee, ON, 376
Tanner's Crossing, MB, 251
Tapscott, AB, 2
Tara, ON, 240
Tarbox Corners, ON, 56
Taronto, Passage de, ON, 408
Taronto, Rivière, ON, 173, 408
Tecumseh, ON, 403
Tedish, NB, 61
Tekemamihouenne, Lac, ON, 322
Témiscaming, QC, xii, 401
Temperance, NS, 63
Temperanceville, ON, 130, 286
Templeton, QC, 141
Templeton-Est, QC, 141
Terra Nova, NF, 267
Terrants Bay, NS, 402
Teskeyville, ON, 12
Tewkesbury, QC, 391-2
The Brook, ON, 44
The Cache, BC, 146
The Coal Banks, AB, 81, 215
The Crossing, AB, 167
The Eddy, BC, 327
The Falls, BC, 444
The Forks, NB, 39
The Narrows, ON, 285
Theodore, Riviere de, NS, 181
The Twelve, ON, 343
The Wedge, NS, 430
Third Crossing, MB, 143
Thlew-ee-choh, NT, 21
Thompson, BC, 336
Thompson's Corners, ON, 119
Thompson's Rapids, ON, 202
Thornloe, ON, 268
Three Creeks, MB, 18
Three Rivers, PE, 142
Three Rivers, QC, 410
Tickle Harbour, NF, 33
Tiefengrund, SK, 202
Tiera del Corte Real, NF, 198
Timagami, ON, 401
Tooley's Corners, ON, 39
Torbolton, ON, 108
Tormentine, Cape, NB, 61
Toronto, ON, 87, 252, 308
Touraine, QC, 141
Tourmentin, Cap de, NB, 61
Tourtes, Rivière aux, ON, 299
Tracadie Bay, PE, 150
Tracadigache, Mont, QC, 63
Tracy, Lac de, ON, 395
Trading Lake, ON, 203
Tranche, Rivière à la, ON, 403
Traverse, Cape, PE, 61
Tring, QC, 366
Trinité, Rivière de la, QC, 22
Trinity, NF, 68
Trois Rivières, PE, 62, 142
Trout Lake, BC, 242, 264
Troy, ON, 2, 20
Tubtown, ON, 32
Tukik, NT, 12
Turgeon, QC, 350
Turks Gut, NF, 238
Turner Bay, NS, 402
Turns Bay, NS, 402
Turtle Cove, NF, 84
Tusket Wedge, NS, 430
Twins Bay, NS, 402
Two Brothers Creek, BC, 408

Ubank, SK, 421
Umingmaktok, NT, 27
Ungava, QC, 1
Union, BC, 92
Union Road, PE, 358
Upper Burton, NB, 54
Upper Chebogue, NS, 12
Upper Garry Lake, NT, 140
Upper Horton, NS, 441
Upper Lawrencetown, NS, 210
Upper Malagash, NS, 231
Upper Musquodoboit, NS, 261
Upper Percy, ON, 427
Upper Rideau Lake, ON, 329
Upper Sackville, NS, 339
Upper St. Bazil, NB, 342
Upper Sussex, NB, 296
Upper Tantallon, NS, 399
Upper Waterton Lakes, AB, 429
Upper West Pubnico, NS, 316
Upper Woods Harbour, NS, 442
Urfé, Rivière d', ON, 150
Utshimassit, NF, 95

Vader's Mills, ON, 444
Valcartier, QC, 352
Valdes Island, BC, 240, 317, 383
Vale Colliery, NS, 404
Valleyfield, QC, 367
Valliantbourg, QC, 206
Val-Saint-Michel, QC, 416
Van Hornes Island, NB, 251
Vanier, Île, NT, 418
Vaughan, ON, 441
Vauuder, SK, 423
Vents, Baie des, NB, 27
Verdonne, Île de, NS, 44
Vermilion Forks, BC, 314
Vermilion River, AB, 325
Vermilion Valley, AB, 168
Vice President, The, BC, 312
Victoria, SK, 325
Victoria, Grand lac, QC, 421
Vil Conomie, NS, 112
Village-Bistodeau, QC, 363
Village-Richelieu, QC, 278
Ville-Marie, QC, 255
Vil Nigeganish, NS, 34
Viner's Island, NT, 4
Virginia, ON, 68
Vixen, Mount, NB, 276

Waghorn, AB, 38
Wakeham Bay, QC, 186
Wales, ON, 176
Wallindale, SK, 321
Walpole, ON, 373
Walsingham, ON, 366
Walterville, ON, 68
Wardsville, ON, 382
Wareham, NF, 68
Wascana, SK, 135
Waswanipi, Lac, QC, 428

Waterford, NS, 316
Waterloo, ON, 67, 131
Waterview, PE, 393
Watford-Ouest, QC, 363
Wawanosh, ON, 112
Waweig, NB, 281
Waweig River, NB, 281
Weatherbie's Corners, PE, 68
Webber, AB, 130
Webster's Creek, NB, 351
Weenusk, ON, 439
Weldford, NB, 160
Welland City, ON, 246
Wellers Bay, ON, 65,
Wellington, NF, 105
Wellington, ON, 186, 379
Wellington Square, ON, 53
Wendigo, ON, 244
West Arichat, NS, 13
West Baccaro, NS, 21
West Chezzetcook, NS, 74
West's Corners, ON, 250
West Coup, NB, 432
West Dublin, NS, 106
West Farnham, QC, 125
West Hants, NS, 159
West Jeddore, NS, 181
West Lynne, MB, 118
Westminster, ON, 203
Westmorland Point, NB, 18
West Niagara, ON, 270
West Oxford, ON, 289
West Port, PE, 250
West Pubnico, NS, 316
Westqua, NB, 432
West River, PE, 115
West River of Pictou, NS, 298
West Thurlow Island, BC, 405
West Winchester, ON, 439
Westwood, ON, 292
West Woolwich, ON, 117
Wheat Belt, AB, 421
White Lake, AB, 424
White Valley, BC, 224
Whitford, AB, 10
Whitles, NF, 441
Whitley's, NF, 441
Whittles, NF, 441
Whitton, QC, 263
Wickwire Station, NS, 248
Wild Bight, NF, 28
William Henry, QC, 384
Williams, ON, 112
Williamsburg, ON, 39, 145
Williamsburg North, ON, 438
Williamsburg West, ON, 258
Williams Siding, BC, 398
Williamstown, ON, 284
Willoughby, ON, 270
Wilmot, ON, 163
Wilmot Corner, NS, 248
Wilson's Corners, ON, 135
Wilson's Mills, ON, 10
Winchester, ON, 50, 74
Windham, ON, 218
Windham River, NS, 143
Winding Ledges, NB, 351
Wind Mountain, AB, 223
Windsor, ON, 435
Windsor, NF, 149
Windy Mountain, AB, 223
Wine River, NB, 28
Winipeke, Baie, QC, 155
Winipeq, QC, 155
Winnipeg River Crossing, ON, 250
Winston, ON, 396
Winthuysen Bay, BC, 263
Wirral, SK, 208
Wiseman's Cove, NF, 433
Witler's Bay, NF, 436
Witless Bay, NF, 436
Wolfe's Cove, QC, 134
Wolstenholme, QC, 179
Woodpecker, AB, 24
Woody Point, BC, 86
Woolastook, NB, 327
Worshcock, NB, 432
Würtele-Moreau-et-Gravel, QC, 126
Wylde's Cove, NS, 260

Yamaska Mountain, QC, 362
Yankee Flats, BC, 274
Yankee Town, BC, 274
Yokasipi, ON, 401
Yonge, ON, 137, 324
York, ON, 112, 408
Yorke River, PE, 276
York Mills, ON, 23
York River, ON, 23
Young's Cove, NS, 102

Zone Mills, ON, 129
Zorra, ON, 112

The Canadian Oxford Dictionary

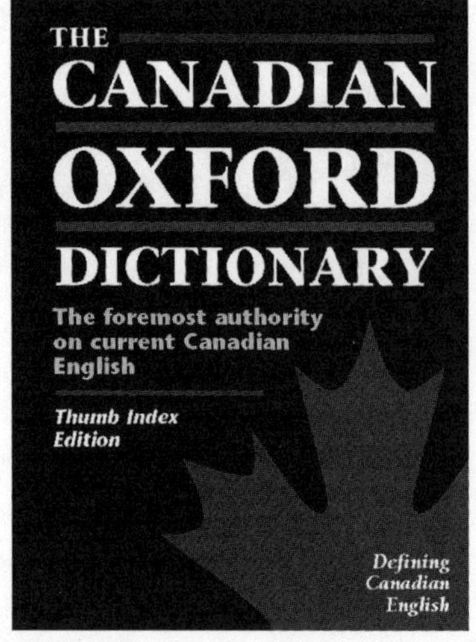

The *Canadian Oxford Dictionary* establishes a new authoritative standard for dictionaries in Canada. This book serves Canadians like no other dictionary, answering their basic questions about the language by giving them advice on Canadian spelling and usage, and defining more Canadian words and senses (almost 2000) than any other dictionary. With about 140,000 words, it combines in one reference book information on English as it is used worldwide and as it is used particularly in Canada. Definitions, worded for ease of comprehension, are presented within entries in order of their familiarity or frequency in Canadian usage, making the dictionary easy to consult.

The dictionary is exceptionally reliable in its description of Canadian English because it is based on thorough research into the language. It represents five years of work by five professionally trained Canadian lexicographers examining almost 20 million words of Canadian text, and another 20 million words of American, British, Australian, and other international English sources. These databases represent over 8000 different Canadian sources. Favoured Canadian pronunciations have been determined by surveying a nationwide group of respondents.

An added feature of this dictionary is its encyclopedic element. It includes short biographies of over 1000 Canadians and 4000 individuals and mythical figures of international significance, and over 5000 place names, more than 1200 of them Canadian.

With the publication of the *Canadian Oxford Dictionary*, Oxford University Press adds another work to its highly respected range of dictionaries.

Publication date: June 1998

ISBN: 0-19-541120-X